Radiological Assessment
Sources and Exposures

Richard E. Faw

J. Kenneth Shultis

Department of Nuclear Engineering
Kansas State University
Manhattan, Kansas

PTR Prentice-Hall, Englewood Cliffs, New Jersey 07632

Library of Congress Cataloging-in-Publication Data

FAW, RICHARD E.
  Radiological assessment : sources and exposures / by Richard E.
Faw, J. Kenneth Shultis.
    p.   cm.
  Includes bibliographical references and index.
  ISBN 0-13-751132-9
  1. Radiation–Safety measures.   2. Radiation dosimetry.
I. Shultis, J. Kenneth.   II. Title.
  [DNLM: 1. Environmental Exposure.   2. Occupational Exposure.
3. Radiation Dosage.   4. Radiation Protection–standards.   WN 650
F278r]
RA569.3.F38   1993
616.9′897–dc20
DNLM/DLC   92-6635
for Library of Congress   CIP

Editorial/production supervision
    and interior design: *Barbara Marttine/Jean Lapidus*
Cover design: *Ben Santora*
Prepress buyer: *Mary Elizabeth McCartney*
Manufacturing buyer: *Susan Brunke*
Acquisitions editor: *Mike Hays*
Editorial assistant: *Dana Mercure*

©1993 by PTR Prentice-Hall, Inc.
A Simon & Schuster Company
Englewood Cliffs, New Jersey 07632

Printed in the United States of America

10 9 8 7 6 5 4 3 2 1

ISBN   0-13-751132-9

PRENTICE-HALL INTERNATIONAL (UK) LIMITED, *London*
PRENTICE-HALL OF AUSTRALIA PTY. LIMITED, *Sydney*
PRENTICE-HALL CANADA, INC., *Toronto*
PRENTICE-HALL HISPANOAMERICANA, S.A., *Mexico*
PRENTICE-HALL OF INDIA PRIVATE LIMITED, *New Delhi*
PRENTICE-HALL OF JAPAN, INC., *Tokyo*
SIMON & SCHUSTER ASIA PTE. LTD., *Singapore*
EDITORA PRENTICE-HALL DO BRASIL, LTDA., *Rio de Janeiro*

To

Beverly *and* Susan

Contents

Appendices

Preface

Readers of this book will soon realize that it is not comprehensive in its treatment of the broad field of radiation protection. That is too much to expect from any one book. The topics emphasized in this book encompass one major aspect of radiation protection—radiological assessment—also a major aspect of each of the fields of health physics, medical physics, and industrial hygiene. What we call radiological assessment deals with evaluation of radiation doses to individuals and population groups from radiation sources both external and internal to the human body, the characterization of radiation sources, and the dispersal of radioactive materials in the environment. Even this restricted scope involves topics in many different disciplines each of which has a vast literature. Our goal has been to bring together the many different ideas and approaches used to assess radiation sources and associated doses. This book is neither a handbook nor a review of all radiological methodologies; rather, we have tried to present basic ideas and concepts in sufficient detail and in a unified fashion so as to allow readers to gain a thorough understanding of the principles and techniques used in radiological analyses.

Two topics are notably absent from the subject matter of this book, operational radiation safety and nuclear measurements. We believe that the former subject is too highly specialized and too dependent on local circumstances and regulations to be included in a book emphasizing principles. We do not minimize the importance of the subject. Rather, we acknowledge our inability to generalize operational radiation safety into a treatment of principles, other than to stress the importance of balancing the costs of radiation exposure with economic and social benefits of its use. Nuclear instrumentation and measurement techniques are essential subjects in the education of radiation protection specialists. There are two reasons for their omission in this text. First, in a formal educational program, the subjects would likely be taken up as a separate course of study. Second, there are, in our opinion, many excellent text and reference books which treat these subjects thoroughly.

In the presentation of the material in this book, we assume that readers possess certain prerequisite knowledge. An understanding of mathematics to the level of algebra and solid geometry is expected, as is a qualitative understanding of the basic nuclear physics of radioactive decay. For certain portions of the text, an understanding of elementary calculus and the solution of ordinary differential equations would be helpful, but not essential. The computational needs for most parts of this book can be provided by simple calculators; however, the methods stressed in this text are ideally suited for small-computer applications.

Although the principles of radiological assessment are emphasized, a relatively large amount of data is provided in the chapters and in the appendices to allow the reader to perform a wide range of realistic calculations. Most chapters in this textbook are supplemented with examples and exercises, and students are urged to study the examples and to complete the exercises. We have also tried to indicate to the reader where more comprehensive data bases and more elaborate radiological models can be found. Throughout the text, and in the examples and exercises, both S.I. and traditional units are used, with preference given to the former.

As with any book of this type, little of the material is original. Most has been gathered from a vast literature, with much of the information coming from reports prepared by committees of various agencies (e.g., UN, ICRP, ICRU, NCRP, and IAEA). To the authors of these reports we extend our appreciation for their important contributions to the discipline of radiological assessment.

Most of the material in the text has been accumulated by us over a period of years for a course on radiation protection. We have attempted to unify notation and to present material in a fashion we have found successful in the classroom. We thank the many students who have made suggestions and who have struggled with and endured earlier drafts of this text. During the preparation of this text, we received help and encouragement from many colleagues and friends. We especially thank J. C. Ryman and K. F. Eckerman of Oak Ridge National Laboratory.

Manhattan, Kansas   *Richard E. Faw*
                                                                     J. Kenneth Shultis

Chapter 1

Introduction to Radiological Assessment

1.1 OVERVIEW OF RADIOLOGICAL ASSESSMENT

Radiological assessment refers to the critical analysis of ionizing radiation sources and associated exposures, usually in the context of radiation protection of humans. Radiological assessment very often involves evaluating individual and societal benefits and detriments of a particular practice or circumstance involving radiation. It very often involves judging whether the conduct of a particular practice can be justified and, if so, designing equipment and developing procedures to assure that no individual is subjected to risks generally deemed unacceptable in normal circumstances, and that any inequities arising from balancing individual risks and societal benefits are minimized.

Radiological assessment requires both judgment and analysis. This text deals directly with analysis, but only indirectly with judgment. It introduces the reader to scientific principles and technical practices applied in characterizing radiation sources, estimating source strengths both external and internal to the human body, examining the transport of radioactivity through the environment, calculating radiation doses, and designing radiation shields. The book also provides information on the consequences of radiation exposure at the individual level and presents procedures for assessment of public-health effects from widespread exposures. It describes typical exposures received by all persons from natural sources and by selected population groups from medical and industrial practices.

Tools from many disciplines are needed in the performance of radiological assessments. The physical principles governing radiation emission from various sources and radiation interactions in matter are important. Geophysical and geochemical concepts are needed to describe the migration of radioactivity through the environment. Mathematical models of the environment are used in estimating the movement of radioactivity through the biosphere and through various organisms in food chains. Physical and metabolic aspects of human biology are applied in mathematical models used in evaluating the retention of radioactivity in the human body and in evaluating radiation exposures.

1

Persons involved with radiological assessment are to be found in many endeavors such as environmental analysis, health physics, nuclear medicine, shielding design, epidemiology, and risk analysis. These persons are engaged professionally in hospitals, academic institutions, research and development laboratories, testing services, government agencies, and industry. Their efforts may be devoted to activities as diverse as planning and analysis of nuclear medicine procedures, design of radiation shields, medical or industrial radiography, planning and monitoring the maintenance and repair of equipment, and inspection and regulation of users of radiation and radioactive materials.

The education and training of persons performing radiological assessments is also diverse. Formal academic preparation may have been in the life sciences, physical sciences, or engineering. Special training may have been acquired during military or industrial service and self-study has no doubt been an important aspect of the specialist's preparation for professional practice.

With all the diversity in preparation, training, and activity, is there the community of purpose and the discipline of a common body of knowledge so that radiological assessment may indeed be called a professional specialty? The authors believe so and, with this book, attempt to set forth the principles supporting that distinguishing body of knowledge.

What is this community of purpose? One might suppose that the purpose is simply protection of humankind from the ill effects of radiation exposure. That is far too simplistic a statement of purpose. It implies a purely mechanistic adherence to set procedures and set goals, not to be dignified as professional practice. Let us say, instead, that the community of purpose is to minimize the risks of radiation exposure in balance with the benefits of the use of radiation and nuclear energy. Judgment in balancing risks and benefits is thus the distinguishing feature in the professional practice of radiological assessment. That judgment requires thorough understanding of both risks and benefits.

Before the principles and analysis techniques used in radiological assessment are discussed in later chapters, this initial chapter deals with the evolution of radiation exposure units and protection standards, including dose limits marking the boundaries between generally acceptable and unacceptable risks.

1.2 HISTORICAL ROOTS

The culmination of decades of study of cathode rays and luminescence phenomena was W. C. Roentgen's 1895 discovery of x rays and A. H. Becquerel's 1896 discovery of what Pierre and Marie Curie in 1898 called *radioactivity*. The beginning of the twentieth century saw the discovery of gamma rays and the identification of the unique properties of alpha and beta particles. It saw the identification of first polonium, then radium and radon, among the radioactive decay products of uranium and thorium.

The importance of x rays in medical diagnosis was immediately apparent and, within months of their discovery, their bactericidal action and ability to destroy tumors were revealed. The high concentrations of radium and radon associated with the waters of many mineral spas led to the mistaken belief that these radionuclides possessed some subtle, broadly curative powers. The genuine effectiveness of radium and radon in treatment of certain tumors was also discovered and put to use in medical practice.

What were only later understood to be ill effects of exposure to ionizing radiation had been observed long before the discovery of radioactivity. Fatal lung disease, later diagnosed as cancer, was the fate of many miners exposed to the airborne daughter products of radon gas, itself a daughter product of the decay of uranium or thorium. Both uranium and thorium have long been used in commerce, thorium in gas-light mantles, and both in ceramics and glassware, with unknown health consequences. The medical quackery and commercial exploitation associated with the supposed curative powers of radium and radon may well have led to needless cancer suffering in later years.

Certain ill effects of radiation exposure such as skin burns were observed shortly after the discovery of x rays and radioactivity. Other effects such as cancer were not soon discovered because of the long latency period, often of many years duration, between radiation exposure and overt cancer expression.

Quantification of the degree of radiation exposure, and indeed the standardization of x-ray equipment, were for many years major challenges in the evaluation of radiation risks and the establishment of standards for radiation protection. Fluorescence, darkening of photographic plates, and threshold erythema (skin reddening as though by first-degree burn) were among the measures early used to quantify radiation exposure. However, variability in equipment design and applied voltages greatly complicated dosimetry. What we now identify as the technical unit of exposure, measurable as ionization in air, was proposed in 1908 by Villard, but was not adopted for some years because of instrumentation difficulties. Until 1928, the threshold erythema dose (TED) was the primary measure of x-ray exposure. Unfortunately this measure depends on many variables such as exposure rate, site and area of exposure, age and complexion of the person exposed, and energy spectrum of the incident radiation.

The first organized efforts to promote radiation protection took place in Europe and America in the period 1913 to 1916 under the auspices of various national advisory committees. After the 1914–1918 world war, more detailed recommendations were made, such as those of the British X-ray and Radium Protection Committee in 1921. Signal events in the history of radiation protection were the 1925 London and 1928 Stockholm International Congresses on Radiology. The first led to the establishment of the International Commission on Radiological Units and Measurements (ICRU). The 1928 Congress led to the establishment of the International Commission on Radiological Protection (ICRP). Both these Commissions continue to foster the interchange of scientific and technical information and to provide scientific support and guidance in the establishment of standards for radiation protection.

In 1929, upon the recommendation of the ICRP, the Advisory Committee on X-ray and Radium Protection was formed in the United States under the auspices of the National Bureau of Standards. In 1946, the name of the Committee was changed to the National Committee on Radiation Protection. In 1964, upon receipt of a Congressional Charter, the name was changed to the National Council on Radiation Protection and Measurements (NCRP). It and its counterpart national advisory organizations maintain affiliations with the ICRP, and committees commonly overlap substantially in membership. A typical pattern of operation is the issuance of recommendations and guidance on radiation protection by both the NCRP and the ICRP, with those of the former being more closely related to national needs and institutional structure. The recommendations of these organizations are in no way mandatory upon government institutions charged with promulgation and enforcement of laws and regulations pertaining to radiation protection. In the United States, for example, the Environmental Protection Agency (EPA) has the responsibility for establishing radiation protection standards. Other federal agencies such as the Nuclear Regulatory Commission or state agencies have the responsibility for issuance and enforcement of laws and regulations.

1.3 BASIC CONCEPTS

This section deals with the basic characterization of radiation sources. Here the discussion is limited to radioisotope sources. X-ray sources are considered in a subsequent chapter. First, the basic nature of biological risk and important features of radiation sources are reviewed. Then, basic radiation and dosimetry quantities are examined in two contexts: (1) quantities of a purely physical nature, and (2) quantities related to biological risk.

Biological risk is the potential consequence of cellular damage either leading to cell death or to disruption of cellular function and reproduction. This disruption results principally from damage to genetic information carried in the cell nucleus. In Chap. 3 it is shown that radiation exposure can lead to a complex range of biological effects and consequent risks.

The type of radiation considered in this book is *ionizing radiation*. Nonionizing radiation such as microwaves (radar), infrared and thermal radiation, and visible and ultraviolet light are not considered. Ionization is a process whereby one or more electrons are liberated from bound states as in atoms or molecules. But, the term is used in a somewhat broader sense, and the ICRU (1980) gives the "official" definition of ionizing radiation.

> *Ionizing radiation* consists of *charged particles* (for example, positive or negative electrons, protons or heavy ions) and/or *uncharged particles* (for example, photons or neutrons) capable of causing ionization by primary or secondary processes. However, the ionization process is not the only process by which energy of the radiation may be transferred to a material. A second important phenomenon is excitation, a process which can also have physical, chemical, or biological consequences. A radiation, such as low energy photons, may be ionizing in one medium but not in another. Hence, the choice of a suitable energy cutoff, below which a radiation is considered as non-ionizing, will depend on circumstances.

1.3.1 The Strength of a Radioisotope Source

A careful distinction must be made between the activity of a source and its strength. Activity is the more precise term, defined as the rate at which nuclear transformations, radioactive decays in the simplest sense, take place. Here, source strength is defined as the rate at which *radiation of a certain type* is released. Both terms have units of (events) per unit time, for example, transformations per second or photons emitted per second. However, the terms are certainly not equivalent.

Two units are in use for activity. The traditional unit is the *curie* (Ci). Originally associated with the rate of decay of one gram of ^{226}Ra, the curie now means precisely 3.7×10^{10} transformations per second. The S.I. unit is the *becquerel* (Bq), meaning precisely one transformation per second. Standard multiples (S.I. prefixes) are appropriate, for example, in the statement that 1 μCi is equivalent to 37 kBq, or that 1 PBq is approximately equivalent to 27 kCi. Two other identifiers of activity are in common parlance, disintegrations per minute (d.p.m.) or disintegrations per second (d.p.s.), but their use is discouraged. Readers may refer to Appendix A for prefixes for S.I. units and for a list of constants and conversion factors.

There are no special units for source strength. One simply states that a source has a strength of so many particles[1] or photons *of a certain type* per unit of time, usually seconds. We are careful here to distinguish between activity and source strength because there is unfortunately an all too common error made in assuming that one transition per second automatically results in the emission of one particle per second. A few examples illustrate this and other points of error, confusion, misunderstanding, and lack of standardization.

The radionuclide ^{137}Cs, along with its daughter ^{137m}Ba, is an important radiation source whose decay scheme is illustrated in Fig. 1.1. It is a fission product and a widely used radiation source for instrument calibration. The parent, with its 30-year half life is much longer lived than the daughter which has a half life of only 2.55 minutes. In nearly all circumstances, the two are in equilibrium; the decay of ^{137}Cs is quickly followed by the decay of ^{137m}Ba. The activities of the two are then the same.

Is ^{137}Cs a gamma-ray emitter? To be precise, it is not. Nevertheless, many data compilations list it as being so, and readers are cautioned accordingly. The nuclide ^{137}Cs decays by beta-particle emission to a *metastable* excited state of ^{137}Ba — designated by the symbol "m" as ^{137m}Ba — a state *which has a finite half life*. The ^{137m}Ba then either emits a gamma ray or an internal-conversion electron. Compare Fig. 1.1 with Fig. 1.2, the decay scheme for ^{60}Co, another important radionuclide. ^{60}Co decays by beta-particle emission to one of two excited states of ^{60}Ni, which then reaches the ground state by gamma-ray emission. The transition from the excited state to the ground state of ^{60}Ni takes place virtually instantly and the gamma ray is thus emitted simultaneously with the beta particle. By contrast, the excited state of ^{137}Ba is *metastable*, existing with a finite half life. ^{60}Co is properly identified as a gamma-ray emitter, ^{137}Cs only ambiguously as such. Although it is by no means universal practice, in this text sources existing in equilibrium are identified as, for example, ^{137}Cs-^{137m}Ba, ^{90}Sr-^{90}Y, or ^{99}Mo-^{99m}Tc.

Return now to the relationship between activity and source strength. In only 94.1% of its transitions does ^{137}Cs decay to ^{137m}Ba by emission of a beta particle of

---

[1]To avoid repeated reference to concepts such as wave-particle duality, the word *particle* will be used to mean x- or gamma-ray photon, beta particle, electron, and so on.

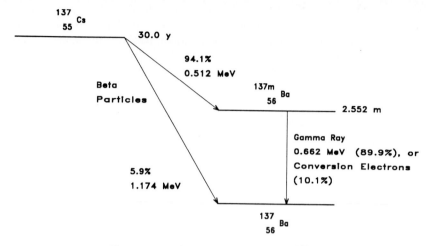

Figure 1.1   Radioactive decay scheme for ^{137} Cs.

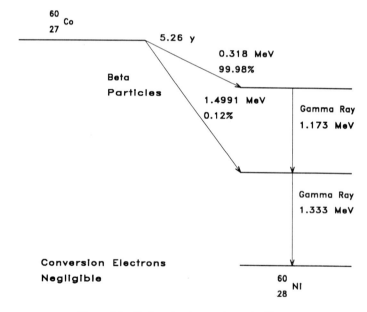

Figure 1.2   Radioactive decay scheme for ^{60} Co.

maximum (end-point) energy 0.512 MeV. ^{137m}Ba emits internal conversion electrons in competition with gamma-ray emission. In only 89.9% of the transitions is a 0.662-MeV gamma ray emitted. Thus, with reference to the activity of ^{137}Cs, per transformation, gamma rays are released with the frequency F of $0.899 \times 0.941 = 0.846$. The source strength S (s^{-1}) for 0.662-MeV gamma rays, in terms of the activity A (Bq) of ^{137}Cs is given by $S = A \times F$.

As another example, consider the radionuclide ^{60}Co which emits two gamma rays. As may be deduced from Fig. 1.2, the frequency of emission of 1.173-MeV gamma rays is 0.9988. That for 1.333-MeV gamma rays is $0.9988 + 0.0012 = 1.0000$. The rough approximation is often made that ^{60}Co emits 1.25-MeV gamma rays (the average of the two energies). If that approximation is made, then it is imperative to assign the frequency 2.0 to the emission process.

1.3.2 Physical Radiation and Dosimetry Quantities

One of the more fundamental and important measures of a radiation field is the number of particles that enter a specified volume, therein perhaps to cause damage. To describe this aspect of a radiation, the concept of *particle fluence* is used. The particle fluence, or simply *fluence*, at any point in a radiation field may be thought of in terms of the number of particles ΔN_p that, during a specified period of time, penetrate a hypothetical sphere of cross sectional area ΔA centered at the given point, as illustrated in Fig. 1.3. The fluence at that point is then defined as

$$\Phi \equiv \lim_{\Delta A \to 0} \left(\frac{\Delta N_p}{\Delta A} \right). \tag{1.1}$$

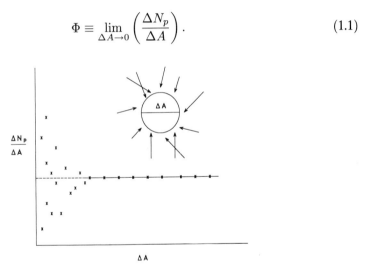

Figure 1.3   Extrapolation of data to zero value of ΔA to obtain a value of fluence.

The limiting process in this definition must be explained carefully. In any radiation field, the number of particles penetrating the test sphere will vary stochastically between repeated measurements because of the "graininess" of the radiation. These fluctuations will increase as the size of the test sphere decreases. Thus, the limiting process of Eq. (1.1) must be performed in a manner so as to smooth or average such statistical fluctuations. This smoothing can be performed by extrapolation from measurements (conceptual or real) made with sufficiently large spheres that avoid the stochastic fluctuations (see Fig. 1.3). However, for very weak radiation fields, this approach would require inordinately large measuring volumes. A better way to smooth the statistical fluctuations is to view ΔN_p as a probable or expected value, the average of many repeated measurements. The extrapolation of the expected ratio $\Delta N_p / \Delta A$ to a point detector, while avoiding the statistical fluctuations of individual measurements, makes the concept of fluence at a point a statistical concept whose value may be experimentally determined only approximately since macroscopic measuring volumes must always be used. In very weak radiation fields, the fluence may become unmeasurable even though conceptually there is a fluence at all positions.

An alternative definition of fluence, very useful conceptually, is as follows: Consider some small volume ΔV of arbitrary shape surrounding a point of interest. The lengths of track segments within the volume are summed for the particles that enter the volume. The fluence is then defined as[2]

$$\Phi \equiv \lim_{\Delta V \to 0} \left(\frac{\text{sum of internal track segment lengths}}{\Delta V} \right), \qquad (1.2)$$

where the limiting process is on the same basis as previously discussed.

The *fluence rate*, or *flux density*, is defined in terms of the incremental fluence $\Delta \Phi$ that occurs in a time interval Δt,

$$\phi \equiv \lim_{\Delta t \to 0} \left(\frac{\Delta \Phi}{\Delta t} \right), \qquad (1.3)$$

where, again, the limiting process is as previously discussed. An alternative interpretation of the flux density is the total path length that all the particles in a unit volume at the point of interest would travel in a unit time. Thus, if n is the number of particles per unit volume, all travelling with speed v, then $\phi = nv$.

It is sometimes more convenient to analyze a radiation field in terms of the energy flux density than in terms of the particle flux density. If all particles were of the same energy E, then the *energy flux density* would be simply $I \equiv E\phi$. But, it is often the case that at least some of the particles are distributed in energy. The energy distribution (spectrum) of the flux density at location $\mathbf{r}$ is described by the function $\phi(\mathbf{r}, E)$ with typical units $\text{cm}^{-2}\ \text{s}^{-1}\ \text{MeV}^{-1}$ and defined so that $\phi(\mathbf{r}, E)dE$ is the flux density at $\mathbf{r}$ of particles with energies within dE about E. The (total) flux density is

$$\phi(\mathbf{r}) = \int dE\ \phi(\mathbf{r}, E), \qquad (1.4)$$

and, correspondingly, the energy flux density is

$$I(\mathbf{r}) = \int dE\ E\ \phi(\mathbf{r}, E), \qquad (1.5)$$

where the range of integration is over all possible values of E.

Consider the question of what properties of a radiation field are best correlated with its hazard. Flux density, or even its energy spectrum, is not closely enough related to hazard to be useful, as a simple query would show. Would you the reader, if faced with exposing all or part of your body in a radiation field of fixed flux density,

---

[2]The equivalence of the two definitions is readily demonstrated for a spherical volume having a mean chord length of 4/3 times the radius.

be indifferent to the energy of the particles? Energy flux density appears to be more closely correlated with radiation hazard than is flux density alone, since the kinetic energy carried by a particle must have some correlation with the damage it can do to biological tissue. Even so, this choice is not entirely adequate — not even for particles of one fixed type. One has good reason to believe that the effect of a thermal neutron, with energy about 0.025 eV, is much greater than one would expect based on its relatively small kinetic energy.

One must examine more deeply the mechanism of the effect of radiation on matter in order to determine what properties of the radiation are best correlated with its hazard. It is apparent that one should be concerned not so much with the passage of particles or energy through a region of the material being exposed to the hazard, but with the creation of certain physical effects in that material. These effects may be, or may result from, the deposition of energy, ionization of the medium, induction of atomic displacements, production of molecular changes, or other phenomena. Major efforts have been made to quantify these phenomena by measurement or calculation as indices of radiation damage. Historically, since medical radiologists were those first concerned with the effects of radiation on biological material, any such quantification is called a "dose" if accumulated over a period of time, or a "dose rate" if the effect per unit of time is of interest.

As is seen, a number of different physical phenomena may be involved and these may be quantified in a variety of ways. Thus, "dose" is not a precise term and is best used in a generic sense to relate to any actual or conceptual measure of physical phenomena involving the effect of radiation on a material. There are a few particular dosimetric quantities which have been precisely defined and which are useful in our study of radiological assessment. To understand these definitions, which follow, one must appreciate the several-stage process in the passage of energy from its original to its final location and form. In general, these stages are as follows for indirectly ionizing radiation:

1. Neutrons or photons, the "primary" radiation, have interactions with the nuclei or the electrons of the material through which they are passing,

2. As a result of the interactions, "secondary" charged particles are emitted from the atoms involved, and each of these starts out with kinetic energy related to the energy of the primary particle and the type of interaction,

3. The secondary charged particles lose energy while traversing the material either (a) through ionization and associated processes (atomic and molecular excitation and molecular rearrangement), or (b) through emission of photons called *bremsstrahlung*. The progress of energy degradation need not be considered further at this point, except to say that the energy removed from the secondary charged particles by the process of ionization and associated mechanisms is distributed along the tracks of the charged particles and, for the most part, is promptly degraded into thermal energy of the medium.

4. The uncharged primary particles may also produce uncharged particles through scattering or other processes. These uncharged particles also carry

off part of the energy of the interaction, but this is not of immediate concern in the definitions which follow.

If an incremental volume of material ΔV around a given point in space has mass Δm and the energy imparted[3] to it by ionizing radiation is $\Delta \epsilon$, the *absorbed dose D* in the given material at this point is defined by

$$D \equiv \lim_{\Delta V \to 0} \left(\frac{\Delta \epsilon}{\Delta m} \right), \tag{1.6}$$

in which the limit process, as before, is such as to ignore statistical variability. The standard unit of absorbed dose is the *gray* (Gy), 1 Gy being equal to an imparted energy of 1 joule per kilogram. A traditional unit for absorbed dose is the *rad*, defined as 100 ergs per gram. Thus, 1 rad = 0.01 Gy = 1 cGy.

It is important in speaking of an absorbed dose to indicate the nature of the absorbing medium. In a measurement process, the element of volume might be a real or conceptual inclusion of one material within another — say a dosimeter in the atmosphere or an air-equivalent ion chamber within a concrete shield. Thus, one may speak of "tissue absorbed dose" within the atmosphere or "air dose" within concrete.

The concept of absorbed dose is very useful in radiation protection. Energy imparted per unit mass in tissue is closely, but not perfectly, correlated with radiation hazard. A closely related quantity, used only in connection with uncharged ionizing radiation, is the *kerma* (an acronym for *k*inetic *e*nergy of *r*adiation produced per unit *ma*ss in matter). It is symbolized by K and defined by

$$K \equiv \lim_{\Delta V \to 0} \left(\frac{\Delta E_k}{\Delta m} \right), \tag{1.7}$$

where the limit process as usual ignores statistical variation, ΔE_k is the energy transferred as kinetic energy *initially* carried by the secondary charged particles released by neutral-particle interactions within ΔV, and the other symbols are as previously defined. That some of the *initial* kinetic energy may be transferred ultimately to bremsstrahlung, for example, is irrelevant.

The use of the kerma requires some knowledge of the material present in the incremental volume, possibly hypothetical, used as an idealized receptor of radiation. Thus, one may speak conceptually of tissue kerma in a concrete shield or in a vacuum, even though the incremental volume of tissue may not actually be present.

Absorbed dose and kerma are sometimes used interchangeably and, if expressed in the same units, are frequently almost equal to each other. Kerma is rather more

---

[3]The energy imparted is precisely defined (ICRU 1980) as the sum of the energies (excluding rest-mass energies) of all charged and uncharged ionizing particles entering the volume minus the sum of the energies (excluding rest-mass energies) of all charged and uncharged ionizing particles leaving the volume, further corrected by subtracting the energy equivalent of any increase in rest-mass energy of the material in the volume. Thus, the energy imparted is that which is involved in the ionization and excitation of atoms within the volume, plus rearrangement of molecules within the volume. This energy is eventually degraded almost entirely into thermal energy.

easily calculated from knowledge of the radiation field, since the accounting of the energy is at the point of primary radiation interaction and the transport of the secondary particles can be ignored; however, it is difficult to measure directly. Absorbed dose, on the other hand, can be measured rather more easily (e.g., by calorimetry, since the energy imparted is eventually almost entirely transformed to thermal energy) but is much more difficult to calculate precisely. Kerma is not as directly related, in principle, to biological effects of radiation and is of value largely by reason of the relative simplicity of its calculation and the hope of approximating the correct absorbed dose by its use.

The quantity called *exposure*, with abbreviation X, is used traditionally to specify the radiation field of gamma or x-ray photons. It is defined as the absolute value of the ion charge (of one sign) produced anywhere in air by the complete stoppage of all negative and positive electrons (except those produced by bremsstrahlung) that are liberated in an incremental volume of air, per unit mass of air in that volume. The exposure is closely related to air kerma, but differs in one important respect. The phenomenon measured by the interaction of the photons in the incremental volume of air is not the kinetic energy of the secondary electrons, but the ionization caused by the further interaction of these secondary electrons with air. The basic unit of exposure is the roentgen, abbreviated R, which is defined as precisely 2.58×10^{-4} coulombs of separated charge per kilogram of air in the incremental volume where the primary photon interactions occur.

It is possible to convert a calculation of air kerma into exposure in two steps. The first involves the change from kinetic energy to ionization by the conversion factor W, which fortunately is almost energy independent for any given material. For air, it is estimated to be 33.85 ± 0.15 electron volts of kinetic energy per ion pair (ICRU 1979). In addition, the result must be reduced somewhat to account for the fact that in general some of the original energy of the secondary electrons may go not into ionization or excitation, but into bremsstrahlung.

Since exposure is a kerma-like concept, there are difficulties in its measurement. The measurement must be such as to ensure that all the ionic charge resulting from interaction in air in a small test volume at a certain location is collected for measurement. However, much of the ionization occurs outside the test volume and some ionization in the test volume takes place as a result of photon interactions just outside the volume. This makes charge collection complicated. The problem can be surmounted, at least for low-energy photons, if in the vicinity of the test volume there is "electronic equilibrium," a concept explained in detail in Chap. 2. Further details concerning measurement of exposure are beyond the scope of this text but may be found in other specialized works (Whyte 1959; ICRU 1962).

The use of exposure as a measure of the photon field is sometimes criticized but has survived because it is a measurable quantity correlated reasonably well with biological hazard. This is due primarily to the fact that, per unit mass, air and tissue are similar in their interaction properties with photons. On the other hand, absorbed dose in tissue is more closely related to hazard and is more fundamentally sound as a radiation quantity for assessing biological effects.

The definitions given in this subsection are for *dose* quantities. *Dose rate* quantities follow naturally. These are symbolized in standard fashion by placing dots over the appropriate symbols specified above, viz., $\dot{D}$, $\dot{K}$, and $\dot{X}$.

1.3.3 Radiation Quantities Related to Biological Hazard

If the energy deposited by ionizing radiation per unit mass of tissue were by itself an adequate criterion of biological damage, absorbed dose would be the best dosimetric quantity to use for radiation protection purposes. However, there are also other factors to consider that are related to the spatial distribution of ionization and excitation at a microscopic level. The charged particles responsible for the ionization may themselves constitute the primary radiation, or they may arise secondarily from interactions of uncharged, indirectly ionizing, primary radiation.

In dealing with the fundamental behavior of biological material or organisms subjected to radiation, one needs to take into account variations in the sensitivity of the biological material to different types or energies of radiation. For this purpose, radiobiologists define a *relative biological effectiveness* (RBE) for each type and energy of radiation, and indeed for each biological effect or endpoint. The RBE is the ratio of the absorbed dose of a reference type of radiation (typically 250-kVp x rays[4]) producing a certain kind and degree of biological effect to the absorbed dose of the radiation under consideration required to produce the same kind and degree of effect. RBE is normally determined experimentally and takes into account all factors relating to biological response to radiation in addition to absorbed dose. The concept of RBE is discussed in more detail in Chap. 3.

RBE depends on many variables: the physical nature of the radiation field, the type of biological material, the particular biological response, the degree of response, the radiation dose, and the dose rate or dose fractionation. For this reason it is too complicated a concept to be applied in the routine practice of radiation protection or in the establishment of broadly applied standards and regulations. In 1964, therefore, a group of specialists established somewhat arbitrarily a related but much more simply defined surrogate called the *quality factor Q*. Unlike the RBE, which has an objective definition in terms of a particular biological endpoint, Q is meant to apply generically to those endpoints consequent to low-level radiation exposure, namely cancer and hereditary illness. For a given type and energy of primary radiation, directly or indirectly ionizing, the quality factor is unique. The product of the quality factor and the absorbed dose is identified as the *dose equivalent* and is recognized as an appropriate measure of radiation risk when applied in the context of establishing radiation protection guidelines and dose limits for population groups. Originally, methods for evaluation of quality factors were somewhat subjective in the magnitudes assigned even though the prescriptions were very precisely defined in terms of macroscopic physical characteristics of the radiation fields. Over the years, subjectivity has been reduced in degree, but not eliminated, and the prescriptions have become more highly reliant on very sophisticated microscopic descriptions of the fields.

---

[4]The designation kVp refers to the peak energy of x rays released when the generating electron beam is produced by a fluctuating voltage.

To understand how the magnitude of the quality factor is determined for various types of radiation, one must first examine the quantity called the *linear energy transfer* for a charged particle of given type and energy. Linear energy transfer (LET or simply L) is the rate per unit distance along a charged particle track at which energy is transferred from the particle to the medium in which it travels. The scope of energy transfer is limited to transfer of energy to electrons via ionization and excitation reactions. Radiative energy loss from the particle is excluded. Most of the energy losses result in low-energy electrons whose energies are ultimately dissipated in the near vicinity of the original charged particle's track, that is, locally. However, some electrons, often called delta rays, have such high energies that their ultimate energy dissipation cannot be considered local. If one restricts the energy loss to an upper limit of δ, for example, one thereby limits the distance from the charged particle track identified as "local." The LET thus defined is called the *restricted linear energy transfer* and is marked by a subscript, e.g., L_δ. The symbol L_∞ denotes that all distances or all energy losses are included. In this case, the LET is equivalent to the collisional stopping power (see Chap. 2). Typical units for LET are keV μm^{-1}.

Since the spatial density of ionization and excitation along particle tracks is believed to be an important parameter in explaining variations in biological effects of radiation of different types and energies, and since the density is clearly proportional to LET, quality factor was defined initially in terms of LET. In particular, since tissue is largely water and has an average atomic number close to that of water, the quality factor was made a mathematical function of the linear collisional stopping power, or L_∞, in water.

The relationship between Q and L_∞, as given in Table 1.1, is that prescribed by the ICRP (1966, 1987). Values of Q are specified only for certain values of L_∞, and any reasonable interpolation procedure is allowed. The ICRP has suggested that, for electrons, a Q-value of unity may be used at all energies, regardless of what Table 1.1 might seem to require. As a charged particle loses energy along its track, L_∞ varies as the particle's energy varies. Quality factor, as a function of L_∞, may be taken as a local value *along the track of a charged particle.* More information is required to ascribe a quality factor to some particular primary radiation, whether that primary radiation be directly or indirectly ionizing. In principle, one must first determine how the absorbed dose is apportioned among particles losing energy at different LETs. One then may account for the variability of Q with L_∞ and determine an average quality factor $\bar{Q}$. Equivalently, one may determine an average LET and, from that, determine an average Q. Thus, the data of Table 1.1 apply to both local and average quality factors and LET values.

As we have indicated, quality factors can be ascribed to uncharged ionizing radiation through a knowledge of the properties of the secondary charged particles they release on interaction with matter. Since secondary electrons released by gamma rays or x rays are always assigned a quality factor of unity, the same factor applies universally to all ionizing photons. The situation for neutrons is not so simple, and average values must be determined as indicated in the following discussion.

The overall hazard of radiation is considered to be closely correlated with a dosimetric quantity called *dose equivalent*. In some regions of tissue, the dose equivalent

TABLE 1.1 Values of Quality Factor as
a Function of Linear Energy Transfer in
Water as Recommended in 1966 by the
ICRP. They Are Not Endorsed by the
1987 NCRP Recommendations.

L_∞ (keV μm^{-1})	Q
3.5 or less	1
7.0	2
23	5
53	10
175	20

Source: ICRP 1966.

is defined as being equal to the product of the absorbed dose in the region and the average quality factor based on the entire fluence of charged particles in the region. The S.I. unit[5] of the dose equivalent H is the *sievert*, abbreviated Sv, and is given by

$$H(\text{sievert}) = D(\text{gray}) \times \bar{Q}, \tag{1.8}$$

where $\bar{Q}$ is the average quality factor, given, in principle, by

$$\bar{Q} \equiv \frac{1}{D} \int_0^\infty dL_\infty D(L_\infty) Q(L_\infty), \tag{1.9}$$

in which $dL_\infty D(L_\infty)$ is the fraction of the absorbed dose D attributed to energy losses by charged particles with LET between L_∞ and $L_\infty + dL_\infty$, and in which the integration is over all possible LET. If it is possible to divide the absorbed dose D into portions D_i to which may be attributed individual (mean) quality factors Q_i, then the overall mean quality factor is given by

$$\bar{Q} \equiv \frac{1}{D} \sum_i D_i Q_i. \tag{1.10}$$

In 1991, the ICRP adopted a change in nomenclature, with the quality factor $\bar{Q}$ being replaced by the *radiation weighting factor* w_R. In this text we retain the quality-factor descriptor.

Because dose equivalent is defined in terms of a quality factor, which is based to some extent on human judgment, and therefore, susceptible to change, the objectively physical nature of the quantity is open to argument. Its direct measurement is not as straightforward as that of the absorbed dose, and it is more common to determine it either by calculation or by conversion from the more elementary quantity, the absorbed dose.

Since the quality factor is defined for use in conventional and routine radiation protection practice, the dose equivalent is firmly established only within this context.

---

[5]The traditional unit for dose equivalent is the *rem*, the product of the quality factor and the tissue absorbed dose in rads. The conversion to S.I. units is: 1 rem = 1 cSv.

TABLE 1.2 Average Quality Factors
Acceptable for Use with Certain Radiations,
as Recommended in 1986 by a Joint Task Force
of the ICRU and the ICRP. Their Use Is Not
Endorsed by the ICRP or the NCRP.

Photons and electrons with $E \geq 30\,\text{keV}$	1
Tritium beta particles	2
Neutrons, protons, alpha particles, other ions	25

Source: ICRU 1986.

It is nevertheless not uncommon for "dose" to be measured in sieverts in the context of very high exposure that might be sustained in emergency peacetime or wartime conditions. One school of thought is that such practice should be deplored and such exposures should be measured in terms of quantities such as absorbed dose; the opposite school of thought is that the use of dose equivalent units is permitted provided the user applies, not the standard quality factor, but a relative biological effectiveness appropriate to the circumstances.

The usual practice is followed in relating a symbol for *dose equivalent rate* to that for dose equivalent, namely $\dot{H}$.

1.3.4 Quality Factors Based on Lineal Energy

As discussed further in Chap. 3, there are shortcomings in the use of the LET as a determinant of the quality factor, especially the fact that LET is not readily measurable and is not directly related to energy deposition in a given volume of irradiated material. Therefore, in 1986 a joint task group of the ICRU and the ICRP (ICRU 1986) issued revised recommendations for evaluation of quality factors. These were recommendations of the task group and were not endorsed by either of the parent organizations. The *lineal energy* rather than the *linear energy transfer* was used as the independent variable in the evaluation. Lineal energy y, a stochastic quantity, is defined as the ratio of the energy imparted to matter in a volume of interest to the mean chord length in the volume. The volume selected was a 1-μm diameter sphere, a volume of tissue for which y can be measured, and a volume thought to be appropriate to radiation damage at the level of the nucleus of the cell. The quality factor may be expressed as a function of y and, for given circumstances of radiation exposure, the absorbed dose may be described as a distribution function in y rather than L. The dose equivalent may thus be evaluated as

$$H = \bar{Q} \times D = \int_0^\infty dy\, D(y)Q(y). \tag{1.11}$$

Suggested values of average quality factors for selected radiations are listed in Table 1.2.

TABLE 1.3 Quality Factor as a Function of Energy for Neutrons[a] as Recommended by the NCRP in 1987.

E (MeV)	Q	E (MeV)	Q
thermal to		7	14
0.001	4	10	13
0.01	5	14	15
0.1	15	20	16
0.5	22	40	14
1.0	22	60	11
2.5	18	100	8
5.0	16	200-400	7

[a]For neutrons of unknown energy, other than thermal neutrons, $Q = 20$.

Source: NCRP 1987 as derived from NCRP 1971a.

TABLE 1.4 Quality Factor (Radiation Weighting Factor) as a Function of LET in Water—a Formulation Adopted by the ICRP (1991).

L_∞ (keV μm^{-1})	$Q(L_\infty)$
< 10	1
10–100	$0.32L_\infty - 2.2$
> 100	$300/\sqrt{L_\infty}$

1.3.5 Re-evaluation of Quality Factors

The original quality factors of Table 1.1, as defined in terms of L_∞, are as of this writing still in use by regulatory agencies in the United States. Nevertheless, organizations such as the ICRP and NCRP have recommended changes in the quality factor formulation which would recognize an RBE for neutrons greater than had been thought appropriate in 1966. The ICRP (1985) accepted an increase in the neutron quality factor by a factor of 2. The NCRP (1987) adopted new neutron quality factors given in Table 1.3, which double the previously accepted values (NCRP 1971b). Basing quality factors on lineal energy (Table 1.2) has not been accepted by the ICRU or the ICRP. The ICRP (1991) has adopted a new formulation of the quality factor in terms of L_∞. The new formulation and the resulting mean quality factors for various types of radiation are summarized in Tables 1.4 and 1.5.

1.4 EVOLUTION OF RADIATION PROTECTION STANDARDS

Lauriston S. Taylor has written at length about the subject of radiation protection and the evolution of standards. He is uniquely qualified to do so, having served as Chairman of the National Council on Radiation Protection and Measurements from its founding in 1929 until 1978 and as a member of the International Commission on Radiological Protection since its founding in 1928. These comments are drawn largely from his history of the two organizations (Taylor 1979) and from his writings for the Health Physics Society (Taylor 1981, 1985). Another excellent source of information on the evolution of radiation protection standards prior to 1967 is the review article by Morgan (1967). An encyclopedic history of research into radiation and health is the book by the pioneering researcher J. Newell Stannard (1988).

TABLE 1.5 Mean Quality Factors (Radiation Weighting Factors) Adopted by the ICRP (1991).

RADIATION	$\bar{Q}$
Gamma rays of all energies	1
Electrons and mesons of all energies	1
Neutrons[a] < 10 keV	5
10–100 keV	10
100 keV–2 MeV	20
2–20 MeV	10
> 20 MeV	5
Alpha particles, fission fragments, heavy nuclei	20

[a]The neutron quality factor may be approximated as (with the neutron energy E in MeV)

$$\bar{Q} = 5 + 17\exp\{-[\ln(2E)]^2/6\}$$

1.4.1 The Tolerance Dose (TD) Concept

The earliest efforts to develop safe practices and the earliest recommendations for radiation protection were directed towards medical radiologists and technicians operating therapeutic x-ray machines. These machines, operating at voltages on the order of 200 kV, were of greater hazard than lower voltage diagnostic machines (NCRP 1954). Handicapped by primitive concepts of dose and by the lack of measurement devices, the pioneer health physicists chose as a "unit of measurement" the threshold erythema dose (TED). Mutscheller (1925) and, independently, Sievert (1925) recommended about 0.01 TED per month [6] as a *tolerance dose*, "... one which an operator can, for a prolonged period of time, tolerate without ultimately suffering injury ..." — simply a dose that could be *tolerated*. By 1924, the roentgen unit of exposure had been established and the TED determined to be equivalent to roughly 300 to 600 R, depending on the x-ray tube voltage. Thus, the TD would equate to about 0.1 to 0.2 R per day exposure. The former figure was adopted as a tolerance dose by the NCRP in 1934, the latter that same year by the ICRP. This was the first major step in establishment of numerical standards for radiation protection. Implied in the concept of tolerance dose is the existence of a threshold value below which no injury was expected to result. Recognition that there may be no threshold for gene mutation effects prompted the NCRP to abandon the tolerance dose concept in 1946.

1.4.2 The Concept of Maximum Permissible Dose

In 1948 the NCRP introduced a new basic radiation standard based on the rem unit of dose equivalent. This standard of 0.3 rem per week was identified as the *maximum permissible dose* (MPD). It was adopted by the ICRP in 1951 and retained in the 1954

---

[6]Sievert recommended 0.1 TED per year.

TABLE 1.6   1954 External-Exposure Recommendations of the NCRP.

Bloodforming organs, gonads, lenses of the eyes	0.3 rem/week
Skin	0.6 rem/week
Hands, forearms, feet, and ankles	1.5 rem/week[a]
Other organs and tissues	0.3–0.6 rem/week, depending on depth

[a] 1.5 rem/week to the skin, with dose equivalent to other tissues corresponding to that resulting from 1.5 R/week skin exposure.

Source: NCRP 1954.

NCRP recommendations (NCRP 1954). At that time, *permissible dose* was defined as ". . . the dose of ionizing radiation that, in the light of present knowledge, is not expected to cause appreciable body injury to a person at any time during his lifetime." Maximum permissible doses were defined as the highest ones permissible *under stipulated conditions of exposure*. The basic 1954 recommendations for long-term exposure to any type of ionizing radiation are given in Table 1.6.

MPDs for any 13 consecutive weeks were set at 10 times the weekly limits, and MPDs for 7 consecutive days were set at 3 times the weekly limit. Recommended limits were also given for one-time emergency exposures, namely 25 R for exposure of the whole body and, in addition, 100 R to hands, forearms, feet, and ankles. The 1954 recommendations contained two instances of dual standards for different exposure groups, the first occurrence of double standards for radiation protection (Taylor 1985). Persons aged 45 years or greater were permitted twice the basic MPDs, except for the lenses of the eyes. MPDs for *non-occupational* exposure of minors were set at 10% of the basic weekly limits.

After continued study of its recommendations, particularly with respect to genetic effects, and in the anticipation of possible shortening of average life expectancy caused by radiation exposure of a larger fraction of the population, the NCRP in 1958 issued a new set of recommendations (NCRP 1958). These and the companion ICRP recommendations (ICRP 1958) formed the basis of the radiation-protection regulations of the U.S. Atomic Energy Commission and Nuclear Regulatory Commission into the 1990s. They are summarized in Table 1.7.

No changes were made in the NCRP basic radiation protection criteria until 1971. By that time, the body of knowledge on biological effects of radiation supported a more generalized system of radiation protection standards, and concepts of maximum permissible dose and acceptable risk had matured. The ICRP (1964) had amplified the permissible-dose concept to the following statement:

The permissible dose for an *individual* is that dose, accumulated over a long period of time or resulting from a single exposure, which, in the light of present knowledge, carries a negligible probability of severe somatic or genetic injuries; furthermore, it is such a

TABLE 1.7   1958 Radiation Protection Recommendations of the NCRP.

BASIC RULES

Accumulated dose (radiation workers)

A. External exposure to critical organs

Whole body, head and trunk, active blood-forming organs, gonads

The maximum permissible dose (MPD), to the most critical organs, accumulated at any age, shall not exceed 5 rems multiplied by the number of years beyond age 18, and the dose in any 13 consecutive weeks shall not exceed 3 rems. MPD = 5(N − 18) rems, where N is the age in years.

B. External exposure to other organs

Skin of whole body

MPD = 10(N − 18) rems and the dose in any 13 consecutive weeks shall not exceed 6 rems.

Lens of the eyes

The dose to the lens of the eyes shall be limited by the dose to the head and trunk.

Hands and forearms, feet and ankles

MPD = 75 rems/year, and the dose in any 13 consecutive weeks shall not exceed 25 rems.

C. Internal exposure

The permissible levels from internal emitters will be consistent as far as possible with the age–proration principles above. Control of the internal dose will be achieved by limiting the body burden of radioisotopes. This will generally be accomplished by control of the average concentration of radioactive materials in the air, water, or food taken into the body.

Emergency dose (radiation workers)

An accidental or emergency dose of 25 rems to the whole body or a major portion thereof, occurring only once in the lifetime of the person, need not be included in the determination of the radiation exposure status of that person.

Medical dose (radiation workers)

Radiation exposures resulting from necessary medical and dental procedures need not be included in the determination of the radiation exposure status of the person concerned.

Dose to persons outside of controlled areas

The radiation or radioactive material outside a controlled area, attributable to normal operations within the controlled area, shall be such that it is improbable that any individual will receive a dose of more than 0.5 rem in any one year from external radiation.

The maximum permissible average body burden of radionuclides in persons outside of the controlled area and attributable to the operations within the controlled area shall not exceed one tenth of that for radiation workers. This will normally entail control of the average concentrations in air or water at the point of intake.

Source: NCRP 1958.

dose that any effects that ensue more frequently are limited to those of a minor nature that would not be considered unacceptable by the exposed individual and by competent medical authorities.

Any severe somatic injuries (e.g., leukemia) that might result from exposure of individuals to the permissible dose would be limited to an exceedingly small fraction of the exposed group; effects such as shortening of life span, which might be expected to occur more frequently, would be very slight and would likely be hidden by normal biological variations. The permissible doses can therefore be expected to produce effects that could be detectable only by statistical methods applied to large groups.

On the basis of the criteria indicated, the Commission has given recommendations with regard to the *maximum* dose which, still fulfilling the above requirements, could be permitted under various circumstances. The Commission has 'balanced' as far as possible the risk of the exposure against the benefit of the practice, and has also considered the possible danger involved in remedial actions once the exposure has occurred. The dose has been called the maximum permissible dose.

TABLE 1.8   1971 NCRP Dose-Limiting Recommendations.

Maximum permissible dose equivalent for occupational exposure	
Combined whole body occupational exposure	
prospective annual limit	5 rems in any one year
retrospective annual limit	10–15 rems in any one year
long-term accumulation to age N	5(N − 18) rems
Skin	15 rems in any one year
Hands	75 rems in any one year (25/qtr)
Forearms	30 rems in any one year (10/qtr)
Other organs, tissues, or systems	15 rems in any one year (5/qtr)
Fertile women (fetus)	0.5 rem in gestation period
Dose limits for the public, or occasionally exposed individuals	
Individual or occasional	0.5 rems in any one year
Students	0.1 rems in any one year
Population dose limits[a]	
Genetic or somatic	0.17 rem average per year
Emergency dose limits—life saving	
Individual (older than 45 years if possible)	100 rems
Hands and forearms	200 rems additional
Emergency dose limits—less urgent	
Individual	25 rems
Hands and forearms	100 rems total
Family of radioactive patients	
Individual (under age 45)	0.5 rem in any one year
Individual (over age 45)	5 rem in any one year

[a]The 1958 ICRP recommendations specified a genetic-dose limit of 5 rems
in 30 years, in addition to background and medical exposure.
Source: NCRP 1971.

The concept of acceptable risk — that appreciable injury manifested in the life-time of the individual is extremely unlikely — had been modified to imply that a risk is acceptable when it is at least offset by a demonstrable benefit. In establishing radiation protection criteria, the NCRP subdivided the general concept of acceptable risk to fit classes of individuals or population groups exposed for various purposes to different quantities and qualities of radiation. Recommendations in numerical terms restricted the term *maximum permissible dose* to occupational situations. Otherwise, as for population exposures, the expression *dose limit* was used. The 1971 criteria are summarized in Table 1.8. The reader is cautioned that application of limits is conditioned by qualifications and comments provided in the report (NCRP 1971a).

1.4.3 Maximum Permissible Body Burden and Related Concepts

By the late 1920s, evidence of the carcinogenic effects of radium as an internal emitter had become apparent. Snyder (1967) and Eisenbud (1987) describe the medical and industrial practices leading to many cases of bone cancer, especially among young women engaged in applying radium-containing paint to luminous dials of timepieces and other instruments. It was found that the lowest *body burden*, that is, total quantity or activity in the entire body, of ^{226}Ra present in a bone-cancer victim was 1.2 μCi.

In 1941, the NCRP recommended a body burden of 0.1 μCi (equivalent to 0.1 μg) of ^{226}Ra as limiting for occupational exposure (NCRP 1941). This criterion and that of 0.1 R per day established in 1934 for external exposure formed the basis for radiation-protection standards established and applied in the wartime Manhattan Project and in the early development of nuclear energy for ship propulsion and electricity generation. By the early 1950s, the NCRP (1953) and the ICRP (1955) had developed limiting body burdens for several dozen radionuclides.

Associated with the concept of limiting or maximum permissible body burden are concepts of *critical body organ, organ burden,* and *maximum permissible concentration* in air and water. Consider a radionuclide of an element such as Sr which is concentrated in bone or I which is concentrated in the thyroid. For any one nuclide, the critical organ is determined by the following considerations (NCRP 1959): (1) the organ that accumulates the greatest concentration of the radioactive material, (2) the essentialness or indispensability of the organ to the well-being of the entire body, (3) the organ damaged by the entry of the radionuclide into the body, and (4) the radiosensitivity of the organ. All these criteria were taken into account in identifying critical body organs, with criterion (1) usually controlling. Given the maximum permissible dose (equivalent) for that organ, it is a conceptually straightforward matter to determine the activity in the critical body organ leading to the MPD, that is, the maximum permissible organ burden, and, with knowledge of the distribution of the element within the body, the maximum permissible body burden. Of course, the data required for that determination were difficult to obtain, and it was necessary, too, to identify the physiological and metabolic characteristics of a *Reference Man* in order to devise broadly applicable radiation protection standards. Given the rates of inhalation and water ingestion of the Reference Man under various circumstances (at work or at rest, for example) it was then possible to determine the maximum permissible concentration (MPC) of the radionuclide in air or water. Inhalation or ingestion, at the MPC, under the specified circumstance, would lead to the maximum permissible body burden and, in turn, the maximum permissible organ dose equivalent.

Many other factors had to be considered in establishing MPCs. Methods had to be devised to account for combined inhalation and ingestion of mixtures of radionuclides, in combination with external exposure. Long-lived radionuclides which were retained for long periods of time in the body posed special problems. MPCs for those radionuclides were based on continuous or routine inhalation or ingestion over a 50-year period, beginning at age 18. Accumulation of the radionuclide in the body would lead to the maximum permissible body burden at the end of the 50-year period. MPDs used in determining the MPCs were as follows: an average annual dose equivalent of 15 rems for most individual organs, 30 rems when the critical organ is the thyroid, and 5 rems when the critical organ is the gonads or whole body. For bone-seeking radionuclides, the maximum permissible limit was based on a comparison of the energy release in the bone with the energy release by a maximum permissible body burden of 0.1 μCi for ^{226}Ra.

By 1960, the NCRP (1959) and ICRP (1960) had issued recommendations for limits on internal exposure for some 240 radionuclides. MPCs and maximum permis-

sible body burdens accompanying the recommendations formed the basis for regulations of the Atomic Energy Commission and Nuclear Regulatory Commission in the
United States into the 1990s.

1.4.4 Risk-Based Dose Limits

The early 1970s saw a transition from standards based on thresholds, safety factors,
and permissible doses to standards based on quantitative risk assessment. The new
line of thinking is exemplified by the 1972 report to the ICRP by the Task Group on
Dose Limits (ICRP 1972). Key portions of the report are summarized in the following discussion. It must be noted that the report is unpublished and not necessarily
reflective of the official ICRP position. The tentative dose limits examined in the report were not based on explicit balancing of risks and benefits, then thought to be an
unattainable ideal. Rather, they were based on the practical alternative of identifying
acceptable limits of occupational radiation risk in comparison with risks in other occupations generally identified as having a high standard of safety and also having risks
of environmental hazards generally accepted by the public in everyday life.

Linear, no-threshold dose-response relationships were assumed for carcinogenic
and genetic effects, namely a 1×10^{-4} probability per rem whole-body dose equivalent
for malignant illness or a 4×10^{-5} probability per rem for hereditary illness within the
first two generations of descendants (ICRP 1977). For other radiation effects, absolute thresholds were assumed.

For occupational risks, it was observed that "occupations with a high standard of
safety" are those in which the average annual death rate due to occupational hazards is
no more than 100 per million workers. An acceptable risk was taken as 50 per million
workers per year, or a 40-year occupational lifetime risk of two fatalities per 1000
workers, that is, 0.002. It was also observed that in most installations in which radiation
work is carried out, the average annual doses are about 10 percent of the doses of the
most highly exposed individuals, with the distribution highly skewed toward the lower
doses. To ensure an *average* lifetime risk limit of 0.002, an upper limit of 10 times
this value was placed on the lifetime risk for any one individual. The annual whole-
body dose-equivalent limit for stochastic effects was thus taken as $(10 \times 0.002)/(40 \times
0.0001) = 5$ rem (5 cSv) per year.

For members of the public, it was observed that risks readily accepted in everyday life, and of a nature that could be assumed to be understood by the public and
not readily avoidable, are typically five deaths per year per million population, or a
70-year lifetime risk of about 4 per 10,000 population. It was observed that some individuals accept risks in everyday life an order of magnitude greater. Thus, the lifetime
risk incurred by a well selected and homogeneous small "critical group" could be considered acceptable if it does not exceed the higher lifetime risk of 4 per 1000. Setting
this limit for the generality of *critical groups* would assure that the lifetime risk to the
average member of the public would be much below 4 in 10000. The whole-body dose
equivalent limit for stochastic risks to individual members of the public would thus be
$0.004/(70 \times 0.0001) = 0.5$ rem (0.5 cSv) per year.

The concept of "risk-based" or "comparable-risk" dose limits provides the ra-
tionale for the 1977 ICRP and the 1987 NCRP recommendations for radiation protection.

1.4.5 1977 ICRP Radiation-Protection Recommendations

In 1977 the ICRP issued recommendations involving major, though still evolutional, changes in the conceptual framework for dose limitation. The 1977 report introduced a system of limits based on the concept of *committed dose equivalent* and on risk coefficients accounting for differing sensitivities of the various organs and tissues of the body. Evaluation of internal doses, following the ICRP methodology (ICRP 1975, 1979), is described at length in Chap. 8.

Before reviewing the recommendations, it is necessary to examine the definitions of some new terms:

Detrimental effects against which protection is required are known as somatic and hereditary; the former if they become manifest in the exposed individual and the latter if they affect descendants.

Stochastic effects comprise malignant and hereditary disease for which the probability of an effect occurring, rather than its severity, is regarded as a function of dose without threshold.

Nonstochastic effects[7] comprise effects such as opacity of the lens and cosmetically unacceptable changes in the skin for which a threshold or pseudothreshold must be exceeded before the effect is induced.

Effective dose equivalent, H_E, the sum of weighted dose equivalents (see following discussion) for irradiated tissues or organs, taking into account the different mortality risks from cancer and the risk of severe hereditary effects in the first two generations associated with irradiation of different organs and tissues.

Committed dose equivalent, H_{50}, to a given organ or tissue from a single intake of radioactive material into the body is the dose equivalent that will be accumulated over 50 years, representing a working life, following the intake. It is the integral over 50 years of the dose equivalent rate in the organ or tissue (ideally, of the person exposed, but, for general application, of Reference Man).

Occupational Exposure Limits.   The ICRP in their 1977 recommendation retained the basic standard for uniform occupational irradiation of the whole body, namely an annual limit of 5 cSv (5 rem) to the whole-body dose equivalent H_{wb} for an acceptable level of occurrence of stochastic effects. The recommended system for limiting exposure is also based on the principle that the limit on risk should be equal whether or not the whole body is irradiated uniformly. This condition is stated in terms of a 5 cSv (5 rem) annual limit on the *effective dose equivalent*, defined as

$$H_E \equiv \sum_T w_T H_T, \tag{1.12}$$

in which H_T is the annual dose equivalent received by tissue T from external exposure, or the annually committed dose equivalent incurred by tissue T as a result of internal

---

[7]An alternate term for "nonstochastic" effects might be "mechanistic effects."

exposure, and w_T is a weighting factor representing the ratio of the stochastic risk resulting from tissue T to the total risk when the whole body is uniformly irradiated. Weighting factors dopted by the ICRP in 1977 are as follows:[8]

Gonads	0.25
Breast	0.15
Red-bone marrow	0.12
Lung	0.12
Thyroid	0.03
Bone surfaces	0.03
Remainder	0.30

As to the "remainder," a weighting factor of 0.06 is applied to each of the five remaining organs or tissues in the body receiving the greatest dose equivalents, the exposure of all other groups being neglected. Extremities, skin, and lens of the eye are excluded from the remainder organs. When the gastrointestinal tract is irradiated, then the stomach, small intestine, upper large intestine, and lower large intestine are treated as four separate organs. The derivation of these factors is explained in Chap. 3, Sec. 3.7.

In their 1990 recommendations, the ICRP (1991) renamed H_T as *dose equivalent* and H_E as *effective dose*. They also changed the symbol H_E to E; but, in this text, the symbol H_E will be retained because the symbol E is reserved for radiation energy.

The ICRP position on nonstochastic effects is that they will be prevented by applying a dose-equivalent limit of 50 cSv (50 rem) in a year to all tissues except the lenses of the eyes, for which the limit is 15 cSv (15 rem) in a year. In their 1990 recommendations, the ICRP recommended that these limits apply to occupational exposure but that, for exposure of the general public, the dose limits for the skin and lens be reduced arbitrarily by a factor of 10. This was done because of the wider range of sensitivity in the general public as compared to the limited population of workers and because the time periods of exposure may be much greater in the general population.

The ICRP prescribes a *secondary limit* and a *derived limit* for the control of internal exposure. These new limits displace the older and no-longer applicable limits expressed in terms of *maximum permissible body burden* and *maximum permissible concentration*.

For a given radionuclide, the *annual limit on intake* (ALI) is a secondary limit designed to meet the basic limits for occupational exposure. ALI is the largest annual intake that would satisfy limits for both stochastic and nonstochastic effects. Separate ALIs may apply to ingestion and inhalation. Procedures are prescribed (see Chap. 7) for dealing with dual modes of intake of mixtures of radionuclides.

For a given radionuclide, there is a derived limit, called the *derived air concentration* (DAC), designed to meet the ALI for inhalation. The DAC is defined as the

---

[8]The weight factors used in the definition of the effective dose equivalent have been revised by the ICRP (1991). The new weight factors, called *tissue weighting factors* are discussed in Chap. 3. The weight factors listed here are those used in the United States in the 1991 revisions of federal radiation-protection standards.

concentration in air that, if breathed by Reference Man for a working year of 2000 hours under conditions of "light activity," would result in the ALI by inhalation. DACs also apply to operations involving exposure to radioactive noble gases and elemental tritium through submersion of reference man in a cloud of gas.

Limits for Planned Occupational Special Exposures.   For situations requiring that a few workers receive dose equivalents in excess of recommended limits, alternatives being unavailable or impractical, the ICRP recommends that effective dose equivalent commitments not exceed twice the relevant annual limit in any single event and, in a lifetime, five times this limit.

Occupational Exposure of Women of Reproductive Capacity.   The ICRP position is that, if occupational exposure is received at an approximately regular rate, it is unlikely that any embryo could receive more than 0.5 cSv (0.5 rem) during the first two months of pregnancy, normal occupational exposure limits thus providing appropriate protection.

Occupational Exposure of Pregnant Women.   The ICRP recommendation is that, when pregnancy has been diagnosed, arrangements be made to ensure that the worker's exposure be limited to no more than 30 percent of the standard effective dose equivalent limits.

Dose-Equivalent Limits for Individual Members of the Public.   The recommendation of the ICRP is that a whole-body dose equivalent limit of 0.5 cSv (0.5 rem) in a year be applied to individual members of the public. A number of qualifications are made with respect to this recommendation. These deal with critical groups within the population, groups representative of those individuals in the population expected to receive the highest dose equivalent and the possibility that some individuals might belong to more than one critical group. The ICRP recommendations would require that, in calculation of dose equivalents incurred by members of the public from intake of radionuclides, account be taken of differences in organ size or metabolic characteristics of children.

Exposure of Populations.   The ICRP makes no specific recommendation as to dose limits for populations, and explicitly withdraws a previously recommended genetic dose limit of 5 rem in 30 years. The point is made by the ICRP that adherence to recommendations for individual members of the public would likely ensure that the average effective dose equivalent to the population would not exceed 0.05 cSv (0.05 rem) per year.

Accidents and Emergencies.   The ICRP makes no specific recommendations for dose levels which would call for remedial actions, the establishment of such being considered the responsibility of national authorities. The ICRP does emphasize, however, the low level of risk associated with its more generally recommended limits and that it should not be obligatory to take remedial action if an effective dose equivalent limit has been or might be exceeded.

1.4.6 1987 NCRP Exposure Limits

The 1987 NCRP recommendations state that "The goal of radiation protection is to limit the probability of radiation induced diseases in persons exposed to radiation (somatic effects) and in their progeny (genetic effects) to a degree that is reasonable and acceptable in relation to the benefits from the activities that involve such exposure." Acceptability implies degrees of risk comparable to other risks accepted in the work-place and by the general public in their everyday affairs. This comparable risk concept requires comparison of the *estimated* risk of radiogenic cancer with *measured* acciden-tal death rates or general mortality rates. Taken by the NCRP as a reasonable basis for estimated risk is the cautious assumption that the dose-risk relationship is without a threshold and is strictly proportional (linear) throughout the range of dose equivalents and dose equivalent rates of importance in routine radiation protection. Specifically, the dose-risk relationship is taken as a lifetime risk of fatal cancer of 10^{-2} $Sv^{-1}(10^{-4}$ $rem^{-1})$ for both sexes and for a normal population age distribution of 18 to 60 years. The genetic component of risk in the first two generations is taken to be 0.4×10^{-2} $Sv^{-1}(0.4 \times 10^{-4}$ $rem^{-1})$. In support of the limits presented in Table 1.9, the following observations are relevant:

- The annual fatal accident rate per million workers ranges from 50 in trade occupations to 600 in mining and quarrying occupations, the all-industry av-erage being 110, that is, about one person per year in 10,000 workers.

- "Safe" industries are taken as those with an annual average fatal accident rate less than one in 10,000, that is, with an annual average risk of less than 1×10^{-4}.

- Among radiation workers, the annual fatal accident rate (nonradiation) is about 0.25×10^{-4}. The annual effective dose equivalent to radiation workers is about 0.23 rem which, based on the estimated risk for radiogenic cancer, results in an annual risk of about 0.25×10^{-4}, comparable to the nonradiation fatal accident rate. Radiation workers thus have a total annual risk of fatality of about 0.5×10^{-4}, well within the range for "safe" industries.

- Overall average mortality risks from accident or disease for members of the public range from about 10^{-4} to 10^{-6} annually. Natural background radi-ation, resulting in an average dose of 0.1 rem annually, results in a risk of mortality of about 10^{-5} per year.

The NCRP recommendations of 1987 are summarized in Table 1.9. Some supplemen-tal information follows, but the interested reader will find it necessary to refer to the NCRP report (1987) for supporting details and qualifications.

Annual Effective Dose-Equivalent Limits versus Dose Commitments. The NCRP adopts the committed effective dose equivalent for radiation protection plan-ning, for use in calculating annual limits of intake and derived air concentrations, and for estimating lifetime risk from a given intake. However, the 1987 recommendations on exposure limits are based on actual received effective dose equivalents (for the most part on an annual basis) and not on committed effective dose equivalents.

Special Occupational Exposures. For those rare occasions requiring annual effective dose equivalents in excess of 50 mSv (5 rem) to selected workers, the NCRP provides the following guidelines:

TABLE 1.9   1987 NCRP Recommendations for Exposure Limits.

Occupational exposures (annual)			
1. Limit for stochastic effects		50 mSv	5 rem
2. Limit for non-stochastic effects			
a. Lens of eye		150 mSv	15 rem
b. All other organs		500 mSv	50 rem
3. Guidance: cumulative exposure	years age ×	10 mSv	1 rem
Planned special occupational exposure		see text	
Guidance for emergency occupational exposure		see text	
Public exposure (annual)			
1. Continuous or frequent exposure		1 mSv	0.1 rem
2. Infrequent exposure		5 mSv	0.5 rem
3. Remedial action levels:			
a. Effective dose equivalent		5 mSv	0.5 rem
b. Radon and decay products		> 2 WLM	
4. Lens of eye, skin and extremities		50 mSv	5 rem
Education and training (annual)			
1. Effective dose equivalent		1 mSv	0.1 rem
2. Lens of eye, skin and extremities		50 mSv	5 rem
Embryo–fetus exposures			
1. Total dose equivalent limit		5 mSv	0.5 rem
2. Dose equivalent limit in a month		0.5 mSv	.05 rem
Negligible individual risk level (annual)			
Effective dose equivalent per source or practice		0.01 mSv	1 mrem

Source: NCRP 1987.

- No worker should receive an effective dose equivalent of more than 100 mSv (10 rem) in a single planned event nor a lifetime effective dose equivalent in excess of 100 mSv incurred in multiple planned events. Older workers with low lifetime effective dose equivalents should be selected for special exposures whenever possible.

- Planned special exposures should be preauthorized in writing by senior management and exposures received should be included, but separately identified, in lifetime exposure records.

The following guidance is provided for emergency occupational exposure:

- Only actions involving life saving justify acute exposures in excess of 100 mSv (10 rem).

- The use of volunteers is desirable, and older workers with low lifetime effective dose equivalents should be selected whenever possible.

- When exposures approach or exceed 1 Gy (100 rad), volunteers should be appraised of the potential for acute radiation effects and the increase in lifetime cancer risk.

1.4.7 The ALAP and ALARA Concepts

With the abandonment of the tolerance dose concept and the assumption of "no-threshold" dose-response relationships, there came the realization that, even within risk-based systems of dose limitation, there was the common-sense desirability of keeping doses "as low as practicable" (ALAP) or "as low as reasonably achievable" (ALARA). For example, the NCRP (1954) stated that ". . . it is strongly recommended that exposure to radiation be kept at the lowest practicable level in all cases." The ICRP (1958) recommended that ". . . doses be kept as low as practicable, and that any unnecessary exposure be avoided." It is perhaps not surprising that no "practicable" way has yet been found to describe quantitatively what is meant by "reasonably achievable." The NCRP (1971a) addressed the issue of practicability in cases for which benefits accrue to one group and risks to another and for which no logical benefit-loss balance sheet could be prepared. As pointed out by Taylor, the key phrase had evolved by 1973 to "as low as reasonably achievable, economic and social considerations being taken into account," and the ICRP (1983) has ventured to explore cost-benefit analysis as a tool in the optimization of radiation protection. Still, quantification of economic and social considerations has proved an elusive goal.

Taylor (1985) has warned of the confusion that may arise with the imposition of secondary numerical standards for radiation protection, and pointed out that avoidance of that confusion was a primary motivation for the NCRP's urging "as low as practicable" exposure within a single standard and not offering specific numerical sublimits. Some of that confusion has no doubt arisen from the introduction in the United States of numerical "guides" to meet the ALAP criterion in the design of light-water-cooled nuclear power reactors [Title 10, U.S. Code of Federal Regulations, Part 50. Appendix I][9].

The current interpretation of the ALARA concept in the United States may be described best by quoting from 1987 federal radiation protection guidance (U.S. 1987).

> This principle is set forth in these recommendations in a simple form: "There should not be any occupational exposure of workers to ionizing radiation without the expectation of an overall benefit from the activity causing the exposure." An obvious difficulty in making this judgment is the difficulty of quantifying in comparable terms costs (including risks) and benefits. Given this situation, informed value judgments are necessary and are usually all that is possible. . .
>
> The principle of reduction of exposure to levels that are 'as low as reasonably achievable' (ALARA) is typically implemented in two different ways. First, it is applied to the engineering design of facilities so as to reduce, prospectively, the anticipated exposure of workers. Second, it is applied to actual operations; that is, work practices are designed and carried out to reduce the exposure of workers. Both of these applications are encompassed by these recommendations. The principle applies both to collective exposures of the work force and to annual and cumulative individual exposures. Its application may therefore require complex judgments, particularly when tradeoffs between collective and individual doses are involved. Effective implementation of the ALARA principle involves

---

[9]These guides, for example, require plant design to limit whole-body dose equivalents to individuals in unrestricted areas to 0.005 cSv (0.005 rem) in one year from gaseous radioactive effluents. The guides also prescribe, in the form of a numerical cost-benefit ratio, introduction of radioactive waste treatment systems to reach a marginal cost of $1000 per reduction of integrated population whole-body dose-equivalent by one person-rem.

most of the many facets of an effective radiation protection program: education of workers concerning the health risks of exposure to radiation; training in regulatory requirements and procedures to control radiation exposure; monitoring, assessment, and reporting of exposure levels and doses; and management and supervision of radiation protection activities, including the choice and implementation of radiation control measures. A comprehensive radiation protection program will also include, as appropriate, properly trained and qualified radiation protection personnel; adequately designed, operated, and maintained facilities and equipment; and quality assurance and audit procedures. Another important aspect of such programs is maintenance of records of cumulative exposures of workers and implementation of appropriate measures to assure that lifetime exposures of workers repeatedly exposed near the limits is minimized.

Secondary numerical standards for radiation protection are, if not endorsed, at least acknowledged in the U.S. federal radiation protection guidance, which identifies as appropriate for certain regulatory authorities:

(1) administrative control levels specifying, for specific categories of workers or work situations, dose levels below the limiting numerical values recommended in this guidance; (2) reference levels to indicate the need for such actions as recording, investigation, and intervention; and (3) local goals for limiting individual and collective occupational exposures.

1.4.8 The ICRP Conceptual Framework for Radiation Protection

In their 1990 recommendations for radiation protection the ICRP sets forth a comprehensive framework for radiation protection. The recommendations are made with the goals of preventing deterministic effects and minimizing stochastic effects of ionizing radiation. They are presented in the context of three general principles, three types of exposure, three exposure situations, and three points of action. The three types of exposure are (1) occupational, (2) medical, and (3) public. The three exposure situations are (1) planned exposure situations, that is, new practices under consideration which, for reasons of radiation risks, may or may not be acceptable, (2) pre-existing exposure situations, that is, situations for which risks would continue even though a practice were discontinued, and (3) potential exposure situations, that is, situations for which personnel exposure, though unlikely, is possible, and for which controls may affect probabilities or magnitudes of exposure. The three points of action are (1) at the source of the radiation or radioactive materials, (2) in the environment through which radionuclides are transported, and (3) at the individual who may be exposed. The three general principles are

justification of a practice: adopt no practice or intervention procedure unless benefits to exposed individuals exceeds radiation induced detriments,

optimization of protection: for any source within a practice, keep exposures as low as reasonably achievable, social factors being taken into account, but constrained by restrictions on risks to individuals so as to limit inequities,

TABLE 1.10   Dose-Limit Recommendations Adopted by the ICRP (1991).

CLASSIFICATION	OCCUPATIONAL LIMIT	PUBLIC LIMIT
Effective dose equivalent	100 mSv (10 rem) in 5 years 50 mSv (5 rem) in any 1 year	1 mSv (0.1 rem) per year averaged over any consecutive 5 years
Annual dose equivalent		
lens of the eye	150 mSv (15 rem)	15 mSv (1.5 rem)
skin (< 100 cm^2) or hands	500 mSv (50 rem)	50 mSv (5 rem)
Mean fetal dose equivalent	5 mSv after diagnosis of pregnancy	

> **setting dose limits**:   ensure that no individual is deliberately exposed to radiation risks judged to be unacceptable in any normal circumstances.

(a) Potential and Planned Exposure Situations.   Once a practice has been justified as being beneficial, it is necessary to optimize the use of resources in reducing radiation risks. The optimization process might well attempt to balance society's risks and benefits in an overall sense. However, certain individuals or groups might well experience inequitably high risks under a contemplated practice unless additional constraints were placed on doses or on risks (in the case of potential exposures).

Dose limits in occupational exposure.   The Commission set out to establish dose limits which would be independent of the age and sex of the worker and which would take into account the following attributes of risk: (1) lifetime attributable probability of death, (2) time lost in the event of attributable death, (3) reduction of life expectancy, (4) annual distribution of attributable probability of death, and (5) increase in age-specific mortality rate. After examining risks for various dose-limit schemes, the Commission concluded that neither lifetime limits nor rigid one-year limits were appropriate. They instead recommended limits on committed or received effective dose equivalent for five-year periods, with only *over-riding limits* for one year periods. They found that prevention of mechanistic effects of radiation required separate limits only for the lens of the eye and the skin. The limits are summarized in Table 1.10. With occupational exposure at an annual rate of 20 mSv beginning at age 18, it was reckoned that the resulting lifetime probability of death is 0.036 — an annual average probability of death of about 1 in 1300.

Dose limits for pregnant workers.   While the dose limits recommended for occupational exposure make no gender distinction, the Commission recommended that, in the event that pregnancy is diagnosed, the dose to the fetus should be limited to 5 mSv during the remainder of the pregnancy, and to 1 mSv during the critical 8 to 15 weeks after conception.

Dose limits for public exposure.   Dose limits for the public are recommended only for planned exposure situations. The considerations taken into account in setting occupational-dose limits were also applied to public-dose limits, but it was more difficult to identify the threshold between tolerable and intolerable exposures. Therefore, as an alternative approach, consideration was also given to the variations in natural background exposure experienced by different segments of the public. The result was an annual limit of 1 mSv which is comparable to natural exposure and to variations in natural exposure. Believing that an annual limit would be too restrictive, the Commission recommended a limit of 1 mSv per year, averaged over any five-year period. This, and related limits, are summarized in Table 1.10.

Dose limits for medical exposure.   While justification and optimization were recognized to be important in the evolution of medical-exposure procedures, the Commission recommended that no limits be placed on exposure of patients undergoing diagnosis or therapy. They did recommend, though, that diagnostic exposure of pregnant or potentially pregnant patients be avoided if possible.

(b) Pre-Existing Exposure Situations.   In such situations, the sources, environmental pathways, and exposed individuals are in place at the time decisions must be made about control measures. Examples include abnormally high natural-radiation exposure and contamination of the public environment from accidents. The principle of justification applies to decisions on whether or not to intervene with control measures. The Commission stressed that dose limits are *not* appropriate for pre-existing exposure situations except for workers dealing with emergencies. They recommend that, except for life-saving actions, effective dose equivalents be limited to 0.5 Sv (5 Sv skin dose equivalent) for those workers conducting immediate and urgent remedial work.

1.4.9 United States Radiation Protection Regulations

In Sec. 1.4.3 it was pointed out that the radiation-protection regulations administered by the Nuclear Regulatory Commission in the 1960s, 1970s, and 1980s were based largely on the 1959 recommendations of the NCRP and the ICRP. Revised regulations, scheduled to go into effect in 1994, are based largely on the methodology of the 1977 and 1990 ICRP recommendations and the limits expressed in the 1987 NCRP recommendations. For example, the new U.S. regulations (Title 10, Code of Federal Regulations, Part 20) give the following annual limits for occupational exposure: the more restrictive of (1) 5 rem (0.05 Sv) for the sum of the committed effective dose equivalent for ingested or inhaled radionuclides and the deep dose equivalent for external exposure, or (2) 50 rem (0.5 Sv) for the sum of the committed dose equivalent to any organ and the deep dose equivalent. The first of these limits (5 rem) is associated with stochastic effects of radiation, the second (50 rem) with mechanistic effects. The doses to the skin and lens of the eye are limited respectively to 50 and 15 rem annually. These dose units — the committed dose equivalent, committed effective dose equivalent, and dose equivalent index — as well as stochastic and mechanistic effects are addressed in subsequent chapters of this text.

1.5 INSTITUTIONAL CONNECTIONS

Persons with professional interest in radiation protection are confronted with an array of statutes, regulations, codes of practice, regulatory guides, recommendations, guidelines, and procedures, not to mention sanctions which may be imposed if bad advice is mistaken for good or requirements are interpreted as recommendations. All this well-intentioned guidance issues from national and international institutions with missions and authorities as confusing as their acronyms. The following is an attempt to help the reader understand the roles of the various institutions. Emphasis is placed on U.S. institutions, not out of national pride, but because the authors are more familiar with the rich traditions of institutional bureaucracy in the United States.

1.5.1 Independent Organizations

Much attention has been given in this chapter to two international and one U.S. organization: the International Commission on Radiation Units (ICRU), the International Commission on Radiological Protection (ICRP), and the National Council on Radiation Protection and Measurement (NCRP). These organizations have enormous influence on radiation-protection standards, not because of any power or authority, but because of their independence and their long-recognized scientific preeminence.

With equal scientific credibility, but with slightly different roles are organizations such as the National Research Council (NRC) of the National Academy of Sciences (NAS) in the United States. Of special note is the NAS/NRC Committee on the Biological Effects of Ionizing Radiation (BEIR) which periodically issues comprehensive reports. While NCRP and ICRP activities are more closely related to standards for protection, those of the BEIR Committee are more closely associated with evaluation and interpretation of scientific research on the biological effects of ionizing radiation.

Independent international and national industrial standards organizations become involved in radiation protection through their publication of specific codes of practice and procedures. Activities of the standards organizations are typically integrated with those of professional and trade organizations.

1.5.2 United Nations Organizations

One UN organization has had far-reaching influence on the understanding of radiation hazards. The United Nations Scientific Committee on the Effects of Atomic Radiation (UNSCEAR), established in 1955, reports yearly to the General Assembly and periodically issues comprehensive reports on radiation hazards.

Several other UN organizations have influence on the practice of radiation protection. Among these are the International Atomic Energy Agency (IAEA), the World Health Organization (WHO) and the Food and Agriculture Organization (FAO). Established in 1957, the IAEA has as its principal objective "to accelerate and enlarge the contribution of atomic energy to peace, health, and prosperity throughout the world." The WHO, created in 1948, has among other goals the promotion of international cooperation in collection and dissemination of health statistics and establishing international health standards. Along with the FAO, and the Joint FAO/WHO Codex Alimentarius Commission, the WHO is involved with establishing codes of practice for the operation of radiation facilities for the preservation of foods.

1.5.3 Governmental Organizations in the United States

The Atomic Energy Act of 1954 gave exclusive authority to an Atomic Energy Commission (AEC) to exercise federal regulation on the use, transportation, and disposal of radioactive materials used in or produced by the nuclear fission process. The authority did not extend to naturally occurring radionuclides, except for uranium and thorium as source materials for nuclear fuels. The first AEC radiation-protection regulations (10CFR20) were published in 1955 and became effective in 1957. The Energy Reorganization Act of 1974 gave rise to a Nuclear Regulatory Commission (NRC) to which were transferred the regulatory functions of the AEC. Other functions of the AEC devolved to the Energy Research and Development Administration (ERDA), later given Cabinet status as the Department of Energy (DOE), which has responsibility for radiation protection and safety at national laboratories.

 The AEC, then NRC, has been permitted since 1959 to delegate to the individual states the responsibility and licensing authority for all regulatory activity save that for nuclear reactors. These "agreement states," which number about half those in the union, are required to maintain radiation protection standards at least as strict and comprehensive as the federal standards.

 During the period 1959 to 1970, the responsibility for establishing standards for radiation protection was assigned to an interagency Federal Radiation Council (FRC). Implementation of the standards, through regulations and enforcement practices was the responsibility of the competent federal agency. The Department of Health, Education and Welfare (HEW) was assigned the primary responsibility for collation, analysis, and interpretation of environmental radiation levels, including medical use of isotopes and x rays. In 1970, the authority of the FRC, along with certain surveillance responsibilities of the Public Health Service (PHS), were transferred to the Environmental Protection Agency (EPA). The Atomic Energy Commission retained the responsibility for implementation and enforcement of radiation standards through its licensing authority. The regulation of radiation from consumer products remained with HEW, now HHS, the Department of Health and Human Services. However, certain functions of the Food and Drug Administration (FDA) Bureau of Radiological Health were transferred to the EPA. As a result of the Orphan Drug Act of 1983 (Public Law 97-414), the National Institutes of Health (NIH) of HHS has had the responsibility for preparation of *probability-of-causation* tables for adjudication of claims of radiation carcinogenesis.

1.6 ADDITIONAL SOURCES OF INFORMATION

Included with this chapter is a bibliography of textbooks, reference books, and handbooks of interest to specialists in radiation protection. In addition, there are the following series of publications:

Standards and Guidelines

Annals of the International Commission on Radiological Protection, available from Pergamon Press (U.S. address: Maxwell House, Fairview Park, Elmsford, N.Y. 10523).

Reports of the International Commission on Radiological Units, available in the United States from ICRU Publications, 7910 Woodmont Ave., Suite, 1016, Bethesda, Md. 20814.

Reports of the National Council on Radiation Protection and Measurements, available in the United States from NCRP Publications, 7910 Woodmont Ave., Suite, 1016, Bethesda, Md. 20814.

NM/MIRD Pamphlets — published by the Medical Internal Radiation Dose Committee of the Society of Nuclear Medicine, 136 Madison Ave., New York, N.Y. 10016.

Regulatory Guides, especially Division 8 dealing with occupational health, published by the U.S. Nuclear Regulatory Commission, Washington, D.C. 20555.

U.S. Code of Federal Regulations, Title 10: Energy, especially Part 20 (10CFR20): Standards for Protection Against Radiation, published by the U.S. Government Printing Office, Washington, D.C. 20402.

American National Standards, published by various professional societies, for example, the American Nuclear Society, 555 North Kensington Ave., La Grange Park, Ill. 60525, and the Health Physics Society, 8000 Westpark Dr., Suite 130, McLean, Va. 22011.

UNSCEAR, IAEA, FAO, and WHO publications, available from United Nations Publications, Room DC2-853, Department 701, New York, N.Y. 10017.

Textbooks and Monographs on Radiation Protection

Cember, H., *Introduction to Health Physics*, (2d ed.), Pergamon Press, Elmsford, N.Y., 1983.

Chilton, A.B, J.K. Shultis, and R.E. Faw, *Principles of Radiation Shielding*, Prentice-Hall, Englewood Cliffs, N.J., 1984.

Eisenbud, M., *Environmental Radioactivity*, (3d ed.), Academic Press, Orlando, Fl., 1987.

Morgan, K., and J.E. Turner, *Principles of Radiation Protection*, John Wiley & Sons, New York, 1967.

Profio, A.E., *Radiation Shielding and Dosimetry*, John Wiley & Sons, New York, 1979.

Shapiro, J., *Radiation Protection*, (3d ed.), Harvard University Press, Cambridge, Mass., 1990.

Turner, J.W., *Atoms, Radiation, and Radiation Protection*, Pergamon Press, Orlando, Fl., 1986.

Textbooks and Monographs on Radiation Detection and Measurement

Attix, F.H., and W.C. Roesch, (eds.), *Radiation Dosimetry*, (2d. ed.), Vols. I-III, Academic Press, New York, 1966.

Chiang, H.H., *Electronics for Nuclear Instrumentation*, Krieger Publishing Co., Malabar, Fl., 1985.

Eicholz, G.G., and J. W. Poston, *Principles of Nuclear Radiation Detection*, Ann Arbor Science Publishers, Ann Arbor, Mich., 1979.

Kowalski, E., *Nuclear Electronics*, Springer-Verlag, New York, 1970.

Knoll, G.F., *Radiation Detection and Measurement*, (2d ed.) John Wiley and Sons, New York, 1989.

Nicholson, P.W., *Nuclear Electronics*, John Wiley & Sons, London, 1974.

Tsoulfanidis, N., *Measurement and Detection of Radiation*, McGraw-Hill (Hemisphere), New York, 1983.

Reference Books with Emphasis on Nuclear Data

Browne, E., and R.B. Firestone, *Table of Radioactive Isotopes*, John Wiley and Sons, New York, 1986.

ICRP Publication 38, *Radionuclide Transformations: Energy and Intensity of Emissions*, Annals of the ICRP, Vols. 11–13, Pergamon Press, Elmsford, N.Y., 1983.

Jaeger, R.G., (ed), *Engineering Compendium on Radiation Shielding*, Vols. I-III, Springer-Verlag, New York, 1968–1975.

Kocher, D.C., *Radioactive Decay Tables*, Report DOE/TIC-11026, National Technical Information Service, Springfield, Va., 1981.

Nuclear Data Tables, published by Academic Press, Orlando, Fl., 32887.

Weber, D.A., K.F. Eckerman, L.T. Dillman, and J.C. Ryman, *MIRD: Radionuclide Data and Decay Schemes*, Society of Nuclear Medicine, New York, 1989.

Handbooks

The following are published by CRC Press, Boca Raton, Florida.

Brodsky, A.B., (ed.), *Handbook of Radiation Measurement and Protection*, Vols. I-II, 1978.

Kereiakes, J.G., and M. Rosenstein, *Handbook of Radiation Doses in Nuclear Medicine and Diagnostic X-Ray*, 1980.

Klement, A.W., (ed.) *Handbook of Environmental Radiation*, 1982.

Waggener, R.G., J.G. Keriakes, and R. J. Shalek, (eds.), *Handbook of Medical Physics*, 1982.

Wang, Y., (ed.), *Handbook of Radioactive Nuclides*, 1969.

The following are published by Adam Hilger, Ltd., Bristol, U.K., as part of a Medical Physics Handbooks series.

Greening, J.R., *Fundamentals of Radiation Dosimetry*, (2d ed.), 1985.

Kathren, R.L., *Radiation Protection*, 1985

McKinlay, A.F., *Thermoluminescence Dosimetry*, 1981.

Mould, R.F., *Radiation Treatment Planning*, (2d ed.), 1985.

Institutions

An institution of enormous benefit to the radiological assessment specialist is the Radiation Shielding Information Center (RSIC). This Center maintains a comprehensive collection of literature and computer programs. It may be addressed at P.O. Box 2008, Oak Ridge National Laboratory, Oak Ridge, Tenn. 37831-6362. Nuclear data may be obtained via on-line computer from the National Nuclear Data Center, Brookhaven National Laboratory, Upton, N.Y. 11973.

Professional societies serving the radiological assessment community include the American Association of Physicists in Medicine, the American Nuclear Society, and the Health Physics Society in the United States, and counterpart societies in other countries. Professional journals of special interest to the radiological assessment community include *Health Physics*, *Journal of Nuclear Medicine*, *Medical Physics*, *Nuclear Science and Engineering*, *Nuclear Applications*, *Radiation Protection Dosimetry*, and *Radiology*.

REFERENCES

EISENBUD, M., *Environmental Radioactivity*, 3d ed., Academic Press, New York, 1987, Chap. 3.

ICRP, "Recommendations of the International Commission on Radiological Protection," *Brit. J. Radiol.* Supplement 6, 1955. [Meeting of the International Congress of Radiology held in Copenhagen, Denmark, July 1953.]

ICRP, "Recommendations of the International Commission on Radiological Protection, 9 Sep 1958," Pergamon Press, London, 1958.

ICRP, *Report of Committee II on Permissible Dose for Internal Radiation*, Publication 2, Recommendations of the International Commission on Radiological Protection, Pergamon Press, Oxford, 1960.

ICRP, *Recommendations of the International Commission on Radiological Protection (As Amended 1959 and Revised 1962)*, Publication 6, International Commission on Radiological Protection, Pergamon Press, Oxford, 1964.

ICRP, *Radiation Protection: Recommendations of the International Commission on Radiological Protection, Adopted Sept. 17, 1965*, Publication 9, International Commission on Radiological Protection, Elmsford, N.Y., 1966.

ICRP, *ICRP Task Group on Dose Limits Report to Main Commission*, 1972. [As reported by Taylor (1979).]

ICRP, *Report of the Task Group on Reference Man*, Publication 23, International Commission on Radiological Protection, Pergamon Press, Oxford, 1975.

ICRP, *Recommendations of the International Commission on Radiological Protection*, Publication 26, International Commission on Radiological Protection, Annals of the ICRP 1, No. 3, Pergamon Press, Oxford, 1977.

ICRP, *Limits for Intakes of Radionuclides by Workers*, Publication 30, International Commission on Radiological Protection, Annals of the ICRP 2, No. 3/4, Pergamon Press, Oxford, 1979.

ICRP, *Cost-Benefit Analysis in the Optimization of Radiation Protection*, Publication 37, International Commission on Radiological Protection, Annals of the ICRP 10, Pergamon Press, Oxford, 1983.

ICRP, *Statement from the 1985 Paris Meeting of the International Commission on Radiological Protection*, Publication 45, International Commission on Radiological Protection, Pergamon Press, Oxford, 1985.

ICRP, *Data for use in Protection against External Radiation,* Publication 51, International Commission on Radiological Protection, Annals of the ICRP 17, No. 2/3, Pergamon Press, Oxford, 1987.

ICRP, *Recommendations of the Commission — 1990,* Draft Report ICRP/90/G-01. International Commission on Radiological Protection, Feb. 1990.

ICRP, *1990 Recommendations of the International Commission on Radiological Protection*, Publication 60, International Commission on Radiological Protection, Annals of the ICRP 21, No. 1-3, Pergamon Press, Oxford, 1991.

ICRU, *Physical Aspects of Irradiation: Recommendations of the International Commission on Radiation Units and Measurements*, Report 10b, International Commission on Radiation Units and Measurements, Washington D.C., 1962. [Published as NBS Handbook 85, National Bureau of Standards, Washington, D.C., 1964.]

ICRU, *Average Energy Required to Produce an Ion Pair*, Report 31, International Commission on Radiation Units and Measurements, Washington D.C., 1979.

ICRU, *Radiation Quantities and Units*, Report 33, International Commission on Radiation Units and Measurements, Washington, D.C., 1980.

ICRU, *The Quality Factor in Radiation Protection,* Report 40, International Commission on Radiation Units and Measurements, Washington D.C., 1986.

MORGAN, K.Z., "Maximum Permissible Exposure Levels — External and Internal," in *Principles of Radiation Protection*, K.Z. Morgan and J.E. Turner (eds.), John Wiley & Sons, New York, 1967, Chap. 14.

MUTSCHELLER, A., "Physical Standards of Protection Against Roentgen-ray Dangers," *Am. J. Roent. Rad. Ther. 13*, 65 (1925).

NCRP, *Safe Handling of Radioactive Luminous Compounds*, Handbook 27, National Bureau of Standards, Washington D.C., 1941.

NCRP, *Maximum Permissible Amounts of Radioisotopes in the Human Body and Maximum Permissible Concentrations in Air and Water*, Handbook 52, National Bureau of Standards, Washington, D.C., 1953.

NCRP, *Permissible Dose from External Sources of Ionizing Radiation, Recommendations of the National Committee on Radiation Protection*, Report 17, National Committee on Radiation Protection, Handbook 59, National Bureau of Standards, Washington, D.C., 1954.

NCRP, *Maximum Permissible Radiation Exposures to Man. A Preliminary Statement of the National Committee on Radiation Protection and Measurement*, Addendum to Handbook 59, National Bureau of Standards, Washington, D.C., 1958.

NCRP, *Maximum Permissible Body Burdens and Maximum Permissible Concentrations of Radionuclides in Air and in Water for Occupational Exposure*, Report 22, National Committee on Radiation Protection, Handbook 69, National Bureau of Standards, Washington, D.C., 1959.

NCRP, *Basic Radiation Protection Criteria, Recommendations of the National Council on Radiation Protection and Measurements*, Report 39, National Council on Radiation Protection and Measurements, Washington, D.C., 1971a.

NCRP, *Protection against Neutron Radiation, Recommendations of the National Council on Radiation Protection and Measurements*, Report 38, National Council on Radiation Protection and Measurements, Washington, D.C., 1971b.

NCRP, *Recommendations on Limits for Exposure to Ionizing Radiation*, Report 91, National Council on Radiation Protection and Measurements, Washington, D.C., 1987.

SIEVERT, R.M., "Einige Untersuchengenüber Vorrichtungen zum Schutz Gegan Rontgenstrahlen," *Acta Radiol. 4*, 61-65 (1925).

SNYDER, W.S., "Internal Exposure," in *Principles of Radiation Protection*, K.Z. Morgan and J.E. Turner (eds.), John Wiley & Sons, New York, 1967, Chap. 10.

STANNARD, J. NEWELL, *Radioactivity and Health, a History*, R.W. Baalman, Jr. (ed), Report DOE/RL/01830-T59, Available from the National Technical Information Service, Springfield, Va., 1988.

TAYLOR, L.S., *Organization for Radiation Protection. The Operations of the ICRP and NCRP, 1928-1974*, Report DOE/TIC-10124, National Technical Information Services, Springfield, Va. 22161, 1979.

TAYLOR, L.S., "The Development of Radiation Protection Standards (1925–1940)," *Health Physics 41*, 227–232 (1981).

TAYLOR, L.S., "The Problems of Radiation Double Standards: Exposure of Potentially Pregnant Persons," *Health Physics 49*, 1043–1052 (1985).

U.S. GOVERNMENT, *Radiation Protection Guidance to Federal Agencies for Occupational Exposure*, Federal Register, Vol. 52, No. 17, 27 Jan. 1987.

WHYTE, G.N., *Principles of Radiation Dosimetry*, Wiley, New York, 1959.

Chapter 2

Radiation Interactions and Response Functions

2.1 INTERACTION COEFFICIENT

The interaction of a given type of radiation with matter may be classified according to the type of interaction and the unit of matter with which the interaction takes place. The interaction may take place with an electron and in many cases the electron behaves as though it were free. Similarly, the interaction may take place with an atomic nucleus which, in many cases, behaves as though it were not bound in a molecule or crystal lattice. However, in some cases, particularly for radiation particles of comparatively low energy, molecular or lattice binding must be taken into account.

The interaction may be a scattering of the incident radiation — deflection accompanied by energy change. A scattering interaction may be elastic or inelastic. Consider, for example, the interaction of a photon with an electron in what is called Compton scattering. In the sense that the interaction is with the entire atom within which the electron is bound, the interaction must be considered as inelastic, since some of the incident photon's energy must compensate for the binding energy of the electron in the atom. However, in most practical cases electron binding energies are orders of magnitude lower than gamma-photon energies, and the interaction may be treated as purely elastic scattering of the photon by a free electron. Neutron scattering by an atomic nucleus may be elastic, in which case the incident neutron's kinetic energy is shared by that of the scattered neutron and that of the recoil nucleus, or it may be inelastic, in which case some of the incident neutron's kinetic energy is transformed to internal energy of the nucleus and thence to a gamma ray emitted from the excited nucleus. It is important to note that for both elastic and inelastic scattering there are unique relationships between energy exchanges and the angles of scattering. These arise from conservation of energy and momentum.

Other types of interactions are absorptive in nature. The identity of the incident particle is lost, and total relativistic momentum and energy are conserved, some appearing as nuclear excitation energy, some as translational, vibrational, and rotational energy. The ultimate result may be the emission of particulate radiation as occurs in the photoelectric effect and in neutron radiative capture.

The interaction of radiation with matter is always statistical in nature, and therefore must be described in probabilistic terms. Consider a particle traversing a homogeneous material and let P denote the probability that this particle interacts in some specified manner (e.g., absorption or scattering) while traveling a distance Δx. It is found that the quantity $\mu \equiv \lim_{\Delta x \to 0}(P/\Delta x)$ is a property of the material for a given interaction. Here the limit process must be interpreted in the manner described in Sec. 1.3.2. In the limit of small path lengths, μ is seen to be the probability per unit differential path length that a particle will undergo a specified interaction. That μ is constant for a given material and for a given type of interaction implies that the probability of interaction per unit differential path length is independent of the path length traveled prior to the interaction.

The interaction probability per unit differential path length is fundamental in describing how radiation interacts with matter and is usually called the *linear attenuation coefficient* (ICRU 1980). It is perhaps more appropriate to use the words *linear interaction coefficient* since many interactions do not "attenuate" the particle in the sense of an absorption interaction. Although this nomenclature is widely used to describe photon interaction, μ is often referred to as the *macroscopic cross section* and given the symbol Σ when describing neutron interactions.

The utility of the linear interaction coefficient in describing interaction of radiation with matter becomes apparent when one interprets the flux density ϕ as the path length traveled by particles in a unit volume in a unit time in the limit of a very small volume. The product $\mu\phi$ is thus seen to be the number of interactions per unit volume per unit time — the volumetric interaction rate. Division by the material density yields $(\mu/\rho)\phi$, the interaction rate per unit mass.

The linear interaction coefficient is a function of the energy of the particle. Depending on the nature of the interaction, it may also be a function of (1) the energy of the particle after scattering, (2) the energy of the recoil atom or electron, (3) the angles of deflection of the scattered radiation and recoil atom or electron, and (4) angles of emission of secondary particles. For example, the scattering interaction coefficient is usually defined in such a way that $\mu(E, E', \theta_s)dE'd\Omega$ is the probability per unit path length for an interaction in which the incident particle of energy E emerges from the interaction with energy between E' and $E' + dE'$ and with scattering angle θ_s, measured with respect to the incident direction, within the differential solid angle $d\Omega$. In this form, μ would have units such as $cm^{-2} \, MeV^{-1} \, sr^{-1}$. Alternatively, the interaction coefficient could be expressed in terms of energies and angles for recoil electrons or atoms, or secondary radiations.

Often it is of interest to deal only with, say, the energy dependence or the angular distribution of scattered radiation. In this case, one or the other of the following forms of the linear interaction coefficient may be used:

$$\mu(E, E') \equiv \int_{4\pi} d\Omega \, \mu(E, E', \theta_s), \qquad (2.1)$$

or

$$\mu(E, \theta_s) \equiv \int_0^\infty dE' \, \mu(E, E', \theta_s). \qquad (2.2)$$

Here $\mu(E, E')dE'$ is the probability per unit differential path length for scattering into dE' at E' without regard to scattering angle and $\mu(E, \theta_s)d\Omega$ is the like probability for scattering into direction range $d\Omega$ without regard to the energy of the exit radiation. Also of interest is

$$\mu(E) \equiv \int_0^\infty dE' \, \mu(E, E'), \qquad (2.3)$$

which is just the total linear interaction coefficient for the scattering of incident radiation of energy E, without regard to energy loss or angle of scattering. To indicate a specific type of interaction, the symbol μ will be be given a subscript, for example, μ_s for scattering.

2.2 MICROSCOPIC CROSS SECTION

The linear interaction coefficient depends on the type and energy of the incident particle, the type of interaction, and the composition and density of the interacting medium. One of the more important quantities that determine μ is the density of target atoms or electrons in the material. It seems reasonable to expect that μ should be proportional to the "target" atom or electron density N in the material, that is, $\mu = \sigma N$, where σ is a constant of proportionality independent of N.[1] The quantity σ is called the *microscopic cross section* and is seen to have dimensions of area. It is often interpreted as being the effective cross-sectional area presented by the target atom or electron to the incident particle for a given interaction. Indeed, in many cases σ has dimensions comparable to those expected from the physical size of the nucleus. However, this simplistic interpretation of the microscopic cross section, although conceptually easy to grasp, leads to philosophical difficulties when it is observed that σ generally varies with the energy of the incident particle and, for a crystalline material, the particle direction. The view that σ is the interaction probability per unit differential path length, normalized to one target atom or electron per unit volume, avoids such conceptual difficulties while emphasizing the statistical nature of the interaction process.

When the microscopic cross section is for the electron, the symbol σ is preceded by the subscript e; otherwise, no subscript is used. Cross sections are usually expressed in units of cm^2. A widely used special unit is the *barn*, having dimensions 10^{-24} cm^2.

Just as the linear interaction coefficients may be differential in nature, so also may be the corresponding microscopic cross sections. In this text, these differential

---

[1]This proportionality is not strictly true for coherent interactions in which the incident particle interacts collectively with multiple target atoms or electrons, as in interactions with orbital electrons of a single atom or with atoms of a crystal lattice. Even for such cases, though, the microscopic cross section is usually defined in this manner.

microscopic cross sections will be denoted by $\sigma(E, E', \theta_s)$, $\sigma(E, E')$, and $\sigma(E, \theta_s)$, which are related by integral expressions analogous to those of Eqs. (2.1) to (2.3). The reader should be aware of alternative forms of notation; for example, $\sigma(E, E', \theta_s) \equiv d^2\sigma/dE'd\Omega$, and the latter form is used in some texts. Sometimes, to emphasize the transition from one energy E to another E', or from one direction Ω to another Ω', the notation $E \to E'$ or $\Omega \to \Omega'$ is used. The following are samples of identical notations: $\sigma(E \to E') \equiv \sigma(E, E')$ and $\sigma(E \to E', \Omega \to \Omega') \equiv \sigma(E, E', \Omega \cdot \Omega') \equiv \sigma(E, E', \omega_s)$, where $\omega_s \equiv \Omega \cdot \Omega' = \cos\theta_s$. All these alternative forms are used in this text.

Data on cross sections and linear interaction coefficients, especially for photons, are frequently expressed as the ratio of μ to the density ρ, called the *mass interaction coefficient*. Since the atomic density N is

$$N = \frac{\rho}{A} N_a, \tag{2.4}$$

in which N_a is Avogadro's number (6.022×10^{23} mol^{-1}), and A is the atomic weight (g mol^{-1}),

$$\frac{\mu}{\rho} = \frac{N\sigma}{\rho} = \frac{N_a}{A}\sigma. \tag{2.5}$$

Thus μ/ρ is an intrinsic property of the interacting medium — independent of its density. This method of data presentation is used much more for photons than for neutrons, in part because, for a wide variety of materials and a wide range of photon energies, μ/ρ is only weakly dependent on the elements of the interacting medium.

For compounds or homogeneous mixtures with, as is the usual case, negligible molecular or crystalline binding effects, the linear and mass interaction coefficients are, respectively,

$$\mu = \sum_i \mu_i = \sum_i N_i \sigma_i, \tag{2.6}$$

and

$$\frac{\mu}{\rho} = \sum_i w_i \left(\frac{\mu}{\rho}\right)_i, \tag{2.7}$$

in which w_i is the weight fraction of component i. In Eq. (2.6), the atomic density N_i and the linear interaction coefficient μ_i are values for the ith material *after* mixing.

2.3 CROSS SECTIONS FOR PHOTON INTERACTIONS

2.3.1 Classification of Types of Interactions

For details of the mechanisms of photon interactions, the reader is referred to the standard reference works of Heitler (1954) and Evans (1955). For many radiological assessment studies, photon energies of 10 keV to 10 MeV are important. For this energy range, the photoelectric effect, pair production, and Compton scattering mechanisms of interaction predominate. Of these three, the photoelectric effect predominates at the lower photon energies. Pair production is important only for higher energy photons. Compton scattering predominates at intermediate energies. In rare instances one may need to account also for coherent (Rayleigh) scattering.

2.3.2 Compton Scattering

The process of Compton scattering is explained with the help of Fig. 2.1. A photon of energy E interacts with an electron initially at rest. After the interaction, the photon has energy E' and moves at angle θ_s measured from its initial direction. The electron moves with kinetic energy T at angle θ_e measured from the initial direction of the photon. Application of the principles of conservation of relativistic total energy and linear momentum leads to the well-known Compton formula,

$$E' = \frac{E}{1 + (E/m_e c^2)(1 - \cos\theta_s)},\qquad (2.8)$$

in which $m_e c^2$ is the rest mass energy of the electron, 8.186×10^{-14} J (0.5110 MeV). Photon transport calculations are frequently carried out using wavelength, rather than energy, as an independent variable. Furthermore, to avoid carrying along various constants, the wavelength is usually measured in multiples of a *Compton unit*, the ratio of the wavelength hc/E to the so-called Compton wavelength $\lambda_c (= hc/m_e c^2 = 2.426 \times 10^{-10}$ cm), where h is Planck's constant, 6.624×10^{-34} J s, and c is the speed of light, 2.998×10^{10} cm s^{-1}. In this unit,

$$\lambda \equiv \frac{m_e c^2}{E}.\qquad (2.9)$$

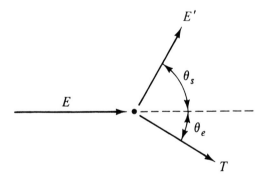

Figure 2.1   Illustration of the Comption scattering process. E and E' are photon energies before and after scattering at angle θ_s. The electron, initially stationary, recoils at an angle θ_e and receives kinetic energy T.

With the use of Compton units, the Compton scattering relationship, Eq. (2.8), is quite simple

$$\lambda' = \lambda + 1 - \cos\theta_s.\qquad (2.10)$$

The kinetic energy T of the recoil electron is given by

$$\frac{m_e c^2}{T} = \lambda \left(1 + \frac{\lambda}{1 - \cos \theta_s}\right). \tag{2.11}$$

The scattering angles of the electron and photon are related by the equation

$$\cot \theta_e = (1 + \lambda^{-1}) \tan(\theta_s/2). \tag{2.12}$$

The maximum transfer of energy from the incident photon to the recoil electron occurs when $\theta_s = \pi$ and $\theta_e = 0$, for which case

$$(\lambda' - \lambda)_{\text{max}} = 2, \tag{2.13}$$

$$\left(\frac{E'}{E}\right)_{\text{min}} = \frac{\lambda}{\lambda + 2}, \tag{2.14}$$

and

$$\left(\frac{T}{E}\right)_{\text{max}} = \frac{2}{\lambda + 2}. \tag{2.15}$$

The relationships among T, E, θ_s, and θ_e are illustrated in Figs. 2.2 and 2.3.

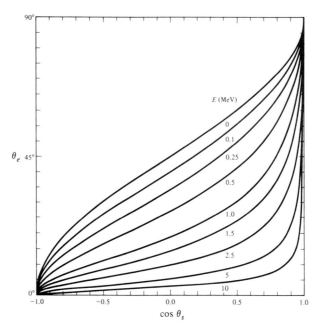

Figure 2.2 Relationship between the angle of electron scattering θ_e (degrees), and the cosine of the angle of photon scattering in the Compton process.

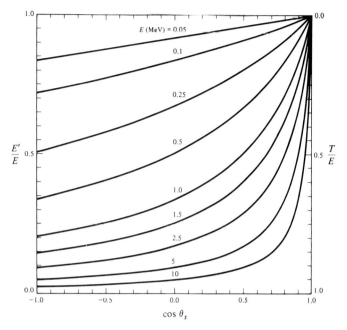

Figure 2.3   Relationship between the energy E' of the scattered photon, the energy T of the scattered electron, and the initial photon energy E as a function of the cosine of the photon scattering angle in the Compton process.

Thomson cross section.   In the limit of zero photon energy, scattering of the photon by a free electron may be treated by the classical theory of radiation. The electron in the electromagnetic field of the incident radiation vibrates with the same frequency as that of the incident radiation, thereby giving rise to the emission of secondary electromagnetic radiation of the same frequency. The total cross section for such scattering, called Thomson scattering, is

$$_e\sigma_T = \frac{8}{3}\pi r_e^2, \tag{2.16}$$

in which r_e is the "classical electron radius." The value of r_e is given by

$$r_e = \frac{e^2}{4\pi\epsilon_o m_e c^2}, \tag{2.17}$$

where e is the electronic charge, 1.602×10^{-19} C, and ϵ_o is the permittivity of free space, 8.854×10^{-14}F cm^{-1}. Thus, $r_e = 2.818 \times 10^{-13}$ cm and $_e\sigma_T = 6.653 \times 10^{-25}$cm^2.

For unpolarized incident radiation, the Thomson cross section per steradian for scattering at angle θ_s is

$$_e\sigma_T(\theta_s) = \frac{1}{2}r_e^2(1 + \cos^2\theta_s). \tag{2.18}$$

Knowledge of the Thomson cross section is important for two reasons. It is the low-energy limit for the incoherent (Compton) scattering cross section. It is also the basis from which are evaluated coherent (Rayleigh) scattering cross sections for collective interactions of photons with atomic electrons.

Incoherent (Compton) scattering cross sections.[2]   Incoherent scattering refers to the interaction of a photon with an individual electron, as distinguished from the coherent interaction of a photon with all electrons of an atom. It is assumed in the discussion that follows that the incident radiation is not polarized.

For free electrons, the incoherent scattering cross section, as a function of the scattering angle, is given by the Klein-Nishina (1929) cross section

$$_e\sigma_c(\lambda, \theta_s) = \frac{1}{2}r_e^2\frac{\lambda^2}{(1 + \lambda - \cos\theta_s)^2}$$

$$\times \left(\frac{\lambda}{1 + \lambda - \cos\theta_s} + \frac{1 + \lambda - \cos\theta_s}{\lambda} - \sin^2\theta_s\right). \tag{2.19}$$

An alternative form is

$$_e\sigma_c(E, \theta_s) = \frac{1}{2}r_e^2 p[1 + p^2 - p(1 - \cos^2\theta_s)], \tag{2.20}$$

in which

$$p \equiv \frac{E'}{E} = \frac{\lambda}{1 + \lambda - \cos\theta_s}. \tag{2.21}$$

This cross section is illustrated in Fig. 2.4. Note that, as λ approaches infinity, that is, E approaches zero, p approaches unity and Eqs. (2.20) and (2.21) reduce to the Thomson formula, Eq. (2.18). A related quantity is the Klein-Nishina *energy scattering cross section*

$$_e\sigma_{ce}(E, \theta_s) \equiv \frac{E}{E'}{_e\sigma_c}(E, \theta_s) = \frac{1}{2}r_e^2 p^2[1 + p^2 - p(1 - \cos^2\theta_s)]. \tag{2.22}$$

The total cross section is obtained from Eq. (2.19) by integration over all directions.

$$_e\sigma_c(\lambda) = 2\pi\int_{-1}^{+1}d(\cos\theta_s){_e\sigma_c}(\lambda, \theta_s)$$

$$= \pi r_e^2\lambda\left[(1 - 2\lambda - 2\lambda^2)\ln\left(1 + \frac{2}{\lambda}\right) + \frac{2(1 + 9\lambda + 8\lambda^2 + 2\lambda^3)}{(\lambda + 2)^2}\right]. \tag{2.23}$$

Incoherent scattering cross sections for bound electrons.   These equations break down when the kinetic energy of the recoil electron would be comparable to its binding energy in the atom. Thus, binding effects might be thought to be an important

---

[2]Here and in all subsequent discussion, the wavelength of the photon is considered to be in Compton units, unless specifically indicated otherwise.

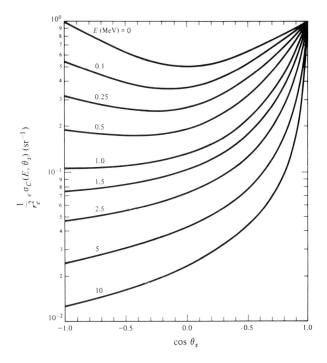

Figure 2.4   Photon incoherent scattering cross section (Compton scattering involving a free electron) as a function of the cosine of the photon scattering angle. The factor r_e is the classical radius of the electron.

consideration for the attenuation of low-energy photons in media of high atomic number. For example, the binding energy of K-shell electrons in lead is 88 keV. However, under these same circumstances, cross sections for the photoelectric interaction of photons greatly exceed incoherent scattering cross sections. Radiation attenuation in this energy region is dominated by photoelectric interactions and, in most attenuation calculations, the neglect of electron binding effects on incoherent scattering causes negligible error. Corrections for electron binding and related data are available in the literature, for example, (Storm and Israel 1967; Biggs and Lighthill 1972a, 1972b; Plechaty, Cullen, and Howerton 1975; Hubbell 1982; Trubey, Berger, and Hubbell 1989). Figure 2.5 shows the relative importance, in lead, of electron binding effects by comparing incoherent scattering and photoelectric cross sections.

2.3.3 Coherent (Rayleigh) Scattering

In competition with the incoherent scattering of photons by individual electrons is coherent scattering by the electrons of an atom collectively. Since the recoil momentum in the Rayleigh interaction is taken up by the atom as a whole, the energy loss of the gamma photon is slight and the scattering angle small. Coherent scattering cross sections may greatly exceed incoherent scattering cross sections, especially for

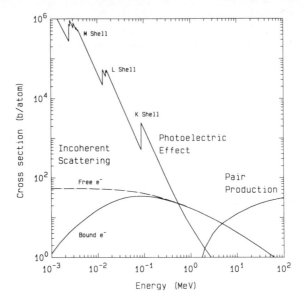

Figure 2.5 Comparison of incoherent-scattering (for both free-electron and bound-electron scattering), photoelectric-effect, and pair-production cross sections for photon interactions in lead. Cross section units are barns per atom.

low-energy photons and high-Z materials. Because of the minimal effect on photon energy or direction, it is common practice to ignore Rayleigh scattering in radiation shielding calculations, especially when electron binding effects mentioned in the preceding paragraph are ignored. Effects of coherent scattering have been addressed in detail by Trubey and Harima (1987). Data are available in an ANSI standard (1991).

2.3.4 Photoelectric Effect

In the photoelectric effect, a photon interacts with an entire atom, resulting in the emission of a photoelectron, usually from the K shell of the atom. Although the difference between the photon energy E and the electron binding energy E_b is distributed between the electron and the recoil atom, virtually all of that energy is carried as kinetic energy of the photoelectron because of the comparatively small electron mass. Thus, $T = E - E_b$.

K-shell binding energies E_k vary from 13.6 eV for hydrogen to 7.11 keV for iron, 88 keV for lead, and 116 keV for uranium. As the photon energy drops below E_k the cross section drops discontinuously. As E decreases further, the cross section increases until the first L edge is reached, at which energy the cross section drops again, then rises once more, and so on for the remaining edges. These "edges" for lead are illustrated in Fig. 2.5. The cross section varies as E^{-n}, where $n \simeq 3$ for energies less than about 150 keV and $n \simeq 1$ for energies greater than about 5 MeV. The atomic cross section varies as Z^m, where m varies from about 4 at E = 100 keV to 4.6 at $E = 3$ MeV. As a very crude approximation

$$\sigma_{ph}(E) \propto \frac{Z^4}{E^3} \tag{2.24}$$

in the energy region for which the photoelectric effect is dominant. Note that the average cross section *per electron* is

$$_e\sigma_{ph}(E) = Z^{-1}\,\sigma_{ph}(E). \tag{2.25}$$

As a general rule, about 80% of photoelectric interactions with heavy nuclei result in ejection of a K-shell electron, while for light nuclei, K-shell electrons are responsible for almost all photoelectric interactions. Consequently, the approximation is often made for heavy nuclei that the total photoelectric cross section is 1.25 times the cross section for K-shell electrons.

As the vacancy left by the photoelectron is filled by an electron from an outer shell, either fluorescence x rays or Auger electrons[3] may be emitted. The probability of x-ray emission is given by the fluorescent yield. For the K shell, fluorescent yields vary from 0.005 for $Z = 8$ to 0.965 for $Z = 90$. Although x rays of various energies may be emitted, as a general rule the approximation can be made that only one x ray or Auger electron is emitted, with energy equal to the binding energy of the photoelectron.

2.3.5 Pair Production

In this process, the incident photon is completely absorbed and in its place appears a positron-electron pair. The phenomenon is induced by the strong electric field in the vicinity of the nucleus and has a photon threshold energy of $2m_ec^2(= 1.02$ MeV). It is possible but much less likely that the phenomenon is induced by the electric field of an electron (triplet production), for which case the threshold energy is $4m_ec^2$. The discussion that follows is limited to the *nuclear* pair production process.

In this process, the nucleus acquires indeterminate momentum but negligible kinetic energy. Thus,

$$T_+ + T_- = E - 2m_ec^2, \tag{2.26}$$

in which T_+ and T_- are the kinetic energies of the positron and electron, respectively. To a first approximation, the total atomic pair production cross section varies as Z^2. The cross section increases with photon energy, approaching a constant value at high energy. The fate of the positron is annihilation in an interaction with an electron, generally after slowing to practically zero kinetic energy. The annihilation process usually results in the creation of two photons moving in opposite directions, each with energy m_ec^2.

---

[3]If an electron in an outer shell, say Y, makes a transition to a vacancy in an inner shell, say X, an x ray may be emitted with energy equal to the difference in binding energy between the two shells. Alternatively, an electron in some other shell, say Y′ which may be the same as Y, may be emitted with energy equal to the binding energy of the electron in shell X less the sum of the binding energies of electrons in shells Y and Y′. This electron is called an Auger electron. If an electron makes a transition from one subshell to a vacancy in another subshell of the same shell, the small difference in binding energies may be transferred to an outer-shell electron, in this case called a Coster-Kronig electron.

2.4 PHOTON ATTENUATION COEFFICIENTS

The photon linear attenuation coefficient μ is, in the limit of small path lengths, the probability per unit distance of travel that a gamma photon undergoes any significant interaction. Thus, for a specified medium,

$$\mu(E) = ZN \,_e\sigma_c(E) + N \,\sigma_{ph}(E) + N \,\sigma_{pp}(E), \qquad (2.27)$$

in which N is the atom density and ZN is the electron density. More common in data presentation is the mass interaction coefficient (see Sec. 2.2)

$$\frac{\mu}{\rho} = \frac{\mu_c}{\rho} + \frac{\mu_{ph}}{\rho} + \frac{\mu_{pp}}{\rho}. \qquad (2.28)$$

Note that Rayleigh scattering and other minor effects are specifically excluded from this definition.[4] In Eq. (2.28),

$$\frac{\mu_c}{\rho} = \frac{Z}{A} N_a \,_e\sigma_c(E), \qquad (2.29)$$

$$\frac{\mu_{ph}}{\rho} = \frac{1}{A} N_a \,\sigma_{ph}(E), \qquad (2.30)$$

$$\frac{\mu_{pp}}{\rho} = \frac{1}{A} N_a \,\sigma_{pp}(E), \qquad (2.31)$$

in which N_a is Avogadro's number. From these results, it is seen that the mass interaction coefficients are independent of the mass density ρ of the material, and it is for this reason that μ/ρ rather than μ values are usually tabulated.

2.5 PHOTON ABSORPTION COEFFICIENTS AND RELATED QUANTITIES

2.5.1 Compton Absorption and Scattering Cross Sections

The cross section per unit wavelength for scattering of a photon into Compton wavelength λ' without regard to angle is, from Eqs. (2.10) and (2.19),

$$_e\sigma_c(\lambda, \lambda') = \pi \, r_e^2 \left(\frac{\lambda}{\lambda'}\right)^2 \left[\left(\frac{\lambda}{\lambda'}\right) + \left(\frac{\lambda'}{\lambda}\right) + (\lambda' - \lambda)(\lambda' - \lambda - 2)\right]. \qquad (2.32)$$

A related quantity is the cross section per unit electron energy for creating a recoil electron with energy T. Here it is convenient to use the ratio $\tau \, (= T/m_e c^2)$. Since $1/\lambda - 1/\lambda' = \tau$,

---

[4]When referring to data tables in other publications, the reader should be aware that occasionally Rayleigh scattering is included.

$$\frac{d\lambda'}{d\tau} = \lambda^2 (1 - \lambda\tau)^{-2}. \tag{2.33}$$

It follows that

$$_e\sigma_c(\lambda, \tau) =_e \sigma_c(\lambda, \lambda') \frac{d\lambda'}{d\tau} \tag{2.34}$$

or

$$_e\sigma_c(\lambda, \tau) = \pi \, r_e^2 \lambda^2 \left[(1 - \lambda\tau) + (1 - \lambda\tau)^{-1} + \frac{(\lambda^2\tau)(\lambda^2\tau + 2\lambda\tau - 2)}{(1 - \lambda\tau)^2} \right]. \tag{2.35}$$

The mean fraction of the photon energy transferred to the recoil electron is designated as f_c and the Compton energy-absorption cross section[5] per electron is defined as

$$_e\sigma_{ca}(\lambda) \equiv f_c \,_e\sigma_c(\lambda). \tag{2.36}$$

The factor f_c is the weighted average of $T/E (= \lambda\tau)$

$$f_c = \frac{1}{_e\sigma_c(\lambda)} \int_0^{2/\lambda(\lambda+2)} d\tau \, \lambda\tau \,_e\sigma_c(\lambda, \tau). \tag{2.37}$$

The energy-scattering cross section is the product of the total cross section and the mean fraction of the photon energy retained by the scattered photon.

$$_e\sigma_{ce}(\lambda) = (1 - f_c) \,_e\sigma_c(\lambda). \tag{2.38}$$

It is evident from the definition of f_c that the factor $(1 - f_c)$ can be evaluated as the mean value of $E'/E (= \lambda/\lambda')$, namely,

$$1 - f_c = \frac{1}{_e\sigma_c(\lambda)} \int_\lambda^{\lambda+2} d\lambda' \frac{\lambda}{\lambda'} \,_e\sigma_c(\lambda, \lambda'). \tag{2.39}$$

Associated with the energy-absorption and energy-scattering cross sections are the mass energy-absorption and energy-scattering coefficients

$$\frac{\mu_{ca}}{\rho} = \frac{Z}{A} N_a \,_e\sigma_{ca}(E), \tag{2.40}$$

$$\frac{\mu_{ce}}{\rho} = \frac{Z}{A} N_a \,_e\sigma_{ce}(E). \tag{2.41}$$

---

[5]The Compton energy absorption cross section thus defined is not a true cross section for photon absorption since, in the interaction, a scattered photon always results. Rather, it is an effective energy absorption cross section with respect to the incident photon energy such that the product $E(\mu_{ca}/\rho)\phi$ is the rate per unit mass at which energy is transferred to kinetic energy of Compton recoil electrons.

2.5.2 Photoelectric Absorption Cross Section

The photoelectric absorption cross section σ_{pha} is the product of σ_{ph} and the average fraction f_{ph} of the incident photon energy appearing as the initial kinetic energy of the photoelectron and, as appropriate, the Auger electron. Thus, to a good approximation,

$$f_{ph} = \frac{1}{E}[(E - E_b) + (1 - w)E_b] = 1 - \frac{wE_b}{E}, \qquad (2.42)$$

in which E_b is the binding energy of the electron, w the fluorescent yield, $E - E_b$ the kinetic energy of the photoelectron, and $(1 - w)E_b$ the average energy transferred to Auger electrons. In a more precise determination, f_{ph} could be refined to weight the relative importance of the photoelectric effect in the K, L, and M electron shells.

The photoelectric linear and mass energy absorption coefficients are

$$\mu_{pha} = N \, \sigma_{pha}, \qquad (2.43)$$

$$\frac{\mu_{pha}}{\rho} = \frac{1}{A} N_a \, \sigma_{pha}. \qquad (2.44)$$

2.5.3 Absorption Cross Section for Pair Production

In a pair production interaction, the fraction of the incident photon energy appearing as the initial kinetic energies of the positron and electron is

$$f_{pp} = 1 - \frac{2m_e c^2}{E}. \qquad (2.45)$$

The pair-production energy-absorption cross sections and coefficients are

$$\sigma_{ppa} = f_{pp} \, \sigma_{pp},$$

$$\mu_{ppa} = N \, \sigma_{ppa},$$

and

$$\frac{\mu_{ppa}}{\rho} = \frac{1}{A} N_a \, \sigma_{ppa}. \qquad (2.46)$$

2.5.4 Corrections for Radiative Energy Loss

The absorption cross sections as defined earlier are based on the transfer of photon energy to initial kinetic energy of electrons and positrons. As these charged particles slow, most of this kinetic energy is dissipated locally by atomic and molecular excitation and ionization processes. Some of the kinetic energy, however, is transferred to bremsstrahlung whose energy may be dissipated far from the point of creation. The fraction radiated may be quite large for charged particles with kinetic energies substantially in excess of their rest-mass energies. Under certain circumstances it is

TABLE 2.1   Radiation Yields for Electrons.

	ELECTRON RADIATION YIELD, $Y(T)^a$						
T (MeV)	C	WATER	AIR	Al	Fe	Pb	U
0.5	0.0018	0.0020	0.0022	0.0043	0.0100	0.0424	0.0491
1.0	0.0034	0.0036	0.0040	0.0764	0.0170	0.0684	0.0792
1.5	0.0050	0.0053	0.0058	0.0110	0.0239	0.0901	0.1035
2.0	0.0067	0.0071	0.0077	0.0145	0.0310	0.1096	0.1249
4.0	0.0142	0.0149	0.0158	0.0292	0.0595	0.1761	0.1955
6.0	0.0222	0.0233	0.0242	0.0444	0.0874	0.2304	0.2518
8.0	0.0304	0.0319	0.0327	0.0596	0.1139	0.2765	0.2990
10.0	0.0387	0.0406	0.0411	0.0745	0.1389	0.3162	0.3394
5.0	0.0595	0.0622	0.0618	0.1105	0.1951	0.3955	0.4193
20.0	0.0798	0.0833	0.0817	0.1438	0.2435	0.4555	0.4790
50.0	0.1856	0.1920	0.1825	0.2959	0.4328	0.6439	0.6635
100.0	0.3181	0.3190	0.3022	0.4448	0.5848	0.7617	0.7766

$^a Y(T)$ is the fraction of the initial kinetic energy T lost as bremsstrahlung. The yields do not account for the minor effects of radiation losses from energetic secondary electrons (delta rays) produced during the deceleration of the primary electron.

Source: ICRU 1984.

necessary to correct the absorption cross sections to account for radiative losses. The correction factor is $(1 - G)$, where G is the fraction of the initial kinetic energies of the charged particles that is lost radiatively, averaged over all types of interactions and all energies. This correction factor may be evaluated for and applied individually to cross sections for Compton scattering, pair production, and the photoelectric effect.

The fraction G may be determined from radiation yields which are listed in Table 2.1. The yield $Y(T)$ is defined as the fraction of the initial electron kinetic energy T that is lost by bremsstrahlung. For the photoelectric effect involving photons of energy E, the kinetic energy of the photoelectron is $T = E - E_b$; thus, neglecting radiative losses from Auger electrons, the value of G is

$$G_{ph} = Y(T). \tag{2.47}$$

For Compton scattering, on the other hand, there is a spectrum of secondary-electron energies, and the value of G in this case is an average over all electron energies, weighted by the energy, namely

$$G_c = \frac{\int dT \, T \, Y(T) \,_e\sigma_c(E, T)}{\int dT \, Y(T) \,_e\sigma_c(E, T)}. \tag{2.48}$$

For pair production, radiative losses by both the positron and the electron must be accounted for in a calculation similar to that of Eq. (2.48).

2.6 CROSS SECTIONS FOR NEUTRON INTERACTIONS

The interaction processes of neutrons with matter are fundamentally different from those for the interactions of photons. Whereas photons interact, almost always, with the atomic electrons, neutrons interact essentially only with the atomic nucleus. Although neutron-electron interactions do occur, this type of interaction is highly improbable, and therefore, negligible when compared to the neutron-nucleus interactions. The cross sections that describe the various neutron interactions are also very unlike those for photons. Neutron cross sections not only can vary rapidly with the incident neutron energy, but also can vary erratically from one element to another and even between isotopes of the same element. The description of the interaction of a neutron with a nucleus involves complex interactions between all the nucleons in the nucleus and the incident neutron, and consequently fundamental theories which can be used to predict neutron cross section variations in any accurate way are still lacking. As a result, all cross-section data are empirical in nature, with little guidance available for interpolation between different energies or isotopes.

Over the years, much effort and money have been expended to measure cross sections for various neutron reactions with many different materials and for wide ranges of neutron energies. Extensive compilations of neutron cross sections have been generated, and the more extensive of these compilations, such as the evaluated nuclear data files (ENDF) (Kinsey, 1979), contain so much information that digital computers are used almost exclusively to process these cross-section libraries to extract cross sections or data for a particular neutron interaction. Even with the large amount of cross-section information available for neutrons, there are still energy regions and special interactions for which the cross sections are poorly known. For example, cross sections which exhibit sudden decreases as the neutron energy changes only slightly or cross sections for interactions which produce energetic secondary photons are of obvious concern to shielding problems and often still are not known to an accuracy sufficient to perform satisfactory deep-penetration shielding calculations.

2.6.1 Classification of Types of Interactions

There are many possible neutron-nuclear interactions, only some of which are of concern in radiation-protection calculations. Ultra-high-energy interactions can produce a myriad of exotic secondary particles; however, the energies required for such reactions are usually well above the neutron energies commonly encountered, and therefore these exotic interactions can be neglected. Similarly, for low-energy neutrons many complex neutron interactions are possible — Bragg scattering from crystal planes, phonon excitation in a crystal, coherent scattering from molecules, and so on — none of which is of particular importance in neutron dosimetry problems. As summarized in Table 2.2, the reactions of principal importance for radiation protection applications are the absorption reactions and the high-energy scattering reactions with their various associated angular distributions.

The total cross section, which is the sum of cross sections for all possible interactions, gives a measure of the probability that a neutron of a certain energy will interact in some manner with the medium. The component of the total cross section for absorption and scattering interactions is usually of primary concern for radiological analysis. Nonetheless, when the total cross section is large, the probability of some type of interaction is great and thus the total cross section, which is the most easily

TABLE 2.2   Principal Nuclear Data Required for Neutron Shielding Calculations.

High-energy interactions (1 eV < E < 20 MeV)
    Elastic scattering cross sections
    Angular distribution of elastically scattered neutrons
    Inelastic scattering cross sections
    Angular distribution of inelastically scattered neutrons
    Gamma-photon yields from inelastic neutron scattering
    Resonance absorption cross sections
Low-energy interactions (< 1 eV)
    Thermal-averaged absorption cross sections
    Yield of neutron capture gamma photons
    Fission cross sections and associated gamma-photon and neutron yields

measured and the most widely reported, gives at least an indication of the neutron energy regions over which one must investigate the neutron interactions in greater detail.

The total cross sections, although they vary from nuclide to nuclide and with the incident neutron energy, have certain common features. For the sake of classification, the nuclides are usually divided into three broad categories: light nuclei (mass number ≤ 25), intermediate nuclei (mass number between 25 and 150), and heavy nuclei (mass number ≤ 150). Example total cross sections for each category are shown in Figs. 2.6 through 2.8.

For light nuclei and some *magic number* nuclei[6] the cross section at low energies (< 1 MeV) often varies as

$$\sigma_t = \sigma_1 + \frac{\sigma_2}{\sqrt{E}}, \tag{2.49}$$

where E is the neutron energy, σ_1 and σ_2 are constants, and the two terms on the right-hand side represent the elastic scattering contribution and the radiative capture (or absorption) reaction, respectively. At energies less than about 0.01 eV, there may be "Bragg cutoffs" representing energies below which coherent scattering from the various crystalline planes of the material is no longer possible. At energies greater than about 0.1 eV, the cross sections are usually slowly varying and "smooth" up to the MeV energy range, at which energies fairly wide (keV to MeV) resonances appear.

Of all the elements, only the lightest (hydrogen and its isotope deuterium) exhibit no resonances (see Fig. 2.6). Notice that for hydrogen, the cross section is almost constant up to the MeV region, where it begins to decrease.

For heavy nuclei (e.g., Fig. 2.8), the total cross section, unless masked by a low-energy resonance, exhibit a $1/\sqrt{E}$ behavior at low energies and usually a Bragg cutoff. The resonances appear at much lower energies than for the light nuclei (usually in the eV region) and have very narrow widths (1 eV or less) with large peak values.

---

[6]A magic number nucleus is one in which the number of neutrons or protons equals 2, 8, 20, 50, 82, or 126. When the nucleus is magic, a particularly stable configuration of the nucleons in the nucleus is achieved analogous to closed electron shells in atomic physics.

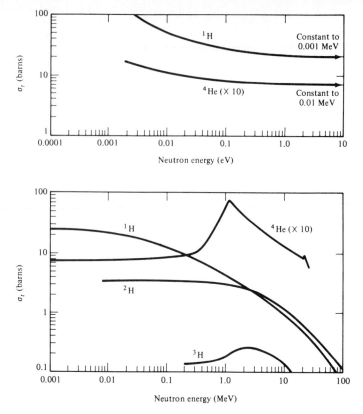

Figure 2.6   Total cross sections for neutron interactions with the lightest nuclei. [Data are from BNL (1958).]

Above 1 keV the resonances are so close together and so narrow that they cannot be resolved and the cross sections appear to be smooth except for a few broad resonances. Finally, the intermediate nuclei (Fig. 2.7), as would be expected, are of intermediate character between the light and heavy nuclei with resonances in the region from 100 eV to several keV. The resonances are not as high nor as narrow as for the heavy nuclei.

For neutron protection purposes, one is usually concerned with fast neutrons (> 500 keV). Neutron penetration studies are somewhat simplified in this energy region compared to the lower-energy region since the heavy nuclides have no resolved resonances and the cross sections vary smoothly with the neutron energy. Only for the light elements must the complicating resonances be considered. Further, the absorption cross sections for all nuclides are usually very small compared to those for other reactions at all energies except thermal energies. Over the fission-neutron energy spectrum, the (n, γ) reaction cross sections seldom exceed 200 mb for the heavy elements, and for the lighter elements this cross section is considerably smaller. Only for thermal neutrons and a few isolated absorption resonances in the keV region for heavy elements is the (n, γ) reaction important.

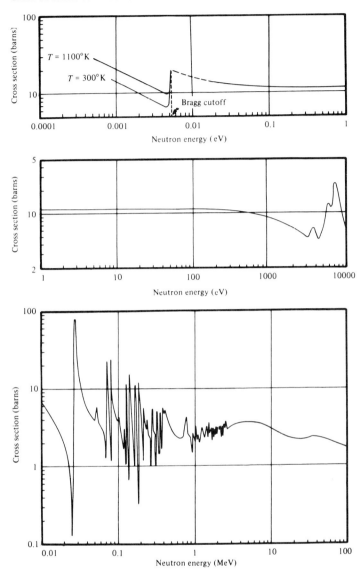

Figure 2.7   Total cross section for neutron interactions with iron. [Data are from BNL (1958).]

In the high-energy region, by far the most important neutron interaction is the scattering process. Generally, elastic scattering is more important, although, when the neutron energy somewhat exceeds the energy level of the first excited state of the scattering nucleus, inelastic scattering becomes possible. For energies greater than about 8 MeV, multiple-particle reactions such as $(n, 2n)$ and $(n, n + p)$ become possible. In most materials of interest for shielding purposes the thresholds for these reactions are sufficiently high and the cross sections sufficiently small that these neutron-producing

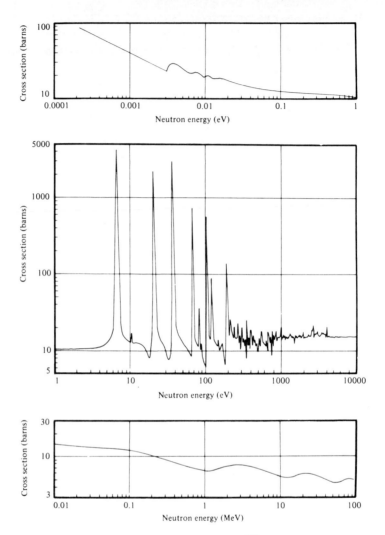

Figure 2.8   Total cross section for neutron interactions with ^{238}U. [Data are from BNL (1958).]

reactions may be ignored compared to the inelastic scattering reactions. However, a few rare reactions producing secondary neutrons should be noted. The threshold for $(n, 2n)$ reactions are particularly low for D and Be (3.3 MeV and 1.84 MeV, respectively) and for these two nuclei there is no inelastic scattering competition. Similarly, for fissionable nuclei, the fast neutrons may cause fission events in which multiple secondary fast neutrons are liberated. In most shielding situations these particular reactions which produce secondary neutrons are not encountered.

Finally, interactions such as (n, p) and (n, α), which produce charged particles, may be of importance when light elements are involved. In the MeV region the (n, α) reaction cross sections for Be, N, and O are appreciable fractions of the total cross sections and may exceed the inelastic scattering contributions. This situation is probably true for most light elements, although only partial data are available. For heavy and intermediate nuclei the charged-particle emission interactions are at most a few percent of the total inelastic interaction cross section and hence are usually ignored.

2.7 NEUTRON SCATTERING INTERACTIONS

The scattering interaction is the most probable interaction of the fast neutron, and is the mechanism on which one relies to slow these neutrons to thermal energies, at which they can be absorbed through (n, γ) reactions. There are two distinct types of scattering processes, both of importance in fast neutron attenuation. In *capture scattering* the incident neutron is absorbed by the scattering nucleus to form a compound nucleus which subsequently decays by the emission of a neutron. If the residual nucleus is left in the ground state, the scattering is called *elastic*. If the residual nucleus is left in an excited state, the scattering is called *inelastic*. The other type of scattering is referred to as *potential scattering*. In this process, which is always elastic, the incident neutron is scattered by the nucleus as a whole — analogous to the diffraction of the incident neutron wave by the entire nuclear potential. Capture-scattering cross sections generally exhibit resonance behavior, while the potential scattering cross sections usually vary slowly with energy.

2.7.1 Kinematics of Neutron Scattering

In all scattering processes the total energy and momentum must be conserved, and consequently, for any given scattering interaction, there are unique relationships among the initial and final neutron energies and the scattering angle. Except for thermal neutron scattering, for which the thermal motion of the target atoms may be comparable to the neutron speed, one can properly neglect the initial kinetic energy of the scattering nucleus in the laboratory coordinate system. If it is further assumed that an amount $-Q$ of kinetic energy[7] is absorbed and retained by the residual nucleus (the excitation energy), then application of the principles of conservation of energy and momentum require that (see Fig. 2.9)

$$E'(\omega_s, E) = \frac{1}{(A+1)^2} \left(\omega_s \sqrt{E} \pm \sqrt{E(\omega_s^2 + A^2 - 1) + A(A+1)Q} \right)^2. \quad (2.50)$$

On physical grounds E' must be real and positive. If $-Q$ is large enough, Eq. (2.50) implies that E' becomes complex, and hence the reaction is not possible. The threshold energy E_t for inelastic scattering with given ω_s is readily obtained by recognizing that, at the threshold energy, the term in the second square root must vanish,

[7]The Q value of any nuclear reaction is the excess of the kinetic energy of the product particles over that of the original particles. For inelastic scattering, the kinetic energy of the scattered neutron and the recoil excited nucleus is less than that of the initial kinetic energies, and thus the Q value is always negative for this reaction.

Laboratory system

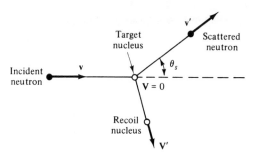

Center-of-mass system

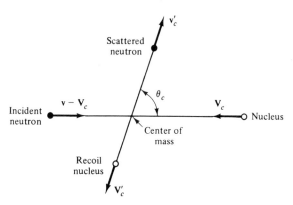

Figure 2.9 Laboratory and center-of-mass systems for the kinematics of neutron scattering. Rather than use the reflection angles θ_s and θ_c in the description of the scattering process, it is more convenient to use their cosines ω_s and ω_c.

that is,

$$E_t(\omega_s) = -\frac{A(A+1)Q}{\omega^2 + A^2 - 1}. \tag{2.51}$$

The minimum threshold energy is observed to occur at $\theta_s = 0$ ($\omega_s = 1$), and thus

$$(E_t)_{min} = -\frac{A+1}{A}Q. \tag{2.52}$$

From this result it is seen that inelastic scattering is impossible if the energy of the incident neutron is less than $(A+1)/A$ times the energy level of the first excited state (minimum value of $-Q$) of the scattering nucleus. In Table 2.3 the energy levels of the first two excited states are listed for selected nuclides. There it is seen that the threshold for inelastic scattering tends to decrease as the atomic mass of the scatterer increases. Notice that the level spacings of the light elements and the magic number nuclides are comparatively large and hence inelastic scattering is generally less signif-

TABLE 2.3   Energies of the First and Second Excited States in MeV for Selected Nuclides.

NUCLIDE	FIRST EXCITED STATE	SECOND EXCITED STATE	NUCLIDE	FIRST EXCITED STATE	SECOND EXCITED STATE
$^{2}_{1}$H, $^{4}_{2}$He[a]	none	none	$^{54}_{26}$Fe[b]	1.408	2.538
$^{6}_{3}$Li	2.185	3.562	$^{56}_{26}$Fe	0.847	2.085
$^{7}_{3}$Li	0.478	4.63	$^{57}_{26}$Fe	0.014	0.136
$^{11}_{5}$B	0.718	1.740	$^{58}_{28}$Fe[b]	0.811	1.675
$^{11}_{5}$B	2.125	4.445	$^{58}_{28}$Ni[b]	1.454	2.776
$^{12}_{6}$C	4.439	7.654	$^{60}_{28}$Ni[b]	1.332	2.159
$^{13}_{6}$C	3.088	3.684	$^{206}_{82}$Pb[b]	0.803	1.167
$^{14}_{6}$C[b]	6.094	6.590	$^{208}_{82}$Pb[b]	0.570	0.898
$^{16}_{8}$O[a]	6.094	6.130	$^{208}_{82}$Pb[a]	2.615	3.198
$^{17}_{8}$O[b]	0.871	3.055	$^{208}_{83}$Bi	0.063	0.510
$^{18}_{8}$O[b]	1.982	3.555	$^{209}_{83}$Bi[b]	0.897	1.609
$^{39}_{19}$K[b]	2.523	2.814	$^{232}_{90}$Th	0.049	0.162
$^{40}_{19}$K	0.030	0.800	$^{233}_{92}$U	0.040	0.092
$^{40}_{20}$Ca[a]	3.352	3.736	$^{234}_{92}$U	0.043	0.143
$^{41}_{20}$Ca[b]	1.943	2.001	$^{235}_{92}$U	75 eV	0.013
$^{45}_{21}$Sc	0.012	0.376	$^{236}_{92}$U	0.045	0.149
$^{55}_{25}$Mn	0.126	0.984	$^{238}_{92}$U	0.045	0.148

[a]Indicates the numbers both of neutrons and protons are magic.
[b]Indicates a magic number of neutrons or protons.
Source: Lederer and Shirley 1978.

icant for these nuclides. Moreover, the odd-even and even-odd nuclides tend to have smaller thresholds than the even-even nuclides.

From Eq. (2.50) it is seen that two discrete values of the final energy E' are possible when $E > E_t(\omega_s)$, $\omega_s > 0$, and when

$$\omega_s\sqrt{E} - \sqrt{E(\omega_s^2 + A^2 - 1) + A(A + 1)Q} \geq 0. \tag{2.53}$$

This condition can be simplified to

$$E \leq -\frac{AQ}{A - 1} \equiv E_c,$$

where E_c is the largest value of the incident neutron energy for which two final neutron energies are possible in the laboratory system and is referred to as the *cutoff energy*. Thus, as the source neutron energy is increased above the threshold energy, but kept below the cutoff energy, the secondary neutron can appear in the forward direction ($\omega_s > 0$) with either of two discrete values of kinetic energy E'. This region of E between $E_t(\omega_s)$ and E_c is called the *double-valued region*.

As the kinetic energy of the incident neutron E is increased above E_c, the value of the left-hand side of the inequality Eq. (2.53) is negative. Thus, only a single value of E is obtained from Eq. (2.50); that is, only the positive sign in front of the square root term of this equation gives a physically realistic result. This region of E above E_c is called the *single-valued region*.

For elastic scattering ($Q = 0$), $E_t = E_c = 0$ and no double energy region occurs. The energy of the scattered neutron E' can be obtained from Eq. (2.50) with $Q = 0$ as

$$E' = \frac{E}{(A+1)^2} \left(\sqrt{\omega_s^2 + A^2 - 1} + \omega_s \right)^2.$$ (2.54)

2.7.2 Cross Sections for Neutron Scattering

Corresponding to each energy level of a nucleus, there is an associated cross section for neutron scattering which may be zero if the incident neutron has insufficient energy to excite the scattering nucleus to that level. The total scattering cross section for an isotope is simply the sum of scattering cross sections associated with all energy levels for which scattering is energetically possible (the ground state is always included). The total inelastic scattering cross section is then the difference between the total scattering and the elastic (ground state) scattering cross sections. The measurement of the inelastic cross section is quite difficult, and although many data have been obtained, there are still many elements and energies for which the inelastic cross sections are unknown. A much easier cross section to measure is the total *nonelastic* cross section, which is the difference between the total cross section and the elastic scattering cross section. When fission and other neutron absorbing cross sections are small, the nonelastic cross section is approximately equal to the sum of all the inelastic scattering cross sections. In Fig. 2.10 nonelastic cross sections for several elements are shown for the MeV energy region and compared to the elastic scattering (total) cross section for hydrogen. Most nuclides show a rapid rise in the nonelastic cross section with increasing neutron energy and then fairly constant values in the MeV range. Finally, it should be emphasized that, although much has been said here about the inelastic scattering cross section, the elastic scattering cross section is usually greater than any other cross section in the MeV energy region. However, as will be seen, elastic scattering for heavy nuclides does little to degrade the energy of a fast neutron and consequently inelastic scattering, although usually less likely, becomes the dominant mechanism for the slowing down of neutrons with energies greater than E_t.

The differential scattering cross section, $\sigma_s(E, \theta)$, is sometimes reported in cross-section libraries in terms of ω_c, the cosine of the scattering angle in the center-of-mass coordinate system. This is done because, in the center-of-mass system, scattering of low- and intermediate-energy neutrons is nearly isotropic, that is,

$$\sigma_s(E, \omega_c) \simeq \frac{\sigma_s(E)}{4\pi}.$$ (2.55)

In fact, for hydrogen, the scattering in the center-of-mass system is isotropic for energies up to about 30 MeV. Generally, the heavier the nuclide, the lower is the energy

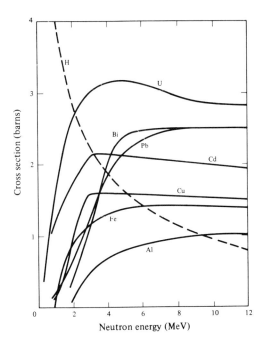

Figure 2.10   Nonelastic cross sections for several elements in the fast-neutron energy range. By constrast the total hydrogen cross section (dashed line) decreases rapidly as the neutron energy increases in this energy region. [Data are from BNL (1958).]

above which elastic scattering becomes anisotropic. Few data are available on the angular distributions of inelastically scattering neutrons. When only one or two levels are involved, the scattering may be anisotropic; however, it has generally been found that it is a good approximation to assume that the inelastically scattered neutrons are emitted isotropically in the center-of-mass system. This is particularly true when multiple levels are involved in the inelastic process. Finally, the potential scattering of fast neutrons is never isotropic, but is highly peaked in the forward directions. In fact, the angular distribution can exhibit several maxima as the scattering angle increases from $\theta_c = 0$. The rapidly changing shape of the angular distribution of scattered neutrons is illustrated in Fig. 2.11 for the case of oxygen. Clearly, such erratic behavior can be expected to greatly complicate detailed calculations of the penetration of a neutron beam into such a material.

2.7.3 Average Energy Transfer in Neutron Scattering

For isotropic inelastic scattering in the center-of-mass system, the average energy of the recoil nucleus is given by

$$\epsilon_{inel} = \frac{1}{2}(1 - \alpha)E\left(1 + \frac{Q}{E}\frac{A+1}{2A}\right), \tag{2.56}$$

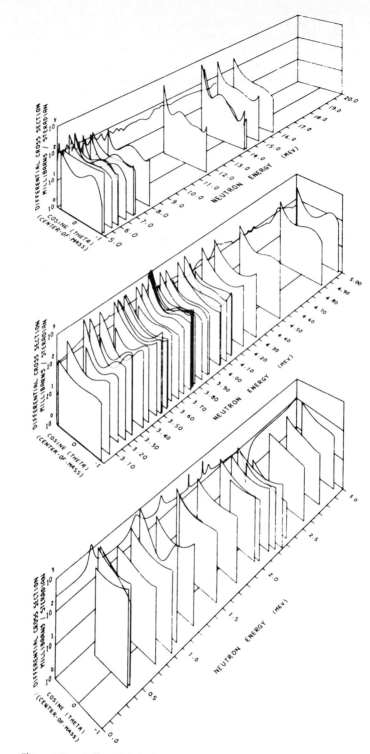

Figure 2.11   Differential elastic scattering cross section of oxygen as a function of neutron energy. [From Garber et al. (1970)]

where $\alpha \equiv (A - 1)^2/(A + 1)^2$. For elastic scattering $(Q = 0)$ Eq. (2.56) reduces to $\epsilon_{elas} = (1 - \alpha)E/2$. Notice that as A becomes large, α approaches unity, and the average energy loss in elastic scattering becomes very small. Only by inelastic scattering can appreciable energy losses be realized. Although for hydrogen the average energy loss is one-half of the initial energy, the total scattering cross section, $\sigma_s(E)$, falls off rapidly in the MeV energy region (see Fig. 2.10) and hydrogen scattering events become relatively improbable. For this reason inelastic scattering by heavy nuclides plays a crucial role in the slowing down of fast neutrons.

The energy of the recoil nucleus from a scattering event caused by a fast neutron is quickly dissipated in solids or liquids, and hence for all practical purposes the recoil energy can be assumed to be deposited locally. For example, a 5 MeV proton travels at most 0.5 mm in aluminum. Heavier recoiling nuclei are stopped in much shorter distances. Only for the case of neutron scattering in a gas does one have to be concerned with the travel of the recoil atoms and then only if very detailed calculations are required. The energy distribution of the recoil nuclei is usually of little concern in shielding analysis. However, if one is interested in precise radiation damage studies, in either biological or nonbiological material, it may be necessary to develop such detailed information for the scattering process.

Of great concern in radiation protection are the secondary gamma photons that are emitted by recoil nuclei formed in inelastic scattering interactions. Such secondary-photon production is a most difficult aspect of fast-neutron shielding since a shield designed only to slow down a fast-neutron beam is usually not adequate to stop the secondary photons. Further, the production of the secondary photons in the shield is far from uniform since, as the energy spectrum of the neutron beam changes, so will the number and type of inelastic reactions vary. The inelastic-scattering cross sections associated with various energy levels in the scattering nucleus (see Fig. 2.12) are known only for a few nuclides and then only for the low-lying energy levels. Secondary-photon cross-section data for neutron scattering has received much attention; nevertheless, much information is still lacking and much design work is performed on the basis of theoretical nuclear models for inelastic scattering.

2.8 RADIATIVE CAPTURE OF NEUTRONS

Most neutrons entering a shield will be absorbed by atoms of the shield material. Any kinetic energy of the neutron plus its binding energy in the resulting compound nucleus (usually around 7 to 9 MeV) leaves the compound nucleus in a highly excited state. The excited nucleus usually decays within 10^{-12} second of the capture, often through several intermediate states, thereby emitting one or more energetic gamma photons. These *capture gamma photons* from (n, γ) reactions can be an important consideration in neutron shielding analyses since they are often produced near the outer surface of a neutron shield and, consequently, the energetic capture photons may have only a small amount of shield to traverse before escaping.

The cross section for radiative capture is very small for high-energy neutrons, typically no more than a few hundred millibarns for neutrons with energies between 20

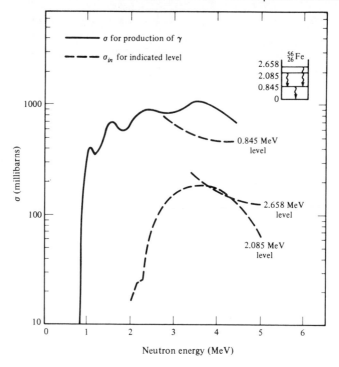

Figure 2.12 Total inelastic scattering cross sections for ^{56}Fe from the first three excited levels (dotted curves). Also shown is the total cross section for the production of the 0.845-MeV photon (solid line). [Data are from BNL (1958).]

keV and 10 MeV. For many nuclides, the (n, γ) capture cross section is poorly known in the keV and MeV energy region. Only for certain important nuclides (e.g., fissionable isotopes) is the cross section known with a good degree of certainty. However, because the cross sections are almost always less than a few percent of the scattering cross sections, (n, γ) reactions are not of particular importance in fast-neutron shielding situations. Of far greater concern are the (n, γ) reactions caused by thermal neutrons which have slowed down by scattering and come into equilibrium with the thermal motion of atoms in the shield. The (n, γ) cross sections may be quite large, up to thousands of barns for nuclides which have a capture resonance near the thermal energy region (e.g., cadmium), and for most isotopes it comprises almost the total absorption cross section. In a material at room temperature, the thermal neutrons will have an average energy of 1/40 eV, corresponding to a speed of 2200 m s^{-1}. In Appendix D, thermal neutron cross sections as well as yields and energies of the capture photons are given for common elements.

The absorption of neutrons in a shielding material can result in a complex secondary-photon energy spectrum characterized by many photons with energies of several MeV. Fortunately, the difficulty in shielding against capture gamma photons can be ameliorated somewhat in certain situations. To minimize the production of these energetic photons, a material can be used that has a high cross section for charged-particle reactions with thermal neutrons. For example, ^{6}Li has a large cross section for (n, α) reactions and thus lithium can be used in a neutron shield to absorb thermal neutrons without secondary photon emission. Usually, less expensive boron is used in neutron shields because ^{10}B also has a large (n, α) cross section for thermal neutrons. Although the product of the reaction, ^{7}Li* (the asterisk denoting an excited state), produces a 0.48-MeV photon by isomeric transition, this photon is of less significance than the capture gamma photons because of its relatively low energy.

2.9 PENETRATION OF CHARGED PARTICLES THROUGH MATTER

Although this text for the most part treats shielding of neutrons and photons, knowledge of the ranges of ionizing charged particles and the rates at which energy is dissipated along their paths is also important for three reasons. First, secondary charged particles resulting from neutron and photon interactions are responsible for the radiation effects of principal concern, namely, biological, chemical, and structural changes. Second, detection and measurement of photons or neutrons is almost always effected through interactions of secondary charged particles. Indeed, the roentgen unit of photon exposure is defined in terms of ionization produced by secondary electrons. Finally, a knowledge of the ranges of charged particles leads directly to the determination of shield thicknesses necessary to stop them or the extents of the regions in which they can cause biological damage.

Two concepts are important: *range* and *stopping power*. Initially, with kinetic energy T_o, a particle is slowed to kinetic energy $T(s)$, after traveling a distance s along its path, as a result of both Coulombic interactions with (atomic) electrons and radiation losses (bremsstrahlung). During deceleration, the stopping power, $-dT/ds$, generally increases until the energy of the particle is so low that charge neutralization or quantum effects bring about a reduction in stopping power. Only for particles of very low energy do collisions with atomic nuclei of the stopping medium become important. The distance the particle travels before being stopped is called the range Λ. Idealized graphs for the range and stopping power are illustrated in Fig. 2.13, which shows the variation of energy and the rate of change of energy with distance along the path of a charged particle.

Heavy charged particles (those with masses greater than or equal to the proton mass), with kinetic energies much less then their rest-mass energies, slow down almost entirely due to Coulombic interactions. A multitude of such interactions take place — so many that the slowing down is virtually continuous and along a straight-line path. These interactions, taken individually, may range from ionization processes producing energetic recoil electrons (*delta rays*) to weak atomic or molecular excitation which may not result in ionization at all. The stopping power resulting from Coulombic interactions, $(-dT/ds)_{coll}$, is called *collisional stopping power*, or the *ionization stopping power*.

Another mechanism, which is important for electrons, is radiative energy loss, characterized by the *radiative stopping power* $(-dT/ds)_{rad}$. Also, a careful treatment of electron slowing down requires accounting for the possibility of delta-ray produc-

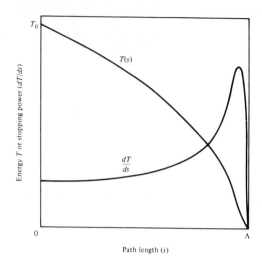

Figure 2.13   Variation of energy and stopping power along the path of a charged particle.

tion and the concomitant deflection of the incident electron from its original direction. In this discussion, electron range will refer to the mean path length rather than the straight-line penetration distance. Information is also available in the literature (Berger 1971) for electron ranges based on straight-line penetration distance as well as for effective ranges and energy loss rates for beta particles from selected radionuclides.

The chapter summarizes the evaluation of range and stopping-power for both electrons and heavy charged particles. A more detailed analysis of electron range and stopping power is presented in Chap. 7, which deals with electron and beta-particle transport and dosimetry.

2.9.1 Collisional Stopping Power

The collisional stopping power depends strongly on the charge number z and speed v of the particle. The speed v is commonly expressed in terms of β, the ratio of v to the speed of light c. If M is the mass of the particle and T its kinetic energy, then $\beta = v/c = \sqrt{T^2 + 2TMc^2}/(T + Mc^2)$. The collisional stopping power also depends on the density ρ of the stopping medium, the ratio Z/A of the medium atomic number to the atomic mass (atomic mass units), and the effective ionization potential I of the medium. Selected values of I are given in Table 2.4. For Z greater than or equal to 13, I for elemental substances, in units of eV, is given approximately by the empirical formula (Barkas and Berger 1964)

$$I = 9.76Z + 58.8Z^{-0.19}. \tag{2.57}$$

For atomic constituents of compounds, I should be increased by a factor of 1.13 (ICRU 1984). For mixtures of elements identified by the index j, the average values of Z/A and I are given by the empirical formulas

TABLE 2.4 Selected Values of the Mean Excitation Energy for Elements, Dry Air, and Atomic Constituents of Compounds in the Condensed State.

MATERIAL		I (eV)
H	saturated bond	19
H	unsaturated bond	16
C	saturated bond	81
C	unsaturated bond	80
C	highly chlorinated	69
N	amines, nitrates, etc	106
N	in rings	82
O	−O−	105
O	=O	94
Al	elemental	166
Air	dry gas	86
Water		75

Source: ICRU 1984.

$$\langle Z/A \rangle = \sum_j w_j \left(\frac{Z_j}{A_j} \right), \tag{2.58}$$

$$\ln I = \langle Z/A \rangle^{-1} \sum_j w_j \left(\frac{Z_j}{A_j} \right) \ln I_j. \tag{2.59}$$

in which w_j is the weight fraction of the jth constituent.

For *heavy charged particles*, an excellent approximation for the collisional stopping power is given by the following result (valid for protons with energies between about 2 and 10 MeV or other heavy particles with comparable speeds) (Evans 1955):

$$\frac{1}{\rho(Z/A)z^2} \left(-\frac{dT}{ds} \right)_{coll} = (4\pi N_a \, r_e^2 m_e c^2)\beta^{-2} \left[\ln \left(\frac{2m_e c^2 \beta^2}{(1 - \beta^2)I} \right) - \beta^2 \right], \tag{2.60}$$

in which N_a is Avogadro's number, z is the charge number of the particle, and r_e is the classical electron radius. The reader will note that $m_e c^2$ and I must be in the same units. The formula has been arranged to emphasize that the stopping power, divided by $\rho(Z/A)z^2$, is a function only of β (or v) and I. For a nonrelativistic particle,[8] $v = \sqrt{2T/M}$. Thus, in a given medium, a 1-MeV proton has the same collisional stopping power as a 2-MeV deuteron, and one-fourth that of a 4-MeV alpha particle. In this way, stopping-power data for protons may be used to evaluate stopping powers for other heavy charged particles. Such data, computed from a slightly more refined version of Eq. (2.60), are presented in Fig. 2.14.

---

[8]By this, we mean a particle whose kinetic energy is very much less than its rest-mass energy — 511 keV for electrons, about 930 MeV for protons or neutrons, and about 4000 MeV for alpha particles.

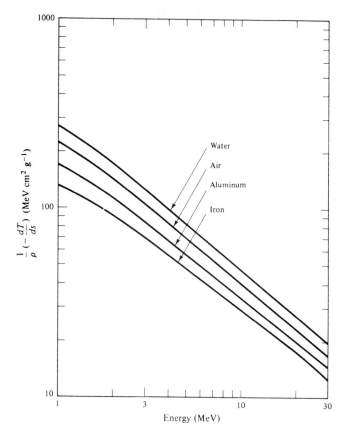

Figure 2.14   Collisional stopping power for protons in air, water, aluminum, and iron. [Data are from Barkas and Berger (1964).]

For *electrons*, the collisional stopping power may be evaluated approximately using the following formula (Evans 1955), which is accurate to within 10% for electrons of energies from 0.01 to 10 MeV and materials ranging from hydrogen through iron

$$\frac{1}{\rho(Z/A)}\left(-\frac{dT}{ds}\right)_{coll} = (2\pi N_a r_e^2 m_e c^2)\beta^{-2}$$

$$\times \left[\ln\left(\frac{\beta^2(m_e c^2)^2}{I^2(1-\beta^2)}[(1-\beta^2)^{-1/2}-1]\right)-\beta^2\right]. \quad (2.61)$$

This formula, which also applies to positrons, has been arranged to emphasize that the stopping power, divided by $\rho(Z/A)$, is again a function only of β and I. Selected stopping-power data, computed using a slightly more refined version of Eq. (2.61), are presented in Fig. 2.15.

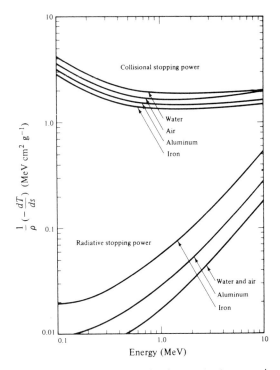

Figure 2.15   Collisional and radiative stopping power for electrons in air, water, aluminum, and iron. [Data are from Berger and Seltzer (1964).]

2.9.2 Radiative Stopping Power for Electrons

No single formula can adequately describe the radiative stopping power over a wide range of electron kinetic energies. It may be said that $(-dT/ds)_{rad}$ is approximately proportional to $\rho(Z^2/A)(T+m_ec^2)$ multiplied by a function strongly dependent on T and weakly dependent on Z. The function varies slowly with T for energies up to about 1 MeV, then increases as T increases, reaching another plateau for energies greater than about 100 MeV. Selected data for the radiative stopping power are presented in Fig. 2.15.

2.9.3 Charged-Particle Range

Because of statistical fluctuations in energy losses along the tracks of charged particles, there is no unique range associated with a particle's initial energy. Furthermore, there is no standard way of defining the *effective* range of a charged particle. It is common to neglect energy-loss fluctuations and assume that particles lose energy continuously along their tracks, with a mean energy loss per unit path length given by the total stopping power. Under this approximation (ICRU 1984), the range is given by

$$\Lambda = \int_0^\Lambda ds = \int_0^{T_o} \frac{dT}{\left(-\frac{dT}{ds}\right)_{total}}. \tag{2.62}$$

Evaluation of the integral is complicated by difficulties in formulating both the radiative stopping power and the collisional stopping power for low-energy particles, particularly electrons, and it is common to assume that the reciprocal of the stopping power is zero at zero energy and increases linearly to the known value at the lowest tabulated energy. The range defined in this way is identified as the CSDA range, that is, the range in the *continuous slowing-down approximation*. Data for heavy charged-particle range and electron range are presented in Figs. 2.16 and 2.17, respectively.

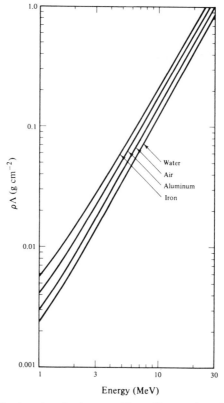

Figure 2.16 Proton range in air, water, aluminum, and iron. [Data are from Barkas and Berger (1964).]

If radiation losses and charge fluctuation may be neglected, interpolation of range data is facilitated by the following rules which may be deduced from Eqs. (2.60) and (2.62):

1. For particles of the *same initial speed* in a given medium, $\rho\Lambda$ is *approximately* proportional to M/z^2.

2. For particles of the *same initial speed*, in different media, $\rho\Lambda$ is *approximately* proportional to $M/z^2(Z/A)$.

Thus, from rule 1, in a given medium a 4-MeV alpha particle has about the same range as a 1-MeV proton.

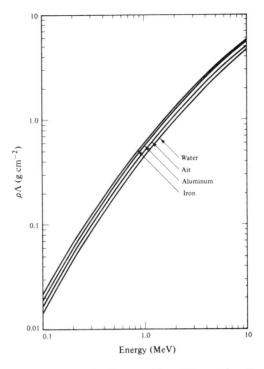

Figure 2.17   Electron range in air, water, aluminum, and iron. [Data are from Berger and Seltzer (1964).]

2.9.4 Approximate Formulas for Range and Stopping Power

Very useful empirical methods are available for making approximate estimates of range and stopping power. Range may be estimated using the formula

$$\rho \Lambda \simeq \delta T^n, \tag{2.63}$$

in which δ and n are empirical constants. Since $(-dT/ds) = (d\Lambda/dT)^{-1}$, it follows that

$$-\frac{dT}{ds} \simeq \frac{\rho T^{1-n}}{n\delta}. \tag{2.64}$$

Values of the parameters n and δ are given in Table 2.5.

2.10 DETECTOR RESPONSE FUNCTIONS

There are many circumstances — for example, source calibration, detector calibration, or external dose evaluation — requiring the relating of the response of a radiation detector, either real or conceptual, to the flux density or fluence of indirectly ionizing radiation. The detector response R can be related to the energy and angular

TABLE 2.5 Constants for the Empirical Energy-Range Formula
$\rho \Lambda$ (g cm^{-2}) = δ [T (MeV)]^n.

		PROTONS		ALPHA PARTICLES		ELECTRONS
MATERIAL	n	δ	n	δ	n	δ
Hydrogen	1.817	8.21($-$4)^a	1.817	6.62($-$5)	1.32	0.155
Carbon	1.787	2.22($-$3)	1.787	1.86($-$4)	1.32	0.356
Aluminum	1.730	3.47($-$3)	1.730	3.15($-$4)	1.32	0.400
Lead	1.680	7.18($-$3)	1.680	7.00($-$4)	1.32	0.640
Water	1.793	1.95($-$3)	1.793	1.62($-$4)	1.32	0.356
Tissue	1.783	2.17($-$3)	1.783	1.83($-$4)	1.32	0.353

^{a}Parentheses indicate powers of ten. Read as 8.21 × 10^{-4}, and so on.

Source: Haffner 1967; by permission of Academic Press.

distribution of either a fluence Φ or a flux density ϕ by a multiplier to be called, generally, the *detector response function* $\mathcal{R}$.

In most general form, the detector response is given by[9]

$$R = \int_0^\infty dE \int_{4\pi} d\Omega \int_{V_d} dV \, \mathcal{R}(\mathbf{r}_d, E, \mathbf{\Omega})\phi(\mathbf{r}_d, E, \mathbf{\Omega}), \qquad (2.65)$$

where the volume integration is over the detector sensitive volume V_d. The function in $\mathcal{R}(\mathbf{r}_d, E, \mathbf{\Omega})$ can be looked upon from a physical point of view as the expected (or stochastically averaged) detector response caused by a particle of energy E traveling in direction $\mathbf{\Omega}$ at point $\mathbf{r}_d$, per differential unit of path length traveled. In general, the explicit representation of the detector response function may be very complicated for a large-volume, directionally dependent detector whose energy dependence varies with location within the volume. For analytical purposes, however, most practical detectors can be well represented by one or another of the simple, idealized detectors having response functions as indicated below

1. Point detector at location $\mathbf{r}_o$, collimated to respond only to radiation with direction $\mathbf{\Omega}_o$

$$\mathcal{R}(\mathbf{r}_d, E, \mathbf{\Omega}) = \mathcal{R}(E)\delta(\mathbf{r}_d - \mathbf{r}_o)\delta(\mathbf{\Omega} - \mathbf{\Omega}_o). \qquad (2.66)$$

2. Point, isotropic detector, at point $\mathbf{r}_o$, responding equally to radiation traveling in any direction

$$\mathcal{R}(\mathbf{r}_d, E, \mathbf{\Omega}) = \mathcal{R}(E)\delta(\mathbf{r}_d - \mathbf{r}_o). \qquad (2.67)$$

---

[9]In this chapter, time-rate response will be discussed rather than the corresponding time-integrated response. For the latter, one uses fluence rather than flux density in Eq. (2.65). The detector response function defined here is a somewhat simplified form of a more general function which may depend, in addition to those factors specified, on time, previous detector history, or even the magnitude of the response. The present formulation is adequate for the idealized types of detector response normally used. Until Sec. 2.14.4 is reached, the discussion assumes that the flux density is known exactly.

3. Point, energy-independent, isotropic detector at point $\mathbf{r}_o$, responding equally to radiation of any energy and traveling in any direction

$$\mathcal{R}(\mathbf{r}_d, E, \mathbf{\Omega}) = \mathcal{R}\,\delta(\mathbf{r}_d - \mathbf{r}_o), \tag{2.68}$$

in which, in this equation only, $\mathcal{R}$ is independent of E.

Most idealized detector types used in radiation-protection practice are of the point, isotropic type; and all the detector concepts described and used in this text are to be considered as such unless a specific statement to the contrary is made. If Eq. (2.67) is substituted into Eq. (2.65), the following somewhat simpler expression for the detector response results

$$R(\mathbf{r}_o) = \int_0^\infty dE\, \mathcal{R}(E)\phi(\mathbf{r}_o, E). \tag{2.69}$$

The concept of the detector response function embraces such simple isotropic detector types as a flux-density detector (in which case $\mathcal{R} = 1$) and an energy flux-density detector [in which case $\mathcal{R}(E) = E$]. Other detectors used for radiation-protection purposes are generally of a dosimetric nature and have response functions related to energy deposition per unit mass. These are discussed later in detail.

2.11 GENERAL FORMULATION FOR DOSIMETRIC DETECTORS

The basic cross-section definition of Secs. 2.1 and 2.2 leads to the fundamental rule that, for a sufficiently small volume ΔV of elemental composition suffused by a uniform field of monoenergetic radiation particles, the number of interactions Δn which occur per unit time is given by

$$\Delta n = \sigma N \phi \Delta V, \tag{2.70}$$

where σ is the microscopic cross section per atom, a function of the type of atom and the energy of the individual particles of radiation, and N is the number of such atoms per unit volume of the medium.

If ϵ is defined as the average amount of energy *transferred* (a concept to be clarified later) from the primary radiation field to the medium in a single interaction, it is easy to see that ΔE, the amount of energy transferred per unit time to the medium due to all interactions within ΔV, is given by

$$\Delta E = \epsilon \sigma N \phi \Delta V. \tag{2.71}$$

Dosimetric concepts commonly deal with the energy deposition rate per unit of mass, so if the mass of the incremental volume is given by $\rho \Delta V$, where ρ is the density of material in ΔV, then the isotropic detector response is

$$R \equiv \frac{\Delta E}{\rho \Delta V} = \epsilon \frac{\sigma N}{\rho} \phi. \tag{2.72}$$

The response R is easily established as a point function, since ΔV can be taken to a zero limit in the stochastic manner discussed in Chap. 1 without any consequent change in formula Eq. (2.72).

Generalization of this formula to take into account a distribution of particle energies and a mixture of types of atoms, as well as the fact that ϵ will differ for different types of interactions, results in the following equation:

$$R(\mathbf{r}) = \int_0^\infty dE \sum_i \frac{N_i(\mathbf{r})}{\rho(\mathbf{r})} \sum_j \sigma_{ji}(E)\epsilon_{ji}(E)\phi(\mathbf{r}, E), \qquad (2.73)$$

where the dependence on atomic species, or even on isotopes of the same element, is characterized by the subscript i and the specific type of interaction involved is indicated by the subscript j. The units on both sides of this equation must be consistent, of course, or a conversion factor must be introduced to account for any inconsistencies.

If the dosimetric response of interest happens to be precisely the energy deposition rate per unit mass of the medium at the point where the idealized detector is located, the corresponding response function is that portion of the integrand in Eq. (2.73) which multiplies ϕ. That is, for this particular circumstance,

$$R = \int_0^\infty dE\, \mathcal{R}(E)\phi(E), \qquad (2.74)$$

where[10]

$$\mathcal{R}(E) = \sum_i \frac{N_i}{\rho} \sum_j \sigma_{ji}(E)\epsilon_{ji}(E). \qquad (2.75)$$

This particular formulation is most directly useful for the type of response called the kerma rate; but it can be used, with precautions, for related responses such as absorbed dose rate and exposure rate.

2.12 RELATIONSHIP OF KERMA RATE AND ABSORBED DOSE RATE

The kerma rate is determined from Eqs. (2.74) and (2.75), as can be seen from its definition in Eq. (1.7), Sec. 1.3.2, as long as ϵ is taken to be the sum of the initial kinetic energies of the secondary charged particles resulting from the interactions of the primary radiation with the atoms in the sensitive volume of the detector.

At first glance, the foregoing equations appear not to be useful for prediction of *absorbed dose rate*, as defined by Eq. (1.6), since this concept requires in principle an accounting of the energy transferred to ionization and related processes along those segments of the tracks of secondary charged particles which lie within the sensitive volume. However, there is frequently a close numerical correspondence between the

---

[10]Note that $N_i/\rho = w_i(N_i/\rho_i)$, where w_i is the weight fraction of the ith constituent. (N_i/ρ_i), which may be written $(N/\rho)_i$, depends only on the identity of the ith constituent.

kerma rate and the absorbed dose rate at a specific location, so that a determination of the kerma rate may be used to estimate the absorbed dose rate to a close approximation.

To appreciate this, one must understand the nature of what is called *charged-particle equilibrium*. This condition is said to exist for a small incremental volume at a given location if, for every charged particle leaving the volume, another of the same type and with the same kinetic energy enters the volume traveling in the same direction. Sometimes, one is concerned only with a specific type of charged particle; thus, one may speak, for example, of *electronic equilibrium*, thus involving only electrons, or *heavy-charged-particle equilibrium*.

This condition is an idealization which is practically impossible to obtain exactly, especially in view of the statistical nature of a radiation field. However, it may be approached under many circumstances, as can be made clear by the following plausible, if not rigorous, analysis.

Consider Fig. 2.18, which depicts all space as being divided up into similar regions, of which the shaded one represents the incremental volume of interest. (The analysis should actually be made on a three-dimensional basis; but the two-dimensional simplification is sufficient to show the principle and is much simpler to draw clearly.) The line marked a–d represents a typical track of a secondary charged particle arising from an interaction by a primary particle within the shaded volume. Ionization occurs along this track. (It is made straight in the figure, but this is not necessary to the argument.) Assume that the primary radiation field is practically constant over this entire region — at least over that portion of it extending away from the shaded volume by a distance equal to the range of the most energetic secondary charged particle likely to be produced. If this is so, one might expect, *on the average*, that for every track $a - d$ starting from the shaded volume, a similar one starts at a similar point in all nearby volumes with the same energy, proceeding in the same direction and with the same trajectory from start to end of its range. Some of these, which start from surrounding volumes, will penetrate the shaded volume. The kinetic energy of the particle starting at a will contribute to the *kerma* in the shaded volume; no other particles shown will make a contribution to this volume. The *absorbed dose*, on the other hand, contributed by the secondary particle from a and all similar particles which penetrate the shaded volume come from energy deposited in ionization and related processes along the track segments a–b, b'–c', and c''–d''. But the energy deposited along b'–c' is the same as the particle kinetic energy lost to the shaded volume in the track segment b–c; similarly, the energy deposited along c''–d'' is the same as the particle kinetic energy lost along the track segment c–d. Thus, as far as the particular tracks shown and corresponding ones nearby are concerned, the contribution to the kerma and the contribution to the absorbed dose are the same in the shaded volume. This compensation for lost secondary particles from the volume of interest is then easily generalized to cover *all* interactions in the shaded volume and in the nearby volumes. When such compensation occurs in the shaded volume, there exists what is called "charged particle equilibrium," and under this condition the kerma and absorbed dose would be equal in this volume.

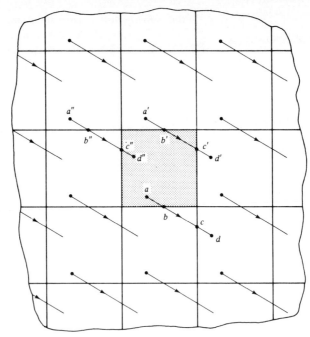

Figure 2.18 All space is divided into similar regions (represented by squares) in the vicinity of the incremental volume of primary interest, shown as the shaded square. In each incremental volume, corresponding interaction events for the primary particles are shown, together with lines representing corresponding secondary charged-particle tracks.

It is easy to see that two separate factors might prevent the establishment of charged-particle equilibrium. In either case, kerma and absorbed dose are not likely to be equivalent.

1. If, within the range of the more energetic charged particles, a boundary exists between the medium containing the shaded volume and another medium of appreciably different properties (which may be the same material but with a substantially different density), the appropriate correspondence between similar tracks may not be established.

2. If the primary radiation field varies appreciably over the region within the range of the secondary particles around the shaded volume, one cannot expect a close correspondence between interactions and secondary charged particles in the shaded volume and in its nearby elements.

At this point the reader is reminded that it has been assumed that $\phi(E)$ in Eq. (2.74) is known exactly. Fluorescence x rays, bremsstrahlung, and annihilation photons are included in the flux density. Even so, and even if the previous two conditions are not violated, there still may be a difference between kerma and absorbed dose. Assume a modification in the depicted figure to the extent that each secondary track, when it reached the point b or b' or b'', was terminated and the remaining kinetic energy in the particle was carried off as a result of bremsstrahlung emission. In such case, the absorbed dose contribution provided by the segments $b'-c'$ and $c''-d''$ in the original case would be lost; but the kerma would remain the same as in the original case.

Subject then to the limitations of charged particle equilibrium and negligible bremsstrahlung production, the kerma rate provides a good approximation to the absorbed dose rate. Since determination of kerma rate does not require examination of the transport of the secondary particles, its analytical determination is much easier than that of absorbed dose and thus there is a strong incentive to use the kerma rate calculation for absorbed dose rate wherever possible. Those who measure absorbed dose by conventional means usually take care to ensure that the measurement process is such as to provide charged-particle equilibrium to a good approximation, so that the measured value can be accurately predicted through calculation of the kerma. This is usually realized by having a sufficiently thick wall around the sensitive volume, with the wall and the sensitive volume having mass interaction coefficients which are very nearly equal over a wide range of photon energies.

2.13 NEUTRON KERMA, ABSORBED DOSE, AND DOSE EQUIVALENT

Even though absorbed dose rate is of more biological and engineering significance than kerma rate, it is quite customary to calculate kerma rate when dealing with neutrons. Except within a distance from a boundary equal to the greatest range of the secondary charged particles (heavy particles, with short ranges, usually), one can expect charged-particle equilibrium to exist to a good approximation (Chen and Chilton 1979a; ICRU 1977). Furthermore, the probability of bremsstrahlung carrying off some of the secondary-particle energy is very small for heavy particles. Thus, for neutrons, kerma rate is usually a very good approximation to absorbed dose rate.

For neutron kerma rate, Eqs. (2.74) and (2.75) apply, modified if one wishes by the notational replacement of $N_i \sigma_{ji}$ with μ_{ji}, the interaction coefficient for the jth type of interaction by a neutron of the given energy with the ith-type atom. For this type of response, ϵ_{ji} refers specifically to the kinetic energy of the charged particle, which may be a recoil nucleus, resulting from this interaction.

The determination of ϵ for use in specifying this response function depends on the interaction involved. For neutron elastic scattering, it can be shown that, on average, the energy per interaction transferred to kinetic energy of the recoil nucleus is given by

$$\epsilon_{elas} = \frac{2EA}{(A+1)^2}[1 - f_1(E)], \tag{2.76}$$

where A is the ratio of the mass of the nucleus involved to that of the neutron[11], and f_1 is the average cosine of the angle of elastic scattering in the center-of-mass system.

---

[11]The ratio of the nuclear and neutron masses is not exactly equal to the atomic mass; nevertheless, the two are so nearly equal that usually no distinction is made, and the same symbol is used.

TABLE 2.6 Values of f_1 for the Neutron Elastic Scattering Cross Section for Various Elements.

	ELEMENT					
E (MeV)	Li	Be	C	O	Si	Fe
6.6282	0.527	0.569	0.220	0.523	0.550	0.790
4.0202	0.340	0.385	0.016	0.372	0.449	0.584
2.4384	0.103	0.017	−0.017	0.022	0.353	0.374
1.4790	0.046	0.222	0.048	0.072	0.287	0.244
0.8970	−0.011	0.166	0.057	0.009	0.217	0.277
0.5441	−0.131	0.045	0.047	0.279	0.152	0.197
0.3300	−0.256	0.036	0.030	−0.109	0.088	0.079
0.2002	0.289	0.024	0.018	0	0.048	0.053
0.1214	0.130	0.015	0.011	0	0.029	0.032
0.0736	0.074	0.010	0.007	0	0.017	0.020
0.0447	0.043	0.006	0.004	0	0.010	0.012
0.0271	0.026	0.004	0.002	0	0.006	0.007
0.0164	0.016	0.002	0.001	0	0.003	0
0.0100	0.010	0.001	0.001	0	0	0

Source: Foderaro, Hoover and Marable 1968; by permission of Springer-Verlag, New York, Inc.

Typical values of f_1 for elastic scattering are given in Table 2.6. For isotropic scattering, $f_1 = 0$, and Eq. (2.76) reduces to $\epsilon_{elas} = 2EA/(A+1)^2$.

For neutron inelastic scattering which excites a level of the nucleus by an amount of energy $|Q|$ above the ground state, the average kinetic energy of the recoil nucleus can be shown to be

$$\epsilon_{inel} = \frac{2EA}{(A+1)^2}\left[1 + \frac{1}{2}\Delta - f_1(E)\sqrt{1+\Delta}\right], \qquad (2.77)$$

where

$$\Delta \equiv -\frac{|Q|}{E}\frac{A+1}{A}. \qquad (2.78)$$

The f_1 coefficients for inelastic scattering are not as well known as those for elastic scattering and are usually taken as zero, in which case the result as shown reduces to Eq. (2.56).

Methods of deriving the energy of charged particles and recoil nuclei in connection with the above, as well as other types of possible reactions, are available in the literature (ICRU 1977). With this information and data on other pertinent variables in Eqs. (2.74) and (2.75), it is possible to determine the response function for various elements and materials of interest. Caswell and Coyne (1980) have done this for various tissue-like materials and produced kerma response functions, which they call "kerma factors." Data for the neutron kerma response function in a four-element approximation to tissue are given in Table 2.7.

TABLE 2.7   Data for the Neutron Kerma Response Function in a Four-Element Approximation to Tissue[a]. For $E < 10^{-5}$ MeV, the Kerma Response Function Can be Calculated Approximately as (NCRP 1971) $\mathcal{R}_K(E) = 3.3 \times 10^{-15}/\sqrt{E}$.

E (MeV)	$\mathcal{R}_K(E)$ (cGy cm)	E (MeV)	$\mathcal{R}_K(E)$ (cGy cm)	E (MeV)	$\mathcal{R}_K$ (cGy cm)
0.110(−04)[b]	0.147(−11)	0.310(+00)	0.128(−08)	0.330(+01)	0.406(−08)
0.200(−04)	0.122(−11)	0.330(+00)	0.132(−08)	0.350(+01)	0.415(−08)
0.360(−04)	0.112(−11)	0.350(+00)	0.137(−08)	0.370(+01)	0.425(−08)
0.630(−04)	0.122(−11)	0.370(+00)	0.132(−08)	0.390(+01)	0.418(−08)
0.110(−03)	0.156(−11)	0.390(+00)	0.148(−08)	0.420(+01)	0.431(−08)
0.200(−03)	0.237(−11)	0.420(+00)	0.162(−08)	0.460(+01)	0.431(−08)
0.360(−03)	0.393(−11)	0.460(+00)	0.164(−08)	0.500(+01)	0.455(−08)
0.630(−03)	0.662(−11)	0.500(+00)	0.160(−08)	0.540(+01)	0.444(−08)
0.110(−02)	0.114(−10)	0.540(+00)	0.165(−08)	0.580(+01)	0.464(−08)
0.200(−02)	0.204(−10)	0.580(+00)	0.171(−08)	0.620(+01)	0.475(−08)
0.360(−02)	0.362(−10)	0.620(+00)	0.177(−08)	0.660(+01)	0.489(−08)
0.630(−02)	0.622(−10)	0.660(+00)	0.183(−08)	0.700(+01)	0.510(−08)
0.110(−01)	0.106(−09)	0.700(+00)	0.189(−08)	0.740(+01)	0.537(−08)
0.200(−01)	0.183(−09)	0.740(+00)	0.194(−08)	0.780(+01)	0.529(−08)
0.360(−01)	0.303(−09)	0.780(+00)	0.199(−08)	0.820(+01)	0.525(−08)
0.630(−01)	0.470(−09)	0.820(+00)	0.204(−08)	0.860(+01)	0.542(−08)
0.820(−01)	0.567(−09)	0.860(+00)	0.210(−08)	0.900(+01)	0.551(−08)
0.860(−01)	0.587(−09)	0.900(+00)	0.217(−08)	0.940(+01)	0.555(−08)
0.900(−01)	0.605(−09)	0.940(+00)	0.227(−08)	0.980(+01)	0.568(−08)
0.940(−01)	0.624(−09)	0.980(+00)	0.245(−08)	0.105(+02)	0.582(−08)
0.980(−01)	0.641(−09)	0.105(+01)	0.248(−08)	0.115(+02)	0.624(−08)
0.105(+00)	0.672(−09)	0.115(+01)	0.246(−08)	0.125(+02)	0.621(−08)
0.115(+00)	0.713(−09)	0.125(+01)	0.256(−08)	0.135(+02)	0.645(−08)
0.125(+00)	0.752(−09)	0.135(+01)	0.265(−08)	0.145(+02)	0.670(−08)
0.135(+00)	0.789(−09)	0.145(+01)	0.269(−08)	0.155(+02)	0.687(−08)
0.145(+00)	0.825(−09)	0.155(+01)	0.277(−08)	0.165(+02)	0.695(−08)
0.155(+00)	0.860(−09)	0.165(+01)	0.287(−08)	0.175(+02)	0.702(−08)
0.165(+00)	0.892(−09)	0.175(+01)	0.291(−08)	0.185(+02)	0.715(−08)
0.175(+00)	0.924(−09)	0.185(+01)	0.303(−08)	0.195(+02)	0.727(−08)
0.185(+00)	0.954(−09)	0.195(+01)	0.304(−08)	0.210(+02)	0.742(−08)
0.195(+00)	0.983(−09)	0.210(+01)	0.313(−08)	0.230(+02)	0.739(−08)
0.210(+00)	0.103(−08)	0.230(+01)	0.318(−08)	0.250(+02)	0.734(−08)
0.230(+00)	0.108(−08)	0.250(+01)	0.331(−08)	0.270(+02)	0.736(−08)
0.250(+00)	0.113(−08)	0.270(+01)	0.346(−08)	0.290(+02)	0.724(−08)
0.270(+00)	0.118(−08)	0.290(+01)	0.360(−08)		
0.290(+00)	0.123(−08)	0.310(+01)	0.373(−08)		

[a]Composition, by mass, is 10.1% H, 11.1% C, 2.6% N, and 76.2% O.

[b]Read as 0.110×10^{-4}, and so on.

Source: Caswell and Coyne 1980.

In order to determine the dose equivalent rate, one must multiply the tissue absorbed dose rate by the average quality factor of all the secondary charged particles within the incremental volume at the specified location.[12] A strict determination of the quality factors demands complete knowledge of the energies of the resulting particles and a careful accounting of the relationship between quality factor and LET as in Table 1.1. It is often satisfactory for practical purposes in determining neutron dose equivalent to assign an overall average quality factor of 10, which is generally a safe estimate (ICRP 1970).

2.14 DOSIMETRIC RESPONSE FUNCTIONS FOR PHOTONS

Equation (2.73) applies to photons, as well as neutrons, but it is usually manipulated so as to appear as

$$R(\mathbf{r}) = \int_0^\infty dE \, \frac{\mu_d(r, E)}{\rho(\mathbf{r})} E \, \phi(\mathbf{r}, E), \tag{2.79}$$

where μ_d, called the *linear energy deposition coefficient*,[13] is given by

$$\mu_d(\mathbf{r}, E) = \sum_i N_i(\mathbf{r}) \sum_j \sigma_{ji}(E) \frac{\epsilon_{ji}(E)}{E}. \tag{2.80}$$

Since N_i is directly proportional to the density of the ith element in the mixture, one sees that the *mass* energy deposition coefficient μ_d/ρ, like μ/ρ, is independent of the density. For a situation in which the attenuating medium is a mixture of elements in a constant proportion to one another, μ_d/ρ is independent of position, even though there may be density variations within the medium.

For the remainder of this section, we shall ignore the variable $\mathbf{r}$ as an argument, since we are concerned only with the response of a point isotropic detector at a given position. Furthermore, we shall consider the situation only for a monoenergetic flux density, since extension to a polyenergetic spectrum is obvious. This restriction will simplify the algebra without compromising the principles to be developed.

2.14.1 Photon Energy Deposition Coefficients

So far, the term "energy deposition," denoted by ϵ, has been kept somewhat vague. As one defines ϵ more precisely, one can develop a number of different energy deposition coefficients; and since several different ones are used, sometimes incorrectly, in the radiation-protection literature, it is necessary to explain the distinction between them carefully.

It is desirable for this purpose to review the energy balance for a small test volume ΔV of a medium interacting with a monoenergetic photon flux density. A slightly

---

[12]The basic dose equivalent concept noted here may be referred to as "primitive," to distinguish it from the more complex phantom-related concept discussed later in Sec. 2.15.

[13]See Sec. 2.2 for a discussion of average coefficients for compounds and mixtures.

simplified picture of the disposition of the energy involved in the many possible photon interactions within the test volume is shown in Fig. 2.19(a). From Eqs. (2.79) and (2.80) and from the definitions of the various photon coefficients provided earlier, one can derive the related picture of the disposition of energy per unit mass of the medium in the test volume as shown in Fig. 2.19(b). One then arrives at the relationship expressed in Fig. 2.19(c) in direct relationship to the diagram in Fig. 2.19(b), by which various energy deposition coefficients are defined schematically. Applications of the various coefficients are discussed in subsequent sections, but their mathematical definitions easily follow:

1. μ_a, called the *linear absorption coefficient* (Hubbell 1968), is given by

$$\mu_a = \mu - (1 - f_c)\mu_c = \mu_{ph} + \mu_{pp} + f_c\mu_c. \tag{2.81}$$

2. μ'_{tr}, although occasionally used (Morris, Chilton and Vetter 1975), has no officially prescribed name. It very well might be called the *linear pseudo-energy-transfer coefficient* since its use is generally as an alternative to μ_{tr} that follows, and is defined by

$$\mu'_{tr} = \mu_a - (1 - f_{pp})\mu_{pp} = \mu_{ph} + f_{pp}\mu_{pp} + f_c\mu_c. \tag{2.82}$$

3. μ_{tr} is called the *linear energy transfer coefficient* (ICRU 1980), and is given by

$$\mu_{tr} = \mu'_{tr} - (1 - f_{ph})\mu_{ph} = f_{ph}\mu_{ph} + f_{pp}\mu_{pp} + f_c\mu_c \tag{2.83}$$

4. μ_{en} is called the *linear energy absorption coefficient* (ICRU 1980) and is given by

$$\mu_{en} = (1 - G)\mu_{tr} = (1 - G_{ph})f_{ph}\mu_{ph} + (1 - G_c)f_c\mu_c$$
$$+ (1 - G_{pp})f_{pp}\mu_{pp}, \tag{2.84}$$

where G is the radiative emission constant defined in Sec. 2.5.4.

Values of the mass energy deposition coefficients μ_{tr}/ρ and μ_{en}/ρ for several elements and mixtures of shielding interest have been extracted from the literature and are provided in Appendix C. Values for μ'_{tr}/ρ are not generally available and must be determined from its basic definition given here. Values for mixtures of elements are determined from either the number fractions or weight fractions for the individual elemental constituents, just as in the case of mass attenuation coefficients.

2.14.2 Photon Kerma, Absorbed Dose, and Dose Equivalent

The response function for photon detectors of the energy deposition type is easily seen from Eq. (2.79) to be

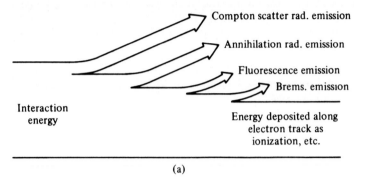

(a)

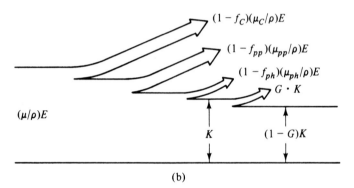

(b)

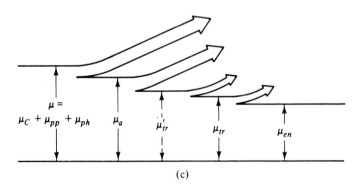

(c)

Figure 2.19   Relationships leading to definitions of various energy deposition coefficients for photons. (a) Energy deposition for photon energy involved in the interactions in an incremental volume of material. (b) Formulas for the energy per unit mass of the material in the incremental volume, corresponding to the various energy increments in (a). (c) Linear coefficients defined by their proportionality to the mass energy relationships in diagrams (a) and (b).

$$\mathcal{R}(E) = E\left(\frac{\mu_d(E)}{\rho}\right), \tag{2.85}$$

where the proper mass energy deposition coefficient $\mu_d(E)$ to be used depends on exactly which detector response is to be determined, and possibly on other related matters to be discussed later in Sec. 2.15.

If ϵ in Eq. (2.80) denotes the average initial kinetic energy of the secondary charged particles (electrons) from photon interactions in the incremental volume, then μ_d is identified as μ_{tr}. With the use of the mass energy transfer coefficient μ_{tr}/ρ in Eq. (2.79), the response then becomes the photon kerma. In Eq. (2.79), as in Eq. (2.74), $\phi(E)$ includes bremsstrahlung as well as fluorescence and annihilation photons. When this is not so, one must resort to one of the approximations discussed in Sec. 2.14.4.

Provided electronic equilibrium exists and there is no bremsstrahlung to remove energy through further photon emission from the secondary charged particles, the photon kerma rate is also a good measure of the absorbed dose rate. This equality between kerma rate and absorbed dose rate for photons is not as nearly universal as it is in the case of neutrons, because the range of the secondary charged particles (electrons) in the photon case may be rather large. In such a case, electronic equilibrium is much more likely to be a poor assumption. Similar considerations relate to the presence of the bremsstrahlung effect; under usual conditions, the secondary electrons arising from photon interactions may be of an energy leading to high secondary-photon yields, especially in high-Z material. For the secondary heavy particles from neutron interactions, the bremsstrahlung photon yield is negligible.

Under conditions in which kerma rate is not a good approximation for the absorbed dose rate, accurate dose-rate assessments require taking into account the transport of the secondary charged particles. The production of bremsstrahlung can be taken into account by the substitution in Eq. (2.79) of μ_{en}/ρ for μ_d/ρ as the mass energy deposition coefficient. Such a substitution is easy and should be done generally.

In case electronic equilibrium does not exist, at least to a close approximation, there is no way around inclusion of secondary-electron transport in the dose assessment, a process that greatly compounds the difficulties inherent in analyzing the radiation field and is rarely undertaken if avoidable.

Since the quality factor for electrons is always taken as unity, the dose equivalent for photon interactions is always numerically given by the absorbed dose, provided that the units are consistent.

The following formulas summarize the response functions, expressed in standard units, for an idealized point isotropic photon detector. The kerma rate response function, in units of Gy cm^2, is

$$\mathcal{R}_K(E) = 1.602 \times 10^{-10}\, E\left(\frac{\mu_{tr}(E)}{\rho}\right). \tag{2.86}$$

Under the assumption of electronic equilibrium, the response function for the absorbed dose rate, in units of Gy cm^2, becomes

$$\mathcal{R}_D(E) = 1.602 \times 10^{-10} \, E \left(\frac{\mu_{en}(E)}{\rho} \right). \tag{2.87}$$

Finally, the response function for the dose equivalent rate, in units of Sv cm^2, is numerically the same as the response function for the absorbed dose rate, in units of Gy cm^2; so that, in the units prescribed,

$$\mathcal{R}_H(E) = \mathcal{R}_D(E). \tag{2.88}$$

In the formulas above, the unit of E is MeV. In order for these response functions to yield a proper result for the detector response when the functions are folded into the flux density, the latter must be expressed in units of cm^{-2}s^{-1}. Units of μ_{tr}/ρ and μ_{en}/ρ must be cm^{2}g^{-1}.

2.14.3 Photon Exposure Rate

The photon exposure is defined in terms of the electric charge (of either sign) able to be created in air as a result of the interactions in a unit mass of air. The factor W, which gives the energy deposited per ion created, is the connecting link between exposure and the energy deposition per unit mass, R; and, because air is the medium of concern, the subscript i in Eq. (2.73) when used for this purpose must relate to air. Thus, the energy deposition coefficient to be used with Eq. (2.84) must be for air only. (The ΔV in such cases represents conceptually the small sensitive volume of an air detector lined with a wall of an air-like substance and embedded in whatever medium the photon exposure field is being determined.) The coefficient to be used should be that related to the kinetic energy of the secondary electrons in air, reduced by a factor which takes into account the fact that some of the kinetic energy may be taken away by bremsstrahlung rather than by causing ionization. This coefficient is μ_{en}/ρ.

Fortunately, W is practically constant with respect to electron energy; and it has a value in air estimated at 33.85 eV per ion pair (ICRU 1979). By this means, the exposure rate can be obtained from the formulas given by Eqs. (2.79) and (2.80), with the use of necessary units conversions and the fundamental definition of the exposure unit of the roentgen (see Chap. 1, Sec. 1.3.2)

$$\dot{X} = \int_E dE \, \mathcal{R}_X(E)\phi(E), \tag{2.89}$$

in which the response function, in units R cm^2, is

$$\mathcal{R}_X(E) = \frac{10^4 \, \text{R kg}}{2.58 \, \text{C}} \times \frac{1.602 \cdot 10^{-19} \, \text{C}}{\text{ion}} \times \frac{10^6 \, \text{ions}}{33.85 \, \text{MeV}}$$

$$\times \frac{10^3 \, \text{g}}{\text{kg}} \times E \left(\frac{\mu_{en}(E)}{\rho} \right)_{air} \tag{2.90}$$

or

$$\mathcal{R}_X(E) = 1.835 \times 10^{-8} \, E \left(\frac{\mu_{en}(E)}{\rho} \right)_{air}. \tag{2.91}$$

This formula requires that E be given in MeV, the mass energy absorption coefficient given in cm^2 g^{-1}, and the flux density spectrum expressed in cm^{-2} s^{-1} MeV^{-1}. In case the exposure rate of Eq. (2.89) is desired in R h^{-1}, as is common, the numerical factor in Eq. (2.91) should be multiplied by 3600 to give 6.606×10^{-5} so that the units for $\mathcal{R}_X$ would be changed to R cm^2 s h^{-1}.

2.14.4 Selection of Proper Mass Energy Deposition Coefficients

In spite of the specificity of the energy deposition coefficients given in the formulas of the preceding section, there are frequent occasions in which other selections may be better. *The previous formulas are valid provided that the photon flux density is known exactly.* On the other hand, many analyses of the flux density are carried out on somewhat less exact assumptions than those presumed for the purpose of these formulas.

Figure 2.19 indicates the following secondary-photon emission processes which may remove energy from a small volume element in which the interactions occur, these photon emissions, of course, contributing to the overall photon field:

1. emission of Compton scattered photons,
2. annihilation radiation following pair production,
3. fluorescence radiation,
4. bremsstrahlung.

Very often analysts will neglect the less important processes, lower on the list, in their analysis of the photon flux density field at the locations of dosimetric interest. In such cases, this is equivalent to making an approximation that the neglected secondary photons are either negligible in number or so weak that they may be considered to be absorbed and to deposit their energy at the site of the original interaction.

It has been found that, in case an approximation of this sort has to be made, the type of mass coefficient best suited for determining the dose rate response is one that is consistent with the approximation made in the flux density determination (Hubbell 1968). One way to rationalize this rule is as follows. If the approximation used is such as to neglect photon emissions from certain reactions, the calculated flux density at most points will be underestimated. However, the value of μ_d consistent with such an approximation, when used to determine the response function, will be too high. This will thus give a response function which is a little high and, when combined with a flux density which is a little low, may provide a fairly accurate answer by the principle of "compensating errors." Figure 2.19 is very useful in identifying the proper energy deposition coefficient to use.

In particular, it is often the practice in a photon transport calculation to account only for photons resulting from Compton scattering and annihilation radiation, implying that fluorescence radiation and bremsstrahlung radiation, if they exist, are absorbed locally. In such case, the coefficient μ'_{tr}/ρ should be used in the response function formulas such as given by Eqs. (2.86) to (2.91).[14] Older calculations have

---

[14] If the fluorescence radiation is negligible, μ_{tr} could be reasonably used here since the distinction between μ'_{tr} and μ_{tr} is correspondingly negligible.

often ignored everything except the Compton scattering for determining the flux density field, and under such circumstances the use of μ_a/ρ for any and all mass energy deposition coefficients is more proper (Goldstein and Wilkins 1954).

Let us illustrate the points just made by way of a series of examples. Suppose that one calculates $\phi(E)$ by successively more sophisticated transport calculations and one then wishes, for each calculation, to make the best estimate of detector response R using Eq. (2.79).

Case 1. In the calculation of $\phi(E)$ only primary and Compton-scattered photons are included. In doing so, one in effect makes the approximation that, when a primary photon suffers a photoelectric interaction, the photon is lost with all its energy being transferred to a photoelectron, no fluorescence takes place, and no bremsstrahlung emission takes place along the path of the photoelectron. One approximation begets another. One is forced to assume that the entire energy of the photon is deposited at the point of the interaction. Similarly, if a primary photon suffers pair production, one is forced to make the approximation that all the photon's energy is deposited at the point of the interaction (including that which otherwise would have appeared as the energies of annihilation photons) with no bremsstrahlung emission along the tracks of the positron and electrons created in the interaction. In other words, one is forced to set both f_{ph} and f_{pp} to unity, and to set G_c, G_{ph}, and G_{pp} to zero. If these approximations are invoked in Eq. (2.84), the equation for μ_{en}, the result is just μ_a, as given in Eq. (2.81). Thus, for Case 1,

$$R \simeq R_1 = \int dE\, \mathcal{R}_1(E)\phi_1(E),\tag{2.92}$$

where

$$\mathcal{R}_1(E) = E\,\frac{\mu_a(E)}{\rho}.\tag{2.93}$$

Case 2. In the calculation of $\phi(E)$ only primary, Compton-scattered, and annihilation photons are included. One treats photoelectric interactions as in Case 1. One still makes the approximation that no bremsstrahlung emission takes place along positron and electron tracks. In other words, one assumes that f_{ph} is unity, and that G_c, G_{ph}, and G_{pp} are all equal to zero. If these assumptions are invoked in Eq. (2.84), the equation for μ_{en}, the result is just μ'_{tr}, as given in Eq. (2.82). Thus, for Case 2,

$$R \simeq R_2 = \int dE\, \mathcal{R}_2(E)\phi_2(E),\tag{2.94}$$

where

$$\mathcal{R}_2(E) = E\,\frac{\mu'_{tr}(E)}{\rho}.\tag{2.95}$$

Case 3. In the calculation of $\phi(E)$ primary, Compton-scattered, fluorescence and annihilation photons are included. In other words, one assumes only that G_c, G_{ph}, and G_{pp} are all equal to zero. If this assumptions is invoked in Eq. (2.84), the equation for μ_{en}, the result is just μ_{tr}, as given in Eq. (2.83). Thus, for Case 3,

$$R \simeq R_3 = \int dE \, \mathcal{R}_3(E)\phi_3(E), \tag{2.96}$$

where

$$\mathcal{R}_3(E) = E \, \frac{\mu_{tr}(E)}{\rho}. \tag{2.97}$$

Case 4.   In this most sophisticated calculation of $\phi(E)$ primary, Compton-scattered, fluorescence and annihilation photons as well as bremsstrahlung are included. No assumptions are need in Eq. (2.84), the equation for μ_{en}, and, for this case,

$$R \simeq R_4 = \int dE \, \mathcal{R}_4(E)\phi_4(E), \tag{2.98}$$

where

$$\mathcal{R}_4(E) = E \, \frac{\mu_{en}(E)}{\rho}. \tag{2.99}$$

The reader will note that, qualitatively,

$$\phi_1 < \phi_2 < \phi_3 < \phi_4, \tag{2.100}$$

while

$$\mathcal{R}_1 > \mathcal{R}_2 > \mathcal{R}_3 > \mathcal{R}_4. \tag{2.101}$$

One may then argue that, by "compensating errors,"

$$R_1 \simeq R_2 \simeq R_3 \simeq R_4. \tag{2.102}$$

2.15 RESPONSE FUNCTIONS FOR EVALUATION OF HAZARDS TO HUMANS

Since radiological assessments are widely used for the protection of humans, the detector responses of particular interest are those related to radiation hazards to people. In order to use any of the response functions thus far discussed in computing the detector response or "dose" by Eq. (2.69), it is necessary for the analyst to determine the flux density energy spectrum at every point where the risk is to be evaluated. It has been the custom, especially for photons, to compute the detector response from the *free-field* flux density, that is, in a radiation field uninfluenced by the actual presence of a human body. Thus, it is still common to examine the hazards of a photon field in terms of free-field values of the exposure rate, air kerma rate, or tissue kerma rate.

Nonetheless, it would appear more proper to establish the risk to a human being at some point in a radiation field by assuming that a human body, or a reasonable semblance of one, is present at the position being evaluated. Of course, the body itself alters the radiation field. The detector response then should be that obtained at some critical point or points within such a body, and clearly the best suited type of detector response is the dose equivalent.

Unfortunately, the determination of dose equivalent inside the human body, or even a simplified representation of the body called a *phantom*, greatly complicates the problem of the analyst. First, the total flux density and its spectral distribution (as perturbed by the presence of the body) must be obtained at each critical point within the body or phantom. Next, one must consider the validity of the assumption regarding the equality of absorbed dose and kerma, so that Eqs. (2.87) and (2.88) can be used in the case of photons or Table 2.7 can be used in case of neutrons. If this assumption is not clearly valid in any specific case, at least to a good approximation, the analyst must follow the transport of the secondary charged particles to see where they dissipate their kinetic energy. Finally, in order to convert the absorbed dose to dose equivalent, one must select a value for the average quality factor, thereby possibly requiring a detailed knowledge of the charged particle spectrum at each point of interest.

Fortunately for the analyst, the foregoing process can be avoided by using an approximate procedure which, although it involves the concept of finding the dose equivalent in the human body or a phantom, is much simpler and gives results which only mildly compromise the accuracy of the hazard determination, most always by overestimating the hazard. This simplified procedure uses a *prescribed response function* which, if folded into the *free-field* flux density spectrum, gives a reasonable estimate of the dose equivalent values at the critical point or points inside the body. Since the free-field flux density is much easier to obtain than that for the radiation inside the actual body, the prescribed response function permits a straightforward use of Eq. (2.83) to determine the dose equivalent in the body.

Phantoms of the following types have been studied to provide information of the interaction of both neutrons and photons with idealizations of the human body:

1. homogeneous cylinders, slabs and spheres, with composition and density representative of the human body as a whole. The more widely used of these is the so-called ICRU sphere, a 30-cm diameter sphere of density $1.0 \, \mathrm{g \, cm^{-3}}$ composed by weight of 76.2% oxygen, 11.1% carbon, 10.1% hydrogen, and 2.6% nitrogen.

2. anthropomorphic phantoms geometrically and physically representative of the human body, including skeletal, lung and soft-tissue regions.

The following is a brief explanation of the procedure, somewhat idealized, used for obtaining the prescribed response function for whatever phantom is selected. Monoenergetic particles are assumed to be directed on the phantom in some prescribed angular distribution. At selected points or regions within the phantom, absorbed-dose, usually approximated by kerma, values are determined. In this determination, contributions by all secondary charged particles at that position are taken into account; and for each type of charged particle of a given energy the L_∞ value in water and, therefore, Q are obtained as in Chap. 1. These are then applied to the absorbed-dose contribution from each charged particle to obtain the dose equivalent contribution at the given location. The resulting distributions of absorbed dose and dose equivalent

TABLE 2.8 Neutron Absorbed Dose Index Response Function and Corresponding Values of the Effective Quality Factor.

NEUTRON ENERGY (MeV)	$\mathcal{R}_{D_p}$ (cGy cm^2)	Q_{eff} [a]
2.5(−08)[b]	0.5(−09)	2.3
1.0(−07)	0.6(−09)	2
1.0(−06)	0.6(−09)	2
1.0(−05)	0.6(−09)	2
1.0(−04)	0.6(−09)	2
1.0(−03)	0.5(−09)	2
1.0(−02)	0.5(−09)	2
1.0(−01)	0.8(−09)	7.4
5.0(−01)	1.8(−09)	11.
1.0	3.1(−09)	10.6
2.0	4.3(−09)	9.3
5.0	5.3(−09)	7.8
10.0	6.0(−09)	6.8
20.0	7.2(−09)	6.0
50.0	9.2(−09)	5.0

[a]These quality factors relate to the dose equivalent for neutrons defined before 1985. They should be multiplied by a factor of 2 to obtain values as redefined by the ICRP for neutrons in 1985 (ICRP 1985).

[b]Read as 2.5×10^{-8}, and so on.

Source: ICRP 1973.

throughout the phantom are then examined to obtain the maximum value, the value otherwise considered to be in the most significant location, or some weighted average dose equivalent for which the weighting factors are based on the radiosensitivity of the various organs and tissues represented by the phantom. The prescribed response function is then that value of either absorbed dose or dose equivalent divided by the free-field fluence of the incident beam postulated. The ratio of the dose equivalent at the point of greatest significance to the absorbed dose at that point is called the *effective quality factor* Q_{eff}.[15]

Two broad classifications of prescribed response functions $\mathcal{R}_{H_p}$ have been studied intensively. One is based on the anthropomorphic phantom and relates the *effective dose equivalent*[16] H_E to the free-field fluence. The other is based on the homogeneous phantom and relates the *absorbed dose index* D_I or the *dose equivalent*

---

[15]The data for the phantom response function for absorbed dose and for the effective quality factor are actually of little direct interest to the analyst. However, information on these are given in Table 2.8 to show the reader the intermediate values in the determination of the phantom dose equivalent. Of particular interest is the variation of effective quality factor with neutron energy.

[16]The effective dose equivalent is defined in Chap. 1, Sec. 1.4.5. In summary, it is an organ-weighted average dose equivalent, with weight factors proportional to the relative sensitivities of the body organs and tissues to radiogenic cancer.

index[17] H_I to the free-field fluence. ICRP and NCRP recommendations (ICRP 1977, NCRP 1987) for radiation protection are based on the effective dose equivalent, but the dose equivalent index may be used as an approximation (ICRP 1977).

2.15.1 Prescribed Response Functions for D_I and H_I

Several organizations, whose recommendations are highly regarded, have provided such response functions for slab and cylindrical phantoms. Prominent among these organizations are the National Council on Radiation Protection and Measurements (NCRP 1971) in the United States, the International Commission on Radiological Protection (ICRP 1973), and the American National Standards Institute (1977). Prescribed response functions provided by these organizations are listed in Tables 2.9 and 2.10. While the response functions given in these tables have been superseded by results of more modern calculations, they are included to assist the reader in the interpretation of the many radiation-protection studies which made use of them.

Prescribed response functions for the dose equivalent index, specific to the maximum value of the dose equivalent in a 30-cm-diameter spherical phantom irradiated with broad, unidirectional, and monoenergetic beams of neutrons or photons have also been studied (Chen and Chilton 1979a; Chilton and Shiue 1981; Shiue and Chilton 1983; Dimbylow and Francis 1979; ICRU 1980). This work has culminated in the response functions given in Tables 2.11 and 2.12 (ICRP 1987). Response functions for the ICRU sphere were calculated for four exposure conditions: (1) a single plane parallel beam, (2) two opposed plane parallel beams, (3) a rotating plane parallel beam, and (4) an isotropic radiation field. The response functions are greatest for case (1) and data for only that case are given in the Tables.

2.15.2 Prescribed Response Functions for H_E

A comprehensive catalog of response functions for the effective dose equivalent was published by the ICRP in 1987. Response functions are presented in Tables 2.13 and 2.14 for five conditions of exposure: (1) anterior-posterior (AP) irradiation with a parallel beam normal to the front and long axis of the body phantom, (2) posterior-anterior (PA) irradiation, (3) lateral (LAT) irradiation, (4) rotational (ROT) irradiation, in which the phantom body is rotated around its long axis while irradiation is in a parallel beam normal to the axis, and (5) irradiation in an isotropic radiation field (ISO). Case 4 (ROT) is thought to be an appropriate choice for the irradiation pattern experienced by a person moving unsystematically relative to the location of a source (ICRP 1987). Included in the ICRP compilation are individual organ doses for the five conditions of exposure.

---

[17]The absorbed dose index at a point is defined as the maximum absorbed dose within a 30-cm diameter sphere centered on the point and consisting of tissue-equivalent material with density 1.0 g cm^{-3}. The dose equivalent index is similarly defined for the maximum dose equivalent, but is further categorized as to the shallow dose equivalent index for the depth of 70 μm to 1 cm and the deep dose equivalent index for depths in excess of 1 cm.

TABLE 2.9 ANSI-Prescribed
Photon Response Function for the
Dose Equivalent Index.

PHOTON ENERGY (MeV)	$\mathcal{R}_{H_p}$ (cSv cm^2)
0.1	7.86(−11)[a]
0.15	1.05(−10)
0.2	1.39(−10)
0.25	1.75(−10)
0.3	2.11(−10)
0.35	2.44(−10)
0.4	2.74(−10)
0.45	3.00(−10)
0.5	3.25(−10)
0.55	3.53(−10)
0.6	3.78(−10)
0.65	4.00(−10)
0.7	4.22(−10)
0.8	4.67(−10)
1.0	5.50(−10)
1.4	6.97(−10)
1.8	8.31(−10)
2.2	9.50(−10)
2.6	1.06(−09)
2.8	1.11(−09)
3.25	1.23(−09)
3.75	1.34(−09)
4.25	1.45(−09)
4.75	1.56(−09)
5.0	1.61(−09)
5.25	1.67(−09)
5.75	1.77(−09)
6.25	1.87(−09)
6.75	1.98(−09)
7.5	2.13(−09)
9.0	2.44(−09)
11.0	2.86(−09)
13.0	3.28(−09)
15.0	3.69(−09)

[a] Read as 7.86×10^{-11}, and so on.

Source: ANSI 1977

2.15.3 Concluding Remarks

   To distinguish detector responses obtained by folding prescribed response functions into the free-field flux density from the more primitive free-field dose equivalent concept discussed in Secs. 2.13 and 2.14, the following convention is adopted. A dose equivalent response or "biological dose" resulting from the folding of prescribed response-function data with the free-field fluence[18] will be called in this text a *pre-*

---

[18] In principle, *phantom-related tissue absorbed dose* is similarly established. However, for most analyses this concept is of much less importance.

TABLE 2.10 Prescribed Neutron Response Functions for the Dose Equivalent Index.

NEUTRON ENERGY (MeV)	$\mathcal{R}_{H_p}$ (cSv cm^2)		
	NCRP	ICRP	ANSI
2.5(−08)[a]	1.02(−09)	1.07(−09)	1.02(−09)
1.0(−07)	1.02(−09)	1.16(−09)	1.02(−09)
1.0(−06)	1.24(−09)	1.26(−09)	1.24(−09)
1.0(−05)	1.24(−09)	1.21(−09)	1.26(−09)
1.0(−04)	1.20(−09)	1.16(−09)	1.16(−09)
1.0(−03)	1.02(−09)	1.03(−09)	1.04(−09)
1.0(−02)	9.9(−10)	9.9(−10)	9.9(−10)
1.0(−01)	6.0(−09)	5.8(−09)	6.0(−09)
5.0(−01)	2.6(−08)	2.0(−08)	2.6(−08)
1.0	3.7(−08)	3.3(−08)	3.7(−08)
2.0	—	4.0(−08)	—
2.5	3.5(−08)	—	3.5(−08)
5.0	4.3(−08)	4.1(−08)	4.3(−08)
7.0	4.1(−08)	—	4.1(−08)
10.0	4.1(−08)	4.1(−08)	4.1(−08)
14.0	5.8(−08)	—	5.8(−08)
20.0	6.3(−08)	4.3(−08)	6.3(−08)
40.0	6.9(−08)	—	—
50.0	—	4.6(−08)	—
100.0	5.0(−08)	5.0(−08)	—

[a] Read as 2.5 × 10^{-8}, and so on.
Source: NCRP 1971; ICRP 1973; ANSI 1977.

scribed dose equivalent, H_p. If there is any need to specify which set of response functions is used in a calculation of H_p, one must qualify the value by giving information on the function used such as, for example, with the phrase "based on the ANSI-prescribed response function."

There still remains some artificiality in the use of response functions based on interaction with body-sized phantoms when the radiation field may be actually highly variable over distances smaller than the dimensions of the phantom or when the point in the field being investigated is so close to a solid object that a human body could not be placed there. The usual procedure is to ignore such conceptual difficulties and accept the artificiality of the calculations for the sake of simplification of the required analysis, in the belief that any realistic analysis would not be likely to improve the accuracy of the risk evaluation to an extent worth the additional effort involved.

For most radiation protection purposes, the specialist can hardly do better than to use one of the response functions for the prescribed dose equivalent given in Tables 2.11 through 2.14, both for neutrons and for photons. Tradition, however, has led to the popular use of the free-field exposure rate for photons as being adequate for practical purposes. Similarly, the free-field kerma rate in a tissue-like detector is sometimes used for neutrons. For shield heating, the absorbed dose rate in the shield

TABLE 2.11   Prescribed Response Functions for Photon Dose Equivalent Indices under Irradiation of a 30-cm Diameter Phantom by a Single Parallel Beam.

	$\mathcal{R}_{H_p}$ $(10^{-10}$ cSv cm$^2)$ FOR DEEP AND SHALLOW INDICES H_I	
ENERGY (MeV)	DEEP DOSE INDEX	SHALLOW DOSE INDEX
0.01	0.0769	6.91
0.015	0.843	3.04
0.02	1.02	1.72
0.03	0.788	0.875
0.04	0.619	0.622
0.05	0.533	0.533
0.06	0.508	0.505
0.08	0.537	0.532
0.10	0.620	0.616
0.15	0.892	0.891
0.20	1.19	1.19
0.30	1.81	1.85
0.40	2.38	2.46
0.50	2.89	3.04
0.60	3.39	3.59
0.80	4.30	4.60
1.0	5.13	5.49
1.5	6.92	7.32
2.0	8.48	8.85
3.0	11.1	11.5
4.0	13.4	13.8
5.0	15.5	15.9
6.0	17.5	17.9
8.0	21.5	22.0
10.0	25.4	26.1

Source: ICRP 1987.

material, or the kerma rate as an approximation, is the appropriate response function for evaluating heat sources in shielding materials (Foderaro, Hoover and Marable 1968; Chilton 1972).

Finally, it should not be forgotten that the proper response function to use is that related to end use of computational results or the purpose of a measurement against which calculated response is to be compared. The detector response more nearly correlated with radiation effects on materials would be the absorbed dose, but this is not universally true. Radiation effects on semiconductors, for example, are related to somewhat more complex physical phenomena than simply energy deposition or ionization. Personnel dosimetry instruments are usually calibrated to read absorbed dose index or dose equivalent index and calculations related to such measurements would make use of the dose index response function. Determination of personnel dose by calculation alone would likely make use of the response function for effective dose equivalent.

TABLE 2.12 Prescribed Response Functions for Neutron Dose Equivalent Indices under Irradiation of a 30-cm Diameter Phantom by a Single Parallel Beam.[a]

| ENERGY (MeV) | $\mathcal{R}_{H_p}$ $(10^{-10}$ cSv cm$^2)$ FOR DEEP AND SHALLOW INDICES H_I | |
	DEEP DOSE INDEX	SHALLOW DOSE INDEX
2.5×10^{-8}	8.00	9.50
1.0×10^{-7}	10.1	11.1
1.0×10^{-6}	11.8	11.2
1.0×10^{-5}	11.6	9.20
1.0×10^{-4}	11.0	7.10
1.0×10^{-3}	9.50	6.20
1.0×10^{-2}	8.60	9.50
2.0×10^{-2}	14.5	19.9
5.0×10^{-2}	35.0	50.6
1.0×10^{-1}	69.0	96.0
2.0×10^{-1}	124	151
5.0×10^{-1}	258	279
1.0	339	364
1.5	363	375
2.0	358	369
3.0	388	418
4.0	418	439
5.0	396	402
6.0	402	415
7.0	416	444
8.0	435	464
10.0	464	481
14.0	520	520
17.0	610	640
20.0	650	660

[a]These values of response functions relate to the dose equivalent for neutrons as defined before 1985. They should be multiplied by a factor of 2 to obtain the dose equivalent as redefined for neutrons in 1985 (ICRP 1985).

Source: ICRP 1987.

REFERENCES

ANSI, *Neutron and Gamma-Ray Flux-to-Dose-Rate Factors*, ANSI/ANS-6.1.1-1991, American National Standards Institute, New York, 1977.

ANSI, *Gamma-Ray Attenuation Coefficients and Buildup Factors for Engineering Materials*, ANSI/ANS-6.4.3, American National Standards Institute, New York, 1991.

BARKAS, W.H. AND M.J. BERGER, *Tables of Energy Losses and Ranges of Heavy Charged Particles*, Report NASA SP-3013, National Aeronautics and Space Administration, Washington, D.C., 1964.

BERGER, M.J. AND S.M. SELTZER, *Tables of Energy Losses and Ranges of Electrons and Positrons*, Report NASA SP-3012, National Aeronautics and Space Administration, Washington, D.C., 1964.

TABLE 2.13   Prescribed Response Functions for Photon Effective Dose Equivalent for Incidence in Various Geometries on an Anthropomorphic Phantom.

ENERGY (MeV)	$\mathcal{R}_{H_p}$ (10^{-10} cSv cm^2) FOR EFFECTIVE DOSE EQUIVALENT H_E				
	AP	PA	LAT	ROT	ISO
0.01	0.062	0.000	0.0200	0.0290	0.0220
0.015	0.157	0.0310	0.0330	0.0710	0.0570
0.02	0.238	0.0868	0.0491	0.110	0.0912
0.03	0.329	0.161	0.0863	0.166	0.138
0.04	0.365	0.222	0.123	0.199	0.163
0.05	0.384	0.260	0.152	0.222	0.180
0.06	0.400	0.286	0.170	0.240	0.196
0.08	0.451	0.344	0.212	0.293	0.237
0.10	0.533	0.418	0.258	0.357	0.284
0.15	0.777	0.624	0.396	0.534	0.436
0.20	1.03	0.844	0.557	0.731	0.602
0.30	1.56	1.30	0.891	1.14	0.949
0.40	2.06	1.76	1.24	1.55	1.30
0.50	2.54	2.20	1.58	1.96	1.64
0.60	2.99	2.62	1.92	2.34	1.98
0.80	3.83	3.43	2.60	3.07	2.64
1.0	4.60	4.18	3.24	3.75	3.27
1.5	6.24	5.80	4.70	5.24	4.68
2.0	7.66	7.21	6.02	6.56	5.93
3.0	10.2	9.71	8.40	8.90	8.19
4.0	12.5	12.0	10.6	11.0	10.2
5.0	14.7	14.1	12.6	13.0	12.1
6.0	16.7	16.2	14.6	14.9	14.0
8.0	20.8	20.2	18.5	18.9	17.8
10.0	24.7	24.2	22.3	22.9	21.6

Source: Reprinted by permission from ICRP 1987, Copyright 1987, Pergamon Press, Inc.

BERGER, M.J., *Distribution of Doses Around Point Sources of Electrons and Beta Particles in Water and Other Media*, NM/MIRD Pamphlet No. 7, Society of Nuclear Medicine, New York, 1971.

BIGGS, F. AND R. LIGHTHILL, *Analytical Approximations for Photon-Atom Differential Scattering Cross Sections Including Electron Binding Effects*, Report SC-RR-72 0659, Sandia Laboratories, Albuquerque, N.M., 1972a.

BIGGS, F. AND R. LIGHTHILL, *Analytical Approximations for Total and Energy Absorption Cross Sections for Photon-Atom Scattering*, Report SC-RR-72 0685, Sandia Laboratories, Albuquerque, N.M., 1972b.

BNL, *Neutron Cross Sections*, Report BNL-325 and Supplements, Brookhaven National Laboratory, Upton, N.Y., 1958.

CASWELL, R.S., J.J. COYNE, AND M.L. RANDOLPH, "Kerma Factors for Neutron Energies below 30 MeV," *Radiat. Res. 83*, 217-254 (1980).

CHEN S.-Y, AND A.B. CHILTON, "Depth-Dose Relationships near the Skin Resulting from Parallel Beams of Fast Neutrons," *Radiat. Res. 77*, 21-33 (1979a).

TABLE 2.14 Prescribed Response Functions for Effective Dose Equivalent for Neutrons Incident in Various Geometries on an Anthropomorphic Phantom[a].

ENERGY (MeV)	$\mathcal{R}_{H_p}$ (10^{-10} cSv cm^2) FOR EFFECTIVE DOSE EQUIVALENT H_E			
	AP	PA	LAT	ROT
2.5×10^{-8}	4.00	2.60	1.30	2.30
1.0×10^{-7}	4.40	2.70	1.40	2.40
1.0×10^{-6}	4.82	2.81	1.43	2.63
1.0×10^{-5}	4.46	2.78	1.33	2.48
1.0×10^{-4}	4.14	2.63	1.27	2.33
1.0×10^{-3}	3.83	2.49	1.19	2.18
1.0×10^{-2}	4.53	2.58	1.27	2.41
2.0×10^{-2}	5.87	2.79	1.46	2.89
5.0×10^{-2}	10.9	3.64	2.14	4.70
1.0×10^{-1}	19.8	5.69	3.57	8.15
2.0×10^{-1}	38.6	8.60	6.94	15.3
5.0×10^{-1}	87.0	30.8	18.7	38.8
1.0	143	53.5	33.3	65.7
1.5	183	85.8	52.1	93.7
2.0	214	120	71.8	120
3.0	264	174	105	162
4.0	300	215	131	195
5.0	327	244	151	219
6.0	347	265	167	237
7.0	365	283	181	253
8.0	380	296	194	266
10.0	410	321	218	292
14.0	480	415	280	365

[a]These values of response functions relate to the dose equivalent for neutrons as defined before 1985. They should be multiplied by a factor of 2 to obtain the dose equivalent as redefined for neutrons in 1985 (ICRP 1985).

Source: Reprinted by permission from ICRP 1987, Copyright 1987, Pergamon Press, Inc.

CHEN S.-Y. AND A.B. CHILTON, "Calculation of Fast Neutron Depth-Dose in the ICRU Standard Tissue Phantom and the Derivation of Neutron Fluence-to-Dose-Index Conversion Factors," *Radiat. Res.* 78, 335-370 (1979b).

CHILTON, A.B. "Methods for Calculating Radiation-Induced Heat Generation," *Nucl. Eng. Des. 18*, 401- 413 (1972).

CHILTON A.B. AND Y.-L. SHIUE, "Progress Report on Studies Concerning Doses in the ICRU Sphere Resulting from Neutron Beams," Proc. 4th Symp. Neutron Dosimetry, *Munich/Neuherberg*, June 1-5, 1981, Comm. Eur. Comm., Luxembourg, EUR 7448 EN, Vol. 1, 1981.

DIMBYLOW, P.J. AND T.M. FRANCIS, *A Calculation of the Photon Depth Dose Distributions in the ICRU Sphere for a Broad Parallel Beam, a Point Source, and an Isotropic Field*, Report NRPB-R92, National Radiation Protection Board, UKAEA Harwell, Oxfordshire, England, 1979.

EVANS, R.D., *The Atomic Nucleus*, McGraw-Hill, New York, 1955.

FODERARO, A., L.J. HOOVER, AND J.H. MARABLE, "Heat Generation by Neutrons," in *Engineering Compendium on Radiation Shielding*, Vol., I., R. G. Jaeger (ed.), Springer-Verlag, New York, Sec. 7.2, 1968.

GARBER, D.E., L.G. STROMBERG, M.D. GOLDBERG, D.E. CULLEN, AND V.M. MAY, *Angular Distribution in Neutron Induced Reactions*, 3rd ed., Report BNL-400, Brookhaven National Laboratory, Upton, N.Y., 1970.

GOLDSTEIN, H., AND J.E. WILKINS, JR., *Calculations of the Penetration of Gamma Rays*, NDA/AEC Report NYO-3075, U. S. Government Printing Office, Washington, D.C., 1954.

HAFFNER, J.W., *Radiation and Shielding in Space*, Academic Press, New York, 1967, *p.* 168.

HEITLER, W., *The Quantum Theory of Radiation*, 3rd ed., Oxford University Press, Oxford, 1954.

HUBBELL, J.H., "Photon Mass Attenuation and Energy Absorption Coefficients," *Int. J. Appl. Radiat. Isot. 33*, pp. 1269-1290 (1982).

HUBBELL, J.H., AND M.J. BERGER, "Attenuation Coefficients, Energy Absorption Coefficients, and Related Quantities," in *Engineering Compendium on Radiation Shielding*, Vol. I., R. G. Jaeger (ed.), Springer-Verlag, New York, Sec. 4.1, 1968.

ICRP, *Recommendations of the International Commission on Radiological Protection, Protection against Ionizing Radiation from External Sources*, ICRP Publication 15, International Commission on Radiological Protection, Pergamon, Elmsford, N.Y., 1970.

ICRP, *Data for Protection Against Ionizing Radiation from External Sources: Supplement to ICRP Publication 15*, Publication 21, International Commission on Radiological Protection, Pergamon, Elmsford, N.Y., 1973.

ICRP, *Recommendations of the International Commission on Radiological Protection*, Publication 26; published as *Annals of the ICRP 1*, No. 3, Pergamon, Elmsford, N.Y., 1977.

ICRP, *Statement from the 1985 Paris Meeting of the International Commission on Radiological Protection*, Publication 45, International Commission on Radiological Protection, Pergamon Press, Oxford, 1985.

ICRP, *Data for Use in Protection Against External Radiation*, Publication 51; published as *Annals of the ICRP 17*, No. 2/3, Pergamon Press, Elmsford, N.Y., 1987,

ICRU, *Neutron Dosimetry for Biology and Medicine,* Report 26, International Commission on Radiation Units and Measurements, Washington, D.C., 1977.

ICRU, *Average Energy Required to Produce an Ion Pair*, Report 31, International Commission on Radiation Units and Measurements, Washington, D.C., 1979.

ICRU, *Radiation Quantities and Units*, Report 33, International Commission on Radiation Units and Measurements, Washington, D.C., 1980.

ICRU, *Stopping Powers for Electrons and Positrons,* Report 37, International Commission on Radiation Units and Measurements, Washington, D.C., 1984.

KINSEY, R., $ENDF/B - V$ *Summary Documentation*, Report BNL-NCS-17541 (ENDF-201), 3rd ed., Brookhaven National Laboratory, Upton, N.Y., 1979.

KLEIN, O., AND Y. NISHINA, *Z. Phys. 52*, 853 (1929).

LEDERER, C.M, AND V.S. SHIRLEY (EDS.), *Table of Isotopes*, 7th ed., Wiley-Interscience, N.Y., 1978.

MORRIS, E.E., A.B. CHILTON, AND A.F. VETTER, "Tabulation and Empirical Representation of Infinite-Medium Gamma-Ray Buildup Factors for Monoenergetic, Point Isotropic Sources in Water, Aluminum, and Concrete," *Nucl. Sci. Eng. 56*, 171-178 (1975).

NCRP, *Protection Against Neutron Radiation*, Report 38, National Council on Radiation Protection and Measurements, Washington, D.C., 1971.

NCRP, *Recommendations on Limits for Exposure to Ionizing Radiation*, Report 91, National Council on Radiation Protection and Measurements, Washington, D.C., 1987.

PLECHATY, E.F., D.E. CULLEN, AND R.J. HOWERTON, Report UCRL-50400, Vol. 6, Rev. 1, National Technical Information Service, Springfield Virginia, 1975. [Data are available as the DLC-139 code package from the Radiation Shielding Information Center, Oak Ridge National Laboratory, Oak Ridge, Tenn.]

SHIUE, Y.-L. AND A.B. CHILTON, "Calculation of Low-Energy Dose Indices and Depth Doses in the ICRU Tissue Sphere," *Radiat. Res. 93*, 421-443 (1983).

STORM, E., AND H.I. ISRAEL, *Photon Cross Sections from 0.001 to 100 MeV for Elements 1 through 100*, Report LA-3753, Los Alamos Scientific Laboratory, Los Alamos, N.M., 1967.

TRUBEY, D.K., M.J. BERGER, AND J.H. HUBBELL, "Photon Cross Sections for ENDF/B-VI," American Nuclear Society Topical Meeting, Advances in Nuclear Engineering Computation and Radiation Shielding, Santa Fe, N. M., 1989. [Data are available as the DLC-136/PHOTX code package from the Radiation Shielding Information Center, Oak Ridge National Laboratory, Oak Ridge, Tenn.]

TRUBEY, D.K. AND Y. HARIMA, "New Buildup Factor Data for Point Kernel Calculations," in *Proceedings of a Topical Conference on Theory and Practices in Radiation Protection and Shielding*, Knoxville, Tenn., April 1987, Vol. 2, p. 503, American Nuclear Society, LaGrange Park, Ill., 1987. [Data are available as the DLC-129 code package from the Radiation Shielding Information Center, Oak Ridge National Laboratory, Oak Ridge, Tenn.]

PROBLEMS

1. A small homogeneous sample of mass m (g) with atomic mass A is irradiated uniformly by a constant flux density $\phi(\text{cm}^{-2}\text{s}^{-1})$. If the total atomic cross section for the sample material with the irradiating particles is denoted by $\sigma_t(\text{cm}^2)$ derive an expression for the fraction of the atoms in the sample that interact during a 1-h irradiation. State any assumptions made.

2. Calculate the linear interaction coefficients in pure air at 20°C and 1 atm pressure for a 1-MeV photon and a thermal neutron (2200 m s^{-1}). Assume that air has the composition 78% nitrogen, 21% oxygen, and 1% argon *by volume*. Use the following data:

Element	Photon μ/ρ cm^2 g^{-1}	Neutron σ_{tot} (b)
Nitrogen	0.0636	11.9
Oxygen	0.0636	4.2
Argon	0.0574	2.2

3. Derive the Compton formula, Eq. (2.8), given that the momentum of a photon of energy E is equal to E/c and the momentum of an electron of kinetic energy T is equal to $(1/c)\sqrt{T(T + 2m_ec^2)}$.

4. Verify Eqs. (2.11) to (2.15).

5. Verify Eq. (2.16); that is, given Eq. (2.18) carry out the integration

$$_e\sigma_t = \int_{4\pi} d\Omega \; _e\sigma_t(\theta_s).$$

6. An air-filled ionization chamber has an effective volume of 500 cubic centimeters and contains air at 20°C and 29.5 in. Hg atmospheric pressure. A constant electrical potential is applied to the chamber. If air in the chamber is exposed to 1 R h^{-1}, and all charge is collected, what is the electrical current (mA) flowing through the chamber?

7. What is the maximum possible kinetic energy (keV) of a Compton electron and the corresponding minimum energy of a scattered photon resulting from scattering of

   (a) a 100-keV photon,
   (b) a 1-MeV photon,
   (c) a 10-MeV photon?

   Estimate for each case the range the electron would have in air of 1.2 mg/cm^3 density.

8. Compute and plot the energy spectrum $F(T)$ of Compton recoil electrons for incident photons of 0.5 and 2.0 MeV. Here $F(T)dT$ is the fraction of electrons with energies between T and $T + dT$.

9. Estimate the production rate of capture gamma photons when a 1-g sample of iron is uniformly irradiated by a thermal neutron flux density of 5×10^{12} cm^{-2} s^{-1}.

10. When an electron moving through air has 5 MeV of energy, what is the ratio, at that energy, of the rate of energy loss by bremsstrahlung to that by collision?

11. What thickness of iron is needed to stop a beam of (a) 10-MeV electrons, (b) 10-MeV alpha particles, and (c) 10-MeV protons? Use Eq. (2.63) and compare your values to ranges shown in Figs. 2.16 and 2.17.

12. Estimate the range of a 10-MeV tritium nucleus in air.

13. Consider the bombardment of a slab of material (bounded by vacuum) by a beam of photons or neutrons. Make a sketch showing qualitatively the ratio of the absorbed dose rate to the kerma rate as a function of depth of penetration into the slab along the direction of the beam. Explain the features of the sketch.

14. On the assumption of isotropic scattering in the center-of-mass system, estimate the response function for 0.1-MeV neutrons, where the dosimetric quantity of interest is the kerma rate in water. How does this compare with the corresponding value for tissue given in Table 2.7? Take scattering cross sections for hydrogen at this energy to be 12.8b and for oxygen to be 3.5b. (Only elastic scattering is involved).

15. A small volume of tissue acting as a detector is located in a beam of 1-MeV neutrons which provides a fluence of 10^{13} cm^{-2}. Compute the kerma using the tissue composition of Table 2.7. Then compute the kerma using the response function of the same table. If the same fluence were incident on a phantom (of the type assumed in Table 2.8) in a broad, parallel-beam situation, what would be the maximum value of the absorbed dose in grays?

How do you account for the differences among these three answers? [*Note*: The scattering cross sections (elastic) for 1-MeV neutrons are to be taken as 4.3 b for H, 2.6 b for C, 2.0 b for N, and 8 b for 0. It may be assumed that scattering is isotropic in the center-of-mass system.]

16. (a) Plot two curves on log-log paper, one for the neutron tissue kerma per unit fluence and the other for the neutron phantom-related dose equivalent per unit fluence based on the ANSI prescribed response functions. What two principal factors account for the differences between the two curves? Cover the energy range 0.001 to 2 MeV. (b) Make a similar plot for photons, assuming that tissue and water have similar characteristics. What accounts for the difference in this case?

17. For an exactly known fluence of 10^{10} cm^{-2} of 1-MeV photons, evaluate the following:

    (a) X (R)
    (b) K (Gy) in air
    (c) K (Gy) in aluminum
    (d) K (Gy) in lead

    State any assumptions or approximations.

18. For a fluence of 10^{10} cm^{-2}, evaluate the following:

    (a) K (Gy) for 1-MeV photons in tissue
    (b) K (Gy) for 1-MeV neutrons in tissue
    (c) H (Sv) for 1-MeV photons
    (d) H_I (Sv) for 1-MeV neutrons (deep dose index)
    (e) H_E (Sv) for 1-MeV neutrons

    In parts (a) and (c) tissue may be approximated as water. In parts (a) to (c) it may be assumed that the fluence is known exactly within the absorbing medium. In parts (d) and (e) the fluence is that in the free field. State any other assumptions or approximations made.

19. At a certain location, the exposure rate due to 1-MeV photons is 100 R h^{-1}. With the same flux density of 1-MeV photons, what would be the absorbed dose rate in concrete (Gy h^{-1})? State any assumptions made.

20. Listed in the following table on page 103 are various coefficients (cm^2 g^{-1}) for gamma-ray interactions in water:

    (a) Using only the data given, fill in the blanks, making a self-consistent data set.
    (b) Explain why $\mu_{en}/\rho < \mu_{tr}/\rho < \mu_a/\rho$.

21. What flux densities of 1 MeV photons would result in

    (a) an exposure rate of 1 R h^{-1}, [air at STP]
    (b) an absorbed dose rate in air (STP) of 1 rad h^{-1},
    (c) a tissue absorbed dose rate of 1 rad h^{-1},
    (d) a tissue kerma rate of 100 ergs g^{-1} h^{-1},
    (e) a dose equivalent rate of 1 rem h^{-1},
    (f) an effective dose equivalent rate of 1 rem h^{-1}?

    Tissue may be approximated as water. State any other assumptions or approximations made.

E(MeV)	μ_c/ρ	μ_{ph}/ρ	μ_{pp}/ρ	μ_{ca}/ρ	μ_{pha}/ρ
0.01		4.720	0.0	0.0029	4.720
0.1	0.163	0.0024	0.0		0.0024
1		0.0	0.0	0.0311	0.0
10	0.0171	0.0			0.0
100		0.0	0.0144	0.0022	0.0

E(MeV)	μ_{ppa}/ρ	μ/ρ	μ_a/ρ	μ_{tr}/ρ	μ_{en}/ρ
0.01	0.0	4.875	4.723	4.723	4.723
0.1	0.0	0.165	0.0248	0.0248	0.0248
1	0.0	0.0707	0.0311	0.0311	0.0311
10		0.0221	0.0167	0.0162	0.0158
100	0.0107	0.0172			0.0129

22. Select at least 10 energies from the first column of Table 2.11. For each of these selected energies, calculate the response function for exposure, in units of R cm^2. From these data, and Table 2.11, determine the ratio of the maximum dose equivalent in a sphere to the free-field exposure under parallel-beam conditions. This ratio is sometimes denoted as C_x and is a function of photon energy. Plot C_x in units of cSv R^{-1} as a function of energy for the range 0.01 to 10 MeV.

23. A "rule of thumb" for exposure from point sources of photons in air at distances over which exponential attenuation is negligible is as follows:

$$\dot{X} = \frac{6CEN}{r^2},$$

where C is the source strength (Ci), E is the photon energy (MeV), N is the number of photons per disintegration, r is the distance in feet from the source, and $\dot{X}$ is the exposure rate (R h^{-1}).

   (a) Re-express this rule in units of Bq for the source strength and meters for the distance.
   (b) Over what ranges of energies is this rule accurate within $\leq 10\%$ and within $\leq 5\%$?

24. (a) For at least six neutron energies between 0.01 and 10 MeV, calculate the ratio of the ICRP-prescribed phantom-related absorbed-dose response function (Table 2.8) to the tissue kerma response function (Table 2.7). Draw a graph of the resulting ratio as a function of neutron energy. (As a check, this can be compared with Fig. 15 of the National Bureau of Standards Handbook 63.) (b) For at least a dozen energies between 2.5×10^{-8} and 50 MeV, calculate the ratio of the ICRP-prescribed phantom-related dose equivalent response function (Table 2.10) to the ICRP-prescribed phantom-related absorbed dose response function (Table 2.8). Draw a graph of the resulting ratio as a function of neutron energy. (As a check, this can be compared with the values for Q_{eff} given in the last column of Table 2.8.)

Chapter 3

Biological Effects of Ionizing Radiation

3.1 INTRODUCTION

In this chapter we examine the mechanisms by which ionizing radiation affects humans, both as individuals and, through hereditary effects, as entire populations. At the most elemental level, exposure to ionizing radiation is thought to result in disruption of the DNA genetic coding in cell nuclei. The consequences of the disruption depend in general on the spatial density of damage loci, which is dependent on radiation dose, and in particular on the proximity of damage loci within a single gene and within a single cell nucleus. Closely spaced loci such as double-strand breaks in the DNA molecule are less likely to be repaired than isolated loci such as single-strand breaks. Consequences are also affected by the rate of creation of damage loci, which is dependent on the radiation dose rate. Two closely spaced loci, otherwise individually reparable, may interact and result in irreparable damage if the second is created before the first has been repaired. In a broader sense, the function of an organ or tissue of the body may be impaired if cell destruction takes place over such a short period of time that natural cell replacement mechanisms cannot compensate. Even if the spatially averaged doses and dose rates are equal, the consequences of exposure to different types of radiation may differ because of nonuniformities in energy dissipation at the microscopic level. For example, energy loss along the tracks of electrons is much more diffuse spatially than is that along the tracks of protons or alpha particles.

Consequences of exposure to ionizing radiation may be classified broadly as *hereditary effects* and *somatic effects*. Damage to the genetic material in germ cells, without effect on the individual exposed, may result in hereditary illness expressed in succeeding generations. Somatic effects are effects on the individual exposed and may be classified by the nature of the exposure, for example, acute or chronic, and by the time scale of expression, for example, short or long term. The short term acute effects on the gastrointestinal, respiratory, and hematological systems are described as the *acute radiation syndrome*.

104

Consequences of exposure to ionizing radiation require examination in many contexts. Acute, life-threatening exposure leads to mechanistic consequences requiring a definite course of medical treatment. Illness is certain, with the scope and degree depending on the radiation dose. On the other hand, minor acute or chronic low-level exposure is thought to result in consequences whose cause-effect relationships are as yet incompletely understood. Hereditary illness may or may not result; cancer may or may not result. Only the probability of illness, not its severity, is dependent on radiation dose. The consequences are *stochastic* as distinct from *mechanistic*.

Much of the material for this chapter has been drawn from the 1980 [BEIR-III] and 1990 [BEIR-V] reports of the U.S. National Research Council Committee on the Biological Effects of Ionizing Radiation (NAS 1980, 1990) and from the 1982, 1986, and 1988 [UNSCEAR] reports of the United Nations Scientific Committee on the Effects of Atomic Radiation (UN 1982, 1986, 1988).

This chapter reviews briefly certain aspects of human cellular reproduction and microdosimetry concepts of importance in characterizing and understanding the biological effects of ionizing radiation. The chapter then examines the nonstochastic effects of exposure to high levels of ionizing radiation. The chapter concludes with examination of stochastic effects of exposure to low levels of ionizing radiation, namely, the induction of cancer and hereditary illness, with special emphasis given to the public-health aspects of exposures of large populations.

3.2 THE HUMAN CELL

In the adult there are some 2×10^{12} cells, each of which, save for the gametes, contain within its nucleus 23 pairs of chromosomes, one in each pair of predominantly maternal and the other of predominantly paternal origin. The nucleus, several μm in diameter, contains some 10^5 genes. Genetic information is carried by the very long deoxyribonucleic acid (DNA) molecule which is described as being in the form of a double (two-stranded) helix, twisted and folded in complex ways to form the individual "arms" of the chromosomes. The backbone of each strand is made up of alternating sugar and phosphate groups. Each sugar group has attached one of four nucleotide bases. The two strands are linked by bonds between these bases, but only in two combinations, adenine-thymine and guanine-cytosine. Genetic sequences of bases along a strand are read in groups of three, each triplet constituting one of 64 possible "words" of information. The spacing between bases is about 0.3 nm and the distance between the two strands is about 2 nm.

In human reproduction, the male and female germ cells divide by meiosis into gametes (sperm and ova), each containing 23 chromosomes. At conception, the male and female gametes join to form a zygote containing 23 pairs of chromosomes. That cell, and its progeny, divide by mitosis into cells containing the same genetic information as the zygote, the divisions leading to some 10^{11} cells in the newborn. Cells of the male and female may be distinguished by the sex chromosomes, an equal (X-X) pair in the female and an unequal (X-Y) pair in the male. The other 22 pairs are said to be *homologous*, meaning that the two chromosomes of each pair are similar in shape, size, and genetic content. Cells also contain some genetic information in the DNA content of the mitochondria. While such DNA, which is always of maternal origin, is also susceptible to radiation damage, there is redundancy of genetic information because of the many mitochondria in a cell, and disruption of mitochondrial DNA is not thought to be a significant effect of exposure to ionizing radiation.

3.3 DOSE-RESPONSE RELATIONSHIPS

In attempting to establish "cause-effect" relationships for biological response to radiation exposure, one no doubt would hypothesize first that the response is somehow related to the average absorbed dose in the irradiated organ or tissue. One might also hypothesize that the extent of a particular response would increase with increasing dose except perhaps at very high doses. A cell can only be killed once; therefore, at very high doses an effect measured by cell destruction per unit dose, for example, might well diminish. One would also expect that a cause-effect relation would have to take into account absorbed dose rate both because the normal mechanism for repair of cellular damage might be overwhelmed by absorption of radiation at a high rate and because certain types of irreparable damage events might result from interaction of two otherwise reparable events. Absorbed dose and dose rate, though, are insufficient to define the extent of a radiation effect. Even if in two cases the dose and dose rate were the same, different types of radiation might lead to effects different in degree or kind. Figure 3.1 contrasts the tracks of heavy charged particles (protons, alpha particles, etc.) and electrons. The differences are striking and these "radiation quality" effects must somehow be taken into account. One powerful methodology for treating not only radiation-quality effects but also dose rate effects involves concepts of *microdosimetry* which are addressed later in this chapter.

Ionizing radiation affects cells either directly or indirectly as a result of ionization and excitation produced through Coulombic interactions along the tracks of charged particles. If the exposure is to indirectly ionizing radiation such as gamma rays or neutrons, then the cellular radiation effect arises from secondary charged particles such as electrons in the case of gamma-ray exposure or recoil atomic nuclei, especially protons, in the case of neutron exposure.

Damage to DNA may result directly from excitation or ionization of constituents of the molecule. Damage may result indirectly from chemical reactions of the DNA molecule with highly reactive atomic, ionic, molecular, and free-radical species produced in the aqueous medium in the cell nucleus through which the radiation passes. For directly ionizing particles such as protons, direct action is thought to predominate; for sparsely ionizing particles such as electrons, indirect action is thought to contribute roughly two thirds of the biological effect (NAS 1990).

Ionizing radiation, in comparison with other cytotoxic agents is extraordinarily efficient in cell killing. Ward et al. (1988) have examined the number and character of DNA lesions in cultures of mammalian cells in which 63 percent of the cells have been killed. They find that, with low-LET ionizing radiation as the agent, there are on average 1000 single-strand DNA breaks, only 40 double-strand breaks, and 440 locally

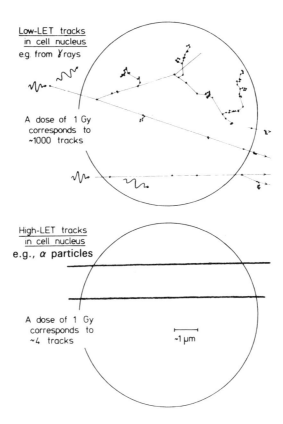

Low-LET tracks in cell nucleus e.g. from γ rays

A dose of 1 Gy corresponds to ~1000 tracks

High-LET tracks in cell nucleus e.g., α particles

A dose of 1 Gy corresponds to ~4 tracks

~1 µm

Figure 3.1   Schematic of a cell nucleus traversed by two electron tracks (upper panel) or by two alpha-particle tracks (lower panel). Ionization clusters at the ends of electron tracks are not to scale. (From Goodhead (1988); by permission of the Health Physics Society.)

multiply damaged sites (LMDS), that is, sites of multiple lesions in close enough proximity to interact in such a way as to cause cell death. For chemical cytotoxic agents, they find predominantly single-strand lesions, but in numbers of tens to hundreds of thousands. This would rule out single-strand breaks, known to be repaired by enzyme action, as being responsible for cell death. Double-strand breaks alone are thought to be too few to account for the efficiency of radiation in cell killing. This leaves the LMDS as being the principal lethal lesion for cellular DNA (NAS 1990).

There are several other radiobiological considerations which bear on dose-response relationships (NAS 1990). Decades of observation have revealed that response is generally a quadratic function of dose and that the sensitivity of cells to damage depends on the phase of the cell cycle. Cells are known to be most sensitive in the stage of mitosis and least sensitive in the stage of DNA synthesis. Synergistic effects of chemicals and radiation are also known to affect cell killing. Oxygen, for example, sensitizes cells, while free-radical scavengers protect cells. Certain chemicals may have no effect on cell killing but may greatly affect radiation carcinogenesis.

3.3.1 Dose and Dose-Rate Effects

The Linear-Quadratic Model.   For some 50 years a simple, straightforward, mechanistic model has provided the framework for interpretation of studies of radiobiological action. Originally the model was applied in studies of radiation effects on simple cellular systems *in vitro* with simple endpoints such as cellular reproductive death. The very robust linear-quadratic model is also applied, if only empirically, to highly complex systems and endpoints such as carcinogenesis in the human.

Under the linear-quadratic model for radiobiological action, a particular effect $\mathcal{E}$ of a dose D is expressed as

$$\mathcal{E}(D) = \alpha D + \beta D^2, \tag{3.1}$$

in which the linear coefficient α and the quadratic coefficient β depend on the type of radiation. Empirically, the model is known to fit the dose-response relationships for many systems and endpoints.

There are physical reasons why, at least for simple systems and endpoints, the response could depend not only on the dose but also on the square of the dose. Consider, for example, a single-strand break of the DNA molecule in the nucleus. Such a break may be called a *sublesion* in that it is likely repaired by normal enzyme action. A double-strand break, not likely to be repaired, and perhaps mutagenic or fatal to the cell, may be called a *lesion*. If closely enough spaced,[1] two sublesions may interact to form another type of lesion, one in which the DNA strand is rejoined, but in error, with some of the genetic information excised or scrambled. Lesions and sublesions, some of the latter of which may interact to form lesions, are created along the individual tracks of ionizing radiations. In a given system the number of these lesions would be proportional to the number of tracks, or to the radiation dose, thus accounting for the linear term in the model. Sublesions created by separate tracks may also combine to form lesions. The probability of this occurring would depend on the square of the number of tracks, thus accounting for the quadratic term in the model. The linear-quadratic dose response can also be explained in terms of a model involving entirely single-track effects, with the ability of the cell to repair lesions being saturated at higher doses (NAS 1990).

With few exceptions, dose-response data, even for simple systems, is available only for low-LET radiation and for doses of about 1 Gy or more delivered at high dose rate. Under these conditions, the quadratic response cannot be ignored and it is extremely difficult to extract from the data the value of the linear-response coefficient, the key information needed for the very important task of assessing the effects of high-LET as well as low-LET radiation at low doses and dose rates. Two exceptions are discussed in the following paragraph. One deals with mutations in stamen hairs of the *Tradescantia* (spiderwort plant), an exquisitely sensitive system with a well-defined endpoint. Another deals with chromosome damage to human lymphocytes in culture, another highly sensitive system with a well-defined endpoint.

---

[1]The distance over which sublesions may combine is thought to be of the order of 1 μm (Kellerer and Rossi 1972).

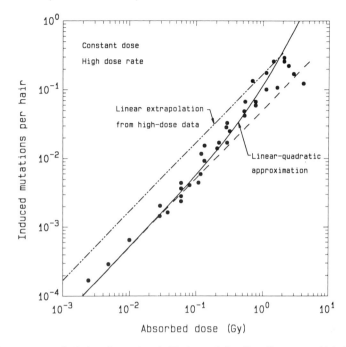

Figure 3.2   Dose response for induced mutations in *Tradescantia* irradiated by x rays at high dose rate using logarithmic scales. [Data are from NCRP (1980).]

Tradescantia Mutations.   Figures 3.2 and 3.3 illustrate composite data from a number of investigations of response to 250 kVp x rays delivered at high dose rate. Figure 3.2 is a full logarithmic plot and in this figure it is clear that at low dose the data fall on a line with unit slope — the linear term in the response model. This line is projected as the dashed line. At doses greater than about 1 Gy the response flattens and then declines due to cell killing or some other effect preventing the expression of the mutation. From about 0.1 to 1 Gy, the response is clearly supralinear, indicative of the quadratic term in the response model. Below the cell-killing dose, the data are well represented by the solid line corresponding to

$$\mathcal{E}(D) = 0.051D + 0.065D^2, \tag{3.2}$$

in which D is the dose in Gy and $\mathcal{E}$ is the mutation incidence. Figure 3.3 displays the same data on a linear plot. In comparison with the logarithmic presentation, the purely linear response at low dose is obscured and the significance of outlying data is magnified. From this view only, one would be unlikely to countenance the linear-quadratic formula for the response.

Rarely in studies of human response to radiation are data available for low dose rates and for doses less than about 1 Gy. Very often too, the assumption is made that the response is purely linear from zero dose to the high doses responsible for the limited data. That such an assumption can be seriously in error is illustrated in the two figures. Shown as broken lines are the linear extrapolations from high dose, with

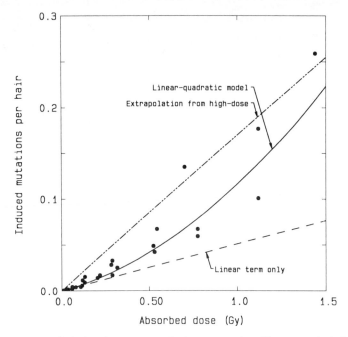

Figure 3.3   Dose response for induced mutations in *Tradescantia* irradiated by x rays at high dose rate using linear scales. [Data are from NCRP (1980).]

slope of 0.17 Gy^{-1}. Suppose one were interested in the response at 0.01 Gy absorbed dose. The linear "model" would yield a mutation response of 1.7×10^{-3} while the linear-quadratic model would yield a response of 5.1×10^{-4} — lower by a factor of 3. This factor has come to be approximated by the *dose rate effectiveness factor* (DREF) and will be discussed further in a later section.

   Chromosome aberrations in human lymphocytes.   The dicentric chromosome structural rearrangement in the lymphocyte has been found to be a very sensitive indicator of human radiation exposure. Because the rearrangement is a type of aberration that can be produced by a single "event" in the nucleus or by interactions of multiple events and occurs spontaneously only rarely, its linear-quadratic dose response has been very thoroughly studied. Table 3.1 summarizes how the model parameters vary with the type of radiation.

   Dose Rate Effects and the Dose Rate Effectiveness Factor.   Studies of radiobiological effects, with endpoints ranging from cell death to tumor induction, in nearly all cases reveal a reduction in the degree of the effect with a reduction in the dose rate.[2] While such studies with human subjects are extremely rare, cellular studies and studies with test animals leave no doubt as to the reality of dose rate effect. For example, in the *Tradescantia* mutation studies, the data shown in the Figs. 3.2 and 3.3 are

---

   [2]By a reduction in dose rate, we include fractionation of the dose, that is, delivery of the dose into a number of fractions of the total dose, separated in time, so that the average dose rate is decreased.

TABLE 3.1 Parameters for the Linear-Quadratic Model for Chromosome Aberrations per Cell in Human Lymphocytes Exposed to Ionizing Radiation of Various Qualities. The Table is a Composite of the Findings of Edwards, Lloyd, Purrott, and Prosser at the UKAEA Research Establishment at Harwell in the U.K. The Response is the Number of Chromosome Aberrations per Cell. The Last Column is Based on ^{60}Co Gamma Rays as the Reference Radiation.

RADIATION	$\alpha\,(Gy^{-1})$	$\beta\,(Gy^{-2})$	α/α_{ref}
15-MeV electrons	0.006	0.057	0.35
15-MeV electrons (pulsed)	0.009	0.061	0.57
^{60}Co gamma rays	0.016	0.050	1.00
250 kVp x rays	0.048	0.062	3.0
14.7 MeV neutrons	0.26	0.088	15
Cyclotron neutrons, $E_{avg} = 7.6$ MeV	0.48	0.064	30
^{252}Cf neutrons, $E_{avg} = 2.3$ MeV	0.60		38
Fission neutrons, $E_{avg} = 0.9$ MeV	0.73		46
Fission neutrons, $E_{avg} = 0.7$ MeV	0.84		53
5.15 MeV alpha particles	0.38		24
4.9 MeV alpha particles	0.29		18

Source: ICRU 1986.

for dose rates on the order of 0.3 Gy min^{-1}. When the x-ray dose rate is reduced to 5×10^{-4} to 5×10^{-3} Gy min^{-1}, the response is linear with dose for doses of 0.15 Gy or less, with the response per unit dose $\alpha \simeq 0.07$ Gy^{-1} (NCRP 1980). At a dose rate of 5×10^{-5} Gy min^{-1}, the response per unit dose is equal to the linear term coefficient 0.051 Gy^{-1} in Eq. (3.2).

Dose and dose rate effects are described symbolically in Fig. 3.4. Let us suppose that the true dose response is given by $\alpha_1 D + \beta D^2$ at doses below which cell killing is significant. When data are available only for high dose and dose rate (the open data points) and it is hypothesized that the response is linear with dose and without threshold, the hypothetical dose response is linear with slope α_L. Now suppose it were possible to make *high-dose* measurements at *low dose rate* (the solid data points), low-dose measurements at any dose rate being beyond the reach of experiment or epidemiolgical study. One may expect that, at sufficiently low dose rate, the dose response would be linear, with slope α_{ex}, and that in the limit of zero dose rate α_{ex} would approach α_1. The dose rate effectiveness factor is defined as

$$DREF = \frac{\alpha_L}{\alpha_{ex}} \simeq \frac{\alpha_L}{\alpha_1}. \tag{3.3}$$

The term α_L/α_1 is called a dose *rate* effectiveness factor because it is based in principle on measurements at high and low dose *rates*. It applies, of course, to the relative effectiveness per unit dose of high and low doses, and the value α_L/α_1, is more properly a dose (magnitude) effectiveness factor.

The NCRP (1980) conducted a thorough study of DREF values for endpoints including mutagenesis and carcinogenesis, with the following conclusion:

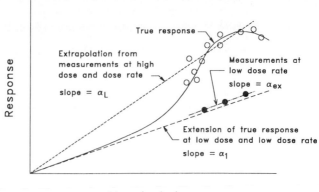

Figure 3.4   Illustration of the derivation of the dose rate effectiveness factor (DREF). [After NCRP (1980)]

For a single exposure to low absorbed doses (between 0 and 20 rads [0.2 Gy]) delivered at any dose rate, and for any total dose delivered at a dose rate of 5 rads y^{-1} [0.05 Gy y^{-1}] or less, it is believed that the DREFs for both mutagenesis and for collective tumors after exposure of the whole body . . . are between 2 and 10.

The BEIR-III Committee (NAS 1980), in their study of radiation carcinogenesis, used a DREF of about 2.25 which was based on data for leukemia and which was applied implicitly in their risk estimates. The UNSCEAR (UN 1988) agreed with the NCRP conclusions. The BEIR-V Committee (NAS 1990) recommended a DREF of 2 for carcinogenesis but did not apply the factor in their risk estimates.

Target Theory and the Linear-Quadratic Model.   Historically, many studies of radiation effects on cells in culture have had biological endpoints characterized by the number of cells surviving in some sense, for example, not killed or not having experienced a mutation. Very often the results have been interpreted in terms of "target theory," described very clearly by Andrews (1974). Central to the theory is the assumption of equal targets and equal-magnitude, independent radiation-induced "hit" events on the targets. Two simplified models are the multihit model and the multitarget model.

In the *multihit model*, any one cell is assumed to be killed or not to survive in its original form if, during the course of the irradiation, it experiences n or more hits. In the simplest case, the probability of a hit is assumed to be proportional to the dose D, namely αD. The probability p_m of m hits on one target is governed by Poisson statistics, that is,

$$p_m = \frac{1}{m!}(\alpha D)^m e^{-\alpha D}. \tag{3.4}$$

The probability of cell survival, $S(D)$, is the probability of less than n hits, namely,

$$S(D) = \sum_{m=0}^{n-1} p_m = e^{-\alpha D} \sum_{m=0}^{n-1} \frac{1}{m!}(\alpha D)^m. \tag{3.5}$$

Survival probability after irradiation by high-LET particles is typically exponential with dose,

$$S(D) = e^{-\alpha D}. \tag{3.6}$$

This result is consistent with cell death following a single hit, as predicted by Eq. (3.5) with $n = 1$. The empirically observed survival function of Eq. (3.6) is also consistent with a linear relationship between dose and response (the probability of a hit). One may generalize the hit probability to be in the form of the linear-quadratic formula for dose response, namely $\mathcal{E}(D) = \alpha D + \beta D^2$. Then, the function for survival from single hits becomes

$$S(D) = e^{-(\alpha D + \beta D^2)}. \tag{3.7}$$

This type of survival function is characteristic of exposure to low-LET radiation.

Survival functions for low-LET radiation may also be interpreted in terms of the *multitarget model*. This model is based on the assumption that any one cell contains n equal targets, each of which must receive at least one hit before the cell is inactivated or killed. The probability of cell survival $S(D)$ is 1 minus the joint probability of one or more hits on each of n targets. If the probability of a single hit is assumed to be proportional to the dose, namely αD, the probability of one or more hits on a single target is $1 - e^{-\alpha D}$, and

$$S(D) = 1 - (1 - e^{-\alpha D})^n. \tag{3.8}$$

Empirically, n is identified as the "extrapolation number." For survival of mammalian cells, n is typically 2 to 20 and α is typically 0.5 to 1 Gy^{-1} (ICRP 1991). At low doses and dose rates, $S(D)$ is often better described by Eq. (3.7).

Survival probabilities for the multihit and multitarget models are illustrated in Fig. 3.5 for $\alpha = 0.5$ and $\beta = 0.01$. These target-theory models have been valuable in providing a framework for correlation of experimental data and have pointed the way toward more comprehensive understanding and modeling of radiation dose-response relationships.

The Microdosimetry Theory of Dual Radiation Action.

Microdosimetric Quantities. Consider first an irradiated medium within which, over the course of the irradiation, many charged-particle tracks have been traversed. Within the irradiated medium are many volumes of special interest. It would be helpful to think of these volumes as encompassing individual cell nuclei or, perhaps, individual chromosomes. Now consider a single ionizing interaction along one of the particle tracks. In that interaction, or event, the *energy deposit* ϵ_i is the energy of the incident particle less the sum of energies of all ionizing particles leaving the interaction, less the energy equivalent of any changes in rest masses. In a given volume of interest, the *energy imparted* to the matter in the volume is $\epsilon = \sum \epsilon_i$. The summation may include multiple events due to one or more statistically independent particle tracks. The *specific energy* z is the ratio of ϵ to the mass m of the volume. The energy imparted and the specific energy are stochastic quantities. After the irradiation, over the many like volumes of interest, there will be many different ϵ or z values, some zero, and others of varying magnitude. There is a distribution function $f(z; D)$, clearly

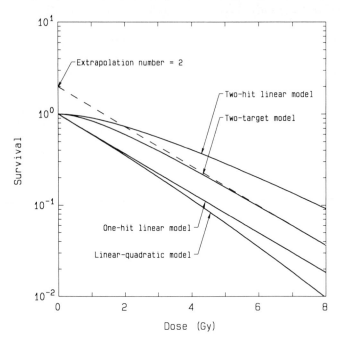

Figure 3.5 Comparison of "survival curves" based on different target theory models. Eq. (3.5) with $n = 2$ is the two-hit linear model; Eq. (3.8) with $n = 2$ is the two-target model; Eq. (3.6) is the one-hit model; and Eq. (3.7) is the linear-quadratic model. For all curves, $\alpha = 0.5$ Gy^{-1} and $\beta = 0.01$ Gy^{-2}.

a function of dose D, defined in such a way that $f(z; D)dz$ is the fraction of volumes receiving specific energy between z and $z + dz$.

In any one volume, part of ϵ may be attributed to overlap of independent particle tracks. As shall be shown, this leads to the dose dependence in $f(z; D)$. Part of ϵ arises from correlated ionization events associated with a single track. For example, multiple delta rays or an Auger electron cascade may collectively contribute to ϵ. If one considers only correlated energy deposits, one may define a single-event distribution $f_1(z)$, independent of D.

The distribution function $f_1(z)$ is more fundamental than $f(z; D)$. It is possible, with some considerable effort, to compute $f(z; D)$ from $f_1(z)$ (Kellerer and Rossi 1978; Kellerer 1985; Zaider and Rossi 1989). The *mean event size* or average specific energy in individual volumes or sites is given by

$$\bar{z} = \int_0^{\infty} dz \, z \, f_1(z). \tag{3.9}$$

The fraction of the absorbed dose with specific energy in the range dz about z, $\mathcal{D}(z)dz$, is given by

$$\mathcal{D}(z) = \frac{z}{\bar{z}} f_1(z). \tag{3.10}$$

If D is the absorbed dose, then the mean number of events per test volume site is

$$\bar{\nu} = \frac{D}{\bar{z}}. \tag{3.11}$$

The probability of ν independent or uncorrelated events occurring in a test volume is given by the Poisson distribution, that is,

$$p(\nu) = \frac{\bar{\nu}^{\nu}}{\nu!} e^{-\bar{\nu}}. \tag{3.12}$$

If $f_{\nu}(z)dz$ is the probability of ν independent events in a volume of interest leading to deposition of specific energy in the range dz about z, then

$$f(z; D) = \sum_{\nu=0}^{\infty} p(\nu) f_{\nu}(z), \tag{3.13}$$

where, for $\nu > 1$,

$$f_{\nu}(z) = \int_{0}^{z} dx \, f_1(x) f_{\nu-1}(z - x), \tag{3.14}$$

and, for $\nu = 0$, $f_0(z) = \delta(z)$, the Dirac delta function.

For very low absorbed dose, with $\bar{\nu} = D/\bar{z} << 1$, it follows that $f(z; D)$ is well approximated by the first two terms in the summation of Eq. (3.13), so that

$$f(z; D) \cong [\delta(z) + f_1(z)(D/\bar{z})] \exp(-D/\bar{z}). \tag{3.15}$$

The function $f(z; D)$ is important in providing a physical explanation for many aspects of the measured relative biological effectiveness. The function plays key roles in explaining observed nonlinear relationships between radiation dose and effect and in supporting the concept of *dual radiation action* (Kellerer and Rossi 1972, 1978).

Dual Radiation Action.   This concept was stressed by the BEIR-III Committee (NAS 1980) in their interpretation of the observed quadratic dose-response relationship for low-LET radiation. It involves hypotheses that (1) the preponderance of stochastic biological effects of ionizing radiation involve the formation of irreparable lesions in autonomous cells, (2) that irreparable lesions result from the interaction of two reparable sublesions, (3) that in any one cell the probability of forming a sublesion is proportional to the specific energy incurred in an irradiation, and (4) that the probability of sublesion interaction is governed by a second order process, that is, is proportional to the mean square of the specific energy. For details, the reader is referred to the work of Kellerer and Rossi (1972, 1978) and Kellerer (1985). Supporting the hypotheses was a substantial body of evidence and opinion that the biological effects of ionizing radiation arise in the main from double-strand rupture of the DNA molecule, single-strand rupture being repaired rapidly by enzyme action. Double rupture may arise from either a single ionizing particle interacting simultaneously with both strands or from two particles separately interacting with the two strands at very nearly the same time and location.

If it is assumed that the extent of effect is proportional to the mean square specific energy, then the effect $\mathcal{E}(D)$ at a given dose D would be given by

$$\mathcal{E}(D) = K \int_0^\infty dz \, z^2 f(z; D), \tag{3.16}$$

in which K is a constant of proportionality. It can be shown using Eq. (3.13) (Kellerer 1985) that

$$\mathcal{E}(D) = K(\zeta D + D^2), \tag{3.17}$$

in which

$$\zeta = \frac{\displaystyle\int_0^\infty dz \, z^2 \, f_1(z)}{\displaystyle\int_0^\infty dz \, z \, f_1(z)}. \tag{3.18}$$

The BEIR-III report suggests that ζ is about 12.5 to 25 cGy for low-LET radiation and about 100 times greater for high-LET radiation. One thus would be justified, on the basis of this model, in relating dose and effect for one type of radiation by Eq. (3.1), in which α and β are to be determined empirically.

One consequence of the specific-energy model of Eq. (3.17) is that the quadratic term would be expected to be of significance only for $D \gg \zeta$ and thus that the quadratic term would be of greater significance, and of importance at lower dose, for low-LET radiation than for high. The reader will find in the discussion to follow later in the chapter, that the dose-effect relationship for neutron irradiation is commonly represented by a linear function, whereas the relationship for gamma irradiation is commonly represented by the linear-quadratic function.

Equation (3.1) has no constant term. The natural, or background, rate of appearance of a biological effect is omitted. Thus $\mathcal{E}$ represents the radiation-induced increment in effect beyond that of background. Nevertheless, the background incidence is a considerable complication in the statistical analysis of data on radiation effect. Nor does Eq. (3.1) account for dose-rate effects. An estimate of the effect can be made for two-track radiation effects. If T is the duration of exposure and τ is the average time between the initiation and repair of a single-track event, then the effect attributable to two-track events may be modified by the multiplicative factor (NAS 1980)

$$G = 2 \left(\frac{\tau}{T}\right)^2 \left(\frac{T}{\tau} - 1 + e^{-T/\tau}\right). \tag{3.19}$$

The factor G approaches the constant value of unity only as $T \ll \tau$. For $T \gg \tau$, G is inversely proportional to T. One must be careful in interpreting this observation. It is only to be expected that radiation damage of the two-track variety becomes negligible at very low dose rate and one-track events may be expected to dominate the radiation effect. Equation (3.1) may be thought of as representing a composite of two-track and single-track events, with the quadratic term being modified as follows:

$$\mathcal{E} = \alpha D + (\beta + \beta' G)D^2, \tag{3.20}$$

in which α, β, and β' are dependent on charged-particle LET.

As Kellerer (1985) has pointed out, survival probability data may be generalized in terms of the specific energy. If $S(z)$ is the probability of cell survival with specific energy z, then for a given radiation at dose D with specific energy distribution $f(z; D)$,

$$S(D) = \int_0^\infty dz\, S(z) f(z; D). \tag{3.21}$$

This formulation can also accommodate a threshold for effect, for example, $S(z) = 1$ for $z < z_c$ and 0 for $z \geq z_c$, in which case

$$S(D) = \int_0^{z_c} dz\, f(z; D). \tag{3.22}$$

3.3.2 Radiation Quality Effects

Factors Characterizing Radiation Quality.

Fluence and Absorbed Dose.   Important quantities used to characterize a radiation field have been introduced in Chap. 1. One is the descriptor called the flux density energy spectrum $\phi(E)$ or its time integral the fluence energy spectrum $\Phi(E)$. These are defined in such a way that $\Phi(E)dE$ is the total path length traversed in a unit volume, in the limit of a small volume, by particles with energies between E and $E + dE$. Another important quantity is the absorbed dose D, defined as the *average* energy imparted to matter per unit mass. The absorbed dose can also be written as a distribution function, that is, in such a way that $D(E)dE$ is that portion of D caused by particles with energies between E and $E + dE$. Figure 3.6 illustrates $\Phi(E)$ for secondary electrons produced in water by gamma and x rays.

Linear Energy Transfer.   The intensity of the ionization and excitation along the track of a charged particle varies as the particle decelerates. Locally, the intensity is described by the linear energy transfer (LET), defined as the spatial transfer rate of energy from a charged particle to a medium in the locality of the particle track through the medium. The "locality" may refer to either a maximum distance from the track or to a maximum value of the discrete energy loss by the charged particle, beyond which losses are no longer considered local. By designating the LET as L_∞, for example, the subscript means that all energies or distances are included. In this case, the LET is equivalent to the collisional stopping power. Here we will drop the subscript and let $L(E)$ denote the collisional stopping power for a particle of energy E. Since $L(E)\Phi(E)dE$ represents the energy locally dissipated per unit volume by electrons with energies in the range dE, it follows that

$$D(E) = \frac{1}{\rho} L(E) \Phi(E). \tag{3.23}$$

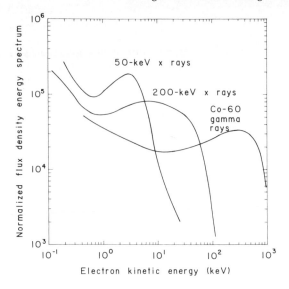

Figure 3.6   Energy spectra of electron fluences $\Phi(E)$, in units of $cm^{-2}\ keV^{-1}$, normalized to 1 rad absorbed dose, for secondary electrons produced in water irradiated by gamma and x rays. [After ICRU (1970)]

For a given type of particle, the fluence energy spectrum may be expressed, not as a function of E, but as a function of L, in such a way that $\Phi(L)dL$ is the fluence of particles with LET between L and $L + dL$. The same can be done for the absorbed dose. Since the product of L and $\Phi(L)dL$ is the energy transferred to the medium per unit volume by particles with LET between L and $L + dL$, then

$$D = \begin{cases} \displaystyle\int_0^\infty dE\ D(E) = \int_0^\infty dL\ D(L), \\[2mm] \displaystyle\frac{1}{\rho}\int_0^\infty dE\ L(E)\Phi(E) = \frac{1}{\rho}\int_0^\infty dL\ L\ \Phi(L). \end{cases} \qquad (3.24)$$

Figure 3.7 illustrates $\Phi(L)$ for the same conditions as in Fig. 3.6, while Figs. 3.8 and 3.9 respectively illustrate $D(E)$ and $D(L)$ for the same conditions. For future reference, let us define here the distribution function $\mathcal{D}(L) \equiv D(L)/D$, defined in such a way that $\mathcal{D}(L)dL$ is the fraction of the absorbed dose delivered by particles with LET in the range dL about L. It follows that the dose-average LET is given by

$$\bar{L} \equiv \int_0^\infty dL\ L\ \mathcal{D}(L) = \frac{1}{\rho D}\int_0^\infty dL\ L^2\Phi(L). \qquad (3.25)$$

3.3.3 The Relative Biological Effectiveness (RBE)

The RBE of one radiation in comparison to another is defined as the ratio of the absorbed doses leading to the same probability of a specific biological effect. Most

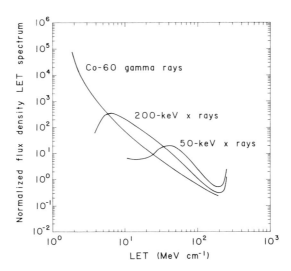

Figure 3.7   LET spectra of electron fluences $\Phi(L)$, in units of cm^{-2} $(keV/cm)^{-1}$, normalized to 1 rad absorbed dose, for secondary electrons produced in water irradiated by gamma and x rays. (Calculated from the energy spectra depicted in Fig. 3.6)

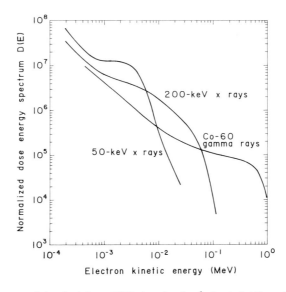

Figure 3.8   Energy spectra of absorbed doses $D(E)$, in units of g^{-1}, that is, keV/g per keV, normalized to 1 rad absorbed dose, for secondary electrons produced in water irradiated by gamma and x rays. (Calculated from the energy spectra depicted in Fig. 3.6)

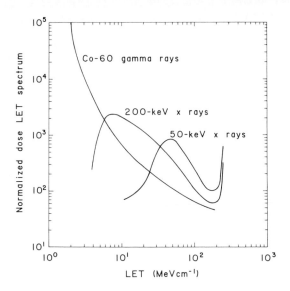

Figure 3.9   LET spectra of absorbed doses $D(L)$, in units of g^{-1}, that is, keV/g per keV, normalized to 1 rad absorbed dose, for secondary electrons produced in water irradiated by gamma and x rays. (Calculated from the energy spectra depicted in Fig. 3.6)

often, one is concerned with the RBE of "high-LET" radiation (e.g., alpha particles or recoil protons arising from fast neutrons) as compared to "low-LET" radiation (e.g., beta particles or secondary electrons arising from gamma or x rays). The LET for an electron is very much less than that for a heavy charged particle of the same energy. For example, LET values in water for 1-MeV electrons and protons are about 2 MeV cm^{-1} and 250 MeV cm^{-1}, respectively. The relative biological effectiveness of the high-LET radiation relative to the low is defined as D_L/D_H, where D_L and D_H are respectively the absorbed doses of the low and high-LET radiations required for equal biological effect. In terms of Eq. (3.17), the relative biological effectiveness is given by

$$\text{RBE} = \frac{\zeta_H D + D^2}{\zeta_L D + D^2}. \tag{3.26}$$

In the limit of very high dose, with $D \gg \zeta_H$, the RBE would approach unity but in the limit of very low dose, $D \ll \zeta_L$, the RBE would approach the maximum value $\text{RBE}_m = \zeta_H/\zeta_L$, or about 100.

In general, the RBE is greater than unity. It is in principle a function of dose and dose rate and, indeed, the nature of the effect. Necessarily, most measurements of RBE have been made at radiation doses for which, at least for the low-LET radiation and perhaps for the high-LET radiation as well, the relationships between dose and effect are nonlinear. For extremely high doses, the RBE is commonly near unity. Only at exceedingly low doses can it be expected that the RBE for high-LET radiation is independent of dose. Measurements at such low doses are rare and are insufficient to support the use of the RBE as a broadly applicable general measure of radiation quality for routine radiation protection purposes. For such purposes a more general and practical concept has been invoked, namely, that of the quality factor.

3.3.4 The Quality Factor

In use since 1964 for radiation protection purposes, the quality factor[3] Q serves as a general measure of the relative risk of equal absorbed doses of different radiations, that is, radiations of different spatial energy deposition patterns or qualities. While the RBE depends on dose, dose rate, radiation qualities of the two radiations, and biological endpoint, the quality factor for a particular radiation is defined to be a function only of its radiation quality. The reference radiation and its quality are fixed. Nevertheless, biological endpoint is certainly a factor to be taken into account in arriving at a relationship for Q as a function of LET. Since the quality factor is designed to be applied only to radiation exposure at low dose and low dose rate, biological endpoints in the classes of genetic mutation and carcinogenesis are the appropriate choices. In selecting RBE values for these classes of effects as a guide to defining the quality factor, one of course would choose values for low doses and low dose rates as being most appropriate. Usually, the RBE is greatest under these exposure conditions. Otherwise, the maximum value, RBE_m, is selected at whatever the dose and dose rate might be. For fission neutrons, for example, with ^{60}Co gamma rays as the reference, RBE_m values range from about 10 to 45 for genetic endpoints in mammals and for life shortening caused by tumors (ICRU 1986). Data for chromosome aberrations in human are given in Table 3.1. Thorough review of all such data, augmented by theoretical analysis using microdosimetry approaches (see the following discussion), support the widely accepted recommendations of organizations such as the International Commission on Radiological Protection for expressing quality factor uniquely as a function of radiation quality.

Two observations support the use of the quality factor as a substitute for the relative biological effectiveness and for expressing Q as a function of linear energy transfer. The first is that the effect of radiation dose rate on RBE is considerably less significant than that of the LET of the radiation. There are sound physical reasons for this observation, and these are taken up subsequently. The second observation is that, for low-LET radiation, the RBE for a given effect is independent of LET. There are also sound physical reasons for this observation. Localized energy deposition at low LET takes place in isolated clusters or "spurs," the effect of each being independent of the distance between clusters. Of course, the distance between clusters varies with LET. By contrast, energy deposition at high LET takes place more uniformly within a columnar region along the particle track. Because of physical and chemical interactions between the products of the deposition, the effect may be expected to be dependent on the "ionization density" along the track, that is, on the LET.

A major advantage allowed by the use of the quality factor as a measure of relative biological effectiveness has been the use of the dose equivalent, the product of the

---

[3]The ICRP in 1991 recommended a name change from quality factor to *radiation weighting factor*, with symbol w_R; however, the quality-factor notation is retained in this text.

absorbed dose and the average quality factor, as an effective measure of the overall risk of radiation exposure, whatever the type of risk.

   Quality Factor versus Linear Energy Transfer.   Since 1966, the ICRP has defined the quality factor in terms of the linear energy transfer L in water for the radiation of concern. If $\bar{L}$ is the mean value of the LET, as given by Eq. (3.25), then the mean quality factor $\bar{Q}$ is expressed as a function of $\bar{L}$. However, the same functional relationship applies for the local value of Q as a function of the local value of L along the track of an individual particle. The functional relationship $Q(L)$ or $\bar{Q}(\bar{L})$, as defined in 1966, is as follows:

L (keV/μm)	$Q(L)$
≤ 3.5	1
7.0	2
23	5
53	10
175	20

Any reasonable interpolation in the table is permissible. In recent years, evidence has come to light indicating that this relationship yields quality factors that are too great to be representative of measured RBE values for very-high LET particles such as recoil atoms, which are produced, for example, in neutron interactions. More important, there is new evidence that the quality-factor relationship yields values too low to be representative of RBE values for intermediate-energy neutrons. As discussed in Chap. 1, the NCRP in 1987 issued proposed mean quality factors for neutrons (Table 1.1c) which are about twice those based on the previously mentioned 1966 $Q(L)$ relationship. However, a new relationship for $Q(L)$ was not suggested. The ICRP in 1985 likewise made an interim recommendation that neutron quality factors then in use be doubled. The ICRP (1991) recommended a new $Q(L)$ formulation, namely

L (keV / μm)	$Q(L)$
< 10	1
$10 - 100$	$0.32L - 2.2$
> 100	$300/\sqrt{L}$

   Even if the radiation is all of one type, evaluation of the mean quality factor or mean LET is difficult, requiring knowledge of the energy spectrum of the flux density of the radiation. If the radiation is of mixed type, then the quality factor must be averaged over the types of radiation, the average being weighted by the absorbed dose attributable to each type. If $D(L)dL$ is the absorbed dose delivered by primary and secondary charged particles with LET between L and $L+dL$, then the average quality factor is given by

$$\bar{Q} = \frac{1}{D} \int_0^\infty dL\, Q(L)\, D(L) = \int_0^\infty dL\, Q(L)\, \mathcal{D}(L), \qquad (3.27)$$

where D is the total absorbed dose and $\mathcal{D}(L) = D(L)/D$.

Rarely in the routine practice of radiation protection can this process of averaging over energy and particle type be justified. Indeed, rarely is the charged-particle flux-density energy spectrum known. Therefore, mean quality factors have been calculated for various particles, using Eq. (3.27), and based on the absorbed dose at a depth of 1 cm in a spherical phantom irradiated by a particle beam. For irradiation by neutrons, mean quality factors based on the 1966 $Q(L)$ formulation are as follows:

Energy (MeV)	$\bar{Q}$
2.5×10^{-8}	2.3
$10^{-7} - 0.01$	2.0
0.1	7.4
0.5	11.0
1.0	10.6
2.0	9.3
4.0	7.8

For all gamma rays, x rays, and beta particles, the quality factor was taken as 1.0. In the absence of information about particle energies, the 1966 quality factors for neutrons and protons were taken as 10 and for alpha particles, 20. Note that, under the 1985 ICRP interim recommendations, the quality factors in the preceding table would be doubled.

Mean quality factors based on the $Q(L)$ formulation recommended by the ICRP (1991) are as follows:

Radiation[a]	$\bar{Q}$
X rays and gamma rays of all energies	1
Electrons and muons of all energies[b]	1
Neutrons, < 10 keV	5
10 − 100 keV	10
100 keV − 2 MeV	20
2 MeV − 20 MeV	10
> 20 MeV	5
Protons, other than recoil protons, energy > 2 MeV	5
Alpha particles, fission fragments, heavy nuclei	20

[a] Values are for radiations incident on the body or, for internal sources, emitted from the sources.
[b] Excluding Auger electrons emitted from nuclei bound to DNA, for which quality factors need be determined from microdosimetry.

Quality Factor versus Lineal Energy.   There are some shortcomings in the use of LET as a determinant of quality factor. Most significantly, LET is not simply related to energy deposition in a given volume of irradiated material because of energy escape via delta rays, straggling in energy loss, and a number of geometrical considerations. These shortcomings may be overcome by expressing the quality factor as a function of the *lineal energy*, the microdosimetric analog of the linear energy transfer and a more easily measured quantity. For these reasons, and others, a joint task group of the ICRP and the ICRU (ICRU 1986) recommended that the lineal energy be used as the determinant of the quality factor.

The *lineal energy* is defined in terms of the specific energy in a volume of interest arising from correlated ionization events associated with individual particle tracks. If $\bar{\ell}$ is the mean chord length in the volume, then the lineal energy y is defined as

$$y = \frac{\epsilon}{\bar{\ell}}. \tag{3.28}$$

For any convex volume, $\bar{\ell}$ is $4V/S$, where V and S are the volume and surface area of the volume. Thus, for a sphere of diameter d, $\bar{\ell} = 2d/3$. The volume of interest chosen by the task force was a sphere of tissue of 1-μm diameter. That diameter was chosen for two reasons: (1) it represents about the smallest region for which y can be measured, and (2) there is empirical evidence (ICRU 1983) that 1 μm is about the average distance over which sublesions in DNA combine to form permanent lesions.

The single-event distribution f_1 is related to the distribution for y by the relationship $f(y)dy = f_1(z)dz$, or

$$f(y) = \frac{\bar{\ell}}{\rho V} f_1(z). \tag{3.29}$$

Analogous to $\bar{z}$ is the *frequency mean lineal energy*

$$\bar{y} = \int_0^\infty dy \, y \, f(y). \tag{3.30}$$

It follows from Eq. (3.29) that the fraction of the absorbed dose delivered with lineal energy y (per unit y) is given by

$$\mathcal{D}(y) = \frac{y}{\bar{y}} f(y). \tag{3.31}$$

Figure 3.10 illustrates the effect of the site diameter on $y\mathcal{D}(y)$ for 1250-keV photons. Figures 3.11 and 3.12 illustrate the spectra $y\mathcal{D}(y)$ for photons and neutrons of various energies.

In order to relate the two distribution functions $f(y)$ and $\mathcal{D}(L)$, one may use an approximation (ICRU 1986) based on the continuous slowing-down approximation (no delta rays) for LET and a spherical reference volume. For a sphere of diameter d, the chord-length distribution function $t(\ell)$ is such that $t(\ell)d\ell = (2\ell/d^2)d\ell$ is the fraction of chords with lengths in $d\ell$. This leads to $\bar{\ell} = 2d/3$. Given L and ℓ, Eq. (3.28) gives $y = \ell L/\bar{\ell}$, whence $dy/d\ell = L/\bar{\ell}$. Since $t(\ell)d\ell$ must equal $f(y)dy$, it follows that,

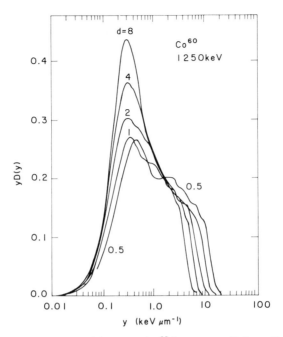

Figure 3.10   Microdosimetric spectra $y\mathcal{D}(y)$ versus y for ^{60}Co gamma radiation, with the diameter $d(\mu m)$ of the sensitive tissue volume as a parameter. [From ICRU (1983); by permission of the ICRU.]

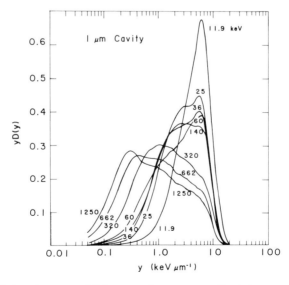

Figure 3.11   Microdosimetric spectra $y\mathcal{D}(y)$ versus y for monoenergetic photons, and a site diameter of 1 μm. [From ICRU (1983); by permission of the ICRU.]

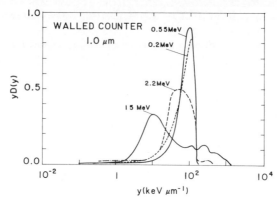

Figure 3.12 Microdosimetric spectra $y\mathcal{D}(y)$ versus y for neutrons and a site diameter of 1 μm. [From ICRU (1983); by permission of the ICRU.]

for fixed L, $f(y) = (8y)/(9L^2)$. With a modest effort, the reader can show that, for a distribution function in L, $\mathcal{D}(L)$, the distribution function for lineal energy is given by

$$f(y) = \frac{8y^2}{9} \int_{2y/3}^{L_{max}} dL \, \frac{\mathcal{D}(L)}{L^3}. \tag{3.32}$$

Having decided to recommend that the quality factor Q be expressed as a function of lineal energy y, the task force was faced with two questions: (1) what should be the reference radiation, and (2) what should be the functional relationship $Q(y)$? Both these questions are discussed later. Once $Q(y)$ is given, the mean quality factor for an irradiation is given by [see Eq. (3.31)]

$$\bar{Q} = \int_0^\infty dy \, Q(y)\mathcal{D}(y) = \int_0^\infty dy \, Q(y)\frac{y}{\bar{y}}f(y). \tag{3.33}$$

The reference radiation was chosen as ^{60}Co gamma rays delivered at low dose and low dose rate. This has been a controversial decision (Eisenhower 1989) because of its deviation from what appears to have been a previously accepted standard of 200 to 250-kVp x rays. The choice of the functional relationship for $Q(y)$ has also been controversial. It was influenced heavily by a comprehensive set of experiments with chromosome aberrations in human lymphocytes as the biological endpoint. The data, as reported in (ICRU 1986), are summarized in the Table 3.1. If the number of chromosome aberrations per cell is expressed as $\alpha D + \beta D^2$, and D is the absorbed dose in units of greys, then the RBE in the limit of low dose, that is, RBE_m is just α/α_{ref}. The reader will note that, compared to ^{60}Co gamma rays, 250-kVp x rays have an RBE_m of 3.0. Of course, had the x rays been chosen as the reference radiation, ^{60}Co gamma rays would have an RBE of only 0.33. According to the LET-based determination of quality factor both radiations are treated as having the same quality factor. This difference set the stage for the controversy over the $Q(y)$ formulation.

     After the lineal energy spectra are determined for each radiation in the table, a statistical method may be employed to find a best fit for a function $Q(y)$. This was done, using a method described by Zaider and Brenner (1985), with the result illustrated in Fig. 3.13. An empirical approximation is

$$Q(y) = \frac{a_1}{y}\left(1 - e^{-a_2 y^2 - a_3 y^3}\right), \tag{3.34}$$

for which, with y in units of keV/μm, $a_1 = 5510$ μm^{-1}, $a_2 = 5 \times 10^{-5}$ (keV/μm)$^{-2}$, and $a_3 = 2 \times 10^{-7}$ (keV/μm)$^{-3}$. Use of Eq. (3.34) results in $\bar{Q} = 1.0$ for photons with energies between 40 and 150 keV (Fig. 3.14). $\bar{Q}$ decreases to about 0.5 for 1-MeV photons and increases to about 1.5 for 10-keV photons. For neutrons, $\bar{Q}$ has a maximum of about 23 for neutrons of energy about 300 keV (Fig. 3.15), about 12 for 50-keV neutrons, and about 7 for 10-MeV neutrons. $\bar{Q}$ is about 25 for 4 to 9-MeV alpha particles. The task force recommends the following approximate values for $\bar{Q}$ for use when $D(y)$ is unknown.

Photons and electrons with $E > 30$ keV	1
Tritium beta particles	2
Neutrons, protons, alpha particles, etc.	25

The use of lineal energy as a determinant of quality factor is still under study. As powerful as microdosimetry concepts may be in the understanding of radiation effects, it appears that, for routine radiation-protection practices and for risk assessment, quality factors will continue to be based on linear energy transfer.

3.4   EFFECTS OF HIGH DOSES OF RADIATION

There are two circumstances under which high doses of ionizing radiation may be received. The first is accidental, and likely to involve a single instance of short duration. The second is medical, and likely to involve doses fractionated into a series delivered over days or weeks, and perhaps administered under conditions designed to modify response in certain organs and tissues. Oxygen, among a group of chemical agents known to modify dose response, promotes the action of radiation presumably through the enhanced production of oxidizing agents along the tracks of the radiation. This section deals only with single, acute exposure of all or part of the body. It does not address issues such as fractionation and effect modification which pertain largely to medical exposure.

3.4.1   Effects on Individual Cells

The likelihood that a cell will be killed or prevented from division as a result of a given radiation exposure depends on factors in addition to the dose-rate and LET effects previously described. One important factor is the state of the cell's life cycle at the time of exposure. Cell death is more likely if the cell is in the process of division than if it is in a quiescent state. Thus, radiation exposure results in a greater proportion of cell death in those organs and tissues with rapidly reproducing cells. Examples include

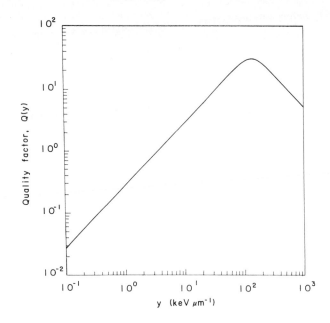

Figure 3.13   The defining relation between quality factor Q and lineal energy y in 1-μm spheres of ICRU tissue. [From ICRU (1986); by permission of the ICRU.]

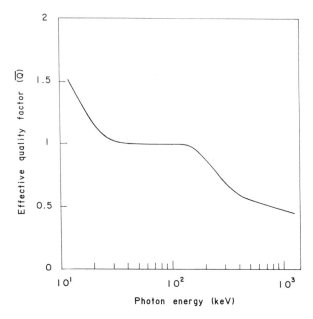

Figure 3.14   Calculated Q values versus photon energy for conditions of charged particle equilibrium. [From ICRU (1986); by permission of the ICRU.]

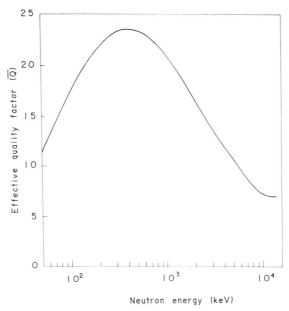

Figure 3.15   Calculated Q values versus neutron energy for conditions of charged particle equilibrium. [From ICRU (1986); by permission of the ICRU.]

the fetus, especially in the early stages of gestation, the bone marrow, and the intestinal lining. Whole-body absorbed doses of several Gy are life-threatening largely because of cell killing in the bone marrow and lining of the intestines. However, in these tissues and in most other tissues and organs of the body, there are ample reserves of cells and absorbed doses of much less than one Gy are tolerable without significant short term effect. Similarly, radiation doses which would be fatal if delivered in minutes or hours may be tolerable if delivered over significantly longer periods of time. Age, general health, and nutritional status are also factors in the course of events following radiation exposure.

For those tissues of the body for which cell division is slow, absorbed doses which might be fatal if delivered to the whole body may be sustained with little or no effect. On the other hand, much higher absorbed doses may lead ultimately to such a high proportion of cell death that, because replacement is so slow, structural or functional impairment appears perhaps long after exposure and persists perhaps indefinitely.

3.4.2 Deterministic Effects in Organs and Tissues

This section deals only with deterministic somatic effects—effects in the person exposed, effects with well-defined patterns of expression and thresholds of dose below which the effects are not experienced, and effects for which the severity is a function of dose. The stochastic carcinogenic and genetic effects of radiation are addressed in later sections.

The risk, or probability of suffering a particular effect or degree of harm, as a function of radiation dose above a threshold, can be expressed in terms of a fiftieth

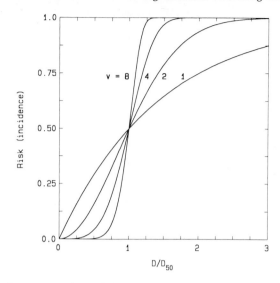

Figure 3.16   The Weibull-type hazard function $(D/D_{50})^v \ln 2$.

percentile dose D_{50}, or median effective dose, which would lead to a specified effect
or degree of harm in half the persons receiving that dose. The D_{50} dose depends, in
general, on the rate at which the dose is received. Furthermore, the range of dose over
which the risk increases from near zero to near unity can generally be described by a
single parameter, for example, a shape factor. This approach to risk characterization
(Scott and Hahn, 1980; Scott 1988) has been applied in a very thorough study of potential health effects from nuclear power plant accidents (Scott and Hahn 1989). The
risk in a particular context is expressed in terms of a hazard function H as follows:

$$\text{Risk } = 1 - e^{-H}. \tag{3.35}$$

The hazard function is a function of the received dose D, and is given by a Weibull-type
function with a single shape parameter v, namely,

$$H = \left(\frac{D}{D_{50}}\right)^v \ln(2). \tag{3.36}$$

Thus, when D is equal to D_{50}, the risk is equal to 0.5. If D is less than a threshold
dose, H is zero. The hazard function is additive for consecutive exposures at different doses and dose rates (with different shape factors) or for several effects with the
same endpoint. For example, with potentially lethal whole body exposure, the hazard functions for lethality because of exposure of the lung, the bone marrow, and the
gastrointestinal system may be combined and used in evaluation of the overall risk
of death. Risk, plotted as a function of D/D_{50}, is illustrated in Fig. 3.16 for hazard
functions governed by shape factors from 1 to 8.

Effects of radiation on individual organs and tissues are described in the paragraphs that follow. A summary, listing median effective doses, shape factors, and
thresholds is provided in Table 3.2.

TABLE 3.2   Median Effective Absorbed Doses D_{50}, Shape Factors v, and Threshold Doses D_{th} for Exposure of Selected Organs and Tissues of the Adult to Low-LET Ionizing Radiation. Central Estimates Are Given, with Upper and Lower Bounds as $\pm$ Values.

ORGAN OR TISSUE	ENDPOINT	BRIEF EXPOSURE[a] D_{50} (Gy)	v^c	D_{th} (Gy)	PROTRACTED EXPOSURE[b] D_{50} (Gy)	v^c	D_{th} (Gy)
Skin	Erythema	6 ± 1	5	3 ± 1	20 ± 10	5	6 ± 2
	Moist desquamation	30 ± 6	5	10 ± 2	80 ± 20	5	40 ± 10
Ovary	Permanent ovulation suppression	3 ± 1	3	0.6 ± 0.4			
Testis	Two-year suppression of sperm count	0.6 ± 0.1	10	0.3 ± 0.1			
Lens	Cataract	3.1 ± 0.9	2	0.5 ± 0.5			
GI System	Vomiting	2 ± 0.5	3	0.5	5 ± 1.2	3	1.5
	Diarrhea	3 ± 0.8	2.5	1	6 ± 1.3	2.5	2.5
	Death	10 ± 5	10	8	35 ± 20	10	18

[a] Absorbed dose rate in excess of 0.06 Gy h^{-1}.

[b] Absorbed dose rate less than 0.06 Gy h^{-1}.

[c] Bounds on shape factor are central value ± 1.

Source: Scott and Hahn 1989.

Skin.   (Langham 1967; Wald 1967; UN 1988) Erythema, equivalent to a first-degree burn, appears about two weeks after an acute absorbed dose of 2 to 3 Gy, although reddening and itching may occur shortly after exposure. Loss of hair may result beginning about two weeks after acute exposure to 2 Gy or more, the effect being complete for doses greater than about 5 Gy and permanent for doses greater than about 6 Gy.

Wet or dry desquamation (transepidermal injury), equivalent to a second-degree burn, may appear one or two weeks after an acute absorbed dose in excess of 10 to 15 Gy. The desquamation results from destruction of basal cells of the epidermis. Dry desquamation is experienced with doses between about 10 and 20 Gy. Greater doses lead to moist desquamation. Doses on the order of 50 Gy result in prompt and extremely painful injury (dermal necrosis) in the nature of scalding or chemical burn.

Repair of sublethal damage to surviving cells is nearly complete in one day, though cellular repopulation is much slower. The radiation dose for a given effect is approximately doubled if the exposure is divided into ten equal daily fractions.

The various layers of skin have differing sensitivity to radiation damage. In terms of both chronic low-level exposure (potentially leading to skin cancer), and transepidermal injury, the basal-cell layer of the epidermis, at a depth range from approximately 0.05 to 0.1 mm, is the most sensitive region because of the mitotic activity that takes place therein. For extremely high doses leading to ulceration, damage to the dermal region, at depth greater than 0.1 mm, is of greater significance than damage to the epidermis.

Lens of the Eye.   (Upton and Kimball 1967; Pochin 1983) The lens of the eye is particularly susceptible to ionizing radiation, as it is to microwave radiation. As little as 1 to 2 Gy absorbed dose from x rays results in detectable opacification of the lens. Effects are cumulative, resulting from the migration of dead cells to a central position at the back of the lens. Dose equivalents in excess of about 10 Sv lead to cataract formation causing detectable impairment of vision.

Blood Forming Tissues.   (Langham 1967; Upton and Kimball 1967; UN 1988) Except for lymphocytes, the mature circulating blood cells are relatively insensitive to radiation. The precursor cells present in bone marrow and lymphoid tissue, being rapidly dividing, are highly sensitive. Exposure of the whole body leads to rapid depletion of lymphocytes and a relatively slow decline in platelets, granulocytes and the long-lived red blood cells. Depression of lymphocytes is central to the impairment of immune response resulting from whole-body exposure to high levels of ionizing radiation. Recovery of the blood-forming tissues occurs through multiplication of surviving precursor cells in the bone marrow, lymphoid organs, spleen and thymus. Blood cell depression may be countered by transfusion. Recovery may be enhanced by bone-marrow transplant, but with attendant immunological difficulties.

Following acute, sublethal whole-body exposure the lymphocyte count reaches its minimum in a few days. Full recovery may take many months or years. For sublethal exposure, the platelet count reaches its minimum 25 to 30 days post exposure and recovery is about 80 to 90 percent complete in about 45 days. Neutrophiles (phagocytes), the major constituent of white blood cells (leucocytes), increase in count the first few days post exposure but then decrease in count, for sublethal exposure, reaching a minimum 40 to 45 days post exposure. The risks of fever and infection are greatest from 20 to 30 days after exposure. Complete recovery requires some six months. For all types of cells, minimum blood counts decrease approximately exponentially with dose for sublethal exposure. A reduction by 50 percent for neutrophiles, lymphocytes, and platelets occurs respectively at 2.0, 1.6, and 1.3 Gy for low-LET radiation.

Gastrointestinal Tract.   (Upton and Kimball 1967) The rapidly dividing cells in the lining of the gastrointestinal tract suffer significant mitotic inhibition at doses on the order of 0.5 Gy for low-LET radiation. For absorbed doses in excess of about 5 Gy, the lining cannot be adequately supplied with new cells. Excessive fluid and salts enter the gastrointestinal system, while bacteria and toxic materials enter the blood stream. The consequences are diarrhea, dehydration, infection, and toxemia.

Urinary, Respiratory, and Neurovascular Systems.   (Upton and Kimball 1967; Ward 1967; UN 1988) The urinary system is relatively radiation resistant. Absorbed doses to the kidney in excess of about 3 Gy of low-LET radiation leads to hypertension. Absorbed doses of greater than 5 to 10 Gy of low-LET radiation lead to atrophy of the kidney and potentially lethal renal failure.

The respiratory system as a whole is relatively radiation resistant as compared to the blood-forming organs and tissues. Very high absorbed doses delivered only to the thorax (as in radiation therapy) may lead to scarring of lung tissue and pulmonary blood vessels, leading to chronic pneumonia-like illness. As discussed in detail in Chap. 4, the bronchial epithelium is very sensitive to the localized exposure resulting from inhaled alpha-particle emitters such as the daughter products of radioactive radon gas.

The nervous system, with little or no mitotic activity, is extremely radiation resistant. Acute whole-body doses in excess of 100 Gy lead in hours or days to death from cerebro-vascular injury. Sublethal exposure may lead to impairment or degeneration of the nervous system as a result of radiation damage to blood vessels of supporting tissues.

Ovaries and Testes.   (Langham 1967; Vogel and Motulski 1979; Pochin 1983; ICRP 1984; UN 1988) In certain stages of cellular life, germ cells and gametes are very susceptible to radiation damage. In other stages, the cells are relatively resistant. There are significant differences between the courses of radiation damage in the male and the female.

Production of ova is completed early during fetal development. Some 2 million ova are present in the ovaries at birth, but the number declines to less than about 10 thousand by age 40. The immature ova, or *oocytes*, begin meiotic division during fetal development. However, the process is suspended in each ovum until a few weeks before that ovum takes part in ovulation. That few-week period is the time of greatest sensitivity for both cell killing and mutation induction. Mature oocytes may be killed by dose equivalents of a few Sv, and complete sterilization requires dose equivalents of 2.5 to 6 Sv, the required dose decreasing with subject age.

The spermatogenesis process which has a duration of about 74 days, occurs in a series of stages. Spermatogonia, produced from spermatogonial stems cells are the most radiosensitive cells. Cytological effects are observable within a few hours after delivery of doses as low as 0.15 Gy. The spermatogonia divide mitotically to spermatocytes. In a process which requires about 46 days, spermatocytes divide by meiosis into spermatids which mature to spermatozoa. Cells in these latter three stages are unaffected by absorbed doses less than about 3 Gy. A 0.15 Gy absorbed dose causes reduction in the sperm count beginning some 46 days later. By 74 days, the count may be reduced by about 80 percent. A 1.0 Gy or greater absorbed dose may lead to complete loss of sperm after the 74-day period. At absorbed doses less than about 3 Gy, fertility is restored completely, but only after many months. Permanent sterility results from greater absorbed doses.

The Embryo.   (NCRP 1971; Pochin 1983; UN 1986; ICRP 1991) In the developing embryo, the loss of but a few cells may risk serious consequences more akin to severe hereditary illness than to cell killing in fully developed somatic tissues.[4] The natural probability for mortality and induction of malformations, mental retardation, tumors, and leukemia during gestation is about 0.06. The combined radiation risk is about 0.2 per Sv delivered throughout gestation. The risk of childhood cancers and leukemia up to age 10 is about 0.02 per Sv. The risk of severe mental retardation is

---

[4]Because of this, the NCRP has issued radiation-protection recommendations specific to occupational exposure of fertile women and the diagnostic radiology of pregnant or potentially pregnant women. The recommendations would limit the dose equivalent to the fetus to no more than 0.005 Sv during the entire gestation period and would minimize radiological examination or treatment during the first two weeks of the menstrual cycle and throughout pregnancy.

greatest during the eighth to fifteenth week of gestation, with a radiation risk of about 0.4 per Sv. The central nervous system is sensitive throughout gestation, but terato-logical (i.e., malformation) risks may be greater by a factor of 2 or 3 during the major organogenesis of the second through eighth weeks.

3.4.3 Potentially Lethal Exposure to Low-LET Radiation

The question of what constitutes a *lethal dose* of radiation has of course received a great deal of study. There is no simple answer. Certainly, the age and general health of the exposed person are key factors in the determination. So too, are the availability and administration of specialized medical treatment. Inadequacies of dosimetry make interpretation of sparce human data difficult. Data from animal studies, when applied to human exposure, are subject to uncertainties in extrapolation. Delay times in the response to radiation, and the statistical variability in response have led to expression of the lethal dose in the form, for example, $LD_{50/60}$, meaning the dose fatal to 50 percent of those exposed within 60 days. The *dose* itself requires careful interpreta-tion. One way of defining the *dose* is the free-field exposure, in roentgen units, for gamma or x rays. A second is the average absorbed dose to the whole body. A third is the *mid-line* absorbed dose, that is, the average absorbed dose near the abdomen of the body, and approximately equivalent to the absorbed dose index (see Chap. 2). For gamma rays and x rays, the mid-line dose, in units of rads, is about two-thirds the free-field exposure, in units of roentgens. Anno et al. (1989) suggest that the lethality of ionizing radiation may be expressed as follows:

Lethality	Mid-line absorbed dose (Gy)
$LD_{5/60}$	$2.0 - 2.5$
$LD_{10/60}$	$2.5 - 3.0$
$LD_{50/60}$	$3.0 - 3.5$
$LD_{90/60}$	$3.5 - 4.5$
$LD_{99/60}$	$4.5 - 5.5$

Scott and Hahn (1989) give median doses D_{50} and shape factors for the risk of death as a function of absorbed dose to the gastrointestinal system (Table 3.2), the lung, and bone marrow (Table 3.3). For whole-body irradiation, the risk of fatality is great-est from the consequences of bone marrow irradiation. As an example of the use of the median doses and shape factors, consider a case in which the bone marrow of a subject receives an absorbed dose of 2.5 Gy. If the subject were to receive mini-mal medical care, the central estimate of the hazard function of Eq. (3.36) would be $H = (2.5/3)^6 \ln 2 = 0.23$. The risk (probability of death) would be $1 - e^{-H} = 0.21$. Upper and lower bounds of risk would be 0.5 and 0.05. However, if the subject were to receive supportive medical care, the central estimate of the hazard function would be $H = (2.5/4.5)^6 \ln 2 = 0.020$. The risk would be $1 - e^{-H} = 0.020$. Upper and lower bounds of risk would be 0.14 and 0.002.

TABLE 3.3   Median Effective Absorbed Doses D_{50}, Shape Factors v, and Threshold Doses D_{th} for Exposure of the Lung and the Bone Marrow of the Adult to Low-LET Ionizing Radiation. Central Estimates are Given, with Upper and Lower Bounds as $\pm$ Values.

ORGAN OR TISSUE	ENDPOINT	ABSORBED DOSE $(Gy\,h^{-1})$	D_{50} (Gy)	v	D_{th} (Gy)
Lung	Death	≥ 100	10 ± 2	12 ± 3	5 ± 2
		10	15 ± 5	12 ± 3	7 ± 2
		1	40 ± 20	12 ± 3	20 ± 10
		0.5	70 ± 30	12 ± 3	40 ± 20
		0.1	310 ± 150	12 ± 3^{a}	160 ± 80
		≤ 0.06	610 ± 300	12 ± 3^{a}	310 ± 150
Bone marrow	Death (with minimal treatment)	≥ 10	3.0 ± 0.5	6 ± 2	1.5 ± 0.3
		1	3.1 ± 0.6	6 ± 2	1.6 ± 0.3
		0.1	3.7 ± 0.6	6 ± 2	1.9 ± 0.3
		0.05	4.4 ± 0.8	6 ± 2	2.2 ± 0.4
		0.01	10 ± 4	6 ± 2	5.0 ± 2.0
Bone marrow	Death (with supportive treatment)	≥ 10	4.5 ± 0.8	6 ± 2	2.3 ± 0.4
		1	4.7 ± 0.9	6 ± 2	2.4 ± 0.5
		0.1	5.6 ± 0.9	6 ± 2	2.8 ± 0.5
		0.05	6.6 ± 1.2	6 ± 2	3.3 ± 0.6
		0.01	15 ± 3	6 ± 2	7.5 ± 1.5

[a]Shape factor 5 ± 1 for internal exposure at the two lower dose rates.

Source: Scott and Hahn 1989.

Representative effects of doses below the threshold for lethality are as follows (NCRP 1971, Anno et al. 1989):

Minimal dose detectable by chromosome analysis	0.05–0.25 Gy
Minimal dose detectable in groups by change in white-blood cell count	0.25–0.50 Gy
Minimal acute dose readily detectable in a specific individual	0.50–0.75 Gy
Mild effects only during first day postexposure 0 with slight depression of blood counts	0.50–1.00 Gy
Minimal acute dose to produce vomiting in 10 percent of exposed individuals	0.75–1.25 Gy
Nausea and vomiting in 20 to 70% of persons exposed fatigue and weakness in 30 to 60%; 20 to 35% drop in blood cell production due to loss of bone marrow stem cells	1.00–2.00 Gy
Acute dose likely to produce transient disability and clear hematological changes in a majority of individuals so exposed.	1.50–2.00 Gy

The lethal effects of radiation exposure may take several courses. For extremely high doses (> 500 Gy), death is nearly instantaneous, resulting from enzyme inactivation or possibly from immediate effects on the electrical response of the heart (Kathren 1985). Lesser but still fatal doses lead promptly to symptoms known collectively as the *prodromal syndrome*. The symptoms, which are expressed within a 48-hour period, are of two main types (Langham 1967):

Gastrointestinal	Neuromuscular
Anorexia	fatigability
Nausea	apathy
Vomiting	sweating
Diarrhea	fever
Intestinal cramps	headache
Salivation	hypotension
Dehydration	hypotensive shock

At doses from about 50 to 500 Gy, most symptoms appear within minutes of exposure, and death results from neurological or cardiovascular failure. At doses on the order of 50 Gy, death is likely to result from cardiac failure within a few days. At lower doses, down to about 6 Gy, death is more likely to result from damage to the gastrointestinal system with attendant dehydration and electrolyte loss.

At doses for which survivability is questionable, the course of radiation illness takes roughly the following pattern (Wald 1967):

Prodromal Stage (0 to 48 hours postexposure).   Symptoms are gastrointestinal and neuromuscular, as described earlier. Reduction in lymphocytes is apparent within 24 hours.

Latent Stage (48 hours to 2 to 3 weeks postexposure).   Remission of prodromal symptoms occurs and there is a period of apparent well being.

Manifest Illness Stage (2 to 3 weeks to 6 to 8 weeks postexposure).   Gastrointestinal and hematological symptoms are sharply defined. Blood elements except for red blood cells are depleted and secondary infection is a threat. Dehydration and loss of electrolytes are life threatening. Skin pigmentation changes and epilation take place.

Recovery Stage (beyond 6 to 8 weeks postexposure).   If the exposed person survives for six weeks, the outlook for ultimate recovery is favorable.

3.5 HEREDITARY RISKS FROM LOW-LEVEL EXPOSURE

Inheritance of radiation-induced abnormalities was discovered by Hermann Muller in 1927 in studies involving high-dose irradiation of fruit flies. Since 1927, studies of plants and animals have revealed approximately linear relationships between mutation frequency and dose, for doses as low as 3 mSv (Pochin 1983). Very little information is available on mutational effects of radiation in humans. Indeed, the only unequivocal evidence relates to chromosomal rearrangement in spermatocytes. However, the evidence from animal studies has left no doubt that heritable mutational effects in the human are possible. Estimation of risks to human populations are based largely on extrapolation of studies of radiation effects in other mammals.

Beginning in the 1950s, radiation protection standards recognized the potential significance of radiation induced hereditary illness. Both the International Commission on Radiation Protection and the National Council on Radiation Protection and Measurements issued recommendations designed to minimize genetic risks in future generations, and which may be interpreted as limiting exposure to human populations to 1.7 mSv average dose equivalent per year in excess of medical and background exposure.

It is often asked whether an increase in mutation rate might be beneficial. This question was addressed and answered in the negative in the BEIR-I and BEIR-III Reports (NAS 1972, 1980). It was argued that, although a genetically diverse population is desirable, there is already ample variability — far greater than that which arises naturally in a single generation. It was argued further that, since virtually all identified mutations are deleterious, any increase in the mutation rate would have net harmful effects on human populations.

This section deals only with exposure to low-LET ionizing radiation, at dose rates comparable to those received from the natural radiation environment. It deals only with heritable genetic effects, and not with radiation induced mutations in somatic cells.

3.5.1 Classification of Genetic Effects

Table 3.4 reports estimates of the incidence of human hereditary or partially hereditary defects and diseases (traits) causing serious handicap at some time during life. Inheritance of a deleterious trait results from mutation(s) in one or both maternal and paternal lines of germ cells. By mutations, we mean either microscopically visible chromosome abnormality or submicroscopic disruption in the DNA making up the individual genes within the chromosomes. Mutations take place in both germ cells and somatic cells, but mutations in germ cells are of concern here.

Regularly inherited traits are those whose inheritance follows Mendelian laws. These are autosomal dominant, X-linked, and recessive traits. Examples of autosomal dominant disorders, that is, those which are expressed even when the person is heterozygous for that trait, are certain types of muscular dystrophy, retinoblastoma, Huntington's chorea, and various skeletal malformations. Examples of recessive disorders, that is, those which are expressed only when the individual is homozygous for the trait, include Tay-Sachs disease, phenylketonuria, sickle-cell anemia, and cystic fibrosis. X-linked disorders, that is, those traits identified with genes in the X chromosome of the X-Y pair[5] and which are expressed mostly in males, include hemophilia, color-blindness, and one type of muscular dystrophy. Chromosome abnormalities are of two types, those involving changes in the numbers of chromosomes (trisomies) and those involving the structure of the chromosomes themselves (unbalanced translocations). Down syndrome is an example of the former. With natural occurrence, numerical abnormalities are more common. Radiation-induced abnormalities are more frequently structural abnormalities.

There is a very broad category comprising what are variously called "irregularly inherited traits," multifactorial diseases, or traits of "complex etiology." This category

---

[5]In the X-Y chromosome pair, otherwise recessive genetic traits carried by the "stronger" maternal X chromosome are expressed as though the traits were dominant.

TABLE 3.4 Natural Incidence of Hereditary Illness as Enumerated by the BEIR and UNSCEAR Committees.

	CURRENT INCIDENCE PER MILLION LIVEBORN	
TYPE OF DISORDER	BEIR	UNSCEAR
Autosomal dominant[a]		10,000[b]
clinically severe[c]	2,500	
clinically mild[d]	7,500	
X-linked	400	
Recessive	2,500	2,500
Chromosomal		
unbalanced translocations[e]	600	400
trisomies[f]	3,800	3,400
Congenital abnormalities	20,000–30,000	60,000

[a]Of the current incidence of 10,000 cases, some 2,000 are thought to be newly introduced from mutations in the parental generation.

[b]Autosomal dominant and x-linked disorders combined.

[c]Survival and reproduction reduced by 20 to 80 percent relative to normal.

[d]Survival and reproduction reduced by 1 to 20 percent relative to normal.

[e]Structural abnormalities.

[f]Numerical abnormalities.

Source: United Nations 1988; NAS 1990.

includes abnormalities and diseases to which genetic mutations doubtlessly contribute but which have inheritances much more complex than result from chromosome abnormalities or mutations of single genes. They are exemplified by inherited predispositions for a wide variety of ailments and conditions. Included in Table 3.4 is a subgroup of irregularly inherited traits, identified as congenital abnormalities. These are well identified conditions such as spina bifida and cleft palate, with reasonably well-known degrees of heritability. One or more other multifactorial disorders, including cancer, are thought to afflict nearly all persons sometime during life; however, the mutational components of these disorders are unknown even as to orders of magnitude (NAS 1990).

3.5.2 Risks for Radiation Induced Hereditary Illness

For the very low doses and dose rates of concern in public-health aspects of radiation exposure, both the BEIR and the UNSCEAR Committees base risk estimates on a linear, no-threshold model for dose response. Three methods have been used in making estimates of radiation risks. One is the *doubling-dose* method, sometimes called the relative-risk method. The second is the *direct* method, sometimes called the absolute-risk method. The third is the *gene-number* method which is sometimes identified as the second of two direct methods. The doubling-dose method makes use of the natural frequency of hereditary illness and arrives at estimates of the degree to which the frequency ultimately would be increased as a result of a sudden sustained increase in radiation exposure of the public. The direct method arrives at an estimate of risk from induced mutations by first determining in test animals the induction rate for a specific class of defects. Then an estimate is made of the proportion of all deleterious dominant disorders in the human that involve similar defects. This estimate, when multiplied by the measured rate in animals, yields an estimate of the rate of inductions of all dominant disorders in humans. The gene-number method arrives at an estimate of risk as the product of three factors (1) the number of genes at which mutations can occur, (2) the radiation dose, and (3) the mutation rate per gene per unit dose. The direct method and the gene-number method apply to first-generation effects, but population-genetic methods can be used to estimate effects in succeeding generations. The doubling-dose method applies to equilibrium effects, but population-genetic methods can be applied to infer first-generation effects.

The doubling-dose method requires far fewer assumptions and approximations and was the favored method of the BEIR-V Committee (NAS 1990). However, the direct method was earlier used by the BEIR-III Committee (NAS 1980) and continues to be used by UNSCEAR (UN 1988). Neither the UNSCEAR nor the BEIR Committees now make use of the gene-number method. However, it has been applied to estimates of X-linked mutations in studies of potential genetic risks from nuclear power accidents (Abrahamson et al. 1989).

The Direct Method for Estimation of First-Generation Genetic Effects.   The direct method is illustrated by using, as a case study, the estimate of first-generation radiation-induced dominant defects in man as was done in the 1977 UNSCEAR Report (UN 1977) and the 1980 BEIR-III Report (NAS 1980). Estimates are based on one observation and three carefully considered assumptions.

Observation: Dominant defects (anomalies) in the skeleton of the mouse appear with a probability of $P = 0.00040$ per Gy absorbed dose to spermatogonia.

Assumption: The proportion of skeletal anomalies in the mouse that might cause serious handicap, should they occur in the human, is given by $A = 1/2$ [UNSCEAR] or 1/2 to 3/4 [BEIR].

Assumption: The proportion of dominant conditions in the human whose main effect is in the skeleton is given by $B = 1/10$ [UNSCEAR] or 1/15 to 1/5 [BEIR].

Assumption: The relative mutational response of mature and maturing oocytes in the human as compared to spermatogonia in the mouse is given by $C = 0$ [UNSCEAR] or 0 to 0.44 [BEIR].

One may estimate that the probability of induction of mutations causing deleterious dominant effects in any of the bodily systems of the human, per Gy gonad dose delivered to *both* parents is given by

$$\text{Probability} = \frac{P \times A \times (1 + C)}{B}, \tag{3.37}$$

which, for UNSCEAR assumptions, is 0.002, and, for BEIR assumptions, is 0.0005 to 0.0065. In other words, using the BEIR estimate, among one million children born into a population which has received an average gonad dose of 1 Gy, one would expect 500 to 6500 children with deleterious dominant defects (500 to 4500 for paternal exposure, 0 to 2000 for maternal exposure).

By 1986, in consideration of additional animal data, the UNSCEAR direct estimates of probabilities of first-generation deleterious mutations (numbers per million liveborn per 0.01 Gy average gonad dose to parents) were revised to

Dominant effects	$10 - 20$ (paternal)	$0 - 9$ (maternal)
Chromosome rearrangements	$1 - 15$ (paternal)	$0 - 5$ (maternal)

The Doubling-Dose Model for Equilibrium Genetic Effects.   The doubling dose D_D is defined as that gonad dose, delivered per generation to all members of a population that, at equilibrium after many generations, would cause an increase in the genetic burden just equal to the spontaneous burden. The genetic burden could also be called the mutation rate or the frequency of hereditary illness. The doubling dose would thus lead to a doubling of the *total* burden. The reciprocal of the doubling dose is sometimes called the *relative mutation risk*. Use of the doubling-dose method allows risks to be expressed in easily understood terms for whole classes of genetic effects. For a given dose D delivered over a generation, the spontaneous burden m_o and the radiation-induced burden m_r are related by

$$m_r = m_o \frac{D}{D_D}. \tag{3.38}$$

For irregularly inherited traits, it is necessary to multiply m_o in Eq. (3.38) by a factor $M \leq 1$, called the mutation component, to account for the fact that expression of the traits does not follow Mendelian inheritance and is not entirely attributable to radiation-induced mutation.

Earliest estimates of D_D, made in the 1950s, and at the time expressed in exposure units, were on the order of 50 R. These estimates were based largely on measurements with *Drosophilia* made at high dose and high dose rate. It was later realized that these estimates were too low because a dose rate effectiveness factor had not been applied to the measured response.

By the 1970s, a large body of information had become available on radiation-induced mutations resulting from chronic exposure of the mouse. The BEIR-I Committee (NAS 1972) based an estimate of D_D on the observations (1) that mutations of any one type (gene locus) in the mouse were induced with a probability or *induction rate* of $a = 2.5 \times 10^{-6}$ per Gy, and (2) that the probability of a spontaneous mutation of any one type in the human, prior to reproduction, was $m_o = 5 \times 10^{-6}$ to 5×10^{-7}. The doubling-dose estimate, given as $D_D = m_o/a$, was 0.2 to 2 Gy. This hybrid estimate, based on both human and mouse data, was revised in the BEIR-III Report (NAS 1980) to an estimate based entirely on mouse data. With $a = 6.6 \times 10^{-6}$ per Gy and $m_o = 7.5 \times 10^{-6}$ (for the mouse), the best estimate of D_D was 1.1 Gy (rounded to 1 Gy). A similar estimate was arrived at by UNSCEAR (UN 1977). Additional evidence has only confirmed the estimate, and 1 Gy as the doubling dose for low-LET radiation delivered at low dose and low dose rate is the currently accepted value (UN 1988, NAS 1990).

It may be the case that the doubling dose for maternal exposure differs from that for paternal exposure and that the spontaneous mutation frequencies differ also for the male and the female. Let the subscripts m and f denote male and female and suppose that both parents receive the same dose D. The total spontaneous burden is $m_o = m_m + m_f$ and the radiation induced burden is $a_m D + a_f D$. The doubling dose for common exposure of both sexes is then

$$D_D = \frac{m_m + m_f}{a_m + a_f}. \tag{3.39}$$

This average doubling dose is referred to as the *gametic* doubling dose. Individual doubling doses m_m/a_m and m_f/a_f are referred to as *sex-specific* doubling doses. In some studies, the doubling dose is defined in terms of the *sum* of the maternal and paternal doses. This is referred to as the *zygotic* doubling dose and is twice the gametic doubling dose.

First-Generation Effects versus Equilibrium Effects.   Suppose that one knows the equilibrium genetic burden arising from continuous radiation exposure and wishes to estimate the burden in the first generation born after commencement of the exposure. Or, suppose one knows the first generation burden and wishes to estimate the equilibrium burden. Or, suppose only one generation receives radiation exposure and one wishes to estimate the total number born with a particular defect in all future generations. All these estimates depend on what is known as the *fitness* of carriers of a particular defect, namely, the probability that a carrier will survive and reproduce. For example, a rule-of-thumb value for the fitness of carriers of deleterious dominant mutations is 80%. Let us define s to be one minus the fitness, the probability that a mutation is lost from the population gene pool as a result of the failure of a carrier to reproduce.

Suppose that in a population remaining static in number, only one generation receives radiation exposure leading to n children born with a particular genetic defect. The next generation of offspring will have among its number $n(1-s)$ children with the defect. The next generation will have $n(1-s)^2$ children with the defect. Let N_{tot} equal the total number of children with the defect born into all future generations.

$$N_{tot} = n + n(1-s) + n(1-s)^2 + n(1-s)^3 + \cdots \tag{3.40}$$

As long as s is not zero, the series converges to yield

$$N_{tot} = \frac{n}{s}. \tag{3.41}$$

Now suppose that each generation receives the same radiation dose, so that the offspring born into each generation include n children carrying the defect. Consider a time far into the future. The offspring of a given generation include n children carrying mutations arising from their parents, $n(1-s)$ children carrying mutations arising from their grandparents, and so on. If N_{eq} is the equilibrium genetic burden, it is given by

$$N_{eq} = n + n(1-s) + n(1-s)^2 + n(1-s)^3 + \cdots \tag{3.42}$$

or

$$N_{eq} = \frac{n}{s}. \tag{3.43}$$

Thus, from the results of Eqs. (3.41) and (3.43), one can say that if a fixed dose is given to one generation, the total number carrying a particular defect born in all future generations is, in magnitude, equal to the equilibrium number born per generation when each generation receives the same dose.

TABLE 3.5 Hereditary Effects per 0.01 Gy (1 rad) Average Gonad Dose to the Entire Population in Each Generation, as Estimated in the BEIR Committee for Low-Dose and Dose Rate, Low-LET Radiation and on the Basis of a Doubling Dose of 1 Gy.

	CASES PER MILLION LIVEBORN	
TYPE OF DISORDER	FIRST GENERATION	EQUILIBRIUM
Autosomal dominant		
clinically severe	5–20[a]	25
clinically mild	1–15[a]	75
X-linked	< 1	< 5
Recessive	< 1	very slow increase
Chromosomal		
structural	< 5	very little increase
numerical	< 1	< 1
Congenital abnormalities	10[b]	10–100[c]

[a]Based on $s = 0.2$ to 0.8 for clinically severe cases and 0.01 to 0.2 for clinically mild cases, where $1 - s$ is the average fitness.

[b]Based on the assumption that the mutational component arises from dominant genes and that $s = 0.1$, where $1 - s$ is the average fitness.

[c]Based on a mutational component of 5 to 35%.

Source: NAS 1990.

3.5.3 Summary of Risk Estimates

Tables 3.5 and 3.6 summarize the 1990 BEIR-V and 1988 UNSCEAR genetic risk estimates. Both are for low-LET radiation, thus the absorbed dose and dose equivalent are the same. Both estimates are based on a population-average gonad absorbed dose 0.01 Gy (1 rad) in a 30-year generation and a doubling dose of 1 Gy (100 rad). There are many other qualifications stated in the reports and the reader is urged to review those qualifications before applying the data to estimates for specific circumstances.

The two tables present risk-estimate data in the context of continuous exposure. In a population static in number, one million born into one generation replace one million in the parental generation. Radiation-induced defects listed in the tables for the one million newborn result from a collective gonad absorbed dose of 10^4 person-Gy to their parents. To illustrate the use of the data in terms of collective dose, use the data from Table 3.6 for total quantifiable risk, namely, 120 cases per million liveborn, at equilibrium, resulting from 10^4 person-Gy of *collective gonad dose to the reproductive population.* This could be restated as a risk of 1.2 percent increase in hereditary illness, at equilibrium, per person-Gy of *collective gonad dose to the reproductive population.*

Data in the tables may also be used to estimate the risk of genetic effects in all future generations resulting from a once-only exposure of the entire population — reproductive and nonreproductive. Again using the data from Table 3.6 for total quantifiable risk, one could say that 120 cases in all future generations would be expected from 10^4 person-Gy of collective gonad dose to the *reproductive population* in one generation. However, it is not correct to say that the same number of cases would arise from 10^4 person-Gy collective gonad dose to the *entire population.* Based on a mean reproductive age of 30 years and a mean life expectancy of 75 years, one would

TABLE 3.6   Hereditary Effects per 0.01 Gy (1 rad) Average Gonad Dose to the Entire Population in Each Generation, as Estimated by the UNSCEAR Committee for Low-dose and Dose Rate, Low-LET Radiation and on the Basis of a Doubling Dose of 1 Gy.

	CASES PER MILLION LIVEBORN PER GENERATION		
TYPE OF DISORDER	FIRST GENERATION[a]	SECOND GENERATION	EQUILIBRIUM
Autosomal dominant and X-linked	15	13	100
Autosomal recessive			
homozygous effects	No increase	No increase	11
partnership effects[b]	Negligible	Negligible	4
Chromosomal			
structural	2.4	1	4
numerical[c]	Not estimated	Not estimated	Not estimated
Total (rounded)	18	14	120

[a] Based on $s = 0.15$ for autosomal dominant and X-linked defects and 0.6 for chromosomal structural anomalies, where $1 - s$ is the average fitness.

[b] Partnership between induced mutations and those already present in the population.

[c] Genetic risks thought to be low.

Source: United Nations 1986, 1988.

expect that in a once-only exposure, the collective (or average) gonad dose to the reproductive population to be 40 percent of the collective (or average) gonad dose to the entire population. Thus, one could say that a collective gonad dose of 10^4 person-Gy to the entire population in one generation would be expected to result in $120 \times 0.4 \simeq 50$ cases of hereditary illness in all future generations.

Expression of genetic risks in terms of population average or collective effective dose equivalent is subject to considerable uncertainty. The weight factor for the gonads in evaluating the effective dose equivalent is 0.2 (ICRP 1991). This means that for an effective dose equivalent of 1 Sv (1 Gy for low-LET radiation), the actual gonad dose could be as little as zero (if the gonads were not exposed) to as great as 5 Gy (if only the gonads were exposed). Thus, one could say that, per 10^4 person-Sv collective effective dose equivalent *to the reproductive population only* in one generation, one could expect $120 \times (0 \text{ to } 5) \simeq 0$ to 600 cases of hereditary illness in all future generations. One could also say that, per 10^4 person-Sv collective effective dose equivalent *to the entire population* in one generation, one could expect $120 \times 0.4 \times (0 \text{ to } 5) \simeq 0$ to 250 cases of hereditary illness in all future generations.

3.6   CANCER RISKS FROM LOW-LET EXPOSURE

A large body of evidence leaves no doubt that ionizing radiation, when delivered in high doses, is one of the many causes of cancer in the human. Excess cancer risk cannot be observed at doses less than about 0.2 Gy, and therefore, risks for lower doses cannot be determined directly (UN 1988). Just how cancer is induced is not understood fully but it is clear that (1) there are no unique cancer types created solely by ionizing radiation, and (2) that induction is a multistep process involving initiation, promotion, and progression. The initiating event is undoubtedly disruption of the genetic coding within a cell nucleus. The mutation must not be so severe that the cell is unable to reproduce and it must somehow confer a selective advantage in multiplication over sister cells in the host organ or tissue. As Pochin (1983) observes, the transformation of a normal cell into one with the potential for abnormal multiplication does not necessarily promote the multiplication. There are many carcinogenic agents, some of which are effective in the initiation phase, some of which are effective in the promotion phase, and some of which, like ionizing radiation, are effective in both phases. There is solid evidence that there is a latent period between radiation exposure and the onset of cancer. While it is not certain that the latent period may be explained simply in terms of initiation-promotion phenomena, the fact that there is a latent period is evidence of the multistep process of carcinogenesis. Tumor progression, known to be enhanced by radiation, denotes either conversion of a benign growth to a malignant growth or the attainment of increasingly malignant properties in an established cancer. In either case, rapid growth of a subpopulation of cancer cells overwhelms less active normal cells.

This section describes first some of the evidence for radiation carcinogenesis. Mechanisms for cancer induction and latency are then examined. Dose-response models are then considered as are age and sex factors in carcinogenesis. Finally, risk estimates, as presented in the 1988 UNSCEAR Report and the 1990 BEIR-V Report are summarized.

3.6.1 Evidence for Radiation Carcinogenesis

Evidence is drawn principally from studies of atomic-bomb survivors, radiation-therapy patients, and various groups subjected to occupational exposure. Some of the main sources of data are described in the following sections.

Atomic Bomb Exposure.   Survivors of the nuclear-weapon attacks on Hiroshima and Nagasaki have been subjects of a sequence of major studies of late somatic effects of ionizing radiation. Some 91,000 survivors and their offspring remain under continuing surveillance. Interpretation of data has been difficult because of the unavailability of a truly representative control group against which to compare incidence of cancer or other effects. Uncertainties in dosimetry and in the relative importance of neutron and gamma-ray exposure have also complicated interpretation. Exposures were delivered under conditions of high dose and high dose rate, and included a high-LET neutron component — circumstances quite different from the chronic, low-dose and low-LET exposures of interest in most risk estimates. Some have argued that the survivors are a select group and perhaps more radiation resistant than the Japanese population at large. Nevertheless, studies of the survivors have provided the bulk of the data used in establishing the basis for risk estimation.

There have been three major studies of individual dosimetry in the populations of Hiroshima and Nagasaki — a 1957 study, identified as T57, a re-evaluation made in 1965, identified as T65, and a re-evaluation made in 1986, identified as DS86. Many important risk analyses, for example, the BEIR-III and UNSCEAR 1982 studies, are based on the T65 dosimetry. Newer UNSCEAR and BEIR risk estimates (UN 1988, NAS 1990) take into account the following differences between the T65 and DS86 dosimetry:

(a) The Hiroshima weapon is presumed to have had a 20% greater yield than had been thought.

(b) Neutron exposures are presumed to be only 10% of previous estimates for Hiroshima and 30 percent for Nagasaki, thus making evaluations of the neutron's relative biological effectiveness very uncertain.

Occupational Exposure.   Several occupational groups have experienced radiation exposure sufficient to induce cancer. One early group, for whom exposures are not well documented, comprises the medical and industrial radiologists and radiology technicians practicing in the first half of the century. Another well documented group consists of the "radium dial painters," mostly young women, engaged in painting instrument dials in the World War I era. ^{226}Ra was use in the paint to cause luminescence, and the practice of tipping paint brushes with the tongue led to ingestion of life-threatening quantities of radium. This was of course not known at the time and it was not until some years later that first anemia and then bone cancer of abnormally high incidence were observed. The group continues to be studied. Through their misfortune has been gained a great deal of knowledge about the distribution and retention of radium in the skeleton of the body and the carcinogenic effects of its presence.

Miners exposed to abnormally high airborne concentrations of radon daughter products constitute a third important occupational group. The inhaled radionuclides, many of which are alpha-particle emitters, are thought to be responsible in part for the abnormally high incidence of bronchial and lung cancer among this group.

Radiation-Therapy Exposure.   Groups of patients given radiation therapy for various reasons have also provided a great deal of information on radiation carcinogenesis. Follow-up studies have been especially valuable for two reasons. Exposures are far better known than is the case for occupational exposure, and control groups for comparison are relatively easy to identify. One important group is British patients who received x-ray therapy for ankylosing spondylitis, a severe form of spinal arthritis. Many thousands of patients were so treated in the 1930s, 40s and 50s. Follow-up studies, continuing to the present, have revealed abnormally high incidence of leukemia among a study group numbering some 14,000. Similar studies have been made of other patient groups with the same disease who were treated in Europe with injections of ^{224}Ra and who later suffered an increased incidence of bone cancer.

Another important group for study of radiation risks consists of 182,000 women from eight countries diagnosed as suffering from cervical cancer. As of 1987, 1.3 million patient years of observation had accumulated, including 624,000 patient years for those treated with radiation (UN 1988).

3.6.2 Factors in Cancer Induction and Latency

Evidence is clear that absorbed doses of ionizing radiation at levels of 1 Gy or greater may lead stochastically to abnormally high cancer incidence in exposed populations. There is no direct evidence that chronic exposure to low levels of ionizing radiation may similarly lead to abnormally high cancer incidence. Risk estimates for chronic, low-level exposure requires extrapolation of high-dose and high dose-rate response data. Methods used for extrapolation are often controversial, any one method being criticized by some as over-predictive and by others as under-predictive. The BEIR-III

Report (NAS 1980) makes the following comments on the complexities involved with interpretation of data on radiation carcinogenesis:[6]

> Cancers induced by radiation are indistinguishable from those occurring naturally; hence, their existence can be inferred only on the basis of a statistical excess above the natural incidence.

> Cancer may be induced by radiation in nearly all tissues of the body.

> Tissues and organs vary considerably in their sensitivity to the induction of cancer by radiation.

> The natural incidence of cancer varies over several orders of magnitude, depending on the type and site of origin of the neoplasm, age, sex, and other factors.

> With respect to excess risk of cancer from whole-body exposure to radiation, solid tumors are now known to be of greater numerical significance than leukemia. Solid cancers characteristically have long latent periods; they seldom appear before 10 years after radiation exposure and may continue to appear for 30 years or more after radiation exposure. In contrast, the excess risk of leukemia appears within a few years after radiation exposure and largely disappears within 30 years after exposure.

> The major sites of solid cancers induced by whole-body radiation are the breast in women, the thyroid, the lung, and some digestive organs.

> The incidence of human breast and thyroid cancer is such that the total cancer risk is greater for women than for men. Breast cancer occurs almost exclusively in women, and absolute-risk estimates for thyroid-cancer induction by radiation are higher for women than for men (as is the case with their natural incidence). With respect to other cancers, the radiation risks in the two sexes are approximately equal.

> There is now considerable evidence from human studies that age is a major factor in the risk of cancer from exposure to ionizing radiation. Both age at exposure and age at cancer diagnosis are important for interpretation of human data. If risks are given in absolute form — i.e., numbers of cancers induced per unit of population and per unit of radiation exposure — then a single value independent of age may be inappropriate. The 1972 BEIR Report (NAS 1972) concluded that the risks of some kinds of cancer was greater after irradiation in childhood and *in utero* than in adult life. It is now apparent that other age groups may also have risks that differ from the average for all ages; e.g., women exposed during the second decade of life have the highest risk of radiation induced breast cancer.

> Various host or environmental factors may interact with radiation to affect cancer incidence in different tissues. These may include hormonal influences, immunological status, exposure to various oncogenic agents, and nonspecific stimuli to cell proliferation in tissues sensitive to cancer induction by radiation.

> The time elapsing between irradiation and the appearance of a detectable neoplasm is characteristically long, i.e., years or even decades. This long latent period must be taken into consideration in all risk calculations, whether these are estimates of risk experienced by populations under study or projections into the future.

---

[6]Reprinted with permission from *The Effects on Populations of Exposure to Low Levels of Ionizing Radiation*: BEIR-III, c. 1980 by the National Academy of Sciences. Published by National Academy Press, Washington, D.C.

The variety of possible biologic mechanisms responsible for human cancer suggests that the dose-response relationship may not be the same for all types of radiation-induced cancer. The fact, however, that epidemiologic studies of widely differing human populations exposed to radiation have given reasonably concordant results for some cancer sites and for a broad range of radiation doses adds considerable strength to the dose-response information now available.

Some of the existing human and animal data on radiation-induced cancers are derived from populations exposed to internally deposited radionuclides for which dose-incidence relationships are influenced by marked nonuniformities in the temporal and spatial distribution of radiation within the body.

Some of the human data concern cancer mortality; others, cancer incidence. It is appropriate to distinguish radiation induced cancers that may not greatly alter the death rate (e.g., skin and thyroid cancer) from others that are generally fatal (e.g., leukemia and lung cancer).

It is not yet possible to estimate precisely the risk of cancer induction by low-dose radiation, because the degree of risk is so low that it cannot be observed directly and there is great uncertainty as to the dose-response function most appropriate for extrapolating in the low-dose region.

3.6.3 Dose-Response Models for Cancer

Models relating cancer incidence or mortality to radiation dose are required both for interpretation of study data using regression analysis and for risk estimation or projection. The projection models described in this chapter are for low-LET radiation only. However, study data may be based in part on high-LET radiation, as is the case for the atomic-bomb casualties. Thus, data interpretation may require consideration of the relative biological effectiveness of the high-LET component. The doses on which projections are based are average whole-body doses that, for low-LET radiation are equivalent to dose equivalent.

Risk estimates for cancer have as the basic elements dose responses that are functions of the cancer site, the age at exposure, the age at which the cancer is expressed or the age at death, and the sex of the subject. Ethnic and environmental factors are not taken into account explicitly, but do affect the estimates. For example, risk estimates based on studies of Hiroshima and Nagasaki nuclear-weapon survivors are strictly applicable only to the Japanese population. Nevertheless, such estimates are widely applied to estimation of risks for other populations.

One may define for a particular type of cancer a dose response $\mathcal{E}(D, a_o, a, s)$, the radiation-induced excess probability of incidence or mortality at age a following exposure to dose D (Gy) at age a_o, for a subject of sex s. The post-exposure time, $t = a - a_o$, may be used instead of a as an independent variable, whence $\mathcal{E}(D, a_o, t, s)$. The response may be expressed in an absolute sense using the *absolute or additive risk* model, or relative to the natural cancer incidence using the *relative or multiplicative risk* model.

The Absolute (Additive) Risk Model.   Both this model and the relative risk model were applied and compared in the 1980 BEIR-III Report (NAS 1980) and the

1988 UNSCEAR Report (UN 1988). The model is based on the supposition that $\mathcal{E}(D, a_o, a, s)$ is independent of the age and sex dependent natural cancer incidence. The BEIR-III Committee expressed the dose response for mortality from a particular type of cancer as

$$\mathcal{E}(D, a_o, a, s) = \alpha(a_o, a, s)D + \beta(a_o, a, s)D^2 \tag{3.44}$$

and used data from many sources to arrive at best estimates of values for parameters α and β.

The Relative (Multiplicative) Risk Model.   In this model, the radiation-induced excess cancer risk (incidence or mortality) is expressed as a multiple of the natural incidence or mortality $\mathcal{E}_o(a, s)$, the cancer-site and sex-specific probability of incidence or mortality in year a. The excess risk is expressed as

$$\mathcal{E}(D, a_o, a, s) = \mathcal{E}_o(a, s) \times \mathcal{R}(D, a_o, a, s), \tag{3.45}$$

in which $\mathcal{R}$ is the relative risk. Values of $\mathcal{E}_o(a, s)$ for cancer *mortality* are given in Table 3.7 for selected types of cancer. Ratios of cancer mortality to cancer incidence are given in Table 3.8. Some of the ethnic and environmental influences on cancer risks are absorbed in the natural-risk factor $\mathcal{E}_o$, thus the the relative risk $\mathcal{R}$ may be applied to varied populations with greater confidence than could be done using the absolute-risk model. The relative risk may be expressed in the form

$$\mathcal{R}(D, a_o, a, s) = \alpha(a_o, a, s)D + \beta(a_o, a, s)D^2 \tag{3.46}$$

or in some other form. For example, relative risks based on the life span study (LSS) of atomic-bomb survivors, summarized in Table 3.9, make use of only linear term in Eq. (3.46). These data are displayed because they represent the principal information analyzed by organizations such as the BEIR-V and UNSCEAR Committees in arriving at estimates of cancer risks.

The BEIR-V Committee (NAS 1990) arrived at relative mortality risks in terms of the years t after exposure. Their results may be expressed in analytical form, as follows, with dose D in Gy and ages and times in years.

Leukemia.   Coefficients are independent of sex. Risk is zero during a two-year latency period. A linear-quadratic fit is used, therefore, a dose-rate effectiveness factor is not required when estimates are made for risks of exposure at low dose and dose rate.

$$\mathcal{R}(D, a_o, t) = (0.243D + 0.271D^2)e^{\gamma(a_o, t)}, \tag{3.47}$$

$$
\begin{aligned}
\gamma &= 4.885, & t &\leq 15, & a_o &\leq 20 \\
\gamma &= 2.380, & 15 &< t \leq 25, & a_o &\leq 20 \\
\gamma &= 2.367, & t &\leq 25, & a_o &> 20 \\
\gamma &= 1.638, & 25 &< t \leq 30, & a_o &> 20
\end{aligned}
$$

TABLE 3.7   United States 1980 Baseline Cancer Mortality per 100,000 per Year, by Sex, Cancer Site, and Age at Death.[a]

AGE AT DEATH	MALES				
	LEUKEMIA	RESPIRATORY	DIGESTIVE	OTHER	NONLEUKEMIA
0– 4	2.7	0.3	0.9	5.0	6.2
5– 9	2.6	0.0	0.1	2.8	2.9
10–14	2.0	0.0	0.1	2.3	2.4
15–19	2.2	0.1	0.2	3.8	4.1
20–24	2.0	0.3	0.8	6.1	7.2
25–29	2.2	0.5	1.2	7.4	9.1
30–34	2.1	1.5	2.6	9.6	13.7
35–39	2.9	6.2	5.7	14.5	26.4
40–44	3.7	20.3	14.2	23.4	57.9
45–49	5.0	52.0	29.8	41.3	123.1
50–54	7.4	106.5	58.2	74.6	239.3
55–59	11.5	179.7	102.3	121.3	403.3
60–64	18.6	275.6	161.1	189.9	626.6
65–69	27.5	382.3	244.7	293.0	920.0
70–74	45.8	476.4	338.0	432.3	1246.7
75–79	62.6	517.7	454.8	621.1	1593.6
80–84	87.7	500.3	572.4	873.5	1946.2
85–	117.1	386.3	705.8	1160.3	2252.4
Total	8.4	71.9	52.7	72.3	196.9

AGE AT DEATH	FEMALES					
	LEUKEMIA	RESPIRATORY	DIGESTIVE	BREAST	OTHER	NONLEUKEMIA
0– 4	2.0	0.2	0.2	0.0	4.0	4.4
5– 9	1.5	0.1	0.1	0.0	2.1	2.3
10–14	1.3	0.0	0.1	0.0	2.0	2.1
15–19	1.5	0.0	0.2	0.0	2.8	3.0
20–24	1.2	0.1	0.4	0.1	3.4	4.0
25–29	1.5	0.3	1.0	1.5	5.3	8.1
30–34	1.8	0.9	2.2	5.3	8.6	17.0
35–39	2.4	3.5	4.2	13.0	14.5	35.2
40–44	2.9	10.8	9.3	23.8	24.8	68.7
45–49	3.9	24.6	19.1	37.9	43.0	124.6
50–54	4.9	44.4	35.9	57.6	69.6	207.5
55–59	7.6	63.7	63.2	75.4	97.6	299.9
60–64	11.1	86.6	97.0	86.1	142.2	411.9
65–69	14.9	104.1	143.4	97.0	183.5	528.0
70–74	23.4	108.5	210.9	106.1	237.6	663.1
75–79	31.9	102.7	283.6	118.1	284.2	788.6
80–84	48.5	90.6	392.0	139.3	359.8	981.7
85–	61.1	96.3	504.3	169.3	424.7	1194.6
Total	6.3	25.2	45.2	30.6	56.3	157.3

[a] International Classification of Diseases (ICD) indices as follows: leukemia 204–208, respiratory 160–165, digestive 150–159, breast 174–175, total 140–208.

Source: HHS 1985.

TABLE 3.8 Ratios of Cancer Mortality to Cancer Incidence Based on Lifetime Expectations at Birth.

	MORTALITY/INCIDENCE RATIO	
SITE	MALES	FEMALES
Leukemia	1.00	1.00
Bone	1.00	1.00
Esophagus	1.00	1.00
Stomach	0.75	0.78
Intestine	0.52	0.55
Pancreas	0.91	0.90
Lung	0.83	0.75
Urinary	0.37	0.46
Lymphoma	0.73	0.75
Breast	—	0.39
Thyroid	0.18	0.20
Liver	1.00	1.00
Average for all except bone cancer and leukemia	0.65	0.50

Source: NAS 1980.

TABLE 3.9 Relative Risk of Death after 1 Gy Shielded Kerma[a] of Low-LET Radiation at High Dose Rate. Basis: the Life Span Study (LSS) of Combined Male and Female[b] Atomic-Bomb Survivors.

EXPOSURE AGE (y)	AGE (y) AT DEATH							
	TOTAL[c]	< 20	20–29	30–39	40–49	50–59	60–69	70+
Leukemia								
< 10	17.05	44.16	3.41	8.64	0.95			
10–19	4.76	54.74		2.45	1.02	0.82		
20–29	5.06		5.33	3.54	43.09	1.02	0.82	
30–39	3.99			0	24.05	10.58	1.47	3.89
40–49	2.55				0.83	3.82	0.82	3.10
50+	6.50					15.63	5.18	6.90
All	4.92	46.47	9.81	4.75	3.68	3.98	1.70	4.40
All cancers except leukemia								
< 10	2.32	70.07	5.89	1.96	1.86			
10–19	1.65	40.90	0.82	1.66	1.39	1.68		
20–29	1.65			1.38	2.09	1.74	1.37	
30–39	1.26			0.84	1.12	1.11	1.23	1.48
40–49	1.24				1.25	1.12	1.13	1.33
50+	1.11					2.58	0.95	1.15
All	1.29	75.32	2.22	1.60	1.58	1.39	1.13	1.29

[a] Shielded kerma accounts for attenuation of structures. Organ doses are lower in magnitude than shielded kermas. Ratios of the two are 0.81 for bone marrow, 0.85 for the breast and thyroid, and 0.72 to 0.76 for other organs.

[b] The ratio of relative risks for male and female are 1.0 for leukemia and 0.41 for the sum of all other cancers.

[c] Totals from (ICRP 1990).

Source: Shimazu, Kato, and Shull 1988.

Respiratory cancer.   Risk is zero during a ten-year latent period. A linear fit is used, therefore, a dose-rate effectiveness factor is required when estimates are made for risks of exposure at low dose and dose rate.

$$\mathcal{R}(D, a_o, t, s) = 0.636 D \; e^{(-1.437 \ln(t/20) + \gamma)}, \qquad (3.48)$$

in which $\gamma = 0$ for the male and 0.711 for the female.

Breast cancer (female only).   Risk is zero during a ten-year latent period. A linear fit is used, therefore, a dose-rate effectiveness factor is required when estimates are made for risks of exposure at low dose and dose rate.

$$\mathcal{R}(D, a_o, t, s) = 1.220 D \; e^{(\gamma_1 - 0.104 \ln(t/20) - 2.212 \ln^2(t/20) - \gamma_2(t-20))}, \qquad (3.49)$$

in which $\gamma_1 = 1.385$, $a_o < 15$, or 0, $a_o \geq 15$, and $\gamma_2 = 0$, $a_o < 15$, or 0.0628, $a_o \geq 15$.

Digestive cancer.   Risk is zero during a ten-year latent period. A linear fit is used, therefore, a dose-rate effectiveness factor is required when estimates are made for risks of exposure at low dose and dose rate.

$$\mathcal{R}(D, a_o, t, s) = 0.809 D \; e^{(\gamma_1 - \gamma_2)}, \qquad (3.50)$$

in which $\gamma_1 = 0$ for males or 0.553 for females, and $\gamma_2 = 0$ if $a_o \leq 25$, $0.198(a_o - 25)$ if $25 < a_o \leq 35$, or 1.98 if $a_o > 35$.

All other cancers.   Risk is zero during a ten-year latent period. A linear fit is used, therefore, a dose-rate effectiveness factor is required when estimates are made for risks of exposure at low dose and dose rate.

$$\mathcal{R}(D, a_o, t, s) = 1.220 D \; e^{-\gamma}, \qquad (3.51)$$

in which $\gamma = 0$ if $a_o \leq 10$, or $0.0464(a_o - 10)$ if $a_o > 10$.

3.6.4 Use of Dose Rate Effectiveness Factors

The dose-response relationships presented in the previous section are for low-LET radiation delivered at high dose and dose rate. Except for the BEIR-V estimates of leukemia risks, the relationships are linear. When such relationships are applied to risks for exposure at low dose and dose rate, estimates are too high by at least a factor of 2 and perhaps by a factor of 10 as agreed by the NCRP and the UNSCEAR Committee (NCRP 1980; UN 1988) [see the discussion of the dose rate effectiveness factor (DREF) in Sec. 3.3.1]. Use of a DREF of 2 for risk estimates based on linear dose-response relationships is recommended but not applied by the BEIR-V Committee. In order to avoid a confusing divergence between tabulated risk estimates in this volume with those in source references, the factor is *not* applied in the risk estimates discussed in this chapter unless specifically stated.

3.6.5 Lifetime Cancer Risk versus Age at Exposure

Evaluation of lifetime risks requires use of *life tables* such as the 1979–81 United States table given in Table 3.10. Use of such tables allows one to determine, for example, the fraction $N(a)/N(a_o)$ of those alive at age a_o surviving to age a. For example, using Table 3.10, the fraction of females alive at age 15 who survive to age 65 is

TABLE 3.12   Excess Cancer Mortality for the U.S. Population by Age at Exposure and Site for 100,000 Persons of Each Age Exposed to a Whole Body or Organ Absorbed Dose of 0.1 Gy (10 rad) from Low-LET Radiation Delivered at High Dose and High Dose Rate. When Applied to Estimation of Risks from Exposure to Radiation at Low Dose and Dose Rate, Dose Rate Effectiveness Factors Need be Applied to All Estimates Except Those for Leukemia.

			MALES			
AGE AT EXPOSURE	TOTAL	LEUKEMIA[a]	NONLEUKEMIA[b]	RESPIRATORY	DIGESTIVE	OTHER
5	1276	111	1165	17	361	787
15	1144	109	1035	54	369	612
25	921	36	885	124	389	372
35	566	62	504	243	28	233
45	600	108	492	353	22	117
55	616	166	450	393	15	42
65	481	191	290	272	11	7
75	258	165	93	90	5	—
85	110	96	14	17	—	—
Average[c]	770	110	660	190	170	300

			FEMALES				
AGE AT EXPOSURE	TOTAL	LEUKEMIA[a]	NONLEUKEMIA[b]	BREAST	RESPIRATORY	DIGESTIVE	OTHER
5	1532	75	1457	129	48	655	625
15	1566	72	1494	295	70	653	476
25	1178	29	1149	52	125	679	293
35	557	46	511	43	208	73	187
45	541	73	468	20	277	71	100
55	505	117	388	6	273	64	45
65	386	146	240	—	172	52	16
75	227	127	100	—	72	26	3
85	90	73	17	—	15	4	—
Average[c]	810	80	730	70	150	290	220

[a]90% confidence limits extend from about one-third the central value to twice the central value except for age 5, for which they extend from one-fifth to four times.

[b]90% confidence limits extend from about one-half the central value to twice the central value.

[c]Weighted average based on the age distribution in a stationary population, as shown in Table 3.10.

Source: Reprinted by permission from NAS 1990, c. 1990 by the National Academy of Sciences. Published by the National Academy Press, Washington, DC.

years, 12.677% are at ages 10 through 19 years, and 6.377% are at ages 80 years or greater. Age dependent leukemia risks for those same age intervals are, from Table 3.12, approximately 75, 72, and 73 fatalities per 100,000 for 0.1 Gy exposure. The average risk for the female population exposed to 0.1 Gy is, therefore,

$$0.12716 \times 75 + 0.12677 \times 72 + \cdots + 0.06377 \times 73 = 80,$$

that is, 80 fatalities per 100,000 females in a static population of all ages. This is the average leukemia risk for females given in Table 3.12. The calculation was made for a population all receiving the same dose. As long as the dose is less than about 0.1 Gy, for which the quadratic term in Eq. (3.47) is negligible, one can apply the same analysis in terms of collective dose. For example, for a collective dose to females of 10^4

TABLE 3.13   Lifetime Cancer Risk Estimates for Low-LET Radiation Expressed as per Capita Lifetime Excess Cancer Deaths for 100,000 Persons (Equal Male and Female Numbers) Receiving 0.1 Gy Organ Absorbed Dose. Risks are Based on the Relative (Multiplicative) Risk Model. When Data are Applied to Estimation of Risks from Exposure to Radiation at Low Dose and Dose Rate, Dose Rate Effectiveness Factors Need be Applied to All Estimates Except Those for Leukemia.

ORGAN/MALIGNANCY	U.S. POPULATION BEIR-V (1990)	JAPANESE POPULATION UNSCEAR (1988)
Red marrow/leukemia	95 (40–235)[a]	97 (71–130)
All cancers except leukemia	695 (485–1030)	610 (480–750)
All cancers	790 (585–1200)	710
Breast	35	30 (14–53)
Digestive	230	
colon		79 (36–134)
stomach		126 (66–199)
bladder		39 (16–73)
Respiratory	170	
lung		151 (84–230)
esophagus		34 (8–72)
Multiple myeloma		22 (6–51)
Ovary		15 (5–34)
Remainder	260	117

[a] 90% confidence intervals given in parentheses.

Source: NAS 1990; United Nations 1988.

person Gy, the expectation is 80 resulting leukemia fatalities. This is true whether 10^5 persons each receive 10^{-1} Gy (10 rad), 10^6 persons each received 10^{-2} Gy (1 rad), or 10^7 persons each receive 10^{-3} Gy (0.1 rad). It is *not* true if 10^4 persons each receive 1 Gy (100 rad) absorbed dose because, for such a dose, the quadratic term in Eq. (3.47) is *not* negligible.

Using life tables, tables of natural cancer mortality, and relative (multiplicative) risks, both the 1988 UNSCEAR Committee and the 1990 BEIR-V Committee computed risks for exposure of Japanese and American populations. These risks are compared in Table 3.13. If one were to apply a DREF of 2 for nonleukemia risks, the expected number of fatalities for exposure at low dose and dose rate would be about 450 fatalities per 10^4 person Gy, that is, a fatality risk of 0.045 per Gy. This risk factor, rounded to 0.05 per Gy may be used as an overall cancer risk factor, as will be done subsequently in our discussion of the *effective dose equivalent*.

The BEIR-V Committee also calculated risks to the United States population under three (low-LET) exposure scenarios: (1) single exposure to 0.1 Gy, (2) continuous lifetime exposure to 1 mSv per year, and (3) exposure to 10 mSv per year from age 18 to age 65. Results are given in Table 3.14. The first scenario is representative of accidental exposure of a large population, the second of chronic exposure of a large population, and the third of occupational exposure.

TABLE 3.14 Excess Cancer Mortality Estimates per 100,000 in the Stationary U.S. Population. Risk Estimates are Based on Organ or Whole Body Absorbed Dose from Low-LET Radiation Delivered at High Dose Rate. When Applied to Estimation of Risks from Exposure to Radiation at Low Dose and Dose Rate, Dose Rate Effectiveness Factors Need be Applied to All Estimates Except Those for Leukemia.

	MALES	FEMALES
SINGLE EXPOSURE TO 0.1 Gy (10 RAD)		
Radiation-induced		
leukemia	110 (50–280)[a]	80 (30–190)
nonleukemia	660 (420–1040)	730 (550–1020)
total	770 (540–1240)	810 (630–1160)
Normal expectation		
leukemia	760	610
nonleukemia	19750	15540
total	20520	16150
CONTINUOUS LIFETIME EXPOSURE TO 1 mGy PER YEAR (100 mRAD PER y)		
Radiation-induced		
leukemia	70 (20–260)	60 (20–200)
nonleukemia	450 (320–830)	540 (430–800)
total	520 (410–980)	600 (500–930)
Normal expectation		
leukemia	790	660
nonleukemia	19760	16850
total	20560	17520
CONTINUOUS EXPOSURE TO 10 mGy PER YEAR (1 RAD PER YEAR) FROM AGE 18 TO AGE 65		
Radiation-induced		
leukemia	400 (130–1160)	310 (110–910)
nonleukemia	2480 (1670–4560)	2760 (2120–4190)
total	2880 (2150–5460)	3070 (2510–4580)
Normal expectation		
leukemia	780	650
nonleukemia	20140	17050
total	20910	17710

[a]90% confidence limits are given in parentheses.

Source: NAS 1990.

3.7 MEASURES OF RISKS FROM LOW-LEVEL EXPOSURE

In the treatment of risks of hereditary illness in terms of collective dose to populations, the use of the *genetically significant dose* has been useful in accounting for the child-bearing potential of different populations. In the treatment of somatic effects in individuals and populations arising from nonuniform radiation exposure, the use of the *somatically significant dose* has been useful in accounting for the relative cancer susceptibilities of the various organs and tissues of the body. The concept of somatically significant dose has evolved to the now widely applied concept of the *effective dose equivalent*. A related concept, used in assessment of medical radiation exposure, is the *cancer detriment index*.

3.7.1 The Genetically Significant Dose (Equivalent)

A measure of biological hazard for hereditary illness is the *genetically significant dose equivalent* (GSD) for a population under examination. The population, for example, might be the entire national population or all workers in a particular industry. Some in the population may receive zero exposure from all but natural sources. The GSD concept originated in the 1950s and is, in essence, a population-average gonad dose equivalent, accounting for the age and sex distribution of the population, and weighted by the child-bearing potential of the recipients of the radiation exposure. A measure of the genetic detriment to a population, the GSD is the dose equivalent to the gonads which, if given to every member of the population, would produce the same genetic detriment as the actual doses received by the various individuals in the population. Explicitly,

$$\text{GSD} \equiv \frac{\sum_i \sum_j N_{ij} \, H_{ij} \, P_j}{\sum_j N_j \, P_j} \tag{3.53}$$

in which N_{ij} represents the number of persons in age/sex class j receiving a gonad dose equivalent H_{ij} within dose range class i, N_j is the total number of persons in age/sex class j, and P_j is the expected future number of children from a person in age/sex class j (see Table 3.15). The total number of cases of hereditary illness in all subsequent generations is obtained by multiplying the GSD by the population size and by a collective risk coefficient 0.012 per person-Sv (UN 1988; see also Table 3.6).[7] If hereditary illness in only the first two generations were of interest, the risk coefficient would be 0.003 per person-Sv. Note that a different risk coefficient is required if the coefficient is to be applied not to the GSD but to the mean gonad dose to the population, that is, the mean dose unweighted by the child-bearing potential. For a mean reproduction age of 30 years, and a life expectancy of 70 to 75 years, the dose received by age 30 is about 40 percent of the total. The risk coefficient for hereditary illness in all future generations is then $0.012 \times 0.4 = 0.005$ per person-Sv.

3.7.2 The Somatically Significant Dose Equivalent

A measure of biological hazard in relation to somatic illness, specifically fatal malignancy, is the *somatically significant dose equivalent* (SSD) for a population. The SSD concept originated in the 1970s and is, in essence, a population-average and tissue/organ average dose equivalent, accounting for the age and sex distribution of the population, and weighted by the age and sex-specific radiation-induced cancer susceptibilities of the various tissues and organs of the body. Explicitly,

$$\text{SSD} \equiv \frac{\sum_i \sum_j \sum_T N_{ij} \, H_{ijT} \, F_{jT}}{\sum_j \sum_T N_j \, H_{jT}}, \tag{3.54}$$

---

[7]The risk coefficient is derived from the subtotal of 115 (rounded to 120) cases per generation per million population, each person receiving 1 rem (0.01 Sv) dose equivalent per generation (nominally, in 30 years), in other words, 0.012 cases per person-Sv. It can be shown (Sec. 3.5.2) that 120 cases would be experienced in all future generations as a result of 1 rem dose equivalent received by one generation in a population of one million.

TABLE 3.15 Expected Number of Future Children, by Sex and Age of Parent, in the United States for 1970.

AGE GROUP	MALES	FEMALES
Fetus	2.5	2.5
0–14	2.6	2.4
15–29	2.7	2.5
30–44	1.1	0.6
45–64	0.1	0.02
65+	0.0	0.0

Source: FDA 1976.

in which N_{ij} is the number of persons in age/sex class j receiving a dose equivalent H_{ijT} in range i to organ or tissue T, N_j is the total number of persons in age/sex class j, and F_{jT} is the number of expected fatal malignancies per unit dose equivalent in tissue T for a person in class j. Values of F_{jT} are given in Table 3.12.

Recognizing that knowledge of F_{jT} values is incomplete, the United Nations Scientific Committee on the Effects of Atomic Radiation (1982) recommend an approximation for the SSD. The approximation is based on the use of values of F_{jT} averaged over the age/sex distribution of a typical population, that is, replacing F_{jT} with an average value F_T (also given in Table 3.12). If this is done, then Eq. (3.54) reduces to

$$\text{SSD} = \frac{1}{N} \sum_i \sum_j \sum_T N_{ij} \, H_{ijT} \left(\frac{F_T}{\Sigma_T \, F_T} \right), \qquad (3.55)$$

in which N is the total size of the population. The term in parentheses is just the fraction of all cancer fatalities associated with organ or tissue T when subjects receive uniform whole-body radiation exposure. The UNSCEAR Committee recommended that, for these factors, one use the weight factors w_T developed by the ICRP (1977) for use in calculation of the effective dose equivalent or dose equivalent commitment. The 1977 weight factors were derived as illustrated in Table 3.16. The 1990 recommendations of the ICRP (1991) contain a new set of weight factors based on current estimates of cancer risks and accounting for the relative severity of various types of cancer in terms of incidence versus mortality and loss of life expectancy. The new set is given in Table 3.17.

Under this approximation, the SSD is replaced by the population average effective dose equivalent. The estimated total number of health effects is given by the product of the SSD, the population size, and an overall risk coefficient.[8] A weakness

---

[8]For the general public, the ICRP (1991) recommends a coefficient of 0.050 per person-Sv for lifetime cancer fatality and 0.075 per person-Sv for lifetime cancer detriment accounting for both cancer fatality and nonfatal cancer incidence. The coefficient of 0.05 may be derived from Table 3.13, as discussed in the previous Sec. For the working population, corresponding factors of 0.040 and 0.060 per person-Sv are recommended. The UNSCEAR ommittee (United Nations 1988) recommends risk coefficients similar in magnitude.

TABLE 3.16 Derivation of the 1977 ICRP Tissue Weighting Factors Using Risk Coefficients Thought Appropriate at the Time.

TISSUE OR ORGAN	TYPE OF RISK	F_T (Gy^{-1})	w_T
Gonads	Two-generation hereditary illness	0.0040	0.25
Bone marrow	Cancer	0.0020	0.12
Bone surface	Cancer	0.0005	0.03
Lung	Cancer	0.0020	0.12
Thyroid	Cancer	0.0005	0.03
Breast	Cancer	0.0025	0.15
Remainder	Cancer	0.0050	0.30[a]
Subtotal	Cancer	0.0125	0.75
Total	ΣF_T	0.0165	1.00

[a] Weight factor 0.06 for each of five other tissues receiving the greatest dose.

Source: ICRP 1977.

TABLE 3.17 The 1990 ICRP Tissue Weighting Factors.

TISSUE OR ORGAN	w_T	TISSUE OR ORGAN	w_T
Bone marrow (red)	0.12	Lung	0.12
Bladder	0.05	Esophagus	0.05
Bone surface	0.01	Skin	0.01
Breast	0.05	Stomach	0.12
Colon	0.12	Thyroid	0.05
Gonads	0.20	Remainder	0.05[a]
Liver	0.05		

[a] The average dose in the adrenals, brain, upper large intestine, small intestine, kidney, muscle, pancreas, spleen, thymus, and uterus. If any one of these organs receives a dose equivalent in excess of that in any one of the twelve organs given weighting factors, factors of 0.025 are applied to it and to the average dose equivalent received by the rest of the remainder organs.

Source: ICRP 1991.

in this approximation is that the effective dose equivalent is heavily weighted by gonad dose, and the estimated health effects include not only fatal malignancies but also hereditary illness in the first two generations.

3.7.3 The Cancer Detriment Index

The Center for Devices and Radiological Health of the U.S. Public Health Service has introduced the *cancer detriment index* (CDI) as a measure of the risk of a diagnostic medical procedure (Rosenstein 1988). This index, which reflects not only the detriment from fatal cancers but also the detriment from cancers that can be treated successfully, is formulated as

TABLE 3.18   Coefficients for Use in Evaluation of the Cancer Detriment Index.

CANCER	$10^5 r_i(f)$ (rad^{-1}) MALE	$10^5 r_i(f)$ (rad^{-1}) FEMALE	$10^5 r_i(c)$ (rad^{-1}) MALE	$10^5 r_i(c)$ (rad^{-1}) FEMALE	s_i
Lung	2.0	2.0	0.1	0.1	0.95
Leukemia[a]	2.4	1.6	0.12	0.08	0.95
Thyroid	0.33	0.67	6.3	12.7	0.05
Breast	—	5.0	—	3.0	0.60
Other[b]	5.0	5.0	1.5	1.5	0.75

[a] Apply coefficients to the absorbed dose in the active bone marrow.

[b] Apply coefficients to the absorbed dose in "trunk tissue."

Source: Rosenstein 1988.

$$\text{CDI} = \sum_i \left[\, r_i(f) \,+\, s_i\, r_i(c) \,\right] \bar{D}_i, \tag{3.56}$$

in which $r_i(f)$ is the lifetime risk coefficient for fatal cancer i (probability per unit absorbed dose), $r_i(c)$ is the lifetime risk coefficient for curable cancer i, s_i is the relative severity associated with successful treatment of cancer i, and $\bar{D}_i$ is the average absorbed dose in the appropriate tissue for cancer i. Coefficients to be applied in the evaluation of the CDI are listed in Table 3.18.

REFERENCES

ABRAHAMSON, S., M. A. BENDER, C. DENNISTON, AND W. J. SHULL, "Genetic Effects," in *Health Effects Models for Nuclear Power Plant Accident Consequence Analysis — Low LET Radiation, Part II: Scientific Bases for Health Effects Models*, Report NUREG/CR–4214, Rev. 1, Part II, U.S. Nuclear Regulatory Commission, Washington, D.C., 1989.

ANDREWS, H. L., *Radiation Biophysics*, 2d. ed., Prentice Hall, Englewood Cliffs, N.J., 1974.

ANNO, G. H., S. J. BAUM, H. R. WITHERS, AND R. W. YOUNG, "Symptomatology of Acute Radiation Effects in Humans After Exposure to Doses of 0.5-30 Gy," *Health Physics, 56*, 821–838 (1989).

EISENHOWER, E. H., "A Review of Neutron Quality Factor Recommendations," *Health Physics Society Newsletter*, June, 1989.

FDA, *Gonad Doses and Genetically Significant Dose from Diagnostic Radiology*, U.S. Department of Health and Human Services, Public Health Service, Food and Drug Administration, Center for Devices and Radiological Health, Rockville, Md., 1976.

GOODHEAD, D. T., "Spatial and Temporal Distribution of Energy," *Health Physics, 55*, 231–240 (1988).

HHS, *Vital Statistics of the United States 1980*, Vol. II, Part A, Publication PHS 85-1101, National Center for Health Statistics, Public Health Service, Department of Health and Human Services, Washington D.C., 1985a.

HHS, *U.S. Decennial Life Tables for 1979-81*, Publication PHS 85-1150-1, National Center for Health Statistics, Public Health Service, Department of Health and Human Services, Washington D.C., 1985b.

ICRP, *Radiation Protection: Recommendations of the International Commission on Radiological Protection, Adopted September, 1965*, Publication 9, International Commission on Radiological Protection, Elmsford, N.Y., 1966.

ICRP, *Recommendations of the International Commission on Radiological Protection,* Publication 26, International Commission on Radiological Protection, *Annals of the ICRP 1*, No. 3, Pergamon Press, Oxford, 1977.

ICRP, *Nonstochastic Effects of Ionizing Radiation*, Publication 41, International Commission on Radiological Protection, Pergamon Press, Oxford, 1984.

ICRP, *1990 Recommendations of the International Commission on Radiological Protection,* Publication 60, International Commission on Radiological Protection, *Annals of the ICRP 2* No. 1–3 (1991).

ICRU, *Linear Energy Transfer*, ICRU Report 16, International Commission on Radiation Units and Measurements, Bethesda, Md., 1970.

ICRU, *Microdosimetry*, ICRU Report 36, International Commission on Radiation Units and Measurements, Bethesda, Md., 1983.

ICRU, *The Quality Factor in Radiation Protection*, ICRU Report 40, International Commission on Radiation Units and Measurements, Bethesda, Md., 1986.

KATHREN, R. L., *Radiation Protection*, Medical Physics Handbook 16, Adam Hilger Ltd., Bristol, 1985.

KELLERER, A. M., "Fundamentals of Microdosimetry," in The Dosimetry of Ionizing Radiation, Vol. 1, K. R. Kase, B. E. Bjarngard, and F. H. Attix (eds.), Academic Press, Orlando, Florida, 1985.

KELLERER, A. M., AND H. H. ROSSI, "The Theory of Dual Radiation Action," *Current Topics in Radiation Research Quarterly, 8,* 85–158 (1972).

KELLERER, A. M., AND H. H. ROSSI, "A Generalized Formulation of Dual Radiation Action," *Radiation Research, 75,* 471–488 (1978).

LANGHAM, W. H., (ed.), *Radiobiological Factors in Manned Space Flight*, Report of the Space Radiation Study Panel, National Academy of Sciences, National Research Council, Washington D.C., 1967.

NAS, National Research Council, Advisory Committee on the Biological Effects of Ionizing Radiations. *The Effects on Populations of Exposure to Low Levels of Ionizing Radiation*, National Academy of Sciences, Washington, D.C., 1972. [The BEIR-I Report]

NAS, National Research Council, Advisory Committee on the Biological Effects of Ionizing Radiations. *The Effects on Populations of Exposure to Low Levels of Ionizing Radiation*, National Academy of Sciences, Washington, D.C., 1980. [The BEIR-III Report]

NAS, National Research Council, Advisory Committee on the Biological Effects of Ionizing Radiations. *Health Effects of Exposure to Low Levels of Ionizing Radiation*, National Academy of Sciences, Washington, D.C., 1990. [The BEIR-V Report]

NCRP, *Basic Radiation Protection Criteria*, NCRP Report 39, National Council on Radiation Protection and Measurements, Washington, D.C., 1971.

NCRP, *Influence of Dose and Its Distribution in Time on Dose-Response Relationships for Low-LET Radiations*, NCRP Report 64, National Council on Radiation Protection and Measurements, Washington, D.C., 1980.

NCRP, *Recommendation on Limits for Exposure to Ionizing Radiation*, NCRP Report 91, National Council on Radiation Protection and Measurements, Bethesda, Md., 1987.

NIH, *Report of the National Institutes of Health ad hoc Working Group to Develop Radioepidemiological Tables*, NIH Publication 85-2748, National Institute of Health, Department of Health and Human Services, Washington, D.C., 1985.

POCHIN, E., *Nuclear Radiation Risks and Benefits*, Clarendon Press, Oxford, 1983.

ROSENSTEIN, M., *Handbook of Selected Tissue Doses in Diagnostic Radiology,* Publication 88-8031, U.S. Department of Health and Human Services, Public Health Service, Food and Drug Administration, Center for Devices and Radiological Health, Rockville, Md., 1988.

SCOTT, B. R., "A Radiation Protection Approach to Assessing Population Risks for Threshold-Type Radiobiological Effects," *Health Physics 55*, 463–470 (1988).

SCOTT, B. R., AND F. F. HAHN, "A Model that Leads to the Weibull Distribution Function to Characterize Early Radiation Response Probabilities," *Health Physics 39*, 521–530 (1980).

SCOTT, B. R., AND F. F. HAHN, "Early Occurring and Continuing Effects," in *Health Effects Models for Nuclear Power Plant Accident Consequence Analysis — Low LET Radiation, Part II: Scientific Bases for Health Effects Models*, Report NUREG/CR-4214, Rev. 1, Part II, U.S. Nuclear Regulatory Commission, Washington, D.C., 1989.

SHIMAZU, Y., H. KATO, AND W. J. SHULL, *Life Span Study Report 11, Part II: Cancer Mortality in the Years 1950-1985 Based on the Recently Revised Doses (DS86)*, Report RERF TR/5-88, Radiation Effects Research Foundation, Hiroshima, 1988.

UN, *Sources and Effects of Ionizing Radiation*, United Nations Scientific Committee on the Effects of Atomic Radiation, United Nations, New York, 1977.

UN, *Ionizing Radiation Sources and Biological Effects*, United Nations Scientific Committee on the Effects of Atomic Radiation, United Nations, New York, 1982.

UN, *Genetic and Somatic Effects of Ionizing Radiation*, United Nations Scientific Committee on the Effects of Atomic Radiation, United Nations, New York, 1986.

UN, *Sources, Effects and Risks of Ionizing Radiation*, United Nations Scientific Committee on the Effects of Atomic Radiation, United Nations, New York, 1988.

UPTON, A. C., AND R. F. KIMBALL, "Radiation Biology," in *Principles of Radiation Protection*, K. Z. Morgan and J. E. Turner (eds), John Wiley & Sons, New York, 1967.

VOGEL, F., AND A. G. MOTULSKY, *Human Genetics*, Springer-Verlag, Berlin, 1979.

WALD, N., "Evaluation of Human Exposure Data," in *Principles of Radiation Protection*, K. Z. Morgan and J. E. Turner (eds), John Wiley & Sons, New York, 1967.

WARD, J. F., C. L. LIMOLI, P. CALABRO-JONES, AND J. W. EVANS, "Radiation vs. Chemical Damage to DNA," in *Anticarcinogenesis and Radiation Protection*, F. Nygaard, M. Simic, and P. Cerutti (eds.), Plenum Press, New York, 1988.

ZAIDER, M. AND D. J. BENNER,, "On the Microdosimetric Definition of Quality Factors," *Radiat. Res. 103*, 302–316 (1985).

ZAIDER, M. AND H. H. ROSSI, "Estimation of the Quality Factor on the Basis of Multi-Event Microdosimetric Distributions," *Health Physics, 56,* 885-892 (1989).

PROBLEMS

1. The following experimental data are for *in vitro* cell survival as functions of absorbed dose for ^{60}Co gamma ray (reference) and fission-neutron exposure. Give your assessment of the neutron relative biological effectiveness for death of the particular cells studied.

Fractions of cells surviving		
Dose (Gy)	Neutrons	Gamma rays
0.5	0.63	0.98
1.0	0.49	0.93
1.5	0.32	0.68
2.0	0.20	0.75
3.0	0.088	0.61
4.0	0.042	0.39
6.0	0.008	0.15
8.0	–	0.036
10.0	–	0.006

2. In the course of radiation therapy, the gastrointestinal system of a patient is exposed to an absorbed dose of 100 rad from low-LET radiation at high dose rate. Other organs and tissues are affected negligibly from the exposure. What is the likelihood that the patient will experience vomiting and/or diarrhea from the exposure?

3. An adult subject receives a uniform whole-body gamma ray dose of 5 Gy at high dose rate. Supportive medical treatment is provided. Give upper and lower bounds on the probability that the subject will die as a result of the exposure.

4. Consider two nuclear plants, each located in such a way that large static populations of the same size are at risk from plant radionuclide releases. Each plant operates for 30 years. One plant continuously releases radionuclides. The other plant has a single accidental release of radionuclides sometime during the 30 years of operation. The collective population doses (person rem) are the same for the two cases. Are the consequences for hereditary illness in all future generations the same? If not, please explain.

5. Suppose a female subject receives at age 22 an accidental occupational whole-body absorbed dose of 3 rad due to low-LET radiation. At age 32 the subject receives a series of diagnostic x rays resulting in a dose of 3 rad to the organs of the digestive system. At age 52 the subject is diagnosed to have stomach cancer expected to be fatal. Give your assessment as to the probability of cancer causation, that is, the likelihood that the cancer is caused by occupational radiation exposure, to medical radiation exposure, or to natural causes. For a general treatment of problems of this type, the reader may wish to refer to radioepidemiological tables (NIH 1985).

6. Revise Table 3.11, correcting the life-table data to account for radiation-induced fatalities.

7. Revise Table 3.11 to apply to a population of all females.

8. You hold a supervisory radiation-protection appointment. A worker, age 20, has received an accidental whole-body radiation exposure of 5 rem. The worker, soon to be married, asks you what the consequences of the exposure might be in terms of (1) her likelihood of developing fatal cancer sometime in the future as a result of the exposure, and (2) the likelihood that any children she might bear would suffer from a hereditary illness caused by the radiation exposure. How do you respond? Prepare a written answer in the form of a memorandum, keeping in mind that your response might be called as evidence in a court of law.

Chapter 4

Exposure to Natural Sources of Radiation

4.1 INTRODUCTION

Life on earth is continually subjected to radiation of natural origin. Exposure is both external and internal, the former arising from cosmic radiation and radionuclides in the environment, the latter arising from radionuclides taken into the body by ingestion or inhalation. Natural sources are the major contributors to human radiation exposure. Study of these sources is important for several reasons. Natural exposure represents a reference against which exposure to man-made sources may be compared, not only for standards-setting purposes, but also in epidemiological studies of the consequences of man-made sources or even of unusually concentrated natural sources in certain areas. The variability in natural exposure introduces uncertainty in the nature of the control population in epidemiological studies. Also, there may be need for mitigation of even natural sources of radiation such as radon daughter products in the home or workplace.

Interest in natural radiation sources has prompted a host of studies and reviews. Among these are the current series of reviews published by the National Council on Radiation Protection and Measurements (NCRP 1987a, 1987b, 1987c). Table 4.1 summarizes average dose equivalent rates to the U.S. population from natural background radiation.

Cosmic radiation at the earth's surface consists mainly of muons and electrons which are debris from cosmic ray showers caused by energetic galactic or intergalactic cosmic rays interacting with atoms in the high atmosphere. The intensity of cosmic radiation varies with atmospheric elevation and with latitude. The 11-year cyclic variation of solar-flare activity affects the earth's magnetic field which, in turn, modulates the intensity of cosmic radiation reaching the earth.

Cosmic radiation reaching the earth's atmosphere consists mainly of high energy hydrogen nuclei, or protons, which interact with constituents of the atmosphere to produce showers of secondary particles including a number of "cosmogenic" ra-

TABLE 4.1   Summary of Average and Effective Dose Equivalent Rates from Various Sources of Natural Background Radiation in the United States.

	DOSE EQUIVALENT RATE (mrem y^{-1} = 10^{-5} Sv y^{-1})				
RADIATION SOURCE	BRONCHIAL EPITHELIUM	OTHER SOFT TISSUES	BONE SURFACES	BONE MARROW	H_E
Cosmic radiation	27	27	27	27	27
Cosmogenic nuclides	1	1	1	3	1
External terrestrial	28	28	28	28	28
Inhaled nuclides	2400				200
Nuclides in body	36	36	110	50	39
Totals (rounded)	2500	90	170	110	300

Source: NCRP 1987b; by permission of the NCRP.

dionuclides. Chief among these are ^{3}H and ^{14}C. These radionuclides are produced at relatively uniform rates and, except as augmented by man-made sources, exist in the biosphere in equilibrium, that is, with equal production and decay rates.

Naturally existing radionuclides not of cosmic-ray origin and not members of decay chains must have half lives comparable to the several billion-year age of the earth. These radionuclides are few and only two, ^{40}K and ^{87}Rb, result in significant portions of the dose rate in humans due to natural sources of radiation.

There are also several decay chains of radionuclides which occur naturally and whose parent radionuclides necessarily have half lives comparable to the age of the earth. Identified by the name of the parent, the only series of significance to human exposure are those of ^{238}U and ^{232}Th. Within these chains or series are subgroups comprising the daughter products of ^{222}Rn and ^{220}Rn. Radon as a source leads to the greatest component of natural radiation exposure and may be responsible for a significant fraction of lung cancer mortality. Health risks from exposure to radon daughter products have been studied intensively (ICRP 1987, NCRP 1984b, NAS 1988) and in many countries efforts are being taken to mitigate the risks. Because of the importance of this topic, special attention is given it in this chapter.

4.2 COSMIC, SOLAR, AND GEOMAGNETICALLY TRAPPED RADIATION

The earth is bombarded continuously by radiation originating from our sun, from sources within our galaxy, and from sources beyond our galaxy. The radiation as it reaches the earth's atmosphere consists of high-energy atomic nuclei. Hydrogen nuclei (protons) constitute the major component, with heavier atoms decreasing in importance with increasing atomic number. Cascades of nuclear interactions in the atmosphere give rise to many types of secondary particles. At the earth's surface, cosmic radiation dose rates are largely caused by muons and electrons. The intensity and angular distribution of galactic radiation reaching the earth is affected by the earth's magnetic field and perturbed by magnetic disturbances generated by solar flare activity. Consequently, at any given location, cosmic ray doses may vary in time by a factor of 3. At any given time, cosmic ray dose rates at sea level may vary with geomagnetic latitude by as much as a factor of 8, being greatest at the pole and least at the equator. The global average cosmic ray dose equivalent rate at sea level is about 240 μSv per year for the directly ionizing component and 20 μSv per year for the neutron component (UN 1988). Cosmic ray dose rates also increase with altitude. At geomagnetic latitude 55° N, for example, the absorbed dose rate in tissue approximately doubles with each 2.75 km (9000 ft) increase in altitude, up to 10 km (33,000 ft). The neutron component of the dose equivalent rate increases more rapidly with altitude than does the directly ionizing component and dominates at altitudes above about 6 km (UN 1988).

Solar cosmic rays associated with flares are mainly hydrogen and helium nuclei. While of too low energy to contribute to radiation doses at the surface of the earth, solar-flare radiation, which fluctuates cyclically with an 11-year period, perturbs earth's magnetic field and thereby modulates galactic cosmic-ray dose rates with the same period. Maxima in solar flare activity lead to minima in dose rates. Solar flare radiation, in comparison to galactic cosmic rays, is of little significance as a hazard in aircraft flight or low orbital space travel. On the other hand, solar-flare radiation presents a considerable hazard to personnel and equipment in space travel outside the earth's magnetic field.

Released continuously from the sun, as an extension of the corona, is the solar wind, a plasma of low energy protons and, presumably, very low energy electrons. The solar wind does not present a radiation hazard, even in interplanetary space travel. However, it does affect the interplanetary magnetic field and the shape of the geomagnetically trapped radiation belts. These radiation belts are thought to be supplied by captured solar-wind particles and by decay into protons and electrons of neutrons created by interactions of galactic cosmic rays in the atmosphere. The trapped radiation can present a significant hazard to personnel and equipment in space missions.

4.2.1 Galactic Cosmic Rays

Cumulative energy spectra for the various components of the galactic cosmic radiation incident on the earth's atmosphere are illustrated in Fig. 4.1. These spectra may be approximated as

$$\phi(E > E_o) \simeq \frac{A}{(M_o c^2 + E_o)^n}, \tag{4.1}$$

in which $\phi(E > E_o)$ is the flux density of particles with energies greater than E_o, $M_o c^2$ is the rest-mass energy equivalent of the particles, A and n are parameters which, generally, depend on the state of the solar cycle.

As a result of nuclear spallation reactions with constituents of the atmosphere, secondary neutrons, protons, and pions, mainly, are produced. Subsequent pion decay results in electrons, photons, neutrons, and muons. Muon decay, in turn, leads to secondary electrons, as do Coulombic scattering interactions of charged particles in the atmosphere. As shown in Fig. 4.2, the number and mix of particles varies with altitude as does the resulting tissue absorbed dose rate. Figure 4.3 illustrates the variation

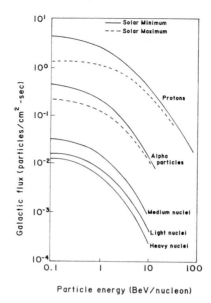

Figure 4.1 Integral energy spectra (i.e., the flux density of particles with energies $> E$) for various components of galactic cosmic radiation. Medium nuclei, $6 \leq Z \leq 9$; light nuclei, $3 \leq Z \leq 5$; heavy nuclei, $Z \geq 10$. [From Haffner (1967); by permission of Academic Press.]

TABLE 4.2 Distribution of U.S. Population with Altitude and Accompanying Dose Equivalent Rates from Cosmic Radiation.

ELEVATION (km)	PERCENT OF POPULATION (%)	RANGE OF DOSE EQUIVALENT RATES[a] $(\text{mrem y}^{-1} = 10\mu\text{Sv y}^{-1})$
0.0–0.2	48.3	26–27
0.2–0.3	35.2	27–28
0.3–0.6	11.0	28–31
0.6–1.2	3.0	31–39
1.2–1.8	2.2	39–52
1.8–2.4	0.35	52–74
2.4–3.0	0.040	74–107
> 3.0	0.0078	107

[a] Assuming 10 percent reduction for structure shielding.
Source: NAS 1980, as derived from NCRP 1975.

with altitude of dose rates from galactic cosmic rays, averaged over geomagnetic latitudes between 43° and 55° and over two periods of solar activity. Table 4.2 describes the distribution of cosmic-ray dose rates among the population of the United States.

Except for short term influences of solar activity, galactic cosmic radiation has been constant in intensity for at least several thousand years. The influence of solar activity is cyclical. The principal variation is on an 11-year cycle, but there are very

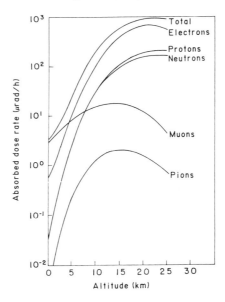

Figure 4.2   Absorbed dose rates at 5 cm depth in a 30-cm thick slab of tissue, from various components of the cosmic radiation during the solar minimum at geomagnetic latitude 55° N. [From NCRP (1987b); by permission of the NCRP.]

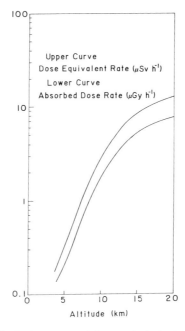

Figure 4.3   Variation of the galactic dose rate and dose equivalent rate with altitude. Doses have been averaged over two periods of solar activity. [Data are from UN (1982).]

small diurnal variations and evidence of a 22-year cycle as well as 27-day quasi-periodic variations (Haffner 1967).

The geomagnetic field of the earth is responsible for limiting the number of cosmic rays which can reach the atmosphere. For particles of atomic number Z vertically incident at geomagnetic latitude λ, the minimum momentum for a cosmic ray to reach the atmosphere is proportional to $Z \cos^4 \lambda$ (Haffner 1967). This accounts for a strong effect of latitude on cosmic-ray dose rates.

4.2.2 Solar Flare Particulate Radiation

Solar flare particles are mostly protons and alpha particles, predominantly the former. Electrons are thought to be emitted as well, but with energies less than those of protons by a factor equal to the ratio of the rest masses. Energy spectra are highly variable, as are temporal variations of intensity. Haffner (1967) describes records of solar flares during the 1956 to 1961 interval which encompasses the 1959 maximum in activity. The large event of July 10, 1959, for example, resulted in the following fluences (cm^{-2}) outside the earth's atmosphere:

PROTONS		ALPHA PARTICLES	
$E > 10\,\mathrm{MeV}$	4.5×10^9	$E > 40\,\mathrm{MeV}$	1.6×10^8
$E > 30\,\mathrm{MeV}$	1.0×10^9	$E > 120\,\mathrm{MeV}$	2.4×10^7
$E > 100\,\mathrm{MeV}$	1.4×10^8	$E > 400\,\mathrm{MeV}$	5.0×10^5

A typical course of events for a flare is as follows. Gamma and x-ray emission takes place over about 4 hours as is evidenced by radio interference. The first significant quantities of protons reach the earth after about 15 hours and peak proton intensity occurs at about 40 hours after the solar eruption (NCRP 1989a).

4.2.3 Trapped Radiation Belts

The earth's geomagnetically trapped radiation belts are also known as Van Allen belts in recognition of James A. Van Allen and his co-workers who discovered their existence in 1958. There are two belts, the inner consisting primarily of protons, the outer of electrons. The particles travel in helical trajectories determined by the magnetic field surrounding the planet. They occur at maximum altitude at the equator and approach the earth most closely near the poles. The spatial distribution of both belts in a plane normal to the solar wind is illustrated in Fig. 4.4. The solar wind compresses the trapped radiation on the sunny side of the earth and the compression is enhanced by solar flare activity. In the earth's shadow, the belts are distended as the solar wind sweeps the magnetosphere outwards. The spatial distribution of the outer (electron) belt in a plane perpendicular to the earth's orbit is illustrated in Fig. 4.5.

Energy spectra of trapped radiation depend in a very complicated way on the position in the belts. However, models have been developed permitting estimates of intensities at any position for both "active" and "quiet" solar conditions. Figures 4.6 and 4.7 illustrate results of such model calculations, the former for a 450-km altitude

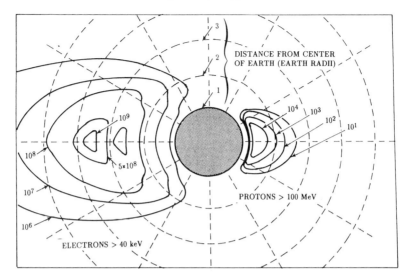

Figure 4.4   Spatial distribution of trapped radiation belts in a plane through the earth and normal to the solar wind. Labels for solid contours are particle flux densities in $cm^{-2} s^{-1}$ for electrons with energies greater that 40 keV (left) and for protons with energies greater that 100 MeV (right). Labels for dashed circular contours are the number of earth radii from the earth's center. [From Haffner (1967); by permission of Academic Press.]

at 28.5° inclination (acute angle from the equator), the latter for a 36,000-km altitude geosynchronous orbit at 160° W longitude.

Because of asymmetries in the geomagnetic field, there is a region identified as the South Atlantic Anomaly (in the vicinity of 30° S, 15° W) wherein proton flux densities are relatively high at altitudes as low as 120 nautical miles. U.S. earth-orbital missions launched from Cape Canaveral are most generally at an inclination of 28.5° and the anomaly accounts for major portions of the doses received by crews of Shuttle missions.

4.2.4 Radiation Doses in Air and Space Travel

For an average altitude for subsonic commercial aviation of 8 km, the average dose equivalent rate from galactic cosmic rays is about 1.35 μSv h^{-1} (Fig. 4.3). Then for an average speed of 600 km h^{-1} and for 1.3×10^{12} passenger-kilometers per year (1984 data, excluding China, from UN 1988), the annual collective dose equivalent to the world population is about 3000 person-Sv from air transportation.[1] For the SST supersonic aircraft, traveling at an average altitude of 16 km, Fig. 4.3 would indicate a dose equivalent rate of 9.5 μSv h^{-1}. Although dose rates in supersonic aircraft exceed those in subsonic, the greater speed compensates. On transatlantic flights, for example, doses incurred in subsonic aircraft are about equal to those incurred in supersonic aircraft (UN 1988).

---

[1]This compares to the NCRP (1987a) estimate of 2500 person Sv annually for U.S. air travel, and the United Nations (1988) estimate of 4300 person Sv annually for world air travel.

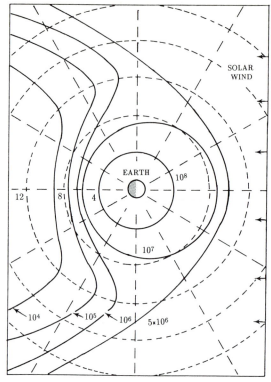

Figure 4.5 Spatial distribution of the outer (electron) trapped radiation belt in the earth's magnetic equatorial plane. Labels for solid contours are solid contours are particle flux densities in cm^{-2} s^{-1} for electrons with energies greater than 40 keV. Circular dashed contours are four earth radii apart. [From Haffner (1967); by permission of Academic Press.]

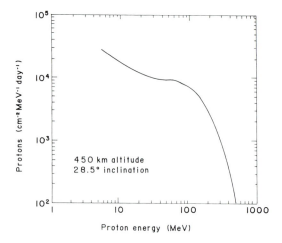

Figure 4.6 Model calculations for trapped-radiation energy spectra for orbital missions at 450-km altitude and 28.5 degree inclination. [Data are from Curtis et al. (1986), as reported in NCRP (1989a); by permission of the NCRP.]

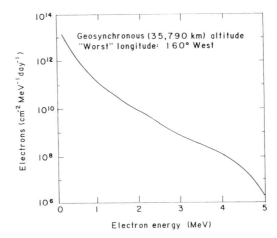

Figure 4.7   Model calculations for trapped-radiation energy spectra for geosynchronous missions and the parking longitude leading to greatest radiation doses. The "altitude" corresponds to nominally 5.7 radii from earth center. [Data are from Curtis et al. (1986), as reported in NCRP (1989a); by permission of the NCRP.]

Solar flare contributions to dose rates in aircraft flight are negligible. Even in the event of a giant flare, there would be ample warning and ample time for evasive action. Such is not necessarily the case in lunar spacecraft missions. Faced with the potential for life-threatening radiation doses from solar flares, the Apollo astronauts were fortunate not to have experienced any significant flares.

Table 4.3 summarizes absorbed doses incurred by astronauts on space missions. Figure 4.8 summarizes dose rates as a function of flight altitude for shuttle missions at the usual 28.5° launch inclination. Durations of missions represented in the figure ranged from two to ten days. While absorbed doses in these missions were well within occupational limits, those for longer-duration, high-altitude missions may well exceed the usual limits associated with occupational exposure. U.S. guidelines for radiation exposure limits in space missions were established by the National Aeronautics and Space Administration (NASA) in 1970 as *career limits* of 4 Sv (400 rem) whole body, 6 Sv to the lens of the eye, 12 Sv to the skin, and 2 Sv to the testes (NCRP 1989). As of this writing, the National Council on Radiation Protection and Measurements (NCRP 1989a) has recommended a revision and broadening of the limits to encompass 30-day and annual limits as well as career limits. If adopted, the limits, in units of Sv (100 rem), would be as follows:

Limit	Blood-Forming Organs	Lens	Skin
30-day	0.25	1.0	1.5
Annual	0.5	2.0	3.0
Career	1 to 4	4.0	6.0

TABLE 4.3   Absorbed Doses Incurred by Astronauts on Space Missions at Various Inclinations and Altitudes.

MISSION OR SERIES	TYPE	INC. (deg)	ALT. (km)	DUR. (h)	ABS. DOSE (mrem)
Apollo VII	Earth Orbital	31.6	229–306	260	120
Apollo VIII	Circumlunar			147	185
Apollo IX	Earth Orbital	32.6	197–249	241	210
Apollo X	Circumlunar			192	470
Apollo XI	Lunar Landing			182	200[a]
Apollo XII	Lunar Landing			236	$\simeq$ 200
Apollo XIV	Lunar Landing			209	$\simeq$ 500
Apollo XV	Lunar Landing			286	$\simeq$ 200
Vostok 1–6	Earth Orbital				2–80
Voskhad 1,2	Earth Orbital				30,70
Soyuz 3–9	Earth Orbital				62–234
Skylab 2	Earth Orbital	50	435	672	1600
Skylab 3	Earth Orbital	50	435	1652	3800
Skylab 4	Earth Orbital	50	435	2160	7700

[a]Corresponding dose equivalent 400 mrem, of which 220 due to protons, 94 to recoil nuclei, 46 due to heavy charged particles, about 12 due to fast neutrons, and about 30 due to electrons and gamma rays.

Source: UN 1982, except for Skylab data (NCRP 1989a) for which are reported mean crew TLD doses (mrad).

Career limits for blood-forming organs depend on sex and age at start of exposure. Lower limits of 1.0 and 1.5 Sv apply respectively to the female and male at age 25. The upper limit of 4 Sv applies to the male, age 55 at start of exposure.

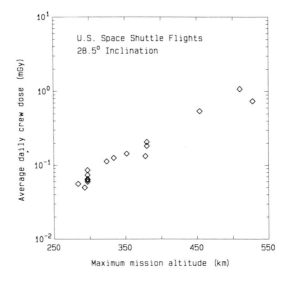

Figure 4.8   Average dose rate to a U.S. space-shuttle crew member, as a function of spacecraft altitude. [Data are from Benton (1986), as reported in NCRP (1989a).]

TABLE 4.4   Global Distribution of Cosmogenic
Radionuclides.

	^{3}H	^{7}Be	^{14}C	^{22}Na
Global inventory (PBq)	1300	37	8500	0.4
Distribution (percent)				
stratosphere	6.8	60	0.3	25
troposphere	0.4	11	1.6	1.7
land surface/biosphere	27	8	4	21
mixed ocean layer	35	20	2.2	44
deep ocean	30	0.2	92	8
ocean sediments			0.4	

Source: UN 1982.

4.3 COSMOGENIC RADIONUCLIDES

Cosmic-ray interactions with constituents of the atmosphere, sea, or earth, but mostly with the atmosphere, lead directly to radioactive products. Capture of secondary neutrons produced in primary interactions of cosmic rays, leads to the formation of many more radionuclides. However, the slow-neutron flux density at sea level is only about 8 cm^{-2} h^{-1} (Morgan 1967) and the total neutron flux density is only about 30 cm^{-2} h^{-1} (UN 1988). Thus, neutron capture in the earth's crust or the sea is of little importance in comparison with capture in the atmosphere, except for production of long-lived ^{36}Cl (NCRP 1975). Of the nuclides produced in the atmosphere, only ^{3}H, ^{7}Be, ^{14}C, and ^{22}Na contribute appreciably to human radiation exposure. Radionuclides borne by the 10^7 kg of "space dust" reaching the earth annually have a total specific activity of less than 450 pCi kg^{-1} and result in only very small atmospheric concentrations (NCRP 1975).

Over the past century, combustion of fossil fuels and the emission of CO_2 not containing ^{14}C has diluted the cosmogenic content of ^{14}C in the environment. Moreover, since World War II, artificial introduction of ^{3}H, ^{14}C, and other nuclides, into the environment by human activity has been significant, especially as a result of atmospheric nuclear-weapons tests. Consequently, these nuclides no longer exist in natural equilibria in the environment.

The tritium ^{3}H nuclide (symbol T) is produced mainly from the ^{14}N(n,T)^{12}C and ^{16}O(n,T)^{14}N reactions. Tritium has a half life of 12.3 years and, upon decay, releases one beta particle with maximum energy 18.6 keV (average energy 5.7 keV). Tritium exists in nature almost exclusively as HTO but, in foods, may be partially incorporated into organic compounds. The global distribution of tritium is described in Table 4.4. In the UNSCEAR (UN 1982) estimate of absorbed dose in the adult due to tritium, it is assumed that ^{3}H exists in the body in the same proportion to stable hydrogen as in continental surface water, namely in the atomic ratio 3.3×10^{-18} (corresponding to 0.0004 Bq cm^{-3} in water). Since the body is 10 percent hydrogen by weight, the average absorbed dose rate in the body is 0.01 μGy y^{-1} (1 μrad y^{-1}).

The nuclide ^{14}C is produced mainly from the ^{14}N(n,p)^{14}C reaction. It exists in the atmosphere as CO_2, but the main reservoir is the ocean. It has a half life of

5730 years and decays by beta particle emission, each decay resulting in a particle of maximum energy 157 keV (average energy 49.5 keV). The natural atomic ratio of ^{14}C to stable carbon is 1.2×10^{-12} (corresponding to 0.226 Bq ^{14}C per gram of carbon (NCRP 1975). For carbon weight fractions of 0.23, 0.089, 0.41, and 0.25 in the soft tissues, gonads, red marrow, and skeleton, annual average absorbed doses in those tissues are respectively 13, 5.0, 23, and 14 μGy.

The ^{7}Be radionuclide, with a half life of 53.4 days, is also produced by cosmic ray interactions with nitrogen and oxygen in the atmosphere. It decays by electron capture, 10.4 percent of the captures resulting in the emission of a 478 keV gamma ray. Environmental concentrations in temperate regions are about 3000 Bq m^{-3} in surface air and 700 Bq m^{-3} in rainwater (UN 1982). Average annual absorbed doses in the adult are 12 μGy to the walls of the lower large intestine, 1.2 μGy to the red marrow, and 5.7 μGy to the gonads.

The nuclide ^{22}Na results from interaction of atmospheric argon with high-energy cosmic-ray secondary neutrons (NCRP 1987b). It has a half life of 2.60 y, decaying by positron emission (90%) and electron capture (10%). The positron has a maximum energy of 546 keV (average energy 216 keV). Essentially all decays are accompanied by emission of a 1.28 MeV gamma ray from the excited ^{22}Ne daughter. The global distribution of ^{22}Na is described in Table 4.4. Annual average absorbed doses to the adult are 0.10 to 0.15 μGy to the soft tissues, 0.22 μGy to red marrow, and 0.27 μGy to bone surfaces.

4.4 SINGLY OCCURRING PRIMORDIAL RADIONUCLIDES

Of the many radioactive species present when the earth formed 4 billion years ago, some 17 very long-lived radionuclides still exist as singly occurring or isolated radionuclides, that is, as radionuclides not belonging to a decay chain, (NCRP 1975). Of these primordial radionuclides, only ^{40}K and ^{87}Rb contribute significantly to human exposure.

The radionuclide ^{87}Rb has a half-life of 4.8×10^{10} y and emits in each decay one beta particle of maximum energy 273 keV (average energy 79 keV) (Kocher 1981). Its natural isotopic abundance is 27.9 percent and the mass concentrations in Reference Man range from 6 ppm in the thyroid to 20 ppm in the testes (ICRP 1975). The greatest annual dose to body tissues is 14 μGy (1.4 mrem) to bone-surface cells (UN 1982). The annual effective dose equivalent is about 6 μSv (UN 1988).

The radionuclide ^{40}K is a major contributor to human exposure from natural radiation. Present in an isotopic abundance of 0.0118% it has a half life of 1.227×10^9 y, decaying both by electron capture (11%) and beta-particle emission (89%). The beta particle has a maximum energy of 1312 keV (average energy 509 keV). Electron capture results in emission of a 1461 keV gamma ray in 10.7 percent of decays. Very low energy Auger electrons and x rays are also released with electron capture (Kocher 1981). The average elemental concentrations of potassium in Reference Man is 2 percent. Annual doses in Reference Man are 140 μGy to bone surface, 170 μGy on average to soft tissue, and 270 μGy to red marrow (UN 1982). ^{40}K also contributes in a major way to external exposure. The average specific activity of the nuclide in soil, $12 \, \text{pCi g}^{-1}$ $(0.44 \, \text{Bq g}^{-1})$, results in an annual whole-body dose equivalent of $120 \, \mu\text{Sv}$ (12 mrem) (UN 1982).

4.5 DECAY SERIES OF PRIMORDIAL ORIGIN

4.5.1 Decay Series Descriptions

Two decay series, identified by the long-lived parents ^{238}U and ^{232}Th contribute appreciably to human exposure to natural radiation. Another series headed by ^{235}U contributes very little and is not discussed here. The two important series are shown schematically in Figs. 4.9 and 4.10, and the principal radiations emitted are given in

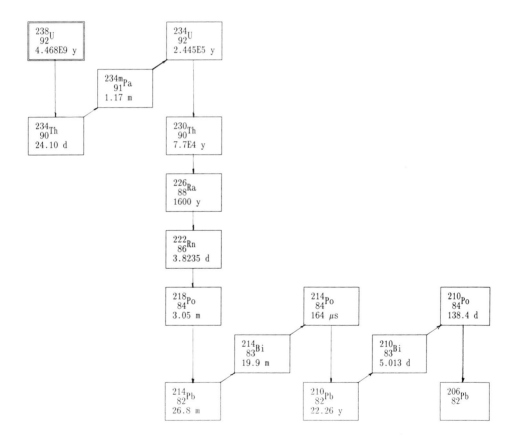

Figure 4.9 Decay scheme for ^{238}U. Alpha decay is depicted by vertically downward lines and beta decay by lines upward to the right. Not shown are (1) the isomeric transition to ^{234}Pa (0.16%) followed by beta decay to ^{234}U, (2) beta decay of ^{218}Po to ^{218}At (0.02%) followed by alpha decay to ^{214}Bi, (3) alpha decay of ^{214}Bi to ^{210}Tl (0.02%) followed by beta decay to ^{210}Pb, and (4) alpha decay of ^{210}Bi to ^{206}Tl (0.00013%) followed by beta decay to ^{206}Pb.

Tables 4.5 and 4.6. Identified in the figures are subseries (NCRP 1975) generally consisting of or headed by a relatively long-lived radionuclide (see Table 4.9). While all the members of a series are not likely to be in radioactive equilibrium in nature for chemical or physical reasons, members of a subseries are more likely to be so. The subseries headed by the gases ^{220}Rn and ^{222}Rn are of special importance and are treated separately in a subsequent section.

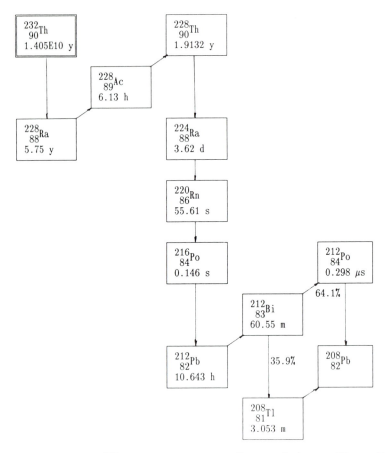

Figure 4.10   Decay scheme for ^{232}Th. Alpha decay is depicted by vertically downward lines and beta decay by lines upward to the right.

4.5.2 Concentrations in Rocks, Soils, and Building Materials

Typical natural radionuclide concentrations in several materials are listed in Tables 4.7 and 4.8. Also listed are dose rates in air 1 m above the nuclide-bearing material, calculated on the basis of radioactive equilibrium throughout the entire series.

TABLE 4.5 Principal Radiation Emitted from the Nuclides of the ^{238}U Decay Series. Not Listed Are X-Rays, Conversion Electrons, Auger Electrons, and Beta or Gamma Rays with Frequencies Less than 0.5%. Certain Nuclides Indicated in Fig. 4.9 are Omitted.

NUCLIDE	ALPHA PARTICLES ENERGY (keV)	ALPHA PARTICLES FREQUENCY (%)	BETA PARTICLES MAX. ENERGY (keV)	BETA PARTICLES AVG. ENERGY (keV)	BETA PARTICLES FREQUENCY (%)	GAMMA RAYS ENERGY (keV)	GAMMA RAYS FREQUENCY (%)
^{238}U	4039	0.23					
	4147	23					
	4196	77					
^{234}Th			75.8	19.5	2.0	63.3	3.8
			95.8	24.8	6.8	92.4	2.72
			96.2	24.9	18.5	92.8	2.69
			188.6	50.6	72.5		
^{234m}Pa			1236	410.2	0.74	1001	0.589
			1471	500.8	0.62		
			2281	825.4	98.6		
^{234}U	4605	0.24					
	4724	27.4					
	4776	72.4					
^{230}Th	4368	0.31					
	4476	0.12					
	4621	23.4					
	4688	76.3					
^{226}Ra	4602	5.55				186.2	3.28
	4785	94.55					
^{222}Rn	5490	99.92					
^{218}Po	6003	99.98					
^{214}Pb			185	50	2.55	53.2	1.11
			490	145	0.83	242.0	7.49
			672	207	48.0	258.8	0.553
			728	227	42.5	295.2	19.2
			1024	337	6.3	351.9	37.2
						785.9	1.10

(continued)

TABLE 4.5 (continued)

NUCLIDE	ALPHA PARTICLES ENERGY (keV)	ALPHA PARTICLES FREQUENCY (%)	BETA PARTICLES MAX. ENERGY (keV)	BETA PARTICLES AVG. ENERGY (keV)	BETA PARTICLES FREQUENCY (%)	GAMMA RAYS ENERGY (keV)	GAMMA RAYS FREQUENCY (%)
^{214}Bi			788	248	1.07	839.0	0.59
			822	261	2.81	609.3	46.3
			977	318	0.56	665.5	1.57
			1066	352	5.61	768.4	5.04
			1077	357	0.89	806.2	1.23
			1151	385	4.43	934.1	3.21
			1253	425	2.50	1120.3	15.1
			1259	427	1.50	1155.2	1.70
			1275	434	1.19	1238.1	5.94
			1380	475	1.59	1281.0	1.48
			1423	492	8.34	1377.7	4.11
			1505	525	17.7	1385.3	0.78
			1540	539	17.9	1401.5	1.39
			1609	567	0.88	1408.0	2.49
			1727	615	3.38	1509.2	2.22
			1855	668	1.01	1583.2	0.72
			1892	684	7.86	1661.3	1.15
			2661	1007	0.6	1729.6	2.97
			3270	1269	17.2	1764.5	15.8
						1847.4	2.09
						2118.6	1.17
						2204.2	4.98
						2447.9	1.56
^{214}Po	7687	99.989					
^{210}Pb			16.5	4.14	80.2	46.5	4.05
			63.0	16.13	19.8		
^{210}Bi			1161.4	389.0	100.0		
^{210}Po	5304	100.0					

Source: Kocher 1981.

180

TABLE 4.6 Principal Radiation Emitted from the Nuclides of the ^{232}Th Decay Series. Not Listed Are X-Rays, Conversion Electrons, Auger Electrons, and Beta or Gamma Rays with Frequencies Less Than 0.5%.

NUCLIDE	ALPHA PARTICLES		BETA PARTICLES			GAMMA RAYS	
	ENERGY (keV)	FREQUENCY (%)	MAX. ENERGY (keV)	AVG. ENERGY (keV)	FREQUENCY (%)	ENERGY (keV)	FREQUENCY (%)
^{232}Th	3830	0.20				59.0	0.19
	3953	23					
	4010	77					
^{228}Ra			38.9	9.9	100		
^{228}Ac			413	118.5	1.59	99.45	1.3
			449	130.0	2.42	105	1.6
			454	131.7	1.54	129.1	2.8
			491	143.8	4.9	154.2	0.9
			494	144.9	0.78	209.3	4.4
			499	146.4	1.30	270.2	3.6
			606	182.1	8	327.6	3.2
			910	290	0.82	338.3	11.4
			969	311	3.3	409.5	2.13
			983	317	7	463.0	4.4
			1014	328	6.6	562.3	0.94
			1115	366	3.4	727.0	0.78
			1168	386	32	755.2	1.05
			1741	611	12	772.2	1.55
			2079	748	8	782.0	0.53
						794.7	4.6
						830.5	0.59
						835.5	1.75
						840.0	0.94
						904.5	0.83
						911.1	27.7
						964.6	5.2
						969.1	16.6
						1246.4	0.54

(*continued*)

TABLE 4.6 (continued)

NUCLIDE	ALPHA PARTICLES		BETA PARTICLES			GAMMA RAYS	
	ENERGY (keV)	FREQUENCY (%)	MAX. ENERGY (keV)	AVG. ENERGY (keV)	FREQUENCY (%)	ENERGY (keV)	FREQUENCY (%)
228Th	5175	0.18				1459.3	1.00
	5212	0.36				1495.8	1.00
	5341	26.7				1501.5	0.55
	5423	72.7				1580.2	0.69
						1588.0	3.5
						1630.4	1.86
						84.4	1.21
224Ra	5449	4.9				241.0	3.95
	5686	95.1					
220Rn	6288	99.90					
216Po	6779	100					
212Pb			158	41.9	5.22	115.2	0.60
			334	94.4	85.1	238.6	44.6
			573	172.7	9.9	300.1	3.41
						39.9	1.02
212Bi	5607	0.402					
	5768	0.600					
	6051	25.22					
	6090	9.63					
212Po	8785	100					
208Tl			1031	340.2	2.92	252.6	0.80
			1072	356.0	0.58	277.4	6.8
			1283	438.7	23.2	510.8	21.6
			1517	532.5	22.7	583.1	84.2
			1794	646.5	49.3	763.1	1.64
						860.4	12.46
						2614.7	99.8

Source: Kocher 1981.

TABLE 4.7   Typical Naturally Occurring Radionuclide Concentrations in Rocks and Soils.

	AVERAGE SPECIFIC ACTIVITY ($pCi\ g^{-1}$)			ABSORBED DOSE RATE IN AIR ($\mu rad\ h^{-1}$)
	^{40}K	^{238}U	^{232}Th	
Igneous Rock				
Acidic	27	1.6	2.2	12
Intermediate	19	0.62	0.88	6.2
Mafic	6.5	0.31	0.30	2.3
Ultrabasic	4.0	0.01	0.66	2.3
Sedimentary Rock				
Limestone	2.4	0.75	0.19	2.0
Carbonate	—	0.72	0.21	1.7
Sandstone	10	0.5	0.21	3.2
Shale	19	1.2	1.2	7.9
Soil Type				
Serozem	18	0.85	1.3	7.4
Gray-brown	19	0.75	1.1	6.9
Chestnut	15	0.72	1.0	6.0
Chernozem	11	0.58	0.97	5.1
Gray forest	10	0.48	0.72	4.1
Sodpodzolic	8.1	0.41	0.60	3.4
Podzolic	4.0	0.24	0.33	1.9
Boggy	2.4	0.17	0.17	1.1
World average	10	0.7	0.7	4.6
Coal				
United States[a]	1.4	0.5	0.6	

[a] United States data from (UN 1982).

Source: UN 1977.

TABLE 4.8   Typical Naturally Occurring Radionuclide Concentrations in Building Materials.

	SPECIFIC ACTIVITY RANGE ($pCi\ g^{-1}$)[a]		
	^{40}K	^{226}Ra	^{232}Th
Brick	16–20	1.4–2.6	1.0–3.4
Concrete	7–19	0.9–2.0	0.9–2.3
Plaster	< 2–10	.09–0.6	< 0.4–2.0
Granite	28–40	2.4–3	2.3–4.5
Limestone/Marble	$\simeq 1$	< 0.5	< 0.5

[a] To convert to units of $Bq\ g^{-1}$, multiply by 0.037.

Source: Eisenbud 1987, UN 1977.

4.5.3 Concentrations in Surface and Ground Water

Environmental mobility of the parent nuclides in the decay series or subseries, of course, affects the redistribution of series daughters. Thorium and radium are relatively insoluble and thus unaffected by water movement. Uranium, on the other hand, forms soluble compounds in an oxidizing or high CO_2 aqueous environment. It may thus be transported great distances, and later reconcentrated in, for example, a reducing environment. Consequently, ^{232}Th and its daughter ^{228}Ra are more evenly distributed in nature than are ^{238}U and its daughter ^{226}Ra.

Concentrations of elements of the ^{232}Th and ^{238}U series in sea water are given in Table 4.9. Concentrations are highly variable in continental surface and ground waters. State-average concentrations of uranium in the United States vary from 0.01 to 7.5 pCi L^{-1} for surface, ground, and domestic water (Hess et al. 1985). The concentration of ^{226}Ra in surface water is low, ranging from 0.1 to 0.5 pCi L^{-1} (Hess et al. 1985). Some 80 percent of 60,000 public water supplies in the United States depend on ground water, within which radium concentrations are highly variable. The greatest concentrations are found in the Piedmont and Coastal Plain regions, comprising parts of New Jersey, North and South Carolina, and Georgia, and in the North Central region, comprising parts of Minnesota, Iowa, Illinois, Missouri, and Wisconsin. Those in aquifers of the Piedmont and Coastal Plain regions are summarized in Table 4.10.

4.5.4 Food Concentrations and Dietary Intake

Estimates of the total dietary intake of certain radionuclides, based on data for Reference Man (ICRP 1975) are shown in Table 4.11. The estimates for ^{238}U and ^{232}Th generally agree with those given by UNSCEAR (UN 1977), which are reported in Table 4.12. Ranges of concentrations in selected foods are reported in Table 4.13.

Dietary intake is highly variable, depending not only on local conditions and dietary choices, but also on the age, sex, and general health of a subject. Certain regions have very high radionuclide concentrations in soil and thus high concentrations in local agricultural products. In the Kerala coast in India, intake of ^{226}Ra is about 0.1 Bq d^{-1}. In the Araxa-Tapira region of Brazil, intake is about 6 Bq d^{-1} (NCRP 1984a). Brazil nuts, for example contain up to 520 Bq kg^{-1} of ^{226}Ra (Eisenbud 1987). In regions within which seafood is a major component of the diet, daily ingestion of ^{210}Pb may average as much as 6.3 Bq (NCRP 1984a). Lichen concentrates a number of radionuclides. In regions of the USSR where caribou or reindeer, lichen feeders, contribute a large portion of the diet, ingestion of ^{210}Pb and ^{210}Po may average as much as 1.6 and 13 Bq d^{-1}, respectively (NCRP 1984a). Cigarettes each contain about 20 mBq of ^{210}Pb and 15 mBq of ^{210}Po, about 10 percent of which enters the lung on inhalation of the tobacco smoke (UN 1988).

4.6 THE ^{222}Rn AND ^{220}Rn DECAY SERIES

4.6.1 Introduction

The noble gas radon diffuses into the atmosphere from rocks, soils, and building materials containing progenitor radionuclides. The gas itself presents little radiation hazard on inhalation, and only minor hazard if ingested in aqueous solution. The

TABLE 4.9   Concentrations of Primordial Radionuclides in Seawater.

SERIES	NUCLIDE	pCi per liter[a]
Singly occurring	^{40}K	342
	^{87}Rb	3.0
^{238}U subseries 1	^{238}U	1.0
	^{234}Th	1.0
	^{234m}Pa	1.0
	^{234}U	1.2
^{238}U subseries 2	^{230}Th	0.0061
^{238}U subseries 3	^{226}Ra	0.098
^{238}U subseries 4	^{222}Rn	0.097
	^{218}Po	0.096
	^{214}Pb	0.095
	^{214}Bi	0.093
	^{214}Po	0.096
^{238}U subseries 5	^{210}Pb	0.084
	^{210}Bi	0.097
	^{210}Po	0.099
^{232}Th subseries 1	^{232}Th	0.000011
^{232}Th subseries 2	^{228}Ra	0.038
	^{228}Ac	0.034
	^{228}Th	0.033
	^{224}Ra	0.003
^{232}Th subseries 3	^{220}Rn	0.0030
	^{216}Po	0.0036
	^{212}Pb	0.0033
	^{212}Bi	0.0032
	^{208}Tl	0.0012
	^{212}Po	0.0021

[a]To convert units to Bq per liter, multiply by 0.037.

Source: NAS 1971.

TABLE 4.10   Summary of ^{226}Ra and ^{228}Ra Concentrations in Ground Water for the Atlantic Coastal Plain and Piedmont Regions of the United States.

AQUIFER TYPE	GEOMETRIC MEAN CONCENTRATION (pCi per liter)[a]	
	^{228}Ra	^{226}Ra
Acidic igneous rocks	1.39	1.80
Metamorphic rocks	0.33	0.37
Sand	1.05	1.36
Arkose	2.16	2.19
Quartzose	0.27	0.55
Limestone	0.06	0.12

[a]To convert units to Bq per liter, multiply by 0.037.

Source: Hess et al. 1985; by permission of the Health Physics Society.

TABLE 4.11   Estimates of Dietary Intake of Certain Radionuclides Based on Element Balances for Reference Man.

ELEMENT	INTAKE (g d^{-1})	NUCLIDE	ISOTOPIC ABUNDANCE	pCi per g OF ELEMENT[a]	INTAKE pCi d^{-1}
K	3.3[b]	^{40}K	0.0118%	8.60×10^2	2800
Rb	2.2×10^{-3}	^{87}Rb	27.85%	2.42×10^4	53
Th	3.0×10^{-6}	^{232}Th	100%	1.09×10^5	0.33
U	1.9×10^{-6}	^{238}U	99.37%	3.34×10^5	0.63

[a] To convert units to Bq per gram or Bq per day, multiply by 0.037.

[b] 1.3 g d^{-1} in the one-year-old, 0.09 g d^{-1} in the breast-fed one-month- old, and 0.14 g d^{-1} in the artificially-fed one-month-old.

Source: ICRP 1975.

TABLE 4.12   Normal Dietary Intake of Selected Radionuclides of the ^{238}U and ^{232}Th series.

NUCLIDE	TYPICAL pCi d^{-1} IN DIET[a]
^{238}U	0.3–0.5
^{232}Th	$\simeq 0.5$
^{226}Ra	0.8–1.7 (mean 0.9)
^{210}Pb	1.2–6.2
^{210}Po	1.3–4.6

[a] To convert units to Bq per day, multiply by 0.037.

Source: NCRP 1984a.

TABLE 4.13   Concentrations of Radionuclides of the ^{238}U and ^{232}Th Decay Series in Foods.

FOOD	CONCENTRATION (pCi kg^{-1})[a]			
	^{40}K[b]	^{226}Ra[c]	^{210}Pb[d]	^{210}Po[d]
Grain products	710–1900	0.80–3.3	2–5	1–5
Meat	2700–3300	0.01–0.02	2–5	1–5
Vegetables	900–6500	0.50–2.8	2–5	1–5
Seafood	930–4500	0.67–1.1		20–500
Fruit (fresh)	620–3700	0.20–0.43		
Eggs	$\simeq 800$	6.1–14		
Dairy products	770–1500	0.19–0.30		
Sugar (white cane)	$\simeq 4$			
Sugar (brown cane)	$\simeq 1900$			

[a] To convert units to Bq per kg, multiply by 0.037.

[b] Source: Klement 1982.

[c] Source: Eisenbud 1987.

[d] Source: NCRP 1984a.

principal hazard associated with radon is due to short-lived daughter products, many of which are alpha-particle emitters causing localized exposure of the basal cells of the bronchial and pulmonary epithelia. The importance of the localized exposure is evident in the data of Table 4.1.

Decay of radon and its daughters in the atmosphere leads first to individual unattached ions or neutral atoms. The ions or atoms may become attached to aerosol particles, the attachment rate depending in a complex manner on the size distribution of the particles. Radioactive decay of an attached ion or atom, because of recoil, usually results in an unattached daughter ion or atom. The size distribution and the unattached fraction both affect lung exposure. Either attached or unattached species may be deposited (plate out) on surfaces, especially in indoor spaces, the rate depending on the surface to volume ratio of the space. Because of plate out, radon daughter products in the atmosphere are not likely to be in equilibrium with the parent. Attachment and lack of equilibrium cause relationships between radon concentrations and lung dose rates to be quite complicated. Special dose quantities such as the *working level* and the *equilibrium-equivalent radon concentration* are used when discussing air-borne alpha emitters.

^{222}Rn and its daughters ordinarily present a greater hazard than ^{220}Rn (thoron) and its daughters, largely because the much shorter half life of ^{220}Rn makes decay more likely prior to release into the atmosphere. Globally, the mean annual effective dose equivalent due to ^{222}Rn daughters is about 1 mSv (100 mrem) while that due to ^{220}Rn daughters is estimated to be about 0.2 mSv (20 mrem) (UN 1982). Relatively little is known about the rates of release, diffusion, inhalation, plate-out, attachment, and so on, for ^{220}Rn and its daughters, and most studies have emphasized the more important ^{222}Rn series. Consequently, this section deals primarily with ^{222}Rn and its daughters; however, the assessment procedures described apply to either decay series.

The 1982 UNSCEAR Report (UN 1982) provides a comprehensive review of what is known about the behavior of the radon decay series and the associated health-hazards. Unless otherwise indicated, data in Sec. 4.6 are drawn from that report.

4.6.2 Special Units and Dosimetry Concepts

Figures 4.9 and 4.10, and Tables 4.5 and 4.6 describe the two radon decay series. Health hazards are presented only by the relatively short-lived daughter products. While only ^{218}Po through ^{214}Po are important for the ^{222}Rn series, all radioactive daughters are important for the ^{220}Rn series.

The need for special dosimetry concepts may be inferred from consideration of the following scenario. A mixture of radon and daughters is inhaled. Radon itself, a noble gas, is exhaled. The daughters, either carried on aerosol particles or present in highly chemically reactive, unattached, atomic or ionic states, are retained in the respiratory system, decaying one into another, and depositing all their alpha-particle energies within the bounds of the respiratory system. Thus, it is not the (parent) radon concentration in the atmosphere which determines the dose to the respiratory system.

Rather, it is the concentration of daughter products. Quantitative interpretation of this situation requires the following new units and concepts.

Potential Alpha Energy Concentrations. Consider now the ^{222}Rn series. Denote the activity concentrations (Bq m^{-3}) of the four daughters ^{218}Po–^{214}Pb–^{214}Bi–^{214}Po by C_i, $i = 1$ to 4 and their decay constants by λ_i (s^{-1}). ^{214}Po releases $E_4 = 7.687$ MeV of alpha particle energy per decay. It thus has a potential alpha particle energy concentration of $E_4 C_4 / \lambda_4$ MeV m^{-3}. ^{214}Bi emits no alpha particles. However, once it decays to ^{214}Po, the alpha particle of the latter is released almost immediately. ^{214}Bi thus has a potential alpha particle energy concentration of $E_4 C_3 / \lambda_3$ MeV m^{-3}. That for ^{214}Pb is similarly $E_4 C_2 / \lambda_2$. ^{218}Po itself emits $E_1 = 6.003$ MeV of alpha particle energy per decay. Its potential alpha energy concentration is thus $(E_1 + E_4) C_1 / \lambda_1$. Thus the total potential alpha energy concentration is

$$E_{tot} = (E_1 + E_4) C_1 / \lambda_1 + E_4 (C_2 / \lambda_2 + C_3 / \lambda_3 + C_4 / \lambda_4). \qquad (4.2)$$

Working Level (WL). A WL is defined as a potential alpha energy concentration of 1.3×10^8 MeV m^{-3} for short lived radon daughters in air. Consider a ^{222}Rn activity concentration of 0.1 μCi m^{-3} (3.7 kBq m^{-3}) in radioactive equilibrium with short-lived daughters. Each daughter thus has the same activity concentration. Using data from Table 4.5 and Fig. 4.9, one finds that the total potential alpha energy concentration is

$$\begin{aligned}
E_{tot}^{equil} &= 3.7 \times 10^3 \left[\frac{6.003 + 7.687}{3.788 \times 10^{-3}} + \frac{7.687}{4.311 \times 10^{-4}} + \frac{7.687}{5.805 \times 10^{-4}} + \frac{7.687}{4.227 \times 10^3} \right] \\
&= 1.3 \times 10^8 \text{MeV m}^{-3}.
\end{aligned}$$

Thus, it is seen that 1 WL corresponds to an activity concentration of 3.7 kBq m^{-3} of ^{222}Rn in equilibrium with its daughters. For ^{220}Rn in equilibrium with its daughters, 1 WL corresponds to an activity concentration of 275 Bq m^{-3}. However, equilibrium is rarely approached, and the activity concentration of radon required to produce 1 WL is often considerably more.

Working Level Month (WLM). The hazard of breathing airborne radon daughters is correlated to the cumulative WL of the air breathed by an individual. This cumulative exposure over some time interval Δt is given by (with time measured in hours)

$$WLh = \int_{\Delta t} dt \, WL(t). \qquad (4.3)$$

Usually, this exposure is expressed in units of *working level months* (WLM) by $WLM = WLh/170$ where 170 is the number of hours in a working month (= 8 hours/day × 5 days/week × 4.25 weeks/month).

Equilibrium Factor (F). To account for the disequilibrium that usually exists between the daughters and the parent radon concentrations, *equilibrium factors* are used. An equilibrium factor may be defined for an individual daughter or for the series as a whole. For an individual daughter, the equilibrium factor f_i is the ratio of the actual daughter activity concentration to that of the parent, that is,

$$f_i = C_i / C_o \, , 1 = 1 \ldots 4. \qquad (4.4)$$

For the series as a whole, the equilibrium factor F may be defined as the ratio of the actual potential alpha energy concentration to that which would exist for daughters in equilibrium with the radon concentration.

$$F = \frac{E_{tot}}{E_{tot}^{equil}}. \tag{4.5}$$

For the ^{222}Rn series, typical individual equilibrium factors f_i are 0.9 for ^{218}Po and 0.7 for ^{214}Pb, ^{214}Bi, and ^{214}Po (NCRP 1984a). The corresponding overall equilibrium factor is $F = 0.712$. However, as a rule of thumb, the equilibrium factor $F = 0.5$ is commonly applied for indoor spaces in the absence of data indicating otherwise (EPA 1986).

Equilibrium Equivalent Concentration (EEC).   This is the activity concentration of radon in equilibrium with short-lived daughters which has the same total potential alpha energy concentration as the actual nonequilibrium mixture. The equilibrium equivalent (activity) concentration EEC and the actual activity concentration C_o of the radon parent are related by

$$EEC = F \times C_o. \tag{4.6}$$

Unattached Fraction.   The distribution of radon daughters lodged in the bronchial-lung system depends on whether or not the inhaled daughters are unattached (free atoms or ions) or attached to dust or aerosol particles. *Unattached fractions* may be defined in terms of activity concentrations of individual daughters or potential alpha energy concentrations for the series. For individual daughters of ^{222}Rn, 1 to 10% of ^{218}Po, ^{214}Pb, and ^{214}Bi atoms are unattached, while the very short-lived ^{214}Po atoms are all unattached. For estimates of respiratory system doses, an important parameter is the fraction f_p of the total potential alpha energy concentration attributable to unattached daughters. Typically, f_p is about 0.05.

4.6.3 Outdoor Radon Concentrations

The exhalation rate of ^{222}Rn from rocks and soils is highly variable, ranging from 0.2 to 70 mBq m^{-2} s^{-1}. An area-weighted average for continental areas, exclusive of Antarctica and Greenland, is 16 mBq m^{-2} s^{-1} (UN 1988). Exhalation from the surface of the sea is only about 1% of that from land areas. Rain, snow, and freezing decrease exhalation rates so that they are generally lower in winter than in summer. Barometric pressure and wind speed also affect exhalation rates, decreasing pressure or increasing wind speed causing the rate to increase.

Radon and daughter products are dispersed in the atmosphere by turbulent diffusion and convection. Extreme conditions of atmospheric stability can lead to ground-level concentrations differing by as much as a factor of 100. Associated with conditions of atmospheric stability are marked diurnal variations in ground-level concentrations, with minima at noon and maxima at midnight being associated respectively with greater and lesser atmospheric instability. Concentrations decrease with altitude, with those of ^{220}Rn decreasing more rapidly because of the shorter half-life of the parent. In the absence of precipitation, parent-daughter equilibrium is approached at elevations exceeding about 100 m. However, rainfall may remove daughter products from the atmosphere, causing absorbed dose rates in air at ground level to be as much as twice normal.

Mean annual concentrations of ^{222}Rn above continental areas range from 1 to 10 Bq m^{-3}, with 5 Bq m^{-3} and an equilibrium factor of 0.8 being typical (UN 1988). Typical mean annual concentrations over the ocean and over island areas are 0.1 Bq m^{-3}. Concentrations of ^{220}Rn daughters are typically 10% of those of ^{222}Rn daughters. Anomalously high levels often exist near coal-fired and geothermal power stations and near uranium-mine tailings.

4.6.4 Sources of Indoor Radon

Sources include exhalation from soil, building materials, water, ventilation air, and natural gas if unvented (as used in cooking). Exhalation from soil can be a significant contribution if a building has cracks or other penetrations in the basement structure or if the building has unpaved and unventilated crawl spaces. Water usage in buildings can present a significant source of Rn if concentrations in the water exceed about 10 kBq m^{-3}. In the United States, concentrations of ^{222}Rn in natural gas average 0.7 kBq m^{-3} but in some states the average is as high as 2 kBq m^{-3}. Radon precursor concentrations in building materials are highly variable. Greater concentrations occur in phosphogypsum and in concrete based on fly ash or alum shale. Sealing concrete surfaces with materials such as epoxy-resin paints greatly reduces radon exhalation.

Surveys of indoor radon concentrations reveal log-normal distributions, such as that shown in Fig. 4.11. A survey in Canada found median and average concentrations of 7.4 and 17 Bq m^{-3} (equilibrium equivalent ^{222}Rn) with an equilibrium factor of 0.52 ± 0.12. Geographic variations were significant, but variations within cities were greater than variations between cities. A survey in homes of U.S. physics professors (Cohen 1986) found geometric and arithmetic mean concentrations of 39 and 54 Bq m^{-3} which, assuming an equilibrium factor of 0.5, correspond to equilibrium concentrations of 20 and 27 Bq m^{-3}. The geometric standard deviation was found to be 2.36. As is indicated in Appendix F-2, for a log-normal distribution, 33 homes in 1000 would be expected to have concentrations in excess of 2 (geometric) standard deviations times the (geometric) mean and 13 homes in 10,000 would be expected to have concentrations in excess of 3 standard deviations times the mean. In the U.S. study, very little correlation was found between concentration and nearly all factors thought to be important in affecting radon levels — basement versus crawl space versus no space, integrity of the barrier between the ground and the house, windiness, draftiness, construction materials, use of natural gas, and, to some extent, even ventilation. Geographical variations were apparently so important as to overwhelm all other factors. A broader based U.S. study (Nero et al. 1986) found a log-normal distribution of ^{222}Rn concentrations in U.S. homes, with a geometric mean of 33 Bq m^{-3} and a geometric standard deviation of 2.8. For the United States as a whole, geometric and arithmetic mean radon concentrations are 1.53 and 3.06 pCi/L for living areas and 2.94 and 4.90 pCi/L for basements (Cohen and Shah 1991). Authors of this 1991 study report that the results may be high by about 20 percent because, among other reasons, high-rise apartments are not represented, urban areas are greatly underrepresented, and people concerned with environmental matters are overrepresented.

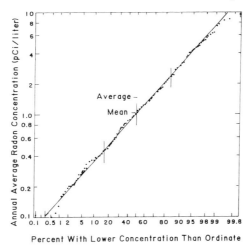

Figure 4.11 Distribution of radon activity concentrations in U.S. residential houses. The ordinate is the annual average radon level, and the abscissa is the probability of having a concentration less than the ordinate. The straight line is a log-normal distribution fitted to the data, and the three vertical lines show the median and ± one standard deviation. [From Cohen (1986); by permission of the Health Physics Society.]

The U.S. Environmental Protection Agency (EPA 1986) has recommended that remedial action be taken when concentrations of ^{222}Rn in the home exceed only 150 Bq m^{-3} (4 pCi L^{-1} or 0.02 WL). Methods for radon remediation, described in an EPA Guide (1987), vary widely in absolute and marginal cost and effectiveness. They range from sealing cracks and covering earth spaces, through various schemes for increasing ventilation, to elaborate house pressurization or subfoundation suction techniques. Moeller and Fujimoto (1984) have analyzed costs of various radon control measures. Measurement and control of radon in the home is also addressed by the NCRP (1988, 1989).

4.6.5 Material Balances for Indoor Radon

Indoor airborne radon and daughter concentrations depend not only on exhalation of the parent gas from surfaces, but also on intake and loss of radionuclides through ventilation and plate-out of daughters on interior surfaces. The rate of plate-out depends on the unattached fraction which is influenced greatly by humidity and by the presence of aerosols (smoke, dust, etc.). The discussion to follow establishes the mathematical framework of a lumped parameter model for the analysis of indoor ^{222}Rn and its daughter concentrations. Example results are given for a representative steady-state case.

Notation. Let C_i represent the indoor airborne activity concentration (Bq m^{-3}) of species i, with $i = 0$ representing the parent ^{222}Rn and $i = 1, \ldots, 4$ the short-lived daughters. Let Q_i similarly represent outdoor activity concentrations present in incoming ventilation air. Let superscripts o and $+$, respectively, identify unattached and attached species. It is assumed that each daughter, when formed by decay of its parent, begins life as an unattached species. Let V (m^3) represent the indoor air volume in question and S (m^2) the surface area through which exhalation of ^{222}Rn takes place (also taken here as the surface area on which plate-out occurs). Mixing of the indoor air is assumed to be sufficiently rapid that any spatial gradients in concentrations can be neglected and a lumped-parameter model used.

Let the average exhalation rate be W_o (Bq m^{-2} s^{-1}), λ_i (s^{-1}) the radioactive decay constant for species i, and λ_v (s^{-1}) the ventilation rate constant, that is, the reciprocal of the time required for complete air exchange (typically 1 to 2 hours). Let λ_a (s^{-1}) represent the rate constant for attachment, that is, the reciprocal of the mean lifetime of an unattached radioactive ion. In clean residential air, a representative unattached lifetime is 100 s. However, in very dusty air, the lifetime may be as short as about 4 s. The diffusion and plate-out of daughters on surfaces is governed by a rate constant $\lambda_d = v_d S/V$, in which v_d is the deposition velocity (m s^{-1}). For unattached daughters, v_d^o is on the order of 0.001 to 0.01 m s^{-1}; for attached daughters, v_d^+ is about 0.00001 to 0.0001 m s^{-1} in clean residential air (UN 1982). Finally, let A_o (Bq s^{-1}) represent an extraneous source of ^{222}Rn from water or natural gas.

Balance Equations.   For the parent ^{222}Rn, which is always unattached,

$$\frac{dC_o^o}{dt} = \frac{W_o S}{V} - \lambda_o C_o^o - \lambda_v C_o^o + \lambda_v Q_o^o + \frac{A_o}{V}. \tag{4.7}$$

For unattached and attached daughters, respectively,

$$\frac{dC_i^o}{dt} = \lambda_i(C_{i-1}^o + C_{i-1}^+) - (\lambda_i + \lambda_a + \lambda_d^o + \lambda_v)C_i^o + \lambda_v Q_i^o, \tag{4.8}$$

$$\frac{dC_i^+}{dt} = \lambda_a C_i^o - (\lambda_i + \lambda_d^+ + \lambda_v)C_i^+ + \lambda_v Q_i^+. \tag{4.9}$$

where $i = 1, \ldots, 4$ and $C_o^+ = 0$ since there is no attached ^{222}Rn.

Steady-State Relationships.   At steady state, time derivatives are zero, so that the previous equations yield

$$C_o^o = \frac{(W_o S + A_o)/V + \lambda_v Q_o^o}{\lambda_o + \lambda_v}, \tag{4.10}$$

$$C_i^o = \frac{\lambda_i\left(C_{i-1}^o + C_{i-1}^+\right) + \lambda_v Q_i^o}{\lambda_i + \lambda_a + \lambda_d^o + \lambda_v}, \tag{4.11}$$

$$C_i^+ = \frac{\lambda_a C_i^o + \lambda_v Q_i^+}{\lambda_i + \lambda_v + \lambda_d^+}. \tag{4.12}$$

Sample Calculation.   A set of sample calculations is presented for the reference house discussed by UNSCEAR (UN 1982) in their review of indoor radon. Characteristics of the house are as follows: $V = 200$ m^3, $S = 350$ m^2, $W_o = 0.002$ Bq m^{-2} s^{-1}, $\lambda_v = 0.5$ h$^{-1} = 0.000139$ s^{-1}, $Q_o^o = 4$ Bq m^{-3}, and $A_o = 0.081$ Bq s^{-1}.

Reference values assumed for other model parameters are as follows: $\lambda_a = 0.01$ $s^{-1}, v_d^+ = 0.000035$ m s$^{-1}, v_d^o = 0.003$ m s^{-1}. For outdoor air, the ratios of activity concentrations are assumed to be 1.0–0.9–0.7–0.7–0.7 for ^{222}Rn–^{218}Po–^{214}Pb–^{214}Bi–^{214}Po. The unattached fractions in outdoor air are assumed to be 0.07 for ^{218}Po, 1.0 for ^{214}Po, and 0.0 otherwise.

Results of steady-state calculations for individual nuclides are as follows:

	^{222}Rn	^{218}Po	^{214}Pb	^{214}Bi	^{214}Po
Concentration (Bq m^{-3})					
unattached	31.6	6.3	0.6	0.4	5.9
attached		15.8	10.1	5.5	0.0
total	31.6	22.0	10.7	5.9	5.9
Equilibrium factor		0.697	0.339	0.186	0.186
Unattached fraction		0.283	0.056	0.066	1.000

Results for the daughter products as a whole are as follows:

Potential alpha energy concentration	3.5×10^5 MeV m^{-3}
Working level	0.00268
Equilibrium equivalent ^{222}Rn concentration	10.1 Bq m^{-3}
Equilibrium factor	0.318
Energy unattached fraction	0.11

4.6.6 Dosimetry for Radon Daughters

Estimates of radiation doses to the respiratory system resulting from inhalation of radon daughters not only are of intrinsic interest, but also are essential elements of risk analysis. We examine here three methods of estimation. In decreasing order of sophistication, these are the methods of the NCRP, the ICRP, and UNSCEAR.

At greatest risk on inhalation of radon daughter products are the basal cells of the tracheobronchial epithelium (bronchial dose). Their minimal depth below the mucus surface is about 35 μm (NCRP 1984b), well within the range of the 7.7 MeV alpha particle emitted by ^{210}Po. Also at risk is the alveolar tissue of the pulmonary region (pulmonary dose). The bronchial dose is very sensitive to the aerosol *activity mean diameter* (AMD), as discussed in Sec. 8.7.1, and the pulmonary dose is very sensitive to the unattached fraction f_p for radon daughter products. Under usual exposure conditions, the pulmonary dose is only about one-eighth the bronchial dose (ICRP 1987).

Many factors affect dose estimates. Factors related to the radiation source are degree of daughter disequilibrium, percent unattachment, particle deposition model, particle size distribution, and method of dose calculation. Factors related to the individual are clearance rates of radioactive materials from the lungs, target-tissue geometry, and subject activity (breathing patterns, smoking habits, etc.).

The NCRP (1984b) has developed a dosimetry model which is sufficiently comprehensive to permit sensitivity studies but which is also readily simplified to accommodate average or representative exposure conditions. This dosimetry model relates lung dose to exposure, that is, to the time integral of activity inhaled or to the WLM measure. The relationship between lung dose and the probability of cancer induction requires additional factors such as age at exposure and age at expression. Both the NCRP and ICRP studies cited previously address dose-exposure relationships as well as cancer risks and population risk estimates.

The NCRP model uses breathing rates adapted from ICRP's Reference Man, namely,

	Liters per Minute		
	Male	Female	Child (age 10)
Light activity	18.75	14.1	14.4
Resting	9.	4.8	4.8

This model is also based on estimates of steady-state activity per unit area on the bronchial tree, accounting for particle deposition, clearance by mucociliary action, and radioactive decay. The model uses a 0.125 μm AMD and dose evaluation at a depth of 22 μm below the surface of the bronchial epithelium (35 μm below the surface of the 13-μm thick mucous layer). Table 4.14 lists calculated dose conversion factors which apply to normal environmental situations. In mines, the AMD is 2 to 4 times greater and the bronchial dose per unit exposure is less. Conversion factors for the child are greater than those for the adult because of the smaller airway area in the child's bronchial tree. Conversion factors for unattached ^{218}Po exceed those for the attached nuclide because of greatly reduced deposition of the very small particles in the nasal and upper tracheal regions.

Dose conversion factors for representative atmospheres may be generated from the data of Table 4.14. The NCRP chose activity concentration ratios for ^{222}Rn–^{218}Po–^{214}Pb–^{214}Bi–^{214}Po of 1.0–0.9–0.7–0.7–0.7 and an activity concentration of unattached ^{218}Po equal to 7 percent of that for ^{222}Rn. The corresponding overall equilibrium factor is 0.712 and 1 pCi m^{-3} of ^{222}Rn is equivalent to an equilibrium-equivalent activity concentration of 1.40 pCi m^{-3}. Dose equivalents given in Table 4.15 are based on a quality factor of 20.

Also listed in Table 4.15 are dose conversion factors based on WLM units. The variability of these conversion factors raises questions about the suitability of the WLM as an exposure standard. It has been suggested (NCRP 1984b) that, while the WL adequately correlates exposure and dose in mine atmospheres (about 0.5 rad per WLM), its use is questionable for environmental atmospheres unless some allowance is made for the unattached fraction in the WL definition.

The ICRP (1987) bases dose conversion factors on assumptions somewhat different from those of the NCRP. Factors are for Reference Man and assume breathing volumes of 1 m^3 h^{-1} outdoors and 0.75 m^3 h^{-1} indoors, an AMD of 0.15 μm, and an

TABLE 4.14 Detailed NCRP Conversion Factors for Bronchial Absorbed Dose Arising from ^{222}Rn Daughter Products in Environmental Atmospheres. ^{218}Po^o Represents the Concentration of Unattached Atoms or Ions. All Other Species Are Assumed to be Present only in Attached Form.

SUBJECT AND ACTIVITY	(mrad y^{-1}) per (pCi m^{-3})^a			
	^{218}Po^o	^{218}Po	^{214}Pb	^{214}Po
Male				
Light activity	0.98	0.029	0.16	0.14
Resting	0.32	0.022	0.12	0.10
Female				
Light activity	0.82	0.029	0.16	0.14
Resting	0.29	0.019	0.10	0.09
10-Year Old Child				
Light activity	2.36	0.060	0.26	0.28
Resting	0.54	0.040	0.17	0.18

^{a}To convert units to (mGy y^{-1}) per (Bq m^{-3}), multiply factors by 0.270.

Source: NCRP 1984b.

TABLE 4.15 NCRP Dose Conversion Factors for ^{222}Rn Daughter Products with Activities in the Following Ratios: ^{222}Rn/^{218}Po/^{214}Pb/^{214}Po = 1./0.9/0.7/0.7. The Activity Concentration of Unattached ^{218}Po is 7 Percent of That of ^{222}Rn. The Quality Factor is Assumed to be 20.

SUBJECT AND ACTIVITY	DOSE PER UNIT EQUILIBRIUM-EQUIVALENT ACTIVITY CONCENTRATION^a		DOSE PER UNIT WLM EXPOSURE^b
	$\dfrac{\text{mrad y}^{-1}}{\text{pCi m}^{-3}}$	$\dfrac{\text{mSv y}^{-1}}{\text{Bq m}^{-3}}$	$\dfrac{\text{rad}}{\text{WLM}}$
Male			
Light activity	0.22	1.2	0.82
Resting	0.14	0.75	0.53
Female			
Light activity	0.21	1.1	0.53
Resting	0.12	0.65	0.46
10-Year Old Child			
Light activity	0.42	2.3	1.6
Resting	0.22	1.2	0.86

^{a}To convert from mrad pCi^{-1} to mGy Bq^{-1}, multiply by 0.270.
^{b}To convert from rad per WLM to Sv per WLM, multiply by 0.01 × 20.
Source: Based on Table 4.14.

unattached fraction $f_p = 0.03$ for the total potential alpha particle energy. Dose conversion factors are summarized in Table 4.16. In the table, the bronchial dose factor of 0.90 mSv y^{-1} per Bq m^{-3}, for example, converts to 0.32 rad per WLM.

TABLE 4.16   ICRP Dose-Equivalent Conversion Factors for Indoor and Outdoor Exposure of Adults to Radon Daughters under Reference Conditions. The Quality Factor is Equal to 20.

	ANNUAL DOSE EQUIVALENT PER UNIT EQUILIBRIUM EQUIVALENT CONCENTRATION (mSv y^{-1} per Bq m^{-3})[a]	
	BRONCHIAL DOSE	PULMONARY DOSE
^{222}Rn daughters		
indoors at home	0.90	0.12
indoors elsewhere	0.23	0.030
outdoors	0.20	0.027
^{220}Rn Daughters		
indoors at home	3.0	0.90
indoors elsewhere	0.75	0.23
outdoors	0.50	0.15

[a]To convert to units of rad per WLM, multiply by 0.359.

Source: ICRP 1987.

The UNSCEAR dose estimation model (UN 1988) is based on the following assumptions: breathing rate = 0.8 m^3 h^{-1} indoors (1.0 m^3 h^{-1} outdoors), AMD = 0.2μm indoors and outdoors, and f_p = 0.025 indoors and outdoors. For both indoor and outdoor exposure, central estimates of absorbed dose rates per unit equilibrium concentration are: 7 nGy h^{-1} per Bq m^{-3} [0.4 rad per WLM] for the bronchial dose, and 0.9 nGy h^{-1} per Bq m^{-3} [0.06 rad per WLM] for the pulmonary dose.

Uncertainties are great in radon-dose estimation. The NCRP (1984b) lists the following percent variations about central dose estimates associated with extremes in the various factors affecting the estimates.

Factor	Percent Variation
Unattached fraction for ^{218}Po	$-20, +10$
Reduction in equilibrium factor	-20
Aerosol activity median diameter	$-20, +100$
Breathing pattern (mouth vs. nose)	$-20, +35$
Child vs. adult	$+60$
Stasis in mucociliary clearance	$+10$
Mucus thickness	$-30, +20$

The dose estimates just reviewed are for absorbed dose in the most sensitive cells of the bronchial and pulmonary compartments of the respiratory system. Conversion to dose equivalent in the same cells simply requires application of a quality factor of 20 for alpha particles. Conversion of dose equivalent to effective dose equivalent is not so straightforward. The weight factor w_T for the "lung" is 0.12 in the ICRP-30 methodology for calculation of the effective dose equivalent (see Chaps. 1 and 3). The

Lung Model Task Group of the ICRP (Masse and Cross 1989) recommends a weight factor of 0.08 to the bronchial dose and 0.02 to the pulmonary dose. Of the total weight factor of 0.12, the remainder is partitioned as follows: nose and nasopharynx 0.003, mouth and oropharynx 0.012, and larynx 0.005. UNSCEAR recommends that weight factors of 0.06 be applied individually to the bronchial and pulmonary dose equivalents, the sum of products yielding the effective dose equivalent.

Sample Calculation.   Suppose that a person is exposed over a very long period of time to an average ^{222}Rn *EEC* of 8.5 Bq m^{-3}. Suppose further that the exposure occurs indoors with the person at rest. What is the resulting steady-state bronchial absorbed dose rate?

The NCRP model (Table 4.15) would estimate $0.14 \times 0.27 \times 8.5 = 0.32$ mGy y^{-1} for the male and 0.27 mGy y^{-1} for the female. This estimate is based on breathing rates of 0.54 and 0.29 m^3 h^{-1}, respectively, for the male and female and on an AMD of 0.125 μm.

The ICRP model (Table 4.16), which does not distinguish male and female, would estimate $0.9 \times 8.5 \div 20 = 0.38$ mGy y^{-1}, in which the divisor of 20 is the quality factor. This estimate is based on a breathing rate of 0.75 m^3 h^{-1} and on an AMD of 0.15 μm. In comparing this estimate with that of the NCRP model, one should note that the greater breathing rate would lead to a higher estimate of absorbed dose rate, while the greater AMD would lead to a lower estimate of the absorbed dose rate.

The UNSCEAR model, unspecific as to sex and activity, would estimate $7 \times 8.5 = 60$ nGy h^{-1} = 0.52 mGy y^{-1}. This is a central estimate, with bounds of about $\pm$ 30% and is based on a breathing rate of 0.8 m^3 h^{-1} and an AMD of 0.2 μm.

4.6.7 Lung Cancer Risks for Radon Exposure

The NCRP, the ICRP, and the U.S. National Academy of Sciences (NCRP 1984b; ICRP 1987; NAS 1988, 1991) have published comprehensive reviews of epidemiological studies of radiation-induced lung cancer, particularly among underground miners. The results are tables of risk factors, some of which are age and sex specific, take into account both age at exposure and age at diagnosis, and take into account concomitant effects of smoking and inhalation of radon daughter products.

Details of the risk estimation methodology are beyond the scope of this text. Here we simply summarize risk factors in terms of exposure rather than absorbed dose or dose equivalent to the tissues of the respiratory system.

The BEIR-IV (NAS 1988) method for long-time chronic exposure employs risk factors expressed as *relative risk coefficients*, R_r/R_o, where R_o is the base-line lung cancer risk (lifetime mortality probability) and R_r is the radiogenic cancer risk (lifetime mortality probability per unit exposure expressed as either WLM or Bq h m^{-3}). Risk coefficients are presented in Table 4.17. When risks are presented in the relative sense, there is agreement between the NAS coefficients (NAS 1988) and those of the ICRP (1987). However, there are substantial differences between the baseline risks adopted by the two bodies. The reader is cautioned that these risk coefficients are based on a linear, no-threshold, dose-response relationship derived for the most part from data for relatively high exposures. For a discussion of the uncertainties in the

TABLE 4.17 Relative Risk Coefficients for Chronic Lifetime Exposure to ^{222}Rn Daughter Products. Coefficients Are Derived from the U.S. National Academy of Sciences Review (NAS 1988) Using Data for 0.1 WLM Annual Exposures. Coefficients Differ Slightly for Other Annual Exposures. Data in Brackets Are from the ICRP Review of Risks (ICRP 1987).

POPULATION	LIFE EXPECTANCY (years)	BASELINE CANCER RISK R_o	RELATIVE EXCESS RISK PER WLM ANNUALLY R_r/R_o
General			
Male	69.7 [70]	0.067 [0.042]	0.52
Female	76.4 [75]	0.025 [0.009]	0.56
Mixed	73.1 [72.5]	0.046 [0.026]	0.54 [0.5]
Nonsmokers			
Male	70.5	0.011	0.92
Female	76.7	0.0060	0.92
Smokers			
Male	69.0	0.123	0.83
Female	75.9	0.058	0.88

estimates, the reader is referred to the previously cited review reports. Reassessment of data (NAS 1991) has taken into account higher breathing rates of miners as well as differences between aerosol-particle sizes in homes and mines. In comparison with previous studies, risks as measured in terms of alpha-particle energy dose per unit exposure, were found to be about 30 percent lower for adults and about 20 percent lower for children.

Table 4.18 summarizes NCRP risk estimates for environmental, not occupational, exposure conditions, taking into account age at commencement of exposure as well as duration of exposure. The data are based on the characteristics of the 1975 population of the United States and do not distinguish between smokers and nonsmokers.

Sample Calculation.   A measure of the possible consequences of exposure to radon daughter products is revealed by the following example. Suppose a large population were exposed 85 percent of the time to ^{222}Rn *EEC*'s of 10 Bq m^{-3}. This translates to annual exposure of $(0.85 \times 8766 \text{ h}) \times (10 \text{ Bq m}^{-3}) = 74{,}500 \text{ Bq h m}^{-3}$. Since 1 WLM corresponds to $(170 \text{ h}) \times (3700 \text{ Bq m}^{-3}) = 629{,}000 \text{ Bq h m}^{-3}$, the average annual population exposure equates to $74{,}500 \div 629{,}000 = 0.12 \text{ WLM}$.

Using data from Table 4.17, and incorporating the BEIR-IV (NAS 1988) baseline cancer risk, one would estimate a relative excess risk of $0.12 \times 0.54 = 0.065$. On the basis of a baseline risk of 0.046, the excess risk is $0.065 \times 0.046 = 0.0030$. For a population of 10^6, this leads to an expectation of $0.0030 \times 10^6 = 3000$ excess lung-cancer fatalities compared to the baseline of 46,000. For a static population of 10^6 with a life expectancy of 73 years, the annual excess fatalities would be $3000 \div 73 = 41$ compared to the baseline annual rate of $46{,}000 \div 73 = 630$ lung-cancer fatalities.

Using data from Table 4.17, but incorporating the ICRP baseline cancer risk and relative excess risk, one would estimate a relative excess risk of $0.12 \times 0.50 = 0.060$.

TABLE 4.18 Lifetime Lung Cancer Risk Per WLM Annual Exposure as a Function of Duration of Exposure and Age at First Exposure. Data Are for the U.S. Population of 1975 and Apply to Environmental Exposure, Not Exposure in Underground Mining.

AGE (y) AT FIRST EXPOSURE	EXPOSURE DURATION (y)				
	1	5	10	30	Life
1	0.000064	0.00034	0.00077	0.0034	0.0091
10	0.000091	0.00050	0.0011	0.0048	0.0091
20	0.00013	0.00069	0.0015	0.0055	0.0077
30	0.00018	0.00098	0.0021	0.0055	0.0077
40	0.00021	0.0010	0.0020	0.0042	0.0045
50	0.00017	0.00084	0.0014	0.0025	0.0027
60	0.00013	0.00055	0.00091	0.0013	0.0013
70	0.000070	0.00028	0.00038	0.00038	0.00038

Source: NCRP 1984b; by permission of the NCRP.

On the basis of a baseline risk of 0.026, the excess risk is $0.060 \times 0.026 = 0.0016$. For a population of 10^6, this leads to an expectation of $0.0016 \times 10^6 = 1600$ excess lung-cancer fatalities compared to the baseline of 26,000.

Using data from Table 4.18 for lifetime exposure beginning at age 1 and an annual exposure of 0.12 WLM, one would estimate a lifetime cancer risk of $0.12 \times 0.0091 = 0.0011$. This estimate—1100 excess fatalities in a population of 10^6 — is more in line with the estimate based on the ICRP data from Table 4.17. The variations among the estimates illustrate not only differences between dosimetry and risk-estimate methodology but also differences in baseline cancer risks.

REFERENCES

BENTON, E.V., "Summary of Radiation Dosimetry Results on Manned Spacecraft,"*Adv. Space. Res. 6*, 315 (1986).

COHEN, B.L, "A National Survey of ^{222}Rn in U.S. Homes and Correlating Factors," *Health Physics 51*, 175–183 (1986).

COHEN, B.L, AND R.S. SHAH, "Radon Levels in United States Homes by States and Counties," *Health Physics 60*, 243–259 (1991).

CURTIS, S.B., W. ATWELL, R. BEEVER, and A. HARDY, "Radiation Environments and Absorbed Dose Estimates on Manned Space Missions," *Adv. Space. Res. 6,* 269 (1986).

EISENBUD, M., *Environmental Radioactivity*, 3d ed., Academic Press, Orlando, Fl., 1987.

EPA, *A Citizen's Guide to Radon*, Report OPA-86-004 of the U.S. Environmental Protection Agency and the Centers for Disease Control of the U.S. Department of Health and Human Services, U.S. Government Printing Office, Washington, D.C., 1986.

EPA, *Radon Reduction Methods*, 2d ed., Report OPA-87-010 of the U.S. Environmental Protection Agency, U.S. Government Printing Office, Washington, D.C., 1987.

HAFFNER, J.W., *Radiation and Shielding in Space*, Academic Press, New York, 1967.

HESS, C.T., J. MICHEL, T.R. HORTON, H.M. PRICHARD, AND W.A. CONIGLIO, "The Occurrence of Radioactivity in Public Water Supplies in the United States," *Health Physics 48*, 552–586 (1985).

ICRP, *Report of the Task Group on Reference Man*, Report 23, International Commission on Radiological Protection, Pergamon Press, Oxford, 1975.

ICRP, "Lung Cancer Risk from Indoor Exposure to Radon Daughters," *Annals of the ICRP 17*, No. 1, ICRP Publication 50, International Commission on Radiological Protection, Pergamon Press, Oxford, 1987.

KLEMENT, A.W., JR., (ED.), *CRC Handbook of Environmental Radiation*, CRC Press, Boca Raton, Fl., 1982.

KOCHER, D.C., *Radioactive Decay Data Tables*, Report DOE/TIC-11026, U.S. Department of Energy, Washington, D.C., 1981.

MASSE, R., AND F.T. CROSS, "Risk Considerations Related to Lung Modeling," *Health Physics 57*, Supp. 1, 283–289 (1989).

MOELLER, D.W. AND K. FUJIMOTO, "Cost Evaluation of Control Measures for Indoor Radon Progeny," *Health Physics 46*, 1181–1193 (1984).

MORGAN, K.Z., "History of Damage and Protection from Ionizing Radiation," in *Principles of Radiation Protection*, K.Z. MORGAN AND J.E. TURNER (EDS.), John Wiley & Sons, New York, 1967.

NAS, *Radioactivity in the Marine Environment*, Report of the Panel on Radioactivity in the Marine Environment, Committee on Oceanography, National Research Council, National Academy of Sciences, Washington, D.C., 1971.

NAS, *The Effects on Populations of Exposure to Low Levels of Ionizing Radiation*, Report of the BEIR Committee [The BEIR-III Report], National Research Council, National Academy of Sciences, Washington, D.C., 1980.

NAS, *Health Risks of Radon and Other Internally Deposited Alpha-Emitters*, Report of the BEIR Committee [The BEIR-IV Report], National Research Council, National Academy of Sciences, Washington, D.C., 1988.

NAS, *Comparative Dosimetry of Radon in Mines and Homes*, National Research Council, National Academy of Sciences, Washington, D.C., 1991.

NCRP, *Natural Background Radiation in the United States*, Report 45, National Council on Radiation Protection and Measurements, Washington, D.C., 1975.

NCRP, *Exposures from the Uranium Series with Emphasis on Radon and its Daughters*, Report 77, National Council on Radiation Protection and Measurements, Washington, D.C., 1984a.

NCRP, *Evaluation of Occupational and Environmental Exposures to Radon and Radon Daughters*, Report 78, National Council on Radiation Protection and Measurements, Washington, D.C., 1984b.

NCRP, *Ionizing Radiation Exposure of the Population of the United States*, Report 93, National Council on Radiation Protection and Measurements, Washington, D.C., 1987a.

NCRP, *Exposure of the Population in the United States and Canada from Natural Background Radiation*, Report 94, National Council on Radiation Protection and Measurements, Washington, D.C., 1987b.

NCRP, *Radiation Exposure of the U.S. Population from Consumer Products and Miscellaneous Sources*, Report 95, National Council on Radiation Protection and Measurements, Washington, D.C., 1987c.

NCRP, *Measurement of Radon and Radon Daughters in Air*, Report 97, National Council on Radiation Protection and Measurements, Washington, D.C., 1988.

NCRP, *Guidance on Radiation Received in Space Activities,* Report 98, National Council on Radiation Protection and Measurements, Washington, D.C., 1989a.

NCRP, *Control of Radon In Houses*, Report 103, National Council on Radiation Protection and Measurement, 1989b.

NERO, A.V., M.B. SCHWEHR, W.W. NAZAROFF, AND K.L. REZVAN, "Distribution of Airborne Radon-222 Concentrations in U.S. Homes," *Science 234*, 992–997 (1986).

UN, *Report of the United Nations Scientific Committee on the Effects of Atomic Radiation*, New York, 1977.

UN, *Report of the United Nations Scientific Committee on the Effects of Atomic Radiation*, New York, 1982.

UN, *Report of the United Nations Scientific Committee on the Effects of Atomic Radiation*, New York, 1988.

PROBLEMS

1. Estimate the change in exposure (mrem y^{-1}) a person would experience in moving from New York to Denver (i.e., from sea-level to 5000 ft elevation). With the data of the previous chapter, what is the increased risk of cancer mortality if a person lived in Denver instead of New York (assuming all other sources of background radiation are the same)?

2. The global inventory of natural tritium is estimated to be 3.5 kg. At what rate (Ci y^{-1}) is tritium introduced globally by natural phenomena?

3. The global inventory of ^{14}C is about 8500 PBq. If all that inventory is a result of cosmic ray interactions in the atmosphere, how many kilograms of ^{14}C are produced each year in the atmosphere?

4. Estimate the equilibrium global mass of ^{22}Na in the biosphere resulting from cosmic rays.

5. What mass of natural potassium is in the average human? Assume Reference Man with a mass of 70 kg. How many radioactive decays of ^{40}K occur per second in this body?

6. For coal with a uranium content of 20 Bq per kg, how much ^{220}Rn (expressed in units of both g and Ci) will be released by the mining and subsequent burning of 14 tons of coal (the annual energy equivalent consumed per capita in the United States)?

7. Consider a two-story house 20×20 m in size with walls 8 m high and a basement 3 m deep. The basement floor and walls are of concrete 30 cm thick, and the outside walls are brick 10 cm thick. Plaster 1 cm thick lines all walls and the ceilings. Estimate the number of Bq of ^{40}K, ^{226}Ra and ^{232}Th in the structural materials of the house.

8. Estimate the increased probability of cancer mortality for a male (as a percent of normal cancer mortality) caused by radiation exposure as a result of the following situations. State all assumptions made.

(a) He becomes a commercial airline pilot at age 35 and continues to make, on the average, four 3000-mile trips a week until retirement at age 65.

(b) He becomes a NASA shuttle pilot at age 40, and in the next four years he makes three 5-day shuttle flights with maximum altitudes of 275, 400, and 500 km.

(c) He receives the average annual internal absorbed dose from cosmogenic radionuclides.

(d) He receives an annual whole-body background dose from natural sources of 1 mSv/y (100 mrem/y).

9. The ^{222}Rn concentration inside buildings is measured at 0.5 pCi per liter with an overall equilibrium factor for its daughters of 0.7. (a) What is the total alpha-particle energy release rate in air (MeV L^{-1} s^{-1})? (b) What is the concentration in working level (WL) units?

10. ^{222}Rn diffuses into a building with an enclosed air volume of 350 m^3. Ventilation of the building results in one complete air exchange per hour. Ventilation air contains 1 Bq m^{-3} of ^{222}Rn. The radon is present at a steady-state specific activity of 0.5 pCi L^{-1}. What is the rate (pCi h^{-1}) at which radon diffuses into the building?

11. The ^{220}Rn (thoron) series (see Fig. 4.9) can be approximated by the following simplified decay scheme by neglecting the short-lived daughters ^{216}Po, ^{212}Po, and ^{208}Tl and assuming all their alpha decay energy is emitted at the time of decay of their parents.

$$^{220}\text{Rn} \xrightarrow{2\alpha} {}^{212}\text{Pb} \xrightarrow{\beta} {}^{212}\text{Bi} \xrightarrow{\alpha} {}^{208}\text{Pb}$$

(a) What is the average energy (MeV) of the alpha particles emitted when a ^{212}Bi nuclide decays to ^{208}Pb?

(b) Derive expressions for indoor activity concentrations of ^{220}Rn and its daughters (using the previous simplified model), in attached and unattached form, analogous to those for ^{222}Rn given in Sec. 4.6.5.

(c) What value would you expect for the attachment fraction of each of the nuclides in this decay chain?

12. Calculate the working level (WL) for a ^{220}Rn (thoron) concentration of 0.1 μCi m^{-3}. Assume equilibrium exists among the parent and daughters.

13. Compute the WL, equilibrium factor and energy unattached fraction for ^{222}Rn daughters as a whole for the air in a basement with a volume of 105 m^3 and a surface area of 210 m^2. Exhalation rate $W_o = .0015$ Bq m^{-2} s^{-1}, $\lambda_v = 0.7$ h^{-1}, $Q_o = 3$ Bq m^{-3}, and $A_o = 0.02$ Bq s^{-1}. For the outdoor air activity concentrations for ^{222}Rn/^{218}Po/^{214}Pb/^{214}Bi/^{214}Po are 1/.92/.75/.65/.65, and the unattached fractions are 1/.09/.01/.01/1. Other model parameters are assumed to have the following values: $\lambda_a = 0.045$, $v_d^+ = .00002$ m s^{-1}, and $v_d^o = .002$ m s^{-1}.

14. The distribution of radon concentration x in U.S. homes, shown in Fig. 4.11, has median and average values of 1.05 and 1.47 pCi L^{-1}, respectively.

(a) Estimate the parameters μ and σ for the log-normal distribution

$$P(x) = \frac{\exp[-(\ln(x) - \mu)^2/(2\sigma^2)]}{x\,\sigma\sqrt{2\pi}}.$$

(b) What fraction of homes have concentrations exceeding 2 pCi L^{-1}? Compare this result to the value obtained from Fig. 4.11.

15. An individual lives in a home with an average indoor *EEC* of ^{220}Rn of 4 pCi L^{-1} and an average outdoor *EEC* of 0.1 pCi L^{-1}. If 14 hours a day are spent indoors, what is the individuals annual exposure in WLM? What is the probability the individual will eventually die of lung cancer as a result of exposure to radon daughters? State all assumptions and approximations.

Exposure to Artificial and Technologically Enhanced Natural Radiation Sources

5.1 INTRODUCTION

The purpose of this chapter is to provide the reader with an appreciation of the significance of human exposure to artificial radiation sources and to natural sources whose concentrations and effects are enhanced technologically. Occupational exposures are reviewed briefly, but primary consideration is given to exposure of the general public.

Sources are considered in the approximate order of their introduction to society. Very shortly after their discoveries, radium and x rays were used for medical purposes, thus medical exposures are examined first. Beginning in 1945, radioactive debris from nuclear weapons tests became an important source of population exposure to radiation. While atmospheric testing of nuclear weapons has greatly diminished since 1962, there remains a substantial legacy of long-lived radionuclides in the environment resulting from the intensive testing carried out in the 1951 to 1962 decade. Nuclear power came into commercial use about 1960, and various activities, from uranium mining to reactor operations, contribute to population exposure to radiation. Natural radioactivity present in coal and other natural resources is dispersed by combustion in power generation and other industrial and domestic activities. These sources of population exposure are also considered in this chapter. Finally, many consumer products contain radioactive materials or emit x rays. While individual exposures from these sources are negligible, collective doses to large populations may be significant, and therefore, deserve examination.

The widespread use of radiation sources and radioactive materials necessarily leads to radioactive wastes whose disposal must be carefully controlled. Detailed descriptions and analysis of the many possible radioactive waste technologies and management schemes for the different types of radioactive wastes produced by human activity is beyond the scope of this text. In this chapter, only the basic principles and

204

general criteria for treating and regulating radioactive wastes are summarized with emphasis being given to wastes from the nuclear power industry.

5.2 MEDICAL SOURCES

There are three broad categories of medical procedures resulting in human radiation exposure: (1) diagnostic x-ray examinations, including mammography and computed tomographic (CT) scans, (2) diagnostic nuclear medicine, and (3) radiation therapy. Collective exposures from diagnostic x rays dominate all other medical exposures. Neither precise estimates of global population exposures and public health consequences nor intercomparison of national practices are at present possible because of differences in reporting practices and methods of averaging doses.

Of all the radiation exposures to the general public arising from human activity, the greatest is due to medical procedures. Medical exposures are delivered at high dose rates and cause the greatest individual organ doses short of those resulting from major accidental exposures. Medical exposures differ from others in that they commonly involve only limited regions of the body. Of course, benefits from medical radiation exposures are high — so high that there is a temptation to ignore the risks. Some may argue that recipients of medical exposure are a small subgroup within the population and that their risks from radiation exposure are far outweighed by the risks of foregoing the radiation procedures. This argument has greatest validity for therapeutic uses of radiation. However, the population subgroup receiving diagnostic x rays is not small. In the United States, some 130 million persons annually receive medical x-ray exposures, of which only 400 thousand persons receive radiation therapy (NAS 1980). The scope for minimizing hazards of medical radiation exposures is indeed great. Risks of both hereditary and somatic illness are of concern, but hereditary illness has received special attention because of the difficulty in balancing benefits to one generation with detriments to succeeding generations. Both technological improvements and procedural refinements are being applied to dose reduction, with the International Commission on Radiological Protection and the World Health Organization being instrumental in effecting changes.

5.2.1 Diagnostic X Rays

The frequencies of diagnostic x-ray procedures vary widely from country to country. In developing countries, the total annual frequency is on the order of 100 to 200 diagnostic examinations per 1000 persons. Of these, dental examinations are but a small component. In the industrial countries, the annual frequency of all diagnostic x rays is on the order of 300 to 900 per 1000 persons, excluding dental and mass-survey examinations (UN 1982). In many countries, mass-survey examinations, largely chest examinations, have declined markedly as the incidence of tuberculosis has declined. Thoracic and skeletal examinations are universally the more common procedures. Other types of examination vary widely in frequency. In Japan, for example, stomach and upper GI fluoroscopy is a common procedure. In Eastern Europe, mass-screening with chest x rays is common. Dental examination frequencies are highly variable, reports are incomplete, and reporting practices differ markedly from country to country. In the United States, the annual examination frequency is about 400 per 1000 persons, and the annual average per capita dose equivalent to the active bone marrow (the critical organ for dental x rays), is about 4 mrad (NCRP 1989a).

When planning or assessing the impact of diagnostic x-ray procedures, the radiation protection specialist is confronted with many factors and variables. Among these, the following are discussed.

Voltage characteristics of the x-ray machine. The machine may employ constant voltage, single phase (rectified) voltage, or three phase voltage. The peak voltage may be as high as 250 kV and is typically 70 to 110 kV. The target may be tungsten, molybdenum, or a combination of the two. Representative energy spectra for x-ray machines are illustrated in Chap. 6, Fig. 6.3.

Source-image distance. This is the distance from the source to the plane of the x-ray image.

Source-skin distance. This is the distance from the source to the nearest skin surface. The SSD and the SID differ by the thickness of the subject and air space between the subject and the image.

Filtration. Total filtration includes inherent filtration within the x-ray tube and added filtration. The total may range from the equivalent of 0.5 mm of Al (inherent) to about 4 mm of Al. The filtration is provided to remove low energy photons which would increase patient dose without contributing to the quality of the x-ray image.

Film size. This is the sensitive area of the image receptor. For films, typical sizes are 7 in. by 17 in. or 14 in. by 17 in.

Image size. This is the x-ray beam area in the plane of the image, as defined by the collimator of the x-ray machine. The ratio of image size to film size is a measure of the care taken in the procedure. Ideally, the ratio is 1.0, but a ratio of 2 is not uncommon (UN 1982).

Entrance exposure. This is the exposure (or exposure rate) at the patient's skin surface nearest the source.

Beam quality. The voltage characteristics and total filtration define a beam quality characterized by the half-value layer (HVL) thickness of additional filtration (aluminum equivalent) required to halve the exposure rate in the x-ray beam.

Projection. This refers to the orientation of the patient during the x-ray exposure. Common designations are posterior-to-anterior (PA), anterior-to-posterior (AP), lateral (LAT), and oblique (OBL).

Assessment of the Radiological Impact of Diagnostic X Rays. The voltage characteristics and total filtration define a reference exposure rate in air at a specified

TABLE 5.1   Half-Value Layer (mm Al Equivalent) as a Function of Filtration and Tube Voltage for X-Ray Diagnostic Units—Full Wave Rectified, Single Phase.

TOTAL FILTRATION (mm Al)	PEAK VOLTAGE (kV)									
	30	40	50	60	70	80	90	100	110	120
0.5	0.36	0.47	0.58	0.67	0.76	0.84	0.92	1.00	1.08	1.16
1.0	0.55	0.78	0.95	1.08	1.21	1.33	1.46	1.58	1.70	1.82
1.5	0.78	1.04	1.25	1.42	1.59	1.75	1.90	2.08	2.25	2.42
2.0	0.92	1.22	1.49	1.70	1.90	2.10	2.28	2.48	2.70	2.90
2.5	1.02	1.38	1.69	1.95	2.16	2.37	2.58	2.82	3.06	3.30
3.0	—	1.49	1.87	2.16	2.40	2.62	2.86	3.12	3.38	3.65
3.5	—	1.58	2.00	2.34	2.60	2.86	3.12	3.40	3.68	3.95

Source: Reprinted with permission from Keriakes and Rosenstein 1980, Copyright CRC Press, Inc.

distance, typically 1 m, per unit electron-beam current in the x-ray machine. Reference exposure rates are illustrated in Fig. 5.1(a) for a constant-voltage machine and in Fig. 5.1(b) for a single-phase full-wave rectified machine. The HVL, as a function of total filtration and tube voltage, is illustrated in Table 5.1.

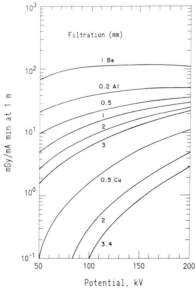

Figure 5.1 (a)   Reference air kerma rate at one meter from a constant-voltage x-ray tube with a tungsten target. The 1 mm beryllium is the tube window. The unit of mGy for air kerma is approximately equivalent to 0.1 R exposure. [From ICRP (1982)]

For a given geometry and entrance exposure, dose equivalents to the various organs and tissues of the body per unit entrance exposure are determined by the type of procedure, the projection, and the beam quality (HVL). An example of this interrelationship is given in Table 5.2. As pointed out by Rosenstein (1988), the increase in the

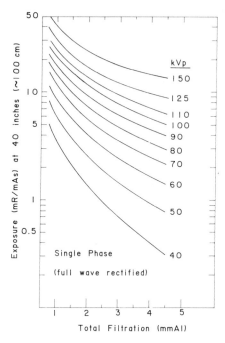

Figure 5.1 (b)   Reference exposure rate in air at 1m from the x-ray source as a function of total filtration for various values of tube potential using a single-phase, full-wave rectified voltage. [From data of McCullough and Cameron (1972), as reported by Nunnaly (1980)]

absorbed dose or cancer detriment index (see Chap. 3, Sec. 3.7.3) with increasing HVL beam quality, when normalized by the entrance exposure, can be misleading. The entrance exposure required to achieve a desired radiographic image is usually lower at higher HVL values. Comprehensive tables are provided by Keriakes and Rosenstein (1980) for adult male and female patients and for pediatric patients (newborn, age 1, and age 5). The data are derived from Monte Carlo x-ray transport calculations for anthropomorphic phantoms.

Tables 5.3 and 5.4 list adult organ dose equivalents H_{Ti} for typical conditions. These are weighted averages accounting for beam quality, entrance exposure, projection, and numbers of images generated during the procedure (Keriakes and Rosenstein 1980). While these organ doses are less than those cited in other reports (NAS 1980), the discrepancies appear to relate mostly to the number of images generated in the procedure. Raw data on which Tables 5.3 and 5.4 are based, namely Monte Carlo x-ray transport calculations, are in use by the U.S. National Center for Devices and Radiological Health (FDA 1973).

A comprehensive study of diagnostic x-ray procedures in the United States was reported by the Bureau of Radiological Health, Food and Drug Administration for 1970 (FDA 1973). Numbers of persons receiving x-ray exposures, and numbers of procedures performed are listed in Table 5.5. Data for adults only and for radiographic procedures, excluding dental, are listed in Table 5.6.

TABLE 5.2 Organ Dose and Cancer Detriment Index (CDI) Per Unit Entrance Exposure (Free-In-Air at Skin Entrance) for Chest X-Rays of the Typical Adult Woman. Source-image distance = 183 cm. Film size = field size = 14 in. × 17 in. (35.6 × 43.2 cm).

BEAM QUALITY HVL (mm Al)		1.5	2.0	2.5	3.0	3.5	4.0
ORGAN/TISSUE		DOSE PER UNIT EXPOSURE (mrad R^{-1})					
Ovaries	AP^a	0.5	1.2	1.9	2.8	3.6	4.3
	PA	0.3	0.7	1.2	1.7	2.3	2.9
	LAT	0.2	0.4	0.6	0.9	1.1	1.4
Thyroid	AP	189	240	283	318	348	372
	PA	11	21	30	40	49	57
	LAT	61	82	100	115	128	138
Active	AP	22	32	42	51	60	69
bone	PA	42	60	79	96	112	127
marrow	LAT	15	22	29	35	41	47
Embryo	AP	0.4	0.9	1.5	2.1	2.7	3.2
(uterus)	PA	0.4	0.8	1.4	2.0	2.6	3.2
	LAT	0.2	0.4	0.6	0.9	1.1	1.3
Lungs	AP	186	256	317	368	412	448
	PA	227	316	394	461	518	566
	LAT	106	153	196	234	266	294
Breasts	AP	580	680	757	819	868	909
	PA	17	31	45	59	72	84
	LAT	180	180	180	280	280	280
Trunk	AP	71	93	113	130	144	157
tissue	PA	70	92	112	129	144	157
	LAT	37	49	59	68	76	83
$10^5 \times$ CDI	AP	5.05	6.10	6.94	7.63	8.20	8.68
(R^{-1})	PA	1.10	1.56	1.99	2.37	2.71	3.01
	LAT	1.77	1.98	2.17	3.02	3.16	3.28

[a] AP—anterior to posterior, PA—posterior to anterior, LAT—lateral.
Source: Rosenstein 1988.

The effective dose equivalent for an examination of type i is given by

$$H_i = \sum_T w_T H_{Ti}, \tag{5.1}$$

where the weight factors are the ICRP factors discussed in Secs. 1.4.5 and 3.7.2. Again, it must be emphasized that the weight factors are averaged over the age and sex distribution of the entire population. Since those persons receiving diagnostic x rays may not be representative in age or cancer susceptibility of the entire population, the effective dose equivalent is but a crude indication of the overall risk to public health or to the individual.

The estimated genetically significant dose equivalent rate in the United States (FDA 1976), based on 1980 practice, is 30 mrem (0.3 mSv) per year (20 to females, 10 to males), up from a total of 20 mrem in 1970 (NCRP 1989a). The somatically

TABLE 5.3 Organ Doses for Common Radiographic Procedures (Adult Females).

| | | | | ORGAN DOSE (mrad)[a] | | | |
EXAMINATION	THYROID	ACTIVE BONE MARROW	LUNGS	BREASTS	OVARIES	EMBRYO OR UTERUS	TOTAL BODY
Chest	6.5	3.0	20	14	0.06	0.06	5.8
Skull	222	31	2.0		< 0.01	< 0.01	37
Cervical spine	404	11	14		< 0.01	< 0.01	23
Ribs	158	42	296	411	0.4	0.5	101
Shoulder (one)	58	6.0	27	77	< 0.01	< 0.01	110
Thoracic spine	75	32	265	276	0.6	0.6	70
Cholecystogram	1.0	66	176		6.0	5.0	85
Lumbar spine	0.3	126	133		405	408	272
Upper G.I.	7.0	114	476	53	45	48	216
KUB	0.01	48	12		212	263	99
Barium enema	0.2	298	48		787	822	396
Lumbosacral sp.	0.05	224	35		640	639	386
IVP	0.01	116	25		636	814	269
Pelvis	< 0.01	27	1.1		148	194	68
Hip (one)	< 0.01	17	< 0.01		78	128	39
Full spine	271	35	117	234	100	128	81
Mammography (xeroradiography)				766			
Mammography (film/screen)				212			

[a] Absence of data implies negligible dose compared to other organs.

Source: Reprinted with permission from Keriakes and Rosenstein 1980, Copyright CRC Press, Inc.

significant dose equivalent, based on data of Tables 5.3, 5.4, and 5.6 and on an adult population of 100 million is about 50 mrem per year. This figure excludes dental, fluoroscopic, and therapeutic exposures. It may be compared to the UNSCEAR SSD estimate of 100 mrem per year for all x-ray exposures in industrial countries (UN 1982). For the United States population in 1980, the population mean annual bone marrow doses due to diagnostic x rays was estimated to be 75 to 115 mrad from medical procedures, including 3 to 4 mrad from dental x rays. It should be noted, though, that over 80 percent of the bone marrow dose occurs in persons over the age of 40, and nearly 60 percent for persons over age 55 (NCRP 1989a). The collective effective dose equivalent to the 1980 U.S. population due to diagnostic x rays amounted to about 73,000 person-Sv. However, because the age distribution of persons receiving diagnostic x rays is more skewed to older ages than is that of the general population, the collective SSD is only 46,000 person-Sv (NCRP 1989a).

Computed Tomography (CT).   CT scans of the head typically result in absorbed doses of 400 to 600 mrem (4 to 6 mSv) to the eyes (UN 1982). However, lens doses may be as great as 12 rad (0.12 Gy), depending on the specific procedure (NCRP 1989a). Scans of the abdomen result in absorbed doses of 1 to 5 mrem to the testes or 1 to 15 mrem to the ovaries (UN 1982). Skin dose may be as great as 56 rad, though 6 rad is more typical (NCRP 1989a). The CT technique, a dramatic advance in diagnostic

TABLE 5.4   Organ Doses for Common Radiographic Procedures (Adult Males).

EXAMINATION	THYROID	ORGAN DOSE (mrad) ACTIVE BONE MARROW	LUNGS	TESTES	TOTAL BODY
Chest	6.5	4.0	19	< 0.01	5.8
Skull	222	31	2.0	< 0.01	37
Cervical spine	404	11	14	< 0.01	23
Ribs	154	49	324	< 0.01	101
Shoulder (one)	58	6.0	39	< 0.01	10
Thoracic spine	75	43	263	< 0.01	70
Cholecystogram	1.0	66	176	< 0.01	85
Lumbar spine	0.3	126	133	7.0	272
Upper G.I.	7.0	117	532	0.4	216
KUB	0.01	48	12	16	99
Barium enema	0.2	298	48	48	396
Lumbosacral sp.	0.05	224	35	43	386
IVP	0.01	116	35	49	269
Pelvis	< 0.01	27	1.1	157	68
Hip (one)	< 0.01	17	< 0.01	368	39
Full spine	271	35	149	10	81

Source: Reprinted with permission from Keriakes and Rosenstein 1980, Copyright CRC Press, Inc.

TABLE 5.5   Estimated Annual Number of Persons Receiving X-Ray Exposures and Annual Number of Procedures in the United States During 1970.[a]

TYPE OF PROCEDURE	MILLIONS OF PERSONS	MILLIONS OF PROCEDURES
Radiographic	76.4	129.1
Photofluoroscopic	9.7	10.4
Non-Photofluoroscopic	66.7	118.7
Dental	59.2	67.5
Fluoroscopic	9.1	12.6
Therapeutic	0.4	2.6

[a]By 1980, the annual number of procedures increased by 32 percent and the rate per 1000 by 18 percent. Both the total number and the rate of dental procedures increased by 40 to 50 percent (NCRP 1989a).
Source: FDA 1973.

radiology, also is thought to reduce radiation risks through reduction in the frequency of angiographic examinations and radioisotope tests (UN 1982).

Mammography.   While mammography does result in significant doses to the breast, when applied with due consideration to the recipient's age and frequency of examination, the benefit from earlier detection of malignancies is thought to outweigh significantly the additional risk of developing fatal malignancies as a result of the radiation exposure (UN 1982). Skin exposure in mammography procedures is typically 1 R, and gland dose at 3-cm depth is typically 270 mrad (NCRP 1989a).

TABLE 5.6 Frequencies of X-Ray Diagnostic Procedures for
Adults in the United States During 1970.[a]

	MILLIONS OF EXAMINATIONS PER YEAR	
EXAMINATION	MALES	FEMALES
Chest	21.3	21.8
Skull	1.5	1.4
Cervical spine	1.6	1.5
Ribs	0.73	0.53
Shoulder (one)	0.86	1.0
Thoracic spine	0.65	0.77
Cholecystogram	1.5	2.3
Lumbar spine	1.9	1.4
Upper G.I.	2.5	2.9
KUB	1.4	1.4
Barium enema	1.5	1.9
Lumbosacral spine	1.0	0.8
IVP	1.7	1.7
Pelvis	0.5	1.2
Hip (one)	0.3	1.0
Full spine	0.1	0.11
Mammography		2.0

[a] By 1980, the annual number of procedures increased by 32 percent and the rate per 1000 by 18 percent. Both the total number and the rate of dental procedures increased by 40 to 50 percent (NCRP 1989a).

Source: Keriakes and Rosenstein 1980.

Measures to Minimize Doses in X-Ray Procedures. The need for dose minimization has been addressed at many levels. Recommendations and proposed methodologies are issued from international bodies such as the World Health Organization, the International Atomic Energy Agency, the European Community, and the International Commission on Radiological Protection. National bodies such as the National Radiological Protection Board in Britain and the National Council on Radiation Protection and Measurements in the United States publish a wide range of recommendations and guidelines dealing not only with operational practices and quality control but also with the structural design of facilities. Licensing bodies at the national level, for example, the FDA's National Center for Devices and Radiological Health (in the United States) and the Health and Safety Executive (in Britain), and authorities at regional level, for example, Agreement States, have inspection and enforcement authority to effect dose minimization.

The International Commission on Radiological Protection (ICRP 1982) recommends a system of dose limitation to ensure that every exposure to ionizing radiation is justified in relation to its benefits. The three main features of the system are as follows:

Justification. The professional judgments of the referring physician and the radiologist that the procedure will produce a net benefit.

Optimization. Balancing the marginal costs of improvements in apparatus and technique with the marginal benefits to the patient.

> *Dose Limitation.* Balancing marginal patient benefits with marginal risks incurred by medical staff and individual members of the public. Dose limitation is discussed in detail in ICRP Publication 26 (1977).

Specific ICRP recommendations on design and operation of radiation facilities (ICRP 1977, 1982) cover structural shielding design, quality assurance, equipment specifications, surveillance and testing, and monitoring of personnel and the workplace. Certain of the detailed recommendations relating to equipment and accessories for normal diagnostic radiology are as follows:

> *X-ray Tube Housing.* Leakage radiation at a distance of 1 m from the focus should not result in an air kerma greater than 1 mGy in one hour (100 mR in one hour).
>
> *Beam Limitation.* Diaphragms, cones or collimators should be used to limit the beam to the area of clinical interest.
>
> *Filtration.* For normal diagnostic work, at least the equivalent of 2.5 mm Al should be used. Exceptions apply to mammography and certain dental procedures.
>
> *Image Intensifiers.* These should be used for all surgical and other procedures requiring mobile fluoroscopy equipment.
>
> *Source-to-Skin Distance.* The distance should not be less than 45 cm and shall not be less than 30 cm.
>
> *Photofluorography.* When using photography of fluorescent screens, the potential for high patient doses is great compared to that from conventional radiography and the selection of camera, film, and fluorescent screen should be carried out very carefully. Satisfactory results for chest surveys should be obtainable with entrance exposures no greater than 100 mR.

Many other recommendations, not repeated here, deal with radiography and darkroom procedures, staff training, protective clothing, and gonad shielding. The NCRP (1989a) stresses that staff training in equipment use, calibration, and quality assurance may reduce diagnostic radiation exposures by as much as 50 percent.

5.2.2 Diagnostic Nuclear Medicine

Internally administered radionuclides are used medically for imaging studies of various body organs and for nonimaging studies such as thyroid uptake and blood volume measurements. Such uses present hazards for both patients and medical staff. Radiopharmaceuticals are also used for *in vitro* studies such as radioimmunoassay measurements, and thus are of potential hazard to medical staff. Frequencies of procedures, while steadily increasing, vary widely from country to country. In industrialized countries, about 10 to 40 examinations involving radiopharmaceuticals are carried out annually per 1000 population. In developing countries, annual frequencies are on the order of 0.2 to 2 examinations per 1000 population.

Frequencies of diagnostic nuclear medicine procedures in the United States are given in Table 5.7. Relative frequencies of liver, lung, and kidney scan procedures remain roughly constant, while bone scans and cardiovascular imaging are increas-

TABLE 5.7 Annual Frequency of Diagnostic Nuclear Medicine Procedures in the United States.

EXAMINATION	NUMBER PER 1000 PERSONS	
	1978	1982
Brain	7.0	3.5
Liver	5.9	6.2
Bone	5.3	7.8
Respiratory	4.8	5.2
Thyroid	3.2	2.9
Urinary	0.9	1.0
Tumor	0.8	0.5
Cardiovascular	0.7	4.1
Other	0.5	0.8
Total	29	32

Source: NCRP 1989a.

TABLE 5.8 Radioisotopes Used in Radiopharmaceuticals and Their Frequency of Administration in Nuclear Medicine Procedures in the United States During 1979.

RADIO-NUCLIDE	CHEMICAL FORM	PERCENT OF PROCEDURES
^{99m}Tc	pertechnate, labelled compounds colloids and particles	81.7
^{131}I	iodide, labelled compounds	8.38
^{133}Xe	gas, saline solution	3.8
^{67}Ga	citrate	2.8
^{123}I	iodide, labelled compounds	0.8
^{68}Ga	citrate	0.4
^{125}I	iodide, labelled compounds	0.19
^{57}Co	cyanocobalamin, bleomycin	0.19
^{111}In	DTPA	0.09
^{14}C	labelled compounds	0.05
^{169}Yb	DTPA	0.05
^{51}Cr	chromate, RBC	0.04
^{32}P	phosphate	0.03
^{58}Co	cyanocobalamin (vit. B-12)	0.01

Source: U.S. Bureau of Radiological Health, as reported in (UN 1982).

ing. While brain scans involving radiopharmaceuticals are decreasing, those involving computed tomography are increasing.

Table 5.8 lists the radiopharmaceuticals in most frequent use in the United States. Most often, the materials are administered by intravenous injection. The list is dominated by ^{99m}Tc (half life 6.02 h), used in some 82 percent of procedures. This radionuclide is ordinarily generated on site by elution (milking) from a ^{99}Mo generator (cow) with a half life of 66.2 h. Emitting almost exclusively low-energy gamma rays, ^{99m}Tc is ideally suited for use with the thin-crystal sodium-iodide gamma scintillation camera, the standard imaging instrument used in nuclear medicine.

Assessment of Radiological Impact.   Detailed evaluation of organ doses from internally administered radionuclides is deferred until Chap. 8. The standard method of evaluation is that of the Society of Nuclear Medicine Committee on Medical Internal Radiation Dose (MIRD 1971, 1985). The MIRD method has been adopted in large measure by the ICRP (1979). For discussion of this and other methods, see (NCRP 1985a, 1985b). Absorbed doses received by the various organs and tissues of the body as a result of intravenous administration depend on the activity administered, the physiological characteristics of the patient, and the chemical form of the radionuclide. The ICRP methodology applies only to large populations, is based on the generic characteristics of Reference Man (ICRP 1975), and, in routine application, treats only broad categories of chemical forms. In addition, the ICRP methodology applies primarily to ingested or inhaled radionuclides. The MIRD technology requires information on uptake and retention highly specific to the chemical form of the radionuclide. Data are available (Keriakes and Rosenstein 1980, ICRP 1987b) for organ dose (equivalents) per unit activity administered for an extensive list of radionuclides and for patient ages of 1, 5, 10, 15 years, newborn and adult. Source-to-target gamma-ray absorbed fractions are also available for subjects of these ages (Cristy and Eckerman 1987). Because of the wide use of ^{99m}Tc, this radionuclide is chosen here for illustration, and Table 5.9 lists organ doses for the five most common procedures of Table 5.7.

In the United States, 10 to 12 million persons annually experience diagnostic nuclear medicine exposures. The most significant tissue dose is to the bone marrow — on the average 300 mrem (3 mSv) per year to the exposed population. Prorated over the entire population, the average is about 14 mrem per year (NAS 1980; NCRP 1989a). Because such procedures are more often performed on older members of the population, the age-weighted effective dose equivalent (i.e., the somatically significant dose, SSD) is only about 6 mrem (NCRP 1989a).

Radiation Protection in Diagnostic Nuclear Medicine.   Radiation protection measures fall into three categories: (1) those for protection of attending medical staff, (2) those for protection of nursing infants, and (3) those for personnel of radiopharmaceutical dispensing laboratories.

Close contact, including feeding, between an infant and a patient mother should be avoided for an appropriate time after radionuclide administration. Under normal practice, external hazards to medical staff from nuclear-medicine patients is minor. An exception might arise in the special case of extreme intensive-care nursing. Body wastes of patients may be disposed of following normal hygienic practices, and contaminated clothing and bedding may be washed using normal laundry procedures, after holding for a sufficient time to allow for radioactive decay.

Because of the large quantities of unsealed radionuclides being processed, there are stringent radiation protection practices required for dispensing laboratories. Monitoring and surveillance are key elements, and means must be at hand for cleanup and decontamination as well as access restriction. Detailed measures are provided in specialized publications (IAEA 1979, ICRP 1987a, WHO 1982, Mould 1985, NCRP 1989c).

TABLE 5.9   Absorbed Doses to Healthy Organs of the Adult Resulting from Intravenous Administration of ^{99m}Tc Radiopharmaceuticals for Diagnostic Imaging.

IMAGE	CHEMICAL FORM	TYP. ACT. (mCi)	ORGAN	DOSE PER UNIT ACTIVITY mGy per MBq[a]
Brain	Pertechnetate	18	testes	0.0027
			ovaries	0.010
			liver	0.0039
			red marrow	0.0061
			bladder	0.019
			stomach	0.029
			ULI[b]	0.062
			LLI[c]	0.022
			effective dose equivalent	0.013
Liver/spleen	Sulfur colloid	5	testes	0.0062
			ovaries	0.0022
			liver	0.074
			red marrow	0.011
			spleen	0.077
			effective dose equivalent	0.014
Skeleton	Pyrophosphate polyphosphate	15	testes	0.0024
			ovaries	0.0035
			liver	0.0013
			kidneys	0.0073
			red marrow	0.0096
			bone surface	0.063
			effective dose equivalent	0.0080
Lung	Macroaggregates microspheres	5	testes	0.0017
			ovaries	0.0026
			kidneys	0.0034
			liver	0.0045
			lungs	0.058
			red marrow	0.0046
			spleen	0.0044
			thyroid	0.0022
			bladder	0.017
			effective dose equivalent	0.011

[a] To convert to mrad per μCi, multiply by 3.70.

[b] Upper large intestine.

[c] Lower large intestine.

Source: ICRP 1987b, Keriakes and Rosenstein 1980, NAS 1980, Mould 1985.

5.2.3 Therapeutic Uses of Radiation

There are three broad categories of radiation therapy — teletherapy, brachytherapy, and therapy using administered radiation sources. *Teletherapy* involves external beams from sources such as sealed ^{60}Co sources, x-ray machines, and accelerators that generate electron, proton, neutron, or x-ray beams. *Brachytherapy* involves sources placed within body cavities (*intracavitary* means) or placed directly within tumor-bearing tissue (*interstitial* means). In the United States, Europe, and Japan, the frequencies for teletherapy and brachytherapy procedures total about 2400 annually per million population (UN 1988).

Thyroid disorders, including cancer, for many years have been treated by ^{131}I, usually by oral administration. Introduced about 1980, in association with the development of techniques for producing monoclonal antibodies, were new cancer diagnosis and treatment methodologies called radioimmunoimaging and radioimmunotherapy. The therapy involves administration of massive doses of antibodies tagged with radionuclides and selected to bind with antigens on the surfaces of tumor cells. Imaging involves administration of very much smaller doses, with the goal of detecting the presence of tumor cells using standard camera and scanner imaging techniques. Imaging requires the use of radionuclides such as ^{99m}Tc which emit low energy gamma rays. Therapy involves the use of radionuclides emitting beta particles and electrons, with minimum emission of gamma rays, thus limiting radiation exposure, to the extent possible, to tumor cells alone. Among radionuclides used in radioimmunotherapy are ^{75}Se, ^{90}Y, ^{111}In, ^{125}I, ^{186}Re, and ^{191}Os. Collective population doses from radiation therapy are very small compared to doses from diagnosis. Furthermore, therapy patients are of a different age distribution than the general population and are otherwise at life-threatening risk. Thus, one would not aggregate therapeutic population doses with those arising involuntarily. Similarly, collective occupational doses in radiation therapy are relatively small. Individual occupational doses are greater for brachytherapy than for teletherapy, in general, but teletherapy with neutrons can lead to long-term gamma-ray occupational exposure arising from neutron-activated accelerator components (UN 1988).

5.2.4 Medical Occupational Exposure

Estimated whole-body exposures for the United States in 1980 are summarized by the National Council on Radiation Protection and Measurements (1989a) as follows: The total number of radiation workers in the medical field amounted to 584,000, including dental, chiropractic, and veterinary practice. Of the total, 277,000 persons were actually exposed occupationally, and the cumulative effective dose equivalent to this population was 416 person-Sv. The average for all workers was 0.7 mSv, and that for exposed workers was 1.5 mSv. Prorated over the population as a whole, the average was only 1.8 μSv (0.18 mrem).

5.2.5 Wastes from Medical Uses of Radionuclides

Radioactive wastes primarily involve short-lived radionuclides. Exceptions are radium sources of various types. Patient excreta containing radionuclides may be released through sanitary sewerage without processing or holdup. Similarly, small quantities of soluble materials may be released through sanitary sewerage. Contaminated clothing and bedding may be held up for radioactive decay and either laundered or, in extreme cases, disposed of as solid waste. Certain contaminated materials, namely scintillation fluors and body parts, are more hazardous chemically or biologically than radiologically and may be disposed of as chemical or biological wastes. Chemical or biological hazards must be eliminated prior to release of solid radioactive wastes for land disposal.

5.3 NUCLEAR EXPLOSIVES

Large fractions of radioactive debris from atmospheric nuclear weapons tests are distributed widely, fallout is global, and the radionuclides remain in the biosphere indefinitely. The hazard is better characterized by the long-term dose commitment than
by the dose rate at any instant and location. Justification for the dose-commitment
concept as a useful single measure of the irradiation consequences for the world population was stated by the UNSCEAR (1977) as follows:

> It is difficult to summarize the doses for the whole world population from nuclear test
> explosions because they arise from a variety of radionuclides which differ widely in their
> behavior in the environment and in their dosimetric characteristics. To state current an
> nual doses would reveal only a small part of an exposure situation, which is known to vary
> not only with time, but also with geographical location, living conditions and age. For a
> given group of individuals, for whom these factors are known, the annual doses from ex
> ternal and internal exposures may be presented as a function of age for some selected
> organs and tissues of interest; that would give a full picture of the annual doses for the
> particular groups. For some long-lived radionuclides that are globally distributed, such
> as ^{137}Cs, this information may be derived for the whole world population as *per caput* an
> nual doses which would be representative of individuals irrespective of age and location.
> For other radionuclides, however, the individual annual doses will vary substantially with
> location and age. That is the case with short-lived radionuclides, such as ^{131}I, as regards
> location, and with bone-seeking radionuclides, such as ^{90}Sr, as regards age ...

> If the variation of individual organ dose rates with time were known for all locations,
> living conditions, and ages, a world population *per caput* dose rate could be calculated.
> Its integral over infinite time would be the global dose commitment from the nuclear
> explosions to date ... [The] integral can be evaluated even though the *per caput* dose
> rate as a function of time is not known or is difficult to derive.

Chapter 11 discusses the modeling used in the committed-dose calculations for fallout.

5.3.1 Radionuclide Source Terms

The fusion and fission energy released in a nuclear-weapon explosion is usually measured in units of megatons (Mt). One megaton refers to the release of 10^{15} calories
of explosive energy — approximately the amount of energy released in the detonation
of 10^6 metric tons of TNT (Glasstone and Dolan 1977). The quantity of fission products produced in a nuclear explosion is proportional to the *weapon fission yield.* For a
1-Mt weapon fission yield, there must be about 1.45×10^{26} fissions, equivalent to the
complete fissioning of about 56 kg of uranium or plutonium. The quantities of ^{3}H and
^{14}C, which are produced in the atmosphere by interactions of high-energy neutrons,
are proportional to the *weapon fusion yield.* There are several fusion reactions used in
thermonuclear devices, with a 1-Mt weapon fusion yield requiring, for example, the
fusion of 7.4 kg of tritium with 4.9 kg of deuterium.

   If the *fission-product yield,* (not to be confused with weapon fission yield) and
radioactive decay constant of a particular nuclide are known, the activity produced
may be calculated as is illustrated in the following example:

^{90}Sr data: fission-product yield $\cong 5.1\%$
decay constant $= 7.54 \times 10^{-10}$ s^{-1} (29.1 y half life)

Activity: $1.45 \times 10^{26} \times 0.051 \times 7.54 \times 10^{-10}$
$\cong 5.6 \times 10^{15}$ Bq Mt^{-1}

Activities, half lives, fission-product yields for ^{235}U induced by fission-spectrum neutrons, and activities produced per unit weapon fission yield for three important fission-product radionuclides are as follows:

Nuclide	Half-Life	Yield (%)	PBq Mt^{-1}
^{90}Sr	29.1 y	5.1	5.6
^{131}I	8.04 d	3.1	4400
^{137}Cs	30.0 y	6.1	6.4

The ^{14}C yield is 1.0 PBq Mt^{-1} (fusion) (UN 1982). Tritium is both a fission product and an activation product. Activities produced per unit weapon yield are as follows: 0.026 PBq Mt^{-1} (fission) and 740 Pbq Mt^{-1} (fusion).

Estimates of nuclear-weapon explosive yields in the atmosphere (cumulative through 1980) are as follows (UN 1982):

Years	Source	No. Tests	Fission (Mt)	Total (Mt)
1945-62	USA	193	72.1	138.6
1949-62	USSR	142	110.9	357.5
1952-53	UK	21	10.6	16.7
1960-74	France	45	10.9	11.9
1964-80	China	22	12.7	20.7
Total		423	217.2	545.4

5.3.2 Atmospheric Dispersal of Weapon Debris

The disposition of weapon debris may be divided into three categories, local fallout, tropospheric fallout, and stratospheric fallout. Local fallout, comprising as much as 50 percent of the debris and consisting of large particles, is defined by UNSCEAR (1982) as that deposited within 100 miles of the test site. Because of the remoteness of test sites, UNSCEAR does not treat this source as contributing to population exposure.

Depending on test altitude and weather conditions, a portion of the test debris is injected into the stratosphere and a portion remains in the troposphere. These two atmospheric regions are separated by the tropopause (about 16 km altitude at the equator and 9 km at the poles). Temperature decreases with elevation in the troposphere. This hydrodynamically unstable condition leads to the development of convective weather patterns superimposed on generally westerly winds. In the stratosphere, temperature is more nearly constant or, in equatorial regions, even rises with elevation. Vertical convective motion is relatively slight and the tropical temperature inversion restricts transfer of material in the stratosphere from hemisphere to hemisphere.

Debris in the troposphere is distributed in longitude but remains within a band of about 30 degrees of latitude (Glasstone and Dolan 1977). The mean lifetime of radioactive debris in the troposphere is about 30 days and tropospheric fallout is important for radionuclides with half lives of a few day to several months (UN 1982). Over the years, the bulk of the radioactive debris from weapons tests has been injected into the stratosphere in the northern hemisphere and at altitudes less than 20 km. Mechanisms for transfer of the debris to the troposphere and thence to fallout on the earth's surface are complex. At elevations less than 20 km, the half-life for transfer of aerosols between hemispheres is about 60 months, while the half-life for transfer to the troposphere is only about 10 months, with little material crossing the tropopause in equatorial regions (Glasstone and Dolan 1977). Consequently, the bulk of the fallout from any one test occurs over the hemisphere of injection and in temperate regions. In terms of the megatons of fission energy, in the period to 1980, 78 percent of the debris was injected into the stratosphere — 70 percent into the northern hemisphere and 8 percent into the southern (UN 1982).

5.3.3 Dose Commitments from Nuclear Weapons Tests

The 1977 and 1982 UNSCEAR reports provide a comprehensive discussion of the distribution of nuclear-weapons fallout, transfer of radionuclides through food chains, and consequent doses and dose commitments to the human population. Some 21 radionuclides are considered, of which only 7 contribute more than 1 percent to the committed effective dose equivalent to the population. These seven, in decreasing order of importance, are ^{14}C, ^{137}Cs, ^{95}Zr, ^{90}Sr, ^{106}Ru, ^{144}Ce, and ^{3}H. Table 5.10 summarizes average organ doses incurred by the world population to the year 2000. Because of its long half-life, 5730 years, the commitment from ^{14}C extends over many human generations. Extending the period for evaluation of the dose commitment only to the year 2000 provides a measure of the radiation hazard presented to those living during the period of intensive atmospheric weapon testing prior to 1962. Extending the accounting period infinitely far into the future provides a measure of the total detriment to the human population resulting from the tests. Doing so adds 2400 μSv to the committed effective dose equivalent due to ^{14}C. The collective effective dose equivalent commitment into the indefinite future due to weapons tests to date is about 3×10^7 person-Sv. This corresponds to about four extra years of exposure of the current world population to natural background radiation (UN 1982).

5.3.4 Wastes from Nuclear Weapons Programs

High-level radioactive wastes generated in the United States in the production of nuclear weapons have accumulated for decades. The wastes are stored at three sites, one in the state of Washington, one in Idaho, and one in South Carolina. The approximately 9000 tonnes of waste has a volume of 380,000 cubic meters (DOE 1988). As discussed in the following section, there are plans to dispose of this U.S. weapons waste in a repository used also for disposal of spent fuel for nuclear power plants.

TABLE 5.10 Summary of Dose Commitments to the Year 2000[a] from Radionuclides Produced in Atmospheric Nuclear Tests Carried out to the End of 1980. The Committed Effective Dose Commitment Is Based on the ICRP Weight Factors Described in Section 1.4.5.

RADIATION SOURCE	ORGAN DOSE COMMITMENT (μGy)							EFFECTIVE DOSE EQUIVALENT (μSv)
	GONADS	BREAST	RED MARROW	LUNGS	THYROID	BONE SURF.	REMAINDER[c]	
External								
Short-lived	310	310	310	310	310	310	310	310
^{137}Cs	370	370	370	370	370	370	370	370
Internal								
^{3}H	47	47	47	47	47	47	47	47
^{14}C	77	203	371	91	91	336	203	179
^{55}Fe	9	9	5	9	9	9	9	9
^{89}Sr			2			3	4	2
^{90}Sr			570	74		1300		116
^{106}Ru				250				30
^{131}I					1100			3
^{137}Cs	170	170	170	170	170	170	170	170
^{144}Ce	310	37						
^{239}Pu[b]	0.03		2	7		25	1	43
^{241}Pu	1.7		10	18		121	5	9
^{241}Am	0.04		0.2			3	.02	5
Total (rounded)	990	1100	1900	1700	2100	2700	1100	1300

[a] The dose commitment to the year 2000 for ^{14}C is 7 percent of the dose commitment over infinite time.

[b] Includes ^{240}Pu.

[c] Whole-body dose commitment except as follows: 5.1 and 15 μGy, respectively, to the upper and lower large intestines for ^{89}Sr, 5.5 μGy to the liver for ^{239}Pu and ^{240}Pu, 25.2 μGy to the liver for ^{241}Pu, and 0.3 μGy to the liver for ^{241}Am.

Source: UN 1982.

5.4 NUCLEAR POWER

As of 1987, there were 300 gigawatts of installed nuclear-electric generating capacity in the world, annually producing 193 GWy of electrical energy. Of the generating capacity, pressurized-water and boiling-water reactors, respectively, accounted for 62 and 23 percent. Gas-cooled reactors (GCR) accounted for 4 percent. Heavy-water reactors (HWR) and light-water graphite reactors (LWGR) each accounted for 5 percent, and fast-breeder reactors (FBR) accounted for 1 percent. Of the energy generated, 52 percent was generated in the United States, 29 percent in France, and 21 percent in the USSR. Leading countries in terms of percent of electricity generated by nuclear means are France (70%), Belgium (66%), Republic of Korea (53%), and Sweden (45%) (UN 1988).

The radiological impact of nuclear power production is examined for the five distinct phases of the nuclear fuel cycle, namely (1) mining and milling, (2) conversion, enrichment and fuel fabrication, (3) reactor operation, (4) fuel reprocessing, and (5) waste storage and disposal.

The radionuclides of concern may be categorized in various ways. One way is by origin: (1) naturally occurring radionuclides at enhanced concentrations, (2) fission products, and (3) activation products. In the first category are the nuclides of the ^{238}U decay chain released in the mining and milling operations. In the second are the radioactive fission products and actinides produced during reactor operation. In the third are products of neutron absorption in structural and fuel-cladding materials during reactor operation.

Another way of categorizing the radionuclides is by their physical-chemical behavior: (1) noble gases, (2) ^{3}H and ^{14}C, (3) halogens, and (4) particulates. These divisions are frequently used in connection with airborne radionuclides and gas streams in plants. They approximate the relative ease of isolation of the radionuclides. The noble gases include the many isotopes of the krypton and xenon fission products as well as the activation product ^{41}Ar. These elements cannot be removed from a gas stream by filtration. Halogens include the many isotopes of the bromine and iodine fission products. If they are present in a gas stream, they are likely to be in a chemical form unsuitable for filtration, and effective removal requires adsorption on a material such as activated charcoal. Other radionuclides and the halogens in ionic form may be removed from a gas stream by filtration. In aqueous liquids, the particulates may be isolated by evaporation, filtration, or ion exchange. The halogens, unless in ionic form, cannot be isolated by evaporation or filtration, nor, of course, can noble gases. Special cases are ^{3}H in the form of tritiated water and ^{14}C as carbon dioxide. The tritium can be isolated only with very great difficulty, and CO_2 removal requires chemical treatment.

Assessment of the global radiological impact of the nuclear fuel cycle as a whole is an extremely complicated task. Principles and techniques described in Chaps. 9 through 11 may be applied to the assessment of the local and perhaps regional impacts of any one nuclear installation for which information is known about the technology in place, local meteorological conditions, local geography and population distributions, and local agricultural practices.

The approach taken in this chapter is a global assessment of radiological impact. It is a summary of the analyses presented in the 1982 and 1988 UNSCEAR reports (UN 1982, 1988), which are themselves summaries of the findings of several international studies. The 1988 report differs from the 1982 report not only in the use of more modern data on generating capacity but also in the accounting for actual practices in the degree to which nuclear fuel is reprocessed. Collective impact on the global population is expressed in terms of the committed effective dose equivalent, averaged over nuclear reactor types according to their contribution to electrical power production and then normalized to unit electrical generation by dividing by the number of gigawatt years of electrical energy produced. Radionuclide releases and dose commitments summarized in the next section are based on "model" installations, chosen to represent the current technology of nuclear power. They are averages over both old and new technology, over normal and abnormal operating conditions, and are based on representative meteorological, agricultural, geographic, and demographic conditions. They do not represent any one installation and should not be applied to assessment of the impact of any one plant.

TABLE 5.11 Normalized Effluent Discharges from Uranium Mining, Milling, Conversion, Enrichment, and Fuel Fabrication.

	MBq per GWy(e)				
	^{238}U^a	^{230}Th	^{226}Ra	^{222}Rn	^{210}Pb^b
Atmospheric Release					
mining				$2.0 \times 10^{+7}$	
milling	$6.6 \times 10^{+2}$	$7.4 \times 10^{+1}$	$4.0 \times 10^{+1}$	$8.8 \times 10^{+5}$	$4.0 \times 10^{+1}$
mill tailings	7.0×10^{-1}	$1.5 \times 10^{+1}$	$1.5 \times 10^{+1}$	$1.0 \times 10^{+6}$	$1.5 \times 10^{+1}$
conversion	$7.4 \times 10^{+1}$	7.4×10^{-1}	7.0×10^{-2}	$8.1 \times 10^{+3}$	
enrichment	$3.7 \times 10^{+1}$	7.4			
fabrication	7.4×10^{-1}				
Liquid Release					
conversion	$8.1 \times 10^{+2}$				
enrichment	$3.7 \times 10^{+2}$				
fabrication	$3.7 \times 10^{+2}$				

^{a}In equilibrium with daughters through ^{234}U.
^{b}In equilibrium with ^{210}Po.
Source: UN 1982.

5.4.1 Radiological Releases

Mining and Milling.   The principal radioactive release associated with uranium mining, underground, or open pit, is natural ^{222}Rn to the atmosphere. Airborne particulates containing natural uranium daughter products also arise from open pit mining and from ore crushing and grinding in the milling process. Mill tailings can also become a long-term source of radioactive contamination due to wind and water erosion, leaching, and radon release, the degree depending on the tailing-stabilization program followed. Mining and milling operations are generally conducted in remote areas and liquid releases containing dissolved uranium daughter products are of little impact on human populations. Activity release rates per unit electrical energy ultimately generated are given in Table 5.11 for combined mining, milling, and fabrication of nuclear fuel.

Uranium Conversion, Enrichment, and Fuel Fabrication.   The product of milling is U_3O_8 "yellow cake" ore concentrate. In this phase of the nuclear fuel cycle, the concentrate is purified and most often converted to UF_6 for enrichment in ^{235}U via gaseous diffusion or centrifuge processes. Prior to fuel fabrication, the uranium is converted to the metallic or the ceramic UO_2 form suitable for use in fuel elements. Large quantities of uranium depleted in ^{235}U are byproducts of the enrichment process. Under current practice the depleted uranium is held in storage as being potentially valuable for use in breeder reactors. In this stage of the nuclear fuel cycle, there are relatively minor liquid and gaseous releases of uranium and daughter products to the environment. These releases are included with the values given in Table 5.11.

Fuel Reprocessing.   As nuclear fuel reaches the end of its useful life in power generation, there remain within the fuel recoverable quantities of uranium and plutonium which may be extracted for re-use in the fuel reprocessing stage of the nuclear fuel cycle. Whether or not the fuel is reprocessed is governed by economic and political considerations. Among the former are costs of reprocessing as compared to costs of mining, milling, conversion, and enrichment of new stocks of uranium. Among the latter are concerns over the potential diversion of plutonium to nuclear-weapons use. At this writing, nuclear fuel used in the power industry is not reprocessed in the United States, despite well-developed technology. There are three commercial reprocessing plants in Europe — at Sellafield in the United Kingdom, and at La Hague and Marcoule in France.

In the reprocessing of oxide fuels, the spent fuel is first dissolved in nitric acid. Plutonium and uranium are extracted into a separate organic phase from which they are ultimately recovered and converted into the oxide form. The aqueous phase containing fission and activation products is then neutralized and stored in liquid form pending solidification and ultimate disposal. Because one reprocessing plant may serve scores of power plants, inventories of radionuclides in process may be very great and extraordinary design features and safety procedures are called for. Because of the time delays between removal of fuel from service and reprocessing, concerns are with only relatively long-lived radionuclides, notably ^{3}H, ^{14}C, ^{85}Kr, ^{90}Sr, ^{106}Ru, ^{129}I, ^{134}Cs, and ^{137}Cs.

During the dissolution step of reprocessing, all the ^{85}Kr, the bulk of the ^{14}C (as CO_2), and portions of the ^{3}H and ^{129}I appear in a gas phase. This gas is cleaned, dried, and released through a tall stack to the atmosphere. All the ^{85}Kr is thus released; however, the major part of the ^{3}H is removed in the drying process and the bulk of the ^{129}I and ^{14}C is removed by reaction with caustic soda. The ^{14}C may then be precipitated and held as solid waste. Depending on the degree of liquid-effluent cleanup, some of the ^{129}I and other fission products subsequently may be released to the environment.

Operational experiences for three reprocessing plants is summarized in Table 5.12. The data for plant releases reveal not only variations in technology, but also lower concentrations of radionuclides in liquid effluents from the Marcoule plant necessitated by that plant's location on a river rather than the seacoast.

Waste Storage and Disposal.   Wastes generated in the nuclear fuel cycle fall into the broad categories of high-level wastes (HLW) and low-level wastes (LLW). The former, comprising unprocessed spent fuel or liquid residues from fuel reprocessing, accounts for only about 1 to 5 percent of the waste volume, but about 99 percent of the waste activity. The latter is comprised of in-reactor components, filter media, ion-exchange resins, contaminated clothing and tools, and laboratory wastes. For the most part, LLW consists of short-lived beta-particle and gamma-ray emitters. Wastes of low specific activity, but containing long-lived alpha-particle emitters, for example, ^{239}Pu, require special handling more in the nature of that required for HLW. Godbee et al. (1986) give a comprehensive analysis of the wastes generated in the nuclear fuel cycle.

In the United States, fuel elements from commercial reactors are presently not processed. By the year 2000, the cumulative spent fuel quantity is expected to reach 16,000 cubic meters, amounting to 40,000 tonnes of uranium and fission products

TABLE 5.12 Normalized Radionuclide Releases from Three Nuclear Fuel Reprocessing Plants During the Period 1980-85.

	TBq per GWy(e)		
	SELLAFIELD	LA HAGUE	MARCOULE
Gaseous Releases			
^{85}Kr	$1.4 \times 10^{+4}$	$1.2 \times 10^{+4}$	$1.4 \times 10^{+4}$
^{3}H	$1.2 \times 10^{+2}$	3.2	$5.7 \times 10^{+1}$
^{14}C	3.5		
particulate (β)	6.3×10^{-2}	4.0×10^{-5}	2.9×10^{-4}
particulate (α)	2.3×10^{-4}	7.5×10^{-6}	
Liquid Releases			
^{3}H	$5.8 \times 10^{+2}$	$2.8 \times 10^{+2}$	$2.9 \times 10^{+2}$
particulate (β)	$9.7 \times 10^{+2}$	$2.6 \times 10^{+2}$	$2.7 \times 10^{+1}$
particulate (α)	8.0	1.6×10^{-1}	6.3×10^{-2}

Source: UN 1988.

(DOE 1988). Most of this spent fuel will be stored at the plant sites where it is generated, which are primarily in eastern states.

Low-level wastes generated in nuclear power plants vary from plant to plant and from year to year, depending on local conditions. On average, in 1981, pressurized-water reactors (PWRs) each generated about 400 m^3 of dry waste and 130 m^3 of wet waste. Boiling-water reactors (BWRs) each generated about 700 m^3 of dry waste and 300 m^3 of wet waste. By 1985, volume reduction techniques had reduced PWR dry and wet volumes to 200 m^3 and 70 m^3, respectively. BWR dry waste volumes had been reduced to about 425 m^3 but BWR wet wastes remained about the same (Taylor 1987).

Storage, as distinct from disposal, refers to waste retained in a retrievable form. Much spent nuclear fuel is stored in unprocessed form. There are three reasons for this practice. First, there is a need to allow radioactive decay of short-lived fission products *in situ* in order to minimize thermal loads and radioactivity inventories during later reprocessing. Second, there are in some jurisdictions unresolved political issues relating to possibly increased nuclear-proliferation risks associated with fuel reprocessing. Third, there are in other jurisdictions no economic incentives to reprocess, owing to the comparatively low costs of fresh fuel. Disposal of waste implies relinquishment of control over the waste, and disposal of HLW has yet to be undertaken. This awkward situation results, in part, from public concerns about safety and consequently from public opposition to the siting of HLW facilities — opposition which deters both appointed and elected officials from reaching decisions on HLW-disposal matters.

It has long been recognized that the favored method of disposal of HLW is placement of the wastes deep in geologically stable formations such as bedded salt or hard rock. Major national programs have been instituted to identify appropriate sites. The HLW technology employed by the nuclear power industry in France provides a good example of a long-term strategy for waste management. A centralized spent-fuel receiving and storage facility has been established where up to 1000 tonnes of fuel may be held in storage. "Prompt"reprocessing takes place after three to five years of storage during which shorter lived fissions products decay. Depending on ^{235}U and ^{239}Pu demand, some fuel may be held for 40 to 50 years prior to "delayed" reprocessing. High-level reprocessing wastes are vitrified, encapsulated in stainless steel, and placed in air-cooled storage vaults. Plans are to hold the vitrified wastes for 20 to 30 years, allowing further radioactive decay, prior to geologic disposal.

The favored method for disposal of LLW has long been shallow land burial. In general, waste is placed as received into trenches excavated in existing soil, with the overburden used to cover the filled trenches in a manner to promote water run-off. Again, the French LLW program serves as an example of a long-term management strategy. Longer-lived wastes and wastes whose packaging does not provide isolation are disposed of in reinforced concrete vaults or monoliths in pits. Two kinds of waste are placed in trenches — those whose activity is low enough to be considered intrinsically safe, and those whose packaging is as secure as concrete monoliths. Metal drums are stacked in concrete enclosures and the stacks are then backfilled and covered with a layer of impermeable clay. An outer layer of topsoil promotes vegetation and blending with the landscape. Permanent inventory records are maintained on the location and contents of each waste container.

The radiological impact on the public from waste management, including transportation, has so far been exceedingly small — no more than 0.1 percent of the total impact of the entire nuclear fuel cycle. Estimates of the possible future impact of geological disposal are very tenuous, as such disposal is intended to isolate wastes from the environment for at least 100,000 years (UN 1982).

5.4.2 Generation and Release of Radionuclides During Nuclear Power Generation

Accumulation of Fission-Product Radionuclides in Fuel.   In the fission process, most often two fragments of the parent atom are created (binary fission) with a distribution in mass as illustrated in Figs. 5.2 and 5.3. Occasionally a third fragment results (ternary fission), usually of low atomic mass and very frequently long-lived tritium, ^{3}H. As is illustrated in the figures, the mass distribution of fission products is bi-modal, with many products having atomic mass numbers around 95 and many others with atomic mass numbers around 140. Among the former are the very important long-lived isotope ^{90}Sr, various isotopes of the halogen bromine, and various isotopes of the noble gas krypton. Among the latter are the very important long-lived isotope ^{137}Cs, the singularly important halogen iodine, especially ^{131}I, and various isotopes of the noble gas xenon. The fission-product nuclei are neutron-rich and decay almost exclusively by beta-particle emission accompanied by gamma-ray emission. Several fission-product decay chains are illustrated in Fig. 5.4. There are characteristically several short-lived fission fragments decaying rapidly into a relatively long-lived daughter. For this reason, yields of fission fragments often are expressed as "chain yields," that is, cumulative yields accounting for precursors. The figure reports both individual and chain yields for a few important products of thermal-neutron induced fission of ^{235}U. An extended table of fission-product yields is given in Appendix H.

Also contained within the fuel are actinides produced by cumulative neutron absorption in uranium, thorium, and plutonium fuels. The actinides are characterized by spontaneous fission in competition with alpha-particle decay, and require sequestration to the same degree as the fission products.

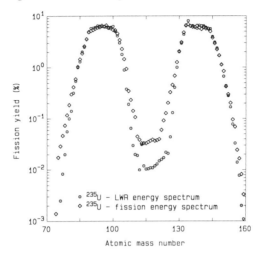

Figure 5.2   Fission product yields for ^{235}U, with fission induced by LWR-spectrum thermal neutrons and by fission spectrum neutrons. [Data are from Ryman (1984).]

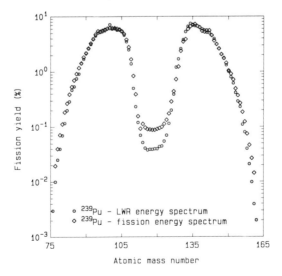

Figure 5.3   Fission product yields for ^{239}U, with fission induced by LWR-spectrum thermal neutrons and by fission spectrum neutrons. [Data are from Ryman (1984).]

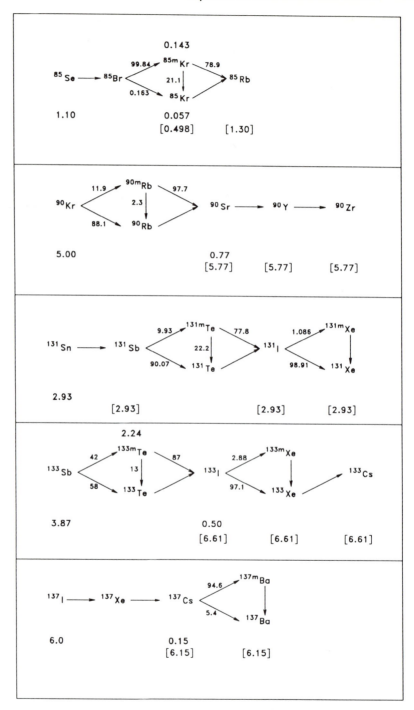

Figure 5.4 Individual and cumulative (chain) yields for selected fission-product chains. See ^{235}U data, App. H. Source: Individual yields from Ryman (1984); branching ratios from Kocher (1981), except for ^{131}Sb (ICRP 1979), and ^{133}Sb (Henry 1974). Chain yields are displayed in brackets.

There are good reasons for keeping track of the "inventory" of radionuclides within individual fuel elements or the entire core of a nuclear reactor. The inventory represents a potential source in the event of an accident and it governs the shielding requirements not only for the reactor in operation but also for spent fuel in storage, transportation, and processing. While procedures for detailed tracking of radionuclides is beyond the scope of this book, there are some approximate methods which are illustrative and which permit the making of rough estimates of inventories.

Suppose $P(t)$ is the thermal power delivered by the reactor core or some portion of the fuel elements, as a function of time t. The power may be related approximately to the fission rate by the factor $k = 3.1 \times 10^{16}$ s^{-1} per thermal MW. Suppose further that the fission yield of product i is Y_i and that the *number of atoms* of the product in the core or fuel elements is $N_{ci}(t)$. This is the *inventory* of radionuclide i, which may readily be converted to units of mass or activity. The following differential equation describes the rate of change of N_{ci} with time, accounting for various source and loss mechanisms.

$$\frac{dN_{ci}}{dt} = \left\{ \begin{array}{l} +kP(t)Y_i \\ +\sum_j \lambda_{ji} N_{cj} \\ +\sum_j \sigma_{ji}\phi N_{cj} \\ -\lambda_i N_{ci} \\ -\sigma_i \phi N_{ci} \\ -f\nu_i N_{ci} \end{array} \right\} \quad \left\{ \begin{array}{l} \text{production by fission} \\ \text{production by decay of precursors} \\ \text{production by transmutation} \\ \text{loss by decay} \\ \text{loss by transmutation} \\ \text{loss by escape from defective fuel} \end{array} \right. \quad (5.2)$$

In this equation, λ_i is the radioactive decay constant for species i, λ_{ji} is the rate of decay of species j into species i, σ_i is the neutron absorption cross section for species i, σ_{ji} is the cross section for neutron absorption in species j that leads to production of species i, and ϕ is average the neutron flux density in the fuel. In this formulation of the balance equation, f is the fraction of the inventory present in defective fuel, and ν_i is the rate constant, called the escape rate coefficient, for movement of species i from the defective fuel into the coolant.

A simpler form of the balance equation of Eq. (5.2) can often be used. The term involving decay of precursors may be omitted, as an approximation, and the individual yield for species i replaced by the chain yield. Furthermore, loss and gain by transmutation are generally of little significance, and, except under extraordinary circumstances, so is the leakage from defective fuel. The approximate balance equation then reduces to

$$\frac{dN_{ci}}{dt} = kPY_i - \lambda_i N_{ci}. \quad (5.3)$$

Were $P(t)$ constant and the initial inventory zero, the time dependence of the inventory would be

$$N_{ci}(t) = N_{ci}^\infty \left(1 - e^{-\lambda_i t}\right). \quad (5.4)$$

in which the equilibrium or steady-state inventory is given by

$$N_{ci}^\infty = \frac{kPY_i}{\lambda_i}. \quad (5.5)$$

As is evident from Eq. (5.4), inventories of long-lived radionuclides such as ^{137}Cs or ^{90}Sr would not approach equilibrium during the usual few years of fuel-element deployment in reactors. On the other hand the inventory of ^{131}I, which has a half-life of 8.05 days, can with reasonable conservatism be assumed to exist at equilibrium throughout power-reactor operation. To gain an appreciation of the magnitude of the equilibrium inventory, use Eq. (5.5) to calculate approximately the full-core, equilibrium inventory of ^{131}I in a LWR reactor operating at a thermal power of 3700 MW. From Fig. 5.4, the chain yield is 0.029. The decay constant is 9.96×10^{-7} s. Thus, the inventory is $N_{ci} = 3.3 \times 10^{24}$ atoms. In other units, this amounts to 5.6 mol, 730 g, 3.3 EBq, or 90 MCi.

Accumulation of Radionuclides in Reactor Coolant. There are two sources of radionuclides in reactor coolant, leakage from defective fuel and activation products produced by neutron interactions in the coolant itself or with fuel and structure in contact with the coolant. Activation product sources are inevitable, and include a number of radionuclides which may be produced in the coolant. For example, ^{16}N is produced as a result of neutron interactions with oxygen, ^{41}Ar as a result of neutron absorption in naturally occurring argon in the atmosphere and ^{3}H as a result of neutron absorption in deuterium and, especially in pressurized-water reactors, by neutron-induced breakup of ^{10}B. Of course, in a sodium-cooled fast reactor, activation of natural sodium to short-lived ^{24}Na is an important consideration for in-plant radiation protection. Other activation products include isotopes of iron, cobalt, chromium, manganese, and other constituents of structural and special-purpose alloys. The radionuclides are leached into the coolant stream. They then may be adsorbed on surfaces or trapped as particulates in the boundaries of coolant streams within the plant, only later to be resuspended in the coolant. These sources can be minimized by carefully specifying the alloy and trace-element concentrations in plant components.

The purification of the reactor coolant will be addressed in a later section. For the moment, assume that the cleanup can be characterized by a first-order rate constant λ_i', the fraction of species i removed per differential unit of time. Typically, λ_i' is on the order of 1 to 3×10^{-5} s^{-1}. By making approximations similar to those invoked in the above treatment of the in-core inventory, the variation of the coolant inventory $N_{wi}(t)$ (atoms of fission-product species i) can be described approximately by the differential equation

$$\frac{dN_{wi}}{dt} = f\nu_i N_{ci} - \lambda_i N_{wi} - \lambda_i' N_{wi}. \tag{5.6}$$

Here again, transmutation and precursor decay have been neglected. However, the leakage term from defective fuel must be included, as it is the only source term. Under steady-state conditions, Eq. (5.5) yields

$$N_{wi}^\infty = \frac{f\nu_i N_{ci}^\infty}{\lambda_i + \lambda_i'} = \frac{kPY_i f\nu_i}{\lambda_i(\lambda_i + \lambda_i')}. \tag{5.7}$$

Not surprising is the fact that the coolant inventory is proportional to the core inventory, and to the fraction of fuel present in defective elements. Escape-rate coefficients ν_i (Horton et al. 1969) are governed by the ability of the nuclide to diffuse through a fuel matrix, such as UO_2 and escape through the fuel cladding into the coolant. Representative values of ν_i, in units of s^{-1}, are as follows:

Noble gases	6.5×10^{-8}
Halogens, Rb, Cs	1.3×10^{-8}
Y, Zr, Nb, Ce, Pr	1.6×10^{-12}
Sr	1.0×10^{-11}
Te	1.0×10^{-9}
Mo	2.0×10^{-9}

As an example, a representative value for the inventory of ^{131}I in the coolant of a light water reactor is estimated. In this example assume $f = 0.0025$, $\lambda_i' \approx 3 \times 10^{-5} \text{ s}^{-1}$, and a coolant mass of 300 tonnes. With these values, $N_{wi}^\infty \approx 1.0 \times 10^{-6} \, N_{ci}^\infty$. In terms of specific activity, the inventory would be 90 MCi $\times 10^{-6} \div 300 \times 10^6$ g $= 0.3 \, \mu$Ci g^{-1}.

Reactor Coolant Purification and Waste-Stream Processing. Methods of coolant purification vary significantly from one reactor plant to another. In a boiling water reactor (BWR) there is only one coolant stream which circulates through the reactor and, as steam, through the turbine. An example coolant cleanup system is illustrated in Fig. 5.5 and determination of λ_i' for ^{131}I is illustrated in problem 6 at the end of the chapter. Noble gases released to the coolant in the reactor core are carried directly into the steam flow to the turbine. However, the weight fraction of iodine in the steam flow from the reactor is only about 1 percent of the weight fraction of iodine in the liquid coolant within the reactor.

Noncondensable gases (primarily air) are extracted from the vapor space in the turbine condenser. Radioactive gases such as those of krypton and xenon are removed along with any air. Halogen compounds are only partially removed in the gas stream, the remainder staying in the liquid. Hydrogen and oxygen formed by radiolysis in the core are converted catalytically to water and returned to the condenser. Some halogen compounds and particulates are then carried back to the reactor with the condensate. Particulate radionuclides in the gas stream are removed by filtration, and halogens are absorbed on activated charcoal, and there allowed to decay. The entire coolant flow passes through a condensate demineralizer before returning to the reactor. Even when the reactor is not in operation, a certain "let-down" coolant flow is extracted from the coolant and passed through a cleanup demineralizer.

For such a BWR system, the removal rate constant for radionuclide species i can be described by

$$\lambda_i' \approx \left(\frac{g}{W}\right) \eta_i + \left(\frac{s}{W}\right) \chi_i \left(\chi_i' + (1 - \chi_i')\eta_i'\right), \tag{5.8}$$

in which g and s are the cleanup and steam mass flow rates, and W is the total coolant mass. Cleanup-demineralizer and full-flow demineralizer efficiencies are denoted, re-

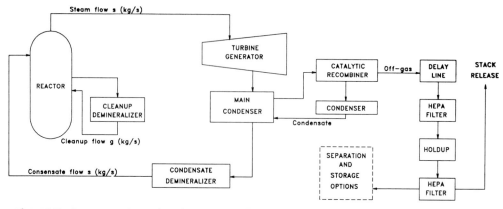

Figure 5.5　A representative coolant cleanup system for a boiling-water reactor. The HEPA filter designation refers to high-efficiency particulate air filter. The separation and storage options may include cryogenic separation of noble gases, with on-site storage of long-lived ^{85}Kr.

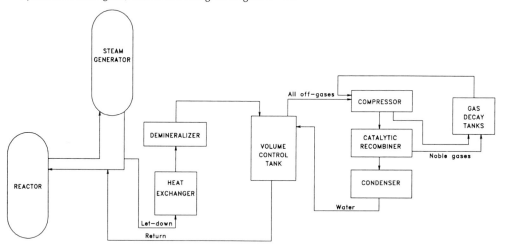

Figure 5.6　A representative primary-coolant cleanup system for a pressurized-water reactor. This system is designed for complete containment of all noble-gases in coolant streams.

spectively, by η_i and η_i' (fraction removed). The ratio of the weight fraction of species i in the steam phase to the weight fraction in the associated liquid phase is given by the partition factor χ_i. The net fraction of species i removed at the condenser air ejector is given by the factor χ_i'. Typical values of these parameters for ^{131}I are $\chi_i = 0.012$, $\eta_i = 0.9$, $\eta_i' = 0.999$, and $\chi_i' = 0.0005$.

In a pressurized water reactor (PWR) the primary coolant circulates in the liquid phase through the reactor and external steam generators. A portion of the coolant stream, about 40 to 80 gallons per minute, is extracted and dissolved gases are removed by sparging with hydrogen. This, among other purification processes, is done in the chemical and volume control system (CVCS) of the plant, as illustrated in Fig. 5.6. Typically, the hydrogen is then recombined with oxygen catalytically to form water, leaving only a small gas volume containing radionuclides. This volume is readily stored and allowed to decay.

There are several other liquid-waste streams that must be treated in a nuclear power plant. General categories are:

High-purity wastes. These streams are of high-chemical purity but, being primary coolant, have potentially high-radioactivity content. They arise from equipment drains associated with valve and pump seals, and so on, and from regeneration of demineralizer resins. In the PWR, the high-purity waste contains ^{3}H, and thus this liquid is ordinarily segregated from other liquid streams that do not containing tritiated water.

Low-purity wastes. These streams are associated with floor drains and similar sources, for which there may be many chemical impurities, but relatively little radioactive contamination.

Detergent wastes. These streams are associated with decontamination operations and the liquids contain chemical detergents. Radioactivity levels are low and these streams are likely segregated and reprocessed for reuse.

Liquid cleanup operations include filtration, demineralization, and evaporation. Efficiencies may be expressed as *decontamination factors*, that is, ratios of concentrations in inlet streams to concentrations in outlet streams. Representative values (Lee et al. 1986) are

Demineralizers	Anion	Cs/Rb	Other
mixed bed	10-100	2-10	10-100
cation bed	1	10	10
anion bed	100	1	1
powdex	10	2	10
Evaporators	100-1000 (iodine)		
	100-10000 (others)		
Reverse osmosis	10-30		

For two demineralizers in series, the decontamination factor for the second unit in the series is reduced by a factor of 10.

Typical Plant Discharges. Average radioactive releases to the environment per unit energy generation are summarized in Table 5.13 for world-wide nuclear reactor operations during the period 1975 through 1979. Typical isotopic compositions are summarized in Tables 5.14 and 5.15.

In light-water cooled reactors (LWR), ^{3}H arises principally from ternary fission and from activation of lithium or boron isotopes which may be exposed to neutrons. In heavy-water cooled or moderated reactors (HWR) ^{3}H arises principally from activation of deuterium. Various mechanisms also are responsible for production of ^{14}C. In LWRs and HWRs, principal sources are ternary fission, (n, α) reactions with ^{17}O present in the oxide fuel, and (n, p) reactions with ^{14}N present in the fuel as an impurity. In gas-cooled, graphite-moderated reactors, the principal sources of ^{14}C are reactions such as (n, γ) with ^{13}C in the moderator and (n, p) with ^{14}N impurities.

Releases to the environment during reactor operations are highly variable owing to variations in operating practices and technology. Variations are so great that

TABLE 5.13 Normalized Environmental Releases of Radionuclides from World-Wide Nuclear Reactor Operations During 1975 to 1984.

	TBq per GWy(e)				
	NOBLE GASES[a]	^{3}H	^{14}C[b]	^{131}I	PARTICULATES[c]
Gaseous Effluents—1975–1979					
PWR	430	7.8	0.22	0.0019	0.0022
BWR	8800	3.4	0.52	0.040	0.053
HWR	460	540	17		0.00004
GCR	3240		1.1		
Gaseous Effluents—1980–1984					
PWR	218	5.9	0.35	0.0018	0.0045
BWR	2150	3.4	0.33	0.0093	0.043
HWR	212	670	6.3	0.00023	0.00004
GCR	2320	5.4		0.0014	0.0014
LWGR	5470		1.3	0.080	0.016
Liquid Effluents—1975–1979					
PWR		38			0.18
BWR		1.4			0.29
HWR		350			0.47
GCR		25			4.8
Liquid Effluents—1980–1984					
PWR		27			0.13
BWR		2.1			0.12
HWR		290			0.026
GCR		96			4.5
LWGR		1.7			

[a]Includes ^{41}Ar.

[b]Data for Argentina, United Kingdom, Federal Republic of Germany, Finland, and the USSR.

[c]All nuclides except ^{3}H for liquid effluents.

Source: UN 1982, 1988.

the historical averages are meaningless for uses other than assessment of the impact of past operation. The reader is cautioned not to apply these data to any one plant, either retrospectively or prospectively. As an example of the extreme variability, consider only the operation of 31 pressurized water reactors in the United States in 1979. The average normalized release rate for ^{133}Xe in airborne effluents was 470 TBq per GWy(e). However, the range was 3 to 5000, and the median was 180. The high extreme was due to abnormal operations in one plant. Excluding that plant from consideration, the average release rate was 350 TBq per GWy(e), with a standard deviation of 330.

5.4.3 Public Dose Commitments from the Fuel Cycle

Assessment of the impact of radionuclides released into the environment require local, regional, and global models for dispersal of the nuclides and their transport through hydrological and food-chain pathways to man. These models are subjects of

Chaps. 9 to 11 of this text. Selected examples of results are provided at this point as illustrations of the use of these models for estimating dose commitments from the nuclear fuel cycle.

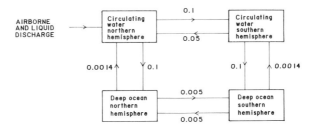

Figure 5.7   Model used for global circulation of ^{3}H and ^{129}I. Factors shown are fractional exchange rates, in units of y^{-1}. [From UN (1982)]

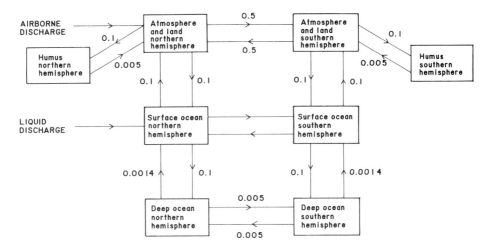

Figure 5.8   Model used for global circulation of ^{14}C. Factors shown are fractional exchange rates, in units of y^{-1}. [From UN (1982)]

Pathway Models.   Figures 5.7 and 5.8 illustrate global models for dispersal of ^{3}H, ^{129}I, and ^{14}C. Application of these models to long-term consequences of releases to the environment in the northern hemisphere is illustrated in Figs. 5.9 to 5.11 The cumulative (time-integral) activities are of use in evaluation of long-term dose commitments to the population.

Dose Commitments.   Estimates of local and regional contributions to the normalized collective effective dose equivalent resulting from world nuclear fuel cycle operations are summarized in Table 5.16. Of the total 4 person-Sv per GWy(e), about 90 percent is delivered within 1 year of release to the environment. About 98 percent is delivered within 5 years. Again, it must be emphasized that these estimates are global averages representing a wide range of technologies and operating experiences.

TABLE 5.14 Isotopic Composition of Airborne Radionuclides Released from U.S. Pressurized Water and Boiling Water Reactors in 1982.

RADIONUCLIDE	ACTIVITY FRACTION	
	PWR	BWR
Noble Gases		
^{41}Ar	0.0050	0.029
^{85m}Kr	0.0042	0.061
^{85}Kr	0.0162	0.013
^{87}Kr	0.0086	0.083
^{88}Kr	0.0039	0.143
^{131m}Xe	0.0063	0.034
^{133m}Xe	0.0059	0.0071
^{133}Xe	0.806	0.198
^{135m}Xe	0.0020	0.056
^{135}Xe	0.139	0.171
^{138}Xe	0.0030	0.195
Other		0.01
Iodine		
^{131}I	0.272	0.065
^{133}I	0.680	0.270
^{135}I	0.043	0.658
Other	0.005	0.007

Source: UN 1988.

TABLE 5.15 Isotopic Composition of Liquid Effluents from Nuclear Power Plants in 1982. U.S. BWRs and PWRs, U.K. GCRs.

NUCLIDE	NORMALIZED RELEASE, GBq per GWh(e)		
	BWR	PWR	GCR
^{54}Mn	6.9	2.5	3.6
^{58}Co	3.3	34	
^{60}Co	30	19	18
^{65}Zn	11	0.11	
^{89}Sr	0.90	1.3	2.1
^{90}Sr	0.15	0.23	290
^{106}Ru		0.20	43
^{110m}Ag	0.82	1.6	
^{125}Sb		1.2	35
^{131}I	3.2	29	
^{134}Cs	6.6	10	450
^{137}Cs	13	18	2100
^{144}Ce	0.57	0.37	123

Source: UN 1988.

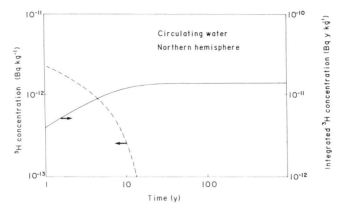

Figure 5.9   The environmental activity concentration of ^{3}H in circulating water in the northern hemisphere caused by release of 1 Bq s^{-1} during 1 year (dashed line). Also shown is the cumulative (time integral) of the activity concentration (solid line). [From UN (1982)]

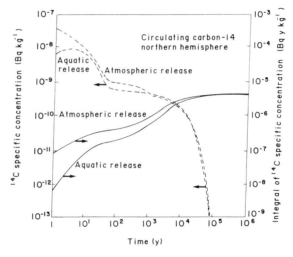

Figure 5.10   Time variation and time integral of environmental activity concentrations of ^{14}C circulating in the northern hemisphere caused by release of 1 Bq s^{-1} during 1 year. [From UN (1982)]

Their application to assessments for any one installation cannot be justified. Nor can their extension very far into the future be justified. Improved technology such as zero-release pressurized-water power plants is already reducing public dose commitments.

To appreciate the variability of dose-commitment assessments, the reader may contrast the data of Table 5.16 with that of an NCRP assessment (NCRP 1987a) of collective effective dose equivalents to regional populations from nuclear fuel cycle operations. The NCRP assessment, given in Table 5.17 is based on 1980 operating records for 47 U.S. nuclear power plants, on operating records for conversion, utilization, and fabrication plants, and on computer modeling for mining and milling operations.

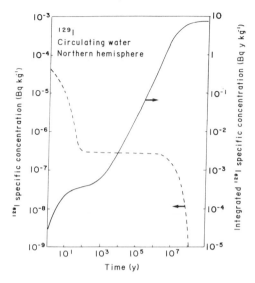

Figure 5.11   Time variation and time integral of environmental activity concentrations of ^{129}I in circulating water in the northern hemisphere caused by release of 1 Bq s^{-1} during 1 year. [From UN (1982)]

Very tenuous estimates of global contributions to the normalized collective effective dose equivalent are summarized in Table 5.18. The estimates are based on a world population reaching the constant level of 10 billion by the year 2000. The estimates are for *cumulative* dose equivalents and, of course, those for 10^6 years, if expressed as per capita annual averages, would appear minuscule. Understanding of this table may be aided by examining the following scenario (UN 1982):

> In order to estimate the maximum *per caput* annual effective dose equivalent in the future as a result of nuclear power production, an incomplete collective dose equivalent commitment truncated over the expected duration of the practice of generation of electricity by fission power, taken here as 500 y, must be used. The releases during the operational stage of the nuclear fuel cycle lead to a local and regional normalized collective effective dose equivalent[1] of 5.7 man Sv per GWy(e) of which 98% is received in the first few years after discharge. For those nuclides which become globally dispersed, the incomplete normalized collective effective dose equivalent commitment to 500 y is about 18 man Sv per GWy(e). The choice of 500 y as a mean duration of the practice of producing power by nuclear fission implies the use of breeder reactors and the rate of mining would decrease. The normalized incomplete collective effective dose equivalent commitment from mining and milling, based on the present fuel cycle, is therefore to 100 y and is likely to be due mainly to radon releases giving 2.5 man Sv per GWy(e). Thus, on the pessimistic assumption that no technological improvements are made and current levels of discharge continue for 500 y, the maximum annual collective effective dose equivalent would be about 25 man Sv per GWy(e). The annual . . . *per caput* effective dose equivalent, . . . assuming that the present released levels are not reduced and that the annual generation of electricity reaches 1 kWy *per caput*, [would reach 25 μSv] in 2500. It can be

---

[1]The 1982 estimate of 5.7 man-Sv was reduced in 1988 to 4 person-Sv (Table 5.17).

TABLE 5.16 Summary of Local and Regional Normalized Collective Effective Dose Equivalent Commitments to the Public from Nuclear Power Production.

SOURCE	person-Sv per GWy(e)
Mining	
radon	0.3
Milling	
uranium, thorium, radium	0.02
radon	0.12
Fuel Fabrication	
uranium	0.003
Reactor Releases	
Atmospheric	
noble gases	0.24
^{3}H	0.53
^{14}C	1.6
iodines	0.003
particulates	0.08
Aquatic	
^{3}H	0.03
other	0.013
Fuel Reprocessing	
Atmospheric	
^{3}H	0.007
^{85}Kr	0.005
^{14}C	0.04
other	0.02
α emitters	0.001
Aquatic	
^{134}Cs, ^{137}Cs	0.8
^{106}Ru	0.4
^{90}Sr	0.03
α emitters	0.005
Transportation	0.1
Total	4

Source: UN 1988.

seen that even with the maximizing assumptions made here, the level of annual *per caput* effective dose equivalent rises to the equivalent of 1% of natural background radiation. The annual *per caput* effective dose equivalent would reduce after the end of the practice, to about 1% of the final value after 100 y.

5.4.4 Occupational Exposures from Nuclear Power

In mining and milling, occupational exposures are due mainly to inhalation of radon daughter products and whole-body gamma-ray irradiation. In other phases, occupational exposures are due mainly to whole-body gamma-ray irradiation. Occupational

TABLE 5.17 Annual Collective Effective Dose Equivalents to Regional Populations Normalized to a 1 GWe Nuclear Power Plant Operating 80% of the Time.

OPERATION	COLLECTIVE EFFECTIVE DOSE EQUIVALENT (person-Sv)
Mining	0.94
Milling	0.25
Conversion	0.0003
Enrichment	0.00004
Fabrication	0.048
Transportation (general)	0.071
Transportation (accidents)	0.054
Total	1.36

Source: NCRP 1987a; by permission of the NCRP.

TABLE 5.18 Summary of Global Normalized Collective Effective Dose Equivalent Commitments to the Public from Nuclear Power Production (Including Releases from the Fraction of Fuel that is Reprocessed).

INTEGRATION PERIOD (y)	person-Sv per GWy(e)				
	10^1	10^2	10^3	10^4	10^6
^{3}H	0.003	0.004	0.004	0.004	0.004
^{85}Kr	0.07	0.12	0.12	0.12	0.12
^{14}C	1.7	6.3	12	63	63
^{129}I	—	0.0008	0.0016	0.0093	1.5

Source: UN 1988.

exposure records are based on personnel dosimetry and on working-level-month radon daughter exposures. As of this writing, records rarely include dose commitments.

Reported average radiation doses require careful interpretation, as they may be based on the total number of workers employed, the number of workers monitored for radiation exposure, or the number of workers actually receiving radiation exposure. Although there is unquestionably a trend in the reduction of individual exposures, such exposures, too, require careful interpretation. It does not necessarily follow that reducing individual doses leads to a reduction in collective doses. Indeed, fostering a balance between individual and collective doses is a major responsibility of the radiological-assessment specialist. Normalized collective doses arising from nuclear power operations in the United States during the period 1973 to 1979 show no discernible trends, ranging from 13 to 22 person-Gy per GWy(e) for boiling water reactors and from 8 to 25 for pressurized water reactors.

The distribution of annual doses to nuclear power workers in the United States is illustrated in Fig. 5.12. This distribution may be compared with that for medical

workers, which is illustrated in Fig. 5.13. Average normalized collective dose equivalents for the various phases of the nuclear fuel cycle are presented in Table 5.19. For
occupational exposure in the fuel cycle as a whole, U.S. data for 1980 (EPA 1984)
indicate that 91 thousand workers actually exposed received an annual average dose
equivalent of 600 mrem (6 mSv).

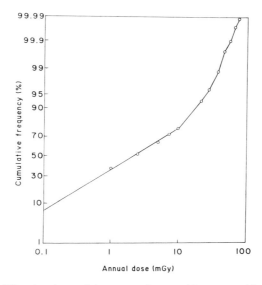

Figure 5.12   Log-probability plot of annual doses to workers receiving measurable doses at light-water reactors in the United States for 1978. [From UN (1982)]

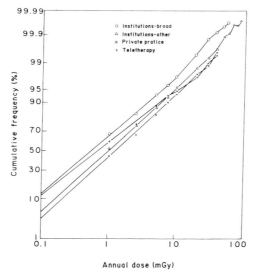

Figure 5.13   Log-probability plot of doses to medical workers with measurable exposure in the United States
for 1975. [From UN (1982)]

TABLE 5.19 World-Average Normalized
Occupational Contributions to Collective Dose
Equivalents from the Nuclear Fuel Cycle.

ACTIVITY	person-Sv per GWy(e)
Mining and milling	0.7
Fuel manufacture	0.5
Reactor operation	10
Reprocessing	0.25
Transportation	0.2
Research and development	5

Source: UN 1982, 1988.

5.5 MISCELLANEOUS SOURCES

Various modern technologies have led to human radiation exposures in excess of those which would have occurred in the absence of the technologies. For example, the mining of coal and other minerals and their use industrially and domestically is responsible for increased releases of naturally occurring radionuclides to the environment.

5.5.1 Utilization of Coal

World production of coal is about 3.7 billion tonnes annually. About 70 percent is used in generation of electricity, the balance mainly in domestic heating and cooking. There are very wide variations in the radioactivity content of coal. The 1982 UNSCEAR report (UN 1982) adopts the following weighted averages, in units of Bq kg^{-1}: ^{40}K, 50; ^{238}U, 20; and ^{232}Th, 20; with daughter products in the decay chains being in equilibrium with the parents. In individual coals, radioactivity content may vary from these averages by as much as a factor of 10.

Partitioning of radionuclides between fly ash and bottom ash during coal combustion is influenced by the flame temperature, by the volatility of the compounds produced, and by the chemical forms of the radioactive materials in the coal. The smaller-sized fly ash particles are more heavily enriched with radionuclides, particularly ^{210}Pb and ^{210}Po. Estimated global average concentrations of radionuclides in fly ash are given in Table 5.20.

The average ash content of coal is about 10 percent by weight, but may be as high as 40 percent. Efficiency of ash removal in power plants is quite variable — from only 80 percent removal to as much as 99 percent removal, the average being about 97.5 percent. In domestic use of coal, as much as 50 percent of the total ash is released to the atmosphere (UN 1982).

In terms of doses to individual tissues, the main impact of atmospheric releases during combustion is the dose to bone surface cells accruing from inhalation of ^{232}Th present in the downwind plumes of particulates from plants. This bone surface dose is also the major component in the effective dose equivalent commitment. A summary of dose commitments from the ^{238}U and ^{232}Th decay chains is given in Table 5.21. Effects of ^{40}K are not relevant, since the potassium content of the human body is

TABLE 5.20   Estimates of Concentrations in Fly Ash and Normalized Atmospheric Discharges of Radionuclides Resulting from Combustion of Coal for Electricity Generation.

NUCLIDE	AVERAGE CONCENTRATION IN FLY ASH	NORMALIZED ATMOSPHERIC RELEASE RATE
	Bq per kg	GBq per GWy(e)^d
^{40}K	500	0.60–12
^{238}U^a	200	0.25–5
^{210}Pb^b	600	0.75–15
^{232}Th^c	200	0.25–5

^{a} In equilibrium with daughters to ^{210}Pb.

^{b} In equilibrium with ^{210}Po.

^{c} In equilibrium with daughters.

^{d} Range for modern and old plants with, respectively, 99.5 and 90 percent fly-ash removal efficiency.

Source: UN 1982, 1988.

under homeostatic control. Most of the dose commitment is realized shortly after radionuclide release, thus the total normalized effective dose equivalent of 2 person-Sv per GWy(e) for electricity generation by coal combustion is comparable to the 4 person-Sv per GWy(e) for electricity generation by nuclear energy.

The total annual committed effective dose equivalent from global coal utilization may be estimated on the basis of 70 percent of the coal production used for electricity generation and the combustion of 3×10^9 kg per GWy(e) of energy. The result is about 2000 person-Sv annually. The same assessment is much more tenuous for the 30 percent of the coal used mainly for domestic purposes. There are few studies on effects of domestic use, but it is apparent that, in domestic use, ash releases are much greater, take place at lower elevation, and occur in more populated areas. Some estimates suggest that the resulting annual total global commitment is even greater than that due to electricity generation.

There are other consequences of the presence of radioactivity in coal. Ash finds use in light-weight aggregates for structural and highway use, and, consequently, may expose populations far from where the coal was burned. However, little information is available on resulting population exposure.

5.5.2 Geothermal Energy Production

The main radiological impact associated with utilization of geothermal energy for electricity production is attributable to ^{222}Rn release to the atmosphere, estimated at 400 TBq per GWy(e). This is expected to result in a normalized collective effective dose equivalent commitment of about 6 person-Sv per GWy(e) (UN 1982). While this commitment exceeds that from coal utilization, geothermal energy production is responsible for only about 0.1 percent of total electricity generation.

TABLE 5.21   Collective Committed Doses and Effective Dose Equivalents Arising Globally from the Combustion of Coal in the Generation of Electricity.

ORGAN/TISSUE	NORMALIZED COLLECTIVE ABSORBED DOSE, person-Gy per GWy(e)			
	INHALATION DURING CLOUD PASSAGE	INTAKE AFTER DEPOSITION	EXTERNAL	TOTAL
Lung	0.210	0.170[a]	0.090	0.470
Bone surface	1.070	0.120	0.090	1.280
Red marrow	0.089	0.017	0.090	0.196
Liver	0.005	0.015	0.090	0.110
Kidneys	0.005	0.030	0.090	0.125
Spleen	0.006		0.090	0.096
GI system	0.0003		0.090	0.090
Other	0.0004	0.011	0.090	0.101
	NORMALIZED COLLECTIVE EFFECTIVE DOSE EQUIVALENT COMMITMENT, person-Sv per GWy(e)			
Average	1.40	0.56	0.09	2.00
Range[b]	0.23–4.6	0.26–1.4	0.015–0.3	0.5–6.2

[a]0.170 to bronchial basal cells and 0.049 to pulmonary epithelium.

[b]Range for modern and old plants with, respectively, 99.5 and 90 percent fly-ash removal.

Source: UN 1982, 1988.

5.5.3 Exploitation of Phosphate Rock

Annually some 1.4 billion tonnes of phosphate rock are mined and processed for use in production of fertilizers (70 percent) and phosphoric acid. Byproduct (phospho)gypsum finds wide use in the construction industry. The U.S. produces about 38 percent of the phosphate rock, the USSR 19 percent, and Morocco 14 percent. While ores in the USSR are primarily magmatic, with low ^{238}U concentration, those from the United States and Morocco are primarily sedimentary, with high concentrations of ^{238}U, generally in equilibrium with daughters. No ores have unusually high concentrations of ^{40}K or ^{232}Th.

Most airborne radioactivity releases are associated with dust produced in strip mining, grinding, and drying of the ore. Utilization of the phosphates leads to both internal and external radiation exposure, the greatest exposure resulting from use of byproduct gypsum in construction. The total annual collective effective dose equivalent commitment from phosphate rock exploitation is estimated at about 3×10^5 person-Sv, of which all but 6000 person-Sv is due to gypsum utilization (UN 1982).

5.5.4 Industrial Uses of Radiation Sources

Radionuclides used in industry contribute very little to collective population doses, although individual occupational exposures may be significant (Table 5.22). Eichholz (1983) catalogs industrial sources and notes that the sources with the greatest activities are those used in radiography, typically comprising 10 to 100 Ci of ^{192}Ir, ^{137}Cs, ^{170}Tm,

TABLE 5.22 Occupational Exposure from Miscellaneous Sources of Radiation in the United States.

	NUMBERS OF RADIATION WORKERS (THOUSANDS)		ANNUAL EFFECTIVE DOSE EQUIVALENT (mSv)[a]	
	ALL	EXPOSED	ALL	EXPOSED
Nuclear power (1984)		98.1		5.6
Other industrial (1970–75)	6.9	2.1	1.6	5.1
Well-logging (1979)	8.7	7.3	3.5	4.2
Dept. of Energy (1983)		80.6		2.0
Public Health Service (1983)	4.6	0.7	0.1	0.5
U.S. Navy Fleet (1984)	30		0.6	
U.S. Navy Shipyard (1984)	21		1.5	
Other Government (1980)	204	105	0.6	1.2
Flight crews/attendants (1980)	97	97	1.7	1.7
Education and transportation (1980)	76	31	0.7	1.6
Other (1980)	115	107	1.7	1.8

[a] To convert to mrem, multiply by 100.

Source: NCRP 1989b.

or ^{60}Co. Borehole logging is accomplished using somewhat lower activity gamma-ray sources and neutron sources such as Pu-Be, Am-Be, or ^{252}Cf. Much lower activity sources, often ^{90}Sr-^{90}Y beta-particle sources, are used for various instrumentation and gauging applications. As with fixed sources used in medical teletherapy, there have been all too frequent cases of industrial sources finding their way into surplus-equipment commerce or municipal wastes.

5.5.5 Consumer Products and Miscellaneous Sources

Table 5.23 is a summary of radiation exposure to the U.S. population from miscellaneous sources. Data are for three categories of sources: (1) sources giving large numbers of people relatively large individual dose equivalents, (2) sources giving large numbers of people relatively small dose equivalents or with dose equivalents to small portions of the body, and (3) sources giving few people small collective dose equivalents.

As a result of these miscellaneous sources, the average person in the United States accrues, in total, an annual effective dose equivalent of from 60 to 130 μSv (6 to 13 mrem). Not included in this total is the annual average dose equivalent of 0.16 Sv (16 rem) to the bronchial epithelium received from ^{210}Po by cigarette smokers. This contributes about 13 mSv to the effective dose equivalent for each smoker. This contribution was not included in the table because it is thought that the main risk of smoking results from effects other than radiation dose.

5.5.6 Occupational Exposure

Table 5.22 summarizes occupational exposures from various sources in the United States. While many workers may be potentially exposed to radiation and may be pro-

TABLE 5.23 Average Annual Effective Dose Equivalents and Effective Dose Equivalent Commitments to the U.S. Population from Exposure to Consumer Products.

SOURCE	MILLIONS EXPOSED	ANNUAL EFFECTIVE DOSE EQUIVALENT (μSv/person) TO EXPOSED POPULATION[a–c]
Domestic water supplies	230	10–60
Building materials	120	70
Mining & agricultural products	200	5–50
Combustible fuels		
Coal	230	0.3–3
Natural gas heaters	16	18
Natural gas cooking ranges	125	4
Dental prostheses	45	0.7
Opthalmic glass	50	< 4
Television receivers	230	< 10
Video display terminals	50	< 10
Radioluminous timepieces	15–20	0.4–1
Airport luggage inspection	30	0.021
Gas and aerosol detectors	100	0.08
Highway construction materials	5	40
Aircraft transport of radionuclides	14	2.4
Elec. tubes and spark gap irradiators	230	0.04
Thorium products gas mantles	50	2
Thorium products		
Welding rods	0.3	160
Check sources	0.8	< 10

[a] To convert to mrem, multiply by 0.1.

[b] To convert to average annual effective dose equivalent for the entire U.S. population, multiply by the ratio of the number exposed to the 230 million total population.

[c] To convert to the annual collective effective population dose equivalent, multiply by the number exposed.

Source: NCRP 1987b.

vided with personnel dosimetry, only a small fraction are measurably exposed. The table, when possible, lists annual effective dose equivalents as averages for all workers in a class and for just those actually receiving radiation exposure from occupational sources.

5.6 REGULATION, STORAGE, AND DISPOSAL OF RADIOACTIVE WASTES

5.6.1 Waste Categories

High-level waste (HLW) includes (1) those highly radioactive materials, containing mainly fission products as well as some actinides that are separated during chemical processing of irradiated fuel, (2) spent fuel itself, if it is declared to be a waste, and (3) any other waste with a radioactivity level comparable to (1) or (2). Low-level waste (LLW) includes all other waste forms. In some jurisdictions, there are various categories of LLW, with different requirements for disposal. Medical and (nonreactor) industrial radioactive wastes are in the category of LLW and, although volumes may be large, generally contain only short-lived radionuclides (ICRP 1985).

5.6.2 U.S. Regulations for LLW Disposal

U.S. federal standards for protection against radiation are found in Title 10, Code of Federal Regulations, Part 20, which is often abbreviated as 10CFR20. These standards apply to licensees of the Nuclear Regulatory Commission which include all civilian nuclear-reactor licensees as well as licensees for use of by-product radionuclides in about half the states. In other states — known as "agreement states" — use of radionuclides is under state jurisdiction whose regulations are very similar to federal regulations.

10CFR20 permits licensees to release small quantities of radionuclides to the atmosphere or to bodies of water, provided (1) concentrations are kept within very strict limits, and (2) radiation doses to individual members of the public and to population groups are kept within very strict limits, as is also required by 10CFR50, the regulations governing licensing of nuclear reactors. Nuclear-fuel-cycle licensees are also required to meet the requirements of 40CFR190 which deal with "environmental radiation protection standards."

Special provisions are provided in 10CFR20 for certain special radioactive wastes characteristic of research laboratories and medical facilities, namely scintillation fluors and animal tissues. ^{3}H or ^{14}C in these materials may be disposed of without regard to radioactivity if the specific activity is less than 0.05 μCi g^{-1} (for tissue, averaged over the weight). Of course, there must be no possibility that contaminated tissue find its way into the livestock or human food chains. 10CFR20 permits unregulated release to sanitary sewerage of all excreta from individuals undergoing medical diagnosis or therapy with radioactive material.

Other special provisions are provided in 10CFR20 for release of radionuclides to sanitary sewerage systems by any one licensee. The material must be readily soluble or dispersible in water. Then, there are criteria for individual radionuclides with limiting concentrations after dilution based on monthly releases and monthly dilution volumes. Gross annual releases cannot exceed 1 Ci total for all nuclides except for ^{3}H and ^{14}C, for which the annual limits are, respectively, 5 Ci and 1 Ci.

In the United States, solid LLW may be transferred to a federally licensed site for near-surface land disposal. Technical criteria for siting and operating a licensed disposal facility are found in 10CFR61. LLW may be accepted in three classes for which there are different disposal criteria. The class depends on the specific activities and the half-lives of the radionuclides contained in the wastes. The shorter-lived, low specific-activity wastes are in Class A. For example, Class A waste may contain up to 700 Ci per m^3 of ^{60}Co but only up to 0.04 Ci per m^3 of longer-lived ^{90}Sr. For ^{90}Sr, specific activities between 0.04 and 44 Ci per m^3 are Class B and specific activities between 44 and 4600 Ci per m^3 are Class C. Greater specific activities cannot be disposed of as LLW. Low-level waste must be stable in volume and not packed in cardboard or fiberboard. There are strict limits on liquid and gas contents, on biological hazards, and on corrosive, pyrogenic, and explosive characteristics. Wastes of different classes may be segregated to take advantage of the different standards for each class in the disposition of the wastes in the repository.

5.6.3 Disposal Strategies for High-Level Waste

Various means have been proposed for the permanent disposal of high-level wastes. In the United States, the decision was reached in 1982 to emphasize construction of a deep-mined geologic repository as the first element of a system for radioactive waste disposal.[2] After a lengthy screening process to select potential sites for a HLW repository, an arid site in Nevada was selected in 1988 for detailed study. On this site HLW would be stored permanently in chambers excavated in volcanic tuff a few hundred meters underground and well above any existing water table. However, to date no decision to begin construction of a permanent HLW repository has been made, and in all likelihood, HLW will probably not begin to be placed in permanent storage at some repository until early in the twenty-first century.

Permanent storage of high-level wastes in deep-mined geologic formations is not the only disposal strategy possible. Some of the alternative disposal means are summarized in the following paragraphs (Flowers 1976, DOE 1989).

Subseabed Disposal.   This disposal strategy involves burial of waste in high integrity containers beneath the ocean floor in tectonically stable, clay-rich sediments of mid-plate regions. While wastes would remain isolated from the biosphere for extremely long times, there are uncertainties in the technology of emplacement of the containers and in the establishment of international structures for regulation and monitoring.

Deep-Hole Disposal.   This method would involve placement of waste containers at depths as great as 10 km, far from the accessible environment and beneath circulating groundwater. There are uncertainties in the technology of drilling and emplacement of wastes and in rock mechanics at the extremes of temperature and pressure associated with this option.

Rock-Melt Disposal or Deep-Well Injection.   Rock-melt disposal would involve injection of liquid or slurry waste into deep underground cavities, with the waste's internal heat generation leading to melting of the rock. Ultimately the rock would resolidify, trapping the material. Deep-well injection would involve pressurized pumping of acidified liquid wastes into deep, hydrofractured formations isolated from the biosphere by impermeable overlying strata. These options were rejected out of hand in the Royal Commission Study (Flowers 1976) and are not considered viable by the DOE (1989) because of the perhaps thousand year delay before resolidification of the waste-rock mix and because of the difficulty in assuring no liquid exchange with aquifers.

Island Geologic Disposal.   Disposal would take place deep beneath the surfaces of remote, uninhabited islands devoid of mineral resources and hydrologically

---

[2]Nuclear Waste Policy Act of 1982: P.L. 97-425 of January 7, 1983 as amended by Title V of P.L 100-203 of December 22, 1987 and P.L. 100-507 of October 18, 1988.

separated from continental land masses. There are uncertainties in potential return pathways to the biosphere possibly arising from seismic and volcanic activity.

Ice Sheet Disposal.   Polar ice sheets are thousands of feet thick and millions of years old. Waste containers placed on the surface would melt the ice and move downward, with ice resealing the passageway of the canisters. There are uncertainties in the long-term stability of ice sheets in view of potential global climate changes, as well as uncertainties in potential return pathways to the biosphere along the interfaces between ice sheets and basement rock.

Space Disposal.   Ejection of waste into the sun or into solar orbit, though possible, would involve the risk of launch failure with potentially disastrous consequences of atmospheric dispersal of the wastes. The method was rejected in the Flower Commission study and jointly by the U.S. Department of Energy and by the National Aeronautic and Space Administration.

Transmutation.   This option is a treatment rather than a disposal method. It could be applied if spent fuel were chemically processed to extract the actinides. The actinides could then be recycled with fresh fuel into reactor operations, with the result of their being transmuted to stable or short-lived radionuclides. The costs would be great and there would be no reduction in the need for long-term storage of fission products such as ^{99}Tc and ^{129}I.

Surface Storage.   In the United States and Western Europe, surface storage is an interim step in waste treatment prior to ultimate disposal. Spent fuel or containers of separated high-level waste would be stored indefinitely in water basins or in dry storage at reactor sites or other collection sites. Monitored surface storage would be continued pending resolution of disputes over permanent disposal methodologies and sites. Waste and fuel elements would be retrievable, allowing separation of by-products and, in the case of spent fuel, unconsumed uranium and plutonium suitable for recycle into power production.

5.6.4 Underground Disposal of High-Level Wastes

The International Atomic Energy Authority (IAEA 1989) has set out the following internationally agreed safety principles and technical criteria for disposal of HLW in underground repositories:

Safety Principles

The burden on future generations shall be minimized by safely disposing of high level radioactive wastes at an appropriate time, technical, social and economic factors being taken into account.

The safety of a high-level waste repository in the post-sealing period shall not rely on active monitoring, surveillance or other institutional controls or remedial actions after the time when the control of the repository is relinquished.

The degree of isolation of high-level radioactive wastes shall be such that there are no predictable future risks to human health or effects on the environment that would not be acceptable today.

As a basic principle, policies and criteria for radiation protection of populations beyond national borders from releases of radioactive substances should not be less stringent than those for the population within the country of release.

For releases from a repository due to 'gradual' processes, the predicted annual dose to individuals of the critical group shall be less than the dose upper bound apportioned by national authorities from the relevant individual dose limits which currently correspond to an annual average dose value of 1 mSv for prolonged exposures.

The level of safety for high-level radioactive wastes shall be such that the predicted risk of a health effect in a year from a repository to an individual of the critical group from disruptive events not covered by [the previous principle] is less than a risk upper bound apportioned by national authorities from an individual limit of risk of health effects of one in a hundred thousand per year.

All radiation exposures that may result from the disposal of high-level radioactive wastes shall be as low as reasonably achievable, economic and social factors being taken into account. The dose and risk upper bounds as defined in [the previous two] principles shall be overriding constraints.

Technical Criteria

The long-term safety of high-level radioactive waste disposal shall be based on the multi-barrier concept,[3] and shall be assessed on the basis of the performance of the disposal system as a whole.

Waste acceptance criteria shall be established for radionuclide content consistent with assumptions made in the repository design.

High-level radioactive wastes to be emplaced in a repository shall be in a solid form with chemical and physical properties favouring the retention of radionuclides and appropriate to the disposal system.

A high-level waste disposal system shall be designed in a way that aims at substantially complete isolation of radionuclides for an initial period of time.

A high-level radioactive waste repository shall be designed, constructed, operated, and closed in such a way that the post-sealing safety functions of the host rock and its relevant surroundings are preserved.

The high-level radioactive waste repository shall be designed and the waste emplaced such that any fissile material remains in a subcritical state.

The repository shall be located at sufficient depth to protect adequately the emplaced waste from external events and processes, in a host rock having properties that adequately

---

[3]Barriers are constituted by waste chemical and physical matrix, containers, backfill material, the repository, host rock, and surrounding geological formations. The safety of waste disposal must not rest on any one barrier. If any one barrier fails, the overall system should still meet safety objectives.

restrict the deterioration of physical barriers and the transport of radionuclides from the repository to the environment.

The repository site shall be selected, to the extent practicable, to avoid proximity to valuable natural resources or material which are not readily available from other sources.

Compliance of the overall disposal system within the radiological safety objectives shall be demonstrated by means of safety assessments based on model that are validated as far as possible.

A quality assurance programme for components of the disposal system and for all activities from site confirmation through construction and operation to closure of the disposal facility shall be established to ensure compliance with relevant standards and criteria.

REFERENCES

CRISTY, M., AND K.F. ECKERMAN, *Specific Absorbed Fractions of Energy at Various Ages from Internal Photon Sources*, 7 vols, Report ORNL/TM-8381, Oak Ridge National Laboratory, Oak Ridge, Tenn., 1987.

DOE, *Integrated Data Base for 1988: Spent Fuel and Radioactive Waste Inventories, Projections, and Characteristics*, Report DOE/RW-0006. Rev. 4, U.S. Department of Energy, Washington, D.C., 1988.

DOE, Studies of Alternative Methods of Nuclear Waste Disposal, Backgrounder Report DOE/RW-0240, U.S. Department of Energy, Washington, D.C., 1989.

EICHHOLZ, G.G., "Source Terms for Nuclear Facilities, and Medical and Industrial Sites," in *Radiological Assessment: A Textbook on Environmental Dose Analysis*, J.E. Till and H.R. Meyer (eds.), Report NUREG/CR-3332, ORNL-5968, U.S. Nuclear Regulatory Commission, Washington, D.C., 1983.

EPA, *Occupational Exposure to Ionizing Radiation in the United States*, Report EPA 520/1-84-005, U.S. Environmental Protection Agency, Office of Radiation Programs, Washington D.C., 1984.

FLOWERS, Brian, *Nuclear Power and the Environment, Sixth Report of the Royal Commission on Environmental Pollution*, H.M. Stationery Office, London, 1976.

FDA, *Population Exposure to X Rays, U.S., 1970*, Publication 73-8047, U.S. Department of Health, Education and Welfare; Public Health Service, Food and Drug Administration, Bureau of Radiological Health, Rockville, Md., 1973.

FDA, *Gonad Doses and Genetically Significant Dose from Diagnostic Radiology*, U.S. Department of Health and Human Services, Public Health Service, Food and Drug Administration, National Center for Devices and Radiological Health, Rockville, Md., 1976.

GLASSTONE, S., AND P.J. DOLAN, (eds.), *The Effects of Nuclear Weapons*, United States Departments of Energy and Defense, Washington, D.C., 1977.

GODBEE, H., A.H. KIBBEY, C.W. FORSBERG, W.L. CARTER, and K.J. NOTZ, "Nuclear Fuel Cycle: An Introductory Overview," in *Radioactive Waste Technology*, A.A. Moghissi, H.W. Godbee, and S.A. Hobart (eds.), American Society of Mechanical Engineers, New York, 1986.

HENRY, E.A., "Nuclear Data Sheets for A = 133," *Nuclear Data Sheets 11*, 510 (1974).

HORTON, H., et al., *Analytical Methods for Evaluating the Radiological Aspects of the GE BWR*, Report APED-5756, General Electric Co., Schenectady, N.Y., 1969.

IAEA, *Preparation and Control of Radiopharmaceuticals in Hospitals*, by K. Kristensen, Technical Report Series 194, International Atomic Energy Agency, Vienna, 1979.

IAEA, *Safety Principles and Technical Criteria for the Underground Disposal of High Level Radioactive Wastes*, Safety Series 99, International Atomic Energy Agency, Vienna, 1989.

ICRP, Report of the Task Group on Reference Man. A Report Prepared by a Task Group of Committee 2 of the ICRP, Publication 23, International Commission on Radiological Protection, Pergamon Press, Oxford, 1975.

ICRP, *Recommendations of the International Commission on Radiological Protection*, *Annals of the ICRP*, Vol. 1, No. 3, Publication 26, International Commission on Radiological Protection, Pergamon Press, Oxford, 1977.

ICRP, *Limits for Intakes of Radionuclides by Workers*, Publication 30 and Supplements, International Commission on Radiological Protection, *Annals of the ICRP*, Pergamon Press, Oxford, 1979.

ICRP, *Protection Against Ionizing Radiation from External Sources Used in Medicine*, *Annals of the ICRP*, Vol. 9, No. 1, Publication 33, International Commission on Radiological Protection, Pergamon Press, Oxford, 1982.

ICRP, *Radiation Protection Principles for the Disposal of Solid Radioactive Waste*, *Annals of the ICRP*, Vol. 15, No. 4, Publication 46, International Commission on Radiological Protection, Pergamon Press, Oxford, 1985.

ICRP, *Protection of the Patient in Nuclear Medicine*, *Annals of the ICRP*, Vol. 17, No. 4, Publication 52, International Commission on Radiological Protection, Pergamon Press, Oxford, 1987a.

ICRP, *Radiation Dose to Patients from Radiopharmaceuticals*, *Annals of the ICRP*, Vol. 18, No. 1–4, Publication 53, International Commission on Radiological Protection, Pergamon Press, Oxford, 1987b.

KERIAKES, J.G., AND M. ROSENSTEIN, *Handbook of Radiation Doses in Nuclear Medicine and Diagnostic X Ray*, CRC Press, Boca Raton, Fl., 1980.

KOCHER, D.C., *Radioactive Decay Tables,* Report DOE/TIC-11026, Technical Information Center, U.S. Department of Energy, Washington, D.C., 1981.

LEE, J.Y., C.A. WILLIS, AND J.T. COLLINS, "Generation of Radioactive Waste in Nuclear Power Reactors, " in *Radioactive Waste Technology,* A.A. Moghissi, H.W. Godbee, and S.A. Hobart (eds.), American Society of Mechanical Engineers, New York, 1986.

McCULLOUGH, E.C., AND J.R. CAMERON, *Exposure Rates from Diagnostic X-Ray Units, Br. J. Radiol.* 43, 448 (1972).

MIRD Committee, Pamphlets, 1–8, Supplements 1–5, Society of Nuclear Medicine, New York, 1968–1971.

MIRD Committee, *Primer for Absorbed Dose Calculations*, Society of Nuclear Medicine, New York, 1985.

MOGHISSI, A.A., H.W. GODBEE, AND S.A. HOBART (eds.), *Radioactive Waste Technology,* American Society of Mechanical Engineers, New York, 1986.

MOULD, R.F., *Radiation Protection in Hospitals*, Adam Hilger, Ltd., Boston, 1985.

NAS, *The Effects on Populations of Exposure to Low Levels of Ionizing Radiation*, Report of the BEIR Committee, National Research Council, National Academy of Sciences, Washington, D.C., 1980.

NCRP, *The Experimental Basis for Absorbed-Dose Calculations in Medical Uses of Radionuclides*, Report No. 83, National Council on Radiation Protection and Measurements, Bethesda, Md., 1985a.

NCRP, *General Concepts for the Dosimetry of Internally Deposited Radionuclides*, Report No. 84, National Council on Radiation Protection and Measurements, Bethesda, Md., 1985b.

NCRP, *Public Radiation Exposure from Nuclear Power Generation in the United States*, Report No. 92, National Council on Radiation Protection and Measurements, Bethesda, Md., 1987a.

NCRP, *Radiation Exposure of the U.S. Population from Consumer Products and Miscellaneous Sources*, Report No. 95, National Council on Radiation Protection and Measurements, Bethesda, Md., 1987b.

NCRP, *Exposure of the U.S. Population from Diagnostic Medical Radiation*, Report No. 100, National Council on Radiation Protection and Measurements, Bethesda, Md., 1989a.

NCRP, *Exposure of the U.S. Population from Occupational Radiation*, Report No. 101, National Council on Radiation Protection and Measurements, Bethesda, Md., 1989b.

NCRP, *Radiation Protection for Medical and Allied Health Personnel*, Report No. 105, National Council on Radiation Protection and Measurements, Bethesda, Md., 1989c.

NUNNALY, James E., "Calibration of Beryllium-Window X-Ray Tubes Operated between 10 kVp and 100 kVp," in *Handbook of Medical Physics*, Vol. 1, R.G. Waggener, J.G. Kereiakes and R.J. Shalek (eds.), CRC Press, Boca Raton, Fl., 1980.

ROSENSTEIN, M., *Handbook of Selected Tissue Doses in Diagnostic Radiology*, Publication 88-8031, U.S. Department of Health and Human Services, Public Health Service, Food and Drug Administration, Center for Devices and Radiological Health, Rockville, Md., 1988.

RYMAN, J.C., "Origen-S Data Libraries," Vol. 3, Section M6 in *Scale-3: A Modular Code System for Performing Standardized Computer Analyses for Licensing Evaluations*, Report NUREG/CR-0200 (ORNL/NUREG/CSD-2), Oak Ridge National Laboratory, Oak Ridge, Tenn., 1984. [Released by the Radiation Shielding Information Center, Oak Ridge National Laboratory, as Code Package CCC-466.]

TAYLOR, G.M., "Volume Reduction: Utilities Caught in the Squeeze," Nuclear News, March 1987, p. 66.

UN, *Report of the United Nations Scientific Committee on the Effects of Atomic Radiation*, New York, 1977.

UN, *Report of the United Nations Scientific Committee on the Effects of Atomic Radiation*, New York. 1982.

UN, *Report of the United Nations Scientific Committee on the Effects of Atomic Radiation*, New York. 1986.

UN, *Report of the United Nations Scientific Committee on the Effects of Atomic Radiation*, New York. 1988.

WHO, *Quality Assurance in Nuclear Medicine*, World Health Organization, Geneva, 1982.

PROBLEMS

1. An adult woman receives a chest x ray involving a single PA view. Particulars are as follows:

   > Voltage = 80 kVP, full-wave rectified, single phase (see Fig. 5.1b).
   > Current = 300 mA
   > Filtration = 3 mm Al
   > Duration = 1/30 second
   > Film size = image size = 14 in. × 17 in.
   > Source-image distance = 72 in.
   > Patient thickness = 26 cm
   > Patient to film distance = 5 cm

   Estimate the resulting cancer detriment index and the absorbed doses to the thyroid, active bone marrow, lungs and breast.

2. A 1-Mt nuclear weapon with 25 percent fission yield is tested in the atmosphere at latitude 40° N. Eighty percent of the radioactive debris is ejected into the stratosphere. Transfer from the stratosphere to the troposphere takes place with a half-life of 10 months. Upon return to the troposphere, debris is relatively rapidly deposited on the surface of the earth as global fallout. Consider now the ^{90}Sr (half life 29.1 y) contained in the debris, assume that all global fallout takes place within latitudes 25° N and 55° N, and assume that the earth is a perfect sphere of radius 6371 km. Estimate the total number of ^{90}Sr atoms deposited per hectare of surface (land and sea) within the prescribed latitude band.

3. The steady-state global inventory of naturally occurring ^{14}C is thought to be 8500 PBq. Estimate the percentage increase in inventory caused by atmospheric nuclear weapons tests through 1980.

4. The average specific activity of ^{238}U in U.S. deposits of coal is 0.5 pCi kg^{-1}. Estimate the concentration of ^{226}Ra in coal in units of (a) pCi kg^{-1} and (b) grams per metric ton.

5. U.S. coal typically contains 400 pCi of ^{238}U per kg. The ash content is typically 10 percent. In combustion of coal for electricity generation, all the ^{238}U is retained in the ash, of which about 2.5 percent is released to the atmosphere. Estimate the quantity of ^{238}U released to the atmosphere per GWy(e) of energy production, expressed as (a) pCi, and (b) kg. You may assume that the thermodynamic efficiency in electricity generation is 33 percent and that the heat of combustion of coal is 28 MJ kg^{-1}.

6. Consider the BWR coolant cleanup system illustrated in Fig. 5.5. Flow data are as follows: coolant mass = 3.4×10^6 lb, steam flow = 1.54×10^7 lb h^{-1}, and cleanup flow = 1.54×10^5 lb h^{-1}. Fuel in defective fuel elements is 0.01 percent of the total. The ratio of the weight fraction iodine in the steam flow from the reactor to the weight fraction iodine in the liquid coolant within the reactor is 0.012. The cleanup demineralizer removes 90 percent of the iodine in the primary coolant passing through it and the condensate demineralizer removes 99.9 percent of the iodine in the primary coolant passing through it. Of the iodine in the steam, 0.5 percent is removed by the condenser air ejector; however, of this quantity, 90 percent is recovered in the air-ejector condenser and returned to the primary coolant flow. The reactor was fueled with fresh fuel and has been operating for one year at a thermal power of 3760 MW. Estimate (1) the cleanup time constant λ'_i (s^{-1}), (2) the ^{131}I specific activity (μCi g^{-1}) in the primary coolant, and (3) the ^{131}I exhaust rate (μCi s^{-1}) from the air-ejector condenser.

Chapter 6

Gamma-Ray and Neutron Shielding and External Dose Evaluation

6.1 INTRODUCTION

This chapter deals with gamma rays, x rays, and neutrons. These radiations are composed of neutral particles and, consequently, are referred to as *indirectly ionizing*. They interact with matter via short-range forces, so that as they pass through matter they travel in series of straight-line segments punctuated by individual scattering interactions. Their paths are terminated by absorption interactions such as neutron radiative capture or neutron-induced fission which may lead to the creation of secondary neutral particles. The scattering interactions ordinarily lead to recoil charged particles and may, as well, lead to secondary neutral particles such as gamma rays produced in neutron inelastic scattering. As secondary charged particles interact with matter they may produce additional charged particles such as electron delta rays or neutral particles such as annihilation gamma rays.

Notwithstanding the production of secondary neutral particles, the attenuation in matter of source neutrons and gamma rays of *fixed* energy is essentially exponential in nature. This exponential attenuation is both predicted by theory and confirmed by observation. It is one of two key elements used in procedures for calculating dose rates arising from shielded radiation sources or for selecting the shielding required for protection from radiations arising from a source. The second key element accounts for the geometric attenuation of radiation, namely, the observation that, in the absence of material attenuation, the intensity of radiation decreases with the inverse square of the distance from a point source to the point of observation.

When dealing with a radiation source in a fixed geometric relationship to a detector or point of observation and with known intervening materials, one is frequently required to predict (1) the dose rate or detector response, given the source strength, or (2) the source strength, given the dose rate. This task requires an analysis or assessment procedure which is usually straightforward though tedious. By contrast, when presented with the task of prescribing the geometry and shielding materials required to limit the dose rate from a known source, one requires a design procedure. While a design procedure makes use of the same concepts and mathematical techniques as the fixed source/detector problem, the design procedure is ordinarily much more difficult, requiring iterative calculations often with the goal of optimizing shield dimensions or costs.

　　This chapter begins by reviewing important basic concepts introduced in Chaps. 1 and 2 and then introduces some new concepts about how neutral particles interact with matter. It next summarizes the characterization of common neutron and gamma-ray sources, and then proceeds with the development of design and analysis methods based solely on geometric and purely exponential attenuation and which are applicable to both neutron and gamma-ray attenuation. Methods of accounting for secondary radiations are then taken up separately for the two radiation types. Finally, some special topics are addressed. These include reflection of radiation, which is an important consideration in instrument calibration, and atmospheric scattering of radiation, called skyshine, which is often an important consideration in design of nuclear facilities.

6.2 REVIEW OF BASIC CONCEPTS

At this point, the reader may wish to review certain portions of Chaps. 1 and 2. The first chapter dwells at some length on the distinction between *source activity* and *source strength*. In attenuation calculations, it is the source strength that is of concern. A radionuclide may emit various radiations. The source strength for a particular radiation, expressed as particles emitted per second, is equal to the product of the activity, expressed as becquerels (or transformations per second) and the frequency of emission of that particular radiation. When a source emits gamma rays of various energies, each energy photon must be accounted for individually in design and analysis calculations.

　　Many sources, however, emit photons or neutrons with a continuous distribution in energy, and for such sources the concept of activity does not apply. Examples of such sources are x-ray machines, and nuclear fission. There are a few design and analysis procedures that treat all source particles collectively, and some will be described in this chapter. Other procedures subdivide the continuous energy spectrum of the source radiation into an energy-group structure, treating each group as though the particles in the group arose with the group-average source energy. These multigroup procedures are extremely powerful and find wide use. Many are based on the same principles that are described in this chapter for monoenergetic radiation. However, some advanced multigroup methods for shield design allow for group-to-group transfer of particles as they lose energy in the attenuation process. These advanced methods are beyond the scope of this text but are discussed in most radiation shielding texts (e.g., Chilton, Shultis, and Faw 1984).

　　Finally there are some radiation sources which, though comprised of individual radionuclides, emit gamma rays of an enormous number of discrete energies — so many that the design and analysis procedures for these sources rely on the multigroup approximation. Examples include fission products in spent nuclear fuel and neutron-capture gamma ray sources in a medium composed of many different elements.

Chapter 1 also introduces the concepts of radiation flux density and fluence, and the concepts of absorbed dose and kerma. These concepts are based on a small hypothetical sphere into and from which radiations travel. Over a time interval long enough to assure statistically representative particle transport, the fluence Φ is the limit, as the sphere volume approaches zero, of the sum of the particle trajectory segments within the sphere divided by the sphere volume, that is, the expected value of the path length per unit volume. The flux density ϕ is the time derivative of the fluence. For neutrons and gamma rays, the kerma, in the same limit, is the total energy transferred within the volume as initial kinetic energy of charged particles, divided by the mass of the volume. The absorbed dose is, in the same limit, the energy imparted to the matter within the volume divided by the mass of the volume.

Chapter 2 describes the energy and momentum conservation laws governing the interactions of neutrons and gamma rays with matter and how the probabilities of interactions can be expressed in terms of cross sections or interaction coefficients. The chapter also shows how the interaction coefficients and reaction energetics are incorporated into response functions which relate fluence and detector response. For example, if Φ is the fluence (cm^{-2}) of gamma rays of energy E (MeV), then the exposure X in roentgen units is given by

$$X = \Phi \mathcal{R}_X, \tag{6.1}$$

in which

$$\mathcal{R}_X = 1.835 \times 10^{-8} E \left(\frac{\mu_{en}}{\rho} \right)^{air}, \tag{6.2}$$

where $(\mu_{en}/\rho)^{air}$ is the mass energy absorption coefficient (cm^2 g^{-1}) in air for photons of energy E and the premultiplier is a conversion factor relating energy absorption to ionization in air. The generic form of Eq. (6.1) is $R = \Phi \mathcal{R}$ in which R is a generalized response and $\mathcal{R}$ is the associated generalized response function. In this chapter, it will be understood that, if the fluence Φ is replaced by the flux density (fluence *rate*), then R becomes the response *rate*.

Chapter 2 also introduces the concept of charged-particle equilibrium. This is an important concept and is of great practical interest in nuclear instrumentation design and application. In this chapter, equilibrium will be assumed, thus permitting approximation of the absorbed dose by the kerma.

6.3 SOME NEW CONCEPTS

6.3.1 Exponential Attenuation

Recall that the interaction coefficient μ is defined in such a way that, in the limit as $\Delta x \to 0$, the product $\mu \Delta x$ is the probability that a particle has an interaction within distance Δx along the particle's original trajectory. Clearly, $1 - \mu\Delta x$ is the probability of traveling elemental distance Δx *without* interaction. Let $p(x)$ be the probability that a particle may travel finite distance x without interaction. It follows that the probability of traveling distance $x + \Delta x$ is the joint probability of traveling distance x without interaction and the probability of traveling the additional distance Δx also without interaction, that is, in the limit as $\Delta x \to 0$,

$$p(x + \Delta x) = p(x) \times (1 - \mu\Delta x). \tag{6.3}$$

In the limit, it thus follows that

$$\frac{dp}{dx} \equiv \lim_{\Delta x \to 0} \frac{p(x + \Delta x) - p(x)}{\Delta x} = -\mu p(x) \tag{6.4}$$

from which, since $p(0) = 1$,

$$p(x) = e^{-\mu x}. \tag{6.5}$$

Thus, for a single interaction mechanism, characterized by μ, unreacted radiation is exponentially attenuated along its path of travel.

6.3.2 The Mean Free Path

The probability that a particle travels a distance between x and $x + \Delta x$ before interaction is the joint probability of traveling distance x without interaction and then having an interaction within distance Δx, that is, $p(x)\mu\Delta x$. In other words, the normalized distribution function for path length x is just $\mu p(x)$. The mean free path is just the path length averaged with this distribution function, or

$$< x > = \int_0^\infty dx \, x\mu e^{-\mu x} = \frac{1}{\mu}. \tag{6.6}$$

In air of density 1.2 mg cm^{-3}, the interaction coefficient for 1.5 MeV photons is 6.2 $\times$ 10^{-5} cm^{-1}. The mean free path is thus 160 meters. In liquid water, the mean free path is 17 cm and in lead it is only about 1.4 cm. Common in practical usage is the *half-value thickness* x_2 for which $p(x_2) = 1/2$ and the *tenth-value thickness* x_{10} for which $p(x_{10}) = 1/10$. It is easily shown that $x_2 = \ln(2)/\mu$ and $x_{10} = \ln(10)/\mu$. The half-value thickness concept is also used with x-ray beams for which there is no unique energy or μ value. In those cases, x_2 is determined empirically.

6.4 CHARACTERIZATION OF RADIATION SOURCES

6.4.1 Individual Radionuclides

Appendix B contains a compilation of the decay characteristics for a number of radionuclides of importance in nuclear medicine, in the research laboratory, and in in-

TABLE 6.1 Fission-Neutron Source Characteristics.

NUCLIDE	T (MeV)	NEUTRONS per FISSION	NEUTRONS per (g s)	HALF LIFE[a]
Spontaneous Fission				
^{238}Pu		2.28	2.6×10^3	87.7 y
^{240}Pu	1.27	2.23	9.1×10^2	6570 y
^{244}Cm	1.33	2.82	1.1×10^7	18.1 y
^{252}Cf	1.47	3.80	2.3×10^{12}	2.64 y
Thermal-Neutron Induced Fission				
^{233}U	1.31	2.48		
^{235}U	1.29	2.43		
^{239}Pu	1.33	2.87		

[a]Half life for combined spontaneous fission and alpha particle decay.
Source: Keepin 1965; Lederer and Shirley 1978; BNL 1969.

dustry. This compilation lists only average energies for beta particles, but lists individually the energies and frequencies for the major gamma and x rays, alpha particles, and electrons released.

6.4.2 Fission Neutrons

Because of the importance of uranium-fueled thermal reactors, more is known about thermal-neutron induced fission of ^{235}U than about fast-neutron induced fission or about fission of other isotopes. While the number of neutrons ν released per fission differs substantially from one type of fission to another, for radiation-protection purposes the energy spectrum of the neutrons from ^{235}U thermal fission is often taken to be representative of that for any other fission, even spontaneous fission. Alternatively, the energy spectrum of fission neutrons is frequently approximated by the Maxwellian distribution

$$\chi(E) = 2\sqrt{\frac{E}{\pi T^3}}\, e^{-E/T}, \qquad (6.7)$$

in which $\chi(E)dE$ is the fraction of fission neutrons with energies between E and $E + dE$, and T is the Maxwellian parameter of Table 6.1 which can be shown to be equal to two-thirds the mean fission-neutron energy.

For ^{235}U, an empirical formula (Cranberg et al. 1956) gives somewhat better agreement with data than does Eq. (6.7). This is a dimensional equation, requiring E to be expressed in units of MeV, and is valid over the range 0 to 20 MeV

$$\chi(E) = 0.4527 e^{-E/0.965} \sinh\sqrt{2.29E}. \qquad (6.8)$$

Another approximation for ^{235}U, useful only over the range 4 to 14 MeV, is (Blizard and Abbott 1962).

$$\chi(E) = 1.75 e^{-0.766E}, \qquad (6.9)$$

in which E is the energy in units of MeV.

TABLE 6.2   Characteristics of (α,n) Neutron Sources.

SOURCE	HALF LIFE	PRINCIPAL α ENERGIES (MeV)	AVERAGE NEUTRON ENERGY (MeV)
^{241}Am/Be	433 y	5.49, 5.44	4.4
^{239}Pu/Be	24100 y	5.16, 5.15, 5.11	4.5
^{210}Po/Ba	138 d	5.30	4.2
^{226}Ra/Be	1599 y	7.69, 6.00, 5.49	3.9
+ daughters		5.30, 4.78	

Source: NBS 1960; Lederer and Shirley 1978.

Many of the transuranic actinide elements fission spontaneously, invariably in competition with alpha-particle emission. Many are responsible for the neutron emission arising from spent nuclear fuel. Important among these are ^{238}Pu, used in power supplies, and ^{252}Cf, used as a neutron source for instrument calibration. Characteristics of some spontaneous fission sources are summarized in Table 6.1.

6.4.3 Neutrons Produced by (α, n) and (γ, n) Reactions

Neutrons were discovered in the 1930s during studies of alpha particle reactions with light elements such as beryllium, lithium, carbon, and fluorine. Such reactions continue to be exploited in the fabrication of compact neutron sources for use in laboratories and for start up of nuclear reactors. Such sources consist of homogeneous mixtures or intermetallic compounds of an alpha-particle source (*emitter*) and a light element (*converter*).

Neutron energy spectra are quite different from those of fission sources. While the spectra are continuous below some maximum energy, there are well defined peaks at energies characteristic of the converter. For example, with ^{9}Be as the converter, neutrons are produced by the reactions ^{9}Be$(\alpha, n)^{12}$C and ^{9}Be$(\alpha, \alpha' + n)^8$Be. Higher energy neutrons are produced by the first reaction, and peaks in the energy spectra are related to the excitation levels of the ^{12}C nucleus. Lower energy neutrons are produced by the second reaction, and the energy spectrum is influenced by the energy of the alpha particle released by the emitter radionuclide.

Table 6.2 lists some characteristic of sources employing beryllium as converter. The ^{239}Pu/Be source is popular because of the long half life of the emitter. It is not useful as a reactor start-up source because of neutron-induced ^{239}Pu fission. The ^{241}Am/Be source is popular as a start-up source because of its long half life and because ^{241}Am has a low cross section for neutron interactions. The ^{226}Ra/Be source, which also relies on radium daughter products for alpha particles, is not widely used because of the relatively intense gamma-ray emission by the radium daughters. Table 6.3 lists threshold alpha-particle energies for a number of (α, n) reactions.

High-energy gamma and x rays can also induce neutron production in light elements, notably deuterium, lithium and beryllium. Laboratory neutron sources relying on the (γ, n) reaction may be fabricated from a mixture of a photo-neutron mate-

TABLE 6.3   Threshold Energies for Selected (α,n) and (γ,n) Reactions.

REACTION	Q (MeV)[a]	THRESHOLD ENERGY (MeV)
$^{6}\mathrm{Li}(\alpha,n)^{9}\mathrm{B}$	−3.975	6.618
$^{7}\mathrm{Li}(\alpha,n)^{10}\mathrm{B}$	−2.791	4.382
$^{9}\mathrm{Be}(\alpha,n)^{12}\mathrm{C}$	5.702	exothermic
$^{10}\mathrm{B}(\alpha,n)^{13}\mathrm{N}$	1.06	exothermic
$^{11}\mathrm{B}(\alpha,n)^{14}\mathrm{O}$	0.157	exothermic
$^{2}\mathrm{H}(\gamma,n)^{1}\mathrm{H}$	−2.225	2.225
$^{6}\mathrm{Li}(\gamma,n+p^{4})\mathrm{He}$	−3.698	3.698
$^{6}\mathrm{Li}(\gamma,n)^{5}\mathrm{Li}$	−5.67	5.67
$^{7}\mathrm{Li}(\gamma,n)^{6}\mathrm{Li}$	−7.251	7.251
$^{9}\mathrm{Be}(\gamma,n)^{8}\mathrm{Bc}$	−1.665	1.665

[a] For a discussion of the Q-value and the threshold energy, please refer to the discussion of neutron kinematics in Chap. 2.

Source: Chilton, Shultis, and Faw 1984.

rial (converter) with an emitter of high energy photons, for example, $^{124}\mathrm{Sb}$. Because $^{123}\mathrm{Sb}$ has a large neutron absorption cross section, the source may be rejuvenated by subjecting it to thermal neutrons within a nuclear reactor. Neutrons emitted in (γ, n) reactions are nearly monoenergetic and it is possible to calculate the neutron energy E_n if one knows the gamma-ray energy E_g and the nuclear properties of the converter, including the threshold energy E_{th} for the reaction. However, it is a good approximation that the photoneutrons are emitted isotropically with energy equal to $E_g - E_{th}$ or, if deuterium is the converter, $(E_g - E_{th})/2$ (Chilton, Shultis, and Faw 1984). Table 6.3 lists threshold energies for a number of emitter-converter (γ, n) reactions. For example, suppose that $^{24}\mathrm{Na}$ is the emitter. It releases two gamma rays upon decay, one of energy 1.37 MeV, the other of energy 2.75 MeV. Only the higher energy gamma ray has sufficient energy to overcome the threshold $-Q$ for the (γ, n) reaction in the converters listed in Table 6.3. If $^{2}\mathrm{H}$ is the converter, the resulting neutron has energy approximately $(2.75 - 2.23)/2 = 0.26$ MeV. If $^{9}\mathrm{Be}$ is the converter, the neutron energy is approximately $2.75 - 1.67 = 1.08$ MeV.

The photoneutrons may be produced in abundance in the targets of electron accelerators as a result of bremsstrahlung interactions. The photoneutrons in nuclear reactors can also be significant, even when the reactor is "shut down," because of the gamma rays emitted from fission products present in the fuel. In reactors cooled or moderated by heavy water or containing beryllium components, the photoneutron production may be so great as to mask criticality and complicate kinetic analyses.

6.4.4 Fusion Neutrons

Many nuclear reactions induced by energetic charged particles can produce neutrons. Most of these reactions require incident particles of very high energies for the reaction to take place, and, consequently, are of little concern in radiological assessment. Only near accelerator targets, for example, would such reaction neutrons be of concern. However, many light elements will fuse exothermically if they are given sufficient kinetic energy, typically more than 10 keV, so that a pair of approaching atoms can overcome the mutual Coulombic repulsion and the nuclei are allow to interact. The two neutron-producing fusion reactions of most interest in the development of thermonuclear power are

$$^2\text{H} + ^2\text{H} \rightarrow ^3\text{He } (0.82 \text{ MeV}) + ^1n(2.45 \text{ MeV})$$
$$^2\text{H} + ^3\text{H} \rightarrow ^4\text{He } (3.5 \text{ MeV}) + ^1n(14.1 \text{ MeV})$$

The 14.1 MeV fusion neutrons are produced copiously in the detonation of a thermonuclear bomb and consequently are of concern in the radiological evaluation of weapons effects. These fusion neutrons are also produced in compact accelerators (called neutron generators) in which deuterium ions are accelerated through a high voltage (100 to 300 kV) and allowed to impinge on a thick deuterium- or tritium-bearing target. Typically in such devices, a 1-mA beam current produces up to 10^9 2.5-MeV neutrons per second from a thick deuterium target and up to 10^{11} 14-MeV neutrons from a thick tritium target.

When these reactions are produced by accelerating one nuclide toward the other, as in a neutron generator, the velocity of the center of mass must be added to the center-of-mass neutron velocity before determining the energy of the neutron in the laboratory system. However, in most fusion devices the velocity of the center of mass is negligible, and one has to contend with monoenergetic 2.45 or 14.1 MeV neutrons.

6.4.5 Fission and Fission-Product Gamma Rays

Gamma and x rays are released in the fission process and as a result of radioactive decay of the products of fission. Fission gamma rays are those released within about 50 ns of the event. In thermal-neutron induced fission of ^{235}U, about 3 photons per fission, with average energy about 0.17 MeV, are released with energies less than 0.3 MeV. In the energy range 0.3 to 7 MeV, about 7 photons are released per fission, with average energy about 1 MeV. For radiation protection calculations the lower energy photons may usually be ignored but it is always necessary to take into account the energy spectrum of the higher energy fission photons. Keepin (1965) gives the following empirical formula for the energy spectrum of the fission gamma rays

$$\mathcal{N}(E) = 6.7e^{-1.05E} + 30e^{-3.8E}, \tag{6.10}$$

where $\mathcal{N}(E)dE$ is the number of photons produced per fission with energies between E and $E + dE$ and in which the energy E must be in units of MeV. This formula is also used as an approximation for photons released in the fission of other nuclides by both thermal and fast neutrons and released in spontaneous fission (Chilton, Shultis, and Faw 1984).

In the fission process, hundreds of different radioactive fission products may be produced. Many decay into other radioactive species leading into very complicated patterns of radioactive decay. Some fission products are very short-lived, some very long-lived. Most decay by beta-particle and gamma-ray emission. These fission products, of course, are responsible for the need for long-term care in the storage and disposal of wastes from the nuclear-power industry. During nuclear reactor operation, decay of previously produced fission products is taking place, and both prompt fission gamma rays and delayed fission-product gamma rays require shielding for purposes of radiation protection.

The energy release from the decay of fission products varies with the fissioning species. The total release (MeV per fission) over all time for selected fissioning species is as follows (Keepin 1965): ^{238}U, 10.9; ^{232}Th, 10.8; ^{235}U, 6.84; ^{239}Pu, 6.15; and ^{233}U, 4.24.

It is possible to take into account fission-product yields, systematics of radioactive decay chains, and decay characteristics of individual fission products and their progeny to calculate the gamma-ray energy spectrum as a function of time after the fission process. As applied to nuclear reactor fission products, such a calculation may be enlarged to account for (1) the thermal energy released by the fission products via both photon and charged-particle release, (2) the reactor-power time dependence during reactor operation prior to removal of fuel for storage, processing, or disposal, (3) the nature of the fuel, for example, mixtures of ^{235}U and ^{239}Pu, and (4), the type of reactor, which determines the energy spectrum of the neutrons responsible for fission. One well-known computer code for accomplishing these objectives is the ORIGEN code, a product of Oak Ridge National Laboratory (Hermann and Westfall 1984). Another such code is the CINDER code, a product of Los Alamos Scientific Laboratory (England, Wilczynski, and Whittemore 1976). Neither code attempts to identify individual gamma rays from fission products. Both use the energy multigroup approach.

LaBauve et al. (1982) analyzed results of detailed calculations to derive an empirical method which can be used quite successfully, even with hand calculations, to determine the energy spectra of gamma and x rays and beta particles and electrons from fission products. In this method the individual gamma rays are collected into six energy groups. Then, t seconds after an individual fission, the energy emission rate $F_j(t)$ (MeV s^{-1}) in energy group j, for t from 10^{-4} s to 10^{+9} s, is expressed approximately as

$$F_j(t) = \sum_{i=1}^{N_j} \alpha_{ij} e^{-\beta_{ij}t}, \qquad j = 1 \text{ to } 6. \tag{6.11}$$

Values of α_{ij} and β_{ij} are given in Table 6.4 for all fission products arising from thermal-neutron induced fission of ^{235}U. The same equation can be used to give the total energy release rate, following a fission, for all gamma and x rays and all beta particles and electrons (parameters for these release rates are also given in Table 6.4). Additional tables are available for fission of ^{239}Pu and, too, for gaseous fission products exclusively and for beta particles (LaBauve et al. 1982).

Suppose that a nuclear reactor has operated with a time-dependent number of fissions per second $P_f(t)$ for time t_o prior to shutdown. Suppose too, that one may neglect transmutation of fission products by neutron absorption — generally a good assumption. Then, at time t_s after shutdown of the reactor the energy release rate (MeV s^{-1}) by fission-product photons with energies in group j is

TABLE 6.4 Constants in the Empirical Approximation for Energy Release by All Fission Products Arising from Thermal-Neutron Induced Fission of ^{235}U. Table Entries Read as, e.g., 1.222E−13 = 1.222 × 10^{-13}.

GROUP 1 (N_1 = 9) 5–7.5 MeV		GROUP 2 (N_2 = 13) 4–5 MeV		GROUP 3 (N_3 = 14) 3–4 MeV		GROUP 4 (N_4 = 13) 2–3 MeV	
α_{ij}	β_{ij}	α_{ij}	β_{ij}	α_{ij}	β_{ij}	α_{ij}	β_{ij}
1.222E−13	6.125E−07	5.629E−20	2.787E−13	2.753E−18	6.788E−14	2.719E−17	8.227E−11
5.409E−12	1.036E−06	7.981E−20	6.023E−08	2.665E−18	3.951E−08	2.632E−11	2.772E−08
6.711E−08	8.330E−04	9.578E−14	5.236E−07	6.126E−16	3.800E−07	5.115E−09	6.458E−07
1.606E−05	3.562E−03	4.786E−11	1.060E−06	1.825E−11	5.905E−07	5.519E−09	2.515E−06
4.867E−05	6.743E−03	2.069E−08	7.083E−05	7.970E−11	9.240E−07	3.111E−06	6.159E−05
1.514E−04	2.279E−02	3.461E−08	1.894E−04	2.066E−07	7.721E−05	1.758E−05	1.356E−04
2.159E−03	1.337E−01	1.045E−06	7.889E−04	2.481E−06	1.396E−04	7.848E−05	5.446E−04
3.220E−03	5.023E−01	2.742E−04	4.142E−03	1.192E−05	6.358E−04	1.872E−04	2.839E−03
4.559E−03	2.539E+00	6.462E−04	1.226E−02	2.584E−04	3.452E−03	2.816E−03	1.199E−02
		1.341E−03	4.521E−02	1.376E−03	1.188E−02	5.830E−03	4.591E−02
		2.204E−03	1.962E−01	2.320E−03	4.160E−02	9.529E−03	1.998E−01
		2.572E−03	7.021E−01	2.799E−03	1.999E−01	1.234E−02	7.130E−01
		2.750E−03	2.953E+00	3.506E−03	6.763E−01	1.240E−02	3.003E+00
				3.559E−03	2.940E+00		

GROUP 5 (N_1 = 14) 1–2 MeV		GROUP 6 (N_6 = 14) 0–1 MeV		GROUP G (N_G = 11) ALL GAMMAS		GROUP B (N_B = 11) ALL BETAS	
α_{ij}	β_{ij}	α_{ij}	β_{ij}	α_{ij}	β_{ij}	α_{ij}	β_{ij}
7.708E−17	2.364E−10	2.817E−11	7.350E−10	2.808E−11	7.332E−10	6.169E−11	7.953E−10
9.292E−12	2.524E−08	4.535E−11	2.139E−08	6.038E−10	4.335E−08	2.249E−09	2.758E−08
7.734E−12	9.229E−08	1.574E−08	1.157E−07	3.227E−08	1.932E−07	2.365E−08	2.082E−07
7.168E−08	6.247E−07	1.794E−07	1.396E−06	4.055E−07	1.658E−06	2.194E−07	1.846E−06
2.206E−08	1.768E−06	1.056E−06	6.652E−06	8.439E−06	2.147E−05	1.140E−05	2.404E−05
3.466E−06	2.719E−05	3.168E−06	2.344E−05	2.421E−04	2.128E−04	1.549E−04	2.337E−04
1.457E−05	8.664E−05	7.944E−05	1.844E−04	1.792E−03	1.915E−03	1.991E−03	1.897E−03
1.033E−04	3.403E−04	2.230E−04	7.375E−04	2.810E−02	1.769E−02	3.256E−02	1.926E−02
2.339E−04	1.119E−03	7.722E−04	3.390E−03	1.516E−01	1.652E−01	2.227E−01	1.573E−01
3.038E−03	7.746E−03	4.001E−03	1.452E−02	4.162E−01	1.266E+00	5.381E−01	1.264E+00
1.112E−02	2.754E−02	2.313E−02	6.692E−02	1.053E−01	5.222E+00	1.282E−01	5.196E+00
2.343E−02	1.156E−01	9.434E−02	2.918E−01				
4.586E−02	5.380E−01	2.007E−01	1.064E+00				
5.000E−02	2.754E+00	1.722E−01	3.665E+00				

Source: LaBauve et al. 1982.

$$\Gamma_j(t_o, t_s) = \int_0^{t_o} dt' P_f(t') F_j(t_s + t_o - t'). \tag{6.12}$$

For the case of a constant P_f, the approximation of Eq. (6.11), when substituted into this expression, yields

$$\Gamma_j(t_o, t_s) = P_f \sum_{i=1}^{N_j} (\alpha_{ij}/\beta_{ij}) e^{-\beta_{ij} t_s} \left(1 - e^{-\beta_{ij} t_o}\right). \tag{6.13}$$

The fission rate P_f is related approximately to thermal power by the factor 3.1×10^{10} fissions per second per watt. Figure 6.1 illustrates $F_j(t)$ for a single fission, and Fig.

6.2 illustrates $\Gamma_j(t_o, t_s)$ for a unit fission rate ($P_f = 1$) and for an operating time $t_o = 30,000$ hours (which is representative of fuel consumption in a nuclear power reactor). The reader will note from Eq. (6.13) that, as t_o becomes very large, $\Gamma_j(t_o, t_s)$ becomes independent of operating time, that is, fission product inventories at the time of shutdown reach equilibrium. The equilibrium approximation is sometimes made in preliminary shielding design calculations, but it is a very conservative approximation leading to an overestimate of gamma radiation from long-lived radionuclides. The reader will also note that, in Eq. (6.13), t_s may be set equal to zero, thereby giving an approximation for the source strength in an operating nuclear reactor of the radiation emitted by the fission products within the reactor fuel.

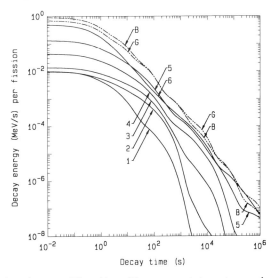

Figure 6.1   Calculated total gamma (G) and beta (B) energy emission rates as a function of time after the fission of ^{235}U. The curves identified by the numbers 1 to 6 are, respectively, rates for photons in the energy ranges (5–7.5), (4–5), (3–4), (2–3), (1–2), and (0–1) MeV. (Calculations based on the data of Table 6.4)

6.4.6 Capture and Inelastic Scattering Gamma Rays

Neutrons produced in a nuclear reactor are rarely absorbed until they are slowed to very low energies, that is, until they reach thermal equilibrium with their environment. They are slowed as a result of both elastic and inelastic scattering. In the elastic scatter process, the kinetic energy of the neutron is partially transferred to kinetic energy of the scattering atomic nucleus. In the inelastic scatter process, some of the kinetic energy is also taken up by excitation of the scattering nucleus. This excitation energy is almost immediately released by emission of one or more gamma rays. Consequently, fast-neutron interactions often produce an associated source of gamma rays. It is seen from Fig. 2.10 that the probability of inelastic scattering decreases rapidly for most materials as the neutron energy falls below 1 or 2 MeV. Fortunately, it is very rare that inelastic-scattering gamma rays exert a major influence on the design of radiation shields.

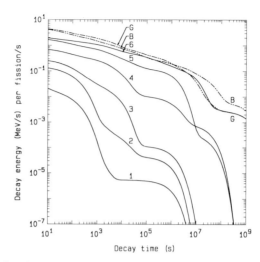

Figure 6.2 Calculated total gamma (G) and beta (B) energy emission rates as a function of time after the fission of f ^{235}U for 30,000 hours of a constant rate. The curves identified by the numbers 1 to 6 are, respectively, rates for photons in the energy ranges (5–7.5), (4–5), (3–4), (2–3), (1–2), and (0–1) MeV. (Calculations based on the data of Table 6.4)

The ultimate fate of a free neutron is capture. Capture in the fuel of a nuclear reactor may result in fission. Otherwise, it almost inevitably results in the prompt emission of one or more capture gamma rays, often of very high energy and penetrating ability. These secondary gamma rays exert a controlling influence on the design of shielding around nuclear reactors. Appendix D contains tables of cross sections for neutron capture in common elements, along with yields of gamma rays in 1-MeV energy groups extending to 11 MeV. In aqueous systems many neutrons are captured by hydrogen atoms. As the tables in Appendix D indicate, each capture results in the emission of a 2.223-MeV gamma ray. Capture of neutrons in hydrogen alone would thus lead to the requirement for gamma ray shielding in addition to neutron shielding.

One important exception to capture gamma-ray production follows absorption of a neutron in ^{6}Li. This nuclide, which has a natural abundance of only 7.5 percent, has a large (n, α) cross section for absorbing a thermal neutron and yielding a helium atom and a tritium atom as the only products (940 b for a neutron energy of 0.0253 eV). Lithium, perhaps enriched in ^{6}Li, is sometimes incorporated into shields to minimize capture gamma-ray production. Natural boron contains 80 percent ^{11}B and only 20 percent ^{10}B, but it is the latter which makes boron so effective in thermal neutron removal through ^{10}B(n, α)^{7}Li reactions which have the exceptionally large cross section of 940 barns. Unfortunately, about 94 percent of the ^{7}Li atoms so created are in an excited nuclear state which is relaxed promptly by emission of a 0.48-MeV gamma ray.

After neutron capture, usually accompanied by prompt gamma-ray emission, the product nuclide may remain radioactive and only later emit gamma rays, beta particles, or other radiations. These so-called activation products may be very useful,

TABLE 6.5   Important Activation Products Produced by Neutron Interactions.

PRODUCT NUCLIDE	REACTION TYPE	PARENT NUCLIDE	PARENT ABUNDANCE	σ_o CROSS SECTION[a]
^{3}H	(n, T)	^{14}N	99.6%	(fast-neutron reaction)
^{3}H	(n, T)	^{16}O	99.8%	(fast-neutron reaction)
^{3}H	(n, α)	^{6}Li	7.5%	940 b
^{14}C	(n, γ)	^{13}C	1.1%	0.9 mb
^{14}C	(n, p)	^{14}N	99.6%	1.81 b
^{16}N	(n, γ)	^{15}N	0.4%	0.02 mb
^{24}Na	(n, γ)	^{23}Na	100.0%	0.93 b
^{28}Al	(n, γ)	^{27}Al	100.0%	0.23 b
^{41}Ar	(n, γ)	^{40}Ar	99.6%	0.66 b
^{56}Mn	(n, γ)	^{55}Mn	100.0%	13.3 b
^{56}Mn	(n, p)	^{56}Fe	91.8%	2.63 b
^{60}Co	(n, γ)	^{59}Co	100.0%	37. b

[a]These are cross sections for 2200 m s^{-1} (0.025 eV) neutrons.
Source: Ryman 1983.

for example, in the production of radiopharmaceuticals or radioisotopes of industrial or research utility. On the other hand, they may be a nuisance, as is often the case when they are produced in the functional materials associated with operation of a nuclear reactor. Some of the more important activation products are described in Table 6.5. Decay characteristics of the products are described in Appendix B. In Table 6.5, as in the tables of Appendix D, the cross section is the so-called 2200 m s^{-1} absorption cross section, $\sigma_a(E_o)$, in reference to the speed of neutrons of energy $E_o = 0.025$ eV, the most probable energy for neutrons in thermal equilibrium with their surroundings at a temperature of $T_o = 293$ K. In a nuclear reactor, the thermal-neutron flux density is ordinarily quoted as the *total* flux density of all thermal neutrons. The appropriate thermal-averaged absorption cross section to use with the total thermal-neutron flux density at absolute temperature T is given by

$$\bar{\sigma}_a \cong \frac{1}{2}\sqrt{\frac{\pi T_o}{T}}\,\sigma_a(E_o). \tag{6.14}$$

6.4.7 X Rays

X rays are produced as a result of rearrangements of atomic electron configurations and as a result of the deceleration of charged particles. The former are so-called characteristic x rays with well defined energies characteristic of the atomic species. The latter are bremsstrahlung, literally "braking radiation," and have energies continuously distributed to a maximum corresponding to the maximum energy of the charged particles. Normally, bremsstrahlung production is significant only in the slowing down of electrons and positrons.

Characteristic x rays arise in many circumstances. When a photon interacts via the photoelectric effect, the atom is left with a vacancy usually in the electron K shell

— a very unstable situation. Rearrangement of the atomic electrons in a process leading to stability results in a cascade of x rays with energies quantized to the differences between electron binding energies in the atomic shells and subshells. The same sequence of events takes place when a radioactive atom decays *via* nuclear capture of an atomic electron (usually from the K-shell) or when an internal-conversion electron is released. The cascade of x rays is accompanied by a cascade of Auger electrons which are released in competition with x rays. This competition, for a particular characteristic x-ray energy, is characterized by the *fluorescence yield*, defined as the probability of x-ray emission rather than Auger-electron emission. Yields approach zero for low-Z elements and unity for high-Z elements. Data tables are available listing K-shell and L, M, and N subshell electron binding energies as well as K-series x-ray energies, frequencies and fluorescence yields (Lederer and Shirley 1978). For example, tungsten ($Z = 74$), the usual target in x-ray machines, has binding energies of K-shell and L_1-, L_2-, and L_3-subshell electrons of, respectively, 69.523, 12.099, 11.542, and 10.205 keV. Transitions from the L_1 to the K shell are *forbidden* (i.e., are not physically allowed by normal transition mechanisms). An electron transition from the L_3 to the K shell thus gives rise to $K_{\alpha 1}$ x rays of energy $69.523 - 10.205 = 59.318$ keV. Similarly, those from the L_2 to the K shell give rise to $K_{\alpha 1}$ x rays of energy 57.981 keV. Generally, the intensity of $K_{\alpha 2}$ x-ray emission is only 50 to 60 percent of that for $K_{\alpha 1}$ x rays.

For radiological assessment applications, the multiplicities of the L, M, N, ... shells are frequently neglected and only the dominant K series is taken into account, with a single representative energy being used. K-series x rays have energies (keV) given approximately as $E_k \cong 0.0065 Z^{2.11}$, where Z is the atomic number (Chilton, Shultis, and Faw 1984). Thus, K-series x rays from tungsten have energy approximately equal to 57 keV. Tables of decay characteristics recorded in Appendix B include energies and frequencies for both characteristic x rays and Auger electrons.

As charged particles slow, the ratio of the rate at which energy is lost by bremsstrahlung to that by ionization and electronic excitation of the stopping medium is given by the ratio of the radiative stopping power to the collisional stopping power. This ratio is approximately (Evans 1955)

$$\frac{(dT/dx)_{rad}}{(dT/dx)_{coll}} \cong \frac{TZ}{700}\left(\frac{m_e^2}{M}\right), \tag{6.15}$$

in which T is the electron energy in units of MeV, m_e is the electron mass, and M is the mass of the charged particle. Clearly, radiative energy loss is most significant for electrons and positrons, the lightest charged particles. It is especially significant for high energy electrons being stopped in high-Z materials. If electrons of energy T_o as great as several MeV impinge on a target thick compared to the electron range, the energy spectrum of the bremsstrahlung per incident electron, is given approximately by

$$\mathcal{E}(E) \cong 2kZ(T_o - E), \tag{6.16}$$

in which $\mathcal{E}(E)$ is the energy released per unit energy (dimensionless), $k = 7 \times 10^{-4}$ MeV^{-1} and Z is the atomic number of the stopping medium (Evans 1955). This function can be used to calculate by convolution the energy spectra of bremsstrahlung produced when a target is bombarded by beta particles or by electrons accelerated through varying voltages.

While it may appear possible, in principle, to estimate the energy spectrum emerging from the target of an x-ray machine, this is not a practical undertaking for several reasons. The high-voltage power supplies of most machines provide a pulsating or fluctuating voltage. This results in a somewhat "softer" spectrum of bremsstrahlung than would result if the voltage were constant at the peak, that is, incident electrons at less than the peak voltage give rise to lower-energy bremsstrahlung. Also, the window of the x-ray machine alone will attenuate preferentially lower energy x rays. Furthermore, filters are deliberately used to attenuate the lower energy x rays thereby minimizing radiation dose to the skin of patients subjected to x-ray diagnosis or therapy. For these reasons, radiations from x ray machines are characterized by measurements made at a reference distance, usually 1 meter, from the machine's target. One measurement is the exposure rate or air kerma at the reference distance per unit beam current in the machine, for a specified peak voltage, voltage form (continuous or pulsed), and amount of filtration (expressed as the thickness of the filter material). Examples are given in Chap. 5, Fig. 5.1.

Another measurement, which broadly characterizes the energy spectrum of the x rays is the *beam quality* expressed as the thickness of aluminum necessary to reduce the exposure rate by half, that is, the half-value-layer (HVL) of aluminum. Table 5.2 in Chap. 5 lists HVLs for rectified, single-phase, diagnostic x-ray units. Measurements of the x-ray energy spectrum are also needed to permit estimates of radiation doses to the various organs and tissues of patients subjected to diagnostic or therapeutic x rays. Such measurements (Fewell and Shuping 1977) were used by the U.S. Public Health Service in developing their handbook of tissue doses for diagnostic radiology (Rosenstein 1988; Peterson and Rosenstein 1989). Representative spectra are illustrated in Fig. 6.3.

6.5 GEOMETRIC AND EXPONENTIAL ATTENUATION

6.5.1 The Point Source and the Point-Kernel Method

With reference to Fig. 6.4, consider a point isotropic source emitting S_p monoenergetic particles per unit time. A receptor or *dose point* is at P, a distance r from the source.

The point source in a vacuum.   In this case, as illustrated in Fig. 6.4(a), all particles move radially outward, and all pass through a spherical surface of radius r. It follows from the definition of the flux density that

$$\phi^o(r) = \frac{S_p}{4\pi r^2}.\tag{6.17}$$

The superscript on ϕ denotes that the flux density applies to particles reaching r without having experienced any interactions. Commonly, ϕ^o is called the *uncollided* or

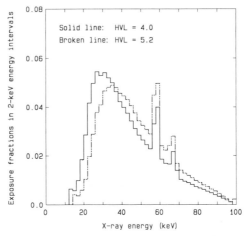

Figure 6.3   Exposure energy spectra at a distance of 1m in air for 100 kVp x-rays with effective filtration HVLs of 4.2 and 5.0 mm aluminum. The spectra are in the form of probability density functions. Each segment of the histogram is the fraction of the exposure rate attributable to x-rays in the 2-keV segment width. [Data are from Peterson and Rosenstein (1989).]

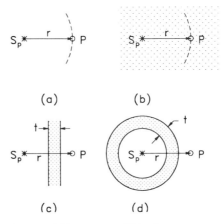

Figure 6.4   The point isotropic source (a) in a vacuum, (b) in a homogeneous medium, (c) with a slab shield, and (d) with a spherical-shell shield.

unscattered flux density. We will adopt the convention that time is measured in seconds and that distances are measured in cm. Thus, the flux density is measured in units of $cm^{-2} s^{-1}$.

   The point source in a uniform attenuating medium.   Now suppose that the source and receptor are present in a homogeneous medium for which the total interaction coefficient for the monoenergetic particles is μ (cm^{-1}) as illustrated in Fig. 6.4(b). Since $e^{-\mu r}$ is the probability that a particle can travel distance r without interaction, it follows that

$$\phi^o(r) = \frac{S_p}{4\pi r^2} e^{-\mu r}. \tag{6.18}$$

Generally, some source particles that scatter one or more times will also reach the receptor a distance r from the source. These scattered particles are not included in $\phi^o(r)$. Their treatment is taken up in Sec. 6.6.

The point source with a shield.   Now suppose that the only attenuating material separating the source and the receptor is in the form of either a slab of material of thickness t with interaction coefficient μ, as illustrated in Fig. 6.4(c), or in the form of a spherical shell of thickness t with interaction coefficient μ, as illustrated in Fig. 6.4(d). In these two cases, $e^{-\mu t}$ is the probability that a particle can reach the receptor without having experienced an interaction. Thus, for these two cases,

$$\phi^o(r) = \frac{S_p}{4\pi r^2} e^{-\mu t}. \tag{6.19}$$

If the slab or the spherical shell is made up of laminates of differing materials each with thickness t_i and interaction coefficient μ_i, then the uncollided flux density is given by

$$\phi^o(r) = \frac{S_p}{4\pi r^2} \exp(-\sum \mu_i t_i). \tag{6.20}$$

The point kernel.   Equations (6.18) to (6.20) are special cases of a more general relationship between the strength S_p of a source at one point, say $\mathbf{r}_A$ and the response of a radiation detector at some other point, say $\mathbf{r}_B$. The general relationship is called the *point kernel* $\mathcal{G}(\mathbf{r}_A, \mathbf{r}_B)$, the response at point $\mathbf{r}_B$ due to a source of *unit* strength at point $\mathbf{r}_B$. Note that the arguments $\mathbf{r}_A$ and $\mathbf{r}_B$ are independent so that $\mathcal{G}$ generally has six independent spatial variables. As will be done in the remainder of this chapter and in the following chapter, point kernels will be used to determine radiation doses from line, plane and volume sources. For example, each differential length of the line source contributes to the response at some point according to the general point kernel. Integration over the length of the line source (the area of the surface source, or the volume of the volume source), weighted by the source strength as it varies with position in the source, leads to the total detector response. Similarly, point-kernel techniques can be used when the detector extends over a region of space as is illustrated in the next chapter when average doses in entire detector regions are computed.

The point-kernel method is widely used in the formulation of radiation shielding and dosimetric calculations. Certain approximations greatly simplify the method, making it one of the most important tools available to the radiation specialist. In certain situations, evaluation of the detector response from a point source does not require treating $\mathbf{r}_A$ and $\mathbf{r}_B$ as independent variables. In an infinite homogeneous medium, the point kernel depends only on the distance between source and detector, namely $r = |\mathbf{r}_A - \mathbf{r}_B|$. This is certainly true for $1/r^2$ geometric attenuation. Even in heterogeneous media, it is often the case that radiation attenuation by matter, as distinct from geometric attenuation, depends primarily on the mass thickness (cumulative ρr) separating source and detector. These circumstances frequently lead to very great simplifications in the point-kernel method. Further, when the point kernel is based on purely exponential attenuation and the inverse square law, as in Eq. (6.18), the application of the point-kernel method to distributed sources is sometimes called *ray theory*. The validity of results obtained with ray theory or other approximations is often difficult to assess *a priori*; nevertheless, ray theory is also a powerful technique, capable of high accuracy when used with discretion (Chilton, Shultis, and Faw, 1984).

6.5.2 The Line Source as a Superposition of Point Sources

An isotropic line source is depicted in Fig. 6.5. The source is of length L and emits S_l particles per unit time per unit length. A receptor is positioned at point P, distance h from the line along a perpendicular to one end of the line. Consider a segment of the line between distance x and distance $x + dx$ along the line. The source within the segment may be treated as an effective point source with strength $S_l dx$. We will first evaluate that portion of the uncollided flux density $d\phi^o$ attributable to the effective point source and then sum, or rather integrate, over all line segments to obtain the total uncollided flux density at P.

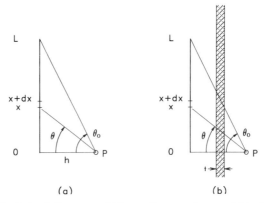

Figure 6.5 The isotropic line source (a) in a uniform medium, and (b) with a slab shield.

Line source in a non-attenuating medium.   In the absence of material interaction,

$$d\phi^o(P) = \frac{1}{4\pi} \frac{S_l\, dx}{x^2 + h^2} \tag{6.21}$$

and thus

$$\phi^o(P) = \frac{1}{4\pi} \int_0^L \frac{S_l\, dx}{x^2 + h^2}. \tag{6.22}$$

As can be seen from the figure, $x = h \tan\theta$ and $x^2 + h^2 = h^2 \sec^2\theta$. It follows that $dx = h \sec^2\theta\, d\theta$ and that one may evaluate the integral as

$$\phi^o(P) = \frac{S_l}{4\pi h} \int_0^{\theta_o} d\theta = \frac{S_l\theta_o}{4\pi h}. \tag{6.23}$$

The angle θ_o must be expressed in radians for this equation to be valid.

Line source in a uniform attenuating medium.   Now suppose that the source and receptor are present in a homogeneous medium with a total interaction coefficient μ. Attenuation along the ray from x to P reduces the uncollided flux density to

$$d\phi^o(P) = \frac{1}{4\pi} \frac{S_l \, dx}{x^2 + h^2} e^{-\mu\sqrt{x^2+h^2}}. \tag{6.24}$$

The total uncollided flux density now is described by the integral

$$\phi^o(P) = \frac{S_l}{4\pi h} \int_0^{\theta_o} d\theta \, e^{-\mu h \sec \theta}. \tag{6.25}$$

This integral cannot be evaluated analytically. It is in the form of the *Sievert integral* or the *secant integral*, defined as

$$F(\theta, b) \equiv \int_0^{\theta} dx \, e^{b \sec x}. \tag{6.26}$$

The function is illustrated in Fig. 6.6. Its evaluation is facilitated by making use of the auxiliary function $\bar{F}(\theta, b) \equiv \theta e^{-b} F(\theta, b)$ as suggested by Shure and Wallace (1975). As may be seen from Table 6.6, accurate interpolation of the tabulated auxiliary function is easier than of the secant integral itself. While the table gives the angle θ in degrees, units of radians must be used in calculations. In terms of the Sievert integral, Eq. (6.25) may then be written as

$$\phi^o(P) = \frac{S_l}{4\pi h} F(\theta_o, \mu h). \tag{6.27}$$

The line source with a slab shield.   Suppose now that the only material separating the line source and the receptor is a parallel slab or concentric cylindrical-shell shield of thickness t and total attenuation coefficient μ_s. For this case, the uncollided flux density at P is

$$\phi^o(P) = \frac{S_l}{4\pi h} F(\theta_o, \mu_s t). \tag{6.28}$$

If the shield is made up of layers of thicknesses t_i and attenuation coefficients μ_{si}, then $\mu_s t$ must be replaced by $\sum \mu_{si} t_i$, the total mean-free-path thickness of the shield.

A superposition procedure for line sources.   Figure 6.7 illustrates two receptor points in relation to a line source in a uniform attenuating medium. Determination of the uncollided flux density at either point may be accomplished by the principle of

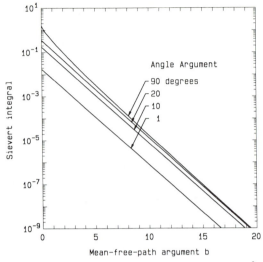

Figure 6.6 Illustration of the Sievert or secant integral $F(\theta, b) = \int_0^\theta dx\, e^{-b \sec x}$.

TABLE 6.6 Values of the Modified Sievert Integral $\tilde{F}(\theta, b)$.

	ANGLE θ (degrees)									
b	1	10	20	30	40	50	60	70	80	90
0.0	1.00000	1.00000	1.00000	1.00000	1.00000	1.00000	1.00000	1.00000	1.00000	1.00000
0.1	0.99999	0.99949	0.99791	0.99511	0.99080	0.98443	0.97492	0.95991	0.93267	0.86443
0.5	0.99997	0.99745	0.98962	0.97600	0.95558	0.92671	0.88677	0.83172	0.75752	0.67563
1.0	0.99995	0.99491	0.97944	0.95306	0.91484	0.86367	0.79870	0.72113	0.63897	0.56810
2.0	0.99990	0.98986	0.95965	0.91012	0.84296	0.76166	0.67274	0.58657	0.51396	0.45686
3.0	0.99985	0.98486	0.94059	0.87078	0.78189	0.68362	0.58826	0.50707	0.44376	0.39445
4.0	0.99980	0.97991	0.92223	0.83468	0.72965	0.62259	0.52802	0.45344	0.39677	0.35268
5.0	0.99975	0.97500	0.90453	0.80149	0.68466	0.57385	0.48282	0.41411	0.36235	0.32209
6.0	0.99970	0.97014	0.88747	0.77092	0.64565	0.53416	0.44747	0.38363	0.33568	0.29838
7.0	0.99964	0.96532	0.87103	0.74273	0.61161	0.50125	0.41891	0.35909	0.31420	0.27929
8.0	0.99959	0.96054	0.85516	0.71667	0.58172	0.47351	0.39522	0.33877	0.29642	0.26348
0.0	0.99954	0.95581	0.83986	0.69255	0.55530	0.44980	0.37516	0.32157	0.28137	0.25011
10.0	0.99949	0.95112	0.82509	0.67018	0.53181	0.42927	0.35789	0.30677	0.26842	0.23860
11.0	0.99944	0.94647	0.81083	0.64941	0.51081	0.41127	0.34282	0.29385	0.25711	0.22855
12.0	0.99939	0.94187	0.79706	0.63008	0.49193	0.39536	0.32951	0.28244	0.24713	0.21967
13.0	0.99934	0.93730	0.78376	0.61206	0.47486	0.38115	0.31765	0.27227	0.23824	0.21177
14.0	0.99929	0.93278	0.77091	0.59524	0.45937	0.36837	0.30699	0.26313	0.23024	0.20466
15.0	0.99924	0.92830	0.75849	0.57951	0.44523	0.35679	0.29733	0.25486	0.22300	0.19822
16.0	0.99919	0.92385	0.74648	0.56478	0.43227	0.34624	0.28854	0.24732	0.21641	0.19236
17.0	0.99914	0.91945	0.73487	0.55096	0.42035	0.33658	0.28049	0.24042	0.21036	0.18699
18.0	0.99909	0.91509	0.72364	0.53797	0.40934	0.32768	0.27307	0.23406	0.20480	0.18205
10.0	0.99904	0.91076	0.71277	0.52575	0.39914	0.31946	0.26622	0.22819	0.19966	0.17748
20.0	0.99899	0.90648	0.70225	0.51423	0.38965	0.31183	0.25986	0.22273	0.19489	0.17324

superposition. The superposition procedure applies as well to a line source in a non-attenuating medium or a line source with parallel slab or cylindrical shell shielding. Point P_1, for example, is on a normal from the end of a projection of the line source.

Were the line source truly of the extended length, then the flux density would be given by Eq. (6.27) with angle argument θ_1. However, that result would be too high by just the amount contributed by a line source of the same strength subtending angle θ_2. Thus, at point P_1

$$\phi^o(P_1) = \frac{S_l}{4\pi h}\left[F(\theta_1, \mu h) - F(\theta_2, \mu h)\right].\tag{6.29}$$

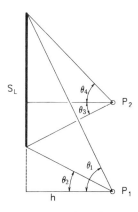

Figure 6.7 The superposition principle applied to the isotropic line source.

By similar reasoning, at point P_2

$$\phi^o(P_2) = \frac{S_l}{4\pi h}\left[F(\theta_3, \mu h) + F(\theta_4, \mu h)\right].\tag{6.30}$$

6.5.3 The Disk Source as a Superposition of Point Sources

While any area source can be represented as a superposition of point sources, there is one special geometry, illustrated in Fig. 6.8, for which the mathematics involved with the superposition is straightforward and the results are very important in radiation shielding design and analysis. This special case involves a disk source with a receptor on the normal to the center of the disk. An annular strip of radius ρ, width $d\rho$, and area $dA = 2\pi\rho d\rho$ is shown in the figure. The width $d\rho$, of course, is infinitesimal and it is clear that all sources within the area dA are equidistant from the receptor point P. If S_a is the area source strength, conventionally in units of $cm^{-2}\,s^{-1}$, then the strength of an effective point source equivalent to the sources between ρ and $\rho + d\rho$ is just $S_a dA$. The distance from this effective point source to the receptor is $r = \sqrt{\rho^2 + h^2}$.

The disk source in a nonattenuating medium.   Only geometric attenuation, through the inverse square law, applies and the flux density at the receptor, accounting for the entire disk source, is given by

$$\phi^o(P) = \frac{S_a}{4\pi}\int_A \frac{dA}{r^2} = \frac{S_a}{2}\int_0^a d\rho\,\frac{\rho}{\rho^2 + h^2}.\tag{6.31}$$

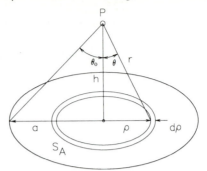

Figure 6.8   The isotropic disk source.

This integral may be evaluated analytically. Since h is fixed, $\rho d\rho = r dr$. One may also note that, since $r = h \sec\theta$, $dr = h \sec\theta \tan\theta d\theta$. It follows that

$$\phi^{o}(P) = \frac{S_a}{2} \int_{h}^{h \sec\theta} dr\, r^{-1} = \frac{S_a}{2} \int_{0}^{\theta_o} d\theta\, \tan\theta = \frac{S_a}{2} \ln(\sec\theta_o). \qquad (6.32)$$

Notice that several choices are available in the integral expression of the flux density. This is often the case in radiation shielding problems. In practical problems, for which numerical evaluation of integrals are frequently required, one of the choices will usually suggest itself as being more efficient for the numerical evaluation.

    The disk source in a uniform attenuating medium.   Suppose that the source and receptor of Fig. 6.8 are present in a uniform medium with total interaction coefficient μ. Exponential attenuation along path r must now be taken into account, and Eq. (6.30) must be modified to read

$$\phi^{o}(P) = \frac{S_a}{2} \int_{h}^{h \sec\theta_o} dr\, r^{-1} e^{-\mu r}. \qquad (6.33)$$

A change of independent variable from r to $x \equiv \mu r$ allows this equation to be written in a form which introduces another special mathematical function useful in shielding analysis.

$$\phi^{o}(P) = \frac{S_a}{2} \int_{\mu h}^{\mu h \sec\theta_o} dx\, x^{-1} e^{-x}. \qquad (6.34)$$

or

$$\phi^{o}(P) = \frac{S_a}{2} \left(\int_{\mu h}^{\infty} dx\, x^{-1} e^{-x} - \int_{\mu h \sec\theta_o}^{\infty} dx\, x^{-1} e^{-x} \right). \qquad (6.35)$$

The special function is the *exponential integral function* $E_n(b)$, defined for integer $n \geq 0$ and positive b as follows:

$$E_n(b) \equiv b^{n-1} \int_{b}^{\infty} dx\, x^{-n} e^{-x} = \int_{1}^{\infty} dx\, x^{-n} e^{-bx} = \int_{0}^{1} dx\, x^{n-2} e^{-b/x}. \qquad (6.36)$$

Examination of Eqs. (6.33) and (6.35) reveals that the uncollided flux density at the receptor is given by

$$\phi^o(P) = \frac{S_a}{2}\left[E_1(\mu h) - E_1(\mu h \sec\theta)\right]. \tag{6.37}$$

As the radius of the disk approaches infinity, the second E_1 function approaches zero.

The exponential integral function is illustrated in Fig. 6.9. Direct interpolation is difficult and evaluation of exponential integrals is facilitated by use of the auxiliary function $\bar{E}_n(b) \equiv be^b E_n(b)$ suggested by Shure and Wallace (1975). This auxiliary function is tabulated in Table 6.7. As b approaches infinity, all $\bar{E}_n(b)$ approach unity. $E_1(b)$ approaches infinity as b approaches 0 and, for very small values of b, a useful approximation is

$$E_1(b) \simeq -\gamma - \ln b, \tag{6.38}$$

in which γ (Euler's constant) is equal to $0.577216 \ldots$ Many special properties of the exponential integral functions, their derivatives and their integrals are discussed by Chilton, Shultis, and Faw (1984).

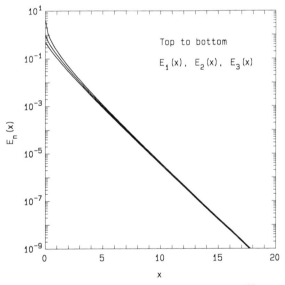

Figure 6.9   The exponential integral function $E_n(x) = x^{n-1}\int_x^\infty du\, u^{-n}e^{-u}$.

The disk source with a slab shield.   Suppose that there is placed between the disk source and the receptor, a slab shield of thickness t and linear interaction coefficient μ_s, and that the medium between source and receptor is otherwise nonattenuating. It is easy to show that, in this case, the uncollided flux density becomes [cf. Eq. (6.37)],

$$\phi^o(P) = \frac{S_a}{2}\left[E_1(\mu_s t) - E_1(\mu_s t \sec\theta)\right]. \tag{6.39}$$

TABLE 6.7 The Modified Exponential Integral Function $\check{E}_n(x)$.

x	$\check{E}_1(x)$	$\check{E}_2(x)$	$\check{E}_3(x)$	x	$\check{E}_1(x)$	$\check{E}_2(x)$	$\check{E}_3(x)$
0.00	0.00000	0.00000	0.00000	1.00	0.59635	0.40365	0.29817
0.01	0.04079	0.00959	0.00495	1.25	0.63879	0.45151	0.34281
0.02	0.06845	0.01863	0.00981	1.50	0.67239	0.49142	0.38143
0.03	0.09148	0.02726	0.01459	1.75	0.69979	0.52537	0.41530
0.04	0.11163	0.03553	0.01929	2.00	0.72266	0.55469	0.44531
0.05	0.12972	0.04351	0.02391	2.50	0.75881	0.60296	0.49630
0.06	0.14623	0.05123	0.02846	3.00	0.78625	0.64125	0.53813
0.08	0.17566	0.06595	0.03736	3.50	0.80787	0.67246	0.57319
0.10	0.20146	0.07985	0.04601	4.00	0.82538	0.69847	0.60306
0.15	0.25522	0.11172	0.06662	5.00	0.85211	0.73945	0.65139
0.20	0.29867	0.14027	0.08597	6.00	0.87161	0.77037	0.68890
0.30	0.36676	0.18997	0.12150	8.00	0.89824	0.81410	0.74359
0.40	0.41913	0.23235	0.15353	10.00	0.91563	0.84367	0.78167
0.60	0.49676	0.30194	0.20942	12.00	0.92791	0.86504	0.80978
0.70	0.52687	0.33119	0.23408	15.00	0.94080	0.88794	0.84047
0.80	0.55300	0.35760	0.25696	20.00	0.95437	0.91258	0.87418

6.5.4 The Rectangular Area Source

Figure 6.10 illustrates a receptor at location P, distance z from the corner of a rectangular area source of dimensions a and b. The source area within differential area $dA = dxdy$ can be treated as an effective differential point source of strength $S_a dA$.

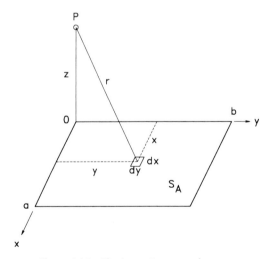

Figure 6.10 The isotropic rectangular source.

The rectangular area source in a nonattenuating medium.   That part of the flux density at P attributable to area dA must account for geometric attenuation along radial distance r, where $r^2 = x^2 + y^2 + z^2$. The total uncollided flux density at P requires integration over both x and y, namely

$$\phi^o(P) = \frac{S_a}{4\pi} \int_0^b dy \int_0^a dx \frac{1}{x^2 + y^2 + z^2}. \tag{6.40}$$

One of the two integrals may be evaluated analytically to obtain

$$\phi^o(P) = \frac{S_a}{4\pi} \int_0^b \frac{dy}{\sqrt{y^2 + z^2}} \tan^{-1}\left(\frac{a}{\sqrt{y^2 + z^2}}\right). \tag{6.41}$$

The remaining integral must be evaluated numerically or by using special techniques such as those developed by Hubbell, Beach, and Lamkin (1960).

Superposition techniques, similar to those for the line source, may be used to evaluate the flux density from any plane source which can be subdivided into rectangular areas.

The rectangular source in an attenuating medium.   While it is easy to modify Eq. (6.40) to account for exponential attenuation, it is not so easy to carry out the integrations. We leave this task to the reader who may instead choose to find the solution published as an appendix in the book by Schaeffer (1973). There are some approximate methods of dealing with this case, or indeed the case of the source in a nonattenuating medium. When the receptor is positioned on a normal to the center of a rectangular area the flux density is often approximated as being equal to that from a disk source of the same strength subtending at the receptor *the same solid angle*, not the same area, as does the rectangular source. Eccentric rectangular areas may be approximated by combinations of disks and annular-sector sources (Spencer 1962).

6.5.5 Surface and Volume Sources

The line, disk, and rectangular sources have been used for purposes of illustration of the superposition technique. This technique is very general and may be applied to surface and volume sources of complex shapes. Many important practical cases have been examined and generalized results have been published. Among the special cases are cylindrical and spherical, surface and volume sources with and without external shields, and with interior as well as exterior receptor locations. In this text, only one of these cases is reviewed, and, for other cases, the reader is referred to the publications of Rockwell (1956), Blizard and Abbott, (1962), Blizard et al. (1968), Hungerford (1966), Schaeffer (1973); Courtney (1975); and Chilton, Shultis, and Faw 1984).

The cylindrical volume source.   This source is important for two reasons: (1) there are many limiting cases of interest, and (2) it is representative of many practical sources such as nuclear fuel elements, nuclear reactor cores, and storage or shipping containers for radioactive materials. Here, receptor locations on the axis and on the surface of the cylinder are considered. The more general problems of off-axis internal receptors, external receptor locations, and external shields are addressed in detail by Blizard et al. (1968).

The source is illustrated in Fig. 6.11. The source strength is uniform of strength S_v (cm^{-3} s^{-1}) in a homogeneous medium with linear interaction coefficient μ (cm^{-1}). Height and radius are, respectively, H and R. The uncollided flux densities at point P_0 at the center of the base and at P_1 on the perimeter of the base are computed as the integrals over the entire volume of the cylinder of contributions from sources within elements of volume $dV = \rho\, d\rho\, d\psi\, dz$ treated as effective point sources. From points P_0 and P_1, the distances r to the volume element dV are determined by the law of cosines and by the Pythagorean theorem, namely

$$r^2 = z^2 + \rho^2, \text{ for } P_0 \tag{6.42}$$

and

$$r^2 = z^2 + R^2 + \rho^2 - 2\rho R \cos\psi, \text{ for } P_1. \tag{6.43}$$

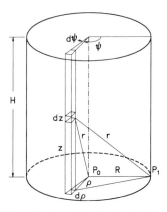

Figure 6.11   The isotropic cylindrical volume source.

The uncollided flux densities, accounting for symmetry in ψ for point P_0, are thus

$$\phi^o(P_0) = \frac{S_v}{2} \int_0^R d\rho\, \rho \int_0^H dz\, r^{-2} e^{-\mu r}, \tag{6.44}$$

and

$$\phi^o(P_1) = \frac{S_v}{2\pi} \int_0^\pi d\psi \int_0^R d\rho\, \rho \int_0^H dz\, r^{-2} e^{-\mu r}. \tag{6.45}$$

These expressions can be converted into forms more suitable for calculations, namely,

$$\phi^o(P_0) = \frac{S_v}{2\mu} G_1(\mu H, \mu R), \tag{6.46}$$

and

$$\phi^o(P_1) = \frac{S_v}{2\mu} G_2(\mu H, \mu R), \tag{6.47}$$

in which the G-functions, dependent only on the total radial and axial mean-free-paths of shielding within the cylinder, are computed from the following equations and are illustrated in Figs. 6.12 and 6.13.

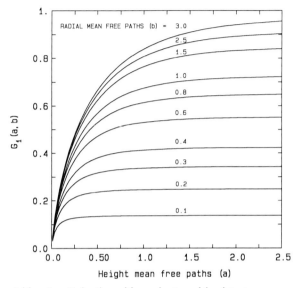

Figure 6.12   The special function $G_1(a, b)$ used for evaluation of the detector response at the center of the base of a cylindrical volume source with height and radius a and b mean-free-paths, respectively.

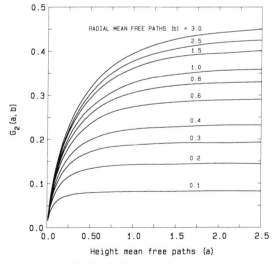

Figure 6.13   The special function $G_2(a, b)$ used for evaluation of the detector response at the perimeter of the base of a cylindrical volume source with height and radius a and b mean-free-paths, respectively.

$$G_1(a, b) = \int_0^a dx [E_1(x) - E_1(\sqrt{x^2 + b^2})], \tag{6.48}$$

and

$$G_2(a, b) = \frac{1}{\pi} \int_0^\pi d\psi \int_0^b dy \, (y/\zeta) F(\tan^{-1}(a/\zeta), \zeta), \tag{6.49}$$

where

$$\zeta^2 = b^2 + y^2 - 2by \cos \psi. \tag{6.50}$$

Superposition procedures may be applied to cylindrical volume sources to obtain the flux density at any point on the surface or the axis. For example, consider Fig. 6.14 and the receptor locations P_2 and P_3. At each location, the flux density may be thought of as arising from two subcylinders of heights H_1 and H_2. Thus, the flux densities are given by

$$\phi^\circ(P_2) = \frac{S_v}{2\mu} [G_1(\mu H_1, \mu R) + G_1(\mu H_2, \mu R)], \tag{6.51}$$

and

$$\phi^\circ(P_3) = \frac{S_v}{2\mu} [G_2(\mu H_1, \mu R) + G_2(\mu H_2, \mu R)]. \tag{6.52}$$

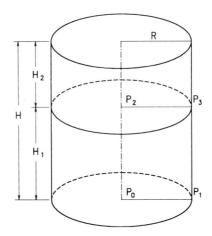

Figure 6.14   Application of the superposition principle to cylindrical volume sources.

The cylindrical volume source in a nonattenuating medium.   Equations (6.44) and (6.45) may be evaluated analytically when $\mu = 0$. The results are

$$\phi^\circ(P_0) = \frac{S_v}{2} \left[R \tan^{-1} \frac{H}{R} + \frac{H}{2} \ln(1 + \frac{R^2}{H^2}) \right], \tag{6.53}$$

and

$$\phi^o(P_1) = \frac{S_v H}{4} \left\{ 1 + 2R/H - \sqrt{1 + 4R^2/H^2} + \ln\left(\frac{1}{2} \left[1 + \sqrt{1 + 4R^2/H^2} \right] \right) \right\}. \tag{6.54}$$

The infinite slab volume source.   Consider a receptor on or above the surface of a homogeneous slab containing a uniform volumetric radiation source of strength S_v. This situation is just a limiting case of cylindrical volume source, namely point P_0 in Fig. 6.11 with $R = \infty$. The flux density is just

$$\phi^o(P_0) = \frac{S_v}{2\mu} G_1(\mu H, \infty) = \frac{S_v}{2\mu} [1 - E_2(\mu H)]. \tag{6.55}$$

As the slab thickness approaches infinity, the uncollided flux density approaches

$$\phi^o(P_0) = \frac{S_v}{2\mu} G_1(\infty, \infty) = \frac{S_v}{2\mu}. \tag{6.56}$$

6.6 ACCOUNTING FOR SCATTERED PHOTONS

The discussion in Sec. 6.5 deals exclusively with the uncollided flux density arising from monoenergetic sources. When material attenuation is very slight, the response of a point isotropic detector to the uncollided flux density, $R^o = \mathcal{R}(E_o)\phi^o$, may approximate well the total response, itself given by

$$R = R^o + R^s = \mathcal{R}(E_o)\phi^o + \int_0^\infty dE\, \mathcal{R}(E)\phi^s(E). \tag{6.57}$$

Here R^s is the response due to scattered radiation, as described by its flux density energy spectrum $\phi^s(E)$. In most cases it is necessary to account for the contribution of scattered radiation to the detector response. There are three main ways of accomplishing this. The first involves calculation of $\phi^s(E)$ directly. The computational methods are quite complicated and beyond the scope of this text. The results of such calculations, however, form the bases for the other two methods of treating scattered gamma rays. The second, which applies to monoenergetic point sources and their superposition into line, area and volume sources, corrects (often very accurately) for scattered radiation with appropriate *buildup factors*. The third method applies to both polyenergetic and monoenergetic sources and deals with the "broad beam" attenuation provided by structural and shielding materials distant from sources such as x-ray machines.

6.6.1 Buildup Factors

Flux density energy spectra have been computed for monoenergetic point sources in infinite homogeneous media as functions of the distance from the source to the receptor location. These spectra are used to calculate buildup factors defined as

$$B \equiv 1 + \frac{R^s}{R^o} = 1 + \frac{1}{\mathcal{R}(E_o)\phi^o(r)} \int_0^\infty dE\, \mathcal{R}(E)\phi^s(r, E). \qquad (6.58)$$

Clearly the buildup factor depends on the source energy E_o, the distance r (or the number of mean free paths, μr), and the nature of the medium, as reflected in $\phi^o(r)$ and $\phi^s(r, E)$. It must be stressed, too, that the buildup factor depends on the nature of the receptor response, as reflected in the response functions $\mathcal{R}(E)$. For example, one may refer to the buildup factor for kerma in iron (as the response) for gamma rays being attenuated in iron (as the medium). One may, as well, refer to the buildup factor for air kerma, exposure, or tissue kerma (as the response) for gamma rays attenuated in iron. The buildup factor is usually denoted as $B(E_o, \mu r)$; its dependence on the composition of the medium and the type of response is implicit.

Infinite medium buildup factors can be used rigorously in calculations only for infinite homogeneous media. Nevertheless, they are very often used for calculations in finite inhomogeneous media, especially for radiation shields made up of layers of different media. They are so used even in some very powerful computer codes for radiation shielding design and analysis. Fortunately, methods are available to correct for the inherently improper application of infinite-medium buildup factors (Kalos 1956; Bunemann and Richter 1968; Chilton, Shultis, and Faw 1984; Trubey and Harima 1987; ANSI 1990). A commonly used rule for two layers of differing materials with thicknesses x_1 and x_2 numbered in the direction from source to detector is as follows: If Z_i and μ_i are the atomic numbers and interaction coefficients for the two media, and if $Z_1 < Z_2$, then the overall buildup factor is approximately equal to the buildup factor B_2 for the higher-Z medium with the use of the total number of mean-free-paths as its argument, that is, $B_2(E_o, \mu_1 x_1 + \mu_2 x_2)$. Otherwise, the overall buildup factor is approximately the product $B_1(E_o, \mu_1 x_1) \times B_2(E_o, \mu_2 x_2)$. Broder et al. (1962) suggest a method for the synthesis of buildup factors for multi-layer shields.

Buildup-factor data.   A limited set of buildup factors is provided in Appendix E (Eisenhauer and Simmons 1975; Chilton, Eisenhauer, and Simmons 1980). They are more broadly applicable than might be thought at first glance. As indicated in Eq. (6.58), it is the ratio $\mathcal{R}(E)/\mathcal{R}(E_o)$ that controls the response-dependence of the buildup factor. For responses such as kerma or absorbed dose in air or water, exposure, or dose equivalent, the ratio is not very sensitive to the type of response. Thus, buildup factors for air kerma or water kerma may be used, for example, for exposure or dose equivalent. Table E-2 (in Appendix E) is for air kerma as the response for gamma rays attenuated in water. These data are useful for internal dosimetry and apply with negligible error to absorbed dose in tissue. They also apply with little error to water kerma as the response. Tables E-1 through E-4 are all for air kerma as the response and for attenuation in air, water, concrete, and lead. Except for attenuation of high-energy gamma rays in lead, they apply with little error to exposure and dose equivalent response.

Figure 6.15 illustrates the buildup factor for concrete, plotted with the photon energy as independent variable and the number of mean free paths as a parameter. That there are maxima in the curves is due to the relative importance of the photoelectric effect, as compared to Compton scattering, in the attenuation of lower energy photons and to the very low fluorescence yields exhibited by the low-Z constituents of concrete. Figure 6.16 illustrates the buildup factor for lead, plotted with the number of mean-free-paths as the independent variable and the photon energy as a parameter. For high-energy photons, pair production is the dominant attenuation mechanism in lead, the cross section exceeding that for Compton scattering at energies above about 5 MeV. The buildup is relatively large because of the production of 0.511-MeV annihilation gamma rays. The extraordinarily large buildup factor for 0.1-MeV photons is due to production of K-shell x rays. At energies below the K-edge, 0.088 MeV, the buildup factors are very small, as indicated in Table E-4.

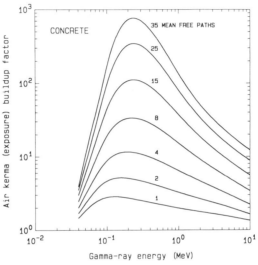

Figure 6.15   Air kerma (exposure) buildup factors for gamma-ray attenuation in concrete. [Data are from Eisenhauser and Simmons (1975).]

Analytical approximations to buildup factors.   A great deal of effort has been directed towards the approximation of buildup factors by mathematical functions which can be used directly in calculations. Two forms of approximation have seen very wide use. One is the *Taylor form*

$$B(E_o, \mu r) \simeq \sum_i A_i e^{-\alpha_i \mu r}, \qquad (6.59)$$

in which the parameters A_i and α_i depend on E_o, the attenuating medium, and the type of response. A second is the *Berger form*

$$B(E_o, \mu r) \simeq 1 + a\mu r e^{+b\mu r}, \qquad (6.60)$$

in which the parameters a and b depend on E_o, the attenuating medium, and the type of response. For purposes of illustration, parameters for the Taylor form of the buildup factor for air kerma in concrete are given in the Appendix in Table E-5 (Chilton 1977). Parameters for the Berger form of the buildup factor for air kerma

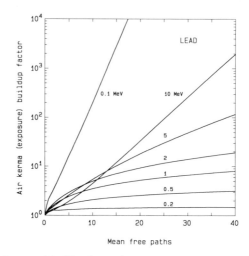

Figure 6.16 Air kerma (exposure) buildup factors for gamma-ray attenuation in lead. [Data are from Simmons and Eisenhauer, as reported in Chilton, Shultis, and Faw (1984).]

in several media are given in Table E-6 (Eisenhauer and Simmons 1975; Chilton 1979; Chilton, Eisenhauer, and Simmons 1980). The data in the latter table were selected to provide the best fit for the buildup factor over the entire span from 0 to 40 mean free paths. For certain energies and distances, errors may be as great as 45 percent.

As would be expected, high accuracy requires an elaborate buildup-factor approximation (i.e., a formula with many parameters). An extraordinarily precise formulation, called the *geometric progression* or *GP* approximation of the buildup factor (Harima et al. 1986), is the favored choice

$$B(E_o, \mu r) \cong \begin{cases} 1 + \dfrac{(b-1)(K^{\mu r} - 1)}{K - 1}, & K \neq 1 \\[2ex] 1 + (b-1)\mu r, & K = 1, \end{cases} \tag{6.61}$$

where

$$K(\mu r) = c(\mu r)^a + d\frac{\tanh(\mu r/\xi - 2) - \tanh(-2)}{1 - \tanh(-2)} \tag{6.62}$$

in which a, b, c, d, and ξ are parameters dependent on the gamma-ray energy, the attenuating medium, and the nature of the response. Example values of the parameters for kerma in air as the response, and for attenuation in air, water, concrete, iron, and lead are listed in the Appendix in Tables E-7 through E-9.

The single-term relaxation-length buildup approximation.   There are circumstances for which a very simple buildup-factor approximation may be applied successfully. Such would be the case when attenuation is known to take place over a limited range of mean free path lengths or when a high degree of conservatism may be satisfactory. To illustrate, consider Fig. 6.17 which depicts the uncollided and the total exposure rates $R^o(\mu r)$ and $R(\mu r)$ as functions of distance in concrete for 0.5-MeV gamma rays. What is plotted is $4\pi r^2 R^o(\mu r)/\mathcal{R}(E_o) = e^{-\mu r}$ for the uncollided response rate and $4\pi r^2 R(\mu r)/\mathcal{R}(E_o) = B(E_o, \mu r)e^{-\mu r}$ for the total response rate. Both functions equal unity for $r = 0$. Superimposed on the curve for the total response is a straight-line approximation which is seen to give good agreement for μr between about 3 and 8 and to be otherwise conservatively in excess of the actual response — very much so for small values of μr. The straight-line approximation is just an approximation of the total response from a point source of unit strength, that is,

$$R = \frac{\mathcal{R}(E_o)}{4\pi r^2}A_1 e^{-\mu' r} \qquad (6.63)$$

in which A_1 and μ' are determined by the slope and intercept of the straight-line approximation. In effect, a single-term Taylor form of the buildup factor is being applied, namely,

$$B(E_o, \mu r) \simeq A_1 e^{-\alpha_1 \mu r}, \qquad (6.64)$$

in which A_1 and $\alpha_1 = \mu'/\mu - 1$ depend on the gamma-ray energy, the attenuating medium, the nature of the response, and the range of mean free paths over which the approximation is applied. Table 6.8 lists values of the parameters for exposure as the response, attenuation in several media, and 4 to 7 mean-free-paths penetration distance.

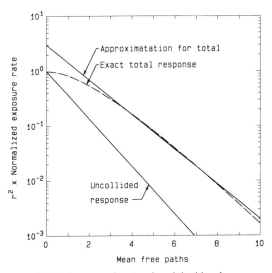

Figure 6.17   The single-term, relaxation-length buildup-favor approximation.

6.6.2 Application of Buildup Factors in Shielding Calculations

In calculations involving extended sources such as line, area, or volume sources, the point-source buildup factor must be applied at the stage of calculation at which effective point sources are being treated. Two examples are illustrated.

TABLE 6.8 Parameters in the Single-Term Taylor Form of the Exposure Buildup Factor for Gamma-Ray Point Sources in Infinite Media. The Approximation is Specifically for 4 to 7 Mean-Free-Paths, but May Be Applied Very Conservatively to Fewer Mean-Free-Paths.

E_o (MeV)	WATER		CONCRETE		IRON		LEAD	
	A_1	$-\alpha_1$	A_1	$-\alpha_1$	A_1	$-\alpha_1$	A_1	$-\alpha_1$
0.2	5.5	0.40	3.8	0.28	2.2	0.15		
0.5	3.7	0.31	3.0	0.27	2.4	0.22	1.4	0.06
1.0	2.9	0.24	2.6	0.23	2.3	0.21	1.5	0.10
1.5	2.6	0.21	2.4	0.20	2.2	0.19	1.5	0.11
2.0	2.4	0.18	2.2	0.18	2.1	0.18	1.5	0.13
5.0	1.8	0.13	1.8	0.14	1.6	0.15	1.1	0.17

Source: Chilton, Shultis, and Faw 1984.

Line source in a uniform attenuation medium. With reference to Fig. 6.5 and Eq. (6.24), the total response rate at detector point P due to photons arising from the differential source length dx is

$$dR(P) = \frac{\mathcal{R}(E_o)S_l dx}{4\pi(x^2 + h^2)}e^{-\mu\sqrt{x^2+h^2}}B(E_o, \mu\sqrt{x^2 + h^2}).   (6.65)$$

In analogy to Eq. (6.25) the response due to the entire line source is given by the integral

$$R(P) = \mathcal{R}(E_o)\frac{S_l}{4\pi h}\int_0^{\theta_o} d\theta\, e^{-\mu h \sec\theta}B(E_o, \mu h \sec\theta).   (6.66)$$

In general, the integral must be evaluated numerically. However, if the Taylor form of buildup-factor approximation, Eq. (6.59), is employed, the integral yields a sum of Sievert integrals. As the reader may verify by substituting the buildup factor from Eq. (6.59) into Eq. (6.66) and changing the order of integration and summation,

$$R(P) = \mathcal{R}(E_o)\frac{S_l}{4\pi h}\sum_i A_i F(\theta_o, [1 + \alpha_i]\mu h).   (6.67)$$

Note the similarity in form between Eq. (6.67) and Eq. (6.27). This illustrates the great utility of the Taylor form of buildup-factor approximation: any analytical solution for the response due to uncollided photons may be converted to a solution for the total response in the form of a series sum with the parameter A_i as a weight factor and with the interaction coefficient μ replaced by the product $(1 + \alpha_i)\mu$.

The disk source in a uniform attenuating medium. With reference to Fig. 6.8 and Eq. (6.33), it is seen that the total response rate at detector location P is given by

$$R(P) = \mathcal{R}(E_o)\frac{S_a}{2}\int_h^{h \sec\theta} dr\, r^{-1}e^{-\mu r}B(E_o, \mu r),   (6.68)$$

or

$$R(P) = \mathcal{R}(E_o)\frac{S_a}{2}\int_{\mu h}^{\mu h \sec \theta} dx\, x^{-1}e^{-x}B(E_o, x), \tag{6.69}$$

Numerical integration is usually required for most approximations of $B(E_o, x)$. However, with the approximation of the buildup factor by the Taylor form, the value of $R(P)$, may be expressed in terms of exponential integral functions, namely,

$$R(P) = \mathcal{R}(E_o)\frac{S_a}{2}\sum_i A_i\left\{E_1[(1+\alpha_i)\mu h] - E_1[(1+\alpha_i)\mu h \sec \theta_o]\right\}. \tag{6.70}$$

The Berger form of the buildup-factor approximation also yields a simple form for the total response. As may be verified by substituting Eq. (6.60) into Eq. (6.69) and carrying out the integration, one obtains

$$R(P) = \mathcal{R}(E_o)\frac{S_a}{2}\left\{[E_1(\mu h) - E_1(\mu h \sec \theta_o)] + \frac{a}{1-b}\left[e^{-(1-b)\mu h} - e^{-(1-b)\mu h \sec \theta_o}\right]\right\} \tag{6.71}$$

6.6.3 Broad-Beam Attenuation — Photon Sources

It is often the case in dealing with the shielding requirements for a radionuclide or x-ray source that the source is located some distance in air from a wall or shielding slab and the concern is with the radiation dose rate on the exterior (cold) side of the wall. Often too, the source is sufficiently far from the wall that the radiation reaches the wall in nearly parallel rays, even though the attenuation in the air is quite negligible in comparison to that provided by the shielding wall. Shielding design and analysis in the circumstances just described, and illustrated in Fig. 6.18, are addressed by the National Council on Radiation Protection and Measurements (NCRP 1976) in their widely used Report 49. Attenuation of photons from both monoenergetic and polyenergetic sources can be established in terms of the following formula:

$$R(P) = R_o(P)A_f, \tag{6.72}$$

in which $R(P)$ is the response at point P, $R_o(P)$ is the response in the absence of the shield wall, accounting only for the inverse-square attenuation, and A_f is an attenuation factor which depends on the nature and thickness of the shielding material, the source energy characteristics, and the angle of incidence θ. The reader will observe that this formula is an extension of Eq. (6.19) incorporating the response function and combining buildup and exponential attenuation into a single factor A_f. For a monoenergetic source of strength S_p, emitting photons of energy E_o,

$$R_o(P) = \frac{S_p}{4\pi r^2}\mathcal{R}(E_o). \tag{6.73}$$

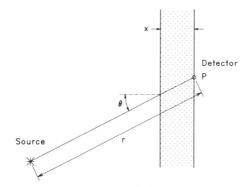

Figure 6.18   Attenuation of gamma and x rays from a point source in air by a shielding wall.

For a radionuclide source emitting photons of multiple energies, $R_o(P)$ would have to account for each energy and its emission frequency. Such accounting has been done, for exposure as the type of detector response, and for a number of radionuclides, in terms of the *specific gamma-ray constant*. This constant Γ is defined as the exposure rate, in vacuum, at unit distance from the source, per unit source activity. A few example values are given in Table 6.9. If $\mathcal{A}$ is the source activity, then the exposure rate at point P is given by

$$R(P) = R_o(P)A_f = \frac{\mathcal{A}}{r^2}\Gamma A_f, \tag{6.74}$$

where, of course, the units of r must correspond to those of Γ. Figures 6.19 and 6.20 display A_f for radionuclide sources and for the gamma rays striking the shield wall perpendicularly. The attenuation factors are for exposure or air kerma as the response, but suffice also for dose equivalent as the response. Figure 6.21, for lead shielding, allows the user to account for obliquely incident gamma rays, in which case the appropriate measure of wall thickness is the oblique thickness traversed along the path from source to detector.

For x-ray sources, the appropriate measure of source strength is the electron-beam current and the appropriate characterization of photon energies, in principle, involves both the peak accelerating voltage (kVp) and the degree of filtration (e.g., mm Al). While the degree of filtration of the x rays would affect their energy spectra, there is only a limited range of filtrations practical for any one voltage and, within that limited range, the degree of filtration has little effect on the attenuation factor (NCRP 1976). If i is the beam current (mA) and r is the source-detector distance (m), then Eq. (6.74), with exposure rate (R min^{-1}) as the response, takes the form

$$R(P) = \frac{i}{r^2}B_f, \tag{6.75}$$

in which B_f is the *normalized shielded output factor*, with units R min^{-1} m^2 mA^{-1}. B_f factors for x rays normally incident on concrete and lead are illustrated in Figs. 6.22 and 6.23.

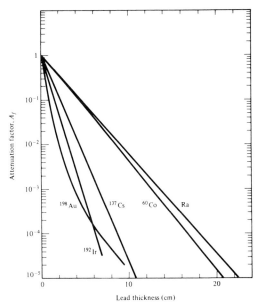

Figure 6.19   Attenuation factor A_f for gamma photons from several radionuclides normally incident on lead slabs. [From NCRP (1976); by permission of the NCRP.]

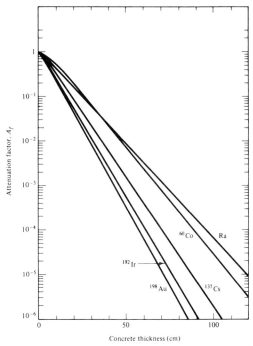

Figure 6.20   Attenuation factor A_f for gamma photons from several radionuclides normally incident on concrete slabs of density 2.25 g/cm^3. [From NCRP (1976); by permission of the NCRP.]

TABLE 6.9   Values of the Specific Gamma-Ray Constant Γ for Certain Radioisotopes.

RADIOISOTOPE	$\dfrac{R\,m^2}{Ci\,h}$	$\dfrac{R\,cm^2}{GBq\,h}$
^{60}Co	1.30	351
^{137}Cs$-^{137m}$Ba	0.32	86
^{192}Ir	0.500	135
^{198}Au	0.232	62.7
^{226}Ra + daughters	0.825^a	223^a

^{a}Source encapsulated in 0.5-mm thick platinum.

Source: NCRP 1976.

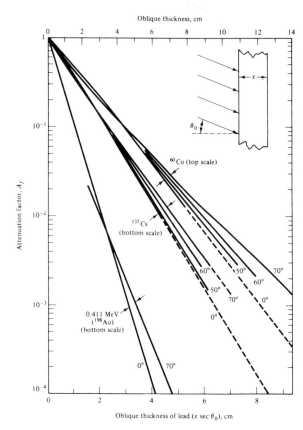

Figure 6.21   Attenuation factors for the transmission of gamma photons from three radionuclide sources through a lead slab. The slab is uniformly illuminated at several angles of incidence with respect to the slab normal. Results for ^{198}Au have been corrected so that only the transmission of the 0.411-MeV photon is shown. [After Kirn, Kennedy, and Wycoff (1954).]

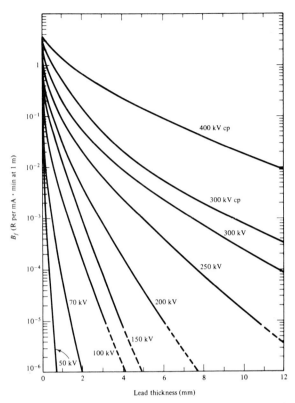

Figure 6.22   Attenuation in lead of x rays produced by various peak potentials (kV$_p$) with a 90° angle between the electron beam and the axis of the x-ray beam. Pulsed waveforms are assumed except for the 300 and 400 kVcp curves which are for constant potential generators. The x-ray beam filtrations are: 0.5 mm of Al for 50 kV, 1.5 for 70 kV, 2.5 for 100 and 150 kV, 3 mm for 200 to 300 kV, and 3 mm of Cu for 300 and 400 kVcp. [From NCRP (1976); by permission of the NCRP.]

   In the design and analysis of shield walls for x-ray installations, it is necessary to account for a number of factors: (1) the *maximum permissible dose* during some prescribed time interval such as one week for an individual situated beyond the shield wall, (2) the *work load*, that is, the cumulative sum during the prescribed time interval of the product of the beam current and the duration of machine operation, (3) the *use factor*, that is, the fraction of machine operation time that the x-ray beam is directed toward the shield wall, and (4) the *occupancy factor*, that is, the fraction of the time during which the x-ray machine is in use and the beam is directed toward the shield wall that the individual at risk is actually present at the location beyond the shield wall. All these factors are taken into account in the methodology of NCRP Report 49. In addition, that methodology also treats leakage radiation from the x-ray machine and radiation scattered from patients or other objects present in the x-ray beam. The methodology is discussed by Chilton, Shultis, and Faw (1984) and a computer code is available for routine x-ray shielding design and analysis (Simpkin 1987). Built into the

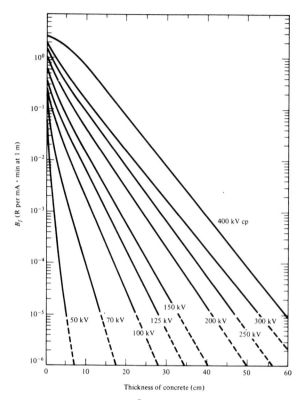

Figure 6.23 Attenuation in concrete (2.35 g/cm^3 density) of x-rays produced by various peak potentials (kVp) with a 90° angle between the electron beam and the axis of the x-ray beam. Pulsed waveforms are assumed except for the 400 kV cp curve which is for a constant potential generator. The x-ray beam filtrations are: 1mm of Al for 50 kV, 1.5 for 70 kV, 2 mm for 100 kV, 3 mm for 100 to 300 kV, and 3 mm of Cu for 400 kV cp. [From NCRP (1976); by permission of the NCRP.]

code are the B_f factors published in NCRP Report 49 and additional factors published elsewhere.

6.7 NEUTRON ATTENUATION AND REMOVAL CONCEPTS

Radiological assessment of the hazards associated with neutron sources is often a complex task. Such an analysis involves not only the primary (source) neutrons but also the production and interaction of secondary photons produced by neutron inelastic scattering and radiative capture. Even secondary neutrons are possible from $(n, 2n)$ reactions and from fission if fissionable material is present. In many cases these secondary radiations are of more radiological concern than are the primary neutrons.

Another complexity of neutron radiological assessments is that few problems are easily solved using elementary techniques. A rigorous calculation of neutron dose rates requires that both the number of neutrons and their energy spectrum be known at the receptor location of interest. Because of the great variation in neutron cross sections both with neutron energy and with material composition and because of the wide ranges of energies neutrons can have, calculation of the neutron dose is much more difficult than for the photon case.

Usually, source neutrons are born at high energies (0.1–10 MeV), slow through the keV and eV energy range by scattering interactions, and then diffuse (again by scattering) as *thermal neutrons* (i.e., neutrons in equilibrium with the thermal motion of the atoms of the material through which they diffuse). Only when the neutrons reach thermal energies (0.0253 eV average energy at room temperature) is it likely that they will be absorbed. Moreover, a slight change in material composition can drastically change the neutron field. To obtain accurate results for neutron dose with errors of only a few percent, it is necessary to use numerical techniques based on precise descriptions of how the neutrons interact in the material of the problem. These techniques, based on transport theory and Monte Carlo simulations are beyond the scope of this text.

In a few special situations, however, there are some simplified approximate methods which have been developed for estimating neutron dose rates. In this section these methods are reviewed. For a discussion of the more sophisticated methods the reader is referred to Chilton, Shultis, and Faw (1984) and Schaeffer (1973).

6.7.1 Differences Between Photon and Neutron Flux Density Calculations

For photon attenuation, the use of exponential attenuation and buildup factors is a very powerful and simple technique. However, the extension of these techniques to neutron attenuation, although attractive because of their simplicity, must be approached cautiously. The correction of the uncollided flux densities for scattered fast neutrons is generally much more difficult than for photons. An approximate technique for fast neutrons, analogous to Eq. (6.63) for photons and developed below for neutrons, is to replace the actual cross sections by an *effective removal cross section* in the uncollided flux density calculation.

A much better technique, in principle, would be to use an appropriate buildup factor to correct the uncollided flux density for scattered neutrons. Unfortunately, the use of neutron buildup factors in this precise way has not been very practical since the buildup of scattered neutrons depends strongly on the material composition, the incident neutron energy spectrum, and above all, the geometry of the particular problem. Because the fast-neutron scattering cross section is greater than the absorption cross section for most materials, and, especially for heavy nuclides which are relatively ineffective at slowing down neutrons, the buildup of scattered neutrons can assume very large values. Moreover, the size and shape of the medium can greatly affect the neutron flux density. For example, a free surface causes a substantial reduction in the scattered neutron flux density near the surface. This reduction, which is far more severe than in the corresponding photon problem, precludes the use of infinite-medium buildup factors for finite size neutron shields. Finally, since neutron cross sections vary dramatically with neutron energy and with material, the buildup of scattered fast neutrons depends sensitively on the initial neutron energy and on the energy dependence of the material cross sections. Furthermore, in contrast to photon scattering, the angular distribution of scattered neutrons can change rapidly with neutron energy. These considerations suggest that buildup factors would have to be calculated for each particular neutron problem, and that the use of infinite-medium buildup factors, which are so successfully used in photon calculations, may introduce serious errors for fast-neutron problems.

Nevertheless, the concept of a buildup correction is well established in shielding methodology, and many simple models for fast-neutron attenuation include a factor to correct, in an approximate manner, for the buildup of scattered fast neutrons. One of these methods is presented below; however, it must be expected in light of the previous discussion that such a model will have a more restricted range of applicability than does the analogous model for photon attenuation.

6.7.2 Fast Neutron Attenuation in Hydrogenous Media

There is one widely encountered situation where the attenuation of a fast neutron beam can be expected to be somewhat insensitive to the buildup of scattered neutrons. When a neutron scatters from hydrogen, one-half of its kinetic energy is lost on average. Consequently, a scattering interaction of a very fast fission neutron with hydrogen effectively removes the neutron from its group of high-energy cohorts. In addition, it is seen from Fig. 2.10 that the cross section for hydrogen in the MeV-energy region increases as the neutron energy decreases. Hence, once a very fast neutron scatters in a hydrogenous medium, subsequent scattering or slowing down interactions will occur relatively near the point of the first scattering interaction.

Thus, the fast neutron dose rate a distance r from a point fission source of strength S_p in a pure hydrogen medium (N_H atoms cm^{-3}) might be expected to be close to the dose rate from uncollided fission neutrons alone, that is,

$$D_H(r) \cong \mathcal{R}_f \, \phi_H^o = \mathcal{R}_f \frac{S_p}{4\pi r^2} \int_{E_f} dE \, \chi(E) \exp[-N_H \sigma_H(E) r], \qquad (6.76)$$

where $\mathcal{R}_f$ is an average detector response function for fast fission neutrons and the integration is over all fission- neutron energies. Above a few MeV, the hydrogen cross section can be approximated by $\sigma_H(E) \cong A \, \exp(-aE)$ and the fission neutron energy spectrum by $\chi(E) \cong BE^{-b}$ where A, B, a, and b are constants. Substitution of these approximations into Eq. (6.76) and evaluation of the integral show that the fast neutron dose rate depends on r as (Chilton, Shultis, and Faw 1984; Albert and Welton 1950)

$$D_H(r) = \frac{S_p}{4\pi r^2} \alpha r^{\gamma/2} \exp(-\beta r^\gamma), \qquad (6.77)$$

where α, β, γ are constants.

The presence of a heavier nonhydrogen component in the attenuating medium (e.g., the oxygen in a water shield) will also degrade fast neutrons in energy, although not nearly as well as hydrogen. Many experiments have indicated that the effect of the oxygen in water is to cause the fast-neutron dose rate to decrease with increasing energy slightly more so than indicated by Eq. (6.77). Specifically, it is found that

$$D_{H_2O}(r) = D_H(r)e^{-\mu_{r,O}r} \tag{6.78}$$

where $\mu_{r,O}$ is an appropriate constant for oxygen. Since the $\exp(-\mu_{r,O}r)$ factor resembles the usual exponential attenuation of uncollided radiation, the constant $\mu_{r,O}$ is called a *removal* coefficient.

The fast neutron dose rate in water arising from a point fission source was the subject of many early investigations. Fits to experimental data have determined values for the parameters α, β, γ, and $\mu_{r,O}$ in Eqs. (6.77) and (6.78) (Casper 1960). Other functional forms have also been fit to experimental results (Grotenhuis 1962, Glasstone and Sesonske 1963). One empirical result for the dose rate in water (in Gy h^{-1}), applicable for $0 < r < 300$ cm, is given by Brynjolfsson (1975) as

$$D_{H_2O}(r) = S_p \frac{9.3 \times 10^{-8}}{4\pi r^2} e^{-b(r)r} \tag{6.79}$$

where

$$b(r) = [0.126 - 0.0001773(r - \frac{r^2}{600})][1 - (2 + \frac{r^4}{5000})^{-1}] \tag{6.80}$$

with r expressed in units of cm. In Fig. 6.24, the fast neutron tissue-absorbed dose in water per source neutron is shown for a ^{235}U fission source as well as for several monoenergetic neutron sources.

6.7.3 Neutron Removal Cross Section

In many situations, fission neutrons will be attenuated not only by an hydrogenous medium but also by an interposed nonhydrogenous shield (see Fig. 6.25). Many experiments have been performed for such situations, and under special circumstances, the effect of the nonhydrogenous shield can be very simply accounted for by an exponential attenuation factor, much as was done for the oxygen correction in the preceding section.

For the situation shown in Fig. 6.25, experimental results reveal that, under certain circumstances, the fast neutron dose rate D' at a detector with the shield in position (i.e., at a distance $r = r_1 + r_2$ of hydrogenous medium plus a thickness t of the nonhydrogenous shield material) is related to the dose rate D at a distance r from the source (without the shield) by

$$D' = D \left(\frac{r}{r+t} \right)^2 e^{-\mu_r t} \tag{6.81}$$

where μ_r, the removal coefficient, is a constant characteristic of the nonhydrogenous shield material for a given fission neutron energy spectrum. Two important restrictions on the experimental arrangement are required for the validity of this result. First, it is important that $\rho_H r_2 \geq 6$ g cm^{-2} where ρ_H is the mass density of hydrogen in the

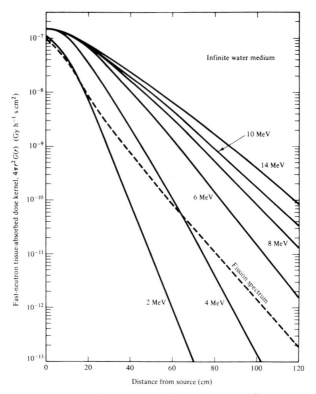

Figure 6.24 Fast-neutron tissue-absorbed dose kernel (multiplied by $4\pi r^2$) expressed in units of Gy h^{-1} for point monoenergetic neutron sources and a point ^{235}U fission source in an infinite water medium. The sources are assumed to be of unit strength (i.e., to emit one neutron per second). [From Chilton, Shultis, and Faw (1984).]

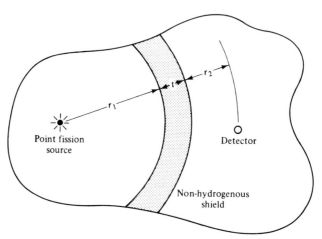

Figure 6.25 Idealized experimental geometry for the measurement of the removal cross section. A non-hydrogeneous shield of thickness t is placed between the source and detector in an infinite hydrogeneous medium.

hydrogenous material, that is, there must be an amount of hydrogen equivalent to about 50 cm of water following the shield. Second, the thickness t must be such that $\mu_r t$ is less than about 5.

Although the factor $\exp(-\mu_r t)$ in Eq. (6.81) appears to indicate that absorption of neutrons is taking place in the shield, the vast majority of fast neutron interactions are scattering interactions in which the fission neutrons are degraded in energy only slightly. However, the hydrogen in the material following the shield (one of the experimental restrictions) will moderate (or remove) the slightly slowed-down neutrons more quickly than those neutrons that pass through the shield without any energy loss since the hydrogen cross section increases with decreasing energy in this energy region. Thus, if there is sufficient hydrogen following the shield to effect the removal of neutrons that are slightly moderated by the shield, it will appear to the detector as if the shield has removed those fast neutrons.

If a series of slabs of different materials is inserted into the hydrogenous medium, the removal term in Eq. (6.81) $\exp[-\mu_r t]$ is replaced by $\exp[-\sum_i \mu_{r,i} t_i]$ where $\mu_{r,i}$ is the removal coefficient for the ith slab of thickness t_i. Similarly, if a slab of a mixture of elements is inserted, the removal coefficient μ_r for the slab is given by $\sum_i N_i \sigma_{r,i}$, where N_i is the atom density of the ith element with microscopic removal cross section $\sigma_{r,i}$. This additive nature of the removal lengths for the nonhydrogen components is a direct consequence of Eq. (6.81) and has generally been supported by experiment, although some deviations have been noted.

There is no firm theoretical reason for the removal cross section to be a material constant, and indeed, it might be expected to vary with the fission neutron energy spectrum, slab thickness, amount of hydrogen material on either side of the shield, and the geometry of the experiment. However, experiment has shown that for most situations (provided the nonhydrogen shield is less than 5 removal-lengths thick, and at least 6 g cm^{-2} of hydrogen follow the shield), μ_r can be taken as a constant for a given fission energy spectrum. In Table 6.10 measured values of some neutron removal cross sections are given. To obtain removal cross sections for other elements, the following empirical formulas (in units of cm^2 g^{-1}) have been obtained (Zoller 1964)

$$\frac{\mu_r}{\rho} = \begin{cases} 0.190\, Z^{-0.43}, & Z \leq 8 \\ 0.125\, Z^{-0.565}, & Z > 8 \end{cases} \tag{6.82}$$

or

$$\mu_r/\rho = 0.206 A^{-1/3} Z^{-0.294} \tag{6.83}$$

where A and Z are the atomic mass and number, respectively, of the element.

6.7.4 Extensions of Removal Theory

The simplicity and effectiveness of removal theory for analyzing the attenuation of fast neutrons by a nonhydrogenous shield in an hydrogenous medium has motivated its extension and modification for use in other applications. For example, removal theory can be used for homogeneous shields (such as concrete shields) provided there is a sufficient hydrogen content in the material to effectively remove the fast fission

TABLE 6.10   Measured Microscopic Removal Cross Sections of Various Elements and Compounds for ^{235}U Fission Neutrons.

MATERIAL	σ_r (b/atom)	MATERIAL	σ_r (b/atom)
Aluminum	1.31 ± 0.05	Oxygen	0.99 ± 0.10
Beryllium	1.07 ± 0.06	Tungsten	3.36
Bismuth	3.49 ± 0.35	Zirconium	2.36 ± 0.12
Boron	0.97 ± 0.10	Uranium	3.6 ± 0.4
Carbon	0.81 ± 0.05	Boric Acid B_2O_3	4.30 ± 0.41[a]
Chlorine	1.2 ± 0.08	Boron Carbide B_4C	4.7 ± 0.3[a]
Copper	2.04 ± 0.11	Fluorothene C_2F_4Cl	6.66 ± 0.8[a]
Fluorine	1.29 ± 0.06	Heavy Water D_2O	2.76 ± 0.11[a]
Iron	1.98 ± 0.08	Hevimet[b]	3.22 ± 0.18
Lead	3.53 ± 0.30	Lithium Fluoride	2.43 ± 0.34[a]
Lithium	1.01 ± 0.05	Oil, CH_2 group	2.84 ± 0.11[a]
Nickel	1.89 ± 0.10	Paraffin $C_{30}H_{62}$	80.5 ± 5.2[a]

[a] Removal cross section is in barns per molecule or per group.

[b] 90 wt% W, 6 wt% Ni, 4 wt% Cu; cross section is weighted average.

Source: Chilton, Shultis, and Faw 1984.

neutrons that are slightly moderated by the nonhydrogen components (Tsypin and Kukhtevich 1968). Corrections have also been obtained for the case in which there is insufficient hydrogen following the nonhydrogenous shield (Shure, O'Brien, and Rothberg 1969). Energy-dependent removal cross sections have also been developed for use with monoenergetic neutron sources in hydrogenous media (Gronroos 1968). Finally, a simple method, analogous to the effective photon attenuation and buildup being described by a single exponential term as in Eq. (6.63), has been developed for fast neutron attenuation in nonhydrogenous shields (Broder and Tsypin 1968).

The attenuation of fast neutrons in a shield necessarily leads to neutrons with intermediate and, eventually, thermal energies. The intermediate energy neutrons can contribute appreciably to the transmitted neutron dose, and the thermal neutrons, which are usually readily absorbed in the shield material, lead to the production of high-energy capture gamma photons. These photons can become the dominate consideration in a neutron shield design. Thus, an important aspect of neutron shield analysis is the calculation of the intermediate and thermal neutron flux densities throughout the shield.

While modern neutronic codes based on rigorous descriptions of how neutrons migrate through shields can calculate very accurately the energy and spatial distributions of neutrons in a shield, these codes are computationally expensive and not suited for routine use or preliminary calculations. Alternative and less costly procedures have been developed based on the simpler diffusion theory and the concept of neutron removal from one energy group to lower energy groups. These simpler removal-diffusion methods can often be used with good effect and far less computational expense. A summary of these methods is provided by Chilton, Shultis, and Faw (1984).

6.8 SPECIAL TOPICS

There are many complex source and detector geometries which cannot be analyzed by the simplified methods presented earlier in this chapter. For several of the most frequently encountered of these situations, specialized simplified techniques have been developed. In this section, two such special cases are considered.

6.8.1 Gamma-Ray Skyshine

In many facilities with intense localized sources of radiation, the shielding against radiation that is directed skywards is usually far less than that for the radiation emitted laterally. However, the radiation emitted vertically into the air will undergo scattering interactions and some radiation will be reflected back to the ground often at distances far from the original source. This atmospherically reflected radiation is referred to as *skyshine* and is of concern both to workers at the facility and to the general population outside the facility site.

A rigorous treatment of the skyshine problem requires the use of computationally expensive methods based on multidimensional transport theory. Alternatively, several approximate procedures have been developed for both gamma-photon and neutron skyshine sources [see Shultis, Faw, and Basset (1991) for a review]. In this section is summarized one approximate method that has been found useful for bare or shielded gamma-ray skyshine sources.

This method, termed the *integral line-beam skyshine method*, is based on the availability of a *line-beam response function* $\mathcal{R}(E, \phi, x)$ which gives the air kerma (cGy per photon = rad per photon) at a distance x from a point source emitting a photon of energy E at an angle ϕ from the source-to-detector axis into an infinite air medium. The air-ground interface is neglected in this method. This response function can be fit over a large range of x to the three-parameter empirical formula (for a fixed value of E and ϕ) (Lampley, Andrews, and Wells 1988).

$$\mathcal{R}(E, \phi, x) = E \, \mathcal{F}(E, \phi, x) \tag{6.84}$$

where

$$\mathcal{F}(E, \phi, x) = \kappa \left(\frac{\rho}{\rho_o} \right)^2 \left[x \frac{\rho}{\rho_o} \right]^b \exp\left(a - cx \frac{\rho}{\rho_o} \right), \tag{6.85}$$

in which ρ is the air density in the same units as the reference density $\rho_o = 0.001225$ g cm^{-3}. When E is measured in MeV, x in meters, $\mathcal{F}$ in cGy MeV^{-1}, and $\mathcal{R}$ in cGy per photon, the constant κ is equal to 1.308×10^{-11}.

The parameters a, b, and c in Eq. (6.85) depend on the photon energy E and the source emission angle ϕ. These parameters have been estimated and tabulated, for fixed values of E and ϕ, by fitting Eq. (6.85) to values of the line-beam response function, at different x distances, obtained by Monte Carlo calculations (Lampley et al. 1988) or by a point-kernel method (Shultis and Faw 1987). Tables of parameters suitable for estimating $\mathcal{R}(E, \phi, x)$ over a source-to-detector range of 20 m $< x < 2600$ m for 12 discrete energies from 0.1 to 10 MeV and for 20 discrete angle ϕ are given by Shultis, Faw, and Basset (1991). Finally, a linear interpolation scheme is used to obtain $\mathcal{R}(E, \phi, x)$ for any E or ϕ in terms of values at the discrete tabulated energies and angles. In this way, the line-beam response function is rendered completely continuous in the x, E and ϕ variables.

To obtain the skyshine dose $R(d)$ at a distance d from a bare collimated source, the line-beam response function, weighted by the energy and angular distribution of the source, is integrated over all source energies and emission directions. Thus if the collimated source emits $S(E, \Omega)$ photons, the skyshine dose is

$$R(d) = \int_0^\infty dE' \int_{\Omega_s} d\Omega \, S(E', \Omega)\mathcal{R}(E', \phi, d), \qquad (6.86)$$

where the angular integration is over all emission directions Ω_s allowed by the source collimation. Here ϕ is a function of the emission direction Ω.

For a shielded source, source photons are exponentially attenuated in the shield. The treatment of photons scattered in the shield, however, is not so straightforward. In the integral line-beam skyshine method, exponential attenuation and an appropriate buildup factor B are used to describe the effect of the shield. However, the energy and angular redistribution of photons scattered in the shield is ignored, that is, the scattered photons are assumed to emerge from the shield with the same energy and direction as the uncollided photons. The skyshine dose rate for a shielded source is thus

$$R(d) = \int_0^\infty dE' \int_{\Omega_s} d\Omega \, e^{-\lambda} B(E', \lambda) S(E', \Omega)\mathcal{R}(E', \phi, d), \qquad (6.87)$$

where $\lambda(\Omega)$ is the mean-free-path length that a photon emitted in direction Ω travels through the shield without collision. Clearly, when there is no source shielding, $\lambda = 0$ this result reduces to the unshielded result of Eq. (6.86).

In the integral line-beam method, the integration over all source emission directions is expressed in terms of a spherical coordinate system with the source at the origin and the polar axis vertically upward. The azimuthal angle ψ is defined with respect to the projection on the horizontal plane of the source-to-detector axis. Thus Eq. (6.87) may be written as

$$R(d) = \int_0^\infty dE' \int_0^{2\pi} d\psi \int_{\omega_{min}(\psi)}^{\omega_{max}(\psi)} d\omega \, e^{-\lambda} B(E', \lambda) S(E', \omega, \psi)\mathcal{R}(E', \phi, d), \quad (6.88)$$

where ϕ and λ are generally functions of ω and ψ, and ω_{min} and ω_{max} define the permissible (cosines of) polar angles for photon emission allowed by any source collimation. For simple geometries and source distributions this result may be reduced analytically (see the following discussion); generally, however, it is necessary to perform the integrations numerically.

Shielded silo example..   The above general result can be reduced to explicit forms suitable for calculation for special geometries and source characteristics. Here we consider the case in which an isotropic, monoenergetic point source, that is,

$S(E', \mathbf{\Omega}) = S_p \, \delta(E' - E)/4\pi$, is shielded from above by a horizontal homogeneous slab shield of thickness t. If ρ_s (g cm^{-3}) is the shield's density and μ/ρ (cm^2 g^{-1}) is the total mass interaction coefficient in the shield material for photons of energy E, then the slab thickness, in units of mean-free-path length, is given by $\tau = t\rho_s(\mu/\rho)$. The source is located on the vertical axis of a cylindrical-shell shield (silo) of inner radius r (see Fig. 6.26). The wall of the silo is assumed to be *black*; that is, no source radiation penetrates it. The source is distance y below the horizontal top of the silo which collimates the emergent radiation into a cone with a polar angle θ_{max} measured from the vertical axis and defined by

$$\omega_o \equiv \cos \theta_{max} = 1 \, / \sqrt{1 + r^2/y^2}. \tag{6.89}$$

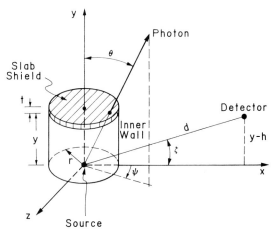

Figure 6.26 Geometry for the shielded silo skyshine problem. A point isotropic source is on the axis of a cylindrical silo with a black wall and shielded from above by a horizontal slab shield.

A detector (receiver or dose point) is located in air of density ρ at radial distance x from the silo axis and at distance h below the horizontal plane of the silo top. If either y or h is above the plane defined by the top of the silo wall, the distance is taken to be negative in sign. Above the source, and generally within the confines of or atop the silo, shielding is permitted in the form of a horizontal slab of material with thickness equivalent to λ mean free paths at photon energy E. The distance from the source to the detector is

$$d = \sqrt{x^2 + (y - h)^2} \tag{6.90}$$

and the angle ζ between the horizontal and the source-to-detector axis is

$$\zeta = \tan^{-1} \frac{(y - h)}{x} \tag{6.91}$$

Consider a ray from the source at polar angle θ (measured from the silo axis) and at azimuthal angle ψ (measured from the vertical plane through the source and detector). The cosine of the angle of emission ϕ between the photon direction and the source-detector axis is the dot product of the unit vector in the emission direction and a unit vector along the source-detector axis, namely,

$$\cos\phi = \sin\theta \, \cos\psi \, \cos\zeta + \cos\theta \, \sin\zeta. \tag{6.92}$$

For a ray at polar angle θ, the slant path through the shield, in units of mean-free-path lengths is given by $\lambda = \tau \, \sec\theta = \tau/\omega$.

Buildup in the shielding slab can be accounted for approximately by application of the air-kerma buildup factor. Using the Berger approximation of Eq. (6.60), for example,

$$B = 1 + \alpha\frac{\tau}{\omega}e^{\beta\tau/\omega}, \tag{6.93}$$

in which the Berger parameters α and β depend on the shield material and the photon energy. In the absence of overhead shielding, the buildup factor is taken as unity.

For this shielded-silo monoenergetic-source problem, the total detector response (rad per photon) of Eq. (6.88), on using the azimuthal symmetry of the geometry and the monoenergetic nature of the source, reduces to

$$\frac{R(d)}{S_p} = \frac{1}{2\pi}\int_0^\pi d\psi \int_{\omega_o}^1 d\omega \, \mathcal{R}(E,\phi,d)e^{-\tau/\omega}[1 + \alpha(\tau/\omega)e^{\beta\tau/\omega}], \tag{6.94}$$

where R depends on the composition and thickness of the overhead shielding material as well as E, r, x, y, and h. The double integral in this result is readily evaluated using standard numerical quadrature.

The integral line-beam method for gamma skyshine calculations has been applied to a variety of other source configurations and found to give generally excellent agreement with benchmark calculations and experimental results (Shultis, Faw, and Basset 1991). It has recently been used as the basis of the microcomputer code *MicroSkyshine* (Grove 1987).

6.8.2 Neutron and Gamma-Ray Reflection

There are circumstances for which the dose rate at some receptor location due to radiation reflected from walls, floors and passageways may be comparable to or even exceed the direct (line-of-sight) dose rate from a radiation source. The term *reflection* in this context does not imply a surface scattering. Rather, gamma rays or neutrons penetrate the surface of a shielding or structural material, scatter within the material, and then emerge from the material with reduced energy and at some location other than the point of entry. Treatment of radiation reflection requires, in principle, accounting for the spatial, energy, and angular distributions of incident and reflected radiation; the nature and thickness of the reflecting material; and the displacement between points of entry and emergence of particles from the reflecting surface. Such a detailed treatment is beyond the scope of this work; however, in many analyses of reflecting surfaces, a simplified method may be used. This method is based on the following approximations:

- Displacement between points of entry and emergence may be neglected. This is a reasonable approximation for reflecting surfaces with dimensions greater than a few mean-free-paths (as measured in the reflecting material) and for incident radiation with only minor variations in intensity over distances of a few mean-free-paths.

- Reflecting media are in the form of half spaces, that is, infinite in thickness and lateral extent, with a plane surface. The neglect of the finite thickness of the reflector is appropriate for thicknesses greater than about two mean-free-paths of the radiation.

- Scattering in air between a source and the reflecting surface and between the reflecting surface and the detector may be neglected.

For discussion of these approximations and for more advanced treatments, the reader is referred to Leimdorfer (1968); Selph (1973); and Chilton, Shultis, and Faw (1984).

The radiation-reflection problem may be discussed in the context of Fig. 6.27. Suppose that incident particles, all of energy E_o and travelling in the same direction, strike area dA in the reflecting surface at angle θ_o measured from the normal to the surface. If ϕ_o is the flux density of the incident particles, the number of incident particles per unit time striking dA is $\phi_o dA \cos \theta_o$. Suppose that the energy spectrum of the angular distribution of the flux density of reflected particles emerging from the surface with energy E and direction characterized by angles θ and φ is $\phi(E, \theta, \varphi)$. The number of particles emerging from dA per unit time with energies in dE about E and with directions in solid angle $d\Omega$ about direction (θ, φ) is $dA \cos \theta \phi(E, \theta, \varphi) dE d\Omega$. The *number albedo* $\alpha_n(E_o, \theta_o; E, \theta, \varphi)$ is defined in such a way that

$$\phi(E, \theta, \varphi) = \phi_o \frac{\cos \theta_o}{\cos \theta} \alpha_n(E_o, \theta_o; E, \theta, \varphi). \tag{6.95}$$

Now suppose that a radiation detector is located at point P, distance r_2 measured along the ray at direction (θ, φ) from the surface. That part $dR^r(P)$ of the response rate at location P due to reflection from area dA is given by

$$dR^r(P) = \frac{dA \cos \theta}{r_2^2} \int_0^\infty dE \, \mathcal{R}(E) \phi(E, \theta, \varphi). \tag{6.96}$$

By substituting Eq. (6.95) into Eq. (6.96) and defining a *response albedo* $\alpha_R(E_o, \theta_o; \theta, \varphi)$ one can see that

$$dR^r(P) = \frac{dA \cos \theta_o}{r_2^2} \mathcal{R}(E_o) \phi_o \, \alpha_R(E_o, \theta_o; \theta, \varphi), \tag{6.97}$$

where

$$\alpha_R(E_o, \theta_o; \theta, \varphi) = \int_0^\infty dE \frac{\mathcal{R}(E)}{\mathcal{R}(E_o)} \alpha_n(E_o, \theta_o; E, \theta, \varphi). \tag{6.98}$$

Equation (6.97) may be used as follows. Suppose that a point isotropic source of strength S_p, emitting particles of energy E_o, is located at a point distance r_1 from

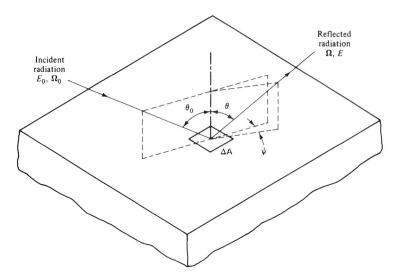

Figure 6.27   Angular and energy relationships in the albedo formulation.

area dA along the incident ray in Fig. 6.27. The flux density of incident radiation at dA is

$$\phi_o = \frac{S_p}{4\pi r_1^2};$$ (6.99)

thus,

$$R^r(P) = S_p \mathcal{R}(E_o) \iint_A dA \frac{\cos\theta_o}{4\pi r_1^2 r_2^2} \alpha_R(E_o, \theta_o; \theta, \varphi).$$ (6.100)

As the spatial variables change during the integration over the reflecting area, the independent variables r_1, r_2, θ_o, θ, and φ also change, as does the albedo function α_R. As is seen in Eq. (6.98), α_R depends on the *ratio* of $\mathcal{R}(E)$ to $\mathcal{R}(E_o)$. One would expect the ratio to be relatively insensitive to the type of response for gamma rays. Therefore, for gamma rays, the exposure albedo α_X is commonly used even when the type of response is tissue kerma, dose equivalent, and so on.

   Gamma-Ray Albedos.   Chilton and Huddleston (1963) devised on theoretical grounds an approximation for the albedo in the form

$$\alpha_X(E_o, \theta_o; \theta, \varphi) = \frac{C(E_o) \, {}_e\sigma_{ce}(E_o, \theta_s) \times 10^{26} + C'(E_o)}{1 + \cos\theta / \cos\theta_o},$$ (6.101)

in which the scattering angle and the energy-scattering Klein-Nishina cross section are given by

$$\cos\theta_s = \sin\theta_o \sin\theta \cos\varphi - \cos\theta_o \cos\theta,$$ (6.102)

TABLE 6.11 Values of the Parameters in the Chilton-Huddleston Formula for the Differential Exposure Albedo for Photons on Various Reflecting Surfaces.

MATERIAL	E_o (MeV)	C	C'
Water	0.2	−0.0187	0.1327
	0.662	0.0309	0.0253
	1.00	0.0470	0.0151
	2.50	0.0995	0.0058
	6.13	0.1861	0.0035
Concrete	0.2	0.0023	0.0737
	0.662	0.0347	0.0197
	1.00	0.0503	0.0118
	2.50	0.0999	0.0051
	6.13	0.1717	0.0048
Iron	0.2	−0.0272	−0.0100
	0.662	0.0430	0.0063
	1.00	0.0555	0.0045
	2.50	0.1009	0.0044
	6.13	0.1447	0.0077
Lead	0.2	−0.0044	−0.0050
	0.662	0.0308	−0.0100
	1.00	0.0452	−0.0083
	2.50	0.0882	0.0001
	6.13	0.1126	0.0063

Source: Chilton, Davisson, and Beach 1965.

and

$$_e\sigma_{ce}(E_o, \theta_s) = \frac{1}{2}r_e^2\, p^2[1 + p^2 - p(1 - \cos^2\theta_s)], \qquad (6.103)$$

where r_e is the classical electron radius (2.818×10^{-13} cm) and

$$p = \frac{1}{1 + (E_o/m_e c^2)(1 - \cos\theta_s)}. \qquad (6.104)$$

Values of the constants C and C' are given in Table 6.11.

A more accurate version of Eq. (6.101), for concrete only, is given by

$$\alpha_X(E_o, \theta_o; \theta, \varphi) = \frac{F(E_o, \theta_o; \theta, \varphi)C(E_o)\, _e\sigma_{ce}(E_o, \theta_s) \times 10^{26} + C'(E_o)}{1 + (\cos\theta/\cos\theta_o)(1 + 2E_o\mathrm{vers}\theta_s)^{1/2}}, \qquad (6.105)$$

in which

$$F(E_o, \theta_o; \theta, \varphi) = A_1(E_o) + A_2(E_o)\mathrm{vers}^2\theta_o + A_3(E_o)\mathrm{vers}^3\theta$$
$$+ A_4(E_o)\mathrm{vers}^2\theta_o\mathrm{vers}^2\theta + A_5(E_o)\mathrm{vers}\theta_o\mathrm{vers}\theta\mathrm{vers}\varphi, \qquad (6.106)$$

where $\mathrm{vers}\theta = 1 - \cos\theta$. Values of the parameters are given in Table 6.12 and the albedo function for 1.25 MeV photons incident on concrete is illustrated in Fig. 6.28 (Chilton 1967). As is evident, the directional dependence of the albedo function is

TABLE 6.12   Parameters for the
Modified Chilton-Huddleston
Formula for the Exposure Albedo
for Photon Reflection from Concrete.

	E_o (MeV)	
PARAMETER	0.662	1.25
C	0.0445	0.0710
C′	0.0161	0.0114
A_1	1.512	1.555
A_2	−0.606	−0.629
A_3	−0.641	−0.605
A_4	0.645	0.539
A_5	−0.157	−0.168

Source: Chilton and Huddleston,
1963.

more pronounced for θ_o approaching 90 degrees, that is, for grazing incidence. In those circumstances, the reflection is greater for θ approaching 90 degrees and for φ approaching 0 degrees, that is, for grazing emergence with little or no change in azimuth.

Neutron Albedos.   Maerker and Muckenthaler (1965) developed the following empirical expression which reproduces results of Monte Carlo calculations within 10%. Source energy is uniformly distributed over the range ΔE_o. The dose unit is the free-field tissue kerma, but the albedo may be used for the dose equivalent index with error not greater than 25% (Cavanaugh 1975).

$$
\alpha_K(\Delta E_o, \theta_o; \theta, \varphi) = \frac{1}{1 + K_1(\cos\theta_o/\cos\theta)} \sum_{m=0}^{8} G_m P_m(\cos\theta_s)
$$

$$
+ \frac{1}{1 + K_2(\Delta E_o, \theta_o, \theta)/\cos\theta} \sum_{k=0}^{4} B_k P_k(\cos\theta_s),
$$

(6.107)

in which θ_s is the scattering angle given by Eq. (6.102), $P_m(\cos\theta_s)$ is the mth order Legendre polynomial of argument $\cos\theta_s$, and

$$
K_2(\Delta E_o, \theta_o, \theta) = \sum_{i=0}^{2}(\cos\theta)^i \sum_{j=0}^{2} \alpha_{ij}(\cos\theta_o)^j.
$$

(6.108)

Legendre polynomials are tabulated by Abramowitz and Stegun (1964) and algorithms for their generation are described by the same authors and by Chilton, Shultis, and Faw (1984). Values of the parameters used in the Maerker and Muckenthaler formula are given in Table 6.13. Albedos calculated by that formula are illustrated in Fig. 6.29 for 0.75 to 1.5-MeV neutrons incident on concrete. As is evident for neutrons as well as gamma rays, the directional dependence of the albedo function is more pronounced for θ_o approaching 90 degrees, that is, for grazing incidence. In those circumstances,

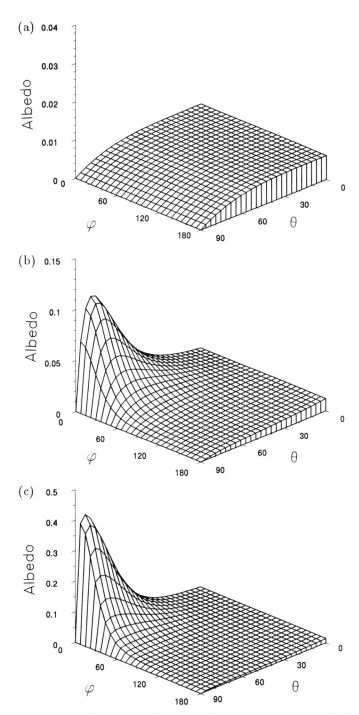

Figure 6.28 Exposure albedo for 1.25-MeV photons incident on concrete, computed using the Chilton-Huddleston formula. Angles are measured in degrees. (a) $\theta_o = 60°$; (b) $\theta_o = 75°$; (c) $\theta_o = 89°$.

TABLE 6.13 Parameters for the Maerker–Muckenthaler Formula for the Differential Dose Albedo for Reflection of Fast Neutrons from Concrete.

| CONSTANT | ENERGY RANGE ΔE_o (MeV) | | | | | |
	0.2–0.75	0.75–1.5	1.5–3	3–4	4–6	6–8
G_0	$6.585(-2)^a$	$7.045(-2)$	$7.211(-2)$	$7.024(-2)$	$6.856(-2)$	$5.899(-2)$
G_1	$5.048(-2)$	$4.393(-2)$	$5.845(-2)$	$7.452(-2)$	$8.294(-2)$	$6.039(-2)$
G_2	$3.710(-2)$	$7.088(-2)$	$5.968(-2)$	$1.000(-1)$	$9.517(-2)$	$7.524(-2)$
G_3	$1.544(-2)$	$1.898(-2)$	$2.729(-2)$	$5.591(-2)$	$7.761(-2)$	$8.140(-2)$
G_4	$7.837(-3)$	$2.408(-3)$	$1.190(-2)$	$2.646(-2)$	$4.292(-2)$	$6.622(-2)$
G_5	0	$-3.589(-3)$	$1.000(-3)$	$-6.908(-4)$	$1.824(-2)$	$3.056(-2)$
G_6	0	0	$4.637(-3)$	$-8.087(-4)$	$5.599(-3)$	$1.595(-2)$
G_7	0	0	$6.490(-3)$	$-1.459(-3)$	$5.288(-3)$	$1.277(-2)$
G_8	0	0	0	$-1.809(-3)$	$1.046(-2)$	$9.380(-3)$
B_0	$6.27(-2)$	$9.00(-2)$	$8.80(-2)$	$9.05(-2)$	$8.744(-2)$	$6.374(-2)$
B_1	$1.50(-2)$	$8.50(-3)$	$1.30(-2)$	$2.15(-2)$	$2.871(-2)$	$1.382(-2)$
B_2	$5.3(-3)$	$9.7(-3)$	$6.0(-3)$	$2.30(-2)$	$2.344(-2)$	$1.178(-2)$
B_3	0	0	0	0	$1.799(-2)$	$1.084(-2)$
B_4	0	0	0	0	$8.517(-3)$	$6.801(-3)$
K_1	1.0	1.0	1.1	0.9	1.1	1.06
α_{00}	0.36	0.51	0.56	0.60	0.43	0.35
α_{01}	1.29	0.32	0.18	0.15	2.02	0.95
α_{02}	0	1.00	1.32	0.48	-0.38	0
α_{10}	0.06	-0.04	-0.14	-0.61	0.05	0.10
α_{11}	-3.06	-2.46	-2.76	-1.08	-9.13	-2.28
α_{12}	0	0	0	0	5.93	1.11
α_{20}	-0.20	0.05	0.05	0.32	0.04	0
α_{21}	1.68	0.95	1.14	0.30	5.97	0
α_{22}	0	0	0	0	-4.39	0

^{a} Read as 6.585×10^{-2}, and so on.

Source: Maerker and Muckenthaler 1965; Selph 1973.

the reflection is greater for θ approaching 90 degrees and for φ approaching 0 degrees, that is, for grazing emergence with little or no change in azimuth.

An empirical fit to results of Monte Carlo calculations for the thermal-neutron number albedo is as follows (Wells 1964):

$$\alpha_n(\theta_o; \theta, \varphi) = 0.21(\cos\theta_o)^{-1/3}\cos\theta. \tag{6.109}$$

Application of the Gamma-Ray Albedo. The use of albedo formulas is illustrated for a series of examples associated with calibration of gamma-ray sources or detectors. Three cases are illustrated in Fig. 6.30. In Case 1, a source is placed on a reflecting slab (say a concrete pad) and a detector is placed directly above the source. In Case 2, the source and detector are both placed on a normal to the reflecting slab. In Case 3, the source and detector are both the same distance above the pad, but separated laterally. Case 1 is a limiting condition of Case 2; therefore, we describe only the latter, the details of which are illustrated in Fig. 6.31.

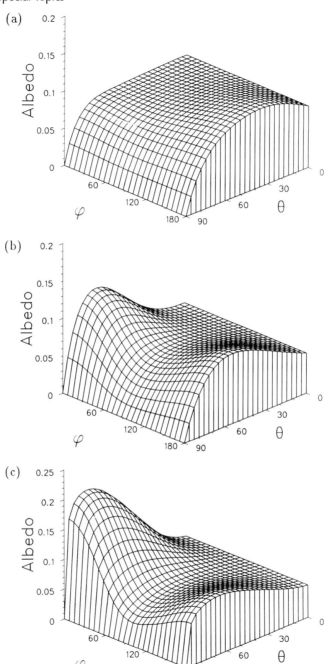

Figure 6.29   Absorbed-dose index albedo for 0.75 to 1.5-MeV neutrons incident on concrete, computed using the Maerker-Muckenthaler formula. Angles are measured in degrees. (a) $\theta_o = 60°$; (b) $\theta_o = 75°$; (c) $\theta_o = 89°$.

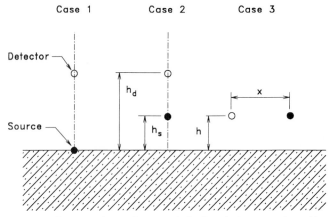

Figure 6.30 Three cases illustrating the application of albedo calculations in the calibration of radiation sources and detectors. Case 1: source on reflecting surface, detector above source. Case 2: source and detector positioned on a normal to a reflecting surface. Case 3: source and detector equally distant from reflecting surface.

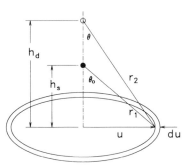

Figure 6.31 A radiation source (black dot) and a radiation detector (open circle) positioned along a perpendicular to a reflecting surface (Case 2).

For Case 2, the *direct* response rate at the detector is just

$$R^o = \frac{S_p \, \mathcal{R}(E_o)}{4\pi(h_d - h_s)^2}. \tag{6.110}$$

The response rate at the detector due to reflected gamma rays is given by Eq. (6.100). In that equation, the element of area (from the figure) is $dA = 2\pi u \, du$. It is apparent that, in order for photons reflected in dA to reach the detector, $\varphi = \pi$. Distances r_1 and r_2, and angles θ and θ_o can be expressed in terms of u. However, the independent variable u can be expressed in terms of angle θ_o, namely $u = h_s \tan\theta_o$. Thus, Eq. (6.100) with integration over u from 0 to ∞ replace by integration over θ_o from 0 to $\pi/2$, leads to

$$R^r = \frac{S_p \, \mathcal{R}(E_o)}{2h_d^2} \int_0^{\pi/2} d\theta_o \frac{\sin\theta_o}{\sec^2\theta} \alpha_R(E_o, \theta_o; \theta, \pi), \tag{6.111}$$

in which $\sec^2\theta = 1 + \tan^2\theta = 1 + (h_s/h_d)^2 \tan^2\theta_o$. The ratio of the reflected response rate to the direct response rate, a measure of the error incurred if the reflected photons are not accounted for, is given by the ratio of R^r to R^o, as given by Eqs. (6.111) and (6.110), namely,

$$\frac{R^r}{R^o} = 2\pi(1 - h_s/h_d)^2 \int_0^{\pi/2} d\theta_o \frac{\sin\theta_o}{\sec^2\theta} \alpha_R(E_o, \theta_o; \theta, \pi). \qquad (6.112)$$

For a given reflecting medium, source energy and type of response, the ratio depends only on the factor h_s/h_d. Indeed, it may be shown that the ratio remains the same if the source and detector positions are reversed. Results are illustrated in Fig. 6.32 for 0.662-MeV and 1.25-MeV photons reflecting from concrete, the response being exposure and the albedo calculated using the Chilton-Huddleston formula. Obviously, if a calibration is to be conducted in Case-2 geometry, uncertainty due to reflection is minimized if the source and detector are in close proximity. However, other considerations such as assurance of electronic equilibrium require that the source and detector be separated as much as possible. Figure 6.32 illustrates the correction factors one may need apply to account for reflected gamma rays.

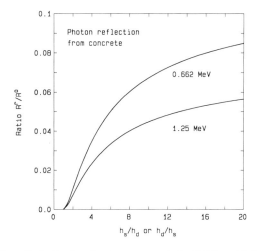

Figure 6.32   Ratio of the reflected to the direct exposure rate for 0.662 and 1.25-MeV gamma rays incident on concrete in Case-2 geometry. The distances of source and detector from the reflecting surface are respectively h_s and h_d.

Case 1 is represented by Case 2 in the limit as h_s goes to zero and the concomitant condition that $\theta = 0$, that is,

$$R^r/R^o = 2\pi \int_0^{\pi/2} d\theta_o \sin\theta_o \alpha_R(E_o, \theta_o; 0, \pi). \qquad (6.113)$$

Note that the ratio is independent of the distance of the detector from the reflecting plane. Both direct and reflected components of the response rate vary inversely with the square of the source-to-detector distance. Sample results are given in Table 6.14 for the exposure response and for the albedo calculated using the Chilton-Huddleston

TABLE 6.14 Ratio of the Exposure Rate Due to Reflected Gamma Rays to the Direct Exposure Rate for Gamma Ray Sources Resting on a Reflecting Surface with the Detector Directly above the Source.

ENERGY (MeV)	REFLECTING MEDIUM			
	WATER	CONCRETE	IRON	LEAD
0.2	0.45	0.34	0.14	0.0072
0.662	0.17	0.11	0.11	0.013
1.0	0.12	0.11	0.079	0.012
1.25	—	0.063	—	—
2.5	0.052	0.049	0.046	0.024
6.13	0.025	0.030	0.041	0.034

formula. Clearly, if a calibration is to be conducted using Case-1 geometry, uncertainty can be minimized if the source rests on a lead surface.

Case 3 represents a common geometry for instrument or source calibration. Source and detector are at the same height h above a reflecting surface and separated by distance x. By calculations similar to those described for Case 2, one can show that, for a given source energy, reflecting medium, and type of response, the ratio of the reflected and direct gamma-ray responses is a function only of the ratio x/h. Ratios are illustrated in Fig. 6.33 for 0.667 and 1.25-MeV gamma rays reflecting from concrete. The response is exposure and the albedos were calculated using the Chilton-Huddleston formula. Reflection is most important for x/h about 10. Unfortunately many calibrations are performed in just about that geometry! Effects of reflection can be minimized only by increasing x/h (thereby increasing the signal-to-noise ratio in the measurement) or decreasing x/h (possibly thereby introducing problems with achieving electronic equilibrium and related difficulties).

Treatment of reflection from walls is also a major complication in the analysis of radiation streaming through ducts and passageways. The interested reader may consult Selph (1973) and Chilton, Shultis, and Faw (1984).

REFERENCES

ABRAMOWITZ, M., AND I. STEGUN (eds.), *Handbook of Mathematical Functions*, NBS Applied Math Series 55, U.S. Government Printing Office, Washington, D.C. 1964.

ALBERT, R.D., AND T.A. WELTON, "A Simplified Theory of Neutron Attenuation and Its Application to Reactor Shield Design," USAEC Report WAPD-15 (Del.), Westinghouse Electric Corp., Atomic Power Division, Pittsburgh, Pa., 1950.

ANSI, *Gamma-Ray Attenuation Coefficients and Buildup Factors for Engineering Materials*, ANSI/ANS-6.4.3, American National Standards Institute, New York, 1991.

BLIZARD, E.P., AND L.S. ABBOTT, (eds.), *Reactor Handbook, Vol. III, Part B: Shielding*, 2nd ed., Interscience, New York, 1962.

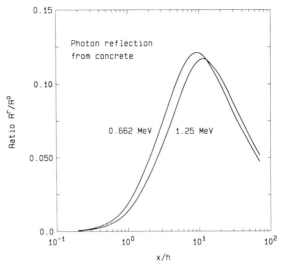

Figure 6.33 Ratio of the reflected to the direct exposure rate for 0.662 and 1.25-MeV gamma rays incident on concrete in Case-3 geometry. Source and detector are separated by distance x and are both distance h above the reflecting surface.

BLIZARD, E.P., A. FODERARO, N.G. GOUSSEV, AND E.E. KOVALEV, "Extended Radiation Sources (Point Kernel Integrations)," in *Engineering Compendium on Radiation Shielding*, Vol. 1, R.G. Jaeger (ed.), Springer Verlag, New York, 1968.

BRODER, D.L., AND S.G. TSYPIN, "Attenuation in Non-Hydrogenous Media," in *Engineering Compendium on Radiation Shielding*, Vol. I, R.G. Jaeger (ed.), p. 325, Springer-Verlag, New York, 1968.

BRODER, D.L., YU.P. KAYURIN, AND A.A. KUTREZOV, *Soviet Journal of Atomic Energy 12*, 26-31 (1962) [as reported in Bunemann and Richter (1968)].

BROOKHAVEN NATIONAL LABORATORY, *Nuclear and Physical Properties of Californium-252*, Report BNL 50168 (T-530), Brookhaven National Laboratory, Upton, N.Y., 1969.

BUNEMANN, D., AND G. RICHTER, "Multilayered Shields," in *Engineering Compendium on Radiation Shielding*, Vol. I, R.G. Jaeger (ed.), Springer Verlag, New York, 1968.

BRYNJOLFSSON, A., "Water," in *Engineering Compendium on Radiation Shielding*, Vol. II, R.G. Jaeger (ed.), Springer-Verlag, New York, 1975.

CASPER, A.W., "Modified Fast Neutron Attenuation Functions," USAEC Report XDC-60-2-76, General Electric Corp., Atomics Products Division, Cincinnati, 1960.

CAVANAUGH, G.P., "A Calculation and Analysis of Albedos Due to 14 MeV Neutrons Incident Upon Concrete," Ph.D. thesis, University of Illinois at Urbana-Champaign, 1975.

CHILTON, A.B., "A Modified Formula for Differential Exposure Albedo for Gamma Rays Reflected from Concrete," *Nucl. Sci. Eng. 27*, 481-482 (1967).

CHILTON, A.B., "Optimized Taylor Parameters for Concrete Buildup Factor Data," *Nucl. Sci. Eng. 64*, 799-800 (1977).

CHILTON, A.B., "Tschebycheff-Fitted Berger Coefficients for Eisenhauer-Simmons Gamma-Ray Buildup Factors in Ordinary Concrete," *Nucl. Sci. Eng. 69*, 436-438 (1979).

CHILTON, A.B., AND C.M. HUDDLESTON, "A Semi-empirical Formula for Differential Dose Albedo for Gamma Rays on Concrete," *Nucl. Sci. Eng. 17*, 419 (1963).

CHILTON, A.B., C.M. DAVISSON, AND L.A. BEACH, "Parameters for C-H Albedo Formula for Gamma Rays Reflected from Water, Concrete, Iron, and Lead," *Trans. Amer. Nucl. Soc. 8*, 656 (1965).

CHILTON, A.B., C.M. EISENHAUER, AND G.L. SIMMONS, "Photon Point Source Buildup Factors for Air, Water, and Iron," *Nucl. Sci. Eng. 73*, 97-107 (1980).

CHILTON, A.B., J.K. SHULTIS, AND R.E. FAW, *Principles of Radiation Shielding*, Prentice-Hall, Englewood Cliffs, N.J., 1984.

COURTNEY, J.C. (ed.), *A Handbook of Radiation Shielding Data*, A Publication of the Shielding and Dosimetry Division of the American Nuclear Society, La Grange Park, Ill., 1975.

CRANBERG, L., G. FRYE, N. NERESON, AND L. ROSEN, "Fission Neutron Spectrum of ^{235}U," *Phys. Rev. 103*, 662-670 (1956).

EISENHAUER, C.M., AND G.L. SIMMONS, "Point Isotropic Buildup Factors in Concrete," *Nucl. Sci. Eng. 46*, 263-270 (1975).

ENGLAND, T.R., R. WILCZYNSKI, AND N.L. WHITTEMORE, *CINDER-7: An Interim Report for Users*, Report LA-5885-MS, Los Alamos Scientific Laboratory, Los Alamos, N.M., 1976.

EVANS, R.D., *The Atomic Nucleus*, McGraw-Hill, New York, 1955.

FEWELL, T.R., AND R.E. SHUPING, "The Photon Energy Distribution of Some Typical Diagnostic X-Ray Beams," *Med. Phys. 4*, 3 (1977).

GLASSTONE, S., AND A. SESONSKE, *Nuclear Reactor Engineering*, D. Van Nostrand, Princeton, N.J., 1963.

GRONROOS, H., "Energy Dependent Removal Cross-Sections in Fast Neutron Shielding Theory," in *Engineering Compendium on Radiation Shielding*, Vol. I , R.G. Jaeger (ed.), p. 305, Springer-Verlag, New York, 1968.

GROTENHUIS, M., "Lecture Notes on Reactor Shielding," Report ANL-6000, Argonne National Laboratory, Argonne, Ill., 1962.

GROVE ENGINEERING, INC., *MicroSkyshine Users' Manual*, 15215 Shady Grove Rd., Rockville, Md. 20950, 1987.

HARIMA, Y., Y. SAKAMOTO, S. TANAKA, AND M. KAWAI, "Validity of the Geometric Progression Gamma-Ray Buildup Factors," *Nucl. Sci. Eng. 94*, 24-35 (1986).

HERMANN, O.W., AND R.M. WESTFALL, *ORIGEN-S: SCALE System Module to Calculate Fuel Depletion, Actinide Transmutation, Fission Product Buildup and Decay, and Associated Radiation Source Terms, Section F7 of SCALE: A Modular Code System for Performing Standardized Computer Analyses for Licensing Evaluation*, Report NUREG/CR-0200, Vol. 2, Oak Ridge National Laboratory, Oak Ridge, Tenn., 1984.

HUBBELL, J.H., R.L. BEACH, AND J.C. LAMKIN, "Radiation Field From a Rectangular Source," *Journal of Research of the National Bureau of Standards, 64C* (No. 2), 121-137 (1960).

HUNGERFORD, H.E., "Shielding," in *Fast Reactor Technology: Plant Design*, J.G. Yevick and A. Amorosi (eds.), MIT Press, Cambridge, Mass., 1966.

KALOS, M.H., *A Monte Carlo Calculation of the Transport of Gamma Rays*, Report NDA 56-7, Nuclear Development Associates, New York, 1956.

KEEPIN, G.R., *Physics of Nuclear Kinetics*. Addison Wesley, Reading, Mass., 1965.

KIRN, F.S., R.J. KENNEDY, AND H.O. Wycoff, "The Attenuation of Gamma Rays at Oblique Incidence," *Radiology 63*, No. 1, 94-104 (1954).

LaBAUVE, R.J., T.R. ENGLAND, D.C. GEORGE, AND C.W. MAYNARD, "Fission Product Analytic Impulse Source Functions," *Nuclear Technology 56*, 332-339 (1982).

LAMPLEY, C.M., C.M. ANDREWS, AND M.B. WELLS, *The SKYSHINE-III Procedure: Calculation of the Effects of Structure Design on Neutron, Primary Gamma-Ray and Secondary Gamma-Ray Dose Rates in Air*, RRA-T8209A, (RSIC Code Collection CCC-289), Radiation Research Associates, Fort Worth, Tex. 1988.

LEDERER, C.M., AND V.S. SHIRLEY, (eds.), *Table of Isotopes,* 7th ed, John Wiley & Sons, New York, 1978.

LEIMDORFER, M., "The Backscattering of Photons," in *Engineering Compendium on Radiation Shielding,* Vol. I, R.G. Jaeger (ed.), Springer Verlag, New York, 1968, pp. 233-245.

MAERKER, R.E., AND F.J. MUCKENTHALER, "Calculation and Measurement of the Fast-Neutron Differential Dose Albedo for Concrete," *Nucl. Sci. Eng. 22*, 455-462 (1965).

NATIONAL BUREAU OF STANDARDS, *Measurement of Neutron Flux and Spectra for Physical and Biological Applications*, Handbook 72, National Bureau of Standards, U.S. Department of Commerce, 1960.

NCRP, *Structural Shielding Design and Evaluation for Medical Use of X Rays and Gamma Rays of Energies up to 10 MeV,* Report 49, National Council on Radiation Protection and Measurements, Washington, D.C., 1976.

PETERSON, L.E., AND M. ROSENSTEIN, *Computer Program for Tissue Doses in Diagnostic Radiology*, Center for Devices and Radiological Health, U.S. Food and Drug Administration, Rockville, Md., 1989.

ROCKWELL, T., *Reactor Shielding Design Manual*, D. Van Nostrand, New York, 1956.

ROSENSTEIN, M., *Handbook of Selected Tissue Doses for Projections Common in Diagnostic Radiology*, HHS Publication (FDA) 89-8031, Center for Devices and Radiological Health, U.S. Food and Drug Administration. Rockville, Md., 1988.

RYMAN, J.C., *ORIGEN-S Data Libraries, Section M6 of SCALE: A Modular Code System for Performing Standardized Computer Analyses for Licensing Evaluation*, Report NUREG/CR-0200, Vol. 2, Oak Ridge National Laboratory, Oak Ridge, Tenn., 1983.

SCHAEFFER, N.M. (ed.), *Reactor Shielding for Nuclear Engineers*, Report TID-25951, U.S. Atomic Energy Commission, 1973. [Available from National Technical Information Service, Springfield, Va.]

SELPH, W.E., "Albedos, Ducts and Voids," in *Reactor Shielding for Nuclear Engineers*, N.M. Schaeffer (ed.), Report TID-25951, U.S. Atomic Energy Commission, 1973. [Available from National Technical Information Service, Springfield, Va.]

SHULTIS, J.K., AND R.E. Faw, *Improved Response Functions for the MicroSkyshine Method*, Report 189, Engineering Experiment Station, Kansas State University, Manhattan, Kans. 1987.

SHULTIS, J.K., R.E. FAW, AND M.S. BASSETT, "The Integral Line-Beam Method for Gamma Skyshine Analysis," *Nucl. Sci. Eng. 107*, 228-245 (1991).

SHURE, K., K.A. O'BRIEN, AND D.M. ROTHBERG, "Neutron Dose Rate Attenuation by Iron and Lead," *Nucl. Sci. Eng. 35*, 371 (1969).

SHURE K., AND O.J. WALLACE, "Compact Tables of Functions for Use in Shielding Calculations," *Nucl. Sci. Eng. 56*, 84 (1975).

SIMPKIN, D.J., "A General Solution to the Shielding of Medical X and γ Rays by the NCRP Report No. 49 Methods," *Health Physics 52*, 431-436 (1987). [Computer Code KUXKVPS released as Packages CCC-509/MICRO and CCC-515/MICRO by the Radiation Shielding Information Center, Oak Ridge National Laboratory, Oak Ridge, Tenn.]

SPENCER, L.V., *Structure Shielding Against Fallout Radiation From Nuclear Weapons*, Monograph 42, National Bureau of Standards, Washington D.C., 1962.

TRUBEY, D.K., AND Y. HARIMA, "New Buildup Factor Data for Point Kernel Calculations," in *Proceedings of a Topical Conference on Theory and Practices in Radiation Protection and Shielding*, Knoxville, Tenn., April 1987, Vol. 2, p. 503, American Nuclear Society, LaGrange Park, Ill., 1987. [Data are available as the DLC-129 code package from the Radiation Shielding Information Center, Oak Ridge National Laboratory, Oak Ridge, Tenn.]

TSYPIN S.G., AND V.I. KUKHTEVICH, "Removal Theory," in *Engineering Compendium on Radiation Shielding*, Vol. I, R.G. Jaeger (ed.), p. 301, Springer-Verlag, New York, 1968.

WELLS, M.B., *Reflection of Thermal Neutrons and Neutron-Capture Gamma Rays from Concrete*, Report RRA-M44, Radiation Research Associates, Fort Worth, Tex., 1964.

ZOLLER, L.K., "Fast Neutron Removal Cross Sections," *Nucleonics 22*, 128 (1964).

PROBLEMS

1. Suppose that a single fuel assembly in a nuclear power plant operates for 30,000 hours at a thermal power of 20 MW. The assembly is then removed from service and placed in storage. After one year, the element is shipped to another site. Estimate, for the time of shipment, the energy release rate (MeV s^{-1}) from the element via (a) gamma and x rays emitted from fission products, and (b) beta particles and electrons emitted from fission products.

2. One gram of natural iron is present in a uniform thermal-neutron flux density of 10^{12} cm^{-2} s^{-1}. Describe quantitatively the resulting emission of capture gamma rays from the iron.

3. A 10^5-GBq point source stored beneath 1 m of water (density 1 g cm^{-3}). At the water surface, evaluate (a) the flux density ϕ^o (cm^{-2} s^{-1}) of uncollided photons, (b) the kerma rate $\dot{K}^o$ (Gy s^{-1}) that would result in water from the uncollided photons, (c) the kerma rate $\dot{K}^o$ (Gy s^{-1}) that would result in air from the uncollided photons, (d) the exposure rate $\dot{X}^o$ (R s^{-1}) that would result from the uncollided photons, (e) the kerma rate $\dot{K}$ (Gy s^{-1}) that would result in water from all photons (computed using the Berger approximation for the buildup factor), and (f) the exposure rate $\dot{X}$ (R s^{-1}) that would result from all photons (computed using the Berger approximation for the buildup factor).

4. A 10^5-GBq source emitting one 1-MeV photon per transformation is uniformly distributed along a hypothetical ring of 2-m diameter entirely within a water medium (density 1 g cm^{-3}). At the center of the ring, evaluate the quantities called for in parts (a) through (f) of problem 6.3.

5. A 10^5-GBq source emitting one 1-MeV photon per transformation is uniformly distributed on the surface of a hypothetical sphere of 2-m diameter entirely within a water medium (density 1 g cm^{-3}). At the center of the sphere, evaluate the quantities called for in parts (a) through (f) of problem 6.3.

6. A 10^5-GBq source emitting one 1-MeV photon per transformation is uniformly distributed within the volume of a hypothetical sphere of 2-m diameter entirely within a water medium (density 1 g cm^{-3}). At the center of the sphere, evaluate the quantities called for in parts (a), (d), and (e) of problem 6.3.

7. A 10^5-GBq source emitting one 1-MeV photon per transformation is uniformly distributed along a hypothetical straight line of 2-m length entirely within a water medium (density 1 g cm^{-3}). At a point one meter from the line, along a perpendicular to the center of the line, evaluate the quantities called for in parts (a), (d), and (e) of problem 6.3.

8. A 10^5-GBq source emitting one 1-MeV photon per transformation is uniformly distributed on the surface of a hypothetical disk of 2-m diameter entirely within a water medium (density 1 g cm^{-3}). At a point one meter from the disk, along a perpendicular to the center of the disk, evaluate the quantities called for in parts (a), (d), and (e) of problem 6.3.

9. Consider a 50-GBq point isotropic source emitting one 1-MeV photon per transformation. The source is to be stored in a lead cask (density 11.5 g cm^{-3}). What minimum cask-wall thickness is required to assure that the exposure rate at the outside surface of the cask is limited to 2.5 mR h^{-1}, (a) neglecting the buildup of scattered photons, and (b) accounting for the buildup of scattered photons?

10. Certain measurements require the placement of a 4000-Ci ^{60}Co source in air on the axis of a concrete cylindrical-shell shield of inside diameter 2.5 m and outside diameter 4.5 m.

Estimate the exposure rate at the outside surface of the shield. Neglect attenuation and buildup in air, but account for both in concrete.

11. Suppose that fallout from a nuclear weapon is approximated as an infinite plane source of 1-MeV photons at the earth-atmosphere interface. Air density is 1.225 mg cm^{-3}. Aerial survey reveals that the exposure rate is 1 mR h^{-1} at an altitude of 1000 m above the interface. Estimate the exposure rate 1 m above the interface (a) neglecting the buildup of scattered gamma rays, and (b) accounting for the buildup of scattered gamma rays.

12. Consider a line source of 1-MeV photons in air, 500-m in length and of strength $S_l = 4 \times 10^{11}$ m^{-1} s^{-1}. A receptor location is 500 m radial distance from the source line and 250 m axial distance beyond the end of the source line. Air is at 20 C, 1 atm pressure, and 45% relative humidity. Estimate the exposure rate (R h^{-1}) at the receptor location (a) neglecting the buildup of scattered gamma rays, and (b) accounting for the buildup of scattered gamma rays.

13. A jogger, travelling along a straight path at a speed of 2 m s^{-1} passes at a distance of 5 meters a bare point isotropic gamma-ray source emitting 10^{12} 1-MeV photons per second. Neglecting exponential attenuation (and buildup) in the atmosphere, estimate the dose equivalent index (rem) received by the jogger.

14. Points A, B, and C are at the vertices of an equilateral triangle 10 m on a side in an air medium. A bare point isotropic source of 1-MeV photons, of strength 10^{13} s^{-1} is to be moved from point A to point B at a uniform speed of 0.5 m s^{-1}. Estimate the effective dose equivalent (rem) accrued by an unshielded observer at point C. Neglect exponential attenuation and buildup in the air.

15. An unshielded 20-m straight section of coolant piping in a nuclear power plant may be treated as a line source of 6-MeV photons of strength $S_l = 10^8$ cm^{-1} s^{-1}. A worker must travel at a distance of 5 m from the source at a speed of 1 m s^{-1}. Because of the presence

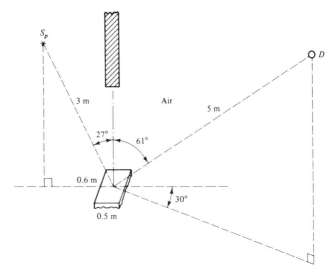

Figure P6.20   Radiation reflection from a concrete surface.

of shielding the worker is exposed only while moving along the 20-m path directly aside the source. Neglecting exponential attenuation (and buildup) in the atmosphere, estimate the dose equivalent index (rem) received by the worker.

16. A 250-kVp x-ray therapy unit operates with a tube current of 20 mA for an average of 18 h per week. How thick should a concrete primary protective barrier (density 2.1 g cm^{-3}) be to reduce the weekly exposure in a controlled area 3 m from the tube focus to less than 100 mR? The wall in question has a use factor of 1/2 assigned to it, and the occupancy factor is unity. What would be the required thickness if the wall were made of lead sheet?

17. A ^{60}Co gamma-ray irradiator containing a 2-TBq source is directed at a 30-cm thick concrete wall (density 2.2 g cm^{-3}). If the wall is 7.5 m from the source, what is the exposure rate behind the wall when the irradiator is on?

18. A beam from a ^{60}Co irradiator containing a 20-TBq source falls obliquely onto a 40-cm-thick concrete wall (density 2.3 g cm^{-3}) 5 m from the source. Estimate how close to perpendicular incidence the beam may be so that the maximum exposure rate behind the wall is less than 100 mR h^{-1}.

19. A ^{242}Cf neutron point source, emitting 10^9 fission neutrons per second is to be stored under water. Make a rough estimate ($\pm 10\%$) of the depth of water required to ensure that the fast-neutron (fission- neutron) dose equivalent index rate does not exceed 1 rem h^{-1} at the water surface. You may assume that the neutron quality factor is 10.

20. A thick concrete slab, 0.5×0.6 m, is placed with reference to a detector and a source as illustrated in Fig. P6.20. A shielding wall prevents any direct radiation reaching the detector. Air scattering is to be ignored and there are no scattering surfaces other than the concrete slab. Estimate the detector response for the following cases:

   (a) A 2-GBq ^{137}Cs-^{137m}Ba source and an exposure-rate response.
   (b) A 1-GBq ^{60}Co source and an exposure-rate response.
   (c) A point isotropic neutron source of energy 2 MeV and strength 10^9 s^{-1}, and response in the form of the dose equivalent index.

   You may make the approximation that the reflecting slab is of infinitesimal area. If you do so, justify the approximation.

21. Repeat problem 6.20 for the source and detector reversed in position.

Chapter 7

Electron Penetration and Dose Evaluation

7.1 INTRODUCTION

This chapter deals with evaluation of the spatial distribution of absorbed dose arising from electron and beta-particle sources in various geometries. The analytical treatment of electron attenuation is distinctly different from that for neutrons and gamma photons. The *indirectly ionizing* neutrons and gamma photons travel along paths composed of straight-line segments punctuated by distinct interactions with atoms or electrons in matter, these interactions being governed by short-range forces. Consequently, the particles have no identifiable "ranges" and their trajectories are best described in terms of concepts such as the interaction coefficient or the mean free path (as discussed in Chap. 6). By contrast, *directly ionizing* electrons are attenuated by a continuous slowing-down process governed largely by long-range Coulombic forces. While the electron range is not precisely defined, the spatial distribution of the energy deposition around a point source of electrons may be described precisely in terms of either a so-called *ninetieth percentile distance* or a fictitious measure of range called the *CSDA range*, that is, the range in the *continuous slowing-down approximation*.

The point-kernel concept introduced in Chap. 6 for neutrons and gamma rays is equally valid for electrons. Moreover, for many electron dosimetry problems, the effects of boundaries or material heterogeneities can be neglected, and the use of infinite-medium point kernel techniques to compute the dose received in one region from an electron source in another results in the dose being expressible as a one-dimensional integral involving an appropriate geometry factor.

Two new dosimetry concepts are introduced in this chapter — the *absorbed fraction* and the *reduction factor*. These concepts deal with relationships between sources distributed throughout one geometric region and average absorbed doses in another (or the same) geometric region. In many applications these quantities are found to be very insensitive to the initial electron energy and are determined primarily by the problem geometry. The absorbed fraction and reduction factor concepts are widely applied in environmental and medical dosimetry, two examples being the evaluation of the dose to the skin arising from beta-particle sources in the atmosphere and the evaluation of the dose to the bone marrow arising from sources present in bone or on bone surfaces.

7.2 BETA-PARTICLE ENERGY SPECTRA

In the beta decay of a radionuclide, there is in effect a transformation, within the nucleus, of a neutron into a proton with the emission of an electron, or beta particle, and an antineutrino. Similarly, in positron decay, there is in effect a nuclear transformation of a proton in the decaying nucleus to a neutron accompanied by emission of a positron and a neutrino. The energy released in the process — the Q value — is shared by the electron and the antineutrino or by the positron and the neutrino. Beta particle or positron decay of a nucleus may occur by any of several different nuclear transitions. For each there is a unique end-point energy, the maximum kinetic energy of the electron or positron. Also, for each transition, there is a unique statistical distribution of energy between the electron and antineutrino or positron and neutrino, and to a very minor extent the residual nucleus. For the balance of this discussion, we will deal with the far more common beta decay, with the recognition that positron decay may be treated very similarly.

While the energy spectrum of the beta particles for any one transition certainly depends on the end-point energy, the spectrum also depends in a very complicated way on the nuclear spin and parity quantum numbers of the nucleus before and after beta-particle emission. These quantum numbers determine whether the transition is referred to as *allowed* or *forbidden*, and, if forbidden, whether the transition is referred to as *unique* or *nonunique*. In a transition, the spin quantum number J may remain constant or it may change by one or more units, that is, $\Delta J = 0, 1, 2, \ldots$. The parity quantum number Π may or may not change, that is, $\Delta \Pi = -1$ (no change) or $+1$ (change). To a first approximation, a single classification index n may be used to characterize the energy spectrum. For $\Delta J \geq 2$, $n = \Delta J - 1$. Otherwise, the various combinations of $\Delta \Pi$ and ΔJ establish the following selection rules which determine the *shape factor* which, in part, governs the beta-particle energy spectrum for a particular transition.

ΔJ	$\Delta \Pi$	Transition Classification	n
0,1	-1	Allowed	0
0,1	$+1$	Nonunique, first-forbidden	0
2	$+1$	Unique, first-forbidden	1
2	-1	Nonunique, second-forbidden	1
3	-1	Unique, second-forbidden	2

As an example, consider the decay of ^{38}Cl to ^{38}Ar. Three beta transitions are possible, with the following characteristics (the maximum and average beta energies are in keV):

Frequency %	E_{max}	E_{avg}	ΔJ	$\Delta \Pi$	n	Classification
32.5	1107	420	1	−1	0	Allowed
11.5	2749	1182	0	+1	0	Non-U, 1st forbidden
56.0	4917	2244	2	+1	1	Unique, 1st forbidden

The energy spectrum of beta particles arising from transition i is conveniently expressed in terms of the electron total energy W (including rest-mass energy), that is, $W = E + m_e c^2$, and the momentum $p = \sqrt{E^2 + 2m_e c^2 E}$. The energy spectrum $N_i(E)$ is defined in such a way that $N_i(E)dE$ is the probability that, in a transformation of the radionuclide, a beta particle is emitted, via transition i, with kinetic energy in the range dE about E. This spectrum can be expressed as

$$N_i(E) = C_i \, p \, W[E_{max,i} - E]^2 F(Z, A, W) \, S_n(Z, A, E_{max,i}, W), \qquad (7.1)$$

in which $E_{max,i}$ represents the maximum (or endpoint) energy, Z and A are the charge and mass numbers of the nucleus after the transition, F is the *Fermi function*, S_n is a *shape factor* determined by the selection-rule classification index n, and C_i is a normalization constant.

The product $pW[E_{max,i} - E]^2$ is a statistical factor and is a major determinant of the energy spectrum. The shape factor S_n is unity for allowed and nonunique first-forbidden transitions; for other transitions it is a complicated function. Suffice to say here that it is an extremely tedious task to evaluate the Fermi function and the shape factor. Procedures are clearly described by Dillman and Von der Lage (1975) and by Dillman (1980) as are procedures required to account for screening by atomic electrons which is especially important for positron decay. Figure 7.1 illustrates individual and composite spectra for the three transitions involved in the beta decay of ^{38}Cl. In Table 7.1 the spectrum for four important beta-emitting radionuclides are given.

If f_i is the frequency for the ith transition, then the spectrum normalization constant C_i is determined by the requirement that

$$\int_0^{E_{max,i}} dE \, N_i(E) = f_i. \qquad (7.2)$$

The average beta-particle energy for a particular transition is

$$\langle E_i \rangle = \frac{1}{f_i} \int_0^{E_{max,i}} dE \, E N_i(E). \qquad (7.3)$$

The average beta-particle energy released per transformation of the radionuclide is given by $\sum_i f_i \langle E_i \rangle$.

TABLE 7.1   Beta Spectra for Four Important Radionuclides. The Spectrum $N(E)$ is Normalized to the Number of Betas per MeV at Energy E that Are Emitted per Decay of the Radionuclide. Data for ^{90}Sr–^{90}Y Are for an Equilibrium Mixture of Parent and Daughter.

^{3}H		^{14}C		^{60}Co		^{90}Sr–^{90}Y	
E (MeV)	$N(E)$	E (MeV)	$N(E)$	E (MeV)	$N(E)$	E (MeV)	$N(E)$
0.00000	82.33	0.00000	9.910	0.00000	6.6370	0.00000	3.2879
0.00010	83.48	0.00010	9.900	0.00040	6.6260	0.00050	3.2800
0.00014	84.19	0.00040	9.867	0.00080	6.6140	0.00100	3.2720
0.00018	84.93	0.00080	9.831	0.00100	6.6090	0.00200	3.2571
0.00020	85.32	0.00100	9.819	0.00500	6.5080	0.00400	3.2272
0.00028	86.85	0.00140	9.821	0.01000	6.4140	0.00600	3.2035
0.00036	88.34	0.00180	9.850	0.01400	6.3420	0.00800	3.1988
0.00040	89.05	0.00200	9.871	0.01800	6.2750	0.01000	3.1950
0.00050	90.74	0.00240	9.925	0.02200	6.2130	0.02000	3.1732
0.00060	92.30	0.00320	10.050	0.02600	6.1520	0.04000	3.1455
0.00070	93.70	0.00400	10.190	0.03000	6.0920	0.06000	3.1300
0.00080	94.98	0.00600	10.520	0.03600	6.0020	0.08000	3.1199
0.00090	96.13	0.00800	10.810	0.04000	5.9410	0.10000	3.1099
0.00100	97.17	0.01000	11.050	0.05000	5.7790	0.12000	3.0975
0.00120	98.95	0.01200	11.240	0.06000	5.6030	0.14000	3.0798
0.00140	100.30	0.01400	11.390	0.07000	5.4120	0.16000	3.0557
0.00160	101.40	0.01600	11.500	0.08000	5.2040	0.18000	3.0230
0.00180	102.20	0.01800	11.580	0.09000	4.9820	0.20000	2.9807
0.00200	102.80	0.02000	11.630	0.10000	4.7460	0.24000	2.8576
0.00240	103.30	0.02200	11.660	0.11000	4.4970	0.28000	2.6751
0.00280	103.00	0.02400	11.660	0.12000	4.2380	0.30000	2.5591
0.00300	102.60	0.02600	11.640	0.13000	3.9690	0.32000	2.4244
0.00320	102.10	0.02800	11.590	0.14000	3.6930	0.36000	2.1035
0.00360	100.70	0.03000	11.530	0.15000	3.4120	0.40000	1.7245
0.00400	98.90	0.03200	11.460	0.16000	3.1270	0.45000	1.2165
0.00450	96.12	0.03600	11.250	0.18000	2.5550	0.50000	0.7806
0.00500	92.86	0.04000	10.990	0.20000	1.9950	0.55000	0.6089
0.00550	89.21	0.04500	10.600	0.22000	1.4660	0.60000	0.6169
0.00600	85.25	0.05000	10.150	0.24000	0.9866	0.65000	0.6226
0.00650	81.04	0.05500	9.648	0.26000	0.5782	0.70000	0.6263
0.00700	76.63	0.06000	9.102	0.28000	0.2624	0.75000	0.6281
0.00750	72.08	0.06500	8.522	0.30000	0.0622	0.80000	0.6282
0.00800	67.42	0.07000	7.917	0.32000	0.0005	0.90000	0.6236
0.00850	62.71	0.07500	7.293	0.36000	0.0005	1.00000	0.6133
0.00900	57.97	0.08000	6.658	0.40000	0.0005	1.10000	0.5973
0.01000	48.56	0.08500	6.019	0.50000	0.0006	1.20000	0.5753
0.01100	39.44	0.09000	5.381	0.60000	0.0005	1.30000	0.5465
0.01200	30.84	0.10000	4.135	0.70000	0.0005	1.40000	0.5101
0.01300	22.95	0.11000	2.971	0.80000	0.0005	1.50000	0.4653
0.01400	15.97	0.12000	1.935	0.90000	0.0005	1.60000	0.4115
0.01500	10.06	0.13000	1.075	1.00000	0.0004	1.80000	0.2790
0.01600	5.37	0.14000	0.438	1.20000	0.0003	2.00000	0.1299
0.01800	0.28	0.15000	0.071	1.40000	0.0000	2.20000	0.0153
0.01857	0.00	0.15640	0.000	1.49112	0.0000	2.28400	0.0000

Source: Obtained with the code by Dillman 1980.

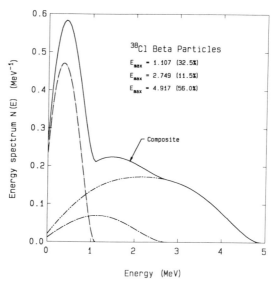

Figure 7.1 Energy spectrum of beta particles released from ^{38}Cl, showing contributions from individual transitions as well as the composite spectrum.

7.3 ENERGY LOSS BY PENETRATING ELECTRONS

As electrons are slowed while passing through matter, their kinetic energy is lost in two ways — as a result of the Coulombic forces exerted primarily by the atomic electrons, and as a result of photon emission (bremsstrahlung) as the electron is decelerated. Interactions with atomic nuclei collectively result in deflection of the electron from a straight path, but account for little energy loss.

7.3.1 Electron Collisional Energy Loss

The kinetic energy loss of an electron as it passes through matter from Coulombic collisions with other electrons is described by a collisional interaction coefficient μ_{coll}.[1] This interaction coefficient is defined in such a way that $\mu_{coll}(E, T)dT$ is the probability, per unit differential distance of travel, that an electron of energy E experiences an interaction in which energy between T and $T + dT$ is lost, that is, a secondary (recoil) electron is produced with energy between T and $T + dT$. By convention, T is always less than $E/2$ and the "original" electron is left with energy greater than or equal to $E/2$. The coefficient is conveniently written in terms of the energies expressed in units of the electron rest-mass energy, that is, $\epsilon = E/m_e c^2$ and $\tau = T/m_e c^2$. Collisional energy loss resulting from free-electron interactions was addressed in 1932 by Møller, who found that

---

[1]Interaction coefficients for electrons are often denoted by the symbol k. However, in keeping with the notation used for interaction coefficients for neutrons and photons, the symbol μ is used here.

$$\mu_{coll}(\epsilon, \tau) = 2\pi N_e r_e^2 \frac{1}{\beta^2} \left[\frac{1}{\tau^2} + \frac{1}{(\epsilon - \tau)^2} + \frac{1}{(\epsilon + 1)^2} - \frac{2\epsilon(\epsilon + 1)}{\tau(\epsilon - \tau)(\epsilon + 1)^2} \right], \quad (7.4)$$

in which N_e is the electron density, r_e is the classical electron radius (2.818×10^{-13} cm), and β is the ratio of the electron speed to the speed of light. Note that $\beta^2 = \epsilon(\epsilon + 2)/(\epsilon + 1)^2$. The Møller interaction coefficient is illustrated in Fig. 7.2. A similar expression for positron energy losses was derived by Bhaba in 1936. The reader will note that, as τ approaches zero, $\mu_{coll}(\epsilon, \tau)$ is singular, varying as τ^{-2}. Indeed, the Møller formula fails for very small τ because atomic electrons cannot then be treated as free, and quantum effects place a nonzero lower limit on τ.

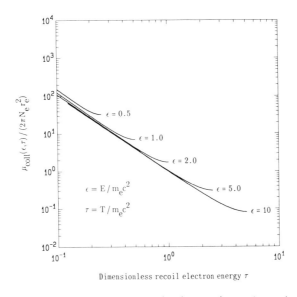

Figure 7.2   The Møller cross section for electron-electron interactions.

7.3.2 Electron Radiation Energy Loss

Classical electromagnetic theory requires that the rate at which electromagnetic energy is radiated from an accelerating charged particle be proportional to the square of the acceleration. In the electric field of a nucleus with atomic number Z, the force on a charged particle with z elemental charges is proportional to the product Zz. The resulting acceleration of the charged particle is proportional to this force and inversely proportional to the particle's mass m. Thus, the rate of radiative energy loss by the charged particle would be expected to be proportional to $(Zz/m)^2$. This inverse-square dependence on particle mass explains why bremsstrahlung from protons or heavier charged particles is negligible in most circumstances compared to that from electrons or positrons. The dependence on the square of Z also explains why, except for the lightest elements, bremsstrahlung in the field of the nucleus far exceeds bremsstrahlung in the fields of atomic electrons.

One may define a differential interaction coefficient for radiative energy loss in such a way that $\mu_{rad}(E, E')dE'$ is the probability per unit differential distance of travel that deceleration of an electron of energy E results in emission of a photon with energy between E' and $E' + dE'$. However, it is not possible to write a simple expression for $\mu_{rad}(E, E')$ or for the associated microscopic cross section $\sigma_{rad}(E, E')$. The interaction coefficient for radiative energy loss by an electron is illustrated in Fig. 7.3 which is based on calculations performed by the PEGS4 computer program (Nelson, Hirayama and Rogers 1985).

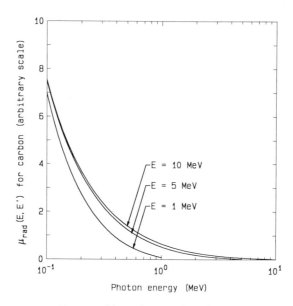

Figure 7.3   Energy spectrum of bremsstrahlung photons released in carbon by radiative energy losses of electrons with initial energy E.

7.4 ELECTRON STOPPING POWER

The stopping power $-dE/ds$ or, equivalently, the LET (linear energy transfer) $L(E)$ is the energy loss per unit differential distance along the track s of a charged particle as it slows through energy E. In principle, the collisional and radiative stopping powers may be written as

$$L_{coll}(E) = \int_0^{E/2} dT\, T\, \mu_{coll}(E, T), \tag{7.5}$$

and

$$L_{rad}(E) = \int_0^E dE'\, E' \mu_{rad}(E, E'). \tag{7.6}$$

7.4.1 Collisional Stopping Power

As indicated in Sec. 7.3, the Møller formula of Eq. (7.4) fails as T approaches zero. Quantum effects place a lower limit on T. These effects are accommodated empirically by using the Møller formula in Eq. (7.5) but imposing a nonzero lower limit T_{min} on the integral. Integration of the first term in the Møller formula gives the result (in dimensionless energies)

$$L_{coll}(\epsilon) = 2\pi N_e m_e c^2 r_e^2 \frac{1}{\beta^2} \int_{T_{min}}^{\epsilon/2} d\tau\ \tau\ [\tau^{-2} + \cdots]$$

$$= 2\pi N_e m_e c^2 r_e^2 \frac{1}{\beta^2} \left\{ \ln\left(\frac{\epsilon}{2T_{min}}\right) + \cdots \right\}. \tag{7.7}$$

In the Bethe-Bloch approximation, for example (Spencer and Fano 1954)

$$T_{min} \simeq \frac{(\bar{I})^2 e^{\beta^2}}{2\epsilon(\epsilon + 2)}, \tag{7.8}$$

in which $\bar{I} = I/m_e c^2$ is a dimensionless *mean excitation energy* and is determined empirically for the stopping medium. This approximation leads to the Rohrlich and Carlson formula used by Berger and Seltzer in their compilations of stopping powers (ICRU 1984):

$$L_{coll}(E) = 2\pi N_e r_e^2 m_e c^2 \beta^{-2} \left[2\ln(\epsilon/\bar{I}) + \ln(1 + \epsilon/2) + F(\epsilon) - \delta \right]. \tag{7.9}$$

The factor F is given by

$$F(\epsilon) = (1 - \beta^2) \left[1 + \epsilon^2/8 - (1 + 2\epsilon) \ln 2 \right]. \tag{7.10}$$

In Eq. (7.9), the *density-effect correction term* δ accounts for polarization of atoms in the stopping medium caused by passage of the electron, the result of which is a reduction of the electric field acting on the moving electron and thus a reduction in the stopping power. Evaluation of δ is quite involved. Procedures are described by the ICRU (1984). In liquid water, for example, the density effect causes a 1.2% reduction in L_{coll} at $E = 1$ MeV, 3.9% at 2 MeV, and 11.5% at 10 MeV. Selected values of mean excitation energies are given in Table 7.2. Stopping powers are illustrated in Fig. 2.15.

For mixtures and compounds with elemental weight fractions w_j, I and δ may be evaluated on the basis of the mean ratio of the atomic number to atomic mass $\langle Z/A \rangle$

$$\langle Z/A \rangle = \sum_j w_j \frac{Z_j}{A_j}, \tag{7.11}$$

$$\ln(I) = \frac{1}{\langle Z/A \rangle} \sum_j w_j \frac{Z_j}{A_j} \ln(I_j), \tag{7.12}$$

$$\delta = \frac{1}{\langle Z/A \rangle} \sum_j w_j \frac{Z_j}{A_j} \delta_j. \tag{7.13}$$

TABLE 7.2 Effective Excitation Energies for Selected Elements, Compounds and Mixtures. Unless Otherwise Indicated, Data Are for Media in Liquid or Solid Form. Data for H, C, N, and O are for Constituents of Compounds.

MEDIUM	I (eV)	MEDIUM	I (eV)
H	19.2	ICRU Tissue	74.7
$H_2(g)$	19.2	ICRU Bone	91.9
C	81.0	Polyethylene	57.4
N	82.0	Polyethylene	
$N_2(g)$	82.0	terephthalate (PTP)	78.7
O	106	Polymethyl	
$O_2(g)$	106	methacrylate (PMMA)	74.0
Air (dry)	85.7	Al	166
H_2O	75.0	Fe	286
$H_2O(g)$	71.6	Ag	470
CO_2 (g)	85.0	Au	790

Source: ICRU 1984.

7.4.2 Stopping Power for Radiative Loss

No simple formula can describe the stopping power or LET for electron radiative energy losses (bremsstrahlung). The total stopping power can be written as the sum of stopping powers associated with radiative losses in the force fields of nuclei and electrons, namely, $L_{rad}(E) = L_{rad,n}(E) + L_{rad,e}(E) = L_{rad,n}(E)(1+\xi/z)$, in which ξ is the "correction term" $L_{rad,e}/L_{rad,n}$. According to ICRU (1984), ξ depends hardly at all on the medium and is less than 1.2 in magnitude for all E. It is about 0.5 at $E = 700$ keV and approaches zero at low E. As is suggested by Eq. (7.6), the radiative stopping power may be written as

$$L_{rad}(E) = N r_e^2 (E + m_e c^2) Z^2 \alpha (1 + \xi/Z) \Phi(E, Z), \qquad (7.14)$$

in which N is the atomic density, $\alpha \simeq 1/137$ is the *fine-structure constant*, and $\Phi(E, Z)$ is the dimensionless scaled radiative energy-loss cross section given by

$$\Phi(E, Z) = \frac{1}{\alpha r_e^2 Z^2 (E + m_e c^2)} \int_0^E dE' \, E' \sigma_{rad,n}(E, E'). \qquad (7.15)$$

The function Φ, too, is not strongly dependent on Z. It is illustrated in Fig. 7.4 for several values of Z. Figure 2.15 illustrates $L_{rad}(E)$ in comparison with $L_{coll}(E)$.

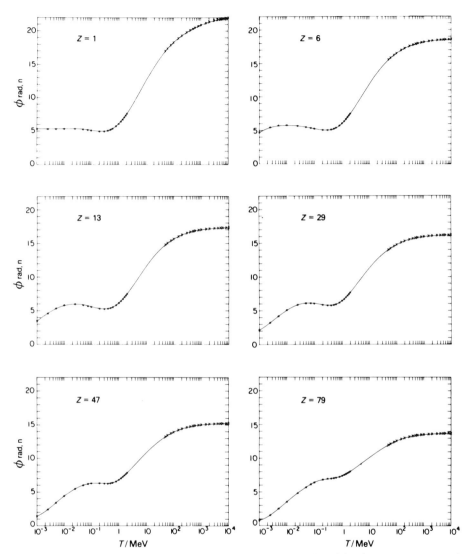

Figure 7.4   Dimensionless scaled radiative energy loss cross sections $\Phi(T, Z)$ for electrons in various media. [From ICRU (1984); by permission of the ICRU.]

7.5 ELECTRON CSDA RANGE

As electrons are slowed, they experience deflections in path and statistical fluctuations in energy loss. Paths are not straight, and, for a group of electrons all starting at the same energy, there are fluctuations in total path length traveled. Thus, one cannot identify an unambiguous electron "range." However, as a *measure* of range one may identify precisely an effective travel distance based on the continuous slowing-down approximation (CSDA), the so-called CSDA range r_o, which is a function of the electron's initial energy E_o.

Under the CSDA approximation it is assumed that an electron slows continuously, with no energy-loss fluctuations, and with a mean energy loss per unit differential path length given by the stopping power $L(E)$ evaluated at the local value of the electron's energy. Thus, the CSDA range is defined as

$$r_o(E_o) = \int_0^{E_o} dE \, [L_{coll}(E) + L_{rad}(E)]^{-1}. \tag{7.16}$$

There are practical difficulties in evaluating the integrand as E approaches zero because electron stopping powers are not well known at low energy. It is usually assumed (ICRU 1984) that the integrand is zero at $E = 0$ and increases linearly with E to the lowest known value. The electron CSDA range in water is illustrated in Fig. 7.5 and Table 7.3.

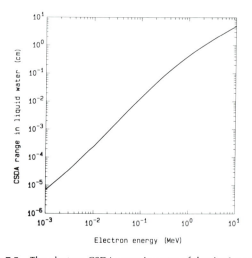

Figure 7.5   The electron CSDA range in water of density 1 g cm^{-3}.

7.6 ELECTRON POINT KERNELS

Prompted by the need for precise medical internal-dosimetry calculations, M.J. Berger (1971, 1973) carried out extensive calculations of the spatial distributions of absorbed dose around point isotropic sources of monoenergetic electrons. He, and others, were then able to use these "point kernel" results to calculate absorbed-dose patterns for beta-particle sources — both point sources and spatially distributed sources.

The point kernel $\mathcal{G}(r, E)$ for an isotropic and monoenergetic source at energy E in a prescribed infinite homogeneous medium may be defined as the mean absorbed dose (MeV g^{-1}) at distance r (cm) from the source per electron emitted. Early published kernels (Berger 1971), still in use, were based on moments-method trans-

TABLE 7.3   Electron CSDA Range r_o in Water of Density 1.0 g cm^{-3}.

E_o (MeV)	ρr_o (g cm^{-2})	E_o (MeV)	ρr_o (g cm^{-2})
0.0005	2.272×10^{-6}	0.0800	9.562×10^{-3}
0.0006	2.897×10^{-6}	0.1000	1.401×10^{-2}
0.0008	4.325×10^{-6}	0.1500	2.760×10^{-2}
0.0010	5.976×10^{-6}	0.2000	4.400×10^{-2}
0.0015	1.092×10^{-5}	0.3000	8.263×10^{-2}
0.0020	1.710×10^{-5}	0.4000	1.264×10^{-1}
0.0030	3.279×10^{-5}	0.5000	1.735×10^{-1}
0.0040	5.268×10^{-5}	0.6000	2.227×10^{-1}
0.0050	7.652×10^{-5}	0.8000	3.248×10^{-1}
0.0060	1.037×10^{-4}	1.0000	4.297×10^{-1}
0.0080	1.689×10^{-4}	1.5000	6.956×10^{-1}
0.0100	2.482×10^{-4}	2.0000	9.613×10^{-1}
0.0150	5.042×10^{-4}	3.0000	1.485
0.0200	8.374×10^{-4}	4.0000	1.997
0.0300	1.715×10^{-3}	5.0000	2.499
0.0400	2.851×10^{-3}	6.0000	2.991
0.0500	4.222×10^{-3}	8.0000	3.950
0.0600	5.807×10^{-3}	10.000	4.880

Source: Berger 1973.

port calculations of Spencer (1959) and were scaled according to the so-called ninetieth percentile distance $r_{90}(E)$, the distance from a source within which 90 percent of the source electron's energy is absorbed on average. This resulted in dimensionless *scaled point kernels* $\mathcal{F}'(r/r_{90}, E)$, independent of the medium density ρ (g cm^{-3}), defined so that[2]

$$\mathcal{G}(r, E) = \frac{E}{4\pi r^2 \rho r_{90}} \mathcal{F}'(r/r_{90}, E), \qquad (7.17)$$

and for which the variation of $\mathcal{F}'$ with E and with the stopping medium was minimal. In water, r_{90} is about 75% of r_o at $E = 0.001$ MeV, and increases to about 90% at $E = 10$ MeV.

More recently published point kernels (Berger 1973) are based on Monte Carlo electron-transport calculations and are scaled according to the more easily determined CSDA range $r_o(E)$. The dimensionless scaled point kernel $\mathcal{F}(r/r_o, E)$ is defined in such a way that

$$\mathcal{G}(r, E) = \frac{E}{4\pi r^2 \rho r_o} \mathcal{F}(r/r_o, E). \qquad (7.18)$$

---

[2]The function $\mathcal{F}'$ is not truly a point kernel since the geometric attenuation factor $1/4\pi r^2$ has been explicitly removed from this function. It might better be termed a *scaled energy-deposition distribution*. However, in this chapter, the term *scaled point kernel*, in accordance with popular usage, will be retained.

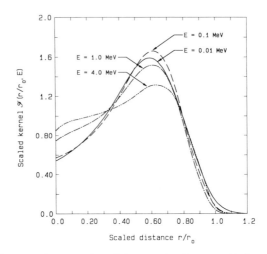

Figure 7.6   Scaled electron point-source kernels in water for electrons with four different initial energies. [From data of Berger (1973).]

An important consideration in the use of scaled point kernels is that the fraction of the electron energy E absorbed within radii from r to $r + dr$ from the source is just $(dr/r_o)\mathcal{F}(r/r_o, E)$. Although the published kernels are for electrons in H_2O (suitable for tissue), procedures are available to convert the kernels for use with other media (Berger 1971). Representative scaled point kernels for water are illustrated in Fig. 7.6. Numerical values of these kernels are given in Table 7.4.

7.6.1 Electron Kernels for Nonaqueous Media

Berger (1971, 1973) described techniques for arriving at approximate point kernels for nonaqueous media. The techniques were originally developed for use with kernels based on ninetieth percentile distances r_{90}, but are applied here without modification for use with kernels based on CSDA ranges. The approximation is most satisfactory for low-Z media such as air, tissue or plastics.

Equation (7.18) can be said to apply approximately to a medium other than water so long as r/r_o is based on the CSDA range r_o for that medium. As noted by Berger (1971), Cross (1968) gives an empirically derived prescription for relating the ninetieth percentile distances of two media. Here we apply the same prescription to CSDA ranges. If $(\rho r_o)_2$ is the mass-thickness (g cm^{-2}) CSDA range in medium 2 and $(\rho r_o)_1$ is the mass-thickness CSDA range in medium 1, then, approximately independent of electron energy, the two are related by a parameter a_{21}, so that

$$(\rho r_o)_2 \simeq \frac{1}{a_{21}}(\rho r_o)_1. \tag{7.19}$$

TABLE 7.4   Scaled Absorbed Dose Distributions $\mathcal{F}(r/r_o, E)$ (Scaled Point Kernels) for Isotropic Sources of Monoenergetic Electrons in Water as a Function of the Scaled Distance r/r_o, where r_o is the CSDA Range of the Electron.

	ELECTRON ENERGY (MeV)								
r/r_o	0.010	0.015	0.020	0.030	0.040	0.050	0.060	0.080	0.100
0.000	0.541	0.541	0.544	0.544	0.550	0.556	0.559	0.563	0.572
0.025	0.563	0.562	0.563	0.564	0.568	0.572	0.575	0.581	0.588
0.050	0.587	0.585	0.584	0.586	0.588	0.591	0.593	0.601	0.607
0.075	0.614	0.609	0.608	0.609	0.610	0.612	0.614	0.622	0.627
0.100	0.642	0.636	0.635	0.634	0.635	0.636	0.638	0.645	0.650
0.125	0.673	0.665	0.664	0.661	0.662	0.663	0.664	0.670	0.676
0.150	0.706	0.696	0.696	0.691	0.691	0.693	0.693	0.698	0.704
0.175	0.742	0.731	0.731	0.724	0.723	0.725	0.724	0.729	0.735
0.200	0.781	0.769	0.768	0.760	0.759	0.760	0.758	0.763	0.769
0.225	0.823	0.810	0.809	0.800	0.798	0.799	0.796	0.802	0.806
0.250	0.869	0.855	0.853	0.843	0.840	0.841	0.837	0.844	0.848
0.275	0.919	0.905	0.901	0.892	0.887	0.888	0.883	0.891	0.893
0.300	0.974	0.959	0.954	0.944	0.939	0.940	0.934	0.943	0.944
0.325	1.033	1.017	1.012	1.002	0.996	0.997	0.991	1.001	1.001
0.350	1.096	1.080	1.075	1.065	1.059	1.061	1.055	1.064	1.063
0.375	1.163	1.147	1.143	1.133	1.127	1.130	1.123	1.131	1.131
0.400	1.231	1.217	1.214	1.203	1.198	1.202	1.196	1.203	1.202
0.425	1.300	1.288	1.287	1.276	1.272	1.277	1.272	1.276	1.276
0.450	1.367	1.358	1.360	1.349	1.346	1.352	1.348	1.350	1.350
0.475	1.430	1.426	1.431	1.420	1.419	1.426	1.423	1.423	1.422
0.500	1.486	1.489	1.497	1.486	1.488	1.494	1.493	1.491	1.490
0.525	1.532	1.544	1.555	1.545	1.550	1.554	1.556	1.553	1.550
0.550	1.567	1.586	1.601	1.593	1.601	1.604	1.608	1.603	1.601
0.575	1.587	1.613	1.631	1.627	1.637	1.640	1.644	1.640	1.637
0.600	1.590	1.622	1.641	1.644	1.655	1.659	1.663	1.659	1.658
0.625	1.576	1.610	1.629	1.641	1.652	1.658	1.660	1.657	1.659
0.650	1.541	1.586	1.593	1.615	1.625	1.635	1.634	1.632	1.638
0.675	1.487	1.519	1.534	1.565	1.575	1.586	1.583	1.582	1.592
0.700	1.413	1.442	1.452	1.492	1.500	1.509	1.509	1.508	1.521
0.725	1.322	1.348	1.352	1.396	1.403	1.406	1.413	1.410	1.422
0.750	1.215	1.238	1.237	1.280	1.287	1.285	1.296	1.292	1.293
0.775	1.097	1.118	1.110	1.149	1.157	1.151	1.169	1.158	1.138
0.800	0.972	0.989	0.977	1.007	1.015	1.007	1.024	1.011	0.969
0.825	0.844	0.855	0.840	0.859	0.863	0.855	0.865	0.856	0.807
0.850	0.717	0.720	0.704	0.711	0.709	0.700	0.697	0.699	0.662
0.875	0.595	0.588	0.573	0.569	0.560	0.550	0.537	0.547	0.533
0.900	0.478	0.464	0.452	0.439	0.427	0.416	0.402	0.408	0.414
0.925	0.368	0.352	0.345	0.326	0.315	0.303	0.296	0.289	0.306
0.950	0.271	0.257	0.253	0.231	0.223	0.211	0.213	0.194	0.211
0.975	0.193	0.179	0.177	0.156	0.150	0.140	0.144	0.124	0.134
1.000	0.133	0.120	0.116	0.101	0.095	0.087	0.089	0.076	0.076
1.025	0.089	0.077	0.072	0.061	0.056	0.051	0.049	0.044	0.039
1.050	0.058	0.047	0.041	0.035	0.030	0.027	0.024	0.022	0.018
1.075	0.037	0.027	0.021	0.019	0.015	0.013	0.010	0.010	0.007
1.100	0.022	0.015	0.010	0.009	0.007	0.006	0.004	0.004	0.003
1.125	0.013	0.008	0.005	0.004	0.003	0.002	0.002	0.001	0.001
1.150	0.007	0.004	0.002	0.002	0.001	0.001	0.000	0.000	0.000
1.175	0.003	0.002	0.001	0.001	0.000	0.000	0.000	0.000	0.000
1.200	0.001	0.001	0.001	0.000	0.000	0.000	0.000	0.000	0.000

(Continued)

TABLE 7.4 (continued)

r/r_o	ELECTRON ENERGY (MeV)								
	0.150	0.200	0.300	0.400	0.500	0.600	0.800	1.000	1.500
0.000	0.588	0.602	0.629	0.653	0.674	0.695	0.719	0.744	0.787
0.025	0.605	0.621	0.650	0.675	0.698	0.719	0.747	0.773	0.816
0.050	0.624	0.641	0.671	0.697	0.721	0.741	0.772	0.798	0.841
0.075	0.645	0.662	0.692	0.718	0.742	0.762	0.795	0.820	0.863
0.100	0.668	0.684	0.714	0.739	0.763	0.783	0.815	0.840	0.881
0.125	0.693	0.708	0.737	0.761	0.783	0.803	0.835	0.858	0.898
0.150	0.721	0.734	0.761	0.784	0.805	0.823	0.854	0.876	0.913
0.175	0.751	0.763	0.787	0.808	0.828	0.845	0.874	0.895	0.929
0.200	0.784	0.794	0.816	0.835	0.854	0.870	0.896	0.915	0.945
0.225	0.819	0.829	0.848	0.866	0.882	0.896	0.920	0.936	0.963
0.250	0.859	0.867	0.884	0.899	0.914	0.926	0.946	0.961	0.983
0.275	0.902	0.910	0.923	0.937	0.949	0.960	0.976	0.988	1.006
0.300	0.951	0.957	0.968	0.979	0.989	0.998	1.010	1.020	1.033
0.325	1.006	1.011	1.019	1.027	1.035	1.041	1.050	1.056	1.063
0.350	1.066	1.070	1.075	1.081	1.086	1.090	1.094	1.096	1.097
0.375	1.133	1.134	1.137	1.140	1.142	1.143	1.143	1.141	1.135
0.400	1.203	1.203	1.203	1.203	1.203	1.201	1.196	1.190	1.175
0.425	1.276	1.274	1.272	1.269	1.266	1.260	1.251	1.242	1.219
0.450	1.350	1.346	1.341	1.336	1.330	1.321	1.307	1.295	1.264
0.475	1.422	1.417	1.410	1.401	1.393	1.381	1.362	1.348	1.310
0.500	1.498	1.484	1.475	1.463	1.452	1.439	1.415	1.397	1.353
0.525	1.548	1.545	1.533	1.518	1.505	1.490	1.462	1.441	1.392
0.550	1.597	1.594	1.580	1.563	1.549	1.532	1.502	1.477	1.426
0.575	1.632	1.630	1.614	1.595	1.579	1.562	1.530	1.503	1.451
0.600	1.651	1.648	1.631	1.611	1.593	1.576	1.545	1.517	1.465
0.625	1.651	1.646	1.628	1.608	1.588	1.572	1.543	1.516	1.466
0.650	1.630	1.620	1.602	1.585	1.562	1.546	1.523	1.499	1.452
0.675	1.583	1.570	1.552	1.537	1.513	1.499	1.481	1.461	1.419
0.700	1.508	1.495	1.477	1.465	1.442	1.430	1.414	1.400	1.364
0.725	1.405	1.397	1.379	1.368	1.349	1.341	1.324	1.313	1.287
0.750	1.281	1.278	1.260	1.249	1.235	1.230	1.210	1.201	1.187
0.775	1.145	1.142	1.125	1.113	1.103	1.097	1.078	1.068	1.066
0.800	1.003	0.995	0.978	0.964	0.956	0.948	0.932	0.920	0.927
0.825	0.849	0.836	0.819	0.804	0.797	0.787	0.778	0.765	0.775
0.850	0.681	0.666	0.652	0.638	0.633	0.624	0.622	0.612	0.623
0.875	0.511	0.499	0.488	0.478	0.473	0.469	0.473	0.472	0.487
0.900	0.364	0.356	0.348	0.341	0.336	0.335	0.340	0.349	0.371
0.925	0.255	0.248	0.240	0.233	0.230	0.228	0.229	0.243	0.263
0.950	0.179	0.171	0.162	0.155	0.152	0.149	0.144	0.153	0.164
0.975	0.120	0.111	0.102	0.095	0.092	0.089	0.081	0.084	0.087
1.000	0.071	0.063	0.055	0.050	0.048	0.045	0.040	0.040	0.040
1.025	0.034	0.029	0.024	0.021	0.019	0.018	0.016	0.016	0.017
1.050	0.013	0.011	0.008	0.007	0.007	0.006	0.006	0.006	0.006
1.075	0.005	0.004	0.003	0.002	0.002	0.002	0.002	0.002	0.003
1.100	0.002	0.001	0.001	0.001	0.001	0.001	0.001	0.001	0.001
1.125	0.001	0.001	0.000	0.000	0.000	0.000	0.000	0.000	0.001
1.150	0.000	0.000	0.000	0.000	0.000	0.000	0.000	0.000	0.000
1.175	0.000	0.000	0.000	0.000	0.000	0.000	0.000	0.000	0.000
1.200	0.000	0.000	0.000	0.000	0.000	0.000	0.000	0.000	0.000

(Continued)

TABLE 7.4 (continued)

	ELECTRON ENERGY (MeV)								
r/r_o	2.000	3.000	4.000	5.000	6.000	8.000	10.00	15.00	20.00
0.000	0.812	0.839	0.849	0.868	0.867	0.865	0.866	0.856	0.841
0.025	0.842	0.871	0.888	0.897	0.901	0.903	0.899	0.886	0.869
0.050	0.868	0.897	0.916	0.921	0.928	0.932	0.926	0.911	0.893
0.075	0.889	0.918	0.936	0.941	0.949	0.952	0.947	0.931	0.912
0.100	0.906	0.935	0.951	0.957	0.964	0.967	0.962	0.946	0.927
0.125	0.922	0.949	0.962	0.970	0.976	0.976	0.974	0.958	0.938
0.150	0.936	0.962	0.971	0.980	0.984	0.984	0.983	0.967	0.946
0.175	0.949	0.973	0.981	0.989	0.992	0.990	0.989	0.974	0.952
0.200	0.963	0.984	0.991	0.997	0.998	0.995	0.994	0.978	0.955
0.225	0.978	0.995	1.002	1.005	1.004	1.001	0.998	0.981	0.958
0.250	0.995	1.007	1.013	1.012	1.011	1.007	1.002	0.984	0.959
0.275	1.014	1.022	1.025	1.021	1.018	1.013	1.006	0.986	0.961
0.300	1.037	1.038	1.038	1.031	1.028	1.020	1.011	0.989	0.964
0.325	1.063	1.057	1.053	1.043	1.038	1.028	1.016	0.991	0.966
0.350	1.091	1.079	1.069	1.056	1.049	1.037	1.022	0.994	0.969
0.375	1.123	1.103	1.086	1.071	1.062	1.046	1.029	0.997	0.971
0.400	1.158	1.131	1.106	1.088	1.076	1.055	1.036	1.001	0.974
0.425	1.196	1.160	1.127	1.107	1.090	1.064	1.043	1.005	0.977
0.450	1.235	1.192	1.151	1.127	1.106	1.074	1.050	1.009	0.979
0.475	1.275	1.224	1.177	1.148	1.122	1.084	1.058	1.013	0.981
0.500	1.313	1.256	1.205	1.170	1.140	1.097	1.067	1.017	0.982
0.525	1.349	1.288	1.236	1.192	1.160	1.111	1.076	1.020	0.983
0.550	1.380	1.317	1.265	1.214	1.180	1.127	1.086	1.022	0.983
0.575	1.404	1.341	1.291	1.235	1.199	1.141	1.096	1.024	0.982
0.600	1.420	1.359	1.308	1.253	1.217	1.154	1.105	1.026	0.980
0.625	1.425	1.367	1.316	1.266	1.230	1.165	1.114	1.027	0.978
0.650	1.416	1.365	1.313	1.274	1.239	1.173	1.121	1.028	0.974
0.675	1.391	1.348	1.301	1.273	1.240	1.176	1.124	1.028	0.969
0.700	1.345	1.312	1.279	1.258	1.228	1.169	1.121	1.025	0.962
0.725	1.277	1.256	1.244	1.225	1.201	1.152	1.111	1.018	0.952
0.750	1.186	1.177	1.183	1.173	1.157	1.123	1.091	1.005	0.940
0.775	1.073	1.077	1.092	1.099	1.092	1.082	1.060	0.986	0.924
0.800	0.943	0.957	0.978	1.005	1.008	1.026	1.013	0.957	0.902
0.825	0.803	0.821	0.856	0.893	0.905	0.949	0.948	0.916	0.872
0.850	0.657	0.682	0.731	0.768	0.790	0.851	0.863	0.862	0.832
0.875	0.511	0.554	0.610	0.639	0.676	0.730	0.761	0.792	0.781
0.900	0.374	0.440	0.490	0.510	0.565	0.597	0.646	0.710	0.719
0.925	0.257	0.326	0.365	0.386	0.446	0.471	0.530	0.615	0.645
0.950	0.166	0.212	0.239	0.274	0.317	0.364	0.420	0.513	0.562
0.975	0.099	0.118	0.135	0.177	0.199	0.268	0.315	0.407	0.472
1.000	0.051	0.057	0.068	0.098	0.111	0.173	0.213	0.307	0.380
1.025	0.022	0.024	0.032	0.045	0.056	0.093	0.126	0.217	0.292
1.050	0.008	0.010	0.014	0.019	0.027	0.044	0.066	0.140	0.209
1.075	0.003	0.004	0.007	0.009	0.013	0.020	0.033	0.082	0.136
1.100	0.001	0.002	0.003	0.004	0.007	0.010	0.017	0.045	0.083
1.125	0.001	0.001	0.002	0.003	0.004	0.006	0.010	0.025	0.051
1.150	0.000	0.001	0.001	0.002	0.003	0.004	0.007	0.017	0.033
1.175	0.000	0.001	0.001	0.002	0.002	0.004	0.006	0.013	0.023
1.200	0.000	0.001	0.001	0.001	0.002	0.004	0.005	0.011	0.019

Source: Berger (1973), as revised in 1990.

TABLE 7.5   The Factor a_{21} for
Medium 2 Relative to Medium
1 (Water) Calculated Using
Eq. (7.20).

MEDIUM	a_{21}
Air	0.887
Muscle	0.989
Bone	0.966
Carbon	0.866
PTP[a]	0.915
LiF	0.871

[a] Polyethylene terephthalate.

If medium 1 is specified as water, then a_{21} is given by

$$a_{21} = \frac{\rho_1 L_2}{\rho_2 L_1}\left[1 + 0.02252(\langle Z\rangle - 6.60)\right], \tag{7.20}$$

in which L_2 and L_1 are, respectively, the collisional stopping powers for 200 keV electrons in medium 2 and medium 1. $\langle Z\rangle$ is the average atomic number of material 2, namely,

$$\langle Z\rangle = \frac{1}{\langle Z/A\rangle}\sum_j w_j \frac{Z_j^2}{A_j}. \tag{7.21}$$

in which w_j is the mass fraction of element j in medium 2 and $\langle Z/A\rangle$ is given by Eq. (7.11). Representative values of the coefficients a_{21} are given in Table 7.5.

7.6.2 Point Kernels for Beta Particles

The scaled kernels may be used as follows for deriving point kernels for beta particle sources. Suppose that the energy spectrum $N(E)$ (MeV^{-1}) for a source is defined in such a way that $N(E)dE$ is the probability that a source transformation results in emission of a beta particle with energy between E and $E + dE$. An example spectrum is shown in Fig. 7.1. If $\mathcal{G}_\beta(r)$ is defined as the mean absorbed dose (MeV g^{-1}) at distance r from the source, per source transformation, then

$$\mathcal{G}_\beta(r) = \frac{1}{4\pi r^2 \rho}\int_0^{E_{max}} dE \frac{E\,N(E)}{r_o(E)}\mathcal{F}(r/r_o, E), \tag{7.22}$$

in which ρ is the medium density (g cm^{-3}) and E_{max} is the maximum energy in the source spectrum $N(E)$. Examples of $\mathcal{G}_\beta(r)$ for beta particles only are illustrated in Fig. 7.7.

   If the source also emits conversion and Auger electrons with discrete energies E_i and frequencies f_i, then the absorbed dose distribution around the source requires an additional term

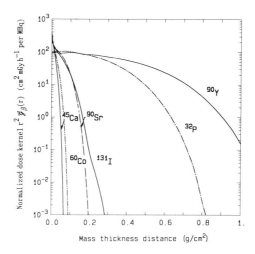

Figure 7.7 Absorbed dose rate in water as a function of distance from point beta particle sources (conversion and Auger electrons are not included). The kernel is multiplied by r^2 and the dose-rate units are converted to mGy h^{-1} per 1 MBq source activity.

$$\mathcal{G}_e(r) = \frac{1}{4\pi r^2 \rho} \sum_i \frac{E_i f_i}{r_o(E_i)} \mathcal{F}(r/r_o, E_i). \tag{7.23}$$

7.7 DOSE EVALUATION IN AN INFINITE MEDIUM

Calculation of the detailed spatial, angular, and energy dependence of a radiation field in realistic geometries is prohibitively difficult. However, there is a common class of dosimetry problems in which the absorbed dose at a point or the dose averaged over some region can be calculated as a simple one-dimensional integral involving only two functions, provided two conditions are satisfied. These conditions are that the source and target regions are part of a medium with uniform interaction coefficients and that the source and target regions are sufficiently far removed from the influence of boundaries that the medium can be modeled as an infinite homogeneous medium (Loevinger 1969; Berger 1971, 1973; NCRP 1991).

7.7.1 Volumetric Sources

For *any* medium and *any* type of radiation, the absorbed dose rate at some target position $\mathbf{r}_t$ due to an isotropic point source of unit activity at location $\mathbf{r}_s$ is given by the point kernel (or Green's function) $\mathcal{G}(\mathbf{r}_t, \mathbf{r}_s)$. This kernel can then be used to calculate the average absorbed dose rate per unit mass (e.g., MeV g^{-1}) in some arbitrary target or receptor volume V_t caused by sources $Q(\mathbf{r}_s)$ (e.g., cm^{-3} s^{-1}) distributed throughout an arbitrary source volume V_s as

$$D(T \leftarrow S) = \frac{1}{m_t} \int_{V_s} dV_s \int_{V_t} dV_t \, Q(\mathbf{r}_s)\rho(\mathbf{r}_t)\mathcal{G}(\mathbf{r}_t, \mathbf{r}_s). \tag{7.24}$$

Here $\rho(\mathbf{r}_t)$ is the density at position $\mathbf{r}_t$ and m_t is the mass of the target volume. The specification of the source and target regions is quite arbitrary; they may be distinct, the same, or even overlap.

If the medium is homogeneous and infinite in extent (or, equivalently, the presence of heterogeneities and boundaries have negligible effect), the point kernel $\mathcal{G}$ depends only on the source-to-detector distance $r = |\mathbf{r}_s - \mathbf{r}_t|$ and the six–dimensional integral of Eq. (7.24) can be reduced to a single one-dimensional integral! Suppose that the origin for the integral over V_t is taken as $\mathbf{r}_s$ and a spherical coordinate system is used (see Fig. 7.8). That part of the volume in a spherical shell of radii r and $r + dr$ that is contained within the target is given by $r^2 dr \Omega_t(r, \mathbf{r}_s)$, where $\Omega_t(r, \mathbf{r}_s)$ is the solid angle subtended at $\mathbf{r}_s$ by all points in V_t that are distance r from $\mathbf{r}_s$. Eq. (7.24) for an infinite homogeneous medium reduces to

$$D(T \leftarrow S) = \frac{1}{V_t} \int_{V_s} dV_s \, Q(\mathbf{r}_s) \int_0^\infty dr \, r^2 \mathcal{G}(r) \Omega_t(r, \mathbf{r}_s), \tag{7.25}$$

where V_t is just m_t/ρ. If the source Q is now assumed to be uniformly distributed in V_s, then changing the order of integration gives the dose in the form of the one-dimensional integral

$$D(T \leftarrow S) = Q V_s \int_0^\infty dr \, \mathcal{G}(r) p(r; T \leftarrow S), \tag{7.26}$$

where the *point-pair distance distribution* $p(r; T \leftarrow S)$ arises formally as

$$p(r; T \leftarrow S) = \frac{1}{V_s V_t} \int_{V_s} dV_s \, r^2 \, \Omega_t(r, \mathbf{r}_s). \tag{7.27}$$

Suppose a point $\mathbf{r}_s$ is selected in the source volume V_s. Of the entire target volume V_t, the fraction at distances between r and $r + dr$ from $\mathbf{r}_s$ is $dr \, r^2 \Omega_t(r, \mathbf{r}_s)/V_t$. Suppose that many points $\mathbf{r}_s$ are selected at random. On average, the probability that a point in the source volume is between distances r and $r + dr$ from a point in the target volume, namely $p(r; T \leftarrow S)dr$, is given by the average of $dr \, r^2 \Omega_t(r, \mathbf{r}_s)/V_t$ over V_s. This average is given by Eq. (7.27). The distribution is independent of which region is identified as the source region and which is the target region. Consequently, p has the symmetry

$$p(r; T \leftarrow S) = p(r; S \leftarrow T). \tag{7.28}$$

A slightly different formulation can be obtained by introducing the *geometry factor*

$$g(r; T \leftarrow S) = \frac{1}{V_s} \int_{V_s} dV_s \frac{\Omega_t(r, \mathbf{r}_s)}{4\pi} = \frac{V_t}{4\pi r^2} p(r; T \leftarrow S), \tag{7.29}$$

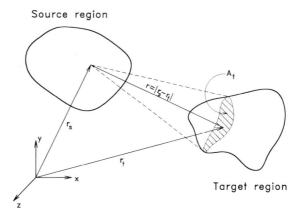

Figure 7.8   Geometry for calculating the dose in a target region T from a uniform volumetric source in region S. The area A_t represents all those points in T that are a distance $r = |\mathbf{r}_s - \mathbf{r}_t|$ from the point $\mathbf{r}_s$ in S. Thus, rays of length r that are emitted isotropically at $\mathbf{r}_s$ within the solid angle $\Omega_t = A_t/r^2$ will terminate in T.

which can be interpreted as the fraction of all rays of length r emitted isotropically in V_s that terminate in V_t. With this geometry factor, the average dose rate in region V_t becomes

$$D(T \leftarrow S) = 4\pi Q \frac{V_s}{V_t} \int_0^\infty dr \; r^2 \mathcal{G}(r) g(r; T \leftarrow S). \tag{7.30}$$

Because of the symmetry of $p(r; T \leftarrow S)$, the geometry factor has the following property for interchanging the source and target regions

$$V_s \, g(r; T \leftarrow S) = V_t \, g(r; S \leftarrow T). \tag{7.31}$$

With this result, the average absorbed dose rate can be expressed alternatively as

$$D(T \leftarrow S) = 4\pi Q \int_0^\infty dr \; r^2 \mathcal{G}(r) g(r; S \leftarrow T). \tag{7.32}$$

This result holds even if the target volume collapses to a point, whereas Eq. (7.30) requires that $V_t > 0$.

The integration in Eqs. (7.30) or (7.32) can usually be performed over a finite range of r because the geometry factor $g(r; T \leftarrow S)$ is nonzero only over a finite range (at least for regions S and T of finite size).

7.7.2 Absorbed Fraction and Reduction Factor

In many dosimetry problems involving point or semi-infinite volumes for the source and target regions, it is more useful to calculate a normalized dose rate which removes the effect of the source strength. Two such normalized dose rates are the *absorbed fraction* and the *reduction factor*.

Absorbed Fraction. The *absorbed fraction* $\phi(T \leftarrow S)$ is defined as the fraction of the energy released from the source that is absorbed in the target, that is,

$$\phi(T \leftarrow S) = \frac{\rho V_t \, D(T \leftarrow S)}{\langle E \rangle \, Q V_s} \tag{7.33}$$

where $\langle E \rangle$ is the average radiation energy emitted per source decay and $Q_{tot} = Q V_s$ is the total activity in the source region. From Eqs. (7.30) and (7.32) the absorbed fraction can be computed as

$$\phi(T \leftarrow S) = \begin{cases} \dfrac{4\pi\rho}{\langle E \rangle} \displaystyle\int_0^\infty dr \, r^2 \mathcal{G}(r) g(r; T \leftarrow S) & \text{(a)} \\[2ex] \dfrac{4\pi\rho}{\langle E \rangle} \dfrac{V_t}{V_s} \displaystyle\int_0^\infty dr \, r^2 \mathcal{G}(r) g(r; S \leftarrow T) & \text{(b)} \\[2ex] \dfrac{\rho V_t}{\langle E \rangle} \displaystyle\int_0^\infty dr \, \mathcal{G}(r) p(r; T \leftarrow S) & \text{(c)} \end{cases} \tag{7.34}$$

The *specific absorbed fraction* $\Phi(T \leftarrow S)$ is equal to $\phi(T \leftarrow S)$ divided by the target mass.

Absorbed fractions for monoenergetic gamma-ray sources and targets among the various organs and tissues of the body have been computed using Monte Carlo techniques and reported by Snyder, Ford, and Warner (1978) and in ICRP Report 23 (1975), and are tabulated in Appendix G. Extensive tables of age and gender-specific absorbed fractions have been computed by Cristy and Eckerman (1987) who also report electron absorbed fractions for bone marrow and trabecular bone as source organs and bone marrow as the target organ.

Reduction Factor. The *reduction factor* $\varphi(T \leftarrow S)$ (Berger 1970) is defined as the ratio of the absorbed dose rate in the target region to the rate at which energy is released per unit mass in the source region, the latter being also the absorbed dose rate in an infinite medium with the same volumetric source strength as is present in the source volume. In terms of the absorbed fraction and the average absorbed dose rate, the reduction factor is

$$\varphi(T \leftarrow S) = \frac{\rho \, V_s}{\langle E \rangle \, Q_{tot}} D(T \leftarrow S) = \frac{V_s}{V_t} \phi(T \leftarrow S). \tag{7.35}$$

When the source and target are the same, the absorbed fraction and the reduction factor are identical. From this definition and the results of Eqs. (7.31) and (7.34), the reduction factor can be computed from

$$\varphi(T \leftarrow S) = \begin{cases} \dfrac{4\pi\rho}{\langle E \rangle} \displaystyle\int_0^\infty dr \, r^2 \mathcal{G}(r) g(r; S \leftarrow T) & \text{(a)} \\[2ex] \dfrac{4\pi\rho}{\langle E \rangle} \dfrac{V_s}{V_t} \displaystyle\int_0^\infty dr \, r^2 \mathcal{G}(r) g(r; T \leftarrow S) & \text{(b)} \\[2ex] \dfrac{\rho V_s}{\langle E \rangle} \displaystyle\int_0^\infty dr \, \mathcal{G}(r) p(r; T \leftarrow S) & \text{(c).} \end{cases} \tag{7.36}$$

TABLE 7.6   Quantities that Have a Non-Zero and Finite Value as Either the Source or Target Region Volume Collapses to Infinite or Zero Size. Values that Are Finite and Non-Zero Are Denoted as "Finite."

QUANTITY	$V_s \to \infty$	$V_t \to \infty$	$V_s \to 0^a$	$V_t \to 0^b$
$p(r; T \leftarrow S)$	0	0	finite	finite
$p(r; S \leftarrow T)$	0	0	finite	finite
$g(r; T \leftarrow S)$	0	finite	finite	0
$g(r; S \leftarrow T)$	finite	0	0	finite
$D(T \leftarrow S)$	finite	0	finite	finite
$\phi(T \leftarrow S)$	0	finite	0	finite
$\varphi(T \leftarrow S)$	finite	0	finite	∞

[a] V_s collapses to a point source with $V_s Q = Q_{tot}$ kept constant.

[b] V_t collapses to a point.

7.7.3 Limiting Source or Target Volumes

So far it has been assumed that the source and target regions are finite in size. However, in many practical dosimetry problems the absorbed dose rate in the vicinity of a point ($V_t \to 0$) is of interest. In the other extreme, the target volume may be effectively infinite (e.g., a slab or cylindrical region). Similarly, the source region may be represented as a point or as a semi-infinite region. For these limiting cases, not all the above measures of the absorbed dose rate or geometric factors yield finite or nonzero values. In Table 7.6, those dosimetric quantities and distributions that yield finite and nonzero values are denoted as "finite." For the indicated limits of each region in this table, the volume of the other region is assumed finite and nonzero.

There is one special case of practical importance in which both the source and target volume become infinite, namely, the case of infinitely long line, cylindrical, or cylindrical shell sources with a parallel (usually co-axial) cylindrical target region. In this case, the absorbed fraction is defined on a per unit axial-length basis. Examples of this special case are presented in Sec. 7.8.6.

7.7.4 Reciprocity Theorem

The symmetry property of Eq. (7.28) for the point-pair distance distribution yields an important relation between the source and target volumes. If the total source activity $Q_{tot} = QV_s$ is kept constant, then it is seen from Eq. (7.26) and the symmetry of $p(r; T \leftarrow S)$ that the average dose rate in T from Q_{tot} distributed uniformly in S is the same as the average dose rate that would occur in S if Q_{tot} were uniformly distributed throughout T. In other words $D(S \leftarrow T) = D(T \leftarrow S)$ when the total source activities are the same.

This reciprocity result also holds even if one of the regions becomes a point. Thus, the mean dose rate in T from a point source S with activity Q_{tot} is the same as the absorbed dose rate at S if the same activity were uniformly distributed throughout T.

If the specific activity Q, rather that the total activity, is kept constant when the source and target regions are reversed, then the total energy absorption rates $V_t D(T \leftarrow S)$ and $V_s D(S \leftarrow T)$ are seen from Eq. (7.26) to be equal.

The above results are a consequence of the reciprocity theorem for radiation transport (Bell and Glasstone 1970) and apply rigorously whenever the point kernel for the absorbed dose rate has the property $\mathcal{G}(\mathbf{r}_s, \mathbf{r}_t) = \mathcal{G}(\mathbf{r}_t, \mathbf{r}_s)$. For most media, this symmetry in the point kernel generally does not apply for scattered radiation. However, it is strictly satisfied in an infinite homogeneous medium or for the unscattered radiation in a heterogeneous medium in which the interaction coefficients are proportional to the local density.

7.7.5 Extension to Nonuniform and Surface Sources

The previous analysis is based on uniform volumetric or point sources. However, the absorbed dose rates arising from other types of sources can also be expressed as one-dimensional integrals involving appropriate kernels and geometry factors. In this section, three extensions are suggested.

Nonuniform Volumetric Sources.   The preceding treatment for a source volume with a uniformly distributed volumetric source can be generalized to allow for a variation of the source strength in the source volume. For such a generalization, the point-pair distance distribution, Eq. (7.27), and the geometric factor, Eq. (7.29), must include a weight function $Q(\mathbf{r}_s)/Q_{tot}$ in the integral where Q_{tot} is the total activity in the source region. The average absorbed dose rate in the target volume $D(T \leftarrow S)$ is then given by Eqs. (7.26) or (7.30) with QV_s replaced by Q_{tot}. The absorbed fraction and reduction factor of Sec. 7.7.2 remain unchanged. However, for nonuniform activity distributions, the symmetry in the point-pair distance distribution and the reciprocity results no longer apply. Consequently, formulas for the absorbed fraction or reduction factor that interchange the source and target regions such as Eqs. (7.34b) and (7.36a) are no longer valid.

Surface Sources.   There is one nonuniform source distribution of special importance. In several dosimetric models, the activity can be considered as being spread uniformly over the surface of the source region (e.g., on a cell membrane or on a bone surface). The use of geometry factors to calculate the absorbed dose in another target region can be extended to this case of a surface source region.

Let the area of the source region S be denoted by A_s and the uniform activity per unit surface area by Q_a (e.g., $cm^{-2} s^{-1}$). The average absorbed dose rate in the target region T is given by Eq. (7.25) with the integral over the source volume replaced by an integral over all points $\mathbf{r}_s$ on the surface of S. The point-pair distance distribution then becomes

$$p_a(r; T \leftarrow S) = \frac{1}{A_s V_t} \int_{A_s} dA_s \, r^2 \, \Omega_t(r, \mathbf{r}_s), \tag{7.37}$$

where $p_a(r; T \leftarrow S) dr$ is the probability that a point randomly picked on the surface of S is within distances r and $r + dr$ from a point randomly selected in T. Similarly, the geometry factor for a surface source becomes

$$g_a(r; T \leftarrow S) = \frac{1}{A_s} \int_{A_s} dA_s \frac{\Omega_t(r, \mathbf{r}_s)}{4\pi} = \frac{V_t}{4\pi r^2} p_a(r; T \leftarrow S) \qquad (7.38)$$

and can be interpreted as the probability a ray of length r that is emitted isotropically from a random point on the surface of S will end in region T.

With these surface source distributions, the absorbed dose rate in T is then given by Eqs. (7.26) or (7.30) with QV_s replaced by the total surface activity $Q_{tot} = Q_a A_s$. While the reduction factor is not defined for a surface source, the absorbed fraction is and is given by Eq. (7.33) or (7.34a). For surface sources, the point-pair distance distribution is no longer symmetric with respect to interchange of S and T, and the relations of Eq. (7.32) and (7.34b) cannot be used.

Finally, the case of surface sources can be further generalized to allow sources of varying strength over the source surface. In this case, evaluation of the geometry factor from Eq. (7.38) must include a weighting factor proportional to the source strength in the integral.

Infinite Cylindrical Sources.   For many dosimetry problems, the source can be modeled by a line or cylindrical volume source of effectively infinite length. The dose in some infinite cylindrical region parallel to the source axis or at some point can again be expressed as a one-dimensional integral involving a scaled line-source kernel and a geometry factor. Explicit treatment of this extension is deferred to Sec. 7.8.6 in which the electron line-source kernel and several geometry factors are derived.

7.7.6 Calculation of Geometric Factors

For source and target volumes with simple geometric shapes, the geometry factor $g(r; T \leftarrow S)$ or, equivalently, the point-pair distance distribution, can often be calculated analytically. In the next section, geometric factors are derived for several simple cases. The determination of these distributions is a standard problem in geometric probability theory (Weil 1983), and many results can be found in the literature (NCRP 1991; Kellerer 1981, 1984; Mäder 1980). Considerable attention has been given to the autologous case (S and T the same region) for which the chord length distribution and the geometry factor are closely related in a convex region (Kellerer 1971).

Many dosimetric problems can be modeled by a few regions with simple shapes imbedded in an infinite medium. For example, a model might consist of several concentric spherical regions for which the absorbed dose in one spherical shell arising from a source in another is sought. In this case, some simple properties of the geometry factor can be used to express the needed geometry factor in terms of geometry factors of a simple problem (e.g., that for a spherical source to a concentric shell). From physical arguments, if A, B, and C represent any three regions (with volumes V_a, V_b, and V_c) in an infinite homogeneous medium, one has the multiple target property

$$g(r; B + C \leftarrow A) = g(r; B \leftarrow A) + g(r; C \leftarrow A). \qquad (7.39)$$

From the reciprocity relation for the geometry factor, that is,

$$V_a\, g(r; B \leftarrow A) = V_b\, g(r; A \leftarrow B), \tag{7.40}$$

and the multiple target property, one can derive the following multiple source property:

$$(V_a + V_b)g(r; C \leftarrow A + B) = V_a\, g(r; C \leftarrow A) + V_b\, g(r; C \leftarrow B). \tag{7.41}$$

The multiple source and target properties can be generalized to any number of regions. In particular, if there are N distinct contiguous regions R_i one has

$$\sum_{j=1}^{N} g(r; R_j \leftarrow R_i) = 1. \tag{7.42}$$

These relations can often be used to obtain analytical results for a multiregion problem in terms of results for simpler problems and thereby to avoid a direct, and often complex, analytic evaluation in the multiregion geometry.

For regions with geometric shapes that are sufficiently complex to preclude easy analytical evaluation of the multiple integrals in the definition of the geometry factor, Monte Carlo techniques can be used to determine the point-pair distance distribution, from which the geometry factor is readily calculated from Eq. (7.29).

7.7.7 Effect of Density Variations

The methods for dose evaluation discussed in this section rigorously apply only when the target region and source regions have the same homogeneous composition and density as the infinite medium in which they are embedded. Of practical importance is the situation in which a single region has a density different from that of the surrounding medium. For example, bone and lung have different densities from that of the surrounding soft tissue. Nevertheless, the previously discussed methods for an infinite uniformly homogeneous medium can be adapted to account for density variations in subregions provided the interaction properties of the different regions are similar, as is often the case in many internal dosimetry problems. For example, soft tissue, bone and air have very similar microscopic cross sections or mass interaction coefficients for charged particles and for photons (except at low energies < 100 keV) because of the similar average atomic numbers of the elemental constituents.

There are some very general theorems for media with density variations that can be derived from the transport equation governing the penetration of radiation through matter. Three of these results are presented here without proof.

Fano's Theorem.   One of the most basic theorems involving density variations is known as Fano's theorem (Fano 1954; Spencer et al. 1980). This theorem, which has important implications in the design of instrumentation for dosimetry, can be stated as follows. In an infinite medium with arbitrary density variations, the radiation field (and, hence, the absorbed dose rate) will be everywhere the same provided (1) the atomic interaction properties are constant (i.e., the mass interaction coefficients are everywhere the same), (2) the energy spectrum and angular distribution of the sources are independent of position, and (3) the volumetric source strengths are everywhere proportional to the local density.

Theorem on Plane-Density Variations.   For the special case in which material properties are everywhere the same except for plane-density variations (i.e., the density $\rho(x)$ varies only with one spatial coordinate x) and in which the strength of the radiation sources also varies only with the distance x, the radiation field at any depth x equals that at a corresponding depth x' in a medium of any *constant* density ρ' with the same mass interaction coefficients. Here corresponding distances x and x' are at the same mass thickness depth, that is, $x'\rho' = \int_0^x \rho(x)dx$, from corresponding reference planes (here taken as $x = x' = 0$). The source strength (per unit mass) in the constant density medium must be the same as that in the variable density medium at corresponding points. This plane-density theorem also holds for semi-infinite media if the boundary conditions of the variable and constant density media are the same.

Figure 7.9 illustrates the attenuation of a beam of 1-MeV electrons normally incident on the plane surface of a half space of water. What is shown is the absorbed dose as a function of the depth of penetration x into the water measured in units of mass thickness ρx. In one case, a water layer between depths of 0.218 and 0.262 g cm^{-2} is replaced by a bone layer of the same mass thickness but of higher density. According to the plane-density variation theorem, if bone and water had the same (not just similar) mass interaction coefficients, then the absorbed dose distribution for both cases should be identical when plotted as a function of mass thickness. As can be seen, the bone produces only a minor distortion of the absorbed-dose distribution, namely a slight enhancement in reflection from the bone layer. If the bone layer were replaced by a metallic layer of the same mass thickness (but whose mass interaction coefficients are quite different from those of water), the distortion would be much more pronounced.

Scaling Theorem.   A very useful theorem for media with density variations is known as the "scaling theorem" (Spencer et al. 1980) which applies even to a finite heterogeneous medium. Consider two media A and B one of which is a scaled version of the other, that is, the coordinates $\mathbf{r}_a$ and $\mathbf{r}_b$ of corresponding points in A and B, respectively, are related by $\mathbf{r}_b = \xi\,\mathbf{r}_a$ where ξ is a constant scaling parameter. The two media are said to be *similar* if (1) all material interfaces occur at corresponding points, (2) the material at corresponding points is the same except that the density at a point in A is ξ times that at the corresponding point in B, and (3) radiation sources at corresponding points are identical in energy and angular distributions and have the same strength per unit mass (the strength per unit volume at a point in B is ξ^2 times that at the corresponding point in A). Corresponding subregions in similar media have identical shapes but different volumes and masses. The scaling theorem then states that if two media are similar then the radiation fields at corresponding points are identical.

As an application of this scaling theorem, consider an important dosimetry problem in which the density ρ' of a single region with volume V' is different from the density ρ of the surrounding infinite medium (for example, a bone in a large mass of soft tissue). The volume V', which may have an arbitrary distribution of sources, is also

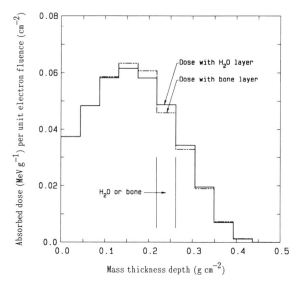

Figure 7.9   Absorbed dose versus depth for 1-MeV electrons normally incident on a water medium. Also shown is the dose profile for the case in which a layer of water is replaced by a layer of bone of the same mass thickness.

the target region of interest. It is supposed that the mass interaction coefficients are the same for both V' and for the remainder of the medium. Then from the above arguments the volume V' can be scaled to an identically shaped volume $V = V'(\rho'/\rho)^3$ with density ρ so that the absorbed fraction ϕ' for volume V' is approximately the same as the absorbed fraction ϕ in a volume V (ICRU 1979). Because the masses contained in the two volumes will be different, the specific absorbed fractions for the two problems would scale as $\Phi = (\rho'/\rho)^2\Phi'$.

If in the above scaling of V' into V the density of the surrounding medium had also been scaled, the equivalence of the absorbed fractions would be exact. However, by keeping the density of the medium outside V at ρ (so as to obtain a completely homogeneous medium), one has neglected the slightly different amounts of energy that would escape from and re-enter the two regions V and V' either from backscatter or because the surfaces wrap around on themselves (i.e., there are areas of local concavity).

Point Kernels for Media with Density Variations.   For infinite media problems in which there are material discontinuities between (or in) the source and target regions, the infinite-medium point-kernel techniques can still be used to good approximation provided all materials have nearly the same electron range measured in terms of mass thickness ρr_o. Such would be the case for constituents of the human body. The medium in and around the regions of interest can thus be homogenized (conceptually), and a point kernel used for the homogenized medium. Alternatively, the point-kernel can be written in terms of the mass-thickness distance z from the source point, for example,

$$\mathcal{G}(r, E) = \frac{E}{4\pi r^2 \rho r_o} \mathcal{F}(z/(\rho r_o), E), \tag{7.43}$$

where ρr_o is an appropriate electron CSDA mass-thickness range (which is assumed nearly equal for all materials involved in the problem). Other methods have also been used to account for discontinuities in material composition in some geometries. An important example is found in the determination of the dose to the skin arising from radioactive gases in the atmosphere (Berger 1974).

7.8 APPLICATIONS OF POINT KERNELS TO BETA DOSIMETRY

In many electron and beta-particle dosimetry problems, the source and target regions can be represented by simple geometric shapes. Further, if the regions are part of an infinite homogeneous medium, the methods of the previous section can be used to find analytical expressions for the absorbed fractions or reduction factors. In this section several such examples are given for electron sources. An important result in electron dosimetry will be seen from these examples: the absorbed fraction or reduction factor, besides being independent of the source strength, is very insensitive to the initial energy of the electron and depends primarily on the problem geometry.

To facilitate the example calculations, the formulas for the absorbed dose rate, absorbed fraction, and reduction factor caused by a monoenergetic electron source of constant volumetric strength Q (e.g., electrons $cm^{-3}\ s^{-1}$) are summarized in terms of the scaled electron point kernel. From Eq. (7.18) and the results of Sec. 7.7.1 and 7.7.2 one has

$$D(T \leftarrow S, E) = \begin{cases} \dfrac{QEV_s}{\rho r_o V_t} \displaystyle\int_0^\infty dr\, \mathcal{F}(r/r_o, E)g(r; T \leftarrow S) \\[2ex] \dfrac{QE}{\rho r_o} \displaystyle\int_0^\infty dr\, \mathcal{F}(r/r_o, E)g(r; S \leftarrow T), \end{cases} \tag{7.44}$$

$$\phi(T \leftarrow S, E) = \begin{cases} \dfrac{1}{r_o} \displaystyle\int_0^\infty dr\, \mathcal{F}(r/r_o, E)g(r; T \leftarrow S) \\[2ex] \dfrac{1}{r_o}\dfrac{V_t}{V_s} \displaystyle\int_0^\infty dr\, \mathcal{F}(r/r_o, E)g(r; S \leftarrow T), \end{cases} \tag{7.45}$$

$$\varphi(T \leftarrow S, E) = \begin{cases} \dfrac{1}{r_o}\dfrac{V_s}{V_t} \displaystyle\int_0^\infty dr\, \mathcal{F}(r/r_o, E)g(r; T \leftarrow S) \\[2ex] \dfrac{1}{r_o} \displaystyle\int_0^\infty dr\, \mathcal{F}(r/r_o, E)g(r; S \leftarrow T). \end{cases} \tag{7.46}$$

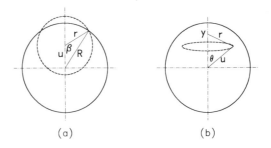

Figure 7.10 Geometry for evaluation of geometry factors from a spherical volume source of electrons (solid circle). (a) Rays of length r emitted a distance u from the source center terminate on the spherical surface represented by the dashed circle. (b) Sources a distance r from an interior dose point at y and also a distance u from the source center are on the dashed circle.

7.8.1 Spherical Volume Source

Average Dose Within a Spherical Volume Source. This example is representative of evaluation of the absorbed dose within a cell, or within the nucleus of the cell, when there is a uniform source of beta particles in the volume of interest. Consider a sphere of radius R in a uniform medium within which is a uniform source emitting Q electrons of energy E (MeV) per cm³ per second. To calculate the mean absorbed dose rate $\langle D(R, E) \rangle$ within the sphere, first find the geometry factor when both the source and target regions are the sphere.

Rays of length r, originating at radial distance u within the sphere, end on a spherical surface illustrated by the dashed sphere in Fig. 7.10(a). The fraction of rays *ending* within the radius-R sphere, that is, $\Omega_t / 4\pi$ in Eq. (7.29), is unity for $r \leq R - u$, and is otherwise just the solid angle fraction subtended by half conical angle β, that is, $(1 - \cos\beta)/2$. By making use of the law of cosines and by averaging over the volume of the sphere as prescribed by Eq. (7.29), one arrives at the geometry factor $g(r; T \leftarrow S)$, here called $g_1(r, R)$, the fraction of all rays of length r originating within the sphere and ending within the sphere, namely

$$g_1(r, R) = \frac{3}{R^3}\left[\int_0^{R-r} du\, u^2 + \frac{1}{2}\int_{R-r}^R du\, u^2\left(1 - \frac{u^2 + r^2 - R^2}{2ur}\right)\right]$$

$$= 1 - \frac{3r}{4R} + \frac{r^3}{16R^3}, \qquad 0 \leq r \leq 2R. \tag{7.47}$$

Because the source and target volumes are the same, the absorbed fraction equals the reduction factor. Thus, from Eqs. (7.45) and (7.46)

$$\phi(R, E) = \varphi(R, E) = \frac{1}{r_o}\int_0^{2R} dr\, g_1(r, R)\mathcal{F}(r/r_o, E). \tag{7.48}$$

This result can be evaluated numerically. Results are shown in Fig. 7.11 as a function of the radius R. Note how insensitive these results are to the initial electron energy.

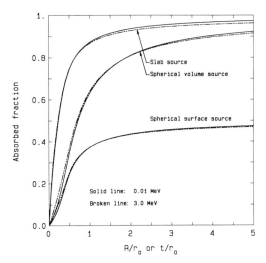

Figure 7.11   Internal absorbed fractions in spherical or slab sources within an infinite water medium. Both volumetric sources contain isotropic, monoenergetic electron sources uniformly distributed within the volume. The spherical surface source has an isotropic, monoenergetic electron source uniformly distributed over its surface. The lines for 0.01 and 3-MeV electrons encompass absorbed fractions for all intermediate energies.

Berger (1973) gives an extended table of absorbed fractions for a wide range of electron energies. Kellerer (1984) extends this analysis to a spheroidal volume.

The *average* absorbed dose rate within the sphere, per unit source strength, is from Eq. (7.35)

$$\langle D(R, E)\rangle = \frac{QE}{\rho}\varphi(R, E),\tag{7.49}$$

in which QE/ρ is the absorbed dose rate which would occur in a homogeneous unbounded medium with the same unit-strength volumetric source.

For a beta-particle source with activity Q per unit volume in the sphere, integration over the source energy spectrum leads to

$$\langle D_\beta(R)\rangle = Q \int_0^{E_{max}} dE\ N(E)\frac{E}{\rho}\varphi(R, E)$$

$$= 4\pi Q \int_0^{2R} dr\ r^2 g_1(r, R)\mathcal{G}_\beta(r).\tag{7.50}$$

Average Dose in a Spherical Shell Surrounding a Spherical Volume Source.
Suppose a sphere of radius $P \geq R$ is concentric with the spherical volume source of radius R. The geometry factor $g(r; T \leftarrow S)$, here called $g_2(r, R, P)$ giving the distribution of rays from the inner source volume to anywhere in the larger sphere can be derived in a manner similar to that used in the derivation of Eq. (7.47). It can

be shown that the fraction of rays of length r originating in the source sphere ending anywhere within the larger sphere of radius P is given, for $P - R \leq r \leq P + R$, by

$$g_2(r, R, P) = \frac{3}{R^3}\left[\frac{R^3 + P^3}{6} - \frac{r(R^2 + P^2)}{8} + \frac{r^3}{48} - \frac{(P^2 - R^2)^2}{16r}\right]. \quad (7.51)$$

For $r \geq P + R$, $g_2 = 0$, and, for $r \leq P - R$, $g_2 = 1$.

The absorbed fraction in the spherical shell between the source sphere and the outer sphere is simply the difference between the absorbed fractions in the source sphere and the outer sphere. Thus, for a monoenergetic electron source (energy E) of constant activity per unit volume in the source sphere, the absorbed fraction in the spherical shell is, from Eq. (7.45),

$$\phi(R, P, E) = \frac{1}{r_o}\int_0^{R+P} dr\,[g_2(r, R, P) - g_1(r, R)]\,\mathcal{F}(r/r_o, E). \quad (7.52)$$

The reduction factor for the shell is then obtained by multiplying this absorbed fraction by the ratio of the source to shell volumes [see Eq. (7.35)] to give

$$\varphi(R, P, E) = \frac{1}{(P/R)^3 - 1}\phi(R, P, E). \quad (7.53)$$

Reduction factors for the shell are illustrated in Fig. 7.12. With superposition techniques, it is an easy matter to obtain absorbed fractions or reduction factors for spherical-shell target regions concentric with, but not contiguous to, source spheres. Problem 7.12 at the end of the chapter extends this analysis to nonconcentric spheres of different radii.

Interior and Exterior Local Doses for the Spherical Volume Source.   Suppose, for the same geometry and source strength, as in Fig. 7.10(b), the radial variation of the absorbed dose rate, both inside and outside the spherical source region is wanted. In such a situation the dose rate is evaluated at a point that is a distance y from the center of the spherical source region. Consider an element of volume $dV = 2\pi u^2 du\,\sin\theta\,d\theta$, in the form of a body of rotation at radius u and polar angle θ, all points of which are distance r from a dose location at radius y, and which contributes $QG(r, E)dV$ to the absorbed dose rate at radius y. Integration over the volume of the source and use of Eq. (7.18) yields

$$D(y, R, E) = \frac{QE}{2\rho r_o}\int_0^R du\,u^2\int_0^\pi d\theta\,\sin\theta\,r^{-2}\,\mathcal{F}(r/r_o, E), \quad (7.54)$$

where

$$\cos\theta = \frac{y^2 + u^2 - r^2}{2uy}. \quad (7.55)$$

A change in the variable of integration from θ to r leads to the following result:

$$D(y, R, E) = \frac{QE}{2y\rho r_o}\int_0^R du\,u\int_{|y-u|}^{y+u} dr\,r^{-1}\,\mathcal{F}(r/r_o, E). \quad (7.56)$$

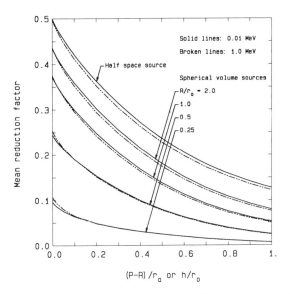

Figure 7.12 Mean reduction factors for (1) source-free spherical shells concentric with spherical volume sources and (2) source free layers bounded by half-space sources, all within infinite water media. The sources are uniform isotropic sources of monoenergetic electrons. The lines for 0.01 and 1-MeV electrons encompass absorbed fractions for all intermediate energies.

Finally, a change in the order of integration and evaluation of the inner integral leads to

$$D(y, R, E) = \frac{QE}{\rho r_o} \int_{r_{min}}^{y+R} \frac{dr}{r_o} g_3(y, r, R) \mathcal{F}(r/r_o, E), \qquad (7.57)$$

where $r_{min} = 0$ if $y \leq R$ and $y - R$ if $y \geq R$. The function $g_3(y, r, R)$ for $y \leq R$ is given by

$$g_3(y, r, R) = 1, \qquad y = 0, \qquad (7.58)$$

$$g_3(y, r, R) = 1, \qquad 0 \leq r \leq R - y, \qquad (7.59)$$

$$g_3(y, r, R) = \frac{R^2 - (r - y)^2}{4yr}, \qquad R - y \leq r \leq R + y, \qquad (7.60)$$

and, for $y \geq R$,

$$g_3(y, r, R) = \frac{R^2 - (y - r)^2}{4yr}, \qquad y - R \leq r \leq y + R. \qquad (7.61)$$

By comparison of Eq. (7.57) to Eq. (7.44b), it is seen that g_3 must be the geometry factor for rays *from* the target point y which end inside the source sphere, that is, $g(r; S \leftarrow T)$.

The *local* reduction factor (i.e., the reduction factor at a point a distance y from the source center) is thus

$$\varphi(y, R, E) = \frac{\rho}{QE} D(y, R, E). \tag{7.62}$$

Local reduction factors $\varphi(y, R, E)$ are illustrated in Figs. 7.13 and 7.14.

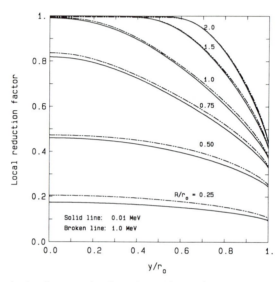

Figure 7.13 Local reduction factors at interior points within uniform isotropic monoenergetic electron sources in spheres of water completely surrounded by the same medium. The lines for 0.01 and 1-MeV electrons encompass absorbed fractions for all intermediate energies.

For a beta particle source, integration over the source energy spectrum leads to

$$
\begin{aligned}
D_\beta(y, R) &= Q \int_0^{E_{max}} dE\, N(E) \frac{E}{\rho} \varphi(y, R, E) \\
&= 4\pi Q \int_{r_{min}}^{R+y} dr\, r^2 g_3(y, r, R) \mathcal{G}_\beta(r).
\end{aligned} \tag{7.63}
$$

Local Dose in a Spherical Volume Surrounded by an Unbounded Source. This case is also illustrated by Fig. 7.10(b) except that the source region is now defined by $u \geq R$. Equation (7.54) is modified as follows:

$$D(y, R, E) = \frac{QE}{2\rho r_o} \int_R^\infty du\, u^2 \int_0^\pi d\theta\, \sin\theta\, r^{-2} \mathcal{F}(r/r_o, E), \tag{7.64}$$

where, again,

$$\cos\theta = \frac{y^2 + u^2 - r^2}{2uy}. \tag{7.65}$$

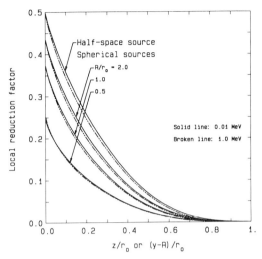

Figure 7.14   Local reduction factors at points exterior to uniform isotropic monoenergetic electron sources in (1) spheres and (2) half spaces of water completely surrounded by the same medium. The lines for 0.01 and 1-MeV electrons encompass absorbed fractions for all intermediate energies.

A change in the variable of integration from θ to r leads to the following result:

$$D(y, R, E) = \frac{QE}{2y\rho r_o} \int_R^\infty du\, u \int_{u-y}^{u+y} dr\, r^{-1}\, \mathcal{F}(r/r_o, E). \tag{7.66}$$

A change in the order of integration leads to

$$D(y, R, E) = \frac{QE}{\rho r_o} \int_{R-y}^\infty \frac{dr}{r_o} g_4(y, r, R)\mathcal{F}(r/r_o, E), \tag{7.67}$$

where, from comparison of this result to Eq. (7.44), $g(r; S \leftarrow T)$, here called $g_4(y, r, R)$, is seen to be the geometry factor for rays from the point y in the sphere to the source region outside, that is, from target to source. This geometry factor is given by

$$g_4(y, r, R) = 1, \qquad r \geq R + y, \tag{7.68}$$

or

$$g_4(y, r, R) = \frac{(y + r)^2 - R^2}{4yr}, \qquad r \leq R + y. \tag{7.69}$$

Finally, the *local* reduction factor $\varphi(y, R, E)$ is related to the dose rate $D(y, R, E)$ by Eq. (7.62). Reduction factors $\varphi(y, R, E)$ for this geometry are illustrated in Fig. 7.15.

7.8.2 Plane Isotropic Source

This example is representative of the evaluation of the skin dose arising from a distributed source on the skin surface. Consider an area source emitting Q_a elec-

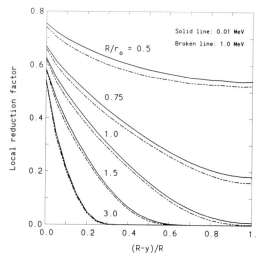

Figure 7.15   Local reduction factors in spherical nonsource water regions surrounded by unbounded, uniform, isotropic and monoenergetic electron sources in water. The lines for 0.01 and 1-MeV electrons encompass absorbed fractions for all intermediate energies.

trons of energy E (MeV) per cm^2 per second. Let $D(z, E)$ be the absorbed dose rate (MeV g^{-1} s^{-1}) at distance z (cm) from the surface into a medium of uniform density ρ (g cm^{-3}). It can be shown that

$$D(z, E) = \frac{EQ_a}{2\rho r_o(E)} \int_z^\infty dr\, r^{-1}\, \mathcal{F}(r/r_o, E). \tag{7.70}$$

If the area source has a surface activity Q_a (e.g., cm^{-2}s^{-1}) of a radionuclide with beta-particle energy spectrum $N(E)$, then evaluation of the absorbed dose rate requires integration over energy, viz.,

$$D_\beta(z) = \frac{Q_a}{2\rho} \int_0^{E_{max}} dE\, \frac{E\, N(E)}{r_o(E)} \int_z^\infty dr\, r^{-1}\mathcal{F}(r/r_o, E)$$

$$= 2\pi Q_a \int_z^\infty dr\, r\mathcal{G}_\beta(r). \tag{7.71}$$

Now suppose the source is a plane isotropic source of the same energy and strength over a disk of radius y and the dose rate is to be calculated at a point distance $z > 0$ along a normal to the center of the disk. Eq. (7.70) evolves to

$$D(z, E) = \frac{EQ_a}{2\rho r_o(E)} \int_z^{\sqrt{z^2+y^2}} dr\, r^{-1}\, \mathcal{F}(r/r_o, E). \tag{7.72}$$

The integration over a beta-particle energy spectrum and the extension to an annular source is straightforward.

7.8.3 Slab Source

Average Dose Within the Slab Source.   Consider a slab source of half-thickness t (cm) embedded within a uniform homogeneous medium and emitting Q electrons of energy E (MeV) per cm³ per second. This is representative of a radionuclide source selectively taken up by some organ or tissue of the body. First evaluate the mean absorbed dose rate within the slab. Following a procedure very similar to that applied in the case of the spherical volume source, one can show that, for rays of length r originating isotropically and uniformly throughout the volume of the slab, the probability that the rays end within the slab is given by

$$g(r,t) = \frac{t}{r} \qquad r \geq 2t, \tag{7.73}$$

or

$$g(r,t) = 1 - \frac{r}{4t} \qquad r \leq 2t. \tag{7.74}$$

It follows from Eq. (7.33), with $V_t/Q_{tot} = V_t/V_s Q = 1/Q$, that the slab-averaged absorbed dose rate is given by

$$\langle D(t,E) \rangle = \frac{EQ}{\rho} \varphi(t,E), \tag{7.75}$$

where the reduction factor, which in this case is equal to the absorbed fraction, is given by Eq. (7.45), namely

$$\varphi(t,E) = \phi(t,E) = \int_0^\infty \frac{dr}{r_o} g(r,t) \mathcal{F}(r/r_o, E). \tag{7.76}$$

Figure 7.11 shows this reduction factors for a slab source.
    For a beta-particle source, integration over the source energy spectrum leads to

$$\langle D_\beta(t) \rangle = 4\pi Q \int_0^\infty dr\, r^2 g(r,t) \mathcal{G}_\beta(r). \tag{7.77}$$

Local Dose Within the Slab Source.   Now consider the local absorbed dose rate within the slab. Let y represent the distance from and normal to the center plane of the slab at which the absorbed dose rate (MeV g⁻¹ s⁻¹) is denoted by $D(y,t)$. By examining the dose rate from a plate of thickness dz within the slab, and treating that thickness as a plane source, one can show that, for $y \leq t$,

$$D(y,t,E) = \frac{QE}{2\rho r_o(E)} \left[\int_0^{t-y} dz \int_z^\infty dr\, r^{-1}\, \mathcal{F}(r/r_o, E) \right.$$

$$\left. + \int_0^{t+y} dz \int_z^\infty dr\, r^{-1} \mathcal{F}(r/r_o, E) \right]. \tag{7.78}$$

One can invert the order of integration and express the result in terms of a local reduction factor, viz.,

$$D(y, t, E) = \frac{QE}{\rho}\varphi(y, t, E),\tag{7.79}$$

in which

$$\varphi(y, t, E) = \frac{1}{r_o}\int_0^\infty dr\, g(y, r, t)\mathcal{F}(r/r_o, E),\tag{7.80}$$

where the geometry factor [identified by comparison of this result to Eq. (7.45)] is given by

$$g(y, r, t) = \frac{[t - y, r]_{min} + [t + y, r]_{min}}{2r}.\tag{7.81}$$

Here the function $[t - y, r]_{min}$ refers to the lesser value of the two arguments $t - y$ and r. Local reduction factors for this case are illustrated in Fig. 7.16.

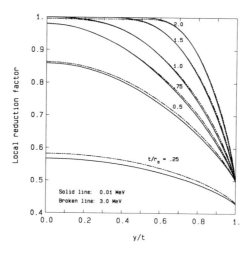

Figure 7.16 Local reduction factors for uniform isotropic monoenergetic electron sources within slabs of water completely surrounded by the same medium. The lines for 0.01 and 1-MeV electrons encompass absorbed fractions for all intermediate energies. Note that for $t/r_o \geq 0.5$, the local absorbed fraction is 0.5 at the slab surface.

For a beta-particle source, integration over the energy spectrum of the source leads to

$$D_\beta(y, t) = 4\pi Q\int_0^\infty dr\, r^2\, g(y, r, t)\mathcal{G}_\beta(r).\tag{7.82}$$

7.8.4 Half-Space Source

Average Dose in a Layer Bounding the Half-Space Source.   Within a half-space source region, Q electrons of energy E (MeV) are emitted per unit volume per unit time (e.g., cm^{-3} s^{-1}) per second. Of interest is the average absorbed dose rate in the layer of thickness h that bounds the plane surface of the half space (see Fig. 7.17). The probability that a ray of length r originating at depth u in the source region ends within the bounding layer of thickness h is given by

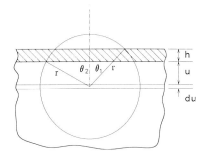

Figure 7.17   Geometry for the non-source layer of thickness h (shaded region) that bounds a uniform isotropic half-space source. Rays of length r that are emitted at a depth u into the source medium terminate on the circle.

$$f(u, h, r) = 0, \qquad r < u, \tag{7.83}$$

$$f(u, h, r) = \frac{1}{2}(1 - \cos \theta_2) = \frac{r - u}{2r}, \qquad u \le r \le u + h, \tag{7.84}$$

$$f(u, h, r) = \frac{1}{2}(\cos \theta_2 - \cos \theta_1) = \frac{h}{2r}, \qquad r \ge u + h. \tag{7.85}$$

The mean-absorbed dose rate (MeV g^{-1} s^{-1}) in the bounding layer is then given by

$$\langle D(h, E) \rangle = \frac{QE}{h\rho} \int_0^\infty du \int_0^\infty \frac{dr}{r_o} f(u, h, r) \mathcal{F}(r/r_o, E). \tag{7.86}$$

The order of integration may be inverted to give

$$\langle D(h, E) \rangle = \frac{QE}{\rho} \int_0^\infty \frac{dr}{r_o} g(h, r) \mathcal{F}(r/r_o, E), \tag{7.87}$$

where

$$g(h, r) = \frac{r}{4h}, \qquad 0 \le r \le h, \tag{7.88}$$

or

$$g(h, r) = \frac{1}{2} - \frac{h}{4r}, \qquad r \ge h. \tag{7.89}$$

By comparison of Eq. (7.87) to Eq. (7.44b), it is seen that $g(h, r)$ must be the geometry factor for rays from the target slab region into the half-space source region.

Results of sample calculations for the corresponding reduction factor $\varphi(h, E) = \rho \langle D(h, E) \rangle / (QE)$ are presented in Fig. 7.12.

Local Dose Exterior to the Half-Space Source. In this problem, the dose is sought at a point that is a perpendicular distance z away from the surface of the half-space source region. This problem was addressed by Berger (1973). The local reduction factor for a uniform monoenergetic and isotropic electron source is given by

$$\varphi(z, E) = \int_z^\infty \frac{dr}{r_o} g(z, r) \mathcal{F}(r/r_o, E), \tag{7.90}$$

where the geometry factor for rays from the target point to the half-space source region is

$$g(z, r) = \frac{1}{2} \left(1 - \frac{z}{r} \right) \qquad r \geq z. \tag{7.91}$$

Representative values of the local reduction factor are illustrated in Fig. 7.13.

7.8.5 Spherical Surface Source

Average Dose Within the Spherical Surface Source. This problem is representative of the evaluation of the average or spatial distribution of the absorbed dose rate within a cell on whose surface is a distributed beta-particle source. Suppose that, on the surface of a sphere of radius R within a uniform medium, there is an isotropic source emitting Q_a electrons of energy E (MeV) per cm^2 per second. First evaluate the average absorbed dose rate in the sphere. Consider rays of length r originating on the surface of the sphere, as illustrated in Fig. 7.18(a). The geometry factor $g_a(r, R)$ for this surface source is the fraction of those rays of length r that are emitted isotropically from the surface and that end in the sphere of radius R. This fraction is just the solid angle fraction subtended by angle β, namely,

$$g_a(r, R) = \frac{(1 - \cos \beta)}{2} = \frac{1}{2} \left(1 - \frac{r}{2R} \right), \qquad 0 \leq r \leq 2R. \tag{7.92}$$

From the general discussion of surface sources in Sec. 7.7.5, average dose rate in the sphere is given by Eq. (7.30) or Eq. (7.44) with QV_s/V_t replaced by $Q_a A_s/V_t$. The area to volume ratio of the sphere is $3/R$, so that the mean absorbed dose rate inside the sphere becomes

$$\langle D(R, E) \rangle = \frac{3E}{\rho R} \frac{Q_a}{r_o(E)} \int_0^{2R} dr \, g_a(r, R) \mathcal{F}(r/r_o, E). \tag{7.93}$$

Although the reduction factor is not defined for a surface source, the absorbed fraction is and is given by Eq. (7.33) in terms of the mean absorbed dose with the total source activity Q_{tot} replaced by $A_s Q_a$. Thus, since $V_t/(Q_a A_s) = R/3$, the absorbed fraction is given by

$$\phi(R, E) = \frac{R\rho}{3E} \frac{1}{Q_a} \langle D(R, E) \rangle = \frac{1}{r_o} \int_0^{2R} dr \, g_a(r, R) \mathcal{F}(r/r_o, E). \tag{7.94}$$

Representative values of $\phi(R, E)$ are illustrated in Fig. 7.11.

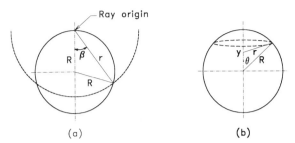

Figure 7.18 Geometry for a spherical surface source (solid circle). (a) Rays of length r that originate from a point on the source surface terminate on a spherical surface denoted by the dashed circle. (b) The dashed circle on the surface represents those source elements that are a distance R from the source center and that are also a distance r from an interior dose location which is a distance y from the source center.

Now, suppose the source is a beta-particle source of strength Q_a cm^{-2} s^{-1} on the surface of the sphere. The average absorbed dose rate within the sphere is then

$$\langle D_\beta(R)\rangle = Q_a \int_0^{E_{max}} dE\ N(E)\frac{3E}{R\rho}\varphi(R,E)$$

$$= Q_a\frac{12\pi}{R}\int_0^{2R} dr\ r^2 g_a(r,R)\mathcal{G}_\beta(r). \tag{7.95}$$

Local Dose Within the Spherical Surface Source.   Now consider the local absorbed dose rate as a function of radius y within the sphere — first for a monoenergetic source. Consider an element of area $dA = 2\pi R^2 \sin\theta\,d\theta$, in the form of a surface of revolution at polar angle θ, as illustrated in Fig. 7.18(b). Each point in dA is at distance r from a dose point at radius y, and contributes $dA\,\mathcal{G}(r,E)$ to the absorbed dose rate at y. Integration over the entire surface of the sphere, expressing $\mathcal{G}(r,E)$ in terms of $\mathcal{F}(r/r_o,E)$, gives the total absorbed dose rate

$$D(y,R,E) = \frac{Q_a E R^2}{2\rho r_o}\int_0^\pi d\theta\ \sin\theta\frac{1}{r^2}\mathcal{F}(r/r_o,E), \tag{7.96}$$

where

$$\cos\theta = \frac{y^2 + R^2 - r^2}{2yR}. \tag{7.97}$$

At the center of the sphere the absorbed dose rate is

$$D(0,R,E) = \frac{Q_a E}{r_o(E)}\mathcal{F}(r/r_o,E). \tag{7.98}$$

A change in the integration variable from θ to r gives, for $y > 0$,

$$D(y,R,E) = \frac{Q_a E R}{2\rho r_o y}\int_{R-y}^{R+y} dr\ r^{-1}\ \mathcal{F}(r/r_o,E), \tag{7.99}$$

or, in terms of a local reduction factor,

$$D(y, R, E) = Q_a \frac{3E}{\rho R} \varphi(y, R, E), \tag{7.100}$$

where

$$\varphi(y, R, E) = \frac{R^2}{6yr_o} \int_{R-y}^{R+y} dr \, r^{-1} \, \mathcal{F}(r/r_o, E). \tag{7.101}$$

The local reduction factor $\varphi(y, R, E)$ is illustrated in Fig. 7.19.

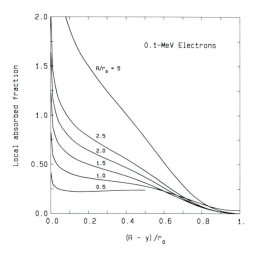

Figure 7.19   Local reduction factors for uniform isotropic monoenergetic electron sources on the surfaces of spheres of water completely surrounded by the same medium. Data for 0.01 to 1-MeV electrons are well represented by these results for 0.1-MeV electrons.

If the source is beta-particles then, at sphere center,

$$D_\beta(0) = 4\pi R^2 Q_a \mathcal{G}_\beta(R). \tag{7.102}$$

Otherwise,

$$D_\beta(y, R) = \frac{2\pi R Q_a}{y} \int_{R-y}^{R+y} dr \, r G_\beta(r). \tag{7.103}$$

If, in Eqs. (7.99), (7.101), and (7.103), the lower limit is changed to $|R - y|$, the equations apply to the local dose exterior to the spherical surface source.

7.8.6 Infinite Line and Cylindrical Sources

Consider an infinitely long line source of monoenergetic electrons of energy E (MeV) isotropically emitting Q_l electrons per unit length in a unit time (e.g., cm^{-1} s^{-1}). The source is embedded in an infinite, homogeneous, water-like medium of density ρ (g cm^{-3}). One can show that the absorbed dose rate $D(r, E)$, in units of MeV g^{-1} s^{-1}, at radial distance r (cm) from the source is given as follows in terms of the point kernel $\mathcal{F}(r/r_o, E)$:

$$D(r, E) = \frac{EQ_l}{2\pi\rho r r_o(E)} \int_0^{\theta_{max}} d\theta \mathcal{F}(r \sec\theta/r_o, E),\tag{7.104}$$

in which, since $\mathcal{F} = 0$ for r greater than about $1.2r_o$ (see Table 7.4), the maximum θ angle is

$$\theta_{max} \simeq \cos^{-1}\left(\frac{r}{1.2r_o}\right).\tag{7.105}$$

Line-Source Kernels.   Equation (7.104) can be recast in terms of a *scaled line-source kernel* $\mathcal{F}_{line}$, defined such that $\mathcal{F}_{line}dr/r_o$ is the fraction of the electron energy E deposited between r and $r + dr$ from the line source, namely $2\pi r\rho D(r, E)dr/E$. From Eq. (7.104), it is seen that this line-source kernel is related to the scaled point-source kernel $\mathcal{F}$ by

$$\mathcal{F}_{line}(r/r_o, E) = \int_0^{\theta_{max}} d\theta \mathcal{F}(r \sec\theta/r_o, E).\tag{7.106}$$

The kernel $\mathcal{F}_{line}(r/r_o, E)$ is illustrated in Fig. 7.20. The absorbed dose rate a distance r from the line source hence can be expressed as

$$D(r, E) = \frac{Q_l E}{2\pi\rho r r_o}\mathcal{F}_{line}(r/r_o, E).\tag{7.107}$$

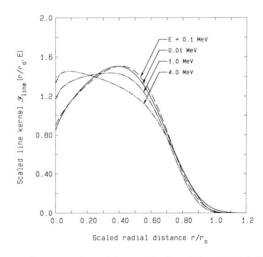

Figure 7.20   Scaled electron line-source kernels in water for four different initial electron energies as calculated numerically from Eq. (7.106) using the data of Table 7.4.

Local Absorbed Dose for an Infinite-Cylinder Volume Source.   Consider an infinitely long cylindrical source of radius R which emits Q electrons of energy E per unit volume per unit time (see Fig. 7.21 for a cross sectional view normal to the source axis). The contribution dD to the dose rate $D(y, R, E)$ at radius y from line sources passing through element of area $dA = u\,du\,d\theta$ is, from Eqs. (7.107),

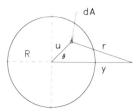

Figure 7.21   Geometry for evaluation of the local reduction factor for the cylindrical volume source of radius R. The source is infinitely long and imbedded in a medium of the same composition as the source material.

$$dD = \frac{Q\,E\,dA}{2\pi\rho r r_o(E)} \mathcal{F}_{line}(r/r_o, E), \tag{7.108}$$

in which

$$r^2 = u^2 + y^2 - 2uy\cos\theta. \tag{7.109}$$

Integration of Eq. (7.108) over the cross sectional area of the source cylinder to obtain the dose rate, and division by QE/ρ, the rate of energy release per unit mass in the source volume, yield the following expression for the local reduction factor:

$$\varphi(y, R, E) = \frac{1}{\pi r_o(E)} \int_0^R du\,u \int_0^\pi d\theta\,r^{-1}\,\mathcal{F}_{line}(r/r_o, E). \tag{7.110}$$

Values of $\varphi(y, R, E)$ are illustrated in Fig. 7.22.

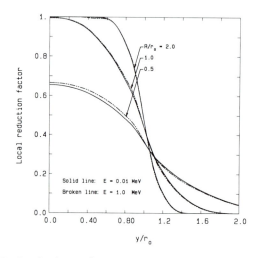

Figure 7.22   Local reduction factor $\varphi(y, R, E)$ for the cylindrical volume source.

Geometry Factors for Infinite-Cylinder Volume Sources.   As has been shown, geometry factors and point kernels can be used to express dose rates and absorbed fractions in some region, arising from volumetric and surface sources, by a simple one-dimensional integral. The same technique can be applied for cylindrical volumetric sources using the scaled line-source kernel (Faw and Shultis 1991).

Figure 7.23 is a cross section normal to the axis of an infinitely long cylindrical source of radius R (cm) in which Q electrons of energy E (MeV) are emitted per cm^3 per second. The geometry factor $g_l(r, R, R')$ for such cylindrical sources is defined in terms of radii in a plane perpendicular to the source axis. Specifically, $g_l(r, R, R')$ is defined as the mean fraction of radii of length r originating within the source volume, $0 \leq u \leq R$, and ending anywhere within the concentric cylinder of radius R', that is, within $0 \leq u \leq R'$.

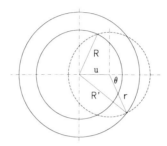

Figure 7.23   Geometry for evaluation of the geometry factor $g(r, R, R')$ and reduction factors $\varphi(R, E)$ and $\varphi(R, R', E)$ for cylindrical volume sources and for annular nonsource shells.

Let $h(u, r, R, R')$ be the fraction of radii with length r that originate at u and end anywhere within the cylinder of radius R'. Clearly, $h = 0$ if $r \geq R' + u$ and $h = 1$ if $r \leq R' - u$ or $r \leq R' - R$ for any u. Otherwise,

$$h(u, r, R, R') = 1 - \frac{\theta}{\pi} = 1 - \frac{1}{\pi} \cos^{-1}\left(\frac{R'^2 - u^2 - r^2}{2ur}\right). \qquad (7.111)$$

The geometry factor $g(r, R, R')$ is just the mean value of h. For $r \geq R + R'$, $g_l = 0$. For $r \leq R' - R$, $g = 1$. Otherwise,

$$g_l(r, R, R') = \frac{1}{\pi R^2} \int_0^R du\, 2\pi u h(u, r, R'),$$

$$= \frac{1}{\pi}\left\{ \cos^{-1}\left(\frac{r^2 + R^2 - R'^2}{2rR}\right) + \frac{R'^2}{R^2} \cos^{-1}\left(\frac{r^2 - R^2 + R'^2}{2rR'}\right)\right. \qquad (7.112)$$

$$\left. - \frac{\left[4r^2 R'^2 - (R^2 - R'^2 - r^2)^2\right]^{1/2}}{2R^2} \right\}.$$

The geometry factor is illustrated in Fig. 7.24 for the special case $R = R'$.

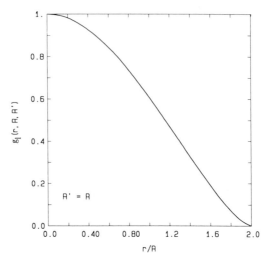

Figure 7.24   Geometry factor $g_\ell(r, R, R')$ for the cylindrical volume source and for the special case $R = R'$.

Mean Absorbed Dose Within an Infinite-Cylinder Source.   In the representation of a cylindrical source of radius R as a bundle of line sources, the fraction of radii r ending within the source is just $g_l(r, R, R)$ and r can be no greater than 2R. The absorbed fraction, that is, the fraction of the energy released within the infinite cylinder absorbed within the same cylinder, is equal to the mean reduction factor, and is just

$$\phi(R, E) = \varphi(R, E) = \int_0^{2R} \frac{dr}{r_o} \mathcal{F}_{line}(r/r_o, E) g_l(r, R, R). \qquad (7.113)$$

This function is illustrated in Fig. 7.25.

Mean Absorbed Dose in a Cylindrical Shell Around a Cylindrical Source.   For a shell of outer radius R', immediately adjacent to a source cylinder of radius R, as illustrated in Fig. 7.23, one may evaluate the absorbed fraction as follows. Per unit distance along the cylinder, the energy absorbed in the source cylinder is $QE\pi R^2\varphi(R, E)$. The rate of energy absorption in the total cylinder of radius R' is, similarly

$$\frac{dE_{abs}}{dt} = QE\pi R^2 \int_0^{R+R'} \frac{dr}{r_o} \mathcal{F}_{line}(r/r_o, E) g_l(r, R, R'). \qquad (7.114)$$

The rate of energy absorbed per unit length along the shell from R to R' is the difference between the expressions in Eqs. (7.113) and (7.114), and the absorbed fraction is just that difference divided by the energy release rate per cm along the source cylinder. Thus,

$$\phi(R, R', E) = \int_0^{R+R'} \frac{dr}{r_o} \mathcal{F}_{line}(r/r_o, E) \left[g_l(r, R, R') - g_l(r, R, R) \right]. \qquad (7.115)$$

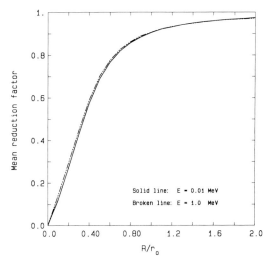

Figure 7.25   The mean reduction factor $\varphi(R, E)$ for a cylindrical volume source.

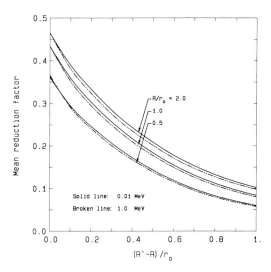

Figure 7.26   The mean reduction factor $\varphi(R, R', E)$ for an annular region of radii R and R' concentric with a cylindrical volume source of radius R.

The absorbed dose rate in the shell is just the difference between the expressions given in Eqs. (7.113) and (7.114) divided by the shell mass per cm along the cylinder, $\pi\rho(R'^2 - R^2)$. The mean reduction factor is this quotient divided by E/ρ, the rate of energy release per unit mass in the source cylinder. To be specific,

$$\varphi(R, R', E) = \frac{\phi(R, R', E)}{(R'^2/R^2 - 1)}. \tag{7.116}$$

This mean reduction factor is illustrated in Fig. 7.26. The methods just illustrated may also be applied to the evaluation of the absorbed fraction for a cylindrical shell concentric with, but not contiguous with, the source cylinder.

REFERENCES

BELL, G. I. AND S. GLASSTONE, *Nuclear Reactor Theory*, Van Nostrand, New York, 1970.

BERGER, M. J., "Beta-Ray Dosimetry Calculations with the Use of Point Kernels," in *Medical Radionuclides: Radiation Dose and Effects*, R. J. Cloutier, C. L. Edwards, and W. S. Snyder (eds.), Proceedings of a 1969 Symposium Held at the Oak Ridge Associated Universities, Oak Ridge, Tenn., 1970.

BERGER, M. J., *Distribution of Absorbed Dose Around Point Sources of Electrons and Beta Particles in Water and Other Media*, NM/MIRD Pamphlet No. 7, *J. Nucl. Med., 12*, Supplement 5 (1971).

BERGER, M. J., *Improved Point Kernels for Electron and Beta-Ray Dosimetry,* Report NB-SIR 73-107, U.S. National Bureau of Standards, Washington, D.C., 1973.

BERGER, M. J., "Beta-Ray Dose in Tissue-Equivalent Material Immersed in a Radioactive Cloud," *Health Physics 26*, 1-12 (1974).

BHABA, H. J., "Scattering of Positrons by Electrons with Exchange on Dirac's Theory of the Positron," *Proc. Royal Soc.* (London) *A154*, 195 (1936).

CRISTY, M., AND K. F. ECKERMAN, *Specific Absorbed Fractions of Energy at Various Ages from Internal Photon Sources,* Report ORNL/TM-8381 (6 vols), Oak Ridge National Laboratory, Oak Ridge, Tenn., 1987.

CROSS, W. G., "Variation of Beta Dose Distribution in Different Media," *Phys. Med. Biol. 13,* 611-618 (1968).

DILLMAN, L. T, AND F. C. VON DER LAGE, *Radionuclide Decay Schemes and Nuclear Parameters for Use in Radiation Dose Estimation*, NM/MIRD Pamphlet No. 10, Society of Nuclear Medicine, 1975.

DILLMAN, L. T., *EDISTR —- A Computer Program to Obtain a Nuclear Decay Data Base for Radiation Dosimetry*, Report ORNL/TM-6689, Oak Ridge National Laboratory, Oak Ridge, Tenn., 1980.

FANO, U., "Note on the Bragg-Gray Cavity Principle for Measuring Energy Dissipation," *Rad. Res. 1*, No. 3, 237-240 (1954).

FAW, R. E., AND J. K. SHULTIS, *Point Pair Distributions for Internal Dosimetry,* Fifth International Symposium on Radiopharmaceutical Dosimetry, Oak Ridge Associated Universities, Oak Ridge, Tenn., 1991.

ICRP, *Report of the Task Group on Reference Man,* Publication 23, International Commission on Radiological Protection, Pergamon Press, Oxford, 1975.

ICRU, *Methods of Assessment of Absorbed Dose in Clinical Use of Radionuclides*, Report 32, International Commission on Radiation Units and Measurements, Washington, D.C. 1979.

ICRU, *Stopping Powers for Electrons and Positrons*, Report 37, International Commission on Radiation Units and Measurements, Washington, D.C. 1984.

KELLERER, A. M., "Considerations on the Random Traversal of Convex Bodies and Solutions for General Cylinders," *Radiation Research 47*, 359–376 (1971).

KELLERER, A. M., "Proximity Functions for General Right Cylinders," *Radiation Research 86*, 264–276 (1981).

KELLERER, A. M., "Chord-Length Distributions and Related Quantities for Spheroids," *Radiation Research 98*, 425–437 (1984).

LOEVINGER, R. "Distributed Radionuclide Sources," in *Radiation Dosimetry*, 2nd ed., Vol. III, F. H. Attix and E. Tochilin (eds.), Academic Press, New York, 1969.

MÄDER, U., "Chord Length Distributions for Circular Cylinders," *Radiation Research 82*, 454–466 (1980).

MØLLER, C., "Passage of Hard Beta Rays Through Matter," *Ann. Physik. 14*, 531 (1932).

NCRP, *Conceptual Basis for Calculations of Absorbed Dose Distributions,* Report 108, National Council on Radiological Protection and Measurements, Bethesda, Md., 1991.

NELSON, W. R., H. HIRAYAMA, AND D. W. O. ROGERS, *The EGS4 Code System*, Report SLAC-265, Stanford Linear Accelerator Center, Stanford, Calif., 1985.

SNYDER, W. S., M. R. FORD, AND G. G. WARNER, *Estimates of Specific Absorbed Fractions for Photon Sources Uniformly Distributed in Various Organs of a Heterogeneous Phantom,* MIRD Pamphlet No. 5 (as revised), Society of Nuclear Medicine, New York, 1978.

SPENCER, L. V., AND U. FANO, "Energy Spectrum Resulting from Electron Slowing Down," *Phys. Rev. 93*, 1172 (1954)

SPENCER, L. V., *Energy Dissipation by Fast Electrons,* NBS Monograph 1, U.S. National Bureau of Standards, Washington, D.C., 1959.

SPENCER, L. V., A. B. CHILTON, AND C. M. EISENHOWER, *Structure Shielding Against Fallout Gamma Rays From Nuclear Detonations*, Special Publication 570, National Bureau of Standards, Washington, D.C., 1980.

WEIL, W., "Sterology — A Survey for Geometers," in *Convexity and its Applications*, P. Gruber and J. M. Wills (eds), Birkenhäuser, Basel 1983.

PROBLEMS

1. From the data given in Table 7.1, calculate the average energy of beta particles emitted by ^{60}Co.

2. Plot the energy and momentum distributions of the beta particles emitted by tritium using the data of Table 7.1.

3. For allowed transitions in beta decay, the shape factor S_n is unity. Show for such beta decay that a plot of $[N(E)/FEp]$ versus E is a straight line. These plots are known as "Kurie plots" and are useful in determining the end-point energy E_o. Construct the Kurie plot for ^{14}C using the data in Table 7.1 and assuming that the Fermi function is constant.

4. What is the relative rate of bremsstrahlung emission by a 1 MeV electron compared to that of a 1 MeV alpha particle in a water medium? What is this ratio if the medium is lead?

5. Estimate and plot the CSDA range in air at standard temperature and pressure as a function of the electron energy from 0.001 to 10 MeV.

6. A small particle containing 10^7 Bq of ^{60}Co becomes attached to the skin of a person working in a nuclear laboratory. Estimate the maximum dose rate (MeV g^{-1} h^{-1}) that would result from the beta particles at a depth of 0.007 cm in the skin (density 1.00 g cm^{-3}) (a) by assuming each beta particle has the average 0.0961 MeV energy, and (b) by using the actual ^{60}Co beta spectrum of Table 7.1.

7. For skin that is uniformly contaminated by ^{60}Co with an activity 10^7 Bq cm^{-2}, what is the beta-particle absorbed dose rate at a mass thickness depth of 7 mg cm^{-2}?

8. Show that the geometry factor for rays of length r from a spherical surface source of radius R_1 to end inside an inner concentric sphere of radius R_2 is given by

$$g_a(r, R_1, R_2) = \frac{1}{2}\left(1 - \frac{r^2 + R_2^2 - R_1^2}{2rR_2}\right)$$

for $R_2 - R_1 \le r \le R_2 + R_1$. For other values of r, $g_a(r, R_1, R_2) = 0$.

9. The infinite-medium absorbed-dose point kernel for alpha particles of initial energy E_o can be estimated by neglecting all delta rays and assuming the alpha particles travel in straight lines of length r_o equal to their range $\Lambda(E_o)$.

   (a) Show that this point kernel can be written as

   $$G_\alpha(r, E_o) = \frac{S(r_o - r)}{4\pi r^2 \rho}, \qquad r < r_o,$$

   where $S(x)$ is the stopping power (MeV cm^{-1}) of an alpha particle with residual range x.

   (b) With the approximations given in Sec. 2.9.4, show that the stopping power as a function of residual range can be approximated by

   $$S(r_o - r) \cong \frac{1}{n}\left(\frac{\rho}{\delta}\right)^{1/n}[(\delta/\rho)E_o^n - r]^{\frac{1}{n}-1}.$$

   (c) What are the limitations of this approximation?

10. Consider a spherical cell with a diameter of 10 μm whose nucleus is also spherical with a diameter of 5 μm and which is concentric with the cell surface. Molecules containing radioactive atoms that emit 5 MeV alpha particles are uniformly attached to the cell surface. (a) What is the range of the alpha particle? (b) With the results of the previous two problems, estimate the absorbed fraction for the cell nucleus.

11. For an infinite homogeneous medium, show that the absorbed dose kernel for a point source of monoenergetic gamma photons can be approximated as

$$G_\gamma(r, E) = \frac{\mu_{en} E}{4\pi r^2 \rho}e^{-\mu r}B(E, \mu r),$$

where μ_{en} and μ are the energy absorption and total interaction coefficients and $B(E, \mu r)$ is the infinite-medium dose buildup factor.

12. Consider a source sphere of radius A and a target sphere of radius B with centers separated by distance $P \geq A + B$.

(a) Show that the point-pair distribution for rays of length r originating in one sphere and ending in the other may be written for $P - A - B \leq r \leq P + A + B$ as

$$p(r; T \leftarrow S) = \frac{9r^2}{4A^3 B^3} \int_0^A du \, u^2 \int_0^\pi d\theta \, \sin\theta \, h(\zeta, r, B)$$

where

$$h = \frac{B^2 - (\zeta - r)^2}{2\zeta r},$$

and

$$\zeta^2 = P^2 + u^2 - 2Pu \cos\theta.$$

(b) In an anthropomorphic phantom of the adult female (Cristy and Eckerman 1987), the spleen and one of two ovaries, respectively, have volumes of 119 and 5.12 cm^3 and centroids separated by a distances of 20.5 cm for the left ovary and 24.8 cm for the right. The two organs and intervening tissue have a density of 1.05 g cm^{-3} and attenuating properties of water for both gamma rays and electrons. Use the results of the previous problem to calculate the absorbed fractions for 1-MeV photons uniformly distributed in the spleen as the source organ and each ovary as target organ and for each ovary as source organ and the spleen as target organ. Use the following approximations:

(1) Treat the two organs as point source and point target, separated by 20.5 and 24.8 cm.

(2) Assume that the two organs are spherical in shape with centers separated by 20.5 and 24.8 cm. For reference, $p(r; T \leftarrow S)$ calculated as described in part (a) is illustrated in Fig. P7.12.

(3) Assume the ovary and spleen are ellipsoids as was done in the calculations of absorbed fractions by Cristy and Eckerman. The spleen is at the origin of Cartesian coordinates, and is described by

$$\left(\frac{x}{2.90}\right)^2 + \left(\frac{y}{1.88}\right)^2 + \left(\frac{z}{5.19}\right)^2 \leq 1.$$

The left ovary is described by

$$\left(\frac{x - 4.31}{1.17}\right)^2 + \left(\frac{y - 2.94}{0.58}\right)^2 + \left(\frac{z - 19.83}{1.80}\right)^2 \leq 1,$$

and the right by

$$\left(\frac{x - 14.67}{1.17}\right)^2 + \left(\frac{y - 2.94}{0.58}\right)^2 + \left(\frac{z - 19.83}{1.80}\right)^2 \leq 1.$$

For reference, $p(r; T \leftarrow S)$ derived from Monte Carlo calculations is illustrated in Fig. P7.12.

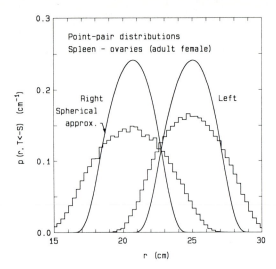

Figure P7.12   Point-pair distributions for the spleen and ovaries of the adult female anthropomorphic phantom (Cristy and Eckerman 1987) as source and target organs. Continuous lines are based on approximations of the organs as spherical in shape. Histograms are based on Monte Carlo calculations for organs ellipsoidal in shape.

(c) Repeat part (b) for the spleen as both source and target and for the ovary as both source and target and for uniformly distributed 1-MeV photon sources.

(d) Repeat part (b) for the ovary as both source and target and for uniformly distributed 1-MeV electron sources.

Chapter 8

Internal Dose Evaluation

8.1 INTRODUCTION

Inhaled or ingested radionuclides generally become distributed throughout the many organs and tissues of the body before they decay or are removed. Such internally distributed radionuclides lead to radiation doses both to the organs or tissues in which they reside and, if gamma or x rays are released, to other parts of the body. Many studies have been conducted to obtain data on the behavior of radionuclides taken into the body. From these bio-kinetic studies, models with varying degrees of sophistication have been developed for estimating internal dose rates to the various body organs and tissues.

This chapter deals with internal dosimetry in two contexts. For purposes of radiation-dose estimation associated with medical administration of radionuclides, one requires bio-kinetic models closely representative of the individual patient and the chemical form of the radionuclide. Indeed, for administration of therapeutic quantities of radionuclides, preliminary studies using tracer quantities of radionuclides may be required to obtain patient-specific parameters of the bio-kinetic model. By contrast, for purposes of establishing radiation-protection programs and conducting radiation-dose estimates associated with public-health consequences of radionuclides in the environment, broadly applicable bio-kinetic models are required for the population as a whole, or for large subsets of the population, and for broad chemical classes of radionuclides.

A popular method for medical internal-dose evaluation is based on the Loevinger and Berman (1976) formalism and is identified as the MIRD method in recognition of the sponsoring organization, the Medical Internal Radiation Dose Committee of the Society of Nuclear Medicine (MIRD 1985). Many aspects of the MIRD method are incorporated in the medical-dose prescriptions and recommendations of the ICRU (1979), ICRP (1987) and NCRP (1985). The internationally sanctioned method of internal dosimetry for radiation-protection and public-health applications is identified here as the ICRP method. The methodology for application of the 1977 ICRP recommendations, which may be identified as the ICRP-30 methodology named for Publication 30 (1979), is very similar to the methodology for application of the

1990 ICRP recommendations, which may be identified as the ICRP-60 methodology, named for ICRP Publication 60 (1991a). In this chapter, the ICRP-30 and ICRP-60 methodologies will be separately identified only when necessary.

The reader will find that the MIRD and ICRP methods are very similar. In most respects, the ICRP method is a simplified and generic version of the MIRD method. Because of the generalization, the ICRP method is strictly applicable only on average to large populations and can be applied only approximately to internal dose evaluation for any individual. Thus, the ICRP method is suitable only approximately for evaluation of internal doses due to radiopharmaceuticals which are especially tailored for selective uptake in specific organs or tissues of the body. Unfortunately the MIRD and ICRP methods have different systems of nomenclature. Despite the differences, in this chapter a parallel description and development of the two methods is attempted. The ICRP method has become the standard for establishment of radiation-protection standards, and recommendations and is widely used in risk estimates related to public health. For these reasons, the ICRP method deserves the considerable attention given it in this chapter.

From the late 1950s and into the 1990s, internal dosimetry methods used in practice, as well as in guides and regulations issued to minimize occupational and public-health radiation risks, were largely based on the 1959 recommendations of the International Commission on Radiological Protection (ICRP 1960, 1964). These recommendations, which appeared in what is called here ICRP Publication 2, were prepared by the Commission's Committee II on Permissible Dose for Internal Radiation. They were accompanied by data and computational procedures suitable for estimating internal doses and dose rates resulting from radionuclide intake. The same procedures were also suitable for establishing maximum permissible radionuclide concentrations in air or water appropriate for use in radiation-protection programs.

Minor revisions of the recommendations were made in 1966 (ICRP 1966) and a detailed compilation of anatomical and physiological characteristics of *Reference Man* were published in 1975 (ICRP 1975). Tabulations of dose-intake conversion factors, derivative of the methods of ICRP Publication 2, have been issued to extend the dose-estimation procedures to age- and sex-specific population groups (NRC 1977; Hoenes and Soldat 1977) and to special exposure circumstances (Kocher 1979). The methods of Publication 2 have also been incorporated into safety regulations such as Part 20 of Title 10 of the United States Code of Federal Regulations (NRC 1960).

In 1977 the ICRP approved a new set of basic recommendations dealing with protection from ionizing radiation. The recommendations were issued in ICRP Publication 26 (1977). Procedures for internal dose evaluation along with tables of limits on annual intakes and air concentrations for occupational safety were issued as ICRP Publication 30 (1979).

In this chapter, the models and calculational procedures recommended by the ICRP in Publications 30 and 60 are summarized using ICRP notation. While the ICRP publications have extensive sets of data, far more than can be presented here, subsets of the data appropriate for many nuclear-power and medical applications are provided in Appendix G. The philosophy, notation, and sophistication of the methods presented in Publications 30 and 60 differ somewhat from those of the older Publication 2. This earlier set of procedures and concepts has become widely incorporated into the dosimetry literature and has served as the basis for many guidelines and regulations. Thus, for historical as well as practical purposes, the principal features of this earlier publication are also reviewed in this chapter and contrasted to the more modern ICRP techniques. The methods recommended in Publications 30 and 60 are incorporated into regulatory rules and guidelines of many countries.

8.2 RADIATION DOSE CONCEPTS

8.2.1 Sources, Targets, and Compartments

Intake of a radionuclide results subsequently in absorption of radiation in varying degrees throughout many if not all organs and tissues of the body. These are identified as *target* organs. The radioactive material itself may exist in the body in well-defined anatomical regions or tissues. These are identified as *source* organs. From a bio-kinetic viewpoint, however, the distribution of the radioactive material in the body might be described more precisely in terms of *compartments* which need not have direct correspondence with anatomical regions. An important example of a compartment is that of "body fluids," a transfer compartment invoked in the ICRP method to account for the delay in time between intake of a radionuclide and its deposition in particular tissues. Other examples include intravascular and interstitial compartments. Compartmental analyses of the respiratory and gastrointestinal systems are taken up later in this chapter. Internal dosimetry in both the MIRD and the ICRP methods is based on source-to-target radiation transport. In the ICRP model, specific organs and tissues are considered instead of compartments. In the more general MIRD method, the tie between compartments and anatomical regions is made through the use of *identification coefficients* (Berman 1977). Such coefficients identify the fraction of the radionuclide in a given compartment which is localized in a given anatomical region, that is, source organ. Use of identification coefficients will be illustrated later in this chapter in the context of bio-kinetic models for radioiodine.

8.2.2 Target Organ Dose and Dose Equivalent

The absorbed dose D_T to a target organ T is the sum of absorbed doses in the target arising from radiations released in radionuclide transformations in all source organs S, namely,

$$D_T = \sum_S D(T \leftarrow S). \tag{8.1}$$

The organ-to-organ absorbed doses $D(T \leftarrow S)$ can be converted to the corresponding dose equivalents $H(T \leftarrow S)$ by multiplying by the appropriate quality factor for each type of radiation contributing to $D(T \leftarrow S)$. Then the total dose equivalent to target organ T is

$$H_T = \sum_S H(T \leftarrow S). \tag{8.2}$$

When the absorbed dose or the dose equivalent is normalized by dividing by the activity taken into the body, the quotient is identified as the *specific* absorbed dose or the *specific* dose equivalent, and is denoted by $\hat{D}_T$ or $\hat{H}_T$.

8.2.3 Effective and Committed Dose Equivalent

The *effective dose equivalent* is a weighted-average dose equivalent, with target-organ weighting factors representative of the fraction of the total stochastic risk resulting from irradiation of those organs. Specifically,

$$H_E = \sum_T w_T H_T. \tag{8.3}$$

The weight factors recommended by the ICRP are derived and their use is explained in Chap. 3, Sec. 3.7.2 and Tables 3.16 and 3.17. Strictly speaking, the weight factors are applicable to radiation protection of occupationally exposed populations. Nevertheless, the effective dose equivalent is useful as an approximate indicator of the risk to either the individual worker or the individual nuclear medicine patient (ICRP 1987).

In this chapter, methods of internal dose evaluation are described for a single intake of radioactive material, through ingestion, inhalation, or injection. The resulting target-organ doses and dose equivalents are *committed* in the sense that they *will* be delivered in the course of time. In the event of a massive intake of radioactive material, measures can be taken to hasten elimination of the material from the body, but discussion of those measures is separate from the issues at hand. In the vast majority of cases, certainly in nuclear medicine, the committed dose is received within days or weeks after the intake. However, there are a few important cases for which the dose delivery may be extended years into the future. These cases involve radionuclides with long half lives *and* in chemical forms leading to avid retention in the body. A good example is ^{226}Ra which has a half life of about 1600 years and which is avidly retained in the skeleton. It makes no sense to evaluate a committed dose equivalent when the period of commitment extends beyond the normal lifespan of the person at risk. For this reason, the ICRP method of internal dose evaluation places a cutoff time η of 50 years on the commitment period. This is a very conservative approach even for commitments accrued early in a working career. An even greater cutoff time, say 70 years, is thought by some to be appropriate for evaluation of dose commitments for lifetime exposure of the general public.

8.2.4 Doses from Multiple Intakes

In most cases, sequential intakes of radionuclides result in committed doses which are simply additive. This is true for both occupational exposure and diagnostic nuclear medicine. It is true for two reasons. First, for low concentrations of radionuclides in the body, bio-kinetics are well described by linear differential equations with constant coefficients. This is the case when the rate of transfer of material from one organ or compartment to another organ or compartment is directly proportional to the concentration of the material in the first compartment. Second, radiation doses are simply additive. To be sure, risks may depend in a nonlinear fashion on total dose and dose rate or dose fractionation; however, the doses themselves are additive.

While this chapter does not explicitly treat the general case of a time-dependent intake of radioactive material, there are well established mathematical techniques for treating the general case for linear bio-kinetic models (Berman 1977). The response from a single intake may be thought of as an impulse response and convolution techniques may be applied to determine the response due to any arbitrary time-dependent intake. For example, if $\hat{R}_i(t)$ is the response of interest (a specific organ dose, say) at time t duc to a unit intake at time 0, then the response $R(t)$ arising from a protracted intake $I(t')$, $-\infty < t' \leq t$, is found by summing (integrating) the responses from an equivalent series of instantaneous intakes over all past time, that is,

$$R(t) = \int_{-\infty}^{t} dt' \, I(t')\hat{R}_i(t - t'). \tag{8.4}$$

With a change of integration variables, the response can be written as the convolution integral

$$R(t) = \int_{0}^{\infty} d\tau \, I(t - \tau)\hat{R}_i(\tau). \tag{8.5}$$

Bio-kinetic models may not be linear when large quantities of radiopharmaceuticals are administered in therapeutic nuclear medicine. A second administration may be applied before the first is eliminated from the body and the doses are not simply additive. A nonradioactive pharmaceutical may be administered prior to the radioactive form and the radiation dose commitment may depend on the quantity and timing of the nonradioactive administration. Bio-kinetics must be examined on a case-by-case basis. An example of nonlinear modeling is described in this chapter in the context of radioiodine dose evaluation.

8.3 THE GENERAL METHOD FOR INTERNAL DOSE EVALUATION

The committed dose equivalent H_T in a particular (target) organ or tissue T is the total dose equivalent which that organ or tissue is committed to receive during the time interval η following the intake of a radionuclide.[1] The value of H_T depends on the mode of intake—injection, ingestion or inhalation—and on the chemical and physical form of the radionuclide. The chemical form affects solubility and metabolic uptake; the physical form of an airborne radionuclide affects initial deposition within the various components of the respiratory system. In evaluating the committed dose equivalent, account must be taken of (1) the partitioning of the radionuclide among the various organs and tissues of the body via body fluids (the *transfer compartment*), (2) the rates of biological elimination and radioactive decay, (3) radioactive daughter products produced, and (4) absorption of radiation in organ or tissue T as a result of radioactive decay (transformations) not only in T but also in other (source) organs and tissues.

---

[1]Clearly the committed dose equivalent depends on η. ICRP Report 30 shows the dependence explicitly by writing $H_{50,T}$ as the 50-year committed dose equivalent. In this chapter's notation, the dependence is implicit. The reader should be aware that, when referring to target-organ dose equivalents arising from external irradiation, H_T refers to the dose equivalent *received* during some interval of exposure.

The committed dose equivalent also depends on the age, sex, and characteristics of the individual exposed as well as the circumstances of the exposure. Data presented here are for general application and average situations, and are thus based on the anatomical and physiological characteristics of Reference Man (ICRP 1975). In certain instances it may be desirable to take into account known parameters of a specific exposed individual when determining committed dose equivalents.

8.3.1 Single Radionuclides

This section deals with the determination, for a particular radionuclide, of the committed dose equivalent to a specified target organ as a result of intake of the nuclide. When normalized to intake of unit activity, the term *specific* committed dose equivalent is applied. The various factors entering the evaluation are discussed in subsequent sections.

Following the unit intake, the radionuclide will be distributed to various source organs and tissues. Upon ingestion, for example, a portion of the material may remain in the gastro-intestinal system causing irradiation of the walls of the stomach (ST), small intestine (SI), upper large intestine (ULI), and lower large intestine (LLI). A portion may enter the body-fluid transfer compartment; of that portion, some may be eliminated directly and some may be taken up by various source organs and tissues. From these tissues, the radionuclide may be eliminated by normal metabolic processes. While present in organ S, the radionuclide experiences U_S nuclear transformations. The symbol U_S, called *source organ transformations*, is equivalent to $\tilde{A}_S$, called *cumulated activity* in the MIRD notation.[2] Evaluation of U_S will be taken up in a later section.

The radionuclide may release a variety of types and energies of radiation, here identified by an index i. Associated with each is a yield Y_i (defined as the fraction of transformations resulting in radiation of type i), the energy of the radiation E_i, and the quality factor Q_i. A fraction of the energy E_i released in organ S may be absorbed in organ T. This fraction is identified as $AF(T \leftarrow S)_i$ or, in the MIRD notation, $\phi(T \leftarrow S)_i$. Per transformation of the nuclide in organ S, the (quality-factor weighted) energy absorbed per unit mass of organ T, called the *specific effective energy* is given by

$$SEE(T \leftarrow S) = \frac{1}{M_T} \sum_i Y_i \, E_i \, AF(T \leftarrow S)_i \, Q_i, \qquad (8.6)$$

in which M_T is the mass of organ T (see Appendix G, Table G-1). Evaluation of $SEE(T \leftarrow S)$ will be taken up in Sec. 8.5. The specific committed dose equivalent $\hat{H}_T$ is thus

$$\hat{H}_T = \sum_S \hat{H}(T \leftarrow S) = \kappa \sum_S U_S \, SEE(T \leftarrow S), \qquad (8.7)$$

---

[2]Clearly the source organ transformations depend on the time interval η, especially for an avidly retained, long-lived radionuclide. In this chapter's notation, as in ICRP Report 30, the dependence is implicit. In this chapter, too, U_S always refers to the *specific* number of transformations, that is, the number of transformations per unit activity taken into the body.

in which κ is a conversion factor relating dose equivalent to specific energy absorption. If H_T is in units of Sv (J kg^{-1}) and SEE is in units of MeV g^{-1}, then $\kappa = 1.6 \times 10^{-10}$. Values of $\hat{H}_T$ as listed in ICRP Publication 30 are reproduced in Appendix G for selected target organs, radionuclides, modes of intake, and physical-chemical forms of the radionuclide.

8.3.2 Radioactive Decay Chains

For a given intake of a parent radionuclide, one may identify both the parent nuclide and the radioactive progeny by index j, with $j = 0$ for the parent. Radionuclide species in the chain have different U_S^j source organ transformations and different specific effective energies. Symbolically,

$$SEE(T \leftarrow S)^j = \frac{1}{M_T} \sum_i Y_i^j E_i^j AF(T \leftarrow S)_i^j Q_i^j, \qquad (8.8)$$

$$\hat{H}(T \leftarrow S) = \kappa \sum_j U_S^j \, SEE(T \leftarrow S)^j, \qquad (8.9)$$

and

$$\hat{H}_T = \kappa \sum_j \sum_S U_S^j \, SEE(T \leftarrow S)^j. \qquad (8.10)$$

We illustrate the evaluation of U_S^j by way of an example presented in the next section.

8.4 SOURCE-ORGAN TRANSFORMATIONS (EXCEPT GI AND RESPIRATORY SYSTEMS)

The calculation of U_S^j, the number of transformations a particular radionuclide species j undergoes in organ S in the time interval η following intake, is one of the key steps needed to evaluate the specific committed dose equivalent $\hat{H}_T$. To calculate the number of source-organ transformations, one must first find the activity of the radionuclide in the specified organ as a function of time after intake. Then by integrating this time-dependent activity, the number of source-organ transformations may readily be found.

To calculate the activity of a radionuclide in an organ, the migration of the radionuclide species to and from the organ as well as radioactive decay must be considered. First presented in this section is the general framework for the MIRD method of calculating source organ transformations. This is followed by a discussion of the ICRP procedures for calculating U_S^j for organs and tissues other than those of the gastrointestinal system and the respiratory system. These systems are considered separately in later sections.

Both methods require first determining the activity $A_S(t)$ of a radionuclide in organ S as a function of time after instantaneous intake of a radionuclide. Assume that the activity is expressed in units of becquerels (Bq) and that the initial intake is A_0. The number of transformations in organ S in time η per unit of intake is then

$$U_S = \frac{1}{A_0} \int_0^\eta dt \, A_S(t). \tag{8.11}$$

The integral's upper limit η is usually taken be equivalent to 50 years. In nearly all cases, the limit may be taken as infinity since most radionuclides decay long before the 50-year limit. A word of caution about units is needed at this point. In Eq. (8.11), A_0 and A_S are in units of Bq, that is, transformations per second. If time is measured in seconds, then U_S has units of seconds, that is, transformations per Bq and $\hat{H}_T$ has units of dose equivalent per Bq. If activities are measured in units other than Bq, or if times are measured in units other than seconds, then conversion factors are required.

8.4.1 The MIRD Method for Source Organ Transformations

Compartmental models are used to represent radionuclide bio-kinetics. Individual compartments may represent distinct anatomical organs or tissues. However, they may also represent apparent (chemical) states of the radionuclide such as, for example, one or more chemical states for the radionuclide present in the bloodstream and one or more states for the radionuclide present in the lymphatic system. Since radiation-dose calculations are based on distinct source and target regions, *identification coefficients* may be used to describe how the materials in the various compartments are apportioned among the various source-organ anatomical regions, as illustrated (later in this section) in Example 8.2. Except for the fact that only singly occurring radionuclides are most generally of interest, there are no general rules or approximations for devising compartmental models. The states and regions of importance depend on the chemical nature of the radionuclide. Rate constants for transfer among compartments are generally patient specific. Berman (1977) presents a comprehensive discussion of the general problem of nonlinear compartment models.

In general, the transfer of radionuclides to and from a compartment depends on the activities in all other compartments. Thus, the activity $A_i(t)$ in compartment i following unit intake of activity is given by the solution of the equation

$$\frac{dA_i(t)}{dt} = f_i(A_1, A_2, \ldots, A_n), \qquad i = 1, \ldots, n, \tag{8.12}$$

in which n is the number of compartments and f_i is the function describing the transfer of the radionuclides to and from compartment i.

For nontherapeutic intakes, the function f_i is usually linear, so that Eq. (8.12) assumes the form

$$\frac{dA_i(t)}{dt} = -A_i(t) \sum_{j=1}^{n} \lambda_{ji} + \sum_{\substack{j=1 \\ j \neq i}}^{n} A_j(t)\lambda_{ij}, \tag{8.13}$$

in which λ_{ji}, $j \neq i$, is the rate constant[3] for transfer of a radionuclide from compartment i into compartment j, and $\lambda_{ii} \equiv \lambda_r$ is the radioactive decay constant.

A general feature of linear systems, discussed in Appendix F-1.4, is that $A_i(t)$ is directly proportional to the initial intake A_0 and a sum of exponential terms, namely,

$$A_i(t) = A_0 \, e^{-\lambda_r t} \sum_{m=1}^{n} B_{mi} \, e^{-C_m t}. \tag{8.14}$$

in which the constants B_{mi} and C_m are determined by the rate constants in the bio-kinetic model. It follows from Eq. (8.11) that

$$U_i = \sum_{m=1}^{n} \frac{B_{mi}}{(\lambda_r + C_m)} \left(1 - e^{-(\lambda_r + C_m)\eta} \right). \tag{8.15}$$

The reader will note that, if λ is measured in units of s^{-1}, then U_i is measured in units of transformations per Bq. The methodology is illustrated in a series of examples.

Example 8.1

This example deals with an elementary model for the bio-kinetics of radioiodine in iodide form. It is the model used in the ICRP method, but it is treated here in the context of the MIRD method. The model is illustrated in Fig. 8.1. If rate constants are defined in such a way that λ_{ij} is the fraction of the radioactive material in compartment j transferred to compartment i per differential unit time, then, for example, $\lambda_{32} = 0.8318 \ \mathrm{d}^{-1}$. In this model, compartment 3, the thyroid, is a distinct anatomical region identified as a source organ. The problem is to evaluate U_3, the source organ transformations in the thyroid subsequent to ingestion of ^{131}I which has a half-life of 8.04 days ($\lambda_r = 0.0862 \ \mathrm{d}^{-1}$).

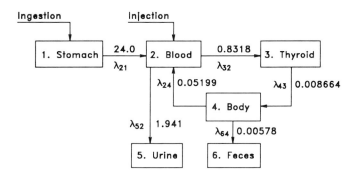

Figure 8.1   The ICRP model for the bio-kinetics of radioiodine in iodide form. Rate constants, in units of d^{-1} represent the fraction of the activity, per unit differential time, transferred from one compartment to another.

Solution   Differential equations describing the activities $A_i(t)$ in the four physical compartments are as follows:

---

[3]While we may say in an abbreviated way that the rate constant λ for an action is the probability per unit time that the action takes place, we must recognize that the rate constant is defined in such a way that, in the *limit* as Δt approaches zero, $\lambda \Delta t$ is the probability that an action takes place in time interval Δt.

$$\frac{dA_1}{dt} = -(\lambda_r + \lambda_{21})A_1$$

$$\frac{dA_2}{dt} = -(\lambda_r + \lambda_{52} + \lambda_{32})A_2 + \lambda_{24}A_4 + \lambda_{21}A_1$$

$$\frac{dA_3}{dt} = -(\lambda_r + \lambda_{43})A_3 + \lambda_{32}A_2$$

$$\frac{dA_4}{dt} = -(\lambda_r + \lambda_{24} + \lambda_{64})A_4 + \lambda_{43}A_3.$$

Initially, $A_1(0) = A_0 = 1.0$ and all other $A_i(0) = 0$. These equations may be solved analytically or numerically. Results for A_2 and A_3 are illustrated in Fig. 8.2. Similarly, U_3 may be obtained analytically or numerically by application of Eq. (8.11), that is, since $A_0 = 1$ and the half-life is much less than 50 years,

$$U_3 \cong \int_0^\infty dt \, A_3(t).$$

The result of this integration is $U_3 = 2.66 \times 10^5$ transformations in the thyroid per becquerel of ingested ^{131}I.

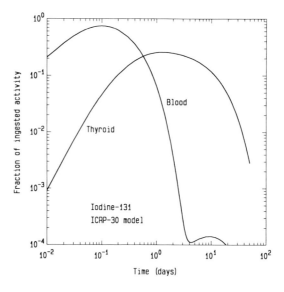

Figure 8.2 Activity of ^{131}I in the thyroid as a function of time after ingestion. Calculations are based on the ICRP model.

Example 8.2

This example deals with an elementary model for the bio-kinetics of radioiodine in ionic (iodide) form. It is an approximate model introduced by Berman (1977) as an example of the MIRD method. The model is illustrated in Fig. 8.3. If rate constants are defined in such a way that λ_{ij} is the fraction of the radioactive material in compartment j transferred to compartment i per differential unit time, then, for example, $\lambda_{21} = 0.727$ d^{-1} which leads to a thyroid uptake of nominally 25 percent of the injected radioiodine. This model illustrates the use of identification coefficients. Only compartment 2, the thyroid, is identified directly as a distinct anatomical region or source organ. Identification coefficients are as follows:

	Compartment			
Source Organ	1	2	3	4
Red blood cells	0.045		0.013	
Salivary glands	0.05			
Plasma	0.099		0.061	0.245
Stomach	0.15			
GI system	0.17			
Extracellular, extravascular	0.423		0.826	0.336
Thyroid		1.0		
Liver	0.063		0.100	0.419

Each entry is the fraction of the content of the compartment associated with the source organ. Of the iodine in the form of extrathyroidal triiodothyronine (compartment 3), for example, the fraction 0.1 is present within the liver as the source organ. Columns, of course, add to unity. The problem is to evaluate the source organ transformations in the thyroid and liver subsequent to injection of ^{131}I which has a half-life of 8.04 days ($\lambda_r = 0.0862$ d^{-1}).

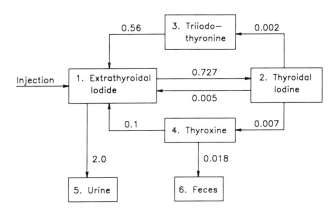

Figure 8.3   The MIRD-12 model for the bio-kinetics of radioiodine in iodide form. Rate constants, in units of d^{-1}, represent the fraction of the activity per unit differential time transferred from one compartment to another.

Solution   Let U_t and U_h represent respectively source organ transformations in the thyroid and liver, and U_1 through U_4 represent the compartment transformations in compartments 1 through 4. From the table of identification coefficients, it is apparent that

$$U_t = U_2,$$
$$U_h = 0.063U_1 + 0.1U_3 + 0.419U_4.$$

Differential equations describing the activities $A_i(t)$ in the four compartments are as follows:

$$\frac{dA_1}{dt} = -(\lambda_r + \lambda_{21} + \lambda_{51})A_1 + \lambda_{12}A_2 + \lambda_{13}A_3 + \lambda_{14}A_4,$$

$$\frac{dA_2}{dt} = -(\lambda_r + \lambda_{12} + \lambda_{32} + \lambda_{42})A_2 + \lambda_{21}A_1,$$

$$\frac{dA_3}{dt} = -(\lambda_r + \lambda_{13})A_3 + \lambda_{32}A_2,$$

$$\frac{dA_4}{dt} = -(\lambda_r + \lambda_{14} + \lambda_{64})A_4 + \lambda_{42}A_2.$$

Initially, $A_1(0) = A_0 = 1.0$ and all other $A_i(0) = 0$. These equations may be solved analytically or numerically. Results for A_1 and A_2 are illustrated in Fig. 8.4. Similarly, U_i, $i = 1$ to 4, may be obtained analytically or numerically by application of Eq. (8.11), that is, since $A_0 = 1$ and the half-life is much less than 50 years,

$$U_i \cong \int_0^\infty dt\, A_i(t).$$

Results, in units of transformations in the compartment per becquerel of injected ^{131}I, are as follows:

$$U_1 = 3.15 \times 10^4 \qquad U_2 = 2.29 \times 10^5$$
$$U_3 = 7.08 \times 10^2 \qquad U_4 = 7.84 \times 10^3.$$

It follows that:

$$U_t = 2.29 \times 10^5 \qquad U_h = 5.34 \times 10^3.$$

Example 8.3

This example illustrates nonlinear effects in the bio-kinetics of radioiodine. It is a highly specialized, patient-specific model for the behavior of a certain ^{123}I-labeled monoclonal antibody used in radiation therapy against B-cell lymphoma. The model, developed by Koizumi et al. (1986), is illustrated in Fig. 8.5. Antibodies (Ab) are introduced intravascularly (Iv) and act against tumor antigens (Ag). Ab-Ag complexes may be released intravascularly and the iodine processed to the iodide form. Both Ab-iodine and iodide equilibrate rapidly between the intravascular and interstitial (Is) spaces. In this model it is assumed that the thyroid is "blocked," that is, that the thyroid has been saturated with iodine prior to administration of the antibodies. The nonlinearity arises as a result of ligand-receptor equilibria in an Ab processor thought to be the liver. Only a certain patient-specific number of receptor sites is available. If F is the ratio of the moles of antibody injected to the moles of available receptors, it can be shown that the equations describing the activities in compartments 1 and 3 are

$$\frac{dA_1}{dt} = -(\lambda_r + \lambda_{21} + \lambda_{61} + \lambda_{91})A_1 - \lambda_{31}(1 - FA_3)A_1 + \lambda_{12}A_2 + \lambda_{13}A_3,$$

$$\frac{dA_3}{dt} = -(\lambda_r + \lambda_{13} + \lambda_{43})A_3 + \lambda_{31}(1 - FA_3)A_1,$$

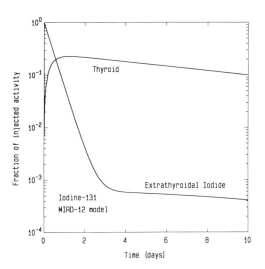

Figure 8.4 Activity of ^{131}I in the thyroid as a function of time after injection. Calculations are based on the MIRD-12 (Berman 1977) model.

where λ_r is the radiological decay constant, 1.27 d^{-1}, and λ_{ij} represents the fraction per differential unit time of the activity in compartment j transferred to compartment i. Equations for activities in other compartments are linear, that is, involve no products such as $A_1 \times A_3$ found in the above pair of equations. A set of rate constants, in units of d^{-1}, for a male patient, 39 years of age, and 108 kg in body weight is as follows:

$$
\begin{array}{lllll}
\lambda_{12} = 1.2 & \lambda_{13} = 15 & \lambda_{21} = 1.6 & \lambda_{31} = 159 & \lambda_{43} = 1.3 \\
\lambda_{45} = 210 & \lambda_{48} = 0.4 & \lambda_{54} = 600 & \lambda_{61} = 0.1 & \lambda_{76} = 0.1 \\
\lambda_{78} = 5.8 & \lambda_{87} = 59.5 & \lambda_{91} = 0.2 & \lambda_{10,7} = 0.5 & \lambda_{11,4} = 17
\end{array}
$$

The problem is to evaluate $A_i(t)$ for compartments 1 and 6 and for values of F ranging from 0.01 to 10. Initial conditions are $A_1(0) = A_0 = 1.0$ and all other $A_i(0) = 0$. $A_1(t)$ will illustrate the rate of clearance of the radioiodine from the body and $A_6(t)$ will illustrate the effectiveness of the radioiodine administration in terms of radiation dose given to tumor cells by the iodinated antibodies bound to tumor antigens.

Solution   For a linear bio-kinetic model, $A_i(t)/A_0$ is independent of A_0. This is clearly not the case for this problem. Figure 8.6 illustrates the fraction of injected iodine-tagged antibodies retained in the intravascular compartment. As F is increased, the radionuclide is cleared more slowly because the Ab processor (liver) acts as a reservoir which is rapidly filled but slowly emptied. As F is decreased, $A_1(t)/A_0$ becomes independent of F, that is, the nonlinearity introduced by the processor becomes irrelevant. What is important to note is that, as F is increased, the fraction of the injected radioiodine bound to tumor cells increases, as is shown in Fig. 8.7. This is desirable from a radiation therapy viewpoint. However, for diagnostic tumor imaging, increasing F may not be desirable because of the higher radiation doses. The way around this contradiction is to administer first an injection of nontagged antibodies in order to saturate liver receptors. A subsequent injection of tagged antibodies would then be bound in greater proportion to tumor cells. For a discus-

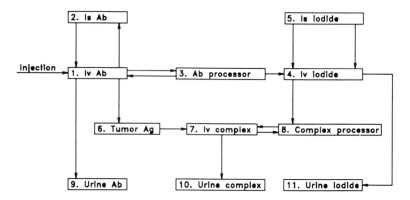

Figure 8.5   A nonlinear bio-kinetic model (Koizumi et al. 1986) for radioiodine-labeled monoclonal antibodies intravascularly administered. AB = antibody, Ag = antigen, IV = intravascular, IS = interstitial, and complex = Ab bound to Ag.

sion of the many additional ramifications of this nonlinear model, the reader is referred to the findings of Koizumi and co-workers (1986).

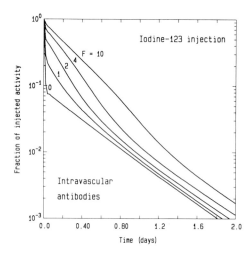

Figure 8.6   Activity of ^{123}I labeled antibodies in the intravascular compartment as a function of time after injection. Calculations are based on the nonlinear bio-kinetics model of Koizumi et al. (1986). *F* is the ratio of the moles of antibodies injected to the moles of receptor in the (liver) antibody processor.

8.4.2 The ICRP Model for Source Organ Transformations

In both Publication 2 and Publication 30, the ICRP employs approximations which constitute a generic simplification of the MIRD model. The major simplification is the introduction of a *body-fluids transfer compartment* which serves first to transport radionuclides from the respiratory and gastrointestinal systems to the various organs

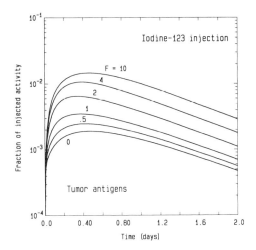

Figure 8.7   Activity of ^{123}I labeled antibodies bound to tumor antigens as a function of time after injection, based on a nonlinear bio-kinetic model (Koizumi et al. 1986). F is the ratio of the moles of antibodies injected to the moles of receptor in the (liver) antibody processor.

and tissues of the body and second to transport radionuclides from the organs and tissues through various paths of biological elimination. The transport is assumed to be so rapid as to be treated as instantaneous. Thus, with few exceptions, the transfer compartment need not be treated as a source compartment for internal dosimetry. The framework of the ICRP model is illustrated in Fig. 8.8. Specific approximations used in the model are as follows:

a. Of the quantity of a radionuclide injected, ingested or inhaled, a fraction f_T is transferred immediately to the body-fluid transfer compartment. For injection, of course, $f_T = 1.0$. For ingestion, f_T is equal to f_{BF}, the fraction of the radionuclide transferred from the GI system to the transfer compartment (see Sec. 8.8.2). If the radionuclide has a half life much greater than two days, f_{BF} is closely approximated by the fraction f_1 of the (nonradioactive) element eventually transferred to body fluid from the GI system (see Appendix G, Table G-2). As will be seen, f_1 depends only on the chemical form or solubility of the material. For inhalation, there are two paths to the transfer compartment. Of the inhaled material, a fraction f'_{BF} is transferred directly to the body-fluid transfer compartment from the respiratory system, while a fraction f_{GI} is transferred to the GI system and thence to the transfer compartment. Thus, for inhalation, $f_T = f'_{BF} + f_{GI}f_{BF}$. The factors f'_{BF} and f_{GI} are derived on the basis of a lung clearance model (see Sec. 8.7.1). For long-lived radionuclides, these factors depend only on the chemical form of the inhaled material and the physical-size characteristics of the aerosol.

b. Materials reaching the transfer compartment are distributed immediately to the various organs and tissues of the body, that is, to the source organs. Distribution fractions, identified as f_2, depend only on the element, not the isotope (see Appendix G, Table G-3). Any one source organ, however, may be subdivided into regions with factors f_2, f'_2, and so on, (see next item). The radionuclides are assumed to be distributed uniformly through the volumes of the sources organs or regions.

c. With certain exceptions, a radionuclide is eliminated biologically in a first-order rate process characterized by a biological decay constant λ_b analogous to the radiological decay constant λ_r. Regions within any one source organ may have separate biological decay constants (see Appendix G, Table G-3).

d. Daughter radionuclides are deposited from the transfer compartment into the source organ in the same fraction as the parent and eliminated biologically from the source organ at the same rate as the parent. Although progeny are chemically distinct from the parent and thus could be expected to have different interaction rates, this assumption that they behave in the same manner as the parent not only simplifies the subsequent analysis but also reflects transfer that is dominated by physical-mechanical mechanisms in which the chemical nature of the matter is not necessarily the controlling factor.

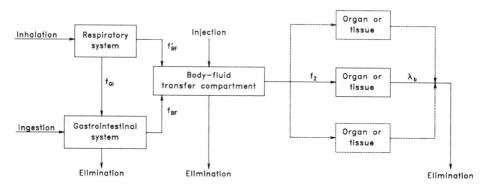

Figure 8.8   Principal features of the ICRP method for internal dosimetry. Transfer from the body-fluid compartment may be to multiple organs and tissues each with a characteristic f_2 and λ_b. Any one anatomical organ or tissue may be divided into multiple regions representing different physical or chemical retention patterns, each region of which has a characteristic f_2 and λ_b.

The ICRP method for evaluating source organ transformations is illustrated for the general case of a radionuclide with a number of radioactive progeny. Suppose that activity A_0 of the parent radionuclide is injected, ingested or inhaled. Let $A_S^0(t)$ represent the activity of the parent radionuclide in some source organ. The initial value of $A_S^0(t)$ is $A_0 f_T f_2$ and the time dependence is governed by the differential equation

$$\frac{dA_S^0}{dt} = -(\lambda_{r0} + \lambda_b)A_S^0. \tag{8.16}$$

Suppose that the radionuclide has a sequence of n daughter products, of activities $A_S^1(t), \ldots, A_S^n(t)$ in the source organ, all initially equal to 0. The time dependence of a daughter-product activity is given by the equation

$$\frac{dA_S^j}{dt} = -(\lambda_{rj} + \lambda_b)A_S^j + \lambda_{rj}A_S^{j-1} , \; j = 1, 2, \ldots, n, \tag{8.17}$$

in which λ_{rj} is the radiological decay constant for the jth daughter product. Note also that λ_b is independent of j, in accord with approximation (d). The solution to this set of equations (see Appendix F, Sec. F.1.3) is as follows:

$$A_S^0(t) = A_0 f_T f_2 e^{-(\lambda_b + \lambda_{r0})t}, \tag{8.18}$$

and, for $j = 1$ to n,

$$A_S^j(t) = A_0 f_T f_2 \left(\prod_{k=1}^{j} \lambda_{rk} \right) \sum_{k=0}^{j} \frac{e^{-(\lambda_b + \lambda_{rk})t}}{\prod_{\substack{p=0 \\ p \neq k}}^{j} (\lambda_{rp} - \lambda_{rk})}, \tag{8.19}$$

provided no two λ_{rk} are equal. Note that this result is in the same form as Eq. (8.14). The total number of disintegrations of radionuclide j in the source organ over interval η is, following Eq. (8.11),

$$U_S^j = \frac{1}{A_0} \int_0^\eta dt\, A_S^j(t). \tag{8.20}$$

For $j = 0$,

$$U_S^0 = f_T f_2 \frac{1 - e^{-(\lambda_b + \lambda_{r0})\eta}}{\lambda_b + \lambda_{r0}}, \tag{8.21}$$

or, for $j = 1$ to n,

$$U_S^j = f_T f_2 \left(\prod_{k=1}^{j} \lambda_{rk} \right) \sum_{k=0}^{j} \frac{1 - e^{-(\lambda_b + \lambda_{rk})\eta}}{(\lambda_b + \lambda_{rk}) \prod_{\substack{p=0 \\ p \neq k}}^{j} (\lambda_{rp} - \lambda_{rk})}. \tag{8.22}$$

Note that this result is in the same form as Eq. (8.15). Values of radiological half lives are given in Appendix B. Values of f_1 and f_2, taken from ICRP Publication 30, are given in Appendix G, Tables G-3 and G-4.

Example 8.4

Consider the ingestion of soluble ^{99}Mo. What are the values of U_S^j for the parent and for the daughter ^{99m}Tc in the liver? Note that only in 87.6 percent of its transformations does ^{99}Mo decay to ^{99m}Tc. The balance of transformations leads to the very long lived ^{99}Tc whose activity is ignored. Assume that $f_T = f_{BF} = 0.785$ (obtained from methods discussed in Sec. 8.8.2).

Solution   From Table G-3, there are two values of f_2 and λ_b, namely, $f_2' = 0.03$ and $\lambda_b' = \ln(2)/1 = 0.693$ d^{-1}, and $f_2'' = 0.27$ and $\lambda_b'' = \ln(2)/50 = 0.0139$ d^{-1}. From tables of radionuclide data in the Appendix, $\lambda_{r0} = 0.251$ d^{-1} for ^{99}Mo and $\lambda_{r1} = 2.76$ d^{-1} for ^{99m}Tc. Thus, for the parent, using Eq. (8.21),

$$U_S^0 = f_T f_2' \left\{ \frac{1 - \exp(-[\lambda_{r0} + \lambda_b']\eta)}{(\lambda_{r0} + \lambda_b')} \right\} + f_T f_2'' \left\{ \frac{1 - \exp(-[\lambda_{r0} + \lambda_b'']\eta)}{(\lambda_{r0} + \lambda_b'')} \right\}$$

$$= 0.83 \text{ d or } 7.4 \times 10^4 \text{ Bq}^{-1},$$

and, for the daughter, using Eq. (8.22),

$$U_S^1 = 0.876 \, f_T f_2' \lambda_{r1} \left\{ \frac{1 - \exp(-[\lambda_{r0} + \lambda_b']\eta)}{(\lambda_{r0} + \lambda_b')(\lambda_{r1} - \lambda_{r0})} - \frac{1 - \exp(-[\lambda_{r1} + \lambda_b']\eta)}{(\lambda_{r1} + \lambda_b')(\lambda_{r1} - \lambda_{r0})} \right\}$$

$$+ \, 0.876 \, f_T f_2'' \lambda_{r1} \left\{ \frac{1 - \exp(-[\lambda_{r0} + \lambda_b'']\eta)}{(\lambda_{r0} + \lambda_b'')(\lambda_{r1} - \lambda_{r0})} - \frac{1 - \exp(-[\lambda_{r1} + \lambda_b'']\eta)}{(\lambda_{r1} + \lambda_b'')(\lambda_{r1} - \lambda_{r0})} \right\}$$

$$= 0.80 \text{ d or } 6.9 \times 10^4 \text{ Bq}^{-1}.$$

8.5 ABSORBED FRACTION AND SPECIFIC EFFECTIVE ENERGY

The general expression for specific effective energy, given by Eq. (8.8) requires a great deal of data. Organ masses are given in Appendix G, Table G-1. Tables of yields and energies are given in Appendix B for selected radionuclides.

Because of the short ranges of charged particles in tissue, a charged particle's kinetic energy is given up to the medium essentially at its place of birth. Thus, for charged particles, $AF(T \leftarrow S)$ or $\phi(T \leftarrow S)$ is unity for $T = S$ and is zero for $T \neq S$. For photons, however, the calculation of AF is not so straightforward.

In developing the supporting technology for both the MIRD and the ICRP methods, irradiation of one organ by radionuclides in another was considered. Monte Carlo radiation transport calculations were performed for a phantom representation of Reference Man. For each organ or tissue, and for selected photon energies, an emitter was assumed to be uniformly distributed in the homogenized phantom source organ. Average energy absorption in each homogenized phantom target organ as well as the source organ were calculated. Results of such calculations have been published in ICRP Publication 23 (ICRP 1975). Spontaneously fissioning radionuclides have received special treatment (Dillman and Jones 1975; Ford et al. 1977). Methodology and results have also been published by the Society of Nuclear Medicine (Loevinger and Berman 1976; Berman 1977; Snyder et al. 1975; Snyder et al. 1978). Representative results are illustrated in Appendix G, Table G-4. Age dependent absorbed fractions for the newborn, ages one, five, ten, and fifteen (or adult female), and adult male have been computed by Cristy and Eckerman (1987).

Example 8.5
Evaluate $SEE(T \leftarrow S)$ for ^{141}Ce if the source and target organs are both the liver.

Solution   Data are given in Appendix B for yields and radiation energies. Values for AF are obtained by interpolation in data of Appendix G, Table G-4. Calculations based on Eq. (8.6) are summarized in the table on page 391.

From Table G-1, $M_T = 1800$ g for the liver. Thus

$$SEE(T \leftarrow S) = \frac{1}{M_T} \sum_i Y_i E_i Q_i AF(T \leftarrow S)_i = 1.0 \times 10^{-4} \text{MeV g}^{-1}.$$

Type of Radiation	Y	E	Q	AF	$Y \cdot E \cdot Q \cdot AF$
Total β^-, e^-	1.000	0.1707	1	1.000	0.1707
γ, x	0.482	0.1454	1	0.163	0.0114
γ, x	0.089	0.0360	1	0.428	0.0014
γ, x	0.049	0.0356	1	0.428	0.0007
					0.1842

8.6 THE SPECIFIC COMMITTED DOSE EQUIVALENT

The specific committed dose equivalent is calculated using Eqs. (8.3) and (8.7) as

$$\hat{H}_E = \sum_T w_T \hat{H}_T = \sum_T w_T \sum_S \hat{H}(T \leftarrow S) = \kappa \sum_T w_T \sum_S U_S \, SEE(T \leftarrow S). \tag{8.23}$$

Values of $\hat{H}_T$ for selected radionuclides are given in Appendix G, Tables G-5 and G-6. These data provide the basis for annual limits of intake and derived air concentrations which are discussed in Sec. 8.12. Table G-5 applies to ingested radionuclides and their radioactive daughters and includes values of f_1, the fraction of the ingested element transferred from the gastrointestinal system to the bloodstream, that is, to the body-fluid compartment. Table G-6 applies to inhaled radionuclides and their radioactive daughters, and values are given for the chemical classification of the inhaled material (see Sec. 8.7). In both tables values are given only for organs and tissues irradiated to a relatively significant extent. Values for "other" organs and "weight" factor given in the last two columns of the tables, are required for evaluation of the (weighted-average) committed effective dose equivalent. Values of $\hat{H}_E$, computed using Eq. (8.23) and values of $\hat{H}_T$ from Appendix G, are given in Tables 8.1 and 8.2. Classification of airborne radionuclides is described in the following section.

8.7 THE ICRP MODEL FOR THE RESPIRATORY SYSTEM

There are two essential features of a mathematical model for the respiratory system. One is a description of how the initial disposition of inhaled and retained materials in the different parts of the system depends on the physical and chemical nature of the airborne materials. The second is a description of how materials are transferred from one part of the respiratory system to another and how they are ultimately cleared from the system by transfer to body fluids or to the gastrointestinal system. When the model is applied to radiation dose assessment, there must be added a third feature, a dosimetry model relating radionuclide transformations in the various parts of the system to radiation doses in the radiosensitive tissues of the system.

TABLE 8.1 Specific Committed Effective Dose Equivalents for Ingestion of Selected Radionuclides.

NUCLIDE	f_1	$\hat{H}_E$ Sv/Bq	$\hat{H}_E$ rem/Ci	NUCLIDE	f_1	$\hat{H}_E$ Sv/Bq	$\hat{H}_E$ rem/Ci
Na-24	1E+00	3.9E−10	1.4E+03	Ru-105	5E−02	2.8E−10	1.0E+03[a]
P-32	8E−01	2.1E−09	7.7E+03	Ru-106	5E−02	5.8E−09	2.1E+04
K-40	1E+00	5.1E−09	1.9E+04	Ag-110m	5E−02	2.9E−09	1.1E+04
Cr-51	1E−01	3.6E−11	1.3E+02	Te-125m	2E−01	9.2E−10	3.4E+03
Cr-51	1E−02	4.0E−11	1.5E+02	Te-127m	2E−01	2.1E−09	7.9E+03
Mn-54	1E−01	7.3E−10	2.7E+03	Te-127	2E−01	1.9E−10	6.9E+02
Mn-56	1E−01	2.6E−10	9.5E+02	Te-129m	2E−01	2.7E−09	9.9E+03
Fe-55	1E−01	1.6E−10	5.8E+02	Te-129	2E−01	5.2E−11	1.9E+02
Fe-59	1E−01	1.8E−09	6.6E+03	Te-131m	2E−01	2.2E−09	8.3E+03
Co-58	5E−02	7.6E−10	2.8E+03	Te-131	2E−01	2.3E−10	8.5E+02
Co-60	5E−02	2.7E−09	1.0E+04	Te-132	2E−01	2.0E−09	7.4E+03
Ni-63	5E−02	1.4E−10	5.4E+02	I-125	1E+00	1.0E−08	3.8E+04
Ni-65	5E−02	1.6E−10	6.1E+02	I-130	1E+00	1.2E−09	4.3E+03
Cu-64	5E−01	1.2E−10	4.3E+02	I-131	1E+00	1.4E−08	5.3E+04
Zn-65	5E−01	3.9E−09	1.4E+04	I-133	1E+00	2.7E−09	1.0E+04
Zn-69	5E−01	2.3E−11	8.5E+01	I-134	1E+00	5.2E−11	1.9E+02
Br-83	1E+00	2.0E−11	7.3E+01	I-135	1E+00	5.4E−10	2.0E+03
Br-84	1E+00	4.1E−11	1.5E+02	Cs-134	1E+00	2.0E−08	7.4E+04
Rb-88	1E+00	4.4E−11	1.6E+02	Cs-136	1E+00	3.1E−09	1.1E+04
Rb-89	1E+00	2.2E−11	8.0E+01	Cs-137	1E+00	1.4E−08	5.0E+04
Sr-89	3E−01	2.2E−09	8.2E+03	Cs-138	1E+00	4.2E−11	1.6E+02
Sr-90	3E−01	3.5E−08	1.3E+05	Ba-139	1E−01	1.1E−10	3.9E+02
Y-90	1E−04	2.7E−09	1.0E+04	Ba-140	1E−01	2.3E−09	8.4E+03
Y-91m	1E−04	1.0E−11	3.9E+01	Ba-141	1E−01	5.5E−11	2.0E+02
Y-91	1E−04	2.4E−09	8.9E+03	La-140	1E−03	2.1E−09	7.7E+03
Y-92	1E−04	5.0E−10	1.9E+03	La-142	1E−03	1.7E−10	6.3E+02
Y-93	1E−04	1.2E−09	4.5E+03	Ce-141	3E−04	7.0E−10	2.6E+03
Zr-95	2E−03	9.2E−10	3.4E+03	Ce-143	3E−04	1.1E−09	4.2E+03
Zr-97	2E−03	2.2E−09	8.0E+03	Ce-144	3E−04	5.3E−09	2.0E+04
Nb-95	1E−02	6.0E−10	2.2E+03	Pr-143	3E−04	1.2E−09	4.5E+03
Mo-99	8E−01	8.1E−10	3.0E+03	Pr-144	3E−04	3.0E−11	1.1E+02
Tc-99m	8E−01	1.6E−11	6.0E+01	Nd-147	3E−04	1.1E−09	3.9E+03
Tc-101	8E−01	1.0E−11	3.8E+01	W-187	3E−01	5.1E−10	1.9E+03
Ru-103	5E−02	7.3E−10	2.7E+03	Np-239	1E−02	8.0E−10	2.9E+03

[a] Read as $1.0 \times 10^{+03}$, and so on.

Source: ICRP 1979.

8.7.1 The ICRP Respiratory Model

The model became part of the ICRP methodology with the release of Publication 30 in 1977. However, the model itself was developed much earlier (ICRP 1966). The model divides the respiratory system into several distinct regions, with rate constants characterizing transfer of radionuclides among the regions and to body fluids directly or to the gastrointestinal system. Deposition of inhaled material into the various regions is dependent on the chemical form and on the particle size.

TABLE 8.2 Specific Committed Effective Dose Equivalents for Inhalation of Selected Radionuclides.

NUCLIDE	CLASS	$\hat{H}_E$ Sv/Bq	$\hat{H}_E$ rem/Ci	NUCLIDE	CLASS	$\hat{H}_E$ Sv/Bq	$\hat{H}_E$ rem/Ci
Na-24	D	2.6E−10	9.5E+02	Br-84	W	2.0E−11	7.5E+01[a]
P-32	W	3.6E−09	1.3E+04	Br-84	D	2.4E−11	8.7E+01
P-32	D	1.5E−09	5.5E+03	Rb-88	D	2.2E−11	8.0E+01
K-40	D	3.4E−09	1.2E+04	Rb-89	D	1.0E−11	3.7E+01
Cr-51	D	2.9E−11	1.1E+02	Sr-89	D	1.6E−09	5.9E+03
Cr-51	W	5.8E−11	2.1E+02	Sr-89	Y	1.0E−08	3.7E+04
Cr-51	Y	7.1E−11	2.6E+02	Sr-90	D	6.1E−08	2.3E+05
Mn-54	D	1.5E−09	5.4E+03	Sr-90	Y	3.5E−07	1.3E+06
Mn-54	W	1.7E−09	6.4E+03	Y-90	W	2.0E−09	7.4E+03
Mn-56	D	8.8E−11	3.3E+02	Y-90	Y	2.2E−09	8.2E+03
Mn-56	W	6.5E−11	2.4E+02	Y-91m	Y	8.4E−12	3.1E+01
Fe-55	D	7.0E−10	2.6E+03	Y-91m	W	5.6E−12	2.1E+01
Fe-55	W	3.3E−10	1.2E+03	Y-91	W	7.9E−09	2.9E+04
Fe-59	W	2.7E−09	9.9E+03	Y-91	Y	1.2E−08	4.4E+04
Fe-59	D	4.0E−09	1.5E+04	Y-92	W	1.6E−10	6.0E+02
Co-58	W	1.2E−09	4.6E+03	Y-92	Y	1.7E−10	6.2E+02
Co-58	Y	1.9E−09	7.1E+03	Y-93	Y	5.6E−10	2.1E+03
Co-60	Y	4.1E−08	1.5E+05	Y-93	W	4.7E−10	1.8E+03
Co-60	W	8.0E−09	3.0E+04	Zr-95	W	3.6E−09	1.3E+04
Ni-63	W	5.1E−10	1.9E+03	Zr-95	Y	4.9E−09	1.8E+04
Ni-63	D	8.4E−10	3.1E+03	Zr-95	D	5.0E−09	1.9E+04
Ni-65	W	4.6E−11	1.7E+02	Zr-97	W	9.7E−10	3.6E+03
Ni-65	D	5.7E−11	2.1E+02	Zr-97	D	6.8E−10	2.5E+03
Cu-64	D	4.4E−11	1.6E+02	Zr-97	Y	1.1E−09	4.0E+03
Cu-64	W	5.8E−11	2.2E+02	Nb-95	W	1.0E−09	3.9E+03
Cu-64	Y	6.2E−11	2.3E+02	Nb-95	Y	1.2E−09	4.5E+03
Zn-65	Y	5.0E−09	1.8E+04	Mo-99	Y	9.8E−10	3.6E+03
Zn-69	Y	9.6E−12	3.6E+01	Mo-99	D	5.3E−10	2.0E+03
Br-83	D	2.1E−11	7.6E+01	Tc-99m	W	5.7E−12	2.1E+01
Br-83	W	2.2E−11	8.0E+01	Tc-99m	D	8.7E−12	3.2E+01

(Continued)

For a given inhalation exposure, analysis of the dynamics of radionuclide transport within the lung is necessary for evaluation of the number of source organ transformations caused by the inhaled radionuclide and its daughters. Although the ICRP model is adequate for evaluation of dose equivalents in the various regions of the respiratory system, ICRP Publication 30 treats the respiratory system as a single source or target organ in evaluation of specific effective energies, $SEE(T \leftarrow S)$, and specific committed dose equivalents H_T.

The ICRP Description of the Respiratory System.   Figure 8.9 illustrates the physical model representing the respiratory system (ICRP 1966). There are three distinct respiratory regions—the naso-pharynx $(N-P)$ region, the trachea and bronchial tree $(T-B)$, and the pulmonary (P) region. Each of these regions is subdivided into

TABLE 8.2 (continued)

NUCLIDE	CLASS	$\hat{H}_E$ Sv/Bq	rem/Ci	NUCLIDE	CLASS	$\hat{H}_E$ Sv/Bq	rem/Ci
Tc-101	W	3.6E−12	1.3E+01	I-125	D	6.6E−09	2.4E+04[a]
Tc-101	D	4.2E−12	1.6E+01	I-130	D	6.7E−10	2.5E+03
Ru-103	W	1.4E−09	5.1E+03	I-131	D	8.7E−09	3.2E+04
Ru-103	D	8.0E−10	3.0E+03	I-133	D	1.5E−09	5.4E+03
Ru-103	Y	2.1E−09	7.8E+03	I-134	D	3.0E−11	1.1E+02
Ru-105	D	9.2E−11	3.4E+02	I-135	D	3.1E−10	1.1E+03
Ru-105	W	9.4E−11	3.5E+02	Cs-134	D	1.3E−08	4.7E+04
Ru-105	Y	1.1E−10	4.1E+02	Cs-136	D	2.0E−09	7.5E+03
Ru-106	W	2.5E−08	9.3E+04	Cs-137	D	8.7E−09	3.2E+04
Ru-106	Y	1.2E−07	4.4E+05	Cs-138	D	2.4E−11	8.8E+01
Ru-106	D	1.5E−08	5.5E+04	Ba-139	D	4.3E−11	1.6E+02
Ag-110m	W	7.2E−09	2.7E+04	Ba-140	D	9.7E−10	3.6E+03
Ag-110m	Y	1.4E−08	5.3E+04	Ba-141	D	2.0E−11	7.4E+01
Ag-110m	D	1.0E−08	3.8E+04	La-140	D	9.2E−10	3.4E+03
Te-125m	W	1.8E−09	6.7E+03	La-140	W	1.2E−09	4.4E+03
Te-125m	D	1.3E−09	4.9E+03	La-142	D	6.0E−11	2.2E+02
Te-127m	D	3.2E−09	1.2E+04	La-142	W	4.2E−11	1.6E+02
Te-127m	W	5.2E−09	1.9E+04	Ce-141	W	1.9E−09	7.1E+03
Te-127	D	6.0E−11	2.2E+02	Ce-141	Y	2.3E−09	8.5E+03
Te-127	W	7.8E−11	2.9E+02	Ce-143	W	7.5E−10	2.8E+03
Te-129m	W	5.5E−09	2.0E+04	Ce-143	Y	8.5E−10	3.2E+03
Te-129m	D	2.2E−09	8.0E+03	Ce-144	Y	9.5E−08	3.5E+05
Te-129	D	2.1E−11	7.7E+01	Ce-144	W	5.2E−08	1.9E+05
Te-129	W	1.8E−11	6.7E+01	Pr-143	Y	2.0E−09	7.3E+03
Te-131m	D	1.1E−09	4.1E+03	Pr-143	W	1.7E−09	6.2E+03
Te-131m	W	1.5E−09	5.5E+03	Pr-144	W	1.1E−11	3.9E+01
Te-131	W	1.2E−10	4.3E+02	Pr-144	Y	1.1E−11	4.2E+01
Te-131	D	1.1E−10	4.0E+02	Nd-147	Y	1.7E−09	6.2E+03
Te-132	W	2.1E−09	7.7E+03	Nd-147	W	1.4E−09	5.3E+03
Te-132	D	1.8E−09	6.5E+03	W-187	D	1.4E−10	5.3E+02
				Np-239	W	5.8E−10	2.2E+03

[a] Read as $2.4 \times 10^{+04}$, and so on.
Source: ICRP 1979.

two or four compartments. All three regions have paths directly to body fluids. Only the P region releases material to the lymphatic (L) system. Both the $N-P$ and the $T-B$ regions have paths to the gastrointestinal system. Connection between the P region and the gastrointestinal system is only through feedback via the $T-B$ region. The factors D_{N-P}, D_{T-B}, and D_P represent the fraction of inhaled material deposited in the three respiratory regions, the balance being the fraction exhaled. These factors, and the D, W, and Y chemical classifications (denoting increasingly avid retention) are discussed in the next section.

For each chemical classification, values are given in Fig. 8.9 for the fraction F of the material deposited in a region that is transferred immediately from the region to each of its compartments. Also given are biological half-lives for transfer of material between compartments. For example, Fig. 8.9 reveals that, for a class W material

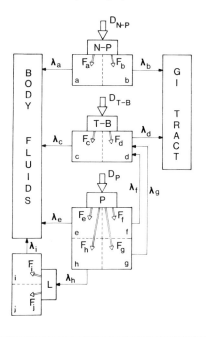

		Class					
		D		W		Y	
Region	Compart-ment	T day	F	T day	F	T day	F
N-P	a	0.01	0.5	0.01	0.1	0.01	0.01
$(D_{N-P} = 0.30)$	b	0.01	0.5	0.40	0.9	0.40	0.99
T-B	c	0.01	0.95	0.01	0.5	0.01	0.01
$(D_{T-B} = 0.08)$	d	0.2	0.05	0.2	0.5	0.2	0.99
	e	0.5	0.8	50	0.15	500	0.05
P	f	n.a.	n.a.	1.0	0.4	1.0	0.4
$(D_P = 0.25)$	g	n.a.	n.a.	50	0.4	500	0.4
	h	0.5	0.2	50	0.05	500	0.15
L	i	0.5	1.0	50	1.0	1000	0.9
	j	n.a.	n.a.	n.a.	n.a.	∞	0.1

Figure 8.9   The mathematical model used to describe clearance from the eight-compartment, four-region, respiratory system. The values for removal half-times T and compartmental fractions F are given for each of the three classes of retained materials. The values given for the three D fractions are the regional depositions for an aerosol with an AMAD of 1 μm. [From ICRP (1979).]

deposited in region P, 15% is transferred to body fluids with a half-life of 50 d; 40% is transferred to the $T - B$ region with a half-life of 1 d; 40% is transferred to the $T - B$ region with a half life of 50 d; and 5% is transferred to the L system with a half life of 50 d. Note that the half-lives are given in units of days. When converting these to decay constants in units of s^{-1}, one must use the factor 86,400 s d^{-1}.

The ICRP Deposition Model.   Deposition of inhaled material depends on the distribution in particle sizes of the carrier (aerosol). Characterizing this distribution is the activity median aerodynamic diameter, AMAD. This quantity is closely approximated by the mass-median aerodynamic diameter, the aerodynamic diameter of a particle being the diameter of a spherical particle with the same settling velocity. Deposition in the different regions of the respiratory system, as a function of AMAD, is illustrated in Fig. 8.10. If AMAD is unknown, then a value of 1 μm should be used, and values of H_T for inhaled materials, as reported in ICRP Publication 30, are based on the 1-μm value. Correction for other values of AMAD, say x μm, may be made as follows:

$$\frac{H_T(x\ \mu m)}{H_T(1\ \mu m)} = f_{N-P}\frac{D_{N-P}(x\ \mu m)}{D_{N-P}(1\ \mu m)} + f_{T-B}\frac{D_{T-B}(x\ \mu m)}{D_{T-B}(1\ \mu m)} + f_P\frac{D_P(x\ \mu m)}{D_P(1\ \mu m)}. \quad (8.24)$$

The fractions f_{N-P}, f_{T-B}, and f_P are, respectively, the proportions of H_T (1 μm) resulting from deposition in the $N-P$, $T-B$, and P regions.

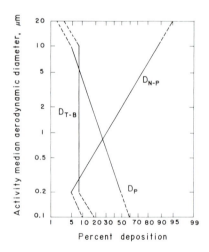

Figure 8.10   Deposition of dust in the respiratory system. The percentages of activity or mass of an aerosol which is deposited in the $N-P$, $T-B$, and P regions is given in relation to the AMAD of the aerosol. The model is intended for AMADs between 0.2 and 10 μm and with geometric standard deviations less than 4.5. Provisional estimates are shown by dashed lines. For AMAD > 20 μm complete deposition in the $N-P$ region can be assumed. The model does not apply to AMAD < 0.1 μm. [From ICRP (1979).]

Retention classifications D, W, and Y are provided in Fig. 8.11 for elements and their compounds. For example, all halides (group 7a) of the alkali metals (group 1a) are in class D, while all halides of the alkaline earths (group 2a) are in class W. All oxides of the lanthanide and actinide elements are in class Y.

Clearance Calculations for the ICRP Model.   As seen in Sec. 8.4, determination of source organ transformations requires evaluation of the time dependence

Class Y — Avid retention: cleared slowly (years)
  Carbides – actinides, lanthanides, Zr, Y, Mo
  Sulfides – none
  Sulfates – none
  Carbonates – none
  Phosphates – none
  Oxides and hydroxides – lanthanides, actinides, Groups 8 (V and VI), 1b, 2b (IV and V), 3b except
  Sc^{3+}, and 6b
  Halides – lanthanum flourides
  Nitrates – none
Class W — Moderate retention: intermediate clearance rates (weeks)
  Carbides – Cations of all Class W hydroxides except those listed as Class Y carbides
  Sulfides – Groups 2a (V + VI), 4a (IV - VI), 5a (IV - VI), 1b, 2b, 6b (V + VI)
  Sulfates – Groups 2a (IV - VII), and 5a (IV - VI)
  Carbonates – lanthanides, Bi^{3+} and Group 2a (IV - VII)
  Phosphates – Zn^{2+}, Sn^{3+}, Mg^{2+}, Fe^{3+}, Bi^{3+}, and lanthanides
  Oxides and hydroxides – Groups 2a (II - VII), 3a (III - VI), 4a (III - VI), 5a (IV - VI), 6a (VI - VI), 8,
  2b (VI), 4b, 5b, and 7b Sc^{+3}
  Halides – lanthanides (except fluorides), Groups 2a, 3a (III - VI), 4a (IV - VI), 5a (IV - VI), 8, 1b, 2b,
  3b (IV - V), 4b, 5b, 6b, and 7b
  Nitrates – all cations whose hydroxides are Class Y and W
Class D — Minimal retention: rapid clearance (days)
  Carbides – see hydroxides
  Sulfides – all except Class W
  Carbonates – all except Class W
  Phosphates – all except Class W
  Oxides and Hydroxides – Groups 1a, 3a (II), 4a (II), 5a (II,III), 6a (III)
  Halides – Groups 1a and 7a
  Nitrates – all except Class W
  Noble Gases – Group 0
Note: Where reference is made from one chemical form to another, it implies that an *in vivo* conversion occurs,
  e.g., hydrolysis reaction.

The following periodic table of the elements is used with the foregoing classification.

Period	1a	2a	3b	4b	5b	6b	7b	8			1b	2b	3a	4a	5a	6a	7a	0
I	H																	He
II	Li	Be											B	C	N	O	F	Ne
III	Na	Mg											Al	Si	P	S	Cl	Ar
IV	K	Ca	Sc	Ti	V	Cr	Mn	Fe	Co	Ni	Cu	Zn	Ga	Ge	As	Se	Br	Kr
V	Rb	Sr	Y	Zr	Nb	Mo	Tc	Ru	Rh	Pd	Ag	Cd	In	Sn	Sb	Te	I	Xe
VI	Cs	Ba	La*	Hf	Ta	W	Re	Os	Ir	Pt	Au	Hg	Tl	Pb	Bi	Po	At	Rn
VII	Fr	Ra	Ac**															

*Lanthanides	Ce	Pr	Nd	Pm	Sm	Eu	Gd	Tb	Dy	Ho	Er	Tm	Yb	Lu
**Actinides	Th	Pa	U	Np	Pu	Am	Cm	Bk	Cf	Es	Fm	Md	No	Lw

Figure 8.11 Pulmonary clearance classification of inorganic compounds. [From ICRP (1966).]

of activity in the organ. Suppose that an activity A_0 is inhaled at time $t = 0$. The
initial activity in each of the compartments a to i of Fig. 8.9 is given by the product of A_0, the D-value for the region containing that compartment, and the F-value
for the compartment. For example, if $A_d(t)$ is the activity in compartment d, then
$A_d(0) = A_0 D_{T-B} F_d$. The rate of clearance depends on the radiological decay constant λ_r of the inhaled radionuclide and the biological decay constant, for example,
λ_d (s^{-1}) $= 7 \ln2/(86400\ T_d)$. Activities in each compartment may be determined by
solution of the following set of differential equations:
For $k = a, b, c, e, f, g$ and h,

$$\frac{dA_k}{dt} = -(\lambda_r + \lambda_k)A_k. \tag{8.25}$$

For $k = d, i$ and j,

$$\frac{dA_d}{dt} = +\lambda_f A_f + \lambda_g A_g - (\lambda_r + \lambda_d)A_d, \tag{8.26}$$

$$\frac{dA_i}{dt} = -(\lambda_r + \lambda_i)A_i + F_i\lambda_h A_h, \tag{8.27}$$

$$\frac{dA_j}{dt} = -\lambda_r A_j + F_j\lambda_h A_h. \tag{8.28}$$

Similar systems of equations may be written for daughter products in a radioactive decay chain. The metabolic behaviors of the daughters are assumed to be the same as that of the parent. Thus, the biological half-lives and partition fractions F can be taken as those of the parent.

Having determined the compartmental activities one may evaluate the rate of transfer to the body fluids, $BF(t)$ Bq s^{-1}, and to the gastrointestinal system, $GI(t)$ Bq s^{-1}, as

$$BF(t) = \lambda_a A_a + \lambda_c A_c + \lambda_e A_e + \lambda_i A_i, \tag{8.29}$$

$$GI(t) = \lambda_b A_b + \lambda_d A_d. \tag{8.30}$$

These two transfer rates lead directly to the evaluation of the fraction f'_{BF} of an inhaled radionuclide transferred directly to the body fluid and the fraction f_{GI} of the inhaled radionuclide transferred to the gastrointestinal system. The evaluation is greatly simplified since holdup times in the respiratory and gastrointestinal systems are very much shorter than the fifty years assumed for η in the calculation of the dose committment.

The fraction f'_{BF} is given by

$$f'_{BF} = \frac{1}{A_0} \int_0^\eta dt \, BF(t)$$

$$\simeq \frac{D_{N-P}F_a\lambda_a}{\lambda_a + \lambda_r} + \frac{D_{T-B}F_c\lambda_c}{\lambda_c + \lambda_r} + \frac{D_P F_e\lambda_e}{\lambda_e + \lambda_r} + \frac{D_P F_h\lambda_h F_i\lambda_i}{(\lambda_h + \lambda_r)(\lambda_i + \lambda_r)}. \tag{8.31}$$

The fraction f_{GI} is given by

$$f_{GI} = \frac{1}{A_0} \int_0^\eta dt \, GI(t)$$

$$\simeq \frac{D_{N-P}F_b\lambda_b}{\lambda_b + \lambda_r} + \frac{D_{T-B}F_d\lambda_d}{\lambda_d + \lambda_r} + D_P \left(\frac{F_f\lambda_f}{\lambda_f + \lambda_r} + \frac{F_g\lambda_g}{\lambda_g + \lambda_r} \right) \frac{\lambda_d}{\lambda_d + \lambda_r}. \tag{8.32}$$

Note that considerable simplification arises if the half life of the radionuclide is long in comparison to half-times in the compartments of the respiratory system.

TABLE 8.3   Approximate Expressions for the Number of Transformations U in the Various Compartments of the Lung Following Inhalation of 1 Bq of Activity.

COMPARTMENT	NUMBER OF TRANSFORMATIONS	COMPARTMENT	NUMBER OF TRANSFORMATIONS
a	$\dfrac{D_{N-P}F_a}{\lambda_a + \lambda_r}$	b	$\dfrac{D_{N-P}F_b}{\lambda_b + \lambda_r}$
c	$\dfrac{D_{T-B}F_c}{\lambda_c + \lambda_r}$	d	$\dfrac{1}{\lambda_d + \lambda_r}\left(D_{T-B}F_d + \dfrac{D_P F_f \lambda_f}{\lambda_f + \lambda_r} + \dfrac{D_P F_g \lambda_g}{\lambda_g + \lambda_r}\right)$
e	$\dfrac{D_P F_e}{\lambda_e + \lambda_r}$	f	$\dfrac{D_P F_f}{\lambda_f + \lambda_r}$
g	$\dfrac{D_P F_g}{\lambda_g + \lambda_r}$	h	$\dfrac{D_P F_h}{\lambda_h \lambda_r}$
i	$\dfrac{D_P F_h \lambda_h F_i}{(\lambda_h + \lambda_r)(\lambda_i + \lambda_r)}$	j	$\dfrac{D_P F_h \lambda_h F_j (1 - e^{-\lambda_r \eta})}{\lambda_r (\lambda_h + \lambda_r)}$

Source: ICRP 1979.

The ICRP Lung Dosimetry Model.   One may evaluate source organ transformations using Eq. (8.20). Combining compartments in each region thus results in values for U_{N-P}, U_{T-B}, U_P, and U_L. In the ICRP model, U_{N-P} is neglected and U_{T-B}, U_P, and U_L are summed to give a single value for the lung as a source organ. The lung is also treated as a single target organ of mass 1000 g, and these conventions have been followed in preparing the data for Appendix G.

U_{N-P}, for example, is given by $U_a + U_b$. Values of U for each compartment are given in Table 8.3. All entries save that for compartment j are based on the approximation that clearance times are much less than the 50-year dose evaluation period. In the entry for compartment j, the symbol η represents 50 years, in units compatible with those of λ_r. For U-values to have units of Bq^{-1}, λ-values must have units s^{-1}. ICRP Publication 30 also provides U-values for daughter products.

Example 8.6

Consider the inhalation of unit activity of ^{137}CsCl. Evaluate U_S, $SEE(T \leftarrow S)$ and $\hat{H}(T \leftarrow S)$ for the lung as the source and target.

Solution   The radioactive half life of ^{137}Cs is so long that radioactive decay may be neglected in determining activities. The inhalation class is D, thus compartments f, g, and j play no roles in lung clearance. With $A_0 = 1$, $A_a(0) = A_b(0) = 0.3 \times 0.5$, $A_c(0) = 0.08 \times 0.95$, $A_d(0) = 0.08 \times 0.05$, $A_e(0) = 0.25 \times 0.8$, and $A_h(0) = 0.25 \times 0.2$. Values of biological decay constants, in units of s^{-1}, are: $\lambda_a = \lambda_b = \lambda_c = 8.02 \times 10^{-4}$, $\lambda_d = 4.01 \times 10^{-5}$ and $\lambda_e = \lambda_h = \lambda_i = 1.60 \times 10^{-5}$.

Solutions of Eqs. (8.25) to (8.28) are as follows. For $k = a$, c, e, and h,

$$A_k(t) = A_k(0)e^{-\lambda_k t}.$$

For compartment i, since $\lambda_h = \lambda_i$,

$$A_i(t) = A_h(0)F_i \lambda_i t e^{-\lambda_i t}.$$

Since all biological half lives are much less than 50 years, Eq. (8.11) yields

$$U_{T-B} = D_{T-B} \left(\frac{F_c}{\lambda_c} + \frac{F_d}{\lambda_d} \right) = 1.94 \times 10^2 \text{ Bq}^{-1},$$

$$U_P = D_P \left(\frac{F_e}{\lambda_e} + \frac{F_h}{\lambda_h} \right) = 1.56 \times 10^4 \text{ Bq}^{-1},$$

$$U_L = \frac{D_P F_h F_i}{\lambda_i} = 3.13 \times 10^3 \text{ Bq}^{-1}.$$

In total, for the lung as the source organ, $U_s = 1.89 \times 10^4$ Bq^{-1}. From the Appendix, each decay of ^{137}Cs, assumed here to be in equilibrium with ^{137m}Ba, results in the following radiations. Values of AF are taken from Appendix G, and all $Q_i's$ are unity.

i	E_i(MeV)	Y_i	AF_i
Total β^-	0.2490	1.0000	1.0
γ, x	0.6616	0.8493	0.049
γ, x	0.0321	0.0371	0.21
γ, x	0.0318	0.0201	0.21

From Eq. (8.6) with $M_T = 1000$ g, $SEE(T \leftarrow S) = 2.8 \times 10^{-4}$ MeV g^{-1}. From Eq. (8.17), $\bar{H}(T \leftarrow S) = \kappa \, U_s SEE(T \leftarrow S) = 8.5 \times 10^{-10}$ Sv Bq^{-1}.

8.7.2 Revision of the ICRP Respiratory Model

An ICRP task group began in 1984 the revision of the model for the respiratory system. Many features of the group's proposals for a new model were described in 1989 (Bair; James et al.; Johnson and Milencoff; Masse and Cross; Roy). The solubility classifications of the ICRP-30 model would be retained in the new model. However, regions of the system would be redefined, and changes would be made in deposition and clearance modeling. Dose calculations would be performed separately for the different regions of the system.

Three major compartments of the respiratory system would be the extrathoracic $(E-T)$ region, the fast-clearing thoracic region, and the slow-clearing thoracic region. The $E-T$ compartment would include the nasal and oral regions, and the pharynx, larynx, and upper trachea. The fast-clearing or tracheo-bronchial $(T-B)$ compartment would include the remainder of the trachea and the bronchi. The slow-clearing or parenchymal-nodular $(P-N)$ compartment would include the bronchioles, alveoli, and thoracic lymph nodes. Deposition modeling would account for differences between nasal and oral breathing.

Clearance calculations for the three compartments would require numerical solution of differential equations because time-dependent coefficients would be used to describe the rates of transfer of material among compartments and to the gastrointestinal tract and body fluids. Dose calculations would be performed separately for

TABLE 8.4   Water Balances for Reference Adults and Children of Age Ten.

	ADULT MALE	ADULT FEMALE	CHILD (10-y)
Gains (cm^3 d^{-1})			
Milk	300	200	450
Water	150	100	200
Other liquid	1500	1100	750
In food	700	450	400
By oxidation	350	250	200
Total	3000	2100	2000
Losses (cm^3 d^{-1})			
In urine	1400	1000	1000
In feces	100	90	70
Insensible[a]	850	600	580
In sweat	650	420	350
Total	3000	2100	2000

[a]Expiration from the lungs and diffusion through the skin.

Source: ICRP 1975.

the three compartments. Instead of applying an overall weight factor of $w_T = 0.12$ for the dose to the lung as a whole when calculating the effective dose equivalent, individual weight factors of 0.02, 0.08, and 0.02 would be applied respectively to the $E - T$, $T - B$, and $P - N$ doses. The doses themselves would be calculated for the cells and tissues most specifically at risk from radiation exposure.

The proposed new model would have distinct advantages over the ICRP-30 model. It would be simpler conceptually even though requiring numerical solution of clearance calculations. It would also have a broader base of supporting data and a more sophisticated dosimetry model.

8.8 DOSIMETRIC MODEL FOR THE GASTROINTESTINAL SYSTEM

The ICRP Model divides the GI system into several distinct organs (see Fig. 8.12), with rate constants characterizing transfer of radionuclides from one organ to the next. Transfer to the bloodstream is assumed to occur only from the small intestine. Values of f_1, the fraction transferred from the gastrointestinal system to the body fluid, are given in Appendix G, Table G-2 for selected elements.

Organs of the GI system are treated as both source and target organs. Analysis of the dynamics of radionuclide transport is necessary for evaluation of the number of source organ transformations experienced by an ingested radionuclide and its daughter products in the various organs of the system. Some problems of radiation dose assessment require information on water balance (see Table 8.4).

8.8.1 Model of the GI System

Figure 8.12 illustrates the physical model of the GI system (Eve 1966). Four organs are involved. Associated with each are reference values for wall mass, content mass,

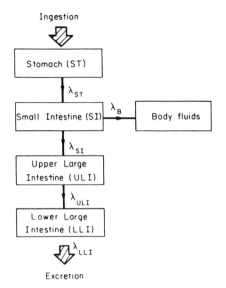

Section of GI tract	Mass of walls (g)	Mass of contents (g)	Mean residence time (day)	λ (day^{-1})
Stomach (ST)	150	250	1/24	24
Small intestine (SI)	640	400	4/24	6
Upper large intestine (ULI)	210	220	13/24	1.8
Lower large intestine (LLI)	160	135	24/24	1

Figure 8.12 Mathematical model used to describe the kinetics of radionuclide behavior in the gastrointestinal system. [From ICRP (1979).]

and mean residence time for the contents. The rate constant λ for transfer of contents from organ to organ is the reciprocal of the mean residence time. It is assumed that the small intestine is the only site for transfer from the GI system to the body fluids.

The rate constant λ_b for transfer from the small intestine to body fluids is related to f_1, the fraction of the ingested *stable* element transferred to body fluids (see Appendix G, Table G-2). From Fig. 8.12 it is clear that, for a stable element, f_1 is just the ratio $\lambda_b/(\lambda_b + \lambda_{SI})$. Thus,

$$\lambda_b = \frac{f_1 \lambda_{SI}}{1 - f_1}. \tag{8.33}$$

A value of unity for f_1 implies an infinite λ_b, that is, instant transfer to body fluids. In this special case, it may thus be assumed that transfer takes place from the *stomach* to body fluids, with none of the element entering the small intestine.

In dealing with radioactive daughter products of an ingested radionuclide, the ICRP model calls for application of the value f_1 (or λ_b) for the parent to all daughter products.

8.8.2 Radionuclide Transport in the GI System

Consider the ingestion of an activity A_0 of a radionuclide at time $t = 0$. Initial values of activities in each organ are: $A_{SI}(0) = A_{ULI}(0) = A_{LLI}(0) = 0$. For ingested radionuclides, $A_{ST}(0) = A_0$. For inhaled radionuclides, $A_{ST}(0) = A_0 f_{GI}$. Time dependencies of these activities are governed by the differential equations

$$\frac{dA_{ST}}{dt} = -(\lambda_r + \lambda_{ST})A_{ST}, \tag{8.34}$$

$$\frac{dA_{SI}}{dt} = -(\lambda_r + \lambda_{SI} + \lambda_b)A_{SI} + \lambda_{ST}A_{ST}, \tag{8.35}$$

$$\frac{dA_{ULI}}{dt} = -(\lambda_r + \lambda_{ULI})A_{ULI} + \lambda_{SI}A_{SI}, \tag{8.36}$$

$$\frac{dA_{LLI}}{dt} = -(\lambda_r + \lambda_{LLI})A_{LLI} + \lambda_{ULI}A_{ULI}, \tag{8.37}$$

in which λ_r is the radiological decay constant. Similar sets of equations may be written for each daughter product in a decay chain. These equations may be solved and then U_S values computed using Eq. (8.21). For example, for an ingested radionuclide

$$A_{ST}(t) = A_0\, e^{-(\lambda_r + \lambda_{ST})t}, \tag{8.38}$$

from which

$$U_{ST} = \frac{1 - e^{-(\lambda_r + \lambda_{ST})\eta}}{\lambda_r + \lambda_{ST}}. \tag{8.39}$$

Because the 50-year dose-evaluation time η is so much longer than the residence time in the GI system, the following approximate expressions for U_S values are appropriate

$$U_{ST} = \frac{1}{\lambda_{ST} + \lambda_r}, \tag{8.40}$$

$$U_{SI} = \frac{U_{ST}\lambda_{ST}}{\lambda_{SI} + \lambda_b + \lambda_r}, \tag{8.41}$$

$$U_{ULI} = \frac{U_{SI}\lambda_{SI}}{\lambda_{ULI} + \lambda_r}, \tag{8.42}$$

$$U_{LLI} = \frac{U_{ULI}\lambda_{ULI}}{\lambda_{LLI} + \lambda_r}. \tag{8.43}$$

For inhaled radionuclides, each quantity represented by Eqs. (8.40) to (8.43) must be multiplied by the factor f_{GI}. Note that for U_S to have units of Bq^{-1}, λ's must have units of s^{-1}. Similar approximations for daughter U_S values are given in ICRP Publication 30.

The fraction of the ingested parent radionuclide transferred to the body fluid is given by

$$f_{BF} = \frac{1}{A_0} \int_0^\eta dt \, \lambda_b \, A_{SI}(t) = \lambda_b \, U_{SI}. \tag{8.44}$$

Note that if the ingested radionuclide is long-lived in comparison with residence time in the stomach and small intestine, $\lambda_r \ll \lambda_{SI} < \lambda_{ST}$ and it follows that $f_{BF} \cong f_1$.

8.8.3 Specific Effective Energies for Organs of the GI System

Some modification of Eq. (8.6) is required to evaluate *SEE* values when target organs are those of the GI system. If the ith radiation is a neutron, x-ray or gamma ray, then the ith term in the summation is as written in Eq. (8.6), with the mass of the target being that of the *wall* of the organ. On the other hand, if the radiation is directly ionizing (β particle, α particle, etc.) and the *source* is the content of the target organ in the GI system, the only radiation dose received is that to the mucosal layer ML of the wall of the organ. The ratio $AF(T \leftarrow T)_i^j / M_T$, that is, the specific absorbed fraction (MeV g^{-1}) in the mucosal layer, is approximated as $0.5\nu / M_T^c$, in which M_T^c is the mass of the *content* of the organ and ν is the factor representing the degree to which the radiation penetrates the mucus, that is, the ratio of the absorbed dose rate in the mucosal layer to that at the interface between the mucus layer and the contents of the GI tract. For β particles, $\nu = 1.0$; for alpha particles and fission fragments, $\nu = 0.01$; and for recoil atoms, $\nu = 0$. The factor 0.5 is introduced because the absorbed dose rate at the interface between the contents and the mucus layer is approximately half that deep within the contents.

Example 8.7

Consider the inhalation of ^{32}P in the Class-D chemical form. The radionuclide has a half life of 14.3 d ($\lambda_r = 5.61 \times 10^{-7}$ s^{-1}) and on each disintegration emits a beta particle of average energy $E = 0.695$ MeV ($Y = 1.0$, $Q = 1.0$). Since ^{32}P emits only beta particles, radiation absorbed in any target organ can arise only in that same organ as a source. Evaluate the specific committed dose equivalent H_T (Sv Bq^{-1}) for each organ of the GI system.

Solution   It may be assumed that the AMAD of the inhaled material is 1 μm. From Fig. 8.9, $D_{N-P} = 0.3$, $D_{T-B} = 0.08$, $F_b = 0.5$, and $F_d = 0.05$. Both F_f and F_g are zero. Also from the figure, $\lambda_b = 8.02 \times 10^{-4}$ and $\lambda_d = 4.01 \times 10^{-5}$ s^{-1}, both much greater than λ_r. From Eq. (8.32),

$$f_{GI} \cong D_{N-P}F_b + D_{T-B}F_d = 0.15;$$

that is, 15 percent of the inhaled ^{32}P is transferred to the GI system. According to the ICRP model, this transfer is assumed to take place immediately upon inhalation for purposes of evaluating dose to the GI system. Values for λ's taken from Fig. 8.12 are listed in the following table. In all cases, the value is much greater than λ_r. Also from Table G-2, $f_1 = 0.8$. Thus, from Eq. (8.33),

$$\lambda_b = \frac{f_1 \lambda_{SI}}{1 - f_1} = 4\lambda_{SI}.$$

For each source (= target) organ, $U_T = U_S$ values may be computed using Eqs. (8.34) to (8.37) modified by the factor f_{GI}. These U values then express the number of disintegrations per inhaled Bq.

In the evaluation of $SEE(T \leftarrow T)$, because only beta particles are emitted,

$$AF(T \leftarrow T)/M_T = 0.5\nu/M_T^c,$$

in which $\nu = 1.0$. Values of M_T^c for contents of the organs may be taken from Fig. 8.12. Eq. (8.2) for the specific committed dose equivalent reduces to

$$\hat{H}_T = \hat{H}(T \leftarrow T) = 1.6 \times 10^{-10}\, U_T SEE(T \leftarrow T).$$

Final results are given in the following table for Class-D inhalation of ^{32}P.

Organ T	$\lambda(\mathrm{s}^{-1})$	$U_T(\mathrm{Bq}^{-1})$	$SEE(T \leftarrow T)(\mathrm{MeV\,g}^{-1})$	$\hat{H}_T(\mathrm{Sv\,Bq}^{-1})$
ST	0.78×10^{-4}	5.54×10^2	1.39×10^{-3}	1.2×10^{-10}
SI	6.94×10^{-5}	4.44×10^2	8.69×10^{-4}	6.2×10^{-11}
ULI	2.08×10^{-5}	1.48×10^3	1.58×10^{-3}	3.7×10^{-10}
LLI	1.16×10^{-5}	2.66×10^3	2.57×10^{-3}	1.1×10^{-9}

8.9 DOSIMETRIC MODEL FOR BONE

Masses of the components of the skeleton of Reference Man are given in Appendix G, Table G-1. Cortical bone (CB) is the compact or dense material of the outside of the bone. Trabecular bone (TB) is the cancellous or spongy inner portion of the bone containing the marrow. Cells at risk from radiation carcinogenesis are hematopoietic stem cells of the red marrow (RM) and the osteogenic cells close to bone surface (BS).

In the ICRP model for bone, there are two source organs, CB and TB, and two target organs, the 1500-g RM and the 120-g osteogenic cells of BS. Absorbed fractions $AF(T \leftarrow S)$ for photon emitters are given in Table G-4. For emitters of nonpenetrating radiation, the following approximations are used in the ICRP model. Radionuclides with half lives less than 15 days are assumed to be distributed on bone surfaces. Isotopes of the alkaline earth elements with half lives greater than 15 days are assumed to be uniformly distributed throughout the volume of bone. For longer lived radionuclides not in the alkaline-earth class, distributions must be obtained from metabolic data for individual elements. Recommended absorbed fractions for alpha-particle and beta-particle emitters are given in Table 8.5.

Further approximations are required in evaluation of source organ transformations, as used in the ICRP model. If a radionuclide is deposited uniformly on all bone surface, then, with MB representing total mineral bone, $U_{TB} \cong U_{CB} \cong 0.5U_{MB}$. If a radionuclide uniformly contaminates all bone mass, then $U_{TB} \cong 0.2U_{MB}$ and $U_{CB} \cong 0.8U_{MB}$. For radioisotopes of the alkaline earths, U_{CB} and U_{TB} can be obtained from data presented in ICRP Publication 20 (1973). Treatment of daughter products is described in ICRP Publications 20 and 30.

TABLE 8.5   Recommended Absorbed Fractions for Dosimetry of Radionuclides in Bone, Based on ICRP Publication 30.

SOURCE ORGAN	TARGET ORGAN	α EMITTER[a] UNIFORM IN VOLUME	α EMITTER[a] ON BONE SURFACE	β EMITTER UNIFORM IN VOLUME	β EMITTER ON BONE SURFACE $E_{avg} \geq 0.2$ MeV	β EMITTER ON BONE SURFACE $E_{avg} < 0.2$ MeV
TB	BS	0.025	0.25	0.025	0.025	0.25
CB	BS	0.01	0.25	0.015	0.015	0.25
TB	RM	0.05	0.5	0.35	0.5	0.5
CB	RM	0.0	0.0	0.0	0.0	0.0

[a] Fission fragments are assumed to have AF values equal to those of alpha particles. Energies of recoil atoms are disregarded for purposes of bone dosimetry.

Source: ICRP 1979.

Example 8.8

Consider the ingestion of ^{32}P. The radionuclide has a half life of 14.3 d ($\lambda_r = 5.61 \times 10^{-7}$ s^{-1}) and on each disintegration emits a beta particle of average energy 0.695 MeV ($Y = 1.0$, $Q = 1.0$). Evaluate the specific committed dose equivalent $\hat{H}_T$ (Sv Bq^{-1}) for bone surface (BS) and red marrow (RM).

Solution   From Example 8.7 and Eq. (8.44), $f_{BF} \cong f_1 = 0.8$. From Table G-3, $f_2 = 0.3$ for mineral bone. Because the half life is less than 15 days, ^{32}P is assumed to be uniformly distributed on bone surfaces. From Table 8.5, $AF(BS \leftarrow TB) = 0.025$, $AF(BS \leftarrow CB) = 0.015$, and $AF(RM \leftarrow TB) = 0.5$. Since phosphorus is permanently retained in bone (Table G-3), Eq. (8.11) reduces to

$$U_{MB} = f_{BF} f_2 \int_0^\eta dt \, e^{-\lambda_r t} \cong \frac{f_{BF} f_2}{\lambda_r} = 4.3 \times 10^5 \text{ Bq}^{-1},$$

and $U_{CB} \cong U_{TB} = 0.5 \, U_{MB} = 2.1 \times 10^5 \, Bq^{-1}$. Equation (8.6) yields, with $M_{BS} = 120$ g and $M_{RM} = 1500$ g,

$$SEE(BS \leftarrow TB) = \frac{Y \, E \, AF(BS \leftarrow TB)}{M_{BS}} = 1.4 \times 10^{-4} \text{ MeV g}^{-1},$$

$$SEE(BS \leftarrow CB) = \frac{Y \, E \, AF(BS \leftarrow CB)}{M_{BS}} = 8.7 \times 10^{-5} \text{ MeV g}^{-1},$$

$$SEE(RM \leftarrow TB) = \frac{Y \, E \, AF(RM \leftarrow TB)}{M_{RM}} = 2.3 \times 10^{-4} \text{ MeV g}^{-1}.$$

From Eq. (8.7),

$$\hat{H}_{BS} = 1.6 \times 10^{-10} [U_{TB} SEE(BS \leftarrow TB) + U_{CB} SEE(BS \leftarrow CB)]$$
$$= 7.6 \times 10^{-9} \text{ Sv Bq}^{-1},$$
$$\hat{H}_{RM} = 1.6 \times 10^{-10} \, U_{TB} SEE(RM \leftarrow TB) = 7.7 \times 10^{-9} \text{ Sv Bq}^{-1}.$$

8.10 DOSIMETRIC MODEL FOR SUBMERSION IN A RADIOACTIVE CLOUD

This discussion is limited to radioisotopes of argon, krypton, and xenon. Radon, which has alpha-active daughter products, poses a substantially different type of risk, and is given a separate appraisal by the ICRP (see Chap. 4). For the noble gases discussed here, exposure due to submersion in a radioactive cloud has been found to be almost exclusively due to external irradiation. Dose equivalents from gases contained in the lungs or absorbed in the body are negligible.

Radiation transport calculations have been carried out to determine dose equivalents in the organs of the body due to external irradiation by photons (including bremsstrahlung from beta particles). Dose rates to the skin and lens of the eye are evaluated over a 0-2 mm depth for photons. For beta particles and electrons, skin doses are evaluated at a depth of 0.07 mm and doses to the lens of the eye are evaluated at a depth of 3 mm.

Factors are reported in Table 8.6 for converting atmospheric concentrations to effective dose equivalent rates for submersion in a semi-infinite cloud, that is, under "2π-geometry" limited by the plane upon which Reference Man stands. Methods are given in Publication 30 (ICRP 1979) for evaluating exposures in room-sized finite clouds of radionuclides.

8.11 ICRP SPECIAL TREATMENT FOR SELECTED RADIONUCLIDES

Internal dosimetry calculations for certain radionuclides have special features requiring considerations beyond those of the standard ICRP methods. Among the radionuclides are tritium which, in water form, easily enters the body through the skin. Uptake and retention of radiocarbon depends strongly on the chemical form. Biokinetic modeling for radioiodine is highly specialized because of the unique role of the thyroid in metabolism. Similarly, highly specialized models are required for the uptake and retention of the bone-seeking alkaline earths.

8.11.1 Tritium

Elemental tritium poses only a minor risk to the lungs, the dose conversion factor being 9.9×10^{-15} Sv h^{-1} lung dose per Bq m^{-3} atmospheric concentration. Inhalation or ingestion of tritiated water results in uniform mixing with the total body water and a specific committed dose equivalent to soft tissues of 1.7×10^{-11} Sv Bq^{-1}. Exposure to an atmospheric concentration C (Bq m^{-3}) results in an intake of $0.02C$ Bq per minute by inhalation and $0.01C$ Bq per minute by absorption through the skin. The biological half life of tritiated water is taken to be 10 days. Analysis of the metabolic behavior of tritiated organic compounds is subject to many uncertainties. Some discussion of this subject is given in ICRP Publication 30.

TABLE 8.6 Dose Conversion Factors (Sv h^{-1} per Bq m^{-3}) for Organs ($\dot{H}_T$) and for the Effective Dose Equivalent ($\dot{H}_E$) for Submersion in Semi-Infinite Clouds of Radioactive Noble Gases, Based on ICRP Publication 30.

	Ar-41	Kr-83m	Kr-85m	Kr-85	Kr-87	Kr-88
Gonads	1.9E−10[a]	6.0E−15	3.3E−11	5.2E−13	1.3E−10	3.5E−10
Breast	2.3E−10	8.0E−15	2.7E−11	4.5E−13	1.5E−10	3.6E−10
RM	2.3E−10	5.5E−15	4.4E−11	5.7E−13	1.5E−10	3.5E−10
Lung	2.2E−10		2.6E−11	4.3E−13	1.4E−10	3.5E−10
Thyroid	2.1E−10		3.0E−11		1.4E−10	3.7E−10
BS	2.5E−10	6.3E−15	4.7E−11	6.2E−13	1.7E−10	3.9E−10
ST				4.3E−13		
ULI	2.3E−10		2.1E−11		1.6E−10	4.2E−10
Kidneys			2.2E−11	4.0E−13		
Liver			2.2E−11	3.8E−13	1.3E−10	3.2E−10
Other	2.2E−10		2.2E−11	4.3E−13	1.5E−10	3.7E−10
(weight)[b]	0.24		0.12	0.12	0.18	0.18
Skin	3.9E−10	1.7E−13	8.3E−11	4.7E−11[c]	5.0E−10	5.2E−10
Lens	2.9E−10	1.7E−13	3.4E−11	5.9E−13	2.7E−10	4.5E−10
$\dot{H}_E$	2.2E−10	3.5E−15	3.0E−11	4.6E−13	1.4E−10	3.6E−10

	Xe-131m	Xe-133m	Xe-133	Xe-135m	Xe-135	Xe-138
Gonads	2.0E−12	6.8E−12	6.3E−12	8.3E−11	5.6E−11	1.7E−10
Breast	1.4E−12	4.9E−12	5.6E−12	7.3E−11	4.2E−11	2.1E−10
RM	2.3E−12	7.4E−12	1.1E−11	8.6E−11	6.2E−11	2.1E−10
Lung	1.0E−12	4.3E−12	4.8E−12	7.0E−11	4.1E−11	2.0E−10
Thyroid			7.1E−12			1.9E−10
BS	2.5E−12	8.0E−12	1.2E−11	9.2E−11	6.6E−11	2.3E−10
ST			3.9E−12	7.3E−11		1.9E−10
ULI						2.2E−10
Kidneys	9.8E−13	3.8E−12	4.3E−12	6.6E−11		
Liver		3.7E−12	4.1E−12		3.5E−11	
Other	9.2E−13	3.9E−12	3.9E−12	7.0E−11	3.7E−11	1.9E−10
(weight)[b]	0.06	0.18	0.12	0.18	0.24	0.18
Skin	1.7E−11[c]	3.7E−11	1.9E−11	1.1E−10	1.2E−10	4.0E−10
Lens	3.0E−12	7.5E−12	9.1E−12	9.3E−11	5.4E−11	2.7E−10
$\dot{H}_E$	1.3E−12	5.2E−12	6.1E−12	7.4E−11	4.6E−11	1.9E−10

[a]Read as 1.9×10^{-10}, and so on.

[b]Weight factor to be applied to the dose conversion factor for other target organs.

[c]Governing factor in determination of the derived air concentration.

Source: ICRP 1979.

8.11.2 Carbon-14

Of carbon monoxide inhaled, 40% is taken to be instantly transferred to the blood (bound to hemoglobin), the remainder being exhaled. The bound carbon is assumed to be uniformly distributed to all organs and tissues of the body and retained with a half life of 200 minutes. Inhaled carbon dioxide is taken to be totally absorbed, with the carbon being distributed uniformly to all tissues and organs of the body. Eighteen percent is retained with a half life of 5 minutes, 81 percent with a half life of 60 minutes, and 1 percent with a half life of 60,000 minutes. Inhaled or ingested ^{14}C in labeled compounds other than CO or CO_2 is assumed to be instantaneously and uniformly distributed throughout all organs and tissues of the body. Of dietary carbon, 90% is assumed to be retained in the body. For other carbon, retention is assumed to be 100%. In either case carbon is assumed to be retained with a biological half life of 40 d. Specific committed dose equivalents (all organs and tissues) of ^{14}C are as follows:

Form	Sv Bq^{-1}
Labeled organics	5.6×10^{-10}
CO gas	7.8×10^{-13}
CO_2 gas	6.4×10^{-12}

8.11.3 Radioiodine

Of iodine entering the body-fluid transfer compartment, 30 percent is assumed to be taken up by the thyroid, the remainder being excreted. Iodine in the thyroid is assumed to be eliminated in organic form with a half life of 80 d. This organic iodine is assumed to be uniformly distributed among all tissues and organs save the thyroid where it is retained with a half life of 12 d. Of this organic iodine, 10 percent is assumed to be excreted and 90 percent returned to body fluids as inorganic iodine. Any radioactive xenon daughter products are assumed to escape immediately from the body. Example 8.1 illustrates this biokinetic model.

8.11.4 Alkaline Earths

The metabolism of bone-seeking Ca, Sr, Ba, and Ra in adult man has been the subject of attention by the ICRP (1973) in Publication 20. A comprehensive mathematical model, supported by extensive experimental observations, allows prediction of the retention of radionuclides in blood, soft tissue, and bone. The compartmentalization of bone is very similar, but not exactly equal, to that used in ICRP Publication 30. Cortical bone and trabecular bone (Publication 30), respectively, may be equated approximately to compact bone and cancellous bone (Publication 20).

The mathematical modeling of metabolism leads to retention functions for the elements in the following compartments (source organs): whole body, blood, bone surface, soft tissue, cortical (compact) bone, and trabecular (cancellous) bone. The retention function $R_S(t)$ is defined as the fraction of the (stable) element retained in compartment S at time t after an initial injection of the element in soluble form into the blood (after transfer to body fluids following ingestion or inhalation). Retention functions for strontium and radium are illustrated in Figs. 8.13 and 8.14. Note that cortical (compact) bone (CB) and trabecular (cancellous) bone (TB) retention functions each include retention in the corresponding bone volume as well as 50 percent of the retention in bone surface.

For ingestion or inhalation of the alkaline earths, source-organ transformations may be computed as follows:

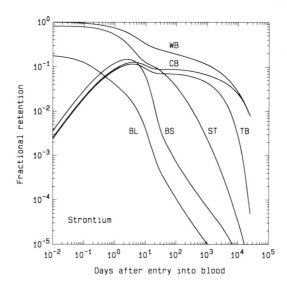

Figure 8.13 Retention factors for stable strontium. WB = whole body, BL = blood, BS = bone surface, ST = soft tissue, CB = cortical (compact) bone, and TB = trabecular (cancellous) bone. [From ICRP (1973).]

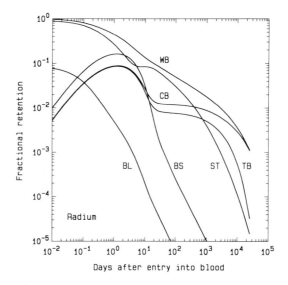

Figure 8.14 Retention functions for "stable" radium. WB = whole body, BL = blook, BS = bone surface, ST = soft tissue, CB = cortical (compact) bone, and TB = trabecular (cancellous) bone. [From ICRP (1973).]

$$U_S = f_T \int_0^\eta dt \, R_S(t) e^{-\lambda_r t}, \tag{8.45}$$

in which f_T is the fraction of the radionuclide transferred to the body-fluid transfer compartment and λ_r is the radioactive decay constant for the radionuclide. Source organ transformations for selected radionuclides are listed in Table 8.7.

TABLE 8.7 Alkaline Earth Source Organ Transformations in 50 Years, U_S/f_T, per Becquerel Transferred to Body Fluids.

NUCLIDE	WHOLE BODY	BLOOD	BONE SURFACE	SOFT TISSUE	CORTICAL BONE	TRABECULAR BONE
^{41}Ca	2.39E+08[a]	8.90E+04	1.24E+06	1.24E+06	1.78E+08	6.26E+07
^{45}Ca	9.54E+06	5.00E+04	6.62E+05	9.16E+05	4.64E+06	3.97E+06
^{80}Sr	1.64E+06	1.85E+04	1.18E+05	5.53E+05	4.81E+05	5.94E+05
^{90}Sr	4.94E+07	2.38E+04	1.61E+05	1.59E+06	1.36E+07	3.44E+07
^{133}Ba	8.16E+06	2.60E+03	3.17E+04	8.51E+05	5.18E+06	2.13E+06
^{140}Ba	2.25E+05	2.26E+03	2.33E+04	1.31E+05	5.05E+04	4.16E+04
^{226}Ra	1.03E+07	2.56E+03	1.03E+05	1.86E+06	6.33E+06	2.19E+06
^{228}Ra	4.30E+06	2.52E+03	1.01E+05	1.40E+06	1.90E+06	1.01E+06

[a]Read as 2.39×10^8, and so on.

Source: ICRP 1973.

8.12 1977 AND 1990 ICRP RECOMMENDATIONS FOR EXPOSURE LIMITS

The procedures recommended in 1977 and 1990 for dose limitation differ from the 1959 recommendations. The earlier recommendations, as described later, were based on limitation of irradiation of a single critical organ at greatest risk from ingestion or inhalation of a given radionuclide. The 1977 and 1990 procedures take account of the combined risk attributable to the exposure of all tissues irradiated. Certain organs and tissues are given explicit recognition because of their susceptibility to radiation damage, the seriousness of such damage, and the extent to which damage is treatable. Tissues at risk, risk natures, and sensitivities[4] are discussed in Chap. 3.

In establishing limits, two broad categories of radiation-induced defects were taken into account, namely,

Stochastic effects—malignant and hereditary disease for which the probability of an effect occurring, rather than its severity, is regarded as a function of dose without threshold, and

Mechanistic effects—effects such as opacity of the lens and cosmetically unacceptable changes in the skin for which an effective threshold of dose must be exceeded before the effect is induced.

1977 Recommended Limits for Occupational Exposure.

Mechanistic Effects. Two tissues are considered to be at risk from mechanistic effects. For the lens of the eye, it is thought that a cumulative dose equivalent in excess of 7.5 Sv can lead to unacceptable opacification. For the skin, it is thought that a cumulative dose equivalent in excess of 25 Sv can lead to unacceptable cosmetic changes. For protection of the lens, the occupational dose-equivalent limit is 0.15 Sv

---

[4]The sensitivity, or risk factor, is the estimated probability per sievert of inducing fatal malignant disease, nonstochastic (mechanistic) change, or substantial genetic defects expressed in liveborn descendants.

Organ or Tissue	w_T	$\hat{H}_T\,(\text{Sv Bq}^{-1})$	
		Ingestion	Inhalation
Gonads	0.25	1.4×10^{-8}	8.8×10^{-9}
Breast	0.15	1.2×10^{-8}	7.8×10^{-9}
RM	0.12	1.3×10^{-8}	8.3×10^{-9}
Lung	0.12	1.3×10^{-8}	8.8×10^{-9}
Thyroid	0.03	1.3×10^{-8}	7.9×10^{-9}
BS	0.03	1.3×10^{-8}	7.9×10^{-9}
SI	0.06	1.4×10^{-8}	9.1×10^{-9}
ULI	0.06	1.4×10^{-8}	9.0×10^{-9}
LLI	0.06	1.4×10^{-8}	9.1×10^{-9}
Remainder	0.12	1.5×10^{-8}	9.5×10^{-9}
$\sum w_T \hat{H}_T$		1.35×10^{-8}	8.67×10^{-9}

Thus, for ingestion, $ALI = 3.7 \times 10^6$ Bq, and, for inhalation, $ALI = 5.8 \times 10^6$ Bq. These limits, for stochastic effects, are more severe than any for mechanistic effects, as would be calculated using Eq. (8.49).

From Eq. (8.50) and the ALI for inhalation, the $DAC = 2.4 \times 10^3$ Bq m^{-3}.

Example 8.10

Evaluate the DAC for submersion in an infinite cloud of ^{83m}Kr based on ICRP-30 weight factors.

Solution   For stochastic effects, Table 8.6 shows that the only organs significantly irradiated are gonads, breast, red marrow, and bone surface. Let $\dot{H}_T$ equal the dose conversion factor (Sv h^{-1} per Bq m^{-3}) from the table. Then,

$$DAC(\text{Bq m}^{-3}) = \frac{0.05 \text{ Sv}}{2000 \text{ h } \sum_T w_T \dot{H}_T}$$

The evaluation is as follows:

Organ	w_T	$\dot{H}_T$
Gonads	0.25	6.0×10^{-15}
Breast	0.15	8.0×10^{-15}
RM	0.12	5.5×10^{-15}
BS	0.03	6.3×10^{-15}
$\sum w_T \dot{H}_T$		3.54×10^{-15}

Thus, for stochastic effects, $DAC = 7.1 \times 10^9$ Bq m^{-3}. For mechanistic effects, Table 8.6 gives $\dot{H}_T = 1.7 \times 10^{-13}$ for both the skin and the lens. Hence, for the lens, $DAC = 0.15$

Sv/(2000 $\dot{H}_{lens}$) = 4.4 × 10^8 Bq m^{-3}. For the skin, DAC = 0.5 Sv/(2000 $\dot{H}_{skin}$) = 1.5 × 10^9 Bq m^{-3}. Mechanistic effects to the lens are limiting and DAC = 4.4 × 10^8 Bq m^{-3} governs.

8.12.2 Occupational Limits for Exposure Combinations

Suppose that workers experience external exposure by penetrating radiation, ingestion, and inhalation of radionuclides, and submersion in radioactive noble gases. Let H_P (Sv) be the dose equivalent index H_I, or, better, the effective dose equivalent H_E experienced in a year by external exposure, and H_{wb}^{max} be the annual limit for whole body dose equivalent. Let I_i be the annual oral intake (Bq) of radionuclide i and I_j the annual inhalation (Bq) of radionuclide j. For submersion in radionuclide k, let U_k' be the annual integral (or sum) of the concentration (Bq m^{-3}) multiplied by the time of exposure (h). ICRP recommended occupational limits will be met if

$$\frac{H_P}{H_{wb}^{max}} + \sum_i \frac{I_i}{(ALI)_i} + \sum_j \frac{I_j}{(ALI)_j} + \sum_k \frac{U_k'}{2000(DAC)_k} \le 1. \qquad (8.51)$$

This inequality may be unduly restrictive if some of the ALIs or DACs are based on mechanistic effects. Then, a more detailed inequality may be derived based on organ-by-organ committed dose equivalents.

8.13 1959 ICRP RECOMMENDATIONS FOR EXPOSURE LIMITS

Publication 2 of the ICRP (1960) and the subsequent Publication 6 (1964) established a widely used data base and set of procedures for internal-dose evaluation. The publication formed the basis of federal radiation-protection regulations used in the United States (NRC 1960) until 1994. The data base of Publication 2 is very limited in comparison with that of Publication 30. The evaluation procedures of the two publications, though similar in certain respects, are significantly different in concept.

Two concepts are central to the methods of Publication 2, namely, the *maximum permissible body burden* A^{max} and the *maximum permissible concentration in air or water*, MPC_a or MPC_w. The total activity of a radionuclide in the body is called the *body burden*. This burden is assumed to be distributed among the organs and tissues of the body according to the physiological characteristics of Reference Man. Each organ or tissue receives a certain dose equivalent rate based on its respective organ burden. Each organ or tissue is assigned a *maximum permissible dose equivalent rate*. The least body burden which results in any one organ or tissue receiving the maximum dose equivalent rate is called the *maximum permissible body burden* and the organ or tissue is called the *critical body organ*.

In determining the maximum permissible concentration, it is assumed that a radionuclide is inhaled or ingested at a uniform weekly-average rate for a (working-life) period of 50 years. The concentration of the radionuclide in air or water leading, after 50 years, to the maximum permissible body burden is called the maximum permissible concentration (MPC_a or MPC_w). In most cases the 50-year period is irrelevant.

Exceptions occur when both the radiological and the biological half lives are longer than a few years.

Publication 2, like Publication 30, employs special models for bone, the respiratory system, and the gastrointestinal system. For other organs and tissues, the data base of Publication 2 provides unique biological decay constants for each element and organ combination. Publication 2 also has a simplified procedure for determination of specific effective energies.

Recommended Limits for Occupational Exposure.   All recommendations are for both internal and external exposure in excess of background and medical exposures. The recommendations as amended in 1964 may be summarized in the following table.

Organ	Limit
Gonads, blood-forming organs, lens	3 rem in 13 weeks
	5 rem in 1 year (to age 18)
	60 rem prior to age 30
	5(N-18) rem through age N years[a]
Skin or thyroid	8 rem in 13 weeks or 30 rem in 1 year
Bone	8 rem in 13 weeks or 30 rem in 1 year[b]
Hands, forearms, feet, ankles	20 rem in 13 weeks or 75 rem in 1 year
Other organs	4 rem in 13 weeks or 15 rem in 1 year

[a] For occupational exposure beginning after age 18.
[b] Equivalent to the bone dose resulting from a body burden of 0.1 μCi of ^{226}Ra.

Maximum permissible concentrations are given in Publication 2 for both continuous (168 hours per week) occupational exposure and for exposure based on 40 hours per week, 50 weeks per year.

There are special limits for adults who work in the vicinity of or might occasionally enter restricted areas and receive radiation exposure in the course of occupational duties, but who are not classed as "radiation workers." Limits are set at 1.5 rem in 1 year to the whole body, gonads, or blood-forming organs and for other organs 10 percent of occupational limits.

Recommended Exposure Limits for Individual Members of the Public.   Limits are 0.5 rem in 1 year to the whole body, gonads, or bloodforming organs and, for other organs, 10 percent of occupational limits. Maximum permission concentrations in areas accessible to the public are to be based on these limits, and 168 hours per week exposure to air or water borne radioactivity. These *MPC*s are thus 10 percent of the *MPC*s for occupational exposure at 40 hours per week.

Recommended Exposure Limits for Population Groups.   The genetic dose to a population is defined as the dose that, if it were received by each person from conception to the mean age of childbearing (assumed to be 30 years), would result in the same genetic burden to the whole population as do the actual doses received by the individuals. The 1964 ICRP recommendation for a permissible genetic dose to a population is 5 rem, or an average of 170 mrem per year to age 30.

8.13.1 The 1959 ICRP Method for Internal Dose Calculations

The method used in ICRP Publication 2 (1960) for internal dose calculations accounts only for gamma and x-ray energy absorption in the organ or tissue within which radiations are emitted, that is, $AF(T \leftarrow S)$ is evaluated only for $T = S$. Each organ is assigned an empirically determined dimension x and, for a photon energy E_i, $AF(S \leftarrow S)$ is calculated from

$$AF(S \leftarrow S) = 1 - \exp[-\mu_a(E_i)x], \tag{8.52}$$

in which $\mu_a(E_i)$ is the linear absorption coefficient for water at energy E_i. Values of x are given in Appendix G, Table G-7, and AF values, computed using μ_a data from Appendix C, are given in Table G-8.

Given here is a sample MPC calculation made using the methods of ICRP Publication 2. The sample is for a radionuclide with no radioactive daughter products. Organs exposed exclude the respiratory system, bone, and the gastrointestinal system, which are treated separately as in Publication 30. The calculations use the notation of ICRP Publication 30.

Suppose that a given radionuclide is taken into the body by ingestion or inhalation at a constant rate $\dot{A}_0$ Bq s^{-1} starting at time zero. Let $A_i(t)$ Bq be the activity in organ i as a function of time. Equation (8.16), with the addition of a source term, is

$$\frac{dA_i}{dt} = -(\lambda_r + \lambda_{bi})A_i + f_T f_{2i}\dot{A}_0. \tag{8.53}$$

Initially A_i is zero, and the solution to the equation is

$$A_i(t) = \frac{f_T f_{2i}\dot{A}_0 \left(1 - e^{-(\lambda_r + \lambda_{bi})t}\right)}{\lambda_r + \lambda_{bi}}. \tag{8.54}$$

The symbol i also can represent the whole body, wb. The fraction of the radionuclide in the body present in organ i at time t is given by the ratio $F_i(t) = A_i(t)/A_{wb}(t)$ computed using Eq. (8.54). At steady state, the ratio is[6]

$$F_i(\infty) = \frac{f_{2i}(\lambda_r + \lambda_{bwb})}{f_{2wb}(\lambda_b + \lambda_{bi})}. \tag{8.55}$$

The quantity $A_i(\eta)$ represents the organ burden at the end of an intake period of $\eta = 50$ years. At that time, the dose equivalent rate to organ i is given by [cf. Eq. (8.7)].

$$\dot{H}_i(\eta) = \kappa \, A_i(\eta)SEE(i \leftarrow i), \tag{8.56}$$

in which the factor $\kappa = 1.6 \times 10^{-10}$ converts MeV g^{-1} to J kg^{-1} and in which the units of $\dot{H}_i(\eta)$ are Sv s^{-1}. The specific effective energy absorption is given by Eq.

---

[6]This fraction is given the symbol f_2 in Publication 2. It should not be confused with the same symbol having different meaning in Publication 30.

(8.6), using the absorbed fraction taken from Table G-8. If, in the same units, $\dot{H}_i^{max}$ is the limiting dose equivalent rate for organ i, then the maximum permissible organ burden is

$$A_i^{\max}(\eta) = \frac{\dot{H}_i^{max}}{\kappa\,SEE(i \leftarrow i)}. \tag{8.57}$$

Associated with the maximum burden calculated for each organ is a corresponding maximum total-body burden $(A^{max})_i = A_i^{max}(\eta)/F_i(\eta)$. The critical organ $(i = c)$ is that organ for which $(A^{max})_i$ is minimum. The corresponding maximum permissible body burden is

$$(A^{max})_c \equiv A^{max} = \frac{A_c^{max}(\eta)}{F_c(\eta)}. \tag{8.58}$$

The maximum permissible concentration may be derived from the value of $\dot{A}_o^{max}$ corresponding to $A_c^{\max}(\eta)$, namely,

$$\dot{A}_o^{max} = \frac{A_c^{\max}(\eta)(\lambda_r + \lambda_{bc})}{f_T f_{2c}\left(1 - e^{-(\lambda_r + \lambda_b c)\eta}\right)}. \tag{8.59}$$

Let $\dot{V}_a$ represent the annual-average rate of inhalation (m^3 s^{-1}) and $\dot{V}_w$ represent the annual-average rate of water intake (m^3 s^{-1}). $MPCs$ are thus given by

$$MPC_a = \dot{A}_0^{max}/\dot{V}_a, \tag{8.60}$$

and

$$MPC_w = \dot{A}_0^{max}/\dot{V}_w. \tag{8.61}$$

ICRP Publication 2 uses the following data for air and water intake rates.

Breathing rate	10 m^3 during 8-hour work day
	20 m^3 during 24 hours
Water consumption	0.0011 m^3 during 8-hour work day
(including food, etc.)	0.0022 m^3 during 24 hours

For example, the annual average breathing rate for 40 hour per week, 50 week per year occupational exposure is $(10 \times 5 \times 50)/(365.25 \times 24 \times 3600) = 7.92 \times 10^{-5}$ m^3 s^{-1}. The average breathing rate for continuous exposure is $20/(24 \times 3600) = 2.31 \times 10^{-4}$ m^3 s^{-1}.

Example 8.11

For ^{137}Cs, including daughter ^{137m}Ba, the whole body is the critical organ. Use the method of ICRP Publication 2 to determine the maximum permissible air concentration for 40 hours per week occupational exposure. Data from Publication 2 are as follows: $\lambda_{bwb} = 1.15 \times 10^{-7}$ s^{-1}, much greater than λ_r, $SEE(\text{wb} \leftarrow \text{wb}) = 6.4 \times 10^{-6}$ MeV g^{-1}, $f_T = 0.75$, and $f_{2wb} = 1.0$. For the whole body, $\dot{H}_{wb}^{max} = 0.05$ Sv y^{-1} or 1.58×10^{-9} Sv s^{-1}.

Solution The maximum permissible body burden is given by Eq. (8.58); thus, since $F_{wb} = 1$ (by definition),

$$A^{max} = A^{max}_{wb}(\eta) = \frac{\dot{H}^{max}_{wb}}{1.6 \times 10^{-10} \; SEE(\text{wb} \leftarrow \text{wb})} = 1.2 \times 10^6 \text{ Bq } (32\mu\text{Ci}).$$

The maximum intake rate is given by Eq. (8.59), or

$$\dot{A}^{max}_0 \cong \frac{A^{max}_{50,wb}\lambda_r}{f_T f_{2wb}} = 0.184 \text{ Bq s}^{-1}(5.8 \times 10^6 \text{ Bq y}^{-1}).$$

By comparison, the corresponding inhalation ALI from Publication 30 is 6×10^6 Bq.

The maximum permissible concentration is given by Eq. (8.60). Since, for 40 hours per week occupational exposure, $\dot{V}_a = 7.92 \times 10^{-5} \text{ m}^3 \text{ s}^{-1}$,

$$MPC_a = \frac{\dot{A}^{max}_0}{\dot{V}_a} = 2.3 \times 10^3 \text{ Bq m}^{-3}(6.3 \times 10^{-8}\mu\text{Ci cm}^{-3}).$$

By comparison, the corresponding DAC from Publication 30 is 2×10^3 Bq m^{-3}.

The ICRP-2 Model for the Respiratory System.   The respiratory system is treated as a single organ, the lung, with mass 1000 g and effective dimension 10 cm. No effects of the particle size of inhaled material are taken into account. Instead, it is assumed that 25 percent of the inhaled material is immediately exhaled. Only two degrees of solubility are recognized. For *soluble* materials, it is assumed that no material is retained in the lung, 50 percent is transferred immediately to the gastrointestinal system, and 25 percent from the lung directly to the bloodstream and from there to other organs of the body. For *insoluble* materials, it is assumed that 12.5 percent of the inhaled material is retained in the lung with a biological half life of 120 days and 62.5 percent is transferred to the GI system with no subsequent transfer to the bloodstream.

The ICRP-2 Model for the Gastrointestinal System.   The physical model used in ICRP Publication 2 is similar to that used in ICRP Publication 30 (Fig. 8.12), except for the contents of the components and mean residence times, which are as follows:

Portion of GI System	Mass of Contents (g)	Mean Residence Time (d)
Stomach	250	1/24
Small Intestine	1100	4/24
Upper Large Intestine	135	8/24
Lower Large Intestine	150	18/24

The dosimetry model for the GI system used in Publication 2 makes use of masses and dimensions of Table G-7 and, for gamma rays, absorbed fractions of Table G-8. For nonpenetrating radiations, absorbed fractions are computed as described in Sec. 8.5.

The ICRP-2 Procedures for Bone Dosimetry.   For bone-seeking radionuclides such as ^{90}Sr which also emit beta and alpha particles, maximum permissible body burdens are based on a comparison with ^{226}Ra and daughters. For ^{226}Ra, $A_{Ra}^{\max}$ has long been established as 0.1 μCi. The assumption is made that 99 percent of the radium in the body is in the skeleton and that, per disintegration of ^{226}Ra, the total energy absorbed in bone is 11 MeV, with a quality factor of 10. With these data, $A_{Ra}^{\max}$ thus produces an average 0.56 rem per week dose equivalent rate to bone. In terms of Publication 30, with the mass of bone taken as 7000 g, SEE(bone$\leftarrow$ bone) $= 11 \times 10/7000 = 0.0157$ MeV g^{-1}. Consider now some other radionuclide x, for which the maximum permissible body burden is A_x^{max}. This radionuclide, with its daughters, has a specific effective energy given by a modified form of Eq. (8.8), namely,

$$SEE^x(\text{bone} \leftarrow \text{bone}) = \frac{1}{M_{bone}} \sum_j \sum_i Y_i^j E_i^j \, AF(\text{bone} \leftarrow \text{bone})_i^j \, Q_i^j \, MF_i^j. \quad (8.62)$$

In this expression, $M_{bone} = 7000$ g, AF is 1.0 for nonpenetrating radiation, or as given in Table G-8 for gamma or X rays. A new factor, MF_i^j, is introduced here. If the ith radiation is a gamma or X ray, $MF = 1$. If the jth radionuclide is a daughter product of ^{226}Ra, then $MF = 1$. Otherwise, $MF = 5$. This modifying factor MF is introduced to account for uncertainty in the effect on bone damage of the physical distribution of radionuclide x as compared to that of ^{226}Ra. Equating the effective dose equivalent rates to bone from $A_x^{\max}$ and $A_{Ra}^{\max}$ leads to

$$A_x^{\max} = A_{Ra}^{\max} \frac{\left[F_{bone}^{Ra}/F_{bone}^x\right]}{\left[SEE^{Ra}/SEE^x\right]}. \quad (8.63)$$

The ICRP-2 Procedures for Submersion Dose.   A very simple energy-balance model is used to evaluate dose conversion factors and maximum permissible concentrations for submersion in semi-infinite clouds of radioactive argon, krypton, and xenon. For a given concentration, the absorbed dose in an infinite volume of air is first calculated. Conservation of energy requires that the energy absorption rate is equal to energy emission rate per unit mass of air. This dose rate (absorption rate) in air is multiplied by a factor of 1.13 to approximate the ratio of the stopping power of charged particles in tissue to the stopping power in air. Then this product is divided by 2 to approximate the tissue dose rate on the ground beneath a large hemispherical cloud. Beta radiation with source energy less than or equal to 0.1 MeV is assumed to contribute only to skin dose. Other beta and gamma radiations are assumed to contribute to both skin dose and whole-body dose.

REFERENCES

Bair, W. J., "Human Respiratory Tract Model for Radiological Protection: A Revision of the ICRP Dosimetric Model for the Respiratory System," *Health Physics 57* (Sup. 1), 249-253 (1989).

BERMAN, M., *Kinetic Models for Absorbed Dose Calculations*, NM/MIRD Pamphlet No. 12, Society of Nuclear Medicine, New York, 1977.

JAMES, A. C., A. BIRCHALL, F. T. CROSS, R. G. CUDDIHY, AND J. R. JOHNSON, "The Current Approach of the ICRP Task Group for Modeling Doses to Respiratory Tract Tissues," *Health Physics 57* (Sup. 1), 271-282 (1989).

CRISTY, M., AND K. F. ECKERMAN, *Specific Absorbed Fractions of Energy at Various Ages from Internal Photon Sources*, 7 Vols., Report ORNL/TM-8381, Oak Ridge National Laboratory, Oak Ridge, Tenn., 1987.

DILLMAN, L. T., AND T. D. JONES, "Internal Dosimetry of Spontaneously Fissioning Nuclides," *Health Physics 29*, 111-123 (1975).

EVE, I. S., "A Review of the Physiology of the Gastrointestinal Tract in Relation to Radiation Dose from Radioactive Materials," *Health Physics 12*, 131-161 (1966).

FORD, M. R., W. S. SNYDER, L. T. DILLMAN, AND S. B. WATSON, "Maximum Permissible Concentration (MPC) Values for Spontaneously Fissioning Radionuclides," *Health Physics 33*, 35-43 (1977).

HOENES, G. R., AND J. K. SOLDAT, *Age-Specific Radiation Dose Commitment Factors for a One-Year Chronic Intake*, Report NUREG-0172, Battelle Pacific Northwest Laboratories, Richland, Wash., 1977.

ICRP, Publication 2, *Recommendations of the International Commission on Radiological Protection Report of Committee II on Permissible Dose for Internal Radiation (1959)*, Publication 2, International Commission on Radiological Protection, Pergamon Press, Oxford. *Health Physics 3* (1960).

ICRP Publication 6, *Recommendations of the International Commission on Radiological Protection (as Amended 1959 and Revised 1962)*, Pergamon Press, Oxford, 1964.

ICRP Task Group on Lung Dynamics, "Deposition and Retention Models for Internal Dosimetry of the Human Respiratory Tract," *Health Physics 12*, 173-207 (1966).

ICRP, *Alkaline Earth Metabolism in Adult Man. A Report Prepared by a Task Group of ICRP Committee 2*, Publication 20, International Commission on Radiological Protection, Pergamon Press, Oxford, 1973.

ICRP, *Report of the Task Group on Reference Man. A Report Prepared by a Task Group of Committee 2 of ICRP*, Publication 23, International Commission on Radiological Protection, Pergamon Press, Oxford, 1975.

ICRP, *Recommendations of the International Commission on Radiological Protection*, Publication 26, International Commission on Radiological Protection, Pergamon Press, Oxford, *Annals of the ICRP 1* (3), (1977).

ICRP, *Limits for Intakes of Radionuclides by Workers*, Publication 30, Parts 1-3, including addenda and supplements, International Commission on Radiological Protection, Pergamon Press, Oxford, *Annals of the ICRP 2* (3-4), 1979, [see also *Annals of the ICRP 3* (1-4), 1979, *4* (3-4), 1980, *6* (2-3), 1981, *7* (1-3), 1982, and *8* (1-4), 1982.]

ICRP, *Radiation Dose to Patients from Radiopharmaceuticals*, Publication 53, International Commission on Radiological Protection, Pergamon Press, Oxford, *Annals of the ICRP 18* (1-4), (1987).

ICRP, *1990 Recommendations of the International Commission on Radiological Protection*, Publication 60, International Commission on Radiological Protection, Pergamon Press, Oxford, *Annals of the ICRP 21* (1-3), (1991a).

ICRP, *1990 Annual Limits on Intake of Radionuclides by Workers Based on the 1990 Recommendations*, Publication 61, International Commission on Radiological Protection, Pergamon Press, Oxford, *Annals of the ICRP 21* (4), (1991b).

ICRU, *Methods of Assessment of Absorbed Dose in Clinical Use of Radionuclides*, Report 32, International Commission on Radiation Units and Measurements, Bethesda, Md., 1979.

JOHNSON, J. R., AND S. MILENCOFF, "A Comparison of Excretion and Retention Between the Current ICRP Lung Model and a Proposed New Model," *Health Physics 57* (Sup. 1), 263-270 (1989).

KOCHER, D. C., *Dose Rate Conversion Factors for External Exposure to Photon and Electron Radiation from Radionuclides Occurring in Routine Releases from Nuclear Fuel Cycle Facilities*, Report NUREG/CR-0494 (ORNL/NUREG/TM-283), Oak Ridge National Laboratory, Oak Ridge, Tenn., 1979).

KOIZUMI, K., G. L. DENARDO, S. J. DENARDO, ET AL., "Multicompartment Analysis of the Kinetics of Radioiodinated Monoclonal Antibody in Patients with Cancer," *J. Nucl. Med. 27*, 1243-1254 (1986).

LOEVINGER, R. AND M. BERMAN, *A Revised Schema for Calculating the Absorbed Dose from Biologically Distributed Radionuclides*, NM/MIRD Pamphlet No. 1 (1968), Revised (1976), Society of Nuclear Medicine, New York, 1976.

MASSE, R., AND F. T. CROSS, "Risk Considerations Related to Lung Modeling," *Health Physics 57* (Sup. 1), 283-289 (1989).

McCULLOUGH, E. C., AND J. R. CAMERON, "Exposure Rates from Diagnostic X-Ray Units," *Br. J. Radiol. 43*, 448 (1972).

MIRD Committee, *MIRD Primer for Absorbed Dose Calculations*, Society of Nuclear Medicine, New York, 1985.

NCRP, *The Experimental Basis for Absorbed Dose Calculations in Medical Uses of Radionuclides*, Report No. 83, National Council on Radiation Protection and Measurements, Bethesda, Md., 1985.

NRC, *Standards for Protection against Radiation*, 10CFR20, 25FR10914, U.S. Nuclear Regulatory Commission (Atomic Energy Commission), Washington D.C., 17 Nov 1960, as amended.

NRC, *Calculation of Annual Dose to Man from Routine Releases of Reactor Effluents for the Purpose of Evaluating Compliance with 10CFR50, Appendix I*, Regulatory Guide 1.109, U.S. Nuclear Regulatory Commission, Washington D.C., 1977.

ROY, M., "Lung Clearance Modeling on the Basis of Physiological and Biological Parameters," *Health Physics 57* (Sup. 1), 255-262 (1989).

SNYDER, W. S., M. R. FORD, G. G. WARNER, AND S. B. WATSON, *Absorbed Dose per Unit Cumulated Activity for Selected Radionuclides and Organs*, NM/MIRD Pamphlet No. 11, Society of Nuclear Medicine, New York, 1975.

SNYDER, W. S., M. R. FORD, AND G. G. WARNER, *Estimates of Specific Absorbed Fractions for Photon Sources Uniformly Distributed in Various Organs of a Heterogeneous Phantom*, NM/MIRD Pamphlet No. 5, Revised, Society of Nuclear Medicine, New York, 1978.

PROBLEMS

1. Consider the accidental intake of ^{131}I (half-life 8.05 days). Over what period of time does the thyroid of Reference Man receive the first 90 percent of its total committed dose equivalent?

2. In a laboratory accident, a worker accidentally ingests ^{125}I. Seven days later, a thyroid scan reveals that the thyroid contains 0.01 μCi (370 Bq) of the radionuclide. Estimate the activity initially ingested and the consequent thyroid committed dose equivalent. Use tabulated $\hat{H}_T$ values. Use f_1 and f_2 values from Appendix G, Tables G-2 and G-3, not the special model of Example 8.1.

3. Repeat Example 8.1 for injection rather than ingestion of ^{131}I.

4. Repeat Example 8.1 for ingestion of ^{125}I rather than ^{131}I.

5. Repeat Example 8.2 for injection of ^{125}I rather than ^{131}I.

6. Repeat Example 8.3 for injection of a mixture within which 50 percent of the activity is in the iodide form and 50 percent is in the form of labeled antibodies. Warning: this is a difficult problem requiring the solution of "stiff" nonlinear differential equations. Nevertheless it is representative of practical problems in medical internal dosimetry.

7. Repeat Example 8.3 for the following case: An initial injection of nonlabeled antibodies ($F = 2$) is followed one hour later by an injection of labeled antibodies ($F = 2$). Warning: this is a difficult problem requiring the solution of "stiff" nonlinear differential equations. Nevertheless it is representative of practical problems in medical internal dosimetry.

8. Consider a hypothetical radionuclide, of 100-day radiological half life, which emits one 2-MeV gamma ray per transformation. The nuclide is in highly soluble form, 100 percent of that ingested being transferred directly to body fluids. Half the activity in body fluids is transferred directly to the skeleton where it is retained with a biological half life of 500 days. Half is transferred to the liver where it is retained with a biological half life of 10 days. For intake by ingestion, estimate $\hat{H}_T$ (Sv Bq^{-1}) for the kidneys as target organ.

9. Thirty days after ingestion of ^{59}Fe (45-day radiological half life), whole body counting reveals the presence of 0.1 mCi of that radionuclide. For Reference Man, estimate the activity ingested and the consequent committed dose equivalent to the liver. Use tabulated $\hat{H}_T$ values.

10. A person ingests ^{59}Fe on a continuous basis at a rate of 0.1 μCi per day. For Reference Man, estimate the equilibrium activity (Bq) present in the liver and the dose equivalent rate (Sv y^{-1}) to that organ. Use tabulated $\hat{H}_T$ values.

11. Reference Man ingests 1 mCi of a radionuclide which has a radiological half life of 1 year and which emits only beta particles with a mean energy release of 1 MeV per transformation. The radionuclide is totally taken up by the body, is uniformly distributed (on a mass basis), and is retained with a biological half life of 1 year. Estimate the committed effective dose equivalent (Sv) arising from the ingestion.

12. The radionuclide ^{64}Cu had a radiological half life of 12.8 hours. Principle radiations are as follows:

Particle	Mean Energy (MeV)	Number per Transformation
β^+	0.279	0.193
β^-	0.188	0.396
γ	1.346	0.006
γ	0.511	0.386

Representative data for oral intake of copper are as follows:

Copper content in the body	72 mg
Copper content in soft tissues	65 mg
Daily dietary intake of copper	3.5 mg
Fraction transferred from GI system to body fluids	0.5
Fraction transferred from body fluids to brain	0.1
Fraction transferred from body fluids to liver	0.1
Fraction transferred from body fluids to pancreas	0.006
Fraction transferred to other tissues/organs	remainder
Biological half life in all tissues	40 d

Estimate (a) the number of source organ transformations (Bq^{-1}) in the liver, (b) SEE(liver ←liver) (MeV g^{-1}), and (c) SEE(kidneys←liver) (MeV g^{-1}).

13. The maximum permitted occupational body burden for ^{226}Ra has long been established as 0.1 μCi. In Reference Man, about 78 percent of the radium in the body resides in the 7 kg skeleton. Suppose that the ^{226}Ra is in equilibrium with daughters to, but not including, ^{210}Pb. For Reference Man, what is the average alpha-particle dose rate (rad y^{-1}) to the skeleton if the body burden is the maximum permitted.

14. Ingestion of ^{90}Sr results in significant exposures to only bone surface and red marrow, for which committed dose equivalents are, respectively, 4.2×10^{-7} and 1.9×10^{-7} Sv Bq^{-1}. For occupational exposure, what is the annual limit of intake?

15. What is the DAC for ^{133}Xe, accounting only for stochastic radiation effects? Suppose a worker is exposed for 1000 hours per year to a ^{133}Xe airborne concentration equal to 50 percent of the DAC. The worker is exposed to no other airborne radionuclides and ingests no radionuclides other than at background levels. If ICRP guidelines are followed, what is the upper limit to the worker's annual effective dose equivalent rate incurred by external exposure to gamma radiation from fixed sources.

16. For submersion in a semi-infinite cloud of ^{133}Xe the weighted dose conversion factor for stochastic effects is 6.1×10^{-12} Sv h^{-1} per Bq m^{-3}. Factors for mechanistic effects on skin and lens are, respectively, 1.9×10^{-11} and 9.1×10^{-12} Sv h^{-1} per Bq m^{-3}. For occupational exposure, what is the DAC, in units of μCi cm^{-3}?

17. A radiation worker is exposed for 1 hour to airborne ^{133}Xe. Personnel dosimetry reveals a deep dose equivalent index of 10 mrem. Estimate (a) the concentration (pCi m^{-3}) of ^{133}Xe in the atmosphere, and (b) the dose equivalent to the skin.

18. In carrying out maintenance on an accelerator, a worker is exposed to airborne tritium in water-vapor form for a period of 30 minutes. This exposure occurred beginning at 2:00 p.m. on a Friday. Starting at 8:00 A.M. on Monday, the worker's urine was collected for 24 hours. The specific activity in the urine was 100 Bq g^{-1}. Estimate the concentration (Bq

m^{-3}) of tritium in the atmosphere to which the worker was exposed and the soft-tissue dose equivalent (Sv) incurred as a result of the exposure.

19. Reference Man ingests 50 μCi (1.85 MBq) of ^{141}Ce. Estimate the committed dose equivalents to each organ of the GI system. Neglect any transfer of the element to body fluids. Neglect any gamma or x-rays with energies less than 0.1 MeV. Plot, as a function of time after ingestion up to 72 h, the activity in each organ of the GI system.

20. Reference Man inhales 1 μg of ^{239}PuO$_2$ in the form of a dust with 1.0 μm AMAD. Prepare a graph of the total activity in the pulmonary region of the lung as a function of time after inhalation. Compute the committed dose equivalent to the lung. ^{239}Pu has a half-life of 2.4×10^4 y and is an alpha-particle emitter. SEE(lung←lung) = 0.10 MeV g^{-1}.

21. ^{51}Cr-labeled denatured erythrocytes are used in the imaging of the spleen. Of the intravenously administered activity, 75 percent is rapidly taken up by the spleen, 15 percent by the liver, and 10 percent uniformly by the remainder of the body. From all organs, 90 percent of the ^{51}Cr is biologically eliminated with a half life of 10 days, 10 percent with a half life of 160 days.

   (a) Verify source organ transformations (Bq^{-1}) for Reference Man, which are approximately as follows: spleen 8.4×10^5, liver 1.7×10^5, and remainder (whole body) 1.1×10^5.

   (b) For Reference Man, evaluate the specific committed dose equivalent (Sv Bq^{-1}) to the spleen as target for this use of ^{51}Cr. (For reference, ICRP Publication 53 lists 5.6 mSv MBq^{-1} as the value to be used in planning the imaging procedure.) Specific effective energies (MeV g^{-1}) may be obtained from the following table:

	Source Organ		
Target organ	Spleen	Liver	Whole body
Spleen	4.10×10^{-5}	1.06×10^{-7}	2.63×10^{-7}
Liver	1.16×10^{-7}	5.39×10^{-6}	2.52×10^{-7}
Whole Body	2.53×10^{-7}	2.54×10^{-7}	2.24×10^{-7}

22. Residents near a metallurgical plant are continuously exposed to airborne cadmium ($Z = 48$). For Reference Man, estimate the length of time after beginning of exposure required for the additional concentration of cadmium in the liver to reach 25 percent of its equilibrium value.

23. Determine the annual limit of intake (ALI by ingestion) for stochastic risk in occupational exposure to ^{32}P. You may use values of specific committed dose equivalents from Appendix G, Table G-5.

24. Persons living and working near a metallurgical plant have been exposed continuously and for a long time to airborne As$_2$O$_3$ in the form of a 1-μm AMAD aerosol (30 percent deposition in the N–P region, 8 percent in the T–B region, and 25 percent in the P region). An autopsy on a person believed to have died under suspicious circumstances reveals in the liver an arsenic concentration of 3 ppm (wet basis), about 10 times the normal. Arsenic, in the trioxide form, has a lung clearance classification of W. Of the arsenic reaching the GI system, 50 percent is transferred to body fluids. From the body fluids, 7 percent is transferred to the liver. Of the arsenic in the liver, 40 percent is retained with a biological half

life of 1 day (decay constant λ_1) and 60 percent with a half life of 10 days (decay constant λ_{10}).

(a) Show that, of the inhaled arsenic, the fraction f'_{BF} transferred directly from the respiratory system to body fluids is about 0.1, and the fraction f_{GI} transferred to the GI system is about 0.5.

(b) If C, as yet to be determined, is the average concentration (μg m^{-3}) of arsenic in the atmosphere, and B is the daily average breathing rate (23 m^3 per day) what is the average *rate* $\dot{A}_0$(μg per day) at which arsenic reaches the liver?

(c) Let m be the *steady-state* mass (μg) of arsenic in the liver. Express m in terms of $\dot{A}_0$, λ_1, and λ_{10}.

(d) Determine the numerical value of C.

Chapter 9

Atmospheric Dispersion of Radionuclides

9.1 INTRODUCTION

The intentional or accidental release of radionuclides to the atmosphere is one of the most important mechanisms for the introduction of radioactivity into the environment. Such atmospheric releases lead to a variety of radiation sources of potential significance to public health. Gamma and beta radiation will be emitted from passing radioactive clouds, radionuclides can be inhaled, and radiation will be emitted from the radioactive particles that are deposited on the ground, nearby structures, and vegetation. The fallout from atmospheric releases of radioactivity may also contaminate both water and food supplies, and thereby lead to internal exposures.

The elevated release of radioactive gases and aerosols or other gaseous pollutants is an effective method for disposing of such effluents while minimizing ground-level concentrations. Nevertheless, atmospheric releases of radioactivity are always potentially hazardous and consequently many studies have investigated how gaseous effluents disperse in the ambient air and are deposited on the ground below (Slinn 1978; Pasquill 1974; Miller 1984).

The exact description of atmospheric dispersion requires a detailed knowledge of all meteorological effects, complete topographic data for the area surrounding the release location, and a description of plume depletion caused by precipitation and interaction with the ground, all combined with elaborate time-dependent fluid mechanical calculations of local and large scale weather patterns. Such exact calculations of plume dispersal from fundamental physical laws are presently beyond the capability of even our largest computers.

Fortunately, several approximate atmospheric dispersion models have been developed over the past 30 years, primarily as a result of the importance of plume dispersal in safety studies for nuclear power plants. These simplified models vary a great deal in sophistication and predictive capabilities. For the most part, they are based on statistical or average behavior under particular meteorological conditions at the site of interest.

In this chapter, the straight-line Gaussian plume model will be described. While this model is based on a time-averaged description of the gaseous discharge, and thus, is unable to account for short-time fluctuations in the wind velocities, it is by far the most frequently used model for the assessment of atmospheric dispersal of radioactive aerosols and gases. It is very simple to use, has been widely tested, and has been shown to give reasonable predictions if a proper selection of model parameters is made.

Before describing this Gaussian model, however, several important properties about meteorological phenomena and their influence on gaseous effluents must be introduced.

9.2 ATMOSPHERIC STABILITY

The dispersion and mixing of gaseous releases in the air of the lower atmosphere depend upon the local winds as well as the resistance of the atmosphere to vertical mixing. In this section, the basic mechanism governing the stability of air to vertical motion is described and its importance to the dispersion of airborne contaminants is presented.

9.2.1 Variation of Pressure with Altitude

It is well known that atmospheric pressure decreases with increasing altitude. The precise description of this relation between pressure P and altitude z is important for the subsequent stability analysis of the atmosphere.

Consider the motionless cylinder of air shown in Fig. 9.1. The downward force on the top surface is denoted by $F(z + \Delta z)$. At equilibrium (i.e., no motion) the upward force, $F(z)$, exerted on the bottom surface must exactly balance the downward force on the top surface plus the weight of the air in the cylinder, that is,

$$F(z) = F(z + \Delta z) + \rho(z)g\Delta zA, \tag{9.1}$$

where ρ is the air density and g is the acceleration due to gravity. In terms of pressure, $P = F/A$, this equation becomes

$$\Delta P \equiv P(z + \Delta z) - P(z) = -\rho(z)g\Delta z. \tag{9.2}$$

Thus, for an increase in altitude of Δz, the atmospheric pressure decreases by an amount $\rho(z)g\Delta z$.

The vertical pressure profile $P(z)$ may be found from Eq. (9.2) by using the ideal gas law

$$PV = MRT = \rho VRT, \tag{9.3}$$

where M is the mass of air in volume V at pressure P and absolute temperature T. The constant R varies among gases but is related to the *universal gas constant* $R_o(= 8.314$ J K^{-1} mol^{-1}) by $R = R_o/M_a$ where M_a is the molecular weight of the gas of interest (29.0 for air). Substitution for $\rho(z)$ from Eq. (9.3) into Eq. (9.2) and taking the limits as ΔP and Δz go to differentials gives

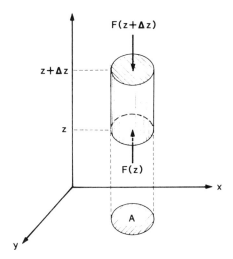

Figure 9.1 A short vertical column of air (of length Δ_z and a horizontal cross section A) showing the pressure forces on the two horizontal faces.

$$\frac{dP}{dz} = -\frac{M_a g}{R_o T} P(z). \tag{9.4}$$

If the temperature is constant with elevation, the solution of this equation is

$$P(z) = P(0)e^{-\alpha z}, \tag{9.5}$$

where $P(0)$ is the pressure at the earth's surface and $\alpha \equiv M_a g/(R_o T)$. For the same isothermal atmosphere, Eqs. (9.3) and (9.5) require the air density also to decrease exponentially with increasing altitude to the ground-level density $\rho(0)$ at zero elevation, that is,

$$\rho(z) = \rho(0)e^{-\alpha z}. \tag{9.6}$$

9.2.2 Stability of Dry Air to Vertical Motion

If a small parcel of air is somehow raised in altitude, it will expand since the pressure of surrounding air decreases. This expansion causes work to be done on the surrounding air. If there is no exchange of energy between the parcel and the surrounding air, this work comes at the expense of the internal energy of the gas in the parcel thereby causing the temperature of the air parcel to decrease. If this lowered temperature is less than that of the temperature of the surrounding air, the density of the air in the parcel will be greater than that of the surrounding air, and the air parcel will tend to sink back towards its initial location. On the other hand, if the temperature in the parcel is greater than that of the surrounding air, the parcel will continue to rise. Thus the resistance of a volume of air to vertical displacements depends on how rapidly the temperature of the atmosphere varies with altitude and how much the temperature inside the volume changes as the volume is displaced.

To obtain a quantifiable limit for the stability of air against vertical motion, consider a volume V of air with surface area S that is somehow raised by an amount Δz (see Fig. 9.2). Because of the decreased pressure at its new elevation, the surface of the parcel will expand through a distance Δd increasing the parcel's volume by $\Delta V = S\Delta d$. From the first law of thermodynamics (conservation of energy), the net heat flowing across the surface into the parcel (ΔQ) must equal the sum of the increase in internal energy of the parcel gas (ΔU) and the net work done by the parcel in expanding against the surrounding gas, (ΔW). In symbols

$$\Delta Q = \Delta U + \Delta W. \tag{9.7}$$

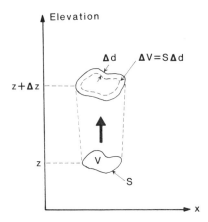

Figure 9.2 An air parcel of volume V and surface area S is displaced upwards a distance Δz. The parcel expands outwards a distance Δd as a result of the decreased air pressure.

The work done by the parcel as it expands against the surrounding gas is just the force exerted by the gas times the displacement Δd, namely, $\Delta W = F\Delta d = PS\Delta d = P\Delta V$. For a unit mass of air at normal atmospheric temperatures and pressures, the internal energy is just proportional to the temperature. Thus, when the air parcel of mass M undergoes a temperature change ΔT the change in internal energy is $\Delta U = Mc_v\Delta T$, where the constant of proportionality c_v is called the *specific heat at constant volume* and is the energy required to raise the temperature of a unit air mass one degree while keeping the volume of the gas constant. Finally, assume the parcel of air is raised to $z + \Delta z$ so quickly that negligible heat is transferred between the parcel and the surrounding air. For such an adiabatic displacement, $\Delta Q = 0$. Hence, the energy balance of Eq. (9.7) becomes

$$0 = c_v M\Delta T + P\Delta V. \tag{9.8}$$

With this relation, the change in temperature of the gas parcel, ΔT, resulting from a change Δz in elevation can now be obtained. From the ideal gas law of Eq. (9.3) it follows that

$$P\Delta V + V\Delta P = MR\Delta T. \tag{9.9}$$

Substitution of $P\Delta V$ from this result into Eq. (9.8) yields

$$(c_v + R)M\Delta T - V\Delta P = 0 \tag{9.10}$$

which, on using Eq. (9.2) and $M = \rho V$, becomes

$$(c_v + R)M\Delta T + Mg\Delta z = 0. \tag{9.11}$$

Rearrangement and use of the limit as ΔT and Δz go to differentials finally yield[1]

$$\left(-\frac{dT}{dz}\right)_a = \frac{g}{c_v + R} = \frac{g}{c_p}, \tag{9.12}$$

where the *specific heat at constant pressure* $c_p = c_v + R$. For dry air near the earth's surface, the evaluation of the right-hand side of Eq. (9.12) gives

$$\left(-\frac{dT}{dz}\right)_a = 0.00977 \text{ C m}^{-1}. \tag{9.13}$$

The rate of *decrease* in temperature with increasing altitude, $-dT/dz$, is termed the *lapse rate*. The lapse rate of almost 0.01 C m^{-1} given by Eq. (9.13) is called the *dry adiabatic lapse rate* and is a critical parameter for atmospheric stability. If the actual atmospheric lapse rate is greater than this value, a rising parcel of air, cooled isentropically, will find itself surrounded by colder and denser air and will, thus, be buoyed further upwards. Similarly, a depressed air parcel will tend to continue sinking. Such an atmospheric lapse rate is termed *superadiabatic* and leads to unstable vertical air motion—a condition that promotes the dilution of airborne contaminants.

By contrast, an ambient lapse rate less than the critical dry adiabatic lapse rate tends to cause vertically displaced air parcels to return to their original elevation. For such cases the atmosphere is said to be *stable*, and for extreme cases in which the lapse rate is negative, that is, $dT/dz > 0$, the atmosphere is strongly stable and is said to be *inverted*. Atmospheric layers with negative lapse rates, so-called *inversion layers*, strongly prevent vertical mixing of air, and are of major concern in the assessment of airborne discharges.

9.2.3 Effect of Humidity on Stability

A rising parcel of moist air will expand and cool. If the temperature decreases past the dew point, moisture will condense if there are sufficient aerosols to promote nucleation. The latent heat of the condensing vapor is then given up to the rising parcel of air so that the air does not cool as much as it would if it were dry. Thus, the wet adiabatic lapse rate $(-dT/dz)$ will be less than that for dry atmospheric air. For warm moist tropical air, the wet adiabatic lapse rate is about one-third that of the dry adi-

---

[1]An alternate derivation of Eq. (9.12) is as follows. For an isentropic (reversible adiabatic) expansion of an ideal gas the product PV^γ is a constant, with γ equal to the ratio of the heat capacities at constant pressure and constant volume, c_p/c_v. Thus, $dV/dz = -(V/P\gamma)dP/dz$. From the ideal gas law, dV/dz can be written in terms of dP/dz and dT/dz. Elimination of dP/dz using Eq. (9.4) yields Eq. (9.12).

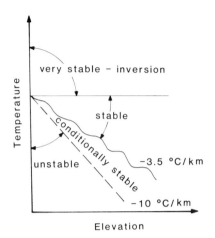

Figure 9.3 Atmospheric stability conditions showing the dry-air adiabatic lapse rate (dotted line) and the lapse rate for warm moist air (wavy line).

abatic lapse rate (Williamson 1973). In polar regions there is little water vapor, and the adiabatic lapse rate is close to that for dry air. In Fig. 9.3 the different regions of stability against vertical mixing are illustrated.

9.2.4 Lapse Rate and Plume Behavior

While many gaseous discharges are invisible to the eye, the effluent of concern is often accompanied by a visible cloud of water condensate or smoke which allows one to follow the subsequent dispersal of the discharge in the air. By observing such visible plumes as they are swept downwind, much has been learned about how gaseous effluents are dispersed in the lower atmosphere. The diffusion models developed later in this chapter are all based on such observations.

The shapes of plumes emitted from smokestacks can vary a great deal depending on local meteorological conditions. One of the most important factors in determining how the plume is dispersed is the local atmospheric lapse rate profile. When the atmosphere is unstable with high winds, the plume is dispersed rapidly. Still conditions under a stable lapse rate, by contrast, permit dispersal to occur only very slowly.

In Fig. 9.4 several typical plume behaviors are shown for different temperature profiles. The unstable lapse rate of Fig. 9.4(a) causes large scale meandering of the plume called *looping*. For a neutral lapse rate (b) the plume spreads out vertically in a uniform manner called *coning*. An inversion (c) inhibits vertical mixing and the plume spreads out only horizontally (due to fluctuations in wind direction) producing what is called *fanning*. Figures 9.4(d) and (e) illustrate conditions for which the lapse rate changes from stable to unstable. In (d) the stable atmosphere below the stack prevents downward mixing of the plume, but the neutral to unstable conditions above the stack allow the plume to diffuse upwards. This condition, known as *lofting*, is ideal for preventing effluents from reaching the ground while allowing the plume to dis-

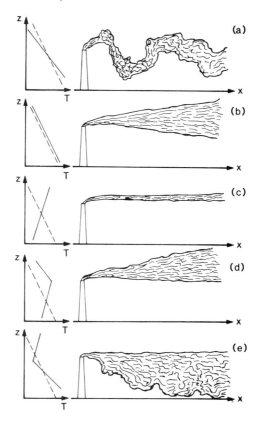

Figure 9.4 Characteristic effluent plumes for idealized atmospheric temperature profiles (solid line) shown to the left (the dashed line represents the dry adiabatic lapse rate): (a) an unstable lapse rate produces *looping*, (b) a neutral lapse rate produces *coning*, (c) strongly stable conditions produce *fanning*, (d) conditions that are stable below and neutral above the emission point produce *lofting*, and (e) unstable conditions below with stable above produce *fumigation*. [After Slade (1968).]

perse. By contrast, a stable condition above and unstable below produces *fumigation*, a situation in which the plume rapidly reaches the ground and stays trapped below the inversion layer.

9.2.5 The Formation of Inversions

The formation of atmospheric inversion layers, which prevent the vertical dispersal of gaseous effluents, is an eventuality of considerable importance for assessing the impact of the gaseous discharges. Inversions can spread over large areas or be quite localized, and can last for many days or be of only a few hours duration. There are three different mechanisms for producing inverted atmospheric layers.

Frontal Inversions.   An advancing warm or cold air mass produces a front in which the more buoyant warm air is layered over the cold air, as shown in Fig. 9.5(a). Although the shape of the fronts are quite different, both can produce inversion layers that extend over a wide area. In fact the inversion helps to sustain the fronts by preventing mixing between the two air masses. Any air pollutants released near the ground will be confined to the cold air, and if the front becomes stagnant, the buildup of pollutants may become severe. Fortunately, the collision of cold and warm air masses is usually accompanied by rain and strong winds which tend to disperse airborne contaminants horizontally.

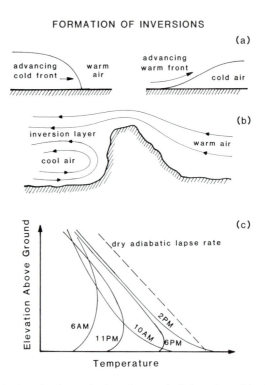

FORMATION OF INVERSIONS

Figure 9.5   Three mechanisms for the production of atmospheric inversions: (a) advancing warm or cold fronts lead to *frontal inversions*, (b) flow of warm air over ground obstacles can cause *advective inversions*, (c) rapid cooling of the ground and the adjacent air during still nights product *radiative inversions*.

Advective Inversions.   The movement of warm air over a cold surface or over a cold air mass also produces inversions. If a warm breeze moves over a cold surface such as the cold ground or a snow field, the lower part of the atmosphere will be cooled by convection, thereby establishing a ground-based inversion. Of more concern to those living near mountains is the deflection of a warm wind by the mountains over cooler air on the lee side of the mountain as shown in Fig. 9.5(b). Such elevated inversion layers are a major cause of air pollution episodes over cities ringed by mountains.

Radiation Inversions.   The most frequently encountered ground-based inversions are caused by the ground cooling rapidly at night. Particularly on cloudless nights, the earth can rapidly radiate its thermal energy into space, and thus, quickly cool the top few centimeters of the ground. The cool ground in turn cools the adjacent air by conduction and convection, and, if there is no wind to mix the air, an inversion layer forms near the ground as shown in Fig. 9.4(c). If the cooling air drops below the dew point, a layer of fog will be observed trapped in the inversion layer next to the ground. More importantly, airborne pollutants emitted during the night will also be trapped in this inversion layer. With the coming of day, the ground is rapidly heated by the sun and the temperature profile returns to the normal lapse rate.

9.3 DIFFUSION OF RADIONUCLIDES IN THE ATMOSPHERE

Radionuclides that are emitted into the atmosphere suffer random collisions with air molecules and experience local fluctuation in the wind that tend to mix the radionuclides with the ambient air. The detailed description of how the radionuclides are increasingly diluted following their release has been the subject of much investigation, but, as yet, no practical description based solely on fundamental principles has been found. Nevertheless, many approximate models have been developed that, in some average sense, describe quite adequately the dissipation of airborne radionuclides and other effluents. By far the more widely used of these simple models are those based on some form of a diffusion approximation. In this section the mechanisms of diffusion and its mathematical description are presented.

9.3.1 The Mechanism of Atmospheric Diffusion

The buffeting received by radionuclides from collisions with the randomly moving air molecules will cause the radionuclides to become increasingly separated in time. This random mixing problem has been well studied, and diffusion theory can describe the dispersion of the radionuclides in still ambient air quite accurately. Unfortunately, this molecular diffusion process turns out to be far too small, usually by several orders of magnitude, to describe the observed dissipation of gaseous effluents in the atmosphere. Only in small, isolated, still-air samples does molecular diffusion become important.

The mechanism responsible for the atmospheric dissipation of radionuclides arises from the normally turbulent nature of air flow in the atmosphere. Wind velocities are seldom if ever steady, but, as in Fig. 9.6, exhibit fluctuations over a wide range of frequencies about some mean value. These random fluctuations are a result of the spontaneous formation of turbulence eddies, and are far more responsible for the mixing of radionuclides into the ambient air than is molecular diffusion.

The size, intensity, and directional dependence of the turbulence eddies varies greatly and depend primarily on atmospheric stability conditions, although terrain features and position in the atmosphere also have influence. For example, near the ground fluctuations in the vertical direction are usually far less than those in the downwind or crosswind directions while at high altitudes turbulence becomes more isotropic. Much research has been devoted to understanding the nature of turbulence, but to

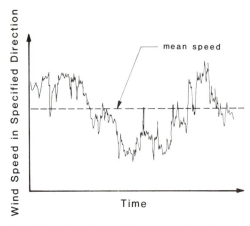

Figure 9.6　A typical wind speed recording showing fluctuations about the mean wind speed. The fluctuations appear with a wide range of frequencies.

date no entirely satisfactory theory has been found, and the description of turbulence mixing must still be based on empirical data.

To see how turbulence promotes the dissipation of airborne pollutants, consider the continuous release of radionuclides from a stack as shown in Fig. 9.7. The turbulence in the atmosphere manifests itself in the form of eddies of all sizes and orientations. Eddies smaller than the size of the plume tend to cause mixing within the plume and to bring air from outside the plume into it, thereby diluting the plume and increasing its size. On the other hand, because the atmosphere is almost infinite (compared to the plume size), there will also be turbulent eddies much larger than the plume. These very large eddies, by contrast, do not disperse the plume's radionuclides but instead deflect the entire plume causing it to meander in a snake-like manner through the atmosphere.

In Fig. 9.7, idealized vertical concentration profiles measured at some downwind location are also shown. The instantaneously measured profile exhibits fluctuations of radionuclide concentration within the plume, and the vertical position of the plume varies with time. However, if measurements are not taken instantaneously, but averaged over some time interval, the plume concentration profile exhibits fewer fluctuations. As the averaging time increases, the plume profile becomes broader and smoother and approaches an ideal Gaussian shape. In effect, the long sampling time averages out the high-frequency components of the wind fluctuations. A similar smoothing occurs for concentration profiles in the cross-wind direction at any elevation.

In lieu of a model for the instantaneous distribution of airborne effluents, a hopelessly complex task, the use of time-averaged concentrations allows one to describe the plume dissipation as if it were a diffusion process averaged over the many fluctuations of the wind. In effect, the turbulence is treated as a diffusion mechanism with some well defined average diffusive property, the so-called *eddy diffusivity*. To see how this characterization of wind turbulence can be used to describe the time-

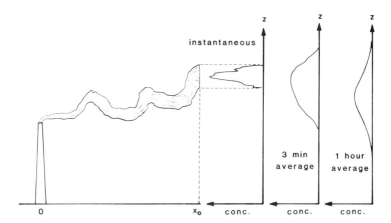

Figure 9.7 An instantaneous view of a plume as it drifts downwind. To the right are veritcal concentration profiles for downwind location x_0 obtained for different sampling times. For sufficiently long sampling times, the concentration profile becomes smooth and bell-shaped as the vertical wind fluctuations are averaged out.

averaged behavior of a discharge plume, it is first necessary to formulate the diffusion process in mathematical terms.

9.3.2 Mathematical Description of Diffusion

The purpose of plume dispersion models is to predict the concentration C of some radionuclide or other pollutant as a function of position and time.[2] To obtain an equation for $C(x, y, z, t)$, consider the small volume of air $dV = dxdydz$, depicted in Fig. 9.8. Assume for the moment that the average wind speeds are zero, but that fluctuations still occur in all directions so that the radionuclides still undergo random motion. For this test volume the following radionuclide balance relation must hold

$$\{\text{rate of increase in } dV\} = \{net \text{ rate of diffusion into } dV\}. \qquad (9.14)$$

This balance relation needs to be expressed mathematically in terms of C. From the definition of C, the left hand side of Eq. (9.14) is simply

$$\text{LHS} = \frac{\partial C(x, y, z, t)}{\partial t} dV. \qquad (9.15)$$

However, expressing the right-hand side in terms of C is not so easy. If the radionuclide were uniformly distributed and the turbulence isotropic, then there would be no net diffusion. On the other hand, if all the radionuclides were initially localized in a small region, it is reasonable to expect that there would be a net migration away

---

[2]In many treatises the symbol χ is used to denote air-born pollutant concentrations; however, the more mnemonic symbol C (which is usually used for concentrations in other media) is used in this book to represent concentrations in all media.

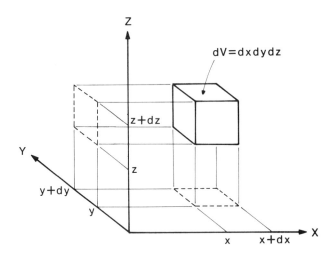

Figure 9.8 Test volume dV used for determining the diffusion of a radionuclide or other gaseous effluent.

from the region of high concentration. The net flow rate from regions of higher concentration to regions of lower concentration can thus be expected to be proportional to the rate at which the concentration varies with position (e.g., $\partial C/\partial x$). If q_x denotes the *net* flow rate in the positive x direction across a unit area oriented perpendicular to the x-axis, then it is reasonable to suppose

$$q_x = -D\frac{\partial C}{\partial x}, \tag{9.16}$$

where D is some positive constant of proportionality called the *diffusion coefficient*. Since $\partial C/\partial x$ is positive (negative) if C increases (decreases) with increasing x, the minus sign in Eq. (9.16) indicates the net flow rate is opposite the direction in which C increases so that, as expected, the radionuclides diffuse away from regions of higher concentration.

Equation (9.16) was proposed in 1855 by Adolf Fick, the German physiologist (Fick 1855), to describe the diffusion of randomly moving molecules in an ambient fluid. Fick's law is often taken as a fundamental relationship for diffusion particles; however, the more precise description of particles in random motion offered by statistical mechanics shows that Fick's law, while not exact, is often an excellent approximation (Duderstadt 1979).

With Fick's law it is now possible to express the right-hand side of the balance relation, Eq. (9.14) in terms of C. The net flow rate *into* dV is simply the sum of the difference in net flow across the three pairs of orthogonal surfaces of dV, namely,

$$\text{RHS} = (q_x - q_{x+dx})dydz + (q_y - q_{y+dy})dxdz + (q_z - q_{z+dy})dxdy. \tag{9.17}$$

Substitution of Fick's law into this expression gives

$$\text{RHS} = \left\{ \left(-D_x \frac{\partial C}{\partial x} \right)\bigg|_{x} - \left(-D_x \frac{\partial C}{\partial x} \right)\bigg|_{x+dx} \right\} dydz$$

$$+ \left\{ \left(-D_y \frac{\partial C}{\partial y} \right)\bigg|_{y} - \left(-D_y \frac{\partial C}{\partial y} \right)\bigg|_{y+dy} \right\} dxdz$$

$$+ \left\{ \left(-D_z \frac{\partial C}{\partial z} \right)\bigg|_{z} - \left(-D_z \frac{\partial C}{\partial z} \right)\bigg|_{z+dz} \right\} dxdy. \qquad (9.18)$$

Here, for generality, the diffusion coefficient is assumed to vary with position and to be different for the three orthogonal directions. Equation (9.18) simplifies to

$$\text{RHS} = \left\{ \frac{\partial}{\partial x} \left(D_x \frac{\partial C}{\partial x} \right) + \frac{\partial}{\partial y} \left(D_y \frac{\partial C}{\partial y} \right) + \frac{\partial}{\partial z} \left(D_z \frac{\partial C}{\partial z} \right) \right\} dV. \qquad (9.19)$$

Finally combining equating Eqs. (9.15) and (9.19), as required by the balance relation Eq. (9.14), one obtains the following differential equation for $C(x, y, z, t)$:

$$\frac{\partial C}{\partial t} = \frac{\partial}{\partial x} \left(D_x \frac{\partial C}{\partial x} \right) + \frac{\partial}{\partial y} \left(D_y \frac{\partial C}{\partial y} \right) + \frac{\partial}{\partial z} \left(D_z \frac{\partial C}{\partial z} \right). \qquad (9.20)$$

If the diffusion coefficients (or *eddy diffusivities*) D_x, D_y and D_z are independent of position (the usual situation for a homogeneous atmosphere) but still, in general, different for each direction, this diffusion equation reduces to

$$\frac{\partial C}{\partial t} = D_x \frac{\partial^2 C}{\partial x^2} + D_y \frac{\partial^2 C}{\partial y^2} + D_z \frac{\partial^2 C}{\partial z^2}. \qquad (9.21)$$

The solution of this equation, subject to appropriate boundary and initial conditions, gives the concentration of the radionuclide of concern as a function of position in the atmosphere and of time. From Eq. (9.20) or (9.21) one can determine how quickly a released radionuclide sample will be diluted in the ambient air. It must be remembered that such solutions are for the time average behavior, as required for the diffusion model to be valid.

9.4 RESULTS OBTAINED FROM THE DIFFUSION MODEL

From the above diffusion equation, the concentrations downwind from gaseous releases of radionuclides or other pollutants can be determined. In the sections that follow, several cases are considered.

9.4.1 Instantaneous Release—Infinite Medium

Suppose some amount Q' {Ci, kg, etc.} of a radionuclide or other pollutant is released at $t = 0$ into an infinite homogeneous air medium that has no steady air flow and none of the pollutant prior to the release. If the release point is at (x_o, y_o, z_o), then the concentration $\{Ci\ m^{-3}, kg\ m^{-3}, etc.\}$ at time t and at position (x, y, z) is given by the solution of Eq. (9.21), namely

$$C(x, y, z, t) = \frac{Q'}{(2\pi)^{3/2}\sigma_x\sigma_y\sigma_z} \exp\left[-\frac{(x - x_o)^2}{2\sigma_x^2} - \frac{(y - y_o)^2}{2\sigma_y^2} - \frac{(z - z_o)^2}{2\sigma_z^2}\right]. \quad (9.22)$$

where the *diffusion parameters* σ_i are defined by

$$\sigma_i(t) = \sqrt{2D_i t}, \qquad i = x, y, \text{ or } z. \quad (9.23)$$

With isotropic diffusion, $D_x = D_y = D_z \equiv D$, Eq. (9.22) reduces to

$$C(r, t) = \frac{Q'}{(2\pi)^{3/2}\sigma^3}e^{-r^2/(2\sigma^2)} = \frac{Q'}{(4\pi Dt)^{3/2}}e^{-r^2/(4Dt)} \quad (9.24)$$

where $r^2 \equiv (x - x_o)^2 + (y - y_o)^2 + (z - z_o)^2$, the square of the distance from the release point to the point of interest.

In Fig. 9.9(a), the spatial distribution of the radionuclide concentration, as specified by Eq. (9.24), is shown for two times after the release. Notice that with increasing time, the radionuclides, while always in greatest concentration at the release point, spread out and become more dilute. Profiles determined by the functional form $\exp[-r^2/2\sigma^2]$ are called *Gaussian*. In Fig. 9.9(b) the temporal behavior of the radionuclide concentration at some fixed position is seen to initially increase in time, reach a maximum, and then decrease towards zero.

In both these figures, a fundamental limitation of diffusion theory is evident. The predicted concentration at any position, even far removed from the source, is nonzero, even for very short times after the release. In other words, some of the radionuclides have apparently moved with speeds far in excess of their thermal speeds. This infinite propagation speed is a characteristic of diffusion theory; and, while this behavior is not physically realistic, the unrealistic nonzero concentrations predicted at very large distances from the release point are so small that, for all practical purposes, they are totally negligible. From many experimental studies, diffusion theory has been found to give an excellent description of the dissipation of radionuclides.

9.4.2 Instantaneous Point Source—Diffusion Plus Advection

Consider the previous problem of an instantaneous release into an infinite medium at $t = 0$ in which there is a steady wind of speed u in the x-direction. For simplicity, take the release point to be the origin. Without the wind, the concentration would be given by Eq. (9.22) (with $x_o = 0$, $y_o = 0$, and $z_o = 0$). The wind simply moves the expanding puff downwind at a constant speed u (see Fig. 9.10). Thus the radionuclide concentration can be obtained directly from Eq. (9.22) by simply moving the center of the puff downwind at a rate u, or,

$$C_{inst}(x, y, z, t) = \frac{Q'}{(2\pi)^{3/2}\sigma_x(t)\sigma_y(t)\sigma_z(t)} \exp\left[-\frac{(x - ut)^2}{2\sigma_x^2(t)} - \frac{y^2}{2\sigma_y^2(t)} - \frac{z^2}{2\sigma_z^2(t)}\right]. \quad (9.25)$$

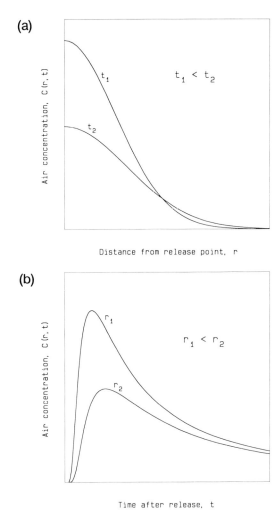

Figure 9.9   Concentration profiles produced by an instantaneous point source in an infinite homogeneous medium as a function of (a) distance from the release point for a fixed time, and (b) time after release for a fixed position.

As the expanding puff is blown past an individual who is at an elevation h from that of the emission point, the amount of radioactivity inhaled will vary in time. Often of most concern is the time-integrated exposure suffered by the individual from the time the radionuclide puff is emitted until it is swept far past and the air returns to its ambient condition. If the individual is directly downwind a distance x and is breathing at a constant rate B (m^3 s^{-1}), the quantity of radionuclides inhaled between t and $t + dt$ is $(Bdt)C_{inst}(x, 0, h, t)$. Thus, the time-integrated exposure, the total amount of radioactivity inhaled, is

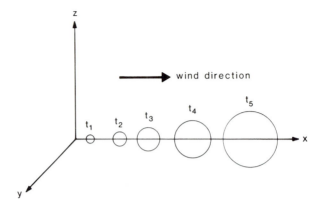

Figure 9.10 An effluent puff at different times after release as it is swept downwind. As it moves, the puff expands as a result of diffusion caused by wind turbulence.

$$B \int_0^\infty C_{inst}(x, 0, h, t)\, dt \equiv B\, \psi(x)$$

where $\psi(x)$ is given by

$$\psi(x) = \frac{Q'}{(2\pi)^{3/2}} \int_0^\infty dt\, \frac{\exp\left[-\frac{(x-ut)^2}{2\sigma_x^2(t)} - \frac{h^2}{2\sigma_z^2(t)}\right]}{\sigma_x(t)\, \sigma_y(t)\, \sigma_z(t)}. \tag{9.26}$$

While this integral cannot be evaluated exactly, it can be well approximated by recognizing that the term $\exp[-(x - ut)^2/2\sigma_x^2]$ makes the integrand narrowly concentrated about the peak at $t = x/u$. Near this narrow peak, if σ_x is sufficiently large, the other terms in the integrand are relatively constant, so that σ_y and σ_z do not change as the puff passes overhead, and

$$\psi(x) \simeq \frac{Q' \exp[\frac{-h^2}{2\sigma_z^2(x/u)}]}{(2\pi)^{3/2}\sigma_x(x/u)\sigma_y(x/u)\sigma_z(x/u)} \int_0^\infty dt\, \exp\left[\frac{-(x - ut)^2}{2\sigma_x^2(x/u)}\right]. \tag{9.27}$$

If the downwind location x is sufficiently far removed from the emission point that the integrand in Eq. (9.27) is negligible at the lower limit compared to the peak, then one can replace the lower limit by $-\infty$. With this replacement, one obtains a definite integral which can be evaluated analytically, namely

$$\int_{-\infty}^\infty \exp\left[-\frac{(x - ut)^2}{2\sigma_x^2(x/u)}\right] dt = \frac{1}{u}\sqrt{2\pi\sigma_x^2(x/u)}. \tag{9.28}$$

Thus Eq. (9.26) becomes

$$\psi(x) = \frac{Q' e^{-h^2/[2\sigma_z^2(x)]}}{2\pi u\sigma_y(x)\sigma_z(x)}. \tag{9.29}$$

Notice that $\psi(x)$ is independent of $\sigma_x(x)$. Exposure results from the entire course of cloud passage and is not affected by the degree of axial dispersion. The diffusion coefficient in the x direction affects the peak concentration but not the quantity of material inhaled during cloud passage. Turbulent diffusion in the y and z directions, of course, is important since it is the only mechanism for dissipating the radionuclides in those directions. Finally, the diffusion parameters, $\sigma_i(x) = \sqrt{2D_i x/u}$, are no longer explicit functions of time, but rather depend on the downwind location x. Later, in Sec. 9.5, procedures for determining $\sigma_y(x)$ and $\sigma_z(x)$ will be reviewed.

9.4.3 Continuous Emission from a Point Source

In many situations, radionuclides are emitted into the atmosphere over an extended period of time rather than as a single instantaneous puff. The airborne concentration resulting from such a continuous source can readily be obtained by treating the emission as a series of instantaneous puffs. Consider an emission in which an amount $Q(t')dt'$ of radionuclides is emitted between t' and $t'+dt'$. If the release point is taken as the origin of an infinite homogeneous atmosphere in which a steady wind of speed u blows in the x direction, the resulting airborne concentration at some point (x, y, z) resulting from the emission in dt' about t' at a later time t is given by Eq. (9.25) as

$$C_{inst}(x, y, z, t; t') = \frac{Q(t')dt'}{(2\pi)^{3/2}\sigma_x(t-t')\sigma_y(t-t')\sigma_z(t-t')} \tag{9.30}$$

$$\times \exp\left(-\frac{[x-u(t-t')]^2}{2\sigma_x^2(t-t')} - \frac{y^2}{2\sigma_y^2(t-t')} - \frac{z^2}{2\sigma_z^2(t-t')}\right)$$

$$\equiv \hat{C}_{inst}(x, y, z, t-t')Q(t')dt' \tag{9.31}$$

The concentration arising from the continuous emission prior to time t is obtained by integrating the above result over all t', namely,

$$C_{cont}(x, y, z, t) = \int_{-\infty}^{t} dt' \, Q(t')\hat{C}_{inst}(x, y, z, t-t') \tag{9.32}$$

or

$$C_{cont}(x, y, z, t) = \int_{0}^{\infty} d\tau \frac{Q(t-\tau)\exp\left[-\frac{(x-u\tau)^2}{2\sigma_x^2(\tau)} - \frac{y^2}{2\sigma_y^2(\tau)} - \frac{z^2}{2\sigma_z^2(\tau)}\right]}{(2\pi)^{3/2}\sigma_x(\tau)\sigma_y(\tau)\sigma_z(\tau)}. \tag{9.33}$$

The integration in Eq. (9.33) can be performed approximately if it is assumed that the emission rate Q is constant, so that C_{cont} no longer is a function of t. Because of the $\exp[-(x-u\tau)^2/2\sigma_x^2]$ term, the integrand is narrowly concentrated around the peak at $\tau = x/u$. Thus Eq. (9.33) may be approximated by

$$C_{cont}(x, y, z) \simeq \frac{Q \exp\left[-\frac{y^2}{2\sigma_y^2(x/u)} - \frac{z^2}{2\sigma_z^2(x/u)}\right]}{(2\pi)^{3/2}\sigma_x(x/u)\sigma_y(x/u)\sigma_z(x/u)} \int_{0}^{\infty} d\tau \exp\left[\frac{-(x-u\tau)^2}{2\sigma_x^2(x/u)}\right]. \tag{9.34}$$

This last integral is approximated in the same way as was used for Eq. (9.27), so that the airborne radionuclide concentration from a continuous source becomes

$$C_{cont}(x, y, z) = \frac{Q}{2\pi u \sigma_y(x)\sigma_z(x)} \exp\left[-\frac{y^2}{2\sigma_y^2(x)} - \frac{z^2}{2\sigma_z^2(x)}\right]. \tag{9.35}$$

The concentration profiles in the crosswind direction y that result from a steady point-emission rate are shown in Fig. 9.11. As would be expected, the maximum concentration in the y direction always occurs directly downwind, at $y = 0$. Moreover, at elevations different from that of the source, the concentration at any cross wind distance y initially increases with downwind distance x, reaches a maximum, and then decreases. Similar profiles are obtained from Eq. (9.35) for profiles in the z direction.

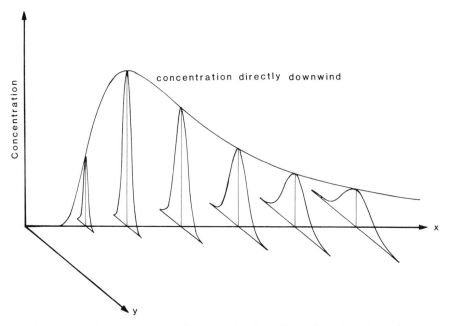

Figure 9.11   Crosswind concentration profiles at an elevation different from that of a continuous point source. The peak concentration initially increases with downwind distance x, and then decreases. The crosswind distribution becomes increasingly spread out as one goes farther from the source.

As the air flow becomes more turbulent because of increasing atmospheric instability or increasing winds, the greater are the eddy diffusivities D_i and hence the greater are the diffusion parameters σ_i. Thus, the Gaussian spread of the plume increases and the radionuclides become less concentrated.

Another implication of Eq. (9.35) is that the concentration of radionuclides is proportional to the amount of radionuclides emitted per unit time and is independent of the concentration of the emitted nuclides at the source. Diluting the effluent before discharge has no effect on plume concentrations! This complete insensitivity to emitted effluent concentration is true provided the volumetric discharge rate is sufficiently small so that the point-source model is a good approximation.

9.4.4 Effect of Ground Interface on Stack Releases

All of the solutions to the diffusion equation obtained so far have been for an infinite homogeneous air medium. Problems of practical importance almost always involve determining the concentration of the effluent at or near the earth's surface. Moreover, the radionuclides of concern often are released at some elevation h above the surface.

Although the air-ground interface may appear to greatly complicate the solution of the diffusion equation, Eq. (9.20) or (9.21), the effect of a flat interface can be readily taken into account for a point source. As the radionuclides or other contaminants reach the ground, they may be reflected if they are nonreactive gases or they may be absorbed or adsorbed to some degree if they are aerosols or reactive gases. For generality, assume some fraction α of the radionuclides that reach the ground are reflected. Of course, atmospheric concentrations may be conservatively overpredicted by setting α to unity.

Rather than solve the diffusion equation directly with the air-ground interface being replaced by a reflecting boundary condition, one can use infinite-air solutions by considering an equivalent problem illustrated in Fig. 9.12 (Gifford 1968; Williamson 1973). In this equivalent, infinite-air problem, a second imaginary point-source is placed a distance h below the location of the air-ground interface. This imaginary source has the same time dependence of the emission rate as does the actual source. However, its strength is only α times the actual source strength. In this manner, as radionuclides from the real source reach the ground and vanish below the interface, they are replaced by radionuclides diffusing upwards from below, thereby simulating the reflection at the ground of the radionuclides from the actual source. Thus, if $C_\infty(x, y, h, t)$ represents the concentration at time t in an infinite medium at location (x, y) on the ground ($z = 0$ plane) due to the point emission at $(0, 0, h)$, the concentration at the same location when the ground reflects a fraction α of the radionuclides is

$$C(x, y, t) = C_\infty(x, y, h, t) + \alpha C_\infty(x, y, -h, t). \tag{9.36}$$

To give an explicit example, consider a continuous, steady emission rate Q from a stack of height h. The effluent concentration at ground level at some location (x, y) downwind is, from Eqs. (9.35) and (9.36),

$$C(x, y) = \frac{Q\,(1 + \alpha)}{2\pi u \sigma_y(x) \sigma_z(x)} \exp\left[-\frac{y^2}{2\sigma_y^2(x)} - \frac{h^2}{2\sigma_z^2(x)} \right]. \tag{9.37}$$

From such a result, one may easily obtain *isopleths*, contours on the ground for equal values of C. As would be expected, the maximum ground-level concentration is seen to occur directly downwind, at $y = 0$. Since the diffusion parameters $\sigma_i(x)$ generally increase with downwind distance x, the ground-level concentration first increases with x, reaches a maximum, and then decreases. The location and value of the maximum downwind concentration depends on the explicit ways in which σ_y and σ_z increase with x. If both σ_y and σ_z have the same x dependence, then from Eq. (9.37) it is easily shown that this maximum occurs when $\sigma_z = h/\sqrt{2}$. For this case, $\sigma_z(x)/\sigma_y(x) = \kappa$, a constant, and the maximum concentration is found to be

$$C_{max} = \frac{(1 + \alpha)Q\kappa}{\pi e h^2 u}. \tag{9.38}$$

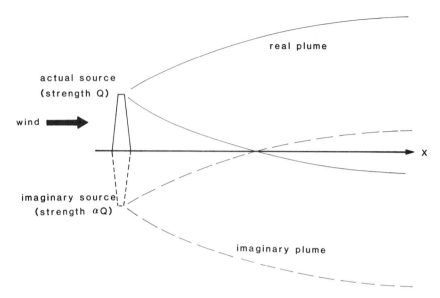

Figure 9.12   Equivalent infinite medium problem to account for the reflection of a fraction α of the effluent that reaches the ground plane ($z = 0$).

From this result for the maximum ground-level concentration, it is seen that C_{max} varies inversely with the square of the stack height, a result first noted by Bosanquet and Pearson (1936). Thus, doubling the stack height reduces the maximum concentration on the ground to one fourth. Also notice that C_{max} varies inversely with the mean wind speed since stronger winds increase the turbulence and hence the diffusion as well as draw out the plume.

Meteorological conditions also affect downwind concentrations since the diffusion parameters σ_y and σ_z are highly dependent on atmospheric stability and wind speed as well as local terrain features. In Fig. 9.13, ground-level concentrations directly downwind are shown for three different release heights for a wide range of atmospheric stability conditions over flat terrain.

9.5 REFINEMENTS TO THE GAUSSIAN PLUME MODEL

The straight-line Gaussian plume model is not only the most widely used model for describing the dissipation of airborne contaminants in the lower atmosphere, but also the most frequently verified atmospheric dispersion model (Miller 1984). The Gaussian plume model developed in the previous section has become the basis for most analyses and experimental investigations.

However, this basic model was developed for a highly idealized situation. For realistic calculations, several corrections need to be made to the basic model to account for the nonidealized nature of the source emission and subsequent behavior of

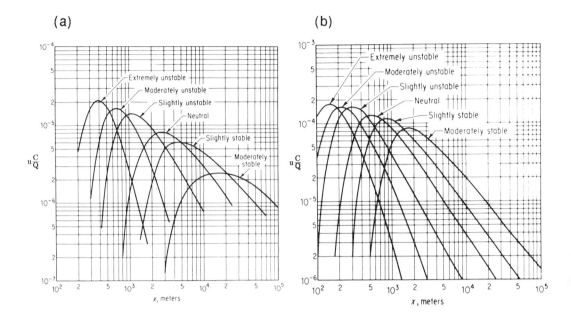

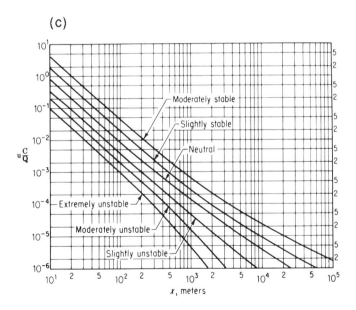

Figure 9.13   Values of uC/Q as a function of downwind distance x for a continuous point source (a) at 100 m elevation, (b) at 30 m elevation, and (c) at ground level. Curves are based on the diffusion parameters given in Fig. 9.15. [From Hilsmeier and Gifford (1962).]

the plume. For example, gaseous effluents may be emitted with some upward momentum and at a temperature greater than ambient. These momentum and buoyancy effects cause the plume to rise higher than the release point. Uneven terrain and hydrodynamic wake effects near the source can also greatly influence the plume development. Moreover, airborne contaminants will settle out of the plume by adhering to the ground as well as by being washed out by precipitation. If the contaminants are radioactive, then radioactive decay must also be considered. Finally, the Gaussian model depends on some appropriate averaging time, and corrections should be made for different averaging times.

The exact treatment of many of these refinements to the basic Gaussian plume model is very complex. Fortunately, several approximate techniques have been developed which allow the analyst to make these corrections easily. In the following sections these refinements are discussed.

9.5.1 Effective Stack Height

In the development of the straight-line Gaussian model, it was assumed that the release point was at the top of a stack of height h above the ground ($z = 0$ plane). In some instances the gaseous effluent is emitted with an upward velocity which tends to displace the plume upwards. Moreover, if the effluent has a higher temperature than that of the ambient air, it will be further buoyed upwards. On the other hand, the wind blowing around and over the stack often creates a downwash that tends to force the plume downwards. Any undulations in the terrain surrounding the stack will also cause the plume to develop differently than it would over flat terrain. None of these factors are included in the basic plume model of Eq. (9.37).

The exact treatment of these effects is complicated and impractical for routine analyses. Fortunately, a simple method has been developed to approximately correct the basic model for all of these effects (USNRC 1982; Briggs 1975). The actual stack height h is simply replaced by an *effective stack height* h_e given by

$$h_e = h + h_{pr} + h_t - c, \tag{9.39}$$

where h_{pr} is the distance the plume rises above the release point as a result of momentum and buoyancy effects, h_t corrects for uneven terrain near the stack, and c corrects for downwash caused by the formation of turbulent eddies in the stack wake.

For most gaseous discharges of radionuclides, the volumetric emission rate is usually small and the discharge is usually close to the ambient temperature. In such instances, the plume rise correction factor h_{pr} can be ignored. However, for those cases involving hot effluents with high discharge velocities (typical of stack releases at fossil-fueled power plants), the plume rise must be considered. Several calculational models and empirical correlations have been developed to describe the plume rise due to momentum and buoyant effects (Smith 1968, Briggs 1969, Carson and Moses 1969, Thomas et al. 1970). One simple formula, suitable for preliminary analyses, is (Moses and Carson 1968)

$$h_{pr} = \frac{\alpha V d + \beta \sqrt{Q_h}}{u}, \tag{9.40}$$

TABLE 9.1   Parameters for Estimating Plume Rise from Eq. (9.40).

ATMOSPHERIC STABILITY	LAPSE RATE ($C\,km^{-1}$)	α	β
Stable	$dT/dz > 1.5$	−1.04	0.071
Neutral	$1.5 > dT/dz > -12.2$	0.35	0.084
Unstable	$-12.2 > dT/dz$	3.47	0.163

Source: Moses and Carson 1968.

where Q_h is the rate of thermal energy discharge (compared to ambient) in watts, V is the discharge speed in m s^{-1} and d is the inside diameter of the stack mouth in meters. By fitting this formula to observed plume rises from various stacks and under various meteorological conditions, Moses and Carson (1968) obtained the values for the empirical constants α and β given in Table 9.1.

The correction factor h_t for uneven ground around the stack is taken as the maximum ground elevation (above the stack base) between the release point and the ground location at which the airborne concentration is desired. The downwash correction factor c is included in Eq. (9.39) only if the effluent exit speed V is less than 1.5 times the horizontal windspeed u. For this case, c is given by (USNRC 1977; Gifford 1972)

$$c = 3d \left(\frac{3}{2} - \frac{V}{u} \right), \tag{9.41}$$

where d is the inside diameter of the stack or other release point.

This procedure is used for elevated releases. Often, however, there are releases near ground level or from vents at elevations below surrounding solid structures. For these cases it is prudent to assume the effective stack height is zero (USNRC 1977).

9.5.2   Correction for Radioactive Decay

As radionuclides drift downwind, some undergo radioactive decay thereby decreasing the concentration in the plume. For long lived isotopes, such radioactive decay is negligible; but if the wind travel time to the downwind location x, namely x/u, is comparable to the radionuclide's half life, then a correction should be made for radioactive decay. To account for radioactive decay in the basic model of Eq. (9.37) for a continuous release, the emission rate is multiplied by the probability the radioactive atom does not decay in time x/u, namely $\exp[-\lambda x/u]$ (Healy and Baker 1968). In this manner, the constant emission rate $Q(0) \equiv Q$ is replaced by an effective spatially dependent source $Q(x)$ given by

$$Q(x) = Q(0)e^{-\lambda x/u}. \tag{9.42}$$

Clearly as one goes farther and farther downwind, radioactive decay becomes an increasingly important mechanism for reducing the concentration in the plume.

9.5.3 Ground Deposition from the Plume

Upon reaching the ground, some fraction of the contaminants in the plume will adhere to the surface and not be reflected back into the plume. Such fallout is of concern to areas surrounding discharge locations, and consequently considerable effort has gone into the development of models for predicting how rapidly airborne contaminants settle out of the plume. The deposition of plume particles onto the ground is usually separated into three categories: (a) dry deposition whereby particles diffusing to the surface are trapped, (b) wet deposition whereby particles are swept from the plume by precipitation, and (c) gravitational settling whereby large particles literally fall out of the plume.

Dry Deposition.   Particles and reactive gases in the plume are deposited on the ground surfaces (e.g., soil, leaves, roofs, etc.) by many physical mechanisms, such as impingement, mechanical trapping, electrostatic attraction, and chemical reactions. It is reasonable to assume that the rate of deposition is proportional to the concentration of the airborne particles and gases near the ground, $C(x, y, 0, t)$. Thus, if C is measured in units of Bq m^{-3}, the total rate ω_{tot} Bq m^{-2} s^{-1} at which particles and gases are removed from the plume and adhere to all surfaces, found in a unit horizontal area of the ground, is given by (Briggs et al. 1968)

$$\omega_{tot}(x, y, t) = v_d\, C(x, y, 0, t) \tag{9.43}$$

where the v_d is an empirical constant of proportionality called the *dry deposition speed* since it must have dimensions of speed.

Besides the rate of radioactivity removal from the plume by dry deposition, the fraction of the removed activity that adheres to vegetation is also of concern since vegetation is one of the means whereby radionuclides can enter the human food chain. This fraction, called the *vegetation retention factor* f_{veg}, is defined as the ratio of the activity deposited on vegetation to the total amount of activity removed from the plume. Thus, the rate ω_{veg} at which effluents in the plume adhere to vegetation in a horizontal unit of ground area is

$$\omega_{veg}(x, y, t) = f_{veg}\, v_d\, C(x, y, 0, t) = v_{veg}\, C(x, y, 0, t), \tag{9.44}$$

where the *vegetation deposition speed* is defined as $v_{veg} = f_{veg} v_d$.

It is important to distinguish between the dry deposition speed v_d and the vegetation deposition speed v_{veg} since they can differ by a factor of 4 or more (Peterson 1983). The former is used to calculated plume depletion rates; the later for deposition rates on vegetation.

The deposition speed and the vegetation retention factor f_{veg} are obtained from experiment by measuring the rate at which a particular pollutant is removed from the plume and deposited on a particular type of vegetation (Hoffman 1977; Miller et al. 1978; Markee 1967). The value of v_{veg} and f_{veg} depend on many factors such as the radionuclide species and its chemical form, the type of vegetation and its density and state of development, the atmospheric conditions, the terrain roughness, and the humidity.

Only limited experimental data for the deposition parameters are presently available. Most data exist for deposition on grasslands and leafy vegetation from which the radionuclides may enter the food chain. Deposition studies have tended to emphasize (simulated) nuclear fallout and various chemical forms of iodine whose radionuclides, as will be seen in Chap. 11, are of paramount concern in the analysis of the food chain. A summary of the results of these studies is given by Peterson (1983).

There is a wide variation in observed deposition speeds, which range from values as large as 0.03 m s^{-1} for molecular iodine on grasses to extremely low values of 10^{-13} m s^{-1} for the noble gas ^{85}Kr. Generally, the more reactive a plume component, the greater will be its deposition speed. The retention factor f_{veg} varies with the type and mass of vegetation on a unit of ground area, increasing as the amount of vegetation increases. For tall grasses such as wheat and pasture grass, f_{veg} ranges from 0.5 to 0.7, while for short grasses such as Bermuda, zoysia, and fescue, f_{veg} is between 0.7 and 0.85. The retention factor also tends to increase with increasing humidity.

As an example, for deposition on grassland, f_{veg} is about 57%, and the following deposition speeds are observed (Miller 1984):

Type of Effluent	v_d (m s^{-1})
Reactive gases (e.g., molecular iodine)	0.035
Small particles (< 4 μm dia.)	0.0018
Unreactive gases (e.g., CH_3I)	0.00018

This concept of deposition speed can be combined with the Gaussian plume models to correct the plume concentration for ground deposition (see Sec. 9.5.4). With this corrected model, plots can be obtained of the fraction of the effluent remaining in the plume versus downwind distance x for different atmospheric stability conditions (USNRC 1977).

Wet Deposition.   The leaching of particles and gases from a plume by falling rain and snow is usually not significant if one is interested in long-term averages resulting from continuous releases. Only when there is a protracted rainy season during grazing or growing seasons or when the effect of an isolated storm is desired should the deposition by precipitation be considered.

Two methods have been developed to predict the rate at which plume contaminants are removed by precipitation and deposited on the ground. The following two methods described assume that the removal rate, and hence, the plume concentrations are averaged over some appropriate time interval—a day, month, season or even a year.

In the first method, the deposition rate by precipitation, per unit area of ground surface (geometric not physical), is given by (Moore 1977; Miller 1984) as

$$\omega(x, y) = \phi \, C_{av} \, L = \phi \int_0^L C(x, y, z) \, dz, \tag{9.45}$$

where C_{av} is the concentration of contaminants vertically averaged over a column of the atmosphere of height L, and ϕ is the fraction of material scavenged from the vertical column of air (per unit time) appropriate for the given precipitation conditions. The length of the column of air, L, should be sufficient to contain the entire plume. Usually, it is taken to be the height of the tropospheric mixing layer.

Another approach (Slinn 1978; Miller 1984), which has been found useful for obtaining long-term average deposition rates, is to use Eq. (9.43) with a suitable *wet-deposition speed* v_w, namely,

$$\omega(x,y) = v_w \, C(x,y,0). \tag{9.46}$$

The wet-deposition speed v_w can be inferred from measured values of the ground-level volumetric radionuclide concentrations in the precipitation and in the air denoted, respectively, by C_p and C_o, both with units such as Bq m^{-3}. Specifically,

$$v_w = \frac{C_p}{C_o} R_p = W_v R_p. \tag{9.47}$$

Here R_p represents the average precipitation rate (e.g., in units of m y^{-1}) and W_v is the *volumetric washout factor* C_p/C_o. The washout factor is sometimes expressed as a *mass washout factor* W_m defined as the ratio of activity concentration in the precipitation to the activity concentration in the air with concentrations expressed on a per unit mass basis. The two washout ratios are related by

$$W_v = W_m \frac{\rho_{rain}}{\rho_{air}} \cong 830 \, W_m \tag{9.48}$$

where ρ_{rain} and ρ_{air} are the mass densities of the rain and air. Typical values of the washout ratio are given in Table 9.2.

Both of these methods give good results for long-term average rates of deposition by precipitation. However, for a single event, Eq. (9.45) is recommended (Miller 1984).

As in dry deposition, the rates given by Eqs. (9.45) and (9.46) are for removal from the plume. If the rate of deposition by precipitation onto vegetation is sought, these values must be multiplied by an appropriate *wet retention factor* f_w which is the fraction of the radionuclides removed from the plume by precipitation that ends up on vegetation surfaces. This retention factor generally decreases with increasing precipitation since heavy rains inhibit accumulation on leave surfaces. For example, the wet retention factor for iodine on grass is about 0.3 in light rain and between 0.1 and 0.2 for heavy rains (Bergström 1967). Peterson (1983) reports initial wet deposition factors of 0.16 for grasses and 0.45 for grains.

Gravitation Settling.   If the effluent contains aerosols or other particulate matter, gravity tends to cause the particles to settle out of the plume. Generally, the larger the particle, the faster will be the *fall velocity* or *settling speed* v_g. The calculation of the fall velocity is generally quite complicated, depending on the particle density, atmospheric pressure, and the particle size and shape. Only for simple cases is it possible to derive the fall velocity.

TABLE 9.2   Washout Factors for Various Plume Effluents. Multiple Entries Refer to Values Obtained from Different Studies.

MATERIAL	MASS W_m (kg_{air}/kg_{rain})	VOLUMETRIC W_v (m^3_{air}/m^3_{rain})
Soluble gases		
water vapor (-10 C)	500	4×10^5
water vapor (-0 C)	240	2×10^5
water vapor (10 C)	110	9×10^4
water vapor (20 C)	60	5×10^4
CO_2	1.0	830
$^{38}Cl_2$	1680	1.4×10^6
SO_2	150	1.2×10^5
Soluble particulates		
generally	1200	1.0×10^6
rhodamine dye	815	6.8×10^5
rhodamine dye	1970	1.6×10^6
NaCl aerosol	1180	9.8×10^5
Insoluble particles		
generally	361	3×10^5
^{238}Pu (fallout)	300 ± 130	$(2.5 \times \pm 1.0) \times 10^5$
^{239}Pu (fallout)	434 ± 132	$(3.6 \times \pm 2.2) \times 10^5$
^{210}Pb (natural)	430	3.6×10^5
lead	290	$2.4 \times \times 10^5$
Fallout radionuclides		
^{137}Cs	230	1.9×10^5
^{137}Cs	560	4.6×10^5
^{137}Cs (old source)	600–800	$(5.0–6.6) \times 10^5$
^{137}Cs (new source)	1050–1100	$(8.7–9.1) \times 10^5$
^{140}Ba	480	4.0×10^5
^{90}Sr	710 ± 370	$(5.9 \pm 3.1) \times 10^5$
^{90}Sr	870 ± 675	$(7.2 \pm 5.6) \times 10^5$
^{106}Ru	675 ± 400	$(5.6 \pm 3.3) \times 10^5$
^{54}Mn	800 ± 410	$(7.5 \pm 3.4) \times 10^5$
^{95}Zr	130	1.1×10^5
^{95}Zr	570 ± 380	$(4.7 \pm 3.1) \times 10^5$
Iodine		
fallout (particulate)	420	3.5×10^5
fallout (particulate)	300–1500	$(2.5–12.4) \times 10^5$
elemental	3000	2.5×10^6
elemental	241	2.0×10^5
elemental (pH 5)	1000	8.3×10^5
methyl (alkyl) iodides	1–50	$(8.3–41.5) \times 10^3$
methyl (alkyl) iodides	5	4.2×10^3

Source: Peterson 1983.

For a small smooth spherical particle of diameter d and density ρ falling in air of density ρ_a, a constant settling speed, v_g, is reached when the gravitational forces are exactly balanced by the aerodynamic drag forces. If slip flow is neglected, this balance is (Briggs et al. 1968; Eisenbud 1963)

$$\frac{\pi}{6}(\rho - \rho_a)d^3 g = \frac{1}{2}\rho_a v_g^2 \left(\frac{1}{4}\pi d^2\right) C_d, \tag{9.49}$$

where g is the gravitational acceleration, and C_d is the dimensionless *drag coefficient*. This equation cannot be solved directly for v_g since the drag coefficient varies with the *Reynolds number* $N_{Re} \equiv \rho_a v_g d/\mu$, where μ is the atmospheric dynamic viscosity.

For Reynolds numbers less than 3, the flow around the sphere is laminar, and the Stokes relation $C_d = 24/N_{Re}$ may be used. Equation (9.49) then yields

$$v_g = \frac{(\rho - \rho_a)g}{18\,\mu}d^2, \qquad N_{Re} < 3. \tag{9.50}$$

If values of μ and ρ_a are substituted for air at standard temperature and pressure, this result becomes

$$v_g = 0.0027\rho d^2, \qquad d < 120/\rho^{1/3}, \tag{9.51}$$

where d is in μm, ρ in g cm^{-3}, and v_g in cm s^{-1}. For larger particles, and higher Reynolds numbers, different approximations for C_d yield (Eisenbud 1963)

$$v_g \cong 0.34\rho^{2/3}d, \qquad 120/\rho^{1/3} < d < 2200/\rho^{1/3} \tag{9.52}$$

and

$$v_g \cong 16\sqrt{\rho d}, \qquad d > 2200/\rho^{1/3}, \tag{9.53}$$

where the same units are used as in Eq. (9.51). For a mineral spherical particle ($\rho = 2.6$ g cm^{-3}), Eq. (9.51), (9.52), or (9.53) is used if the particle diameter is in the range $d < 85$ μm, 85 μm $< d < 1.5$ mm, or $d > 1.5$ mm, respectively.

The effect of *slip flow* increases the settling speed over the nonslip result, and occurs when the mean-free-path length of the air molecules is greater than the particle size. It is, thus, important only for very small particles at normal air pressures, say $d < 0.5$ μm for $\rho = 2.6$ g cm^{-3}. Also, particles with nonspherical shapes generally fall at as little as half the speed of a spherical particle of the same density and volume.

For very small particles with fall velocities of less than 1 cm s^{-1} (typically for $d < 15$ μm diameter), vertical turbulence dominates any tendency for gravitational settling, and such particles diffuse together with other nonparticulate effluents. However, the sedimentation of particles with fall velocities greater than 1 cm s^{-1} becomes an important plume depletion mechanism. A simple method for describing a plume of particles falling with speeds between 1 and 100 cm s^{-1} is to use the standard Gaussian diffusion model with the particle plume centerline tilted downwards to correct for the settling of the particles. Thus at a distance x downwind, the effective plume centerline height h_e is replaced by $h_e - v_g x/u$.

If it is assumed that the settling particles remain on the ground once they have fallen, the amount removed from the plume and deposited per unit area on the ground per unit time is

$$w(x, y, t) = v_g C(x, y, 0, t), \tag{9.54}$$

where $C(x, y, 0, t)$ is the ground-level concentration of particles in the tilted plume at time t.

9.5.4 Correction of Plume Model for Effluent Depletion

As the plume travels downwind from a continuous source, some of the effluents are deposited on the ground thereby reducing effluent concentrations in the plume. Similarly, radioactive decay further reduces plume concentration. To account for such plume depletion at a downwind distance x, the emission rate $Q \equiv Q(0)$ in Eq. (9.37), with $\alpha = 1$, is most frequently decreased to some smaller value $Q(x)$.

Consider some downwind location (x, y) at which the rate of effluent deposition per unit area of the ground is given, for a steady release, by [cf. Eqs. (9.43), (9.46), and (9.54)]

$$w(x, y) = v\, C(x, y, 0) = \frac{v\, Q(x)}{\pi u \sigma_y(x) \sigma_z(x)} \exp\left[-\frac{y^2}{2\sigma_y^2(x)} - \frac{h_e^2}{2\sigma_z^2(x)} \right]. \tag{9.55}$$

where $v = v_d + v_w + v_g$ is the total deposition speed for the plume. The rate at which the source strength $Q(x)$ decreases with increasing x must equal the rate at which effluent is deposited at the downwind distance x, that is,

$$\frac{dQ(x)}{dx} = -\int_{-\infty}^{\infty} dy\, w(x, y) = -\sqrt{\frac{2}{\pi}} \frac{v Q(x)}{u \sigma_z(x)} \exp\left[-\frac{h_e^2}{2\sigma_z^2(x)} \right]. \tag{9.56}$$

This first order differential equation is readily solved for $Q(x)$ to give

$$Q(x) = Q(0) \exp\left\{ -\sqrt{\frac{2}{\pi}} \frac{v}{u} \int_0^x dx'\, \frac{1}{\sigma_z(x')} \exp\left[-\frac{h_e^2}{2\sigma_z^2(x')} \right] \right\}. \tag{9.57}$$

Once $\sigma_z(x)$ is determined as an explicit function of x (see Sec. 9.6), the integration in Eq. (9.57) can be performed to give the effective source strength $Q(x)$.

For wet deposition described by a scavenging coefficient ϕ rather than by a deposition velocity, the plume depletion rate is given by Eq. (9.45), namely

$$w(x, y) = \phi \int_0^L dz\, C(x, y, z). \tag{9.58}$$

Generally, L is taken to be well above the top of the plume so that the upper limit can be replaced by ∞. Hence the source strength is given by

$$\frac{dQ(x)}{dx} = -\phi \int_{-\infty}^{\infty} dy \int_0^{\infty} dz\, C(x, y, z). \tag{9.59}$$

If the Gaussian model for an elevated plume [Eq. (9.35)] is substituted into this equation [after replacing Q by $Q(x)$], one obtains upon integrating

$$\frac{dQ(x)}{dx} = -\frac{\phi}{u}Q(x). \qquad (9.60)$$

Integration from 0 to x gives the effective source strength as

$$Q(x) = Q(0)\exp(-\phi x/u). \qquad (9.61)$$

Finally, if radioactive decay is also an important plume depletion mechanism, the effective source strength of Eq. (9.57) or (9.61) is multiplied by $\exp[-\lambda x/u]$ (see Sec. 9.5.2). Thus, the airborne concentration at ground level due to a constant emission rate $Q(0)$ at an effective height h_e is

$$C(x, y, 0) = \frac{Q(x)\exp[-\lambda x/u]}{\pi u \sigma_y(x)\sigma_z(x)}\exp\left[-\frac{y^2}{2\sigma_y^2(x)} - \frac{h_e^2}{2\sigma_z^2(x)}\right], \qquad (9.62)$$

where $Q(x)$ is calculated from Eq. (9.57) or (9.61).

9.5.5 Wake Effects

Implicit in the Gaussian straight-line model is the assumption that the motion of the ambient air moves in straight streamlines parallel to each other and the ground. However, uneven ground or large buildings near the evolving plume often disturb this ideal flow pattern, and thus, produce airborne concentrations that are quite different from those predicted by simple diffusion models.

The air flow around an obstruction is generally very complicated, as is illustrated in Fig. 9.14 which shows the flow around a large reactor building. The normal parallel air flow is displaced upwards over and laterally around the building, producing the so-called *displacement zone*. In the lee of the building there is a wake characterized by highly turbulent flow. In the wake, close to the obstruction, there is usually a region in which two large counter-flowing eddies are formed. Clearly, any plume which encounters such an obstruction is going to be perturbed significantly.

The calculation of airborne concentrations in the vicinity of such flow obstructions is complicated and well beyond the scope of this text. Because of the importance of such obstructions on the dispersal of gaseous effluents, there have been many studies performed on plume behavior near various obstructions (Briggs et al. 1968). Generally, the lower the release point, the more important are the effects of flow obstructions. For some specialized situations very approximate methods have been proposed (Smith 1968; USNRC 1977). Unfortunately, no general simple technique has been developed for predicting these effects, and site-specific studies must usually be performed using complex analysis programs.

9.6 ESTIMATION OF THE DIFFUSION PARAMETERS

To use the straight-line Gaussian models developed in the previous section to predict airborne concentrations, it is first necessary to obtain numerical values for the diffu-

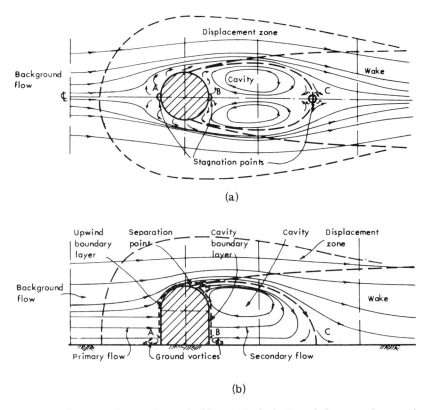

(a)

(b)

Figure 9.14   Air flow around a containment building: (a) in the horizontal plane near the ground, and (b) in the longitudinal center plane. [From Briggs et al. (1968).]

sion parameters that determine how rapidly the plume dissipates. The diffusion parameters σ_y and σ_z can vary widely and change with the atmospheric lapse rate, wind speed, downwind location, height of release, and local terrain features. Whenever possible, one should use values actually measured at the location of interest under the atmospheric conditions of concern.

Unfortunately, measured values are rarely available, and one is forced to use values that have been obtained at some other location under slightly different or idealized conditions. Over the past several decades many studies have attempted to define values for the diffusion parameters. It is from these studies that general guidelines for evaluating the diffusion parameters have evolved which make the Gaussian plume model a useful tool for predicting concentrations of airborne effluents.

Since the primary mechanism for plume dispersion is atmospheric turbulence (a problem beyond our present abilities to treat from first principles), it is not surprising that no entirely satisfactory theoretical approach has been developed to predict values for σ_y or σ_z. Some success has been obtained, however, in correlating these diffusion parameters to various fluctuations in the wind velocity (Pasquill 1974). Of more practical interest are those investigations that have measured plume dispersion under a wide variety of meteorological conditions. Through such measurements and semi-empirical fits to the resulting data, general guidelines have been established to help the analyst in selecting appropriate values for the diffusion parameters needed in a particular situation.

In the subsections that follow, some of the more widely used results for evaluating diffusion parameters are summarized. It should be emphasized that these values are for idealized circumstances, and that whenever possible locally derived values are preferable.

9.6.1 Pasquill's Diffusion Parameters

In the late 1950s Pasquill suggested a practical method for estimating the diffusion parameters $\sigma_y(x)$ and $\sigma_z(x)$ (Pasquill 1961). Guided by theory and the results of extensive smoke-plume experiments, a set of curves (Fig. 9.15) were developed which gave σ_y and σ_z versus x for six atmospheric stability classes (Pasquill 1974; Gifford 1961, 1968). This formulation of the diffusion parameters is extensively used and has been found to provide reasonable estimates of airborne concentrations for many situations.

The six atmospheric classifications (labeled A through F in increasing order of stability), as originally used by Pasquill, are related to average wind speed and the solar radiation balance. These classifications are defined in Table 9.3. To the original six atmospheric stability categories, a seventh G class for extremely stable conditions has recently been added with $\sigma_y^G = 0.6\,\sigma_y^F$ and $\sigma_z^G = 0.6\,\sigma_z^F$. These stability categories may also be defined in terms of the atmospheric lapse rate or the standard deviation of the horizontal (azimuthal) wind direction, σ_a, as shown in Table 9.4.

The Pasquill-Gifford curves of Figs. 9.15 are based on experimental data for ground releases made over flat smooth terrain and measured over short time intervals (6–10 minutes). This set of diffusion parameters, which is by far the most commonly used set, has been found to give reasonable estimates for long-term concentrations arising from ground, or near-ground, continuous releases over flat ground. The Pasquill-Gifford parameters are less reliable for elevated releases over complex terrain, and the slope of the σ_z curve for extremely unstable conditions (class A) at large distances has been questioned (Brenk et al. 1983).

Although graphical representations for $\sigma_y(x)$ and $\sigma_z(x)$ are satisfactory for hand calculations, they are not easily incorporated into computer programs. Empirical fits have to be made to the data of Fig. 9.15 and the resulting formulas are far easier to use in machine calculations. One such approximating formula has the form (Brodsky 1982)

$$\sigma_i = a_i x (1 + b_i x)^{-c_i}, \qquad i = y \text{ or } z. \tag{9.63}$$

Values of the constants a_i, b_i and c_i are given in Table 9.5.

A somewhat more complicated approximation is based on Pasquill's results corrected for averaging times and surface roughness. This approximation, which is recommended (Hanna et al. 1977) for surface releases, has the form (Miller 1984)

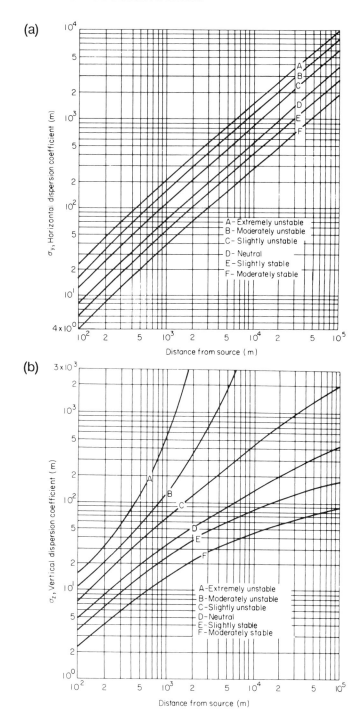

Figure 9.15   Pasquill-Gifford horizontal (a) and vertical (b) diffusion parameters as a function of downwind distance from the source for different atmospheric stability categories. [From Gifford (1968).]

TABLE 9.3   Relation of Turbulence Types to Weather Conditions

A—EXTREMELY UNSTABLE CONDITIONS			D—NEUTRAL CONDITIONS[a]	
B—MODERATELY UNSTABLE CONDITIONS			E—SLIGHTLY STABLE CONDITIONS	
C—SLIGHTLY UNSTABLE CONDITIONS			F—MODERATELY STABLE CONDITIONS	

SURFACE WIND SPEED ($m\,s^{-1}$)	DAYTIME INSOLATION			NIGHT TIME CONDITIONS	
	STRONG	MODERATE	SLIGHT	THIN OVERCAST OR $\geq$ 4/8 CLOUDINESS[b]	$\leq$ 3/8 CLOUDINESS
< 2	A	A–B	B		
2	A–B	B	C	E	F
4	B	B–C	C	D	E
6	C	C–D	D	D	D
> 6	C	D	D	D	D

[a] Applicable to heavy overcast, day or night.

[b] The degree of cloudiness is defined as that fraction of the sky above the local apparent horizon which is covered by clouds.

Source: Gifford 1968.

TABLE 9.4   Classification of Atmospheric Stability Categories.

STABILITY CLASSIFICATION	PASQUILL CATEGORIES	σ_a (deg)	TEMPERATURE CHANGE WITH HEIGHT ($^\circ$C per 100m)
Extremely unstable	A	25.0°	< −1.9
Moderately unstable	B	20.0°	−1.9 to −1.7
Slightly unstable	C	15.0°	−1.7 to −1.5
Neutral	D	10.0°	−1.5 to −0.5
Slightly stable	E	5.0°	−0.5 to 1.5
Moderately stable	F	2.5°	2.5 to 4.0
Extremely stable	G	1.7°	> 4.0

Source: Gifford 1968; USNRC 1972.

TABLE 9.5   Coefficients for Eq. (9.63) That Approximate Pasquill's Diffusion Parameters. Downwind Distance x is Measured in Meters.

STABILITY CLASS	$\alpha_y(x)$ (m)			$\sigma_z(x)$ (m)		
	a_y	b_y	c_y	a_z	b_z	c_z
A	0.22	10^{-4}	1/2	—	—	—
B	0.16	10^{-4}	1/2	—	—	—
C	0.11	10^{-4}	1/2	0.08	2×10^{-4}	1/2
D	0.08	10^{-4}	1/2	0.06	1.5×10^{-4}	1/2
E	0.06	10^{-4}	1/2	0.03	3×10^{-4}	1
F	0.04	10^{-4}	1/2	0.02	3×10^{-4}	1
G	0.024	10^{-4}	1/2	0.012	3×10^{-4}	1

Source: Brodsky 1982.

TABLE 9.6   Coefficients for the Approximation of the Pasquill–Gifford Diffusion Parameters, Eqs. (9.63) and (9.64), for Six Atmospheric Stability Classes. Downwind Distance x and σ_i are in Meters.

	ATMOSPHERIC STABILITY CATEGORY					
COEFFICIENT	A	B	C	D	E	F
a_1	−0.023	−0.015	−0.012	−0.0059	−0.0059	−0.0029
a_2	0.35	0.25	0.18	0.11	0.0881	0.0541
b_1	0.88	−0.99	−1.19	−1.35	−2.88	−3.80
b_2	−0.15	0.82	0.85	0.79	1.26	1.42
b_3	0.15	0.017	0.0045	0.0022	−0.0421	−0.0551

Source: Miller 1984.

$$\sigma_y(x) = x(a_1 \ln x + a_2) \tag{9.64}$$

and

$$\sigma_z(x) = \frac{1}{2.15} \exp(b_1 + b_2 \ln x + b_3 \ln^2 x), \tag{9.65}$$

where values of the approximation parameters are given in Table 9.6.

9.6.2 Other Diffusion Parameter Results

Many plume dispersion experiments to measure value of σ_y and σ_z have been performed over the past few decades. From such measurements other empirical formulas have been derived. Most of these results have been expressed in terms of the power functions

$$\sigma_i(x) = a_i x^{b_i}, \qquad i = y, z. \tag{9.66}$$

where the approximation parameters a_i and b_i change with the stability classification of the atmosphere.

Sutton (1953) proposed, on theoretical grounds, that the diffusion parameters in the transverse direction (σ_y and σ_z) vary with downwind distance x according to Eq. (9.66). For releases close to the ground $b_y > b_z$ since fluctuations in the vertical direction are usually less than those in the horizontal crosswind direction. However, for a release elevation $h > 25$ m, it is found that $b_y \cong b_z$. Consequently, σ_y and σ_z both have the same x-dependence, so that the analytical expression for the maximum, downwind, ground-level concentration given by Eq. (9.38) is valid. Experimentally, determined values of the Sutton approximation parameters are given by Eisenbud (1963).

Klug (1969) proposed diffusion parameters (the "Klug" set), expressed by Eq. (9.66), for ground-level releases over terrain with low surface roughness and for source distances of less than 3 km. From tracer studies of elevated release (108 m) over terrain of medium roughness with source distances up to 60 km, Singer and Smith (1966) have proposed parameters (the "Brookhaven" set) for a power function description

of σ_y and σ_z. For ground level releases in metropolitan areas or for sites with extreme surface roughness, McElroy and Pooler (1968) have provided parameters for
the power function approximation (the "St. Louis" set). Based on experiments for
elevated releases (50 and 100 m) and source distances out to 11 km, power function
approximations have been generated (the "Jülich" set) which are applicable to sites
with medium- to high-surface roughness caused by vegetation and settlements (Geiss
et al. 1978).

9.6.3 Diffusion Parameters Based on Meteorological Data

The suggested values of the diffusion parameters presented in the previous sections
are all for specific site and release conditions (e.g., given release elevations, terrain
roughness, measurement time, etc.). Whenever possible actual data specific to the
site under study should be used. From smoke-plume experiments where the visible
plume is observed, site-specific values for the diffusion parameters can be deduced
(Islitzer and Slade, 1968).

   Often, however, such extensive experiments are impractical. It is usually fea-
sible only to obtain site-specific information on wind speeds and their fluctuations.
Methods have been developed to use such wind measurements for estimating σ_y and
σ_z. One such simple method, the ASME method, is based on the following power
function formulas (Smith 1968):

$$\left.\begin{array}{l} \sigma_y(x) = 0.15\,\sigma_a\,x^{0.71} \\ \sigma_z(x) = 0.15\,\sigma_e\,x^{0.71} \end{array}\right\} \text{ stable } \left(\frac{dT}{dz} > 0\right), \qquad (9.67)$$

or

$$\left.\begin{array}{l} \sigma_y(x) = 0.045\,\sigma_a\,x^{0.86} \\ \sigma_z(x) = 0.045\,\sigma_e\,x^{0.86} \end{array}\right\} \text{ unstable } \left(\frac{dT}{dz} < 0\right), \qquad (9.68)$$

where σ_a and σ_e are the standard deviations (in degrees) of, respectively, the azimuthal and elevation angles of the wind velocity. Such standard deviations are readily
measured with a bivane recording indicator, and usually σ_a and σ_e are taken as one-
sixth the maximum hourly range. Besides measuring the vertical and horizontal wind
fluctuations, estimates of the atmospheric lapse rate must be obtained.

   In case complete meteorological data to use Eqs. (9.67) or (9.68) is not available,
the following approximations are suggested (Smith 1968):

Lapse rate unknown:	assume stable atmosphere if $\sigma_e < 1.5^o$
	otherwise assume atmosphere is unstable
σ_e **unknown:**	if stable, let $\sigma_e = \sigma_a$
	if unstable, let $\sigma_e = 0.7\,\sigma_a$
σ_e **and** σ_a **unknown:**	if stable, let $\sigma_a = 2^o$ (any elevation)
	if unstable, let $\sigma_a = 23/u + 4.75^o$ at 100 m,
	where u is the mean-wind speed (m/s) at 100 m

   Sometimes, available wind data are for a height h'' different from the effluent
release elevation h'. In such cases, it is necessary to correct the measured values of

σ_e'' and σ_a'' to obtain values for the release elevation. At slightly elevated positions from the ground, the lateral and vertical fluctuations, $\sigma_v \equiv u\,\sigma_a$ and $\sigma_w \equiv u\,\sigma_e$, respectively, are approximately independent of elevation even through u, σ_a and σ_e change with height. Thus,

$$\sigma_a' = \frac{\sigma_a''\,u''}{u'}, \text{ and } \sigma_e' = \frac{\sigma_e''\,u''}{u'}. \tag{9.69}$$

The mean wind speed u is observed to vary with height approximately as

$$\frac{u'}{u''} = \left(\frac{h'}{h''}\right)^n, \tag{9.70}$$

where $n \cong 0.25$ or 0.5 for an unstable or stable atmosphere, respectively, (Smith 1968). Substituting this result into Eq. (9.68), one can readily estimate σ_a' and σ_e' in terms of values measured at a different elevation.

9.7 AVERAGING TIMES

The Gaussian diffusion models describe the spatial variation of the mean airborne concentration. These concentrations are time-averaged values corresponding to observed concentrations that are averaged over some sampling time. In the diffusion models this averaging time is reflected in the data used to obtain values for the diffusion parameters σ_y and σ_z. If averaging time different from that used to generate the diffusion parameters is desired, then a correction must be made. The modification for different averaging time falls into two categories: (1) times comparable to experimental measurements, and (2) averaging times over some long time interval, such as a year.

9.7.1 Short Averaging Times

In Sec. 9.3.1 it was seen how wind turbulence causes the plume from a steady source to fluctuate so that the instantaneous concentration profile in the crosswind direction y, at any fixed downwind location x, would appear to meander about the average downwind direction. If the profile were averaged over longer and longer times, the average profiles would appear to spread out, decrease in amplitude, and become more symmetric about the $y = 0$ downwind axis (see Fig. 9.7).

The need to describe the concentrations in terms of time averages, while allowing use of diffusion theory, precludes information about the short term fluctuations in the concentrations. As the averaging time becomes longer, the maximum concentration at any particular location decreases. For averaging times τ between 10 minutes and 5 hours, the peak concentration is found to vary as $1/\sqrt{\tau}$ (Williamson, 1973). Thus, quadrupling the averaging time would approximately halve the observed peak concentration. For averaging times τ of less than 10 minutes, the ratio of the short-term peak concentration to the 1-hour mean concentration equals $\tau^{-\alpha}$ where τ is the averaging time expressed as a fraction of an hour, and α is a constant that depends on the atmospheric stability. For neutral, unstable, and very unstable conditions, α equals 0.35, 0.52, and 0.65, respectively (Smith 1968).

The expressions for the diffusion parameter presented in the previous section are based on plume concentration profiles that were averaged over some specific observation time. For example, the Pasquill values of Fig. 9.15 are based on 3 minute averages while other studies have shorter or longer averaging times. It is possible to correct the diffusion model for different averaging times by modifying the diffusion parameters. If τ' and τ'' represent two averaging times, the corresponding crosswind diffusion parameters are related approximately by (Hanna et al. 1977; Miller 1984)

$$\frac{\sigma_y'}{\sigma_y''} = \left(\frac{\tau'}{\tau''} \right)^\alpha . \tag{9.71}$$

The constant α is usually between 0.25 and 0.3 for $1\text{ h} < \tau' < 100\text{ h}$, while for shorter times α is about 0.2 if $3\text{ min} < \tau' < 1\text{-hour}$. A similar correction can be made to the vertical diffusion parameter σ_z if one is within a few kilometers of the release point.

9.7.2 Long Averaging Times

For many environmental studies, one is interested not in the airborne concentrations under a given meteorological condition but in concentrations averaged both over a long period (e.g., a month or a year) and over all meteorological conditions that are likely to occur during the averaging period. Over such long averaging intervals, the weather conditions vary greatly. It is thus necessary to average the concentrations over the expected frequency with which different wind speeds, wind directions, and atmospheric stabilities arise.

To obtain the necessary statistical meteorological data, wind and atmospheric stability measurements must be taken throughout each day for several averaging periods, preferably at or near the site under study. From these data, the distributions of wind velocities under various stability conditions can then be estimated. It is standard practice to classify wind direction (the compass heading from which the wind is blowing) as belonging to one of n (usually 16) compass sectors, each of which spans an angle of $2\pi/n$ radians at the source location. By classifying all wind speeds and directions by sector over a sufficiently long period of time, one can then obtain a *wind rose* which gives the distribution of velocities over some specified time of the year (e.g., monthly, seasonally, or yearly). A typical wind rose is shown in Fig. 9.16.

To calculate the long-term average of the effluent concentration field that is produced by a steadily emitting point source, the equivalent of a wind rose for each atmospheric stability class must first be generated. In practice, a matrix f_{ijk} is generated that gives the probability (fraction of the averaging time) that the wind is in sector k, with speed u_j, and the atmosphere is in stability class i. Clearly,

$$\sum_i \sum_j \sum_k f_{ijk} = 1. \tag{9.72}$$

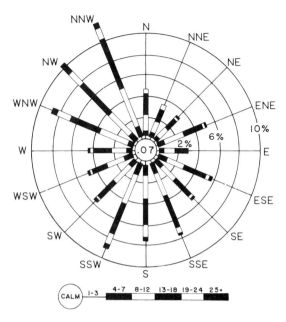

Figure 9.16 A typical wind rose showing the frequency with which the wind, in different wind speed categories, comes from a particular compass sector. All speeds are in miles h^{-1}. [After NSP (1960).]

Rather than to calculate effluent concentrations C directly, it is more usual to calculate *dilution factors*[3] (C/Q) where Q is the emission rate of the source. Thus, the dilution factor is independent of the source strength. When the atmosphere is in stability class i and the wind has speed u_j, the dilution factor at a location (x, y) on the ground is found from Eq. (9.37), with $\alpha = 1$, to be

$$(C/Q)_{ij} = \frac{1}{\pi u_j \sigma_{yi}(x) \sigma_{zi}(x)} \exp\left[-\frac{h_e^2}{2\sigma_{zi}^2(x)} - \frac{y^2}{2\sigma_{yi}^2(x)}\right], \tag{9.73}$$

where σ_{yi} and σ_{zi} are, respectively, the horizontal and vertical diffusion parameters under stability condition i, and where α is taken here to be 1, corresponding to complete ground reflection.

At the downwind distance x, the sector crosswind range is

$$y = \pm x \tan\left(\frac{\pi}{n}\right) \cong \pm \frac{\pi}{n} x. \tag{9.74}$$

Thus, the average over the sector's crosswind width is

$$\overline{\left(\frac{C}{Q}\right)}_{ij} = \frac{n}{2\pi x} \int_{-\pi x/n}^{\pi x/n} dy \left(\frac{C}{Q}\right)_{ij}. \tag{9.75}$$

---

[3]The dilution factors for airborne pollutants are often referred to as (χ/Q)-values.

If the downwind distance x is sufficiently large so that $\pi x/n \geq 2\,\sigma_{yi}(x)$, the limits on the above integral can be replaced by $\pm\infty$. Hence,

$$\overline{\left(\frac{C}{Q}\right)}_{ij} = \sqrt{\frac{2}{\pi}}\left(\frac{n}{2\pi x}\right)\frac{e^{-h_e^2/2\sigma_{zi}^2(x)}}{u_j\,\sigma_{zi}(x)}. \tag{9.76}$$

Notice that this sector-averaged result depends only on the vertical diffusion parameter σ_z.

Finally, the long-term average dilution factor for sector k is obtained by taking a weighted average of Eq. (9.76) over all wind speeds and atmospheric conditions for that sector, namely,

$$\overline{\overline{\left(\frac{C}{Q}\right)}}_k = \sum_i \sum_j \overline{\left(\frac{C}{Q}\right)}_{ij} f_{ijk}. \tag{9.77}$$

Diffusion Parameters for Sector Averages. To estimate the sector-averaged dilution factors from Eq. (9.77), special empirical formulas have been developed for the vertical diffusion parameter $\sigma_z(x)$ (Watson and Gammertsfelder 1963; Fuquay et al. 1964). For moderately or very stable atmospheres

$$\sigma_z^2(x) = a\left(1 - e^{-c^2 x^2/u^2}\right) + \frac{bx}{u}, \tag{9.78}$$

while for neutral or unstable conditions

$$\sigma_z^2(x) = \frac{1}{2}C_z^2 x^{2-n}. \tag{9.79}$$

The parameters a, b, c, C_z^2, and n in these formulas are given in Table 9.7. Dilution factors calculated with these particular formulas are considered to be quite realistic and are widely used in environmental impact analyses. It should be emphasized that these results apply only to sector-averaged calculations. As an alternative, one may also use the σ_z values discussed in Sec. 9.5.

9.8 LIMITATIONS OF THE GAUSSIAN DIFFUSION MODEL

The Gaussian straight-line diffusion model developed in this chapter has several inherent limitations. The model is based on the assumption of homogeneous turbulence in the atmosphere which then makes the diffusion coefficients D_i independent of position. At best, this condition only approximately holds in real atmospheres. Horizontal homogeneity is approached if the terrain is flat and uniform. Vertical homogeneity is harder to achieve since the wind speed tends to increase with elevation, thereby, altering the vertical distribution of turbulent eddies.

Another assumption in the development of the Gaussian diffusion model requires there to be a constant, nonzero, wind speed. This nonzero wind speed condition allows neglect of diffusion in the downwind direction. The constant wind speed condition allows all parts of the plume to move advectively with the same speed. In

TABLE 9.7   Constants in Eqs. (9.78) and (9.79) for σ_z^2 Used for Sector-Averaged Calculations. Downwind Distance x is in Meters and the Temperature Gradient dT/dz (Negative Lapse Rate) is in Units of $°C$ per 100 m.

	ATMOSPHERIC STABILITY			
	VERY STABLE	MODERATELY STABLE	NEUTRAL	UNSTABLE
PARAMETER	$\frac{dT}{dz} \geq 1.5$	$1.5 > \frac{dT}{dz} \geq -0.5$	$-0.5 > \frac{dT}{dz} \geq -1.5$	$-1.5 \geq \frac{dT}{dz}$
a (m^2)	34	97	—	—
b (m^{2}s^{-1})	0.025	0.33	—	—
c (s^{-2})	8.8×10^{-4}	2.5×10^{-4}	—	—
C_z^2 (u = 1–3 m s^{-1})	—	—	0.15	0.30
(u = 4–8 m s^{-1})	—	—	0.12	0.26
(u $\geq$ 9 m s^{-1})	—	—	0.11	0.24
n	—	—	0.25	0.20

Source: Watson and Gammertsfelder 1963; Fuquay et al. 1964.

reality, in the atmospheric boundary layer, (z less than about 150 m), the upper parts of the plume spread downwind faster than do the lower portions of the plume.

The Gaussian model also requires that air concentrations be averaged over a sufficiently long time interval (at least a few minutes) so as to smooth out spatial fluctuations in the plume. For routine continuous releases, this requirement presents no limitation. However, for an instantaneous puff release (arising from an accident, for example), the model is unable to predict instantaneous air concentrations. Only in a statistical sense when the *expected* value (or *ensemble average*) of the instantaneous concentration at some downwind location is sought (i.e., an average over many similar instantaneous releases) can the Gaussian model be used.

The time-averaged plume behavior, even for continuous releases, is quite different from the instantaneously observed plume, especially during unstable conditions, when the plume is seen to fluctuate considerably. To describe these plume fluctuations, a modified diffusion model has been developed (Gifford 1959, 1968) in which the centers of the expanding disk elements (cross section slices of the plume) are distributed randomly about their mean position.

The standard Gaussian model also is limited to regions within several tens of kilometers of the release point. To predict dispersion of gaseous effluents over regional or continental scales, the variation in wind speed and directions must be taken into account. Trajectory models have been developed (Heffter et al. 1975; Feber et al. 1982) which use measured wind velocity data to construct particle trajectories, usually for 3-hour time segments, at various elevations up to about 5 km. These trajectories are then combined with diffusion and deposition models to predict concentrations and effluent fallout over regional and even global areas.

The Gaussian model developed in this chapter is also applicable only to releases occurring within a few hundred meters of the surface. For debris introduced into the troposphere or stratosphere (for example, by a thermonuclear explosion), predictions of ground level deposition become very complex. Residence times for debris in the troposphere range from a few days for dust in the lower rain-bearing elevations up to 40 days if the debris is injected at the upper elevations (Eisenbud 1963, 1987). In the stratosphere, dust can remain for many months or even several years. High altitude jet streams and long residence times cause gaseous effluents injected high into the atmosphere to become widely spread around the world. Modeling of such high altitude releases is exceedingly difficult.

Despite the limitations of the Gaussian diffusion models, it is by far the most widely used model for analyzing the dispersion of radionuclides emitted from nuclear facilities. It has the great advantage of being computationally simple, and, for many practical applications or preliminary analyses, it has proven to be quite satisfactory.

REFERENCES

BERGSTRÖM, S. O. W., "Transport of Fallout ^{131}I into Milk," in *Radiological Concentration Processes*, B. Aberg and F. P. Hungate (eds.), 337–354, Proc. of Inter. Sympos., April 1966 in Stockholm, Sweden, Pergamon, Oxford (1967).

BOSANQUET, C. H., AND J. L. PEARSON, "The Spread of Smoke and Gases from Chimneys," *Trans. Faraday Soc. 32,* 1249, (1936).

BRENK, H. D., J. E. FAIROBENT, AND E. H. MARKEE, "Transport of Radionuclides in the Atmosphere," in *Radiological Assessment: A Textbook on Environmental Analysis*, J. E. Till and H. R. Meyer, (eds.), NUREG/CR-3332, U.S. Nuclear Regulatory Commission, Washington, D.C., 1983.

BRIGGS, G. A., I. VAN DER HOVEN, R. J. ENGLEMANN, AND J. HALITSKY, "Processes Other Than Natural Turbulence Affecting Effluent Concentrations," in *Meteorology and Atomic Energy 1968*, D. H. Slade (ed.), Report TID-24190, pp. 189–255, U.S. Atomic Energy Comm., Washington, D.C., 1968.

BRIGGS, G. A., *Plume Rise*, AEC Critical Review Series, TID-25075, 1969.

BRIGGS, G. A., "Plume Rise Predictions," in *Lectures on Air Pollution and Environmental Impact Analyses*, Workshop Proceedings, American Meteorological Society, Boston, Mass., pp. 59–111, 1975.

BRODSKY, A., "Models for Calculating Doses from Radioactive Materials Released to the Environment," in *Handbook of Radiation Measurement and Protection*, Vol. II, A. Brodsky (ed.), pp. 367–422, CRC Press, Boca Raton, Fla., 1982.

CARSON, J. E., AND H. MOSES, "The Validity of Several Plume Rise Formulas," *J. Air Poll. Control Assoc. 19*, 862 (1969).

DUDERSTADT, J. J., AND W. R. MARTIN, *Transport Theory*, Wiley-Interscience, New York, 1979.

EISENBUD, M., *Environmental Radioactivity*, 1st ed., McGraw-Hill, New York, 1963.

EISENBUD, M., *Environmental Radioactivity*, 3rd ed., Academic Press, New York, 1987.

FEBER, G. J., J. L. HEFFTER, AND A. W. KLEMENT, JR., "Meteorological Dispersion of Released Radioactivity," in *Handbook of Environmental Radiation*, A. W. Klement (ed.), pp. 29–42, CRC Press, Boca Raton, Fla., 1982.

FICK, A., 1855, "Uber Diffusion," *Ann. Physik Chem. 2[94]*, 59–86 (1855).

FUQUAY, J. J., C. L. SIMPSON, AND W. T. HINDS, "Prediction of Environmental Exposures from Sources Near the Ground Based on Hanford Experimental Data," *J. Appl. Material 3,* 6 (1964).

GEISS, G. A., K. J. VOGT, H. S. EHRLICH, AND G. POLSTER, "Recent Results of Dispersion Experiments at Emission Heights of 50 and 100 m," in *Proc. 12th Ann. Symp. Fachverband für Strahlungschutz*, H. Kellerman (ed.), Norderney, Germany, 1978.

GIFFORD, F. A., "Statistical Properties of a Fluctuating Plume Dispersion Model," in *Advances in Geophysics*, Vol. 6, pp. 117–138, F. N. Frenkiel and P. A. Sheppard (eds.), Academic Press, New York, 1959.

GIFFORD, F. A., "Use of Routine Meteorological Observations for Estimating Atmospheric Dispersion," *Nuclear Safety 2(4)*, 44–57 (1961).

GIFFORD, F. A., "An Outline of Theories of Diffusion in the Lower Layers of the Atmosphere," in *Meteorology and Atomic Energy 1968*, Slade, D. H. (ed.), pp. 65–116, Report TID-24190, U.S. Atomic Energy Commission, Washington, D.C.,1968.

GIFFORD, F. A., "Atmospheric Transport and Dispersion Over Cities," *Nuclear Safety 13,* 391–402 (1972).

HANNA, S. R., G. A. BRIGGS, J. DEARDORFF, B. A. EAGAN, F. A. GIFFORD, AND F. PASQUILL, "AMS Workshop on Stability Classification Schemes and Sigma Curves—Summary of Recommendations," *Bull. Am. Meteorol. Soc. 58(12),* 1305–1309 (1977).

HEALY, J. W., AND R. E. BAKER, "Radioactive Cloud-Dose Calculations," in *Meteorology and Atomic Energy 1968*, Slade, D. H. (ed.), pp. 301–377, Report TID-24190, U.S. Atomic Energy Commission, Washington, D.C., 1968

HEFFTER, J. L., A. D. TAYLOR, AND G. J. FERGER, *A Regional-Continental Scale Transport, Diffusion, and Deposition Model*, Envir. Research Lab., NOAA Tech. Memo ERL-ARL-50, Air Research Lab., Nat. Oceanic and Atm. Admin., Silver Spring, Md., 1975.

HILSMEIER, W. F., AND F. A. GIFFORD, *Graphs for Estimating Atmospheric Dispersion*, Report ORO-545, USAEC, Washington, D.C., 1962.

HOFFMAN, F. O., "A Reassessment of the Deposition Velocity in the Prediction of the Environmental Transport of Radioactive Releases," *Health Physics 32,* 437–441 (1977).

ISLITZER, N. F., AND D. H. SLADE, "Diffusion and Transport Experiments," in *Meteorology and Atomic Energy 1968*, Slade, D. H. (ed.), pp. 117–188, Report TID-24190, U.S. Atomic Energy Commission, Washington, D.C., 1968.

KLUG, W., "Ein Verfahren zur Bestimmung der Ausbreitungsbedingungen aus synoptischen Beobachtunger," *Staub. 29*, 143 (1969).

MARKEE, E. H., "A Parametric Study of Gaseous Plume Depletion by Ground-Surface Absorption," in *Proc. of USAEC Meteorological Information Meeting*, C. A. Mawson (ed.), AECL-2787, pp. 602–613, 1967.

McELROY, J. L., AND F. POOLER, *St. Louis Dispersion Study: Vol. II—Analysis*, U.S. Dept. of Health, Education and Welfare, Arlington, Va., 1968.

MILLER, C. W., F. O. HOFFMAN, AND D. L. SHAEFFER, "The Importance in the Variation of the Deposition Velocity Assumed for the Assessment of Airborne Radionuclide Releases," *Health Physics 34,* 730–374 (1978).

MILLER, C. W., "Atmospheric Dispersion and Deposition," in *Model and Parameters for Environmental Radiological Assessments*, C. W. Miller (ed.), pp. 11–20, DOE/TIC-11468, U.S. Dept. of Energy, Washington, D.C., 1984.

MOSES, H., AND J. E. CARSON, "Stack Design Parameters Influencing Plume Rise," *J. Air Poll. Control Assoc. 18,* 456 (1968).

MOORE, R. E., *The AIRDOSE-II Computer Code for Estimating Radiation Dose to Man from Airborne Radionuclides in Areas Surrounding Nuclear Facilities*, DOE Report ORNL-5245, Oak Ridge Nat. Lab., Oak Ridge, Tenn., 1977.

NSP (Northern States Power Co.), *Final Safety Analysis Report*, Vol. I, Sect. II, NSP, Minneapolis, Minn., 1969.

PASQUILL, F., "The Estimation of the Dispersion of Windborne Material," *Meteorol. Mag. 90,* 33 (1961).

PASQUILL, F., *Atmospheric Diffusion*, 2nd ed., Halsted Press, Wiley, New York, 1974.

PETERSON, H. T. JR., "Terrestrial and Aquatic Food Chain Pathways," in *Radiological Assessment: A Text Book on Environmental Dose Analysis*, J. E. Hill and H. R. Meyer (eds.), NUREG/CR-3332, U.S. Nuclear Regulatory Commission, Washington, D.C., 1983.

SINGER, I. A., AND M. E. SMITH, "Atmospheric Dispersion at Brookhaven National Laboratory," *Air and Water Poll. Int. J., 10,* 125–135 (1966).

SLADE, D. H. (ed.), *Meteorology and Atomic Energy 1968*, Report TID-24190, U.S. Atomic Energy Commission, Washington, D.C., 1968.

SLINN, W. G. N., "Parameterizations for Resuspension and for Wet and Dry Deposition of Particles and Gases for Use in Radiation Dose Calculations," *Nuclear Safety 19,* 205–219 (1978).

SMITH, M. (ed.), *Recommended Guide for the Prediction of the Dispersion of Airborne Effluents*, Am. Soc. of Mech. Eng., New York, NY, 1968.

SUTTON, O. G., *Micrometeorology*, McGraw-Hill, New York, 1953.

THOMAS, F. W., S. B. CARPENTER, AND W. C. COOLBAUGH, "Plume Rise Estimates for Electric Generating Stations," *J. Air Poll. Control Assoc. 20,* 170 (1970).

USNRC (U.S. Nuclear Regulatory Commission), *Safety Guide 23*, USNRC, Washington, D.C., 1972.

USNRC (U.S. Nuclear Regulatory Commission), *Methods for Estimating Atmospheric Transport and Dispersion of Gaseous Effluents in Routine Releases from Light-Water-Cooled Reactors*, Regulatory Guide 1.111, U.S. Nuclear Regulatory Comm., Washington, D.C., 1977.

USNRC (U.S. Nuclear Regulatory Commission), *Atmospheric Dispersion Models for Potential Accident Consequence Assessments at Nuclear Power Plants*, Regulatory Guide 1.145, U.S. Nuclear Regulatory Commission, Washington, D.C., 1982.

WATSON, E. C., AND C. C. GAMMERTSFELDER, *Environmental Radioactive Containment as a Factor in Nuclear Plant Siting Criteria*, HW-SA-2809, Hanford, Wash., 1963.

WILLIAMSON, S. J., *Fundamentals of Air Pollution*, Addison-Wesley, Reading, Mass., 1973.

PROBLEMS

1. By what percent does the atmospheric pressure change in an isothermal atmosphere per 100 m increase in elevation? Dry air at standard temperature and pressure has an effective molecular weight of 28.97 and a density of 1.293 mg cm^{-3}.

2. If the pressure on a parcel of dry air is increased reversibly and adiabatically by 0.001 MPa, what is the resulting change in the air temperature?

3. Verify that Eq. (9.22) is a solution of Eq. (9.21). Show that $C(x, y, z, t) = 3x + y + 10z$ also satisfies the diffusion equation. Why is Eq. (9.22), and not this latter solution, the physically realistic solution to the problem of an instantaneous release into an infinite medium?

4. Estimate the inhalation exposure received by an individual who is standing underneath the release point when an amount Q' of radionuclides is emitted into an infinite medium. A wind of speed u is blowing perpendicularly to the vertical line between the person and the source point.

5. Consider a continuous stack release of a long-lived radionuclide noble gas from an elevation $h = 75$ m. The wind speed is 1.2 m s^{-1}, and the measured lapse rate is large and negative, indicating extremely unstable conditions. Measurement of σ_a and σ_e are not available. Estimate and plot on a log-log scale, C/Q at ground level versus downwind distance (100 – 5000 m) for $y = 0$. Use (a) ASME guidelines of Sec. 9.6.3, and (b) Pasquill's results.

6. Suppose that the radionuclide of problem 9.5 has a decay constant of 0.001 s^{-1}. Repeat problem 9.5(a) for this case and illustrate the comparison.

7. Suppose that conditions of problem were "stable." Repeat problem 9.5(a) for this case and illustrate the comparison. Use downwind distances from 3000 to 50,000 m.

8. For problem 9.5(b) compute and illustrate the cross-wind variation of C/Q at x = 2000 m, over the crosswind range from $y = 0$ to 1800 m.

9. Sutton's empirical formulas for the diffusion coefficients are expressed in the form

$$\sigma_y = ax^n \text{ and } \sigma_z = bx^n,$$

where a, b, and n are constants. Thus, the ratio $\sigma_z/\sigma_y = \kappa$ is also a constant. Show that, for this choice of the diffusion coefficients, the maximum downwind ground concentration (resulting from a continuous emission Q (Ci s^{-1}) from a stack of height h above flat terrain) occurs at a downwind distance

$$x_{max} = (h^2/2b^2)^{1/2n}$$

and has the value

$$C_{max} = [(1 + \alpha)Q\kappa/\pi euh^2].$$

10. Two 75-m high stacks, 100 m apart, both emit 20 nCi s^{-1} of ^{134}I on a clear day during which the atmospheric lapse rate is 1.6 C per 100 m. If the mean wind velocity is 5 m s^{-1} perpendicular to the line between the stacks, estimate the rate at which ^{134}I is deposited on a pasture 20 km directly downwind of the stacks. Use Pasquill's diffusion coefficients and assume an iodine deposition velocity of 0.04 m s^{-1}. The half life of ^{134}I is 52.5 minutes.

11. Consider a long, straight, heavily traveled motorway on which traffic emits a particular gaseous pollutant at a rate Q_l kg m^{-1} s^{-1}. Derive an expression for the airborne concentration of this pollutant at ground level a distance x downwind when the wind is blowing perpendicularly to the motorway.

12. Calculate the effective stack height for an effluent that is discharged from a circular chimney of 2-m inside diameter and 50-m height with a speed of 10 m s^{-1}. The temperature of the effluent is 75 C above ambient, and the mean wind speed is 7 m s^{-1} in an unstable atmosphere. Assume flat terrain around the stack.

13. Estimate the rate at which ^{131}I is deposited on a unit area of pasture land 1.5 km from a nuclear power plant on a day with winds of 2 m s^{-1} under neutral conditions. The iodine is released continuously from a 75-m stack at a rate of 55 nCi s^{-1}. Assume that 95% of the iodine is in a nonreactive gaseous form while the rest is molecular iodine.

14. Smoke particles with a mean size of 20 μm and density 2.0 g cm^{-3} are emitted from a 50-m tall stack into a stable atmosphere with a mean wind speed of 1.5 m s^{-1}. Plot the normalized deposition rate directly downwind, $w(x,0)/Q$, as a function of downwind distance x.

15. The annual gaseous radionuclide releases from a 100-m high stack at a BWR power plant are listed in the following table:

Noble Gases	Ci y^{-1}	Halogens	Ci y^{-1}	Particulates	Ci y^{-1}
^{85}Kr	5.79×10^3	^{131}I	0.576	^{137}Cs-^{137m}Ba	$< 3.7 \times 10^{-4}$
^{87}Kr	1.03×10^5	^{133}I	1.16	^{140}Ba-^{140}La	$< 1.3 \times 10^{-3}$
^{88}Kr	1.40×10^5	^{135}I	2.26	^{90}Sr	4.4×10^{-5}
^{133}Xe	1.14×10^5			^{134}Cs	$< 2.0 \times 10^{-4}$
^{135}Xe	1.85×10^5			^{89}Sr	5.5×10^{-5}
^{138}Xe	6.34×10^4				

At this plant the prevailing wind direction is in the SSE sector 10.6% of the time. In that direction the plant boundary is 1000 m downwind and a dairy farm is adjacent to the plant. Representative atmospheric conditions are "slightly unstable" in the Pasquill classification scheme with a mean wind speed of 1.5 m s^{-1}. Only the following nuclides are of significance: ^{131}I, ^{87}Kr, ^{88}Kr, ^{135}Xe and ^{138}Xe.

(a) What is the appropriate annually averaged dilution factor for the dairy farm (ignore radioactive decay)?

(b) Correcting for radioactive decay, what are the ground-level yearly-average concentrations of each of the significant nuclides at the dairy farm?

(c) What is the annual whole-body exposure due to noble gases that an individual who lived at the farm would receive?

(d) What is the annual thyroid dose rate an infant would receive who drinks milk from the cows on the farm?[*4]

16. Consider the BWR plant described in problem 9.15. Noble gases are assumed to be released at an elevation of 100 m. Halogens and particulates are assumed to be released at ground level. The prevailing wind direction is in the SSE sector. In that direction, the

[4]Subproblems marked by asterisk require use of data and methods from Chap. 11.

plant boundary is 1000-m downwind and dairy cows are pastured adjacent to the boundary. Wind speeds in the SSE sector are divided into 6 groups (subscript k). Average wind speeds for each group at both 0 and 100-m elevations are listed in the following table:

Mean Wind Speeds (m s^{-1}) in Groups $k = 1$–6		
k	u_k (0 elevation)	u_k (100-m elevation)
1	0.5	1.3
2	1.5	3.5
3	2.5	6.3
4	4.0	9.5
5	5.5	13.3
6	6.5	15.5

There are four wind stability conditions: very stable ($j = 1$), moderately stable ($j = 2$), neutral ($j = 3$), and unstable ($j = 4$). Bivariate wind-speed stability frequencies are listed below. f_{kj} is defined as the percentage of the time that the wind is in the SSE sector with speed in the kth group and stability in the jth group.

Bivariate Wind-Speed Stability Frequencies, f_{kj}				
k	$j = 1$	$j = 2$	$j = 3$	$j = 4$
1	0.073	0.000	0.0158	0.0744
2	0.396	0.148	0.238	0.389
3	1.02	0.891	0.758	1.19
4	0.846	0.126	1.16	1.56
5	0.000	0.192	0.474	0.595
6	0.000	0.000	0.000	0.433

Your assignment is to assess whether the plant meets the following design guidelines for gaseous radionuclide releases per reactor.

Gaseous Effluents to Atmosphere

10 mrad y^{-1} gamma-ray air dose at plant boundary
20 mrad y^{-1} gamma-ray plus beta-particle air dose at boundary
5 mrem y^{-1} (total body) at potentially occupied point
15 mrem y^{-1} (skin) at potentially occupied points

Radioactive Iodine and Particulates to Atmosphere

15 mrem y^{-1} from inhalation at potentially occupied points
15 mrem y^{-1} from food pathways at potentially occupied points*

17. An accident at a nuclear power station resulted in the release, over a 24-hour period, of 15 Ci of ^{131}I and 7.5×10^6 Ci of ^{133}Xe. During and subsequent to the accident, generally southern winds varied between 120° and 200° (directions from which winds were blowing). Average wind speed was 2.5 mph and conditions were slightly unstable (Pasquill category C). Releases were at an elevation of 30 m above grade. The effective exclusion area of the station has a radius of 2 miles. The nearest dairy farm is 4 miles north of the plant. A city of population 100,000 is located 11 miles from the plant at an azimuth of 320°. Aside from the city, the area within 2 to 20 miles radially from the plant is rural, with an average population density of 80 persons per square mile. The area between 20 and 50 miles of the plant is rural and small-town with an average population density of 270 persons per square mile. Estimate:

(a) The maximum potential instantaneous dose equivalent rate off site from ^{133}Xe (mrem h^{-1} whole body).

(b) The maximum potential dose equivalent off site, averaged over all wind directions, during the entire course of the accident from ^{133}Xe (mrem whole body).

(c) The ^{131}I concentration, as a function of time, in milk from the nearest dairy farm. For iodine, assume that the average quantity deposited over the 24-hour period is, instead, instantly deposited.*

(d) The maximum potential thyroid dose equivalent to the infant due to consumption of milk from the nearest dairy farm.*

Chapter 10

Dispersion of Radionuclides in Surface and Ground Waters

10.1 INTRODUCTION

The movement of liquid water is one of two primary mechanisms, atmospheric dispersion being the other, by which radionuclides, chemical or biological contaminants, or other pollutants are spread throughout our ecosystem and introduced into its many food chains. Unlike atmospheric dispersion, the distances involved in describing the migration of radionuclides dissolved or suspended in liquid water can vary considerably. Migration of radionuclides in soil, subsoil and uptake in vegetation is typically quite local, involving distances of centimeters to a few meters. Movement of radionuclides in underground aquifers are of concern for distances measured in meters to several tens of kilometers. Finally, the dispersal of radionuclides in surface waters can involve distances of hundreds of meters in ponds and lakes, to hundreds of kilometers in rivers, and even to global dimensions in oceans.

The modeling of radionuclide dispersal in the environment by surface and ground waters is a formidable challenge. The hydrodynamic description of the movement of surface water is made difficult by very irregular geometries (e.g., estuaries and mountain streams), variable flow rates (e.g., tides and precipitation), complex flow patterns (e.g., currents and eddies), density variations (e.g., thermal stratification), interactions with the atmosphere (e.g., wave motion), and usually by turbulent flow conditions. Movement of ground water, while usually not turbulent, is also complicated as a result of complex geometries (e.g., rock crack structures) and strong interactions between the water and the matrix in which water is moving (e.g., porous flow). The dispersion and transport of radionuclides in surface and ground waters is furthered complicated by the need to consider such effects as adsorption and desorption of radionuclides by sediments and channel surfaces and the need to account for radioactive decay and the buildup of radioactive progeny.

477

With the introduction of supercomputers, many advances have been made in the development of computational models capable of describing many complex aquatic dispersion problems (see, for example, Falconer et al. 1989). These models require considerable computer resources since they solve numerically the governing transient, multidimensional, coupled, flow equations which can account for the many flow variables, including, for example, the effects of unsteady sediment movement. Such powerful computer-based techniques, combined with field measurements, are always needed to determine accurate dispersion profiles in geometrically complex aquatic systems.

However, for many preliminary analyses, much simpler approximate models may be used effectively to estimate the spatial and temporal distribution of radionuclides following their release into surface or ground waters. These analytical models, which are mostly based on diffusion theory and empirical formulas, are rigorously applicable only to relatively simple geometries. Nevertheless, with proper discretion in their application, they often give insight and useful estimates of radionuclide concentrations for many important environmental applications. Such analytic dispersion models for aquatic systems are the subject of this chapter. The dispersal of radionuclides entering surface waters is first discussed, and then the problem of radionuclide migration in soil and ground water is addressed. No attempt is made to be complete in either case; rather, only relatively simple and idealized dispersion models that illustrate the essential features of the different hydrospheric systems are considered.

10.2 INITIAL (NEAR-FIELD) MIXING IN SURFACE WATERS

If liquid effluent is discharged as a high-speed jet or a buoyant plume into a large body of water, the resulting turbulent interaction between the discharge and the ambient water causes rapid dilution of any radionuclides in the discharge. This initial mixing occurs very near the discharge point, typically within 10 to 100 times the characteristic discharge dimension, half the square root of the discharge cross-sectional area. This rapid, near-field dilution of the effluent concentration, often by factors of up to 100, is a very effective means of reducing the concentrations of discharged radionuclides. While later dilution in the receiving water further reduces concentrations, this far-field mixing occurs much more slowly and over far greater distances than does the initial near-field mixing of the discharge plume as a result of the greatly reduced turbulence levels that naturally occur in the receiving water.

Many factors affect the initial dilution and the region over which it occurs. These factors include the momentum and buoyancy of the discharge, the configuration of the discharge structure, the discharge location in the receiving water, and the depth and currents of the receiving water near the discharge location. There are many different types of discharges, each with its own near-field mixing characteristics. One encounters surface and submerged discharges, single port and multiport outfalls, discharges into deep and shallow receiving waters, discharges to stagnant and flowing waters, and discharges with varying degrees of buoyancy. For nuclear plants, buoyant discharges are of primary concern since routine radionuclide releases are usually mixed with the cooling water discharges for once-through systems or with the blowdown for closed-cycle systems.

Many initial mixing models have been developed to describe different discharge situations. These models, usually in the form of computer codes, are generally too involved to include here. Rather, in this section some simple but useful empirical results are presented which allow rapid estimation of the amount of initial mixing for several common discharge configurations. Specifically, the following two quantities are estimated: (1) the *dilution S*, which is the ratio of the radionuclide concentration in the discharge to that at some point of interest (usually at the end of the near-field mixing zone), and (2) a measure of the size of the near-field mixing zone. The importance of determining the near-field dilution S is that it determines the concentrations to be used as the starting conditions for the far-field calculations discussed in later sections. The near-field mixing zone is usually very small compared to the distances involved with the subsequent far-field calculations. Hence, the source for the far-field calculations can usually be approximated as an effective point source whose concentration is determined from the near-field mixing.

10.2.1 Point Surface Discharges

For an outfall near the surface, for example, an open channel or a slightly submerged pipe, the initial mixing depends on the nature of the receiving water and the buoyancy and momentum of the discharge. This momentum dependence is usually expressed in terms of the dimensionless Froude number

$$F_o \equiv u_o / \sqrt{(\Delta\rho/\rho)g\ell_o}, \tag{10.1}$$

where u_o is the mean discharge velocity, ρ the ambient water density, $\Delta\rho$ the difference between the ambient density and the density of the discharge, g the acceleration of gravity (9.8 m s^{-2}), and ℓ_o a characteristic dimension of the outfall that is related to the discharge cross-sectional area A_o by $\ell_o = \sqrt{A_o/2}$.

Buoyant Discharge into Still Waters. The discharged plume entrains ambient water as it moves in the discharge direction. Eventually, the turbulence induced by the shear forces between the plume and the ambient water becomes less than the ambient turbulence, marking the end of the initial mixing zone at which point the effluent will have a bulk dilution S. This distance from the discharge point at which the jet becomes stabilized and entrainment of ambient water ceases is called the *transition distance* x_t and marks the end of the initial near-field mixing phase.

The receiving water can be considerd deep when the maximum vertical depth to which a buoyant plume descends h_{max} is considerably less than the water depth H. When the deep receiving water is stagnant or only has weak crosscurrents, it is observed that (Jirka et al. 1983)

$$S_{deep} \cong 1.4 \, F_o. \tag{10.2}$$

and

$$x_t \cong 15\,\ell_o\,F_o. \tag{10.3}$$

For this case the maximum vertical penetration of the surface plume is found to be $h_{max} \cong 0.42\,\ell_o F_o$ which occurs approximately at a distance $5.5\,\ell_o F_o$ from the outfall.

For shallow receiving water $(h_{max} > 0.75\,H)$ the dilution is inhibited by the limited depth of ambient water available for mixing into the plume. The bulk dilution for such shallow water discharges, after initial mixing, can be estimated by multiplying the result from Eq. (10.2) by an empirical factor to give

$$S_{shallow} = 1.4 \left\{ \frac{0.75H}{h_{max}} \right\}^{0.75} F_o. \tag{10.4}$$

Buoyant Discharge into a Strong Crossflow. When effluent is discharged from the shore into a strong crosscurrent (e.g., into a fast flowing river), it is swept downstream, and if the current is sufficiently strong the discharge plume may become attached to the shore. Such shoreline attachment prevents the plume from entraining ambient water along one side, thereby inhibiting near-field dilution. This reduced dilution from shoreline attachment is further aggravated in shallow water where the plume is in contact with the bottom and the crossflow is prevented from passing under the plume. Jirka et al. (1981) have proposed a condition for the occurrence of shoreline attachment for perpendicular discharges into a straight channel with crossflow speed u and depth H. Attachment will occur when

$$u > 0.05\,u_o \left(\frac{h_{max}}{H} \right)^{-3/2}. \tag{10.5}$$

No simple model exists for calculating the initial dilution for attached plumes. However, it is conservative to assume the dilution for attached plumes is one-half of that for unattached plumes as given by Eq. (10.4). for shallow water.

Discharges with Neutral or Negative Buoyancy. The results in the previous subsection apply to buoyant discharges $(\Delta\rho > 0)$. As the density difference between the effluent and ambient water becomes small, F_o becomes very large indicating large dilutions and large transition distances before the discharge jet becomes stabilized (i.e., entrainment of ambient water from the jet action ceases).

For a truly neutral jet discharged into an infinite stagnant medium, the dilution (averaged over the jet cross-sectional area) is given by (Fischer et al. 1979)

$$S(x) = \frac{0.32\,x}{\ell_o}. \tag{10.6}$$

While this result indicates that dilution would continue to increase without limit as the jet moves further into the medium, in practice this result is valid only as long as the jet-induced turbulence dominates the ambient turbulence, that is, until the jet becomes stabilized.

A plume discharged at the surface with negative buoyancy will sink to the bottom, entraining ambient water as it sinks. This sinking plume is just the opposite of a rising buoyant plume discharged at the bottom, and can be analyzed using the methods presented in the next section.

10.2.2 Submerged Point Discharges

Effluent discharges into water are often injected well below the surface of the water, usually near the bottom. Discharge structures can range from simple open pipes to elaborate arrays of diffusers oriented at different angles so as to promote increased dilution during the near-field mixing phase. In this section, the simplest case of a point discharge on the bottom and directed vertically upward is considered. More elaborate discharge geometries are treated by Roberts (1977) and Jirka (1982).

The initial mixing of a submerged discharge is strongly influenced by the depth of the receiving water. Two fundamentally different cases are encountered. First, in deep water a distinct buoyant plume can rise to the surface entraining ambient water during its rise and thereby diluting the radioactive effluent (see Fig. 10.1a). If the water is thermally stratified, the plume may rise only to the equilibrium level before beginning to disperse horizontally. In the second situation, encountered in shallow waters, the discharge momentum may be sufficiently large to cause a general breakdown in the rising thermal plume and, thereby, to create complex recirculation zones (see Fig. 10.1b). This unstable shallow water case is commonly encountered in heated water discharges from power plants. Unfortunately, it is also more difficult to analyze since mixing in the recirculation zones must be considered and simple plume models cannot be used.

The distinction between these two basic types of submerged discharges is important. The transition from a stable thermal plume rise (deep receiving water case) to the unstable counterflow situation (shallow receiving water case) occurs when (Jirka et al. 1983)

$$H/d_o < 0.22 F_o, \tag{10.7}$$

where H is the depth to the discharge point, d_o is the diameter of the circular discharge orifice, and F_o is the Froude number of Eq. (10.1) with the characteristic length ℓ_o replaced by d_o. This result is insensitive to the discharge angle, although for most of the results presented in the following subsections, vertical discharges are assumed.

Deep Receiving Water.   If the shallow-water condition of Eq. (10.7) is not satisfied, then models based on buoyant plumes can be used to estimate the radionuclide concentration in the plume as it rises to the surface. For discharges into stagnant waters or waters with only weak crosscurrents, the centerline dilution of the plume, after rising a vertical distance z, can be approximated by a fit of data given by Roberts (1977):

$$S_c(z) = \frac{5}{9} F_o \left(\frac{z}{d_o F_o} \right)^{\alpha(z)/8}, \qquad 0.1 \leq \frac{z}{d_o F_o} \leq 100. \tag{10.8}$$

(a)

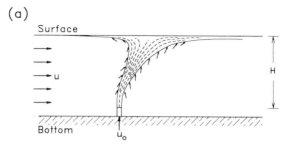

(b)

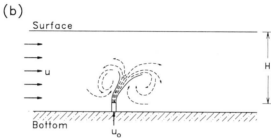

Figure 10.1 Two types of bottom discharges: (a) a buoyant plume rises in deep water and spreads out as ambient water is entrained from the sides and produces a mixed region on the surface, and (b) a high-speed jet or buoyant plume discharged into shallow water becomes unstable before reaching the surace and produces regions of counter flow. [After Jirka et al. (1983).]

Here $\alpha(z) \equiv 6 + \ln[z/(d_o F_o)]$, and d_o is the effective diameter of the discharge port (including any contraction produced by a sharp edged nozzle). Because a mixed layer forms on the water surface, the total distance the plume can rise is somewhat less than the discharge depth H. Usually in calculating surface concentrations, a rise distance $z = 0.8\,H$ is used. Finally, a near-field bulk dilution averaged over the whole surface plume can be estimated from

$$S \cong 1.4\,S_c \times 0.8\,H. \tag{10.9}$$

For stratified receiving waters, the rising thermal plume can become trapped below the free surface. For an analysis of such situations see Jirka (1975) and Fischer (1979).

Shallow Receiving Water.   When Eq. (10.7) is satisfied, the strong momentum effects of the discharge jet causes complex flow patterns to be established in the receiving water. For a vertical jet discharged into waters with only weak crosscurrents, the bulk dilution in the near field can be estimated from (Lee and Jirka 1981)

$$S \cong 0.9 \left(\frac{H}{d_o}\right)^{5/3} F_o^{-2/3}. \tag{10.10}$$

If the jet is oriented horizontally, it is conservative to assume that the discharge quickly rises to the surface and behaves as a surface jet in shallow water. For this case, the near-field bulk dilution can then be estimated from Eq. (10.4), with ℓ_o in the definition of F_o replaced by d_o.

Moderate to Strong Crossflows.   When submerged discharges are made into waters with moderate to strong currents (especially into deep waters), the crossflows provide an additional mechanism to dilute the discharged radionuclides. Hence, it is conservative to apply the above results that were derived for discharges into stagnant waters.

10.3 DISCHARGES TO SMALL LAKES AND RESERVOIRS

In small shallow ponds and reservoirs, there is usually rapid mixing of any influx of dissolved or suspended radionuclides. Winds cause rapid lateral mixing and the lack of thermal stratification in the pond permits vertical mixing. Such water bodies are typical of many power-plant cooling ponds in which plant discharges are in the form of buoyant surface jets which rapidly mix with the ambient pond water. In them, the radionuclides can be considered uniformly distributed in the pond water.

A simple model to describe the time variation of radionuclide concentrations can be developed for these well-mixed ponds. Consider a pond with a volume V of water. To keep this volume constant, the influx of water into the pond must equal the sum of evaporative water loss and the rate of discharge from the pond, q_o (m^{-3} s^{-1}). Let $Q(t)$ be the time-dependent rate at which a particular radionuclide enters the pond from, for example, atmospheric fallout, plant discharges, or surface water runoff. If none of the radionuclides becomes attached to the sediment on the pond bottom, then the rate of change in the concentration $C(t)$ of the radionuclide in the pond water equals the rate of input less the rates of decay and outflow, namely,

$$\frac{dC(t)}{dt} = \frac{Q(t)}{V} - \left(\lambda + \frac{q_o}{V}\right) C(t). \tag{10.11}$$

in which λ is the radioactive decay rate constant and $Q(t)$ may be in units of Bq s^{-1}, atoms s^{-1}, and so on.

The general solution of this equation for an arbitrary initial concentration $C(t_o)$ is, from Appendix F, Sec. F-1.1,

$$C(t) = e^{-\lambda' t} \left\{ C(t_o) e^{\lambda' t_o} + \frac{1}{V} \int_{t_o}^{t} dt' Q(t') e^{\lambda' t'} \right\}, \qquad t > t_o, \tag{10.12}$$

where $\lambda' \equiv \lambda + q_o/V$. For the special case of a constant input rate Q_o that begins at $t_o = 0$ into an uncontaminated pond, this result shows that the subsequent radionuclide concentration increases as

$$C(t) = \frac{Q_o}{q_o + \lambda V} \left\{ 1 - e^{-(\lambda + q_o/V)t} \right\}, \tag{10.13}$$

which, after a sufficiently long time, leads to a steady-state concentration

$$C_\infty = \frac{Q_o}{q_o + \lambda V}.$$ (10.14)

For a constant influx, it is seen that the steady-state concentration is determined by two rate constants: the decay rate λ and the volumetric fraction discharge rate q_o/V. For a long lived radionuclide (λ small) and a small outflow (q_o small), the concentrations can buildup to very large values. This observation stresses the importance of maintaining a sufficiently large outflow from such limited bodies of water to avoid high concentrations of radionuclides. On the other hand, if the purpose of the reservoir is to hold radionuclides until most decay and thereby to limit the discharge of radioactivity to other surface waters, then the outflow should be minimized.

For the case of an instantaneous accidental release of an amount Q_i of radionuclide at $t = 0$ into a small pond, $Q(t) = Q_i\delta(t)$, Eq. (10.12) shows that the concentration subsequently decreases exponentially as

$$C(t) = \frac{Q_i}{V}e^{-(\lambda+q_o/V)t}.$$ (10.15)

This model is appropriate for many small shallow ponds in which the lifetime of the radionuclides is comparable to or longer than the mean impoundment time for the water (typically a few days or weeks). For very short lived radionuclides, the assumption of uniform instantaneous mixing is not valid, and a spatially dependent model must be used. Similarly, for ponds with vertical thermal stratification, which inhibits vertical mixing, or for wide irregularly shaped ponds in which uniform mixing is invalid, more complex hydrological models must be employed [see Johnson (1980) for a review of such models].

10.4 DISCHARGES TO RIVERS

The dispersal of radionuclides introduced into a river (see Fig. 10.2) has three distinct phases. First, the momentum and thermal energy of the discharge plume become rapidly mixed vertically near the point of discharge. This near-field mixing of the discharge plume is common to many water discharge problems and was discussed separately in Sec. 10.2. Once the vertical near-field mixing is complete, the effluent spreads laterally, the rate of spreading determined by the turbulence and currents of the river flow. Eventually, the radionuclides become uniformly distributed laterally, that is, transverse to the flow, and the radionuclide concentration is finally governed by longitudinal dispersion which smooths out concentration gradients caused by variations in the effluent discharge rate or river flow.

10.4.1 Transverse Mixing

Instantaneous Releases. Consider a very wide, shallow, straight river[1] with an average depth H flowing with a speed u in the x direction in which an amount Q'

---

[1]Although the river models presented here assume a straight river, one can use these models for a meandering and bending river by using a curvilinear coordinate system with x directed along the river centerline.

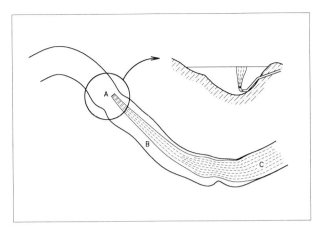

Figure 10.2   Discharge into a river has three stages of mixing: (A) initial momentum and buoyance of the discharge plume determine mixing near outlet; (B) turbulence and currents in the river cause lateral mixing; (C) after lateral mixing is complete, longitudinal diffusion erases concentration gradients caused by unsteady discharge or river flow rates. [After Fischer et al. (1979).]

of a radionuclide is discharged at some point in the river at time $t = 0$. The natural turbulence in the river causes the discharge to diffuse outward from its point of release and the river flow sweeps the expanding diffusing cloud downstream. The mixing in the vertical z direction is extremely rapid, and for the sake of this analysis, will be assumed to be instantaneous. Thus, an instantaneous, vertically uniform, line source of radionuclides with strength Q'/H (Bq m^{-1}) is effectively introduced at time 0 at the discharge point (here taken as the origin of the coordinate system).

The subsequent diffusion of radionuclides laterally in the y-direction and longitudinally in the x-direction is governed by the time-dependent diffusion equation of Sec. 9.3.2 in two dimensions. First consider the case in which the river flow is negligible, that is, $u \cong 0$, and no appreciable radioactive decay occurs. For this case the radionuclide concentration $C(x, y, t)$ determined from

$$\frac{\partial C(x,y,t)}{\partial t} = D_x \frac{\partial^2 C(x,y,t)}{\partial x^2} + D_y \frac{\partial^2 C(x,y,t)}{\partial y^2} \tag{10.16}$$

where D_y and D_x are the diffusion coefficients in the transverse (y) and longitudinal (x) directions, respectively. These coefficients are, in general, different since the distribution of eddies and currents in the x and y directions are dissimilar. The solution of this equation for $t > 0$ is

$$C(x, y, t) = \frac{Q'/H}{2\pi\sigma_x\sigma_y} \exp\left(-\frac{x^2}{2\sigma_x^2} - \frac{y^2}{2\sigma_y^2}\right), \tag{10.17}$$

where the diffusion parameters $\sigma_i = \sqrt{2D_i t}$, $i = x$ and y.

To account for the advective motion of the river water which sweeps the expanding region of radionuclides downstream, x in the previous result is replaced by the corresponding distance from the center of the moving radionuclide region, that is, x is replaced by $x - ut$. Thus in the flowing river

$$C(x, y, t) \equiv Q'\hat{C}(x, y, t) = \frac{Q'}{2\pi\sigma_x\sigma_y H} \exp\left(-\frac{(x - ut)^2}{2\sigma_x^2} - \frac{y^2}{2\sigma_y^2}\right), \quad (10.18)$$

where $\hat{C}(x, y, t)$ is the concentration at x, y at time t resulting from a unit release at time 0.

Continuous Discharges.   The radionuclide concentration in an infinitely wide river arising from a protracted release at the rate $Q(t)$ (Bq s^{-1}) is obtained by integrating the previous result for an instantaneous release over all earlier release times, that is,

$$C(x, y, t) = \int_{-\infty}^{t} dt' \, Q(t')\hat{C}(x, y, t - t'), \quad (10.19)$$

or equivalently

$$C(x, y, t) = \int_{0}^{\infty} d\tau \, Q(t - \tau)\hat{C}(x, y, \tau). \quad (10.20)$$

For a constant release rate Q, this integral can be approximated analytically. Substitution from Eq. (10.18) for $\hat{C}$, and approximation of the resulting integral in the same manner used to reduce Eq. (9.26) to Eq. (9.29) yields the steady-state concentration distribution

$$C(x, y) = \frac{Q}{\sqrt{4\pi D_y x u H}} \exp\left[\frac{-y^2 u}{4D_y x}\right]. \quad (10.21)$$

To correct for the effect of radioactive decay in the time interval x/u required for the radionuclides to reach a distance x downstream, simply multiply the previous result by the factor $\exp[-\lambda x/u]$, where λ is the decay constant of the radionuclide under consideration.

Effect of River Banks.   The earlier results are valid as long as the dispersing patch of radionuclides is far from the river banks. The river banks, however, eventually limit further lateral dispersion of the radionuclides. At the banks $\partial C/\partial y = 0$. In effect, the banks cause the incident radionuclides to be "reflected" back into the river. The solution of the diffusion equation, subject to river bank boundary conditions, can be obtained from the previous results using the method of virtual images (as was used in Sec. 9.4.4 to account for the effect of the ground on a diffusing gas cloud). Consider the case in which the source is at $x = 0$ and a distance y_o from the stream centerline (the solid circle in Fig. 10.3). When the diffusing plume from this source reaches a bank, a compensating plume from a virtual source (open circles) enters the river channel, exactly compensating for the plume leaving the channel. By using an infinite array of virtual sources as shown in Fig. 10.3, every time a plume leaves the river a

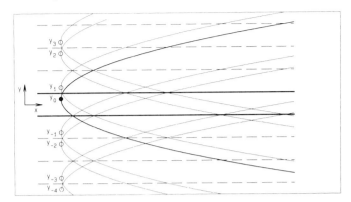

Figure 10.3 Geometry for constrained flow between two river banks. The discharge point (solid circle) is a distance y_o from the stream centerline. Virtual sources (open circles) are placed so that whenever a plume leaves the river channel, a compensating plume from a virtual sources enters the channel so that there is no net flow across the channel boundaries.

compensating plume enters, thereby maintain a zero net flow of radionuclides across the river bank boundary.

The y coordinates of the sources in Fig. 10.3 (measured from the river centerline) can be expressed as

$$y_n = nW + (-1)^n y_o, \qquad n = 0, \pm 1, \pm 2, \ldots \tag{10.22}$$

The concentration at some point (x, y) in the river, is then found by summing the concentrations produced by each source in this infinite array, that is,

$$C(x, y) = \frac{Q}{\sqrt{4\pi D_y x u H}} \sum_{n=-\infty}^{\infty} \exp\left[\frac{-(y - y_n)^2 u}{4 D_y x}\right]. \tag{10.23}$$

In practice, only a summation over the sources nearest to the river (e.g., $n = 0, \pm 1, \pm 2$) is necessary to evaluate this expression.

As the downstream distance x increases, Eq. (10.23) reduces to a constant value, independent of y, namely

$$C(x, y) \to C_o \equiv \frac{Q}{HWu}. \tag{10.24}$$

This is precisely the value one would expect if the constant discharge rate Q were fully mixed in the volumetric flow rate HWu of the river. An example of the lateral concentration profiles, as calculated by Eq. (10.23), at various downstream distances is shown in Fig. 10.4. The downstream distance required for the lateral profile to have less than 5% variation from the mean lateral concentration (i.e., the *lateral mixing length L*) is (Fischer et al. 1979)

$$L = \frac{\delta u W^2}{D_y}, \tag{10.25}$$

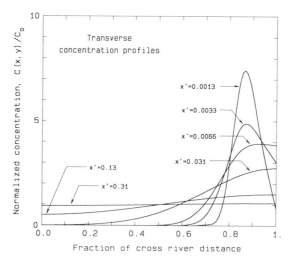

Figure 10.4   Transverse concentration profiles at various downstream distances from a discharge point positioned one-eight of the channel width from the right bank. Down stream distance is measured in dimensionless distance $x^1 = (xD_y)/(uW^2)$ where x is the actual distance. The concentration is normalized by dividing by the fully-mixed uniform concentration.

where δ is between 0.1 (for discharge at midstream) to 0.4 (for discharge at the river bank).

Finally, Eq. (10.23) can be corrected for the radioactive decay of the radionuclides as they disperse laterally by multiplying Eq. (10.23) by the probability a radionuclide does not decay while traveling a distance x downstream, namely, by $\exp(-\lambda x/u)$.

10.4.2 Longitudinal Dispersion

Once the radionuclides become fully mixed laterally, their further dispersal is controlled by the advective flow of the river. If the radionuclides undergo appreciable decay during the times of interest or for unsteady-discharges, it is important to include the effects of longitudinal dispersion and radioactive decay following the lateral mixing zone. Two special cases will be considered here.

For a steady release rate Q, the concentration $C(x)$ following the lateral mixing zone is given by the steady-state diffusion equation (now including an advective and a decay term), namely

$$u \frac{dC(x)}{dx} = D_x \frac{d^2C(x)}{dx^2} - \lambda C(x). \tag{10.26}$$

The solution of this equation, subject to the boundary condition that $C(x)$ vanish as $x \to \infty$, is

$$C(x) = A \, \exp\left(\frac{xu}{2D_x}(1 - \sqrt{1+\alpha}) \right), \tag{10.27}$$

where A is a constant of integration and $\alpha = 4\lambda D_x/u^2$. If the lateral mixing occurs relatively quickly, the value of this constant can be determined from the condition that the release rate must equal the rate at which all radionuclides downstream are decaying, that is,

$$Q = \int_0^\infty dx \, \lambda C(x) HW. \tag{10.28}$$

From this condition, it is found that

$$A = \frac{2Q}{HWu\alpha} \left[\sqrt{1+\alpha} - 1 \right]. \tag{10.29}$$

For many rivers, α is much less than unity and this result reduces to the simple "plug-flow" result in which longitudinal dispersion is negligible, namely,

$$C(x) = \frac{Q}{HWu} \exp\left[\frac{-\lambda x}{u} \right]. \tag{10.30}$$

Another important river discharge problem is to determine the far-field concentration arising from a single accidental point release Q' (Bq) at time $t = 0$. Following the lateral mixing zone, the far-field concentration $C(x,t)$ is obtained by integrating Eq. (10.18) (corrected for radioactive decay) over all lateral distances and by uniformly distributing the radionuclides across the river channel. The result is

$$C(x,t) = \frac{Q'}{2\pi\sigma_x\sigma_y H} \frac{1}{W} \int_{-\infty}^\infty dy \, \exp\left(-\frac{(x-ut)^2}{2\sigma_x^2} - \frac{y^2}{2\sigma_y^2} - \lambda t \right), \tag{10.31}$$

which, on integrating and substituting $\sigma_x = \sqrt{2D_x t}$, gives

$$C(x,t) = \frac{Q'}{WH\sqrt{4\pi D_x t}} \exp\left(-\frac{(x-ut)^2}{4D_x t} - \lambda t \right). \tag{10.32}$$

10.4.3 Estimation of Diffusion Coefficients

Before any of the previous models can be used, it is necessary to obtain values for the diffusion coefficients D_x and D_y. These coefficients depend on the turbulence and secondary currents that arise from the energy-dissipation characteristics of the channel as a result of shear stresses on the river bed. These shear stresses are often expressed in terms of the shear velocity $u^* = \sqrt{\tau_o/\rho}$ where τ_o is the shear stress on the channel bottom and ρ is the water density. In a uniform open channel, the shear velocity is found to be (Henderson 1966)

$$u^* = \sqrt{gHs} \tag{10.33}$$

where g is the gravitational acceleration constant (9.8 m s^{-2}) and s is the slope of the channel (i.e., the rate of decrease in channel elevation with downstream distance).

For lateral diffusion, it is found that D_y is proportional to this shear velocity and to the average river depth H, that is,

$$D_y = \beta u^* H. \tag{10.34}$$

The parameter β varies with the type of channel (Fischer et al. 1979). For straight uniform channels, β is in the range 0.1 to 0.2. Curves and irregularities along the banks increase the turbulence, so that in natural waterways β is seldom less than 0.4. If the river only slowly meanders and bank irregularities are moderate, β falls in the range of 0.4 to 0.8. Sharp curves and rapid meandering cause β to be even larger, and for such cases, field tests are usually required to determine appropriate values.

Generally, the longitudinal diffusion coefficient D_x is two or three orders of magnitude greater than the lateral coefficient. The longitudinal coefficient D_x is much more difficult to determine from field studies and far less data are available. Fischer et al. (1979) gives the following formula for estimating this coefficient:

$$D_x = \frac{0.011 u^2 W^2}{H u^*}.\tag{10.35}$$

10.5 DISCHARGES TO ESTUARIES

Rivers flowing into the sea form estuaries whose waters are a combination of fresh and salt waters. Estuaries vary considerably in structure, ranging from deep, narrow, stratified fjord estuaries to multiple-branched, shallow, bar-built estuaries. The calculation of concentrations of discharged radionuclides in estuaries is much more difficult than for rivers. The difficulty arises from the complex manner in which effluent is mixed in the estuary waters before eventually finding its way to the open sea.

In estuaries, there is turbulent diffusive mixing, as in rivers, but many other large-scale factors affect the dispersal of effluents. Daily tidal flows to and from the sea promote mixing in two ways. First, friction from the tidal flows of the heavier salt water along the bottom generates turbulence which leads to mixing. Then, the flow of the tides themselves induces oscillatory currents in the estuary thereby causing large-scale mixing.

The rivers entering an estuary provide the source of water which flushes the estuary and causes an overall net flow into the sea. In some nearly tideless estuaries, the fresh river water flows over the heavier sea water directly into the ocean and the dispersal is dominated by density driven currents. In other estuaries with strong tidal actions, the waters are well mixed and density effects can be neglected.

Winds, which are important for mixing in oceans and large lakes, may or may not be important for the mixing of estuary waters. For wide estuaries or estuaries with many long bays, winds may cause significant surface currents which are important mixing mechanisms for surface-attached discharges such as oil spills or buoyant plumes. In sheltered narrow estuaries, by contrast, winds may be only a minor factor in mixing.

Finally, seasonal and unusual events can greatly change the mixing characteristics of a estuary. An estuary may be well stratified in wet seasons but vertically mixed in dry seasons. A flood can lead to strong stratification in a normally well mixed estuary. By contrast, a passing hurricane or sea storm can cause a normally stratified estuary to become suddenly well mixed. These catastrophic changes in estuary structure can take weeks to relax to normal conditions. Thus, an analyst must be aware that estuaries are rarely in steady state and that an analysis for one season may not be applicable for another or for unusual weather conditions.

10.5.1 Estuary Modeling

In some estuaries, the geometry and currents are so irregular that no simple analytical approach can be used. For these cases, only field studies or elaborate multidimensional computer models can be used to predict the mixing of effluents. See, for example, the computer models presented by Falconer et al. (1989). However, in many assessments of radionuclide concentrations in estuaries, there is little need for the fine spatial and temporal resolution afforded by these computer models. For example, in a continuous release of long lived radionuclides, the long-term estuary concentrations, averaged over many tidal cycles, can be estimated from the mean residence time of the estuary water. This residence time is determined by the freshwater flow through the estuary and the mean estuary volume, in the same manner as was used for the pond model described in Sec. 10.3. For such long-term concentration assessments, spatial variations are often relatively minor, and the use of more elaborate models is not justified.

An intermediate approach to estuary modeling can be called generically one-dimensional analytical modeling. In this approach, all the mixing mechanisms are lumped together to give a model that describes the mixing in only one direction (e.g., transverse or longitudinal). The main difficulty with using such simplified models is the need to obtain appropriate values for the model parameters. Nevertheless, with appropriate field tracer studies, parameters for these models can often be estimated so that useful quantitative results can be obtained.

Because of the variation of water and currents in an estuary throughout the day, primarily as a result of the tides, it is difficult to estimate the spatial distribution of radionuclides at any instant of time. However, at some position x in a vertically well-mixed estuary (where x is measured along the estuary "centerline" towards the sea), one could average the flow laterally over the cross sectional area A of the estuary. The variation between tidal cycles of this laterally averaged concentration $C(x,t)$ is then given by

$$A\frac{\partial C}{\partial t} + Q_f\frac{\partial C}{\partial x} = \frac{\partial}{\partial x}\left(DA\frac{\partial C}{\partial x}\right) - \lambda C + S, \qquad (10.36)$$

in which the time derivative represents the change per tidal cycle. Here λ is the radionuclide decay constant, Q_f is the volumetric flow rate of fresh water (averaged over a tidal cycle), D is a dispersion coefficient representing all the mixing processes that occur during a tidal cycle, and S is the discharge rate of radionuclides averaged over a tidal cycle.

For the case of a steady discharge Q (Bq s^{-1}) at some point in the estuary, here taken as $x = 0$, Eq. (10.36) can be solved for the steady-state radionuclide concentration in an estuary with a constant cross section, dispersal coefficient, and freshwater flow rate. In this case, Eq. (10.36) can be written as

$$u\frac{dC(x)}{dx} = D\frac{d^2C(x)}{dx^2} - \lambda C(x) + \frac{Q}{A}\delta(x), \tag{10.37}$$

where $u = Q_f/A$ is the effective speed of the freshwater through the estuary. This equation is the same equation used to describe the longitudinal dispersion in a river [cf., Eq. (10.26)]. The general solution of this equation is

$$C(x) = C_1\,e^{\omega_1 x} + C_2\,e^{\omega_2 x}, \qquad x \neq 0 \tag{10.38}$$

where C_1 and C_2 are constants of integration and the relaxation constants ω_1 and ω_2 are

$$\omega_1 \equiv [1 + \sqrt{1+\alpha}]\frac{u}{2D} > 0, \tag{10.39}$$

and

$$\omega_2 \equiv [1 - \sqrt{1+\alpha}]\frac{u}{2D} < 0, \tag{10.40}$$

in which $\alpha \equiv 4D\lambda/u^2$. Unlike longitudinal dispersion in a river, for which α is often much less than unity, α for an estuary is often quite large because of the typically large values of D and small values of u encountered. Typical values of D range from 30 to 300 m^2 s^{-1} and effective freshwater speeds can range down to zero for an estuary without a river.

To determine values for the constants C_1 and C_2 in Eq. (10.38), boundary conditions must be specified. Here it is assumed that $C(x)$ vanishes as $x \to -\infty$ (i.e., far inland) and also at $x = L$ (the distance to the estuary mouth from the discharge location). For these boundary conditions, the solution for $x < 0$ is

$$C(x) = \frac{Q}{Q_f\sqrt{1+\alpha}}\left[1 - \exp(-\sqrt{1+\alpha}\,uL/D)\right]\exp(\omega_1 x), \tag{10.41}$$

and for $x > 0$

$$C(x) = \frac{Q}{Q_f\sqrt{1+\alpha}}\{\exp[\omega_2(x-L)] - \exp[\omega_1(x-L)]\}\exp(\omega_2 L). \tag{10.42}$$

In Fig. 10.5, two example concentration profiles are shown for steady discharges made at two different distances from the estuary mouth. Notice that significant radionuclide concentrations occur above the discharge location. It must be emphasized that such profiles (or any result obtained from this model) do not represent the actual concentration that would be observed at any instant of time. During the slack tide, radionuclides build up near the discharge location. This peak is then dispersed during the following flood or ebb tides. The example profiles are obtained only after averaging actual concentrations over a tidal cycle.

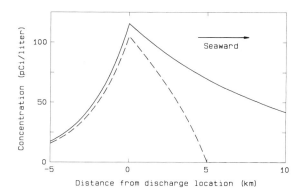

Figure 10.5   Concentration profiles, averged over a tidal cycle, along the length of an estuary as predicted by Eqs. (10.41) and (10.42). Here a radionuclide with a decay constant of 0.2 d^{-1} is discharged at a rate of 2 μ Ci s^{-1} at a location 5 km (dashed line) and 30 km (solid line) from the mouth of the estuary. The estuary has a cross sectional area of $A = 300$ m^2, a tidal averged dispersion coefficient $K = 60$ m$^2 s^{-1}$, and a freshwater flow $Q_f = 10$ m$^3 s^{-1}$.

10.6 DISCHARGES TO LARGE LAKES AND OCEANS

The potential for dilution of effluents in large lakes and oceans, unlike rivers and reservoirs, is almost without limit. For such discharges, the near-field dilution and subsequent dispersal of the near-field concentrations are of interest. Submerged multiport diffusers often are used to ensure large near-field dilutions for routine discharges from coastal plants. The sudden drop in temperature (and salinity increase for oceans) with depth in the thermocline (see Fig. 10.6) inhibits radionuclides in the warmer surface waters from mixing with the denser deep water. This stratification in oceans and great lakes causes buoyant effluents, after near-field mixing, to be effectively trapped in the surface layer where their dispersal is governed by surface currents and turbulent diffusion. On a much longer time scale, some radionuclides eventually migrate from the upper mixed layer through the deep water into the sediment on the bottom. In this section, only the near-term dispersal of effluents confined to the upper layer is discussed.

Unlike turbulent diffusion in rivers, diffusion in very large bodies of water is governed by eddies of ever increasing size as the effluent spreads out. Consequently, the diffusion coefficient is not a constant, but, as in atmospheric diffusion processes, increases with the time the effluent has been diffusing. As a rule of thumb, the eddy diffusivity (diffusion coefficient) increases as the four-thirds power of the eddy size, the so-called *four-thirds law*.

The dispersal of surface effluents is also affected by generally complex currents, especially near coasts. The advective movement, that is, motion caused by the bulk movement of the water in which the radionuclides are diffusing, often deviates significantly from the straight-line trajectories used for river and atmospheric models. For

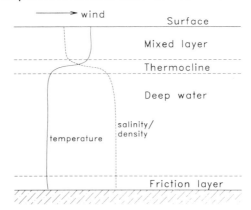

Figure 10.6 Typical vertical distribution of salinity and temperature in deep stratified lakes or oceans. The large change in water density across the thermocline inhibits mixing of the warm upper waters with the colder and denser deep waters.

cases in which the advective motion is in a known direction, diffusion-advective models can be applied (see the following discussion). For complex flow fields, an alternative Lagrangian approach is often used whereby the flow field is first determined, either computationally or experimentally, and then a diffusion calculation is superimposed on the advective motion. This approach is discussed in Sec. 10.6.2.

10.6.1 Ocean Advective-Diffusion Models

Consider a submerged buoyant discharge of radionuclides at a constant rate Q (Bq s^{-1}) into an infinitely wide ocean with a steady surface current of speed u (see Fig. 10.7). The plume rises, spreading as it does so, eventually reaching the surface where it is confined in a surface layer of thickness H, the actual water depth for coastal waters or the depth to the thermocline in deep waters. When the plume reaches the surface (at $x = 0$) it has an effective width $2W$, which increases further as the surface-constrained plume moves with the current. The concentration in the surface layer is governed by the steady-state advective-diffusion equation

$$u\frac{\partial C(x,y)}{\partial x} = \frac{\partial}{\partial x}\left[D_x\frac{\partial C(x,y)}{\partial x}\right] + \frac{\partial}{\partial y}\left[D_y\frac{\partial C(x,y)}{\partial y}\right] - \lambda C(x,y). \quad (10.43)$$

In the following analysis, it is assumed that the diffusion in the x-direction is small compared to the advective term, that is, the first term on the right in Eq. (10.43) can be neglected.

　　As a first approximation, assume the diffusion coefficient D_y is constant. The plume when it reaches the surface can be treated as a plane source at $x = 0$ with width $2W$ and depth H. The strength of this vertical plane source is $Q/(2WH)$ (Bq m^{-2} s^{-1}). The solution of Eq. (10.43) for this source can be found by treating the plane source as a superposition of a series of vertical line sources [width dy', depth H, strength $Q\,dy'/(2WH)$], that is,

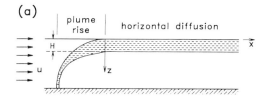

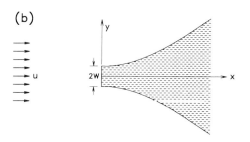

Figure 10.7   *Geometry for a submerged buoyant discharge into stratified water. The top figure shows the rising plume which eventually reaches the surface and is trapped in the top layer. In the plan view (bottom), the surface confined plume continues to diffuse laterally.* [After Jirka et al. (1983).]

$$C(x, y) = \int_{-W}^{W} dy'\, C_{line}(x, y; y') \tag{10.44}$$

where $C_{line}(x, y; y')dy'$ is the differential concentration at x, y arising from the effective line source of width dy' about y'. Equation (10.21) gives the solution for a line source of strength Q/H at $y = 0$. From this result (with Q/H replaced by $Qdy'/(2WH)$, with the line source placed at y', and with a radioactive decay correction), one obtains

$$C_{line}(x, y; y')dy' = \frac{Qdy'}{2WH\sqrt{4\pi D_y x u}} \exp\left[-\frac{(y - y')^2}{4D_y x} - \lambda x/u \right]. \tag{10.45}$$

Substitution of this result into Eq. (10.44) and integration gives the desired solution for the plane source, namely

$$C(x, y) = \frac{Q}{4WHu} e^{-\lambda x/u} \left\{ \mathrm{erf}\left(\frac{y + W}{2}\sqrt{\beta} \right) - \mathrm{erf}\left(\frac{y - W}{2}\sqrt{\beta} \right) \right\}, \tag{10.46}$$

where $\beta \equiv u/(D_y x)$ and $\mathrm{erf}(x)$ is the *error function* (see Appendix F, Sec. F-3.1) defined as

$$\mathrm{erf}(x) \equiv \frac{2}{\sqrt{\pi}} \int_0^x e^{-\tau^2} d\tau. \tag{10.47}$$

Eq. (10.43) can also be solved for the case in which the diffusion coefficient varies as the four-thirds power of the plume width (Brooks 1960). The result is given by Eq. (10.46) with (Jirka et al. 1983)

$$\beta = \frac{3}{2}\left[\left(\frac{1 + 2D_{yo}x}{uW^2}\right)^3 - 1\right]^{-1}.$$ (10.48)

Here D_{yo} is the initial value of the diffusion coefficient at $x = 0$.

For an instantaneous release Q' (Bq) spread uniformly to depth H at $x = 0$ and $y = 0$, the resulting time-dependent concentration for the case of constant turbulent diffusion is given by Eq. (10.18) (corrected for radioactive decay). The result may be written as

$$C(x,y,t) = \frac{Q'/H}{4\pi t\sqrt{D_x D_y}}\exp\left[-\frac{(x - ut)^2}{4D_x t} - \frac{y^2}{4D_y t} - \lambda t\right].$$ (10.49)

10.6.2 Decoupled Advective-Diffusion Methods

An alternative to using the combined diffusion–advection model is to separate the dispersal problem into two parts. First, the velocity field is determined, either from experimental data or from sophisticated hydrodynamic calculations which can accurately treat coastal geometries, upwellings, tidal effects, and so on. Once the velocity field is determined, the dispersal of a radionuclide discharge can be analyzed by considering its diffusion with respect to the moving release point whose motion is known from the velocity field. Thus, in this Lagrangian approach, a pure diffusion calculation is superimposed on a predetermined flow (current) field.

This separation of the velocity calculation (or advective component) from the diffusion component is valid as long as the releases are small and have negligible buoyancy. In such a case, the release has little influence on the velocity field. The advantage of this approach is that very complicated geometries and current patterns can be treated and that non-constant diffusion parameters can be accommodated.

Because of the almost unlimited size of turbulent eddies which can be formed in great lakes and oceans, the diffusion coefficient generally increases as the size of the dispersing region increases. Hence, the standard deviation of a diffusing region can be expected to increase with time, that is,

$$\sigma_r^2 = \beta t^\alpha,$$ (10.50)

where t is the time the patch has been diffusing. If the size ℓ of the circular diffusing region is defined arbitrarily as $3\sigma_r$ (thereby containing 95% of the Gaussian diffusing radionuclides), the corresponding diffusion coefficient D_r, which is related to σ_r by $D_r = \sigma_r^2/(4t)$, is

$$D_r = \frac{\beta}{4}t^{\alpha-1} = \frac{\beta}{4}(3\sqrt{\beta})^{-\gamma}\ell^\gamma,$$ (10.51)

where $\gamma = 2(1 - 1/\alpha)$. Values of α and β have been found by fitting Eq. (10.50) to experimental data obtained in several ocean studies. For t measured in seconds and σ_r in cm, Okubo (1971) found $\beta = 0.0108$ and $\alpha = 2.34$ (so that $\gamma = 1.15$) for t between 10^3 and 10^6 seconds. Over a longer time range of 10^3 to 10^8 seconds,

Pritchard et al. (1971) report values of $\beta = 0.006$ and $\alpha = 2.54$ (giving $\gamma = 1.21$). For both correlations, the implied γ values are slightly lower than the commonly assumed four-thirds power law (i.e., D varies as $\ell^{4/3}$).

With this result for σ_r, the concentration $C(r)$ at a distance r from the moving release point (i.e., in a coordinate system whose origin is the center of the diffusing region) is readily found. If an instantaneous release of Q' (Bq) at $t = 0$ is spread over a depth H (the actual water depth for coastal waters or the depth to the thermocline in deep waters) then (Jirka 1983)

$$C(r,t) = \frac{Q'e^{-\lambda t}}{H\pi\sigma_r^2(t)} \exp\left[\frac{-r^2}{\sigma_r^2(t)}\right]. \tag{10.52}$$

Finally, by expressing r in terms of the fixed coordinates x and y and by superimposing the diffusion solution of Eq. (10.52) onto the known velocity field, one can obtain the concentration $C(x, y, t)$ in the fixed coordinate system. This Lagrangian approach has been generalized for continuous releases to variable velocity fields by Adams et al. (1975).

10.7 SEDIMENTATION EFFECTS IN SURFACE WATERS

Both suspended sediment and bedded sediment can greatly affect the radionuclide concentrations in surface waters. Adsorption of radionuclides on sediments decreases the dissolved concentration and makes the radionuclides less available to various members of aquatic food chains including humans. For example, 90% of the ^{137}Cs discharged into a river at Oak Ridge National Laboratory was found to be adsorbed by suspended sediments within 10 miles of the release point (Churchill et al. 1965). Conversely, when bedded sediment is resuspended during high-turbulence conditions, the sediment can act as a source of radionuclides long after the dissolved radionuclides have flowed away from the region of concern. Clearly, consideration must be given to sediment-radionuclide interactions if accurate predictions of spatial and temporal distributions of radionuclides in aquatic environments are to be obtained.

Suspended sediments, because of their large surface area per unit mass, are usually responsible for most of the adsorption of radionuclides. The physical and chemical mechanisms involved in radionuclide adsorption/desorption interactions with sediment are many and complex. The extent to which radionuclides are adsorbed by suspended sediments is characterized by a dimensionless *equilibrium distribution coefficient* K_d defined as the ratio of the amount of bound radionuclide per unit volume of dry sediment to the amount of radionuclide dissolved in a unit volume of water. This distribution coefficient, also known as the *solid/liquid partition coefficient*, is sometimes defined on a weight basis, namely the ratio of the amount of bound radionuclide per unit mass of dry sediment to the dissolved amount in a unit volume of water. This later definition, denoted here by $\hat{K}_d$, has units such as (mL g^{-1}) and is related to K_d by

TABLE 10.1 Values for the equilibrium distribution coefficient $\hat{K}_d$ {L kg^{-1}} for the adsorption of selected elements on sediment with emphasis on oxidizing conditions.

ELEMENT	FRESH WATER		SALINE WATER	
	RANGE	MEDIAN	RANGE	MEDIAN
Am	85–40,000	5×10^3	97–650,000	10^4
Ce	7,800–140,000	10^4	9,700–10^7	5×10^4
Cr[a]	0–10^3	low	0–10^3	low
Cs	50–8×10^4	10^3	17–10^4	3×10^2
Co	1,000–71,000	5×10^3	7,000–300,000	10^4
Cm	100–70,000	5×10^3		
Eu	200–800	5×10^2	5,000–130,000	10^4
Fe[a]	10^3–10^4	5×10^3	20,000–450,000	5×10^4
I	0–75	10	0–100	10
Mn[a]	10^2–10^4	10^3	10^2–10^4	10^3
Np[a]	0.2–127			
P[a]		high		high
Pu	10^2–10^7	10^5	10^2–10^5	5×10^4
Pm	10^3–10^4	5×10^3	10^3–10^5	10^4
Ra	10^2–10^3	5×10^2	10^1–10^3	10^2
Sr	8–4,000	1,000	6–400	50
Tc[a]	0–10^2	5	0	0
Th	10^3–10^6	10^4	10^4–10^5	5×10^4
^{3}H	0	0	0	0
U[a]	16			
Zn	10^2–10^3	5×10^2	10^3–10^4	5×10^3
Zr	10^3–10^4	10^3	10^3–10^5	10^4

[a] Highly dependent on oxidation-reduction conditions.

Source: Onishi et al. 1981 as reported by Jirka et al. 1983.

$$\hat{K}_d = \frac{K_d}{\rho_s} \quad (10.53)$$

where ρ_s is the mass density of dry sediment (typically between 2.0 and 2.5 g cm^{-3}).

Values of the distribution coefficient depend on many factors including radionuclide type, chemical state, and concentration, as well as on the sediment concentration and type, the flow and water quality, and the contact time. Consequently, there is not a single value of K_d for a particular radionuclide that can be used in all situations. Onishi (1981) has reviewed the adsorption/desorption process on sediment and summarized observed $\hat{K}_d$ values for many elements (see Table 10.1).

The amount of radioactivity adsorbed on solid sediment particles suspended in water can be described by the activity C_{susp} adsorbed to suspended sediment per unit volume of the sediment or by $\tilde{C}_{susp}$ adsorbed to the sediment in a unit volume of liquid (water plus dissolved matter). Although both concentrations have the same units (e.g., Bq cm^{-3}), it is important to distinguish between them. If a suspension has a volume V_{susp} of suspended sediment solids in a unit volume of liquid, these two concentrations are related by

$$\tilde{C}_{susp} = V_{susp}C_{susp}. \tag{10.54}$$

In a unit volume of liquid (i.e., in a volume $1 + V_{susp}$ of suspension), let the activity concentration of radionuclides dissolved in the water be denoted by C_w. If the adsorption and desorption to and from the sediment are in equilibrium, then, from the definition of the distribution coefficient K_d, the dissolved and sediment-bound concentrations are related by

$$C_w = \frac{C_{susp}}{K_d} = \frac{1}{K_d}\frac{\tilde{C}_{susp}}{V_{susp}}. \tag{10.55}$$

Finally, under equilibrium conditions, the total activity concentration in a unit volume of liquid, C_{tot}, is given by

$$C_{tot} = C_w + \tilde{C}_{susp} = C_w(1 + V_{susp}K_d). \tag{10.56}$$

10.7.1 Sediment Models

The sorption of radionuclides on sediment, the settling and resuspension of sediments, and the varying rates at which sediment is transported make a detailed analysis of radionuclide concentrations in surface waters quite complex. The spatial and temporal movement of both sediment and water must be determined as well as the movement of radionuclides between the water and the sediment.

Many computer models have been developed for performing such analyses. Two dimensional models include SERATRA (Onishi et al. 1976, 1982) for vertical and longitudinal analyses, and FETRA (Onishi et al. 1976; Onishi 1981) for longitudinal and transverse analyses. The three dimensional FLESCOT model (Onishi and Trent 1982) has been applied to estuaries and coastal regions. For river modeling there are the CARICHAR (Belleudy et al. 1987) and SEDICOUP (Holly 1988) models.

While such computer models can provide detailed estimates of radionuclide distributions in water and their sediments, they require considerable input data and computational effort. For preliminary estimates, several simple one-dimensional or lumped parameter models have been developed which can yield analytical results (Jirka et al. 1983; USNRC 1978). One such model is presented in the following subsection for lake analyses.

Lake Model with Sediment Interactions.   To determine the rate at which radionuclides are removed from the water in a lake, both by sediment attachment and by flushing, Codell (USNRC 1978) has developed a simple two-compartment lake model. The lake geometry and the radionuclide exchange mechanisms between the two model compartments are shown in Fig. 10.8.

The lake volume V (m^3) and the rate of outflow q_o (m^3 y^{-1}) are assumed constant. The radionuclides are introduced at a rate $Q(t)$ (Bq y^{-1}) and are assumed to be rapidly mixed so that there is no spatial variation of radionuclide concentrations in the water. The lake-bed area is A (m^2), so that the effective lake depth is $d_1 = V/A$.

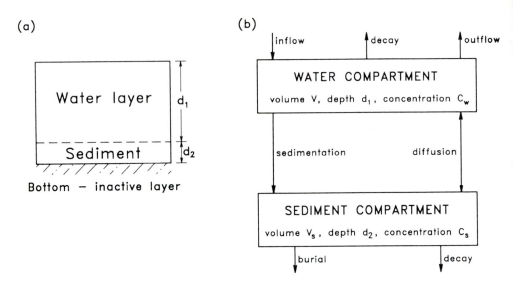

Figure 10.8 The two compartment model for analyzing the radionuclide concentrations in a lake with sediment. (a) Sediment settles through the water scavenging radionuclides and forming a sediment layer at the bottom. Only the top thickenss d_2 of the sediment layer is assumed to exchange radionuclides directly with the water. (b) The two compartments (water and sediment) have several mechanisms by which radionuclides are moved into and out of the compartments. [After USNRC (1978).]

The rate of sediment accumulation on the lake bottom, v_s, is also assumed constant (typically 10^{-3} to 10^{-2} m y^{-1}). The thickness of the "active" sediment layer d_2 (m) remains constant (typically 0.1 m) so that, as sediment falls, the bottom portion of the sediment bed becomes inactive, thereby maintaining a constant active thickness d_2 that exchanges radionuclides with the water column above, which has a thickness d_1 m. The volume of bedded sediment is $V_s = Ad_2$. The volume of suspended solids is negligible in comparison to the lake volume. Finally, the radionuclides sorbed onto bedded sediment are always assumed to be in equilibrium with the dissolved radionuclides in the interstitial water within the sediment bed. Similarly, the radionuclides sorbed onto suspended sediment are assumed to be always in equilibrium with the radionuclides dissolved in the supernatant water within the lake.

The equations describing the time-dependence of the activity concentrations of dissolved radionuclides $C_w(t)$ (per unit volume of liquid) and of radionuclides adsorbed on bedded sediment $C_s(t)$ (per unit volume of sediment) are obtained by summing the gain and loss rates (see Fig. 10.8) for each compartment. The rates of inflow and outflow of dissolved radionuclides, in units of Bq y^{-1}, are respectively $Q(t)$ and $q_o C_w(t)$. The rates of loss by radioactive decay of dissolved radionuclides and adsorbed radionuclides on bedded sediment, in units of Bq y^{-1}, are respectively $\lambda V C_w(t)$ and $\lambda V_s C_s(t)$. The rate (Bq y^{-1}) at which radioactivity leaves the bedded sediment/compartment via inactivation (burial) is $A v_s C_s(t)$.

There are two mechanisms by which radionuclides are exchanged between the lake water and the sediment bed, namely, (1) the adsorption of radionuclides dissolved in the lake water by the settling suspended sediment that, in turn, becomes incorporated into the sediment bed (*sedimentation transfer*), and (2), the diffusion (or *direct exchange*) between the radionuclides in the bedded sediment and those dissolved in the lake volume. The rate of radionuclide exchange, in units of activity per unit time, can be expressed as follows:

sedimentation: The concentration (Bq m^{-3}) in suspended sediment is $C_w K_d$. The rate (m^3 y^{-1}) of settling is $v_s A$. Therefore, the rate of transfer by settling sediment to the sediment bed (Bq y^{-1}) is $v_s A C_w K_d$.

diffusion: The concentration of radionuclides in the interstitial water in the sediment bed is assumed to be in equilibrium with that of adsorbed radionuclides in the sediment, that is, C_s/K_d (Bq m^{-3}). The rate of radionuclide transfer from this interstitial water to the lake volume is assumed to be proportional to the interfacial area A and to the difference between concentrations in the two zones, with a constant of proportionality K_f (m y^{-1}) called the *coefficient for direct transfer.* The rate of exchange (Bq y^{-1}) is thus $K_f A(C_s/K_d - C_w)$.

The rates of activity transfer, when divided by the lake volume V or the sediment volume V_s, represent, respectively, the rates of change of concentrations C_w and C_s. Thus, the balance equations are

$$\frac{dC_w(t)}{dt} = \frac{Q(t)}{V} - \lambda C_w(t)4 - \frac{q_o}{V}C_w(t) - \frac{v_s}{d_1}K_d C_w(t)$$
$$+ \frac{K_f}{d_1}\left(\frac{1}{K_d}C_s(t) - C_w(t)\right) \tag{10.57}$$

and

$$\frac{dC_s(t)}{dt} = -\lambda C_s(t) - \frac{v_s}{d_2}C_s(t) + \frac{v_s}{d_2}K_d C_w(t) - \frac{K_f}{d_2}\left(\frac{1}{K_d}C_s(t) - C_w(t)\right). \tag{10.58}$$

These balance equations can be rearranged into a more convenient form for solution, namely,

$$\frac{dC_w(t)}{dt} = \lambda_1 C_s(t) - \lambda_2 C_w(t) + \frac{Q(t)}{V} \tag{10.59}$$

and

$$\frac{dC_s(t)}{dt} = \lambda_3 C_w(t) - \lambda_4 C_s(t), \tag{10.60}$$

where

$$\lambda_1 \equiv \frac{K_f}{d_1 K_d}, \tag{10.61}$$

$$\lambda_2 \equiv \frac{q_o}{V} + \lambda + \frac{v_s K_d}{d_1} + \frac{K_f}{d_1}, \tag{10.62}$$

$$\lambda_3 \equiv \frac{v_s K_d}{d_2} + \frac{K_f}{d_2}, \tag{10.63}$$

and

$$\lambda_4 \equiv \lambda + \frac{v_s}{d_2} + \frac{K_f}{d_2 K_d}. \tag{10.64}$$

For the case of an instantaneous input of Q' (Bq) into the lake water at time $t = 0$, the solution of this equation can be written as

$$C_w(t) = \frac{Q'}{V(\omega_1 - \omega_2)} \left[(\lambda_2 + \omega_1)e^{\omega_2 t} - (\lambda_2 - \omega_2)e^{\omega_1 t}\right] \tag{10.65}$$

and

$$C_s(t) = \frac{\lambda_3 Q'}{V(\omega_1 - \omega_2)} \left\{e^{\omega_1 t} - e^{\omega_2 t}\right\}, \tag{10.66}$$

where ω_1 and ω_2 are the constants

$$\omega_{1,2} = \frac{1}{2}\left\{-(\lambda_4 + \lambda_2) \pm \sqrt{(\lambda_4 + \lambda_2)^2 - 4(\lambda_2\lambda_4 - \lambda_3\lambda_1)}\right\}. \tag{10.67}$$

Results are illustrated in Fig. 10.9 for a test problem representative of the behavior of long-lived ^{90}Sr in a lake similar to one of the Great Lakes in the United States.

For a continuous release $Q(t)$ which begins at $t = 0$, the water concentration is found by convolution of $Q(t)$ with the impulse response, namely,

$$C_w(t) = \int_0^t d\tau \, C_w^{inst}(t - \tau)Q(\tau) \tag{10.68}$$

where $C_w^{inst}(t)$ is the response for an instantaneous unit input as given by Eq. (10.65) with $Q' = 1$.

10.8 RADIONUCLIDE MIGRATION IN GROUNDWATER

Another important pathway for movement of radioactivity through the biosphere is provided by the flow of underground water. Radionuclides can seep into groundwater from a variety of surface sources, for example, defective radioactive waste repositories, mine-tailing ponds, or fallout. Because groundwater tends to move very slowly, many years may pass before any contamination is found in wells near the input site.

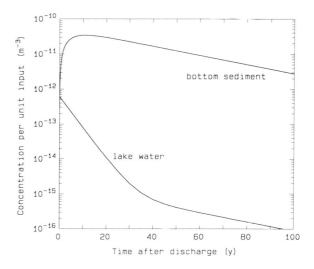

Figure 10.9 Concentrations of ^{90} Sr in lake water and bedded sediment as functions of time after the instantaneous introduction of unit activity in a lake representative of one of the Great Lakes. Parameters for the calculations (USNRC 1978) in units of meters and years are as follows: $\lambda = 0.0239$, $q_o = 2.09 \times 10^{11}$, $V = 1.64 \times 10^{12}$, $K_d = 2400$, $K_f = 0.4$, $d_1 = 30$, $d_2 = 0.1$, and $v_s = 5 \times 10^{-4}$.

Similarly, it may take many years for contaminated groundwater to be restored to its uncontaminated state after the source of contamination is removed.

Unlike atmospheric and surface water dispersion, the movement of radionuclides in underground waters is highly dependent on the chemical form of the radionuclides and the nature of the geological matrix through which groundwater is flowing. As radionuclides are carried along in groundwaters, they adsorb to the surface of soil and rock particles thereby slowing down their migration speed. For many radionuclides the effective migration speed can be many orders of magnitude slower than that of the groundwater itself.

The rate of adsorption (and subsequent desorption) depends strongly on the nature of the soil (chemically as well as structurally) and on the chemical state of the particular radionuclide being considered. Positive ions are generally more strongly sorbed than negative ions because most soil particles have a preponderance of negative surface charges. Also, smaller multivalent ions are usually more strongly sorbed than are large univalent ions. For porous media such as soil, the fraction of the dissolved radionuclides sorbed is inversely proportional to the soil particle size. It is for this reason that clays, which have particle sizes less than 2 μm in diameter, are extremely effective sorbents.[2]

Because of the strong tendency with which many radionuclides are sorbed in soils and geological formations, the speed at which they move (even through satu-

---

[2]The small particle size produces a very large surface area on which ions can be sorbed. For example, in 1 cm^3 of loam (50% sand, 25% silt, and 25% clay) it is estimated that there is a surface area of about 18,000 cm^2 (Eisenbud, 1987).

rated matrices) can be very slow, often only small fractions of a meter in a year. In waste repositories at which radioactive liquids have leaked into the soil, it has been found that many of the radionuclides have migrated only a few meters over several years. Nevertheless, when radionuclides with long lifetimes reach the water table of an aquifer beneath such a facility, they can then move several kilometers over century time scales and enter the biosphere through water extracted from wells or by entering surface waters if the aquifer intersects a surface waterway.

The description of how radionuclides discharged on the earth's surface are dispersed by groundwater is approached in several stages. First, and usually the most difficult, is a description of the vertical movement of the radionuclides down through the soil into an underlying water table or aquifer. The difficulty arises primarily because of the variable amount of percolating water (e.g., precipitation) which seeps into the soil and carries the radionuclides downwards. Such unsaturated dispersion calculations are extremely difficult, and almost always a simplified and hopefully conservative analysis is made as in Sec. 10.8.2. Once radionuclides reach an aquifer, in which the pores of the geological matrix are usually filled with water, saturated flow calculations are performed to determine the spatial and temporal spread of the radionuclides as they are carried along by the aquifer flow as in Sec. 10.8.3. Finally, many aquifers reach the surface as springs or intersect lakes and rivers, and the calculation of the rate at which radionuclides leave the aquifer and enter such surface waters is the third stage of a groundwater analysis (see Sec. 10.8.4).

In the following sections, some simple groundwater models are presented. These models, because of their simplicity, must be used cautiously. In many real groundwater problems, the geometry may be far more complicated than can be accommodated by these simple models. However, the greatest weakness in such models is the large uncertainty in many of the model parameters such as water flow rates and adsorption coefficients. To use these models effectively, field data usually must be obtained to refine estimates of model parameters.

10.8.1 Description of Groundwater Dispersion

In this section equations are developed to describe the migration of radionuclides through porous media. For the most part, saturated conditions are assumed, that is, all voids in the geological matrix are filled with water, and the medium is assumed isotropic and homogeneous. The theory developed below can be extended to more general conditions (ANS 1980; Li 1972; and Lappala 1981). However, such generalizations preclude the derivation of simple analytical results, and they are not considered here.

Equation for Radionuclide Concentrations in Groundwater. Consider a geological formation composed of porous and granular material with water filling all pores and space between the grains, that is, a *saturated* medium. The amount of space available for water is called the *total porosity* n defined as the total volume available for liquid in a unit volume of the formation. Not all the water is free to move; some will be trapped in closed-in voids and dead-end cracks. The *effective porosity* $n_e(\leq n)$ is

TABLE 10.2 Typical Values of Porosity n and Effective Porosity n_e for Various Aquifer Materials.

AQUIFER MATERIAL	NUMBERS OF ANALYSES	RANGE	ARITHMETIC MEAN
Porosities			
Igneous Rocks			
weathered granite	8	0.34–0.57	0.45
weathered gabbro	4	0.42–0.45	0.43
basalt	94	0.03–0.35	0.17
Sedimentary Materials			
sandstone	65	0.14–0.49	0.34
siltstone	7	0.21–0.41	0.35
sand (fine)	245	0.25–0.53	0.43
sand (coarse)	26	0.31–0.46	0.39
gravel (fine)	38	0.25–0.38	0.34
gravel (coarse)	15	0.24–0.36	0.28
silt	281	0.34–0.51	0.45
clay	74	0.34–0.57	0.42
limestone	74	0.07–0.56	0.30
Metamorphic Rocks			
schist	18	0.04–0.49	0.38
Effective Porosities			
Sedimentary Materials			
sandstone (fine)	47	0.02–0.40	0.21
sandstone (medium)	10	0.12–0.41	0.27
siltstone	13	0.01–0.33	0.12
sand (fine)	287	0.01–0.46	0.33
sand (medium)	297	0.16–0.46	0.32
sand (coarse)	143	0.18–0.43	0.30
gravel (fine)	33	0.13–0.40	0.28
gravel (medium)	13	0.17–0.44	0.24
gravel (coarse)	9	0.13–0.25	0.21
silt	299	0.01–0.39	0.20
clay	27	0.01–0.18	0.06
limestone	32	$\sim$ 0–0.36	0.14
Wind-Laid Materials			
loess	5	0.14–0.22	0.18
eolian sand	14	0.32–0.47	0.38
tuff	90	0.02–0.47	0.21
Metamorphic Rocks			
schist	11	0.22–0.33	0.26

Source: McWhorter and Sunada 1977 as reported by Codell and Duguid 1983.

thus defined as the volume of moving water in a unit volume of the formation. Typical values of porosities and effective porosities are shown in Table 10.2.

Consider a unit volume at point $\mathbf{r}(x, y, z)$ of the formation at time t. In this unit volume, radionuclides are dissolved in the flowing water with a concentration $C(\mathbf{r}, t)$ per unit liquid volume, dissolved in the trapped stagnant water with a concentration $C_t(\mathbf{r}, t)$ per unit liquid volume, and sorbed onto the surface of the solid matrix material with a concentration $C_s(\mathbf{r}, t)$ per unit volume of solid. The concentrations can be measured in any appropriate units such as $(Bq\ cm^{-3})$ or $(atoms\ m^{-3})$. Thus the total amount of radioactivity in the unit volume of the formation is

$$C_{tot} = n_e C + (n - n_e)C_t + (1 - n)C_s. \tag{10.69}$$

To obtain an equation that describes the spatial and time dependence of the radionuclide concentration in the flowing water, one may perform a mass balance on the total radioactivity. This balance equation for the total radioactivity in the unit volume V of the formation can be expressed in words as

    (a) rate of increase of radioactivity in V =
               + (b) net rate at which flowing water brings radionuclides into V
               + (c) net rate at which nuclides diffuse into V
               − (d) rate at which nuclides in V decay.

Term (a) is simply $\partial C_{tot}/\partial t$, and term (d) is $-\lambda C_{tot}$ where λ is the decay constant for the radionuclide species being considered. If the flowing water has a velocity u (with components u_x, u_y, and u_z in the three coordinate directions), then the rate at which radioactivity is swept *into* volume V by water flowing in the x direction is $-u_x(\partial n_e C/\partial x)$. Similar expressions hold for the advective flow in the y and z directions.

At first it may seem surprising that the dispersal of radionuclides in groundwater also has a diffusive component. Since groundwater almost always flows in a laminar fashion (unlike flow in the atmosphere and surface waters), there can be no turbulence diffusion processes in groundwater flow. Also molecular diffusion is far too small to be an important contributor. However, the tortuous paths the water follows while flowing through pores and around grains acts as an effective diffusion mechanism (see Fig. 10.10). This diffusion mechanism is often referred to as *mechanical dispersion*. Thus the net rate at which radioactivity diffuses into the unit volume under consideration is expressed, in analogy to Eq. (9.19), as

$$\text{Term (c)} = n_e \frac{\partial}{\partial x}\left(d_x \frac{\partial C}{\partial x}\right) + n_e \frac{\partial}{\partial y}\left(d_y \frac{\partial C}{\partial y}\right) + n_e \frac{\partial}{\partial z}\left(d_z \frac{\partial C}{\partial z}\right), \tag{10.70}$$

in which d_x, d_y, and d_z are appropriate diffusion coefficients.

Substituting these expressions into the balance relation for a homogeneous medium (i.e., one in which d_x, d_y, d_z, n, and n_e are constants) gives

$$n_e \frac{\partial C}{\partial t} + (n - n_e)\frac{\partial C_t}{\partial t} + (1 - n)\frac{\partial C_s}{\partial t} = -n_e \left\{ u_x \frac{\partial C}{\partial x} + u_y \frac{\partial C}{\partial y} + u_z \frac{\partial C}{\partial z} \right\}$$

$$+ n_e \left\{ d_x \frac{\partial^2 C}{\partial x^2} + d_y \frac{\partial^2 C}{\partial y^2} + d_z \frac{\partial^2 C}{\partial z^2} \right\} - \lambda \left\{ n_e C + (n - n_e)C_t + (1 - n)C_s \right\}. \tag{10.71}$$

It is usually assumed that the radionuclide concentrations in the flowing and trapped water are equal (i.e., $C = C_t$). If it is also assumed that the dissolved radionuclides are in equilibrium with the radionuclides adsorbed on the grain and pore

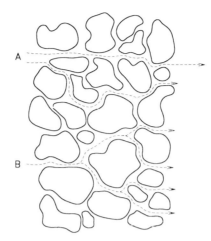

Figure 10.10   The tortuous paths used by groundwater flowing through an aquifer lead to a diffusion-like phenomenon called *hydraulic dispersion*. Longitudinal dispersion (A) occurs because water in the upper path travels farther in a given time interval than in the lower path. Transverse dispersion (B) results from the repeated splitting and deflection of the flowing water. [After Bouwer (1978).]

surfaces, then from Eqs. (10.53) and (10.55) the adsorbed concentration C_s can be expressed in terms of the dissolved concentration C as

$$C_s = K_d C = \hat{K}_d \rho_s C. \tag{10.72}$$

Here, the mass density of matrix material ρ_s is related to the bulk or partial density ρ'_s (mass of solid matrix per unit volume of the formation) by

$$\rho'_s = (1 - n)\rho_s. \tag{10.73}$$

Typical values of the distribution coefficient $\hat{K}_d$ in geological material are shown in Tables 10.3 and 10.4.

With the previously discussed assumptions, the concentrations C_t and C_s can be eliminated from Eq. (10.71) to give

$$n_e R_d \frac{\partial C}{\partial t} = -n_e \left\{ u_x \frac{\partial C}{\partial x} + u_y \frac{\partial C}{\partial y} + u_z \frac{\partial C}{\partial z} \right\}$$

$$+ n_e \left\{ d_x \frac{\partial^2 C}{\partial x^2} + d_y \frac{\partial^2 C}{\partial y^2} + d_z \frac{\partial^2 C}{\partial z^2} \right\} - \lambda n_e R_d C. \tag{10.74}$$

where the *retardation factor* R_d is defined as

$$R_d = \frac{n}{n_e} + \frac{1-n}{n_e} \rho_s \hat{K}_d = \frac{n}{n_e} + \rho'_s \frac{\hat{K}_d}{n_e}. \tag{10.75}$$

Finally, if *effective advective speeds* U_i are defined as

TABLE 10.3 Distribution Coefficients $\hat{K}_d(\text{L kg}^{-1})$ for Strontium and Cesium in Various Geological Materials.

MATERIAL	Sr	Cs
basalt, 32–80 mesh	16–135	792–9520
basalt, 0.5–4 mm, 300 ppm TDS	220–1220	39–280
basalt, 0.5–4 mm, sea water	1.1	6.5
basalt-fractured in situ measurement	3	
sand, quartz, pH 7.7	1.7–3.8	22–314
sands	13–43	100
carbonate, greater than 4 mm	0.19	13.5
dolomite, 4000 ppm TDS	5–14	
granite, greater than 4 mm	1.7	34.3
granodiorite, 100–200 mesh	4–9	8–9
granodiorite, 0.5–1 mm	11–23	1030–1810
Hanford sediments	50–300	300
tuff	45–4000	800–17800
soils	19–282	189–1053
shaley siltstone greater than 4 mm	8	309
shaley siltstone greater than 4 mm	1.4	102
alluvium, 0.5–4 mm	48–2454	121–3165
salt, > 4 mm saturated brine	0.19	0.027

Source: Isherwood 1981 as reported by Codell and Duguid 1983.

$$U_i = u_i/R_d, \qquad i = x, y, z, \tag{10.76}$$

and *effective diffusion coefficients* D_i as

$$D_i = d_i/R_d, \qquad i = x, y, z, \tag{10.77}$$

then Eq. (10.74) reduces to

$$\frac{\partial C}{\partial t} + U_x \frac{\partial C}{\partial x} + U_y \frac{\partial C}{\partial y} + U_z \frac{\partial C}{\partial z} = D_x \frac{\partial^2 C}{\partial x^2} + D_y \frac{\partial^2 C}{\partial y^2} + D_z \frac{\partial^2 C}{\partial z^2} - \lambda C. \tag{10.78}$$

This equation has the same form as the dispersion equation used earlier to describe the dispersal of radionuclides in the atmosphere and surface waters. Here, however, the flow velocity has been greatly reduced since the retardation factor is typically very large, a reasonable result since the sorption of radionuclides greatly slows their effective migration speed. In later sections, solutions to this equation for simple geometries will be presented.

One final note. The typically large values of the retardation factor R_d found for most radionuclides in many geological formations implies that only a small amount of total activity in a unit volume of the formation, C_{tot}, is dissolved in the flowing groundwater. With the assumed equilibrium of Eq. (10.72) and the equality of concentrations in the flowing and trapped water (i.e., $C = C_t$), the concentration in the flowing water is found from Eq. (10.69) to be

$$C = C_{tot}/(n_e R_d). \tag{10.79}$$

TABLE 10.4   Geometric Mean Values of the Distribution Coefficients $\hat{K}_d$ (L kg^{-1}) for Various Elements in Different Soil Types.

ELEMENT	SAND	LOAM	CLAY	ORGANIC
Ac	450	1500	2400	5400
Ag	90	120	180	15000
Am	1900	9600	8400	1500
Be	250	800	1300	3000
Bi	100	450	600	1500
Br	15	50	75	180
C	5	20	1	70
Ca	5	30	50	90
Cd	80	40	560	800
Ce	500	8100	20000	3300
Cm	4000	18000	6000	6000
Co	60	1300	550	1000
Cr	70	30	1500	270
Cs	280	4600	1900	270
Fe	220	800	165	600
Hf	450	1500	2400	5400
Ho	250	800	1300	3000
I	1	5	1	25
K	15	55	75	200
Mn	50	750	180	150
Mo	10	125	90	25
Nb	160	550	900	2000
Ni	400	300	650	1100
Np	5	25	55	1200
P	5	25	35	90
Pa	550	1800	2700	6600
Pb	270	16000	550	22000
Pd	55	180	270	670
Po	150	400	3000	7300
Pu	550	1200	5100	1900
Ra	500	36000	9100	2400
Rb	55	180	270	670
Re	10	40	60	150
Ru	55	1000	800	66000
Sb	45	150	250	550
Se	150	500	740	1800
Si	35	110	180	400
Sm	245	800	1300	3000
Sn	130	450	670	1600
Sr	15	20	110	150
Ta	220	900	1200	3300
Tc	0.1	0.1	1	1
Te	128	500	720	1900
Th	3200	3300	5800	89000
U	35	15	1600	410
Y	170	720	1000	2600
Zn	200	1300	2400	1600
Zr	600	2200	3300	7300

Source: Sheppard and Thibault 1990.

TABLE 10.5   Typical Values of the Hydraulic Conductivity K for Various Porous Materials.

MATERIAL	NUMBER OF ANALYSES	RANGE $(cm\ s^{-1})$	ARITHMETIC MEAN $(cm\ s^{-1})$
Igneous rocks			
weathered granite	7	$(3.3\text{–}52) \times 10^{-4}$	1.64×10^{-3}
weathered gabbro	4	$(0.5\text{–}3.8) \times 10^{-4}$	1.89×10^{-4}
basalt	93	$(0.2\text{–}4250) \times 10^{-8}$	9.45×10^{-6}
Sedimentary materials			
sandstone (fine)	20	$(0.5\text{–}2270) \times 10^{-6}$	3.31×10^{-4}
siltstone	8	$(0.1\text{–}142) \times 10^{-8}$	1.90×10^{-7}
sand (fine)	159	$(0.2\text{–}189) \times 10^{-4}$	2.88×10^{-3}
sand (medium)	255	$(0.9\text{–}567) \times 10^{-4}$	1.42×10^{-2}
sand (coarse)	158	$(0.3\text{–}6610) \times 10^{-4}$	5.20×10^{-2}
gravel	40	$(0.3\text{–}31.2) \times 10^{-1}$	4.03×10^{-1}
silt	39	$(0.09\text{–}7090) \times 10^{-7}$	2.83×10^{-5}
clay	19	$(0.1\text{–}47) \times 10^{-8}$	9.00×10^{-8}
Metamorphic Rocks			
schist	17	$(0.002\text{–}1130) \times 10^{-6}$	1.9×10^{-4}

Source: McWhorter and Sunada 1977 as reported by Codell and Duguid 1983.

Calculation of Flow Speeds.   Before Eq. (10.78) can be used to calculate the concentration of radionuclides at some point in a groundwater medium, the effective seepage or pore velocity U $(= u/R_d)$ of the flowing water must first be determined. This velocity, which depends on the pressure field and the viscous forces between the water and the geological formation, is often difficult to calculate, especially for nonsteady flows in heterogeneous anisotropic media.

For an isotropic medium with steady flow, the volumetric rate of water flowing through a unit area perpendicular to the x-axis, v_x (which has dimensions of speed), has been found empirically to be given by

$$v_x = -K \frac{\partial H}{\partial x}. \tag{10.80}$$

In this relation, known as Darcy's law, K is the *hydraulic conductivity* and is a property of the medium (Table 10.5 gives typical values). The hydraulic head H is defined as

$$H(x, y, z) = z + \frac{p(x, y, z)}{\rho g}, \tag{10.81}$$

where z is the depth into the formation, p the fluid pressure, ρ the water density, and g the gravitational acceleration constant.

The speed v_x is known as the *Darcy speed* in the x direction. However, it is not equal to the actual macroscopic water speed u_x (the seepage or pore flow speed).

Rather, v_x is the speed the water would have to have if it were moving through the entire unit area normal to the x-axis, solids as well as pores and cracks, to produce the actual volumetric flow rate. Thus, v_x is considerably smaller than u_x. These two speeds are related by $v_x = f_e u_x$, where f_e is the fraction of the unit area normal to the x-axis that is occupied by pores and capillaries through which water is flowing. This fraction for an isotropic medium is just the effective porosity n_e, so that the seepage speed in the x-direction is $u_x = v_x/n_e$.

In a three dimensional isotropic medium, Darcy's law can be written as

$$\mathbf{v} = -K \left(\mathbf{i}\frac{\partial H}{\partial x} + \mathbf{j}\frac{\partial H}{\partial y} + \mathbf{k}\frac{\partial H}{\partial z} \right). \tag{10.82}$$

The Darcy velocity $\mathbf{v}$ (with components v_x, v_y, and v_z) is the volumetric water flow rate across a unit area perpendicular to the flow direction (the direction of $\mathbf{v}$). This velocity is related to the actual seepage or flow velocity by

$$\mathbf{u}(x, y, z) = \frac{\mathbf{v}(x, y, z)}{n_e}. \tag{10.83}$$

To find the water velocity $\mathbf{u}$ from Eqs. (10.82) and (10.83), it is first necessary to know the derivatives of the hydraulic head at the point of interest. The calculation of $H(x, y, z)$ is usually the difficult part of groundwater calculations, especially for unsaturated flows in heterogeneous, anisotropic media. Differential equations can be derived for the hydraulic head (ANS 1980), and, in some simple cases, analytical solutions can even be obtained (Li 1972). In most cases, however, these equations must be solved numerically (Willis and Yeh 1987).

For steady-flow in homogeneous, isotropic, saturated media, in which the pressure varies only slowly with position, the derivative of the hydraulic head may be approximated by the rate of change in elevation in the flow direction. For example, in an aquifer with a horizontal bed that has a water table slope S_x in the x direction, the pore or seepage speed in the x direction is

$$u_x = -\frac{K}{n_e}\frac{\partial H}{\partial x} \cong -\frac{K}{n_e}S_x. \tag{10.84}$$

Similarly, for a vertically downward flow ($u_x = u_y = 0$) through a saturated water table, the soil water pressure is approximately constant. From Eq. (10.81) the seepage speed can then be approximated by

$$u_z = -\frac{K}{n_e}\frac{\partial H}{\partial z} \cong -\frac{K}{n_e}. \tag{10.85}$$

For flows and transport in unsaturated zones, other approximate methods must be used. For the percolation of water through surface soils, the downward speed u_z can be estimated from the rate of recharge of water at the surface, R_r, and from the mean volumetric water content θ as (Codell and Duguid 1983)

$$u_z = \frac{R_r}{\theta}. \tag{10.86}$$

The recharge rate is usually estimated from the water budget of the soil which balances the rainfall with evaporation and root uptake (Gupta et al. 1978). The mean volumetric water content of the unsaturated soil θ can be conservatively estimated (i.e., to yield an overestimate of u_z) from the differences between the total and effective porosities, $n - n_e$.

The Diffusion Coefficient. The diffusion coefficients needed for Eq. (10.78) are often expressed in terms of the dispersivities of the medium and the Darcy speed of the flow. If molecular diffusion effects are ignored and if the medium is assumed to be isotropic, the diffusion coefficients can be expressed as (Scheidegger 1961; Codell and Duguid 1983)

$$\theta d_i = \alpha_T |\mathbf{v}| + \frac{(\alpha_L - \alpha_T) v_i^2}{|\mathbf{v}|}, \qquad i = x, y, z, \tag{10.87}$$

where θ is the volumetric water content, α_L and α_T are, respectively, the *longitudinal* and *transverse dispersivities*, and $|\mathbf{v}|$ is the magnitude of the Darcy velocity $[= (v_x^2 + v_y^2 + v_z^2)^{1/2}]$. Typical values of dispersivities in different geological formations are listed in Table 10.6.

For saturated media, the volume fraction θ can be estimated as the effective porosity n_e. Thus, from Eq. (10.83), the diffusion coefficients can be expressed in terms of the seepage speeds as

$$d_i = \alpha_T |\mathbf{u}| + \frac{(\alpha_L - \alpha_T) u_i^2}{|\mathbf{u}|}, \qquad i = x, y, z. \tag{10.88}$$

Finally, upon division by the retardation factor R_d, the effective radionuclide diffusion coefficients become

$$D_i = \alpha_T |\mathbf{U}| + \frac{(\alpha_L - \alpha_T) U_i^2}{|\mathbf{U}|}, \qquad i = x, y, z. \tag{10.89}$$

For the special case in which the flow is in the direction of one of the coordinate axes (the x-axis, say), $|\mathbf{U}| = U_x$ and $U_y = U_z = 0$. The effective diffusion coefficients from the above expression then simplifies to

$$D_x = \alpha_L U_x \tag{10.90}$$

and

$$D_y = D_z = \alpha_T U_x. \tag{10.91}$$

10.8.2 Migration from the Surface

One task a groundwater analyst often encounters is the determination of the rate at which radionuclides enter an underground aquifer after migrating from a surface release point downwards into the ground. In this section two idealized vertical migration problems are considered. The first problem deals with an accidental release of

TABLE 10.6   Values of Dispersivities α_L and α_T Obtained by (1) Analysis of Tracer Break Through Curves in Groundwater Solute Transport (Left Half), and (2) by Calibration of Various Numerical Transport Models against Observed Groundwater Solute Transport (Right Half).

TRACER MEASUREMENTS			FIT TO MODELS		
SITE	α_L(m)	α_T(m)	SITE	α_L(m)	α_T(m)
Chalk River, Ontario alluvial aquifer	0.034		Rocky Mtn. Arsenal alluvial sediments	30.5	30.5
Chalk River, strata of high velocity	0.034–0.1		Arkansas River Valley coalluvial sediments	30.5	9.1
Alluvial aquifer	0.5		California	30.5	9.1
Alluvial, strata of high velocity	0.1		alluvial sediments Long Island	21.3	4.3
Lyons, France alluvial aquifer	0.1–0.5		glacial deposits Brunswick, Georgia	61	20
Lyons (full aquifer)	5		limestone		
Lyons (full aquifer)	12.0	3.1–14	Snake River, Idaho	91	136.5
Lyons (full aquifer)	8	0.015–1	fractured basalt		
Lyons (full aquifer)	5	0.145–14.5	Idaho, fractured	91	91
Lyons (full aquifer)	7	0.009–1	basalt		
Alsace, France alluvial sediments	12	4	Hanford Site, Wash. fractured basalt	30.5	18
Carlsbad, N. Mexico fractured dolomite	38.1		Barstow, Calif. alluvial deposits	61	18
Savannah River, SC fractured schistgneiss	134.1		Roswell Basin, NM limestone	21.3	
Barstow, Calif. alluvial sediments	15.2		Idaho Falls, lava flows and sediments	91	137
Dorset, England			Barstow, Calif.	61	0.18
chalk (fractured)	3.1		alluvial sediments		
chalk (intact)	1.0		Alsace, France	15	1
Berkeley, California sand/gravel	2–3		alluvial sediments Florida (SE)	6.7	0.7
Mississippi limestone	11.6		limestone		
NTS, carbonate aquifer	15		Sutter Basin, Calif. alluvial sediments	80–200	8–20
Pensacola, Florida limestone	10				

Source: Evenson and Dettinger 1980 as reported by Codell and Duguid 1983.

radionuclides falling over a large ground area. The second problem is concerned with a constant surface source which slowly leaks into the ground.

In both of these problems the soil is assumed to be homogeneous, isotropic, and infinitely deep. The flow is assumed to be steady, constant, and directed vertically downward, that is, $U = kU_z$. Moreover, the surface source is assumed to be infinite and uniform laterally so that the radionuclide concentration in the ground depends only on the depth z (measured from the surface) and the time t after the surface source is imposed (assumed to be at $t = 0$ and $z = 0$). Under these conditions, Eq. (10.78) reduces to

$$\frac{\partial C(z,t)}{\partial t} + U\frac{\partial C(z,t)}{\partial z} = D\frac{\partial^2 C(z,t)}{\partial z^2} - \lambda C(z,t), \quad z, t > 0, \qquad (10.92)$$

where $D = \alpha_L U$ and $U = U_z$.

Instantaneous Fallout Field.   Consider a situation in which a radionuclide is uniformly deposited at time $t = 0$ over a wide surface area ($z = 0$ plane). Each unit area of the surface is assumed to receive the same amount Q_o of radioactivity (Bq, Ci, atoms, etc.). Prior to the fallout, the concentration in the soil is zero, that is, $C(z,t) = 0$ for $t \leq 0$, $z > 0$. The resulting concentration in the soil is governed by Eq. (10.92) subject to

$$\lim_{z \to \infty} C(z,t) = 0, \qquad t > 0, \tag{10.93}$$

and, from the conservation of $Q_o/(n_e R_d)$ radionuclides initially placed in the flowing water [see Eq. (10.79)],

$$\int_0^\infty dz\, C(z,t) = \frac{Q_o e^{-\lambda t}}{n_e R_d}, \qquad t > 0. \tag{10.94}$$

The solution can be written as

$$C(z,t) = \frac{Q_o e^{-\lambda t}}{n_e R_d \sqrt{D}} \left\{ \frac{1}{\sqrt{\pi t}} \exp\left[-\frac{(z - Ut)^2}{4Dt} \right] \right.$$
$$\left. - \frac{U}{\sqrt{4D}} \exp\left(\frac{Uz}{D} \right) \mathrm{erfc}\left(\frac{z + Ut}{\sqrt{4Dt}} \right) \right\} \tag{10.95}$$

where $\mathrm{erfc}(x)$ is the *complementary error function* (see Appendix F, Sec. F-3.1). In Fig. 10.11 concentration profiles are shown for various times after release for a radionuclide with a very long lifetime (i.e., $\lambda \cong 0$).

Constant Surface Source.   The analysis of radioactivity seepage into the ground from a shallow cooling or mill-tailing pond can also be modeled by Eq. (10.92) if the pond is assumed to be very wide. In this case, a radionuclide is introduced into the pond at $t = 0$ and maintained at a constant concentration C_o for $t > 0$. It is assumed that there is no radioactivity present for $t < 0$ (i.e., $C(z,t) = 0$, $t < 0$). As the water enters the soil, the concentration in the flowing water will decrease on account of adsorption by the soil particles and entrapment by the water in closed pores. Thus, from Eq. (10.79), the boundary condition at $z = 0$ becomes $C(0,t) = C_o/(n_e R_d)$. Also as $z \to \infty$, $C(z,t)$ must vanish. The solution of Eq. (10.92) subject to these two boundary conditions is

$$C(z,t) = \frac{C_o}{2n_e R_d} \left\{ \exp\left[\frac{z(U - b)}{2D} \right] \mathrm{erf}\left(\frac{z - bt}{\sqrt{4Dt}} \right) \right.$$
$$\left. + \exp\left[\frac{z(U + b)}{2D} \right] \mathrm{erf}\left(\frac{z + bt}{\sqrt{4Dt}} \right) \right\}, \tag{10.96}$$

where $b = \sqrt{U^2 + 4\lambda D}$ and $\mathrm{erf}(x)$ is the error function (see Appendix F, Sec. F-3.1).

10.8.3 Migration in Underground Aquifers

Once radionuclides are introduced into an underground aquifer, either by seepage from the surface or by being released into a well which intersects the formation, they

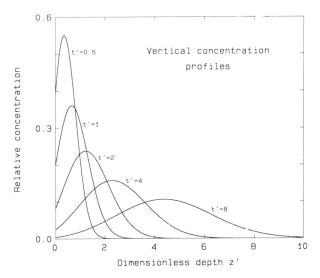

Figure 10.11 Vertical concentration profiles in moving groundwater resulting from an instantaneous deposition of radionuclides on the surface at $t = 0$. The relative concentration $C(z, t)D/(uQ_o)$ is plotted against dimensionless depth $z' = zU/(2D)$ for several dimensionless times $t' = u^2 t/D$ after the surface deposition.

can migrate large distances given sufficient time. In this section, results are presented for several types of source configurations for instantaneous releases at $t = 0$ into a horizontally confined saturated aquifer.

The aquifer, with thickness h, is assumed to be parallel to the $x - y$ plane and to be infinite in extent. The effective seepage speed $U = u/R_d$ of the radionuclide being considered is assumed to be constant and in the x-direction. The radionuclide concentration in the flowing aquifer water is given by Eq. (10.78), which for the present problem assumes the form

$$\frac{\partial C}{\partial t} + U\frac{\partial C}{\partial x} = D_x \frac{\partial^2 C}{\partial x^2} + D_y \frac{\partial^2 C}{\partial y^2} + D_z \frac{\partial^2 C}{\partial z^2} - \lambda C. \tag{10.97}$$

The origin is assumed to be at the top of aquifer, so that the bottom of the aquifer is at $z = h$. Because the aquifer flow is horizontally constrained, there is no flow of radionuclides vertically upward or downward at the aquifer boundaries, that is,

$$\left.\frac{\partial C(x, y, z, t)}{\partial z}\right|_{z=0} = \left.\frac{\partial C(x, y, z, t)}{\partial z}\right|_{z=h} = 0. \tag{10.98}$$

The boundary conditions in the other directions are that $C(x, y, z, t)$ vanishes as x or y approaches $\pm\infty$.

For all of the source configurations presented below, it is assumed that a unit of activity is released instantaneously at $t = 0$. Some of this activity will quickly be adsorbed on the aquifer solids and some will be dissolved in water trapped in dead-end pores and cracks. Of the unit activity released, the amount that is dissolved in the

TABLE 10.7   Green's Functions Used in the Solutions for Aquifer Concentrations.

$$X_1(x, t) = \frac{1}{\sqrt{4\pi D_x t}} \exp\left[-\frac{(x - Ut)^2}{4D_x t} - \lambda t\right]$$

$$X_2(x, t) = \frac{1}{4a}\left\{\text{erf}\left[\frac{x + a - Ut}{\sqrt{4D_x t}}\right] + \text{erf}\left[\frac{x - a - Ut}{\sqrt{4D_x t}}\right]\right\}$$

$$Y_1(y, t) = \frac{1}{\sqrt{4\pi D_y t}} \exp\left[-\frac{y^2}{4D_y t}\right]$$

$$Y_2(x, t) = \frac{1}{2b}\left\{\text{erf}\left[\frac{b + y}{\sqrt{4D_y t}}\right] + \text{erf}\left[\frac{b - y}{\sqrt{4D_y t}}\right]\right\}$$

$$Z_1(z, t) = \frac{1}{h}\left\{1 + 2\sum_{m=1}^{\infty} \exp\left[-\frac{m^2 \pi D_z t}{h^2}\right] \cos(m\pi z_s/h)\cos(m\pi z/h)\right\}$$

$$Z_2(z, t) = \frac{1}{\sqrt{4\pi D_z t}}\left\{\exp\left[-\frac{(z - z_s)^2}{4D_z t}\right] + \exp\left[-\frac{(z + z_s)^2}{4D_z t}\right]\right\}$$

flowing water and that is subsequently dispersed throughout the aquifer is, from Eq. (10.79), $(n_e R_d)^{-1}$.

Solutions of Eq. (10.97), subject to the above boundary conditions, are presented below for several source configurations (Codell et al. 1982). These solutions are given in terms of the Green's functions X_i, Y_i, and Z_i defined in Table 10.7.

Point Source.   Source at $x = y = 0$ and $z = z_s$, where $0 \leq z_s \leq h$

$$C(x, y, z, t) = \frac{1}{n_e R_d} X_1(x, t) Y_1(y, t) Z_1(z, t). \tag{10.99}$$

For an infinitely deep aquifer (i.e., $h \to \infty$) replace $Z_1(z)$ by $Z_2(z)$.

Vertical Line Source. $(0 < z < h)$ at $x = y = 0$

$$C(x, y, z, t) = \frac{1}{n_e R_d h} X_1(x, t) Y_1(y, t). \tag{10.100}$$

Horizontal Line Source. $(-b < y < b)$ centered at $x = 0$, $y = 0$, $z = z_s$

$$C(x, y, z, t) = \frac{1}{n_e R_d} X_1(x, t) Y_2(y, t) Z_1(z, t). \tag{10.101}$$

For an infinitely deep aquifer (i.e., $h \to \infty$) replace $Z_1(z)$ by $Z_2(z)$.

Vertical Area Source. $(-b < y < b, 0 < z < h)$ at $x = 0$

$$C(x, y, z, t) = \frac{1}{n_e R_d h} X_1(x, t) Y_2(y, t). \tag{10.102}$$

Volumetric Source. $(-a < x < a, -b < y < b, 0 < z < z_s)$

$$C(x, y, z, t) = \frac{1}{n_e R_d h} X_2(x, t) Y_2(y, t). \qquad (10.103)$$

For continuous releases at a rate $Q(t)$ beginning at $t = 0$, the radionuclide concentration C in the flowing aquifer water can be found in terms of the above instantaneous solutions (denoted by $\hat{C}$) as

$$C(x, y, z, t) = \int_0^t d\tau \, \hat{C}(x, y, z, t - \tau) Q(\tau). \qquad (10.104)$$

10.8.4 Migration Through Aquifers to Surface Waters

Migrating radionuclides in underground water present little direct hazard to the biosphere. However, underground waters can reach the surface where their dissolved radionuclides can enter various food chains through water extracted from wells, for example. Many aquifers also are intercepted by surface waters such as rivers and lakes, thereby allowing radionuclides in the underground water to enter the biosphere.

Simple flux models have been developed to estimate the discharge rate of radionuclides entering surface waters from intercepted aquifers (Codell et al. 1982). As illustrated in Fig. 10.12, radionuclides are introduced into an aquifer a distance x upflow from where the aquifer empties into a body of surface water. In the flux model, all radioactivity introduced into the aquifer is assumed to eventually reach the surface water, except that which decays before reaching the surface water. This model only gives the total rate of radionuclide input to the surface water. No information is provided about the spatial distribution of the radioactivity entering the surface water.

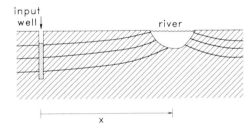

Figure 10.12 Radionuclides enter an aquifer through a contaminated well and eventually emerge into a river a distance x away. The acquifer flow is assumed to be in the x-direction, and the acquifer is infinite in lateral extent and has a depth H. [After Codell et al. (1982).]

For unidirectional flow in the x direction, the flow of radionuclides (e.g., $Bq\,s^{-1}$) crossing a differential area $dA = dx\,dy$ at some point (x, y, z) in the aquifer and at time t is

$$dF(x, y, z, t) = n_e \left[uC(x, y, z, t) - d_x \frac{\partial C(x, y, z, t)}{\partial x} \right] dz\,dy, \qquad (10.105)$$

where C is the radionuclide concentration in the dissolved flowing phase. If the entire flow of the aquifer (depth h and infinite width) is assumed to enter the surface water a distance x from the point at which radionuclides are introduced into the aquifer, the rate at which radionuclides enter the aquifer is thus, using Eqs. (10.76) and (10.77),

$$F(x,t) = n_e R_d \int_0^h dz \int_{-\infty}^{\infty} dy \left(UC(x,y,z,t) - D_x \frac{\partial C(x,y,z,t)}{\partial x} \right). \quad (10.106)$$

Instantaneous Point Release.   For a unit activity released at a point in the aquifer at time $t = 0$, the subsequent concentration in the aquifer water is given by Eq. (10.99). Substitution of this result into Eq. (10.106) and performance of the integration yield

$$\frac{F(x,t)}{nR_d} = \frac{x + Ut}{2t} X_1(x,t), \quad (10.107)$$

where $X_1(x,t)$ is given in Table 10.7.

Instantaneous Volumetric Release.   For a unit activity introduced instantaneously over a vertically averaged volume of the aquifer at time $t = 0$, the resulting concentration is given by Eq. (10.103). Then from Eq. (10.106), the subsequent rate of discharge into the surface waters a distance x from the source is found to be

$$\frac{F(x,t)}{nR_d} = UX_2(x,t)$$

$$-\frac{1}{4a} \sqrt{\frac{D_x}{\pi t}} \left\{ \exp\left[-\frac{(x + a - Ut)^2}{4D_x t} \right] \exp\left[-\frac{(x - a - Ut)^2}{4D_x t} \right] \right\},$$

$$(10.108)$$

where $X_2(x,t)$ is defined in Table 10.7.

REFERENCES

ADAMS, E.E., K.D. STOLZENBACH, AND, D.R. HARLEMAN, *Near and Far Field Analysis of Buoyant Surface Discharges into Large Bodies of Water*, Report 205, Parsons Lab. Water Resour. and Hydrodynamics, Dept. Civil Eng., Mass. Inst. of Technology, Cambridge, Mass., 1975.

ANS (American Nuclear Society), *Evaluation of Radionuclide Transport in Ground Water for Nuclear Power Sites*, Standard 2.17, LaGrange Park, Ill., 1980 (reaffirmed 1989).

BELLEUDY, P., J.L. RAHUEL, AND YANG GUOLU, "CARICHAR — Mobile Bed Modeling of Graded Sediments in Unsteady Flow," Proc. IAHR 22nd Congress, Lausanne, Switzerland, 1987.

BOUWER, H., *Groundwater Hydrology*, McGraw-Hill, New York, 1978.

BROOKS, N.H. "Diffusion of Sewage Effluent in an Ocean Current," in *Proc. of 1st Intern'l Conf. on Waste Disposal in the Marine Environment*, E. Pearson (ed.), Pergamon, New York, 1960.

CHURCHILL, M.A., J.A. CRAGWALL, R.W. ANDREWS, AND S.L. JONES, *Concentrations, Total Sediment Loads, and Mass Transport of Radionuclides in the Clinch and Tennessee Rivers*, Report ORNL-3721, Supplement 1, Oak Ridge National Laboratory, Oak Ridge, Tenn., 1965.

CODELL, R.B., K.T. KEY, AND G. WHALEN, *A Collection of Mathematical Models for Dispersion in Surface Water and Groundwater*, Report NUREG-0868, U.S. Nuclear Regulatory Commission, Washington, D.C., 1982.

CODELL R.B., AND J.D. DUGUID, "Transport of Radionuclides in Groundwater," in *Radiological Assessment: A Textbook on Environmental Analysis*, J.E. Till and H.R. Meyer (eds.), Report NUREG/CR-3332, U.S. Nuclear Regulatory Commission, Washington, D.C., 1983.

EISENBUD, M., *Environmental Radioactivity*, 3rd ed., Academic Press, New York, 1987.

EVENSON, D.E., AND M.D. DETTINGER, *Dispersive Processes in Models of Regional Radionuclide Migration*, Univ. of California, Lawrence Livermore Laboratory, Livermore, Calif., 1980.

FALCONER, R.A., P. GOODWIN, AND R.G.S. MATTHEW (eds.), *Hydraulic and Environmental Modeling of Coastal, Estuarine and River Waters*, Grower Technical Publ., Aldershot, England, 1989.

FISHER, H.B., E.J. LIST, R.C.Y. KOH, J. IMBERGER, AND N.H. BROOKS, *Mixing in Inland and Coastal Waters*, Academic Press, New York, 1979.

GUPTA, S.K., K. TANJI, D. NIELSON, J. BIGGAR, C. SIMMONS, AND J. MACINTYRE, *Field Simulation of Soil-Water Movement with Crop Water Extraction*, Water Science and Engineering Paper 4013, Dept. Land, Air and Water Resour., University of California-Davis, 1978.

HENDERSON, F.M., *Open Channel Flow*, Macmillan, New York, 1966.

HOLLY, F.M., JR., *CHARIMA and SEDICOUP Codes for Riverine Mobile-Bed Simulation*, 2nd Interagency Seminar on Stream Sedimentation Models, Denver, Colo., Oct. 19–20, 1988.

ISHERWOOD, D., *Geoscience Data Base Handbook for Modeling a Nuclear Waste Repository*, Report NUREG/CR- 0912, Vols. 1 and 2, U.S. Nuclear Regulatory Commission, Washington, D.C., 1981.

JIRKA, G.H., E.E. ADAMS, AND K.D. STOLZENBACH, "Properties of Buoyant Surface Jets," J. Hydr. Div., Am. Soc. Civ. Eng., 107, 1467–88, 1981.

JIRKA, G.H., "Multiport Diffuser for Heat Disposal: A Summary," *J. Hydraul. Div. Am. Soc. Civil Eng. 108*, 1025–1068 (1982).

JIRKA, G.H., A.N. FINDIKAKIS, Y. ONISHI, AND P.J. RYAN, "Transport of Radionuclides in Surface Waters," in *Radiological Assessment: A Textbook on Environmental Dose Analysis*, J. E. Till and H. R. Meyer (eds.), Report NUREG/CR-3332 (ORNL-5968), U.S. Nuclear Regulatory Commission, Washington, D.C., 1983.

JOHNSON, B.H., *A Review of Numerical Reservoir Hydrodynamic Modeling*, U.S. Army Corps of Engineers, Waterways Experiment Station, Vicksburg, Miss., February, 1980.

LAPPALA, E.G., "Modeling of Water and Solute Transport Under Variably Saturated Conditions," in *State of the Art, Modeling and Low-Level Waste Management: An Interagency Workshop*, C.A. Little and L.E. Stratton (eds.), National Tech. Info. Serv., Springfield, Va., June 1981.

LEE, J.H.W., AND G.H. JIRKA, "Vertical Round Buoyant Jet in Shallow Water," *J. Hydraul. Div. Am. Soc. Civil Eng. 107*, 1651–1675 (1981).

LI, W-H., *Differential Equations of Hydraulic Transients, Dispersion, and Groundwater Flow*, Prentice-Hall, Englewood Cliffs, N.J., 1972.

McWHORTER, D.B., AND D.K. SUNADA, *Ground-Water Hydrology and Hydraulics*, Water Resources Publications, Fort Collins, Colo., 1977.

OKUBO, A., "Oceanic Diffusion Diagrams," *Deep-Sea Res. 18*, 789–802 (1971).

ONISHI, Y., P.A. JOHNSON, R.G. BACA, AND E.L. HILTY, *Studies of Columbia River Water Quality–Development of Mathematical Models of Sediment and Radionuclide Transport Analysis*, Report BNWL-B-452, Pacific Northwest Laboratory, Richland, Wash., 1976.

ONISHI, Y., R.J. SERNE, E.M. ARNOLD, C.E. COWAN, AND F.L. THOMPSON, *Critical Review: Radionuclide Transport, Sediment Transport, and Water Quality Mathematical Modeling; and Radionuclide Adsorption/Desorption Mechanisms*, Report NUREG/CR-1322, PNL-2901, Pacific Northwest Laboratory, Richland, Wash., 1981.

ONISHI, Y., AND D.S. TRENT, *Mathematical Simulation of Sediment and Radionuclide Transport in Estuaries*, Report NUREG/CR-2423, PNL-4901, Pacific Northwest Laboratory, Richland, Wash., 1982.

ONISHI, Y., S.B. YABUSAKI, AND C.T. KINCAID, "Performance of the Sediment-Contaminant Transport Model SERATRA," Proc. Conf. *Applying Research to Hydraulic Practical*, P.E. Smith (ed.), American Society of Civil Engineers, pp. 623–632, 1982.

PRITCHARD, D.W., R.O. REID, A. OKUBO, AND H.H. CARTER, "Physical Processes of Water Mixing Movement and Mixing," in *Radioactivity in the Marine Environment*, National Academy of Sciences, Washington, DC, 1971.

ROBERTS, P.J.W., *Dispersion of Buoyant Wastewater Discharged from Outfall Diffusers of Finite Length*, Report KH-R-35, W.M. Keck Lab. Engineering Materials, California Institute of Technology, Pasadena, Calif., March 1977.

ROBERTS, P.J.W., "Line Plume and Outfall Dispersion," *J. Hydraul. Div. Am. Soc. Civil Eng., 105*, 313–331 (1979).

SCHEIDEGGER, A.D., "General Theory of Dispersion in Porous Media," *J. Geophysical Research, 66* (10), 3273–3278 (1961).

SHEPPARD, M.I., AND D.H. THIBAULT, "Default Soil Solid/Liquid Partition Coefficients, K_dS, for Four Major Soil Types: A Compendium," *Health Physics, 59*, 471–482 (1990).

USNRC (U.S. Nuclear Regulatory Commission), *Liquid Pathway Generic Study. Impacts of Accidental Radioactivity Releases to the Hydrosphere from Floating and Land-Based Nuclear Power Plants*, Report NUREG-0040, Washington, D.C., 1978.

WILLIS, R., AND W. YEH, *Groundwater Systems Planning and Management*, Prentice-Hall, Englewood Cliffs, N.J., 1987.

PROBLEMS

1. Water containing dissolved ^{137}Cs with a concentration of 0.5 mBq m^{-3} is discharged at a steady rate of 0.75 m^3 s^{-1} from a square channel of width 0.5 m that is positioned 1.5 m beneath the surface of a large lake which has negligible currents. The warm discharge is 0.2% less dense than the receiving lake water. Estimate the size of the near field zone and the ^{137}Cs concentration at the edge of the near field if the water depth around the discharge point is (a) 200 m deep, and (b) 2.5 m deep.

2. The blowdown from a wet cooling tower at a nuclear power plant is discharged from the shore and near the surface of a run-of-the-river reservoir. The blowdown has a temperature of 20 C and is emitted through a pipe 0.5 m in diameter into water that has a depth of 2 m and that is 7 C colder than the effluent. The river is very wide and the river velocity near the outfall varies from near zero to 0.3 m s^{-1} depending on the operation of the dam which controls the river flow. If the blowdown has an initial activity concentration of 1 kBq cm^{-3}, estimate the activity concentration after near-field mixing is completed for the two extreme river flow speeds.

3. Liquid waste from a radwaste processing plant is discharged vertically from a single pipe at the bottom of a shallow coastal bay where the water has a depth of 4 m. The waste is discharged at the rate of 1.5 m^3 s^{-1} and has a density close to that of fresh water while the ocean water has a density of 1.025 g cm^{-3}. If the effluent has an initial activity of 1 nCi m^{-3} estimate the activity concentration at the surface of the unstratified ocean if the waste is discharged at a speed of (a) 20 m s^{-1} and (b) 0.5 m s^{-1}.

4. A radionuclide with a half life of 8 days is accidentally leaked into a small reservoir at a rate of 1 mCi d^{-1} for a period of 2 weeks. The reservoir has a constant water volume of 1.8×10^4 m^3 and a normal outflow of 75 m^3 h^{-1}. (a) Plot the activity concentration in the reservoir water over the ten week period following the start of the leak. (b) To minimize contamination downstream of the reservoir, the outflow is stopped when the accidental leak is discovered (i.e., after leaking for 2 weeks) until the reservoir doubles its size and water has to again be released from the reservoir at its normal rate. Plot the activity concentration in the reservoir water for this situation.

5. 500 liters of water containing 15 TBq of a radioisotope with a half life of 10 h is accidentally discharged into a river. The river has a width of 50 m, an average depth of 1.5 m, a river-bed slope of 0.0004, and a flow of 7×10^4 m^3 h^{-1}. (a) Find the maximum activity concentration in the river water at a water intake 20 km downstream from the release point. (b) Over what time interval (measured from the time of release) should the water intake be closed to avoid extracting water with more than 1% of the maximum concentration?

6. If the discharge described in problem 10.5 occurs near one shore of the river, estimate the downstream distance from the discharge point for the radioactivity to becomes uniformly mixed transversely (i.e., the transverse concentration profile is uniform to within 5%). What is this distance if the release was made in the center of the river?

7. Verify that Eq. (10.27) is a solution of Eq. (10.26). Show that the condition of Eq. (10.28) leads to the constant of integration being given by Eq. (10.29).

8. A radioisotope with a half life of 50 days is discharged at a rate of 2 μCi d^{-1} into an unstratified estuary with an average width of 500 m and an average depth of 5 m. The discharge point is 50 km above the mouth of the estuary. The longitudinal dispersion coefficient is estimated to be 240 m^2 s^{-1} and the tidal averaged flow speed of the water through the estuary is 0.01 m s^{-1}. Estimate the activity concentrations (averaged over a tidal cycle) 30 km below and above the release point.

9. Of the several parameters in problem 10.8 (release rate, isotope half life, diffusion coefficient, average flow speed, estuary dimensions, discharge location, and tidal averaged diffusion coefficient), which parameter has the most effect on the tidal-averaged concentration 30 km upstream from the release point?

10. Derive Eq. (10.46).

11. If 10 TBq of a radioisotope with a half life of 75 d are instantaneously discharged far from land on the surface of an ocean 50 m above the thermocline, estimate the activity concentration in the surface water 4 km from the release point (in the Lagrangian coordinate system) at the following times after the release: 1 h, 5 h, 1 d, 5 d, 2 weeks, and 2 months.

12. Fallout deposits 10 mCi of ^{90}Sr on a large lake with a water volume of 1.2×10^6 m^3 in the which the outflow and inflow of water are 10 L s^{-1}. The sediment builds up on the bottom at the rate of 1 cm y^{-1} and the coefficient K_f for direct (diffusion) exchange between the sediment bed and the lake water is 0.02 m^2 s^{-1}. The average lake depth is 75 m and the thickness of the active sediment bed is in 0.075 m. The half life and equilibrium distribution coefficient for ^{90}Sr are 29.1 y and 450 mL g^{-1}, respectively. Calculate the activity concentration dissolved in the lake water and attached to the bottom sediment 3 years after the fallout incident.

13. Derive Eqs. (10.65) and (10.66).

14. A continuous discharge of a radionuclide with a half life of 30 d at the rate 3 pCi y^{-1} is made into a large lake with a water volume of 0.74×10^6 m^3. The sediment bed builds up at the rate of 0.5 cm y^{-1} and the equilibrium distribution coefficient K_d between the sediment and the dissolved radionuclide is 1300. The lake depth averages 50 m and the thickness of the active sediment layer is 0.1 m which has a coefficient for direct exchange with the lake water $K_f = 0.05$ m^2 s^{-1}. Thirty percent of the lake volume flows from the lake every year. Estimate the eventual steady-state activity concentration in the lake water and in the active sediment layer.

15. Consider a freshwater aquifer of coarse gravel with a slope of 0.001. (a) Estimate the seepage speed of the groundwater. (b) Estimate an effective advective speed of a cesium contaminant that has leaked into this aquifer.

16. Calculate the retardation factor for strontium in quartz sand with a density of 2.93 g cm^{-3} and an equilibrium coefficient $K_d = 60$.

17. At a particular site, precipitation infiltrates the ground at a rate of 20 cm y^{-1} to recharge the water table. Estimate the average downward speed of water in the unsaturated zone which is composed of fine sand.

18. The surface of a large sandy field is contaminated uniformly with 0.4 mCi of ^{90}Sr per m^2. This field experiences an average rainfall of 30 cm y^{-1} and has negligible runoff. The unsaturated sandy soil has the following properties: dispersivities $\alpha_L = \alpha_T = 30$ m; porosity $n = 0.39$; effective porosity $n_e = 0.30$; hydraulic conductivity $K = 0.002$ cm s^{-1}; distribution coefficient $K_d = 110$; bulk density $\rho_s = 2.9$ g cm^{-3}. Plot the activity concentration in the soil at depths of 10 cm, 1 m, and 10 m as a function of time after the contamination event.

19. Ten curies of a radionuclide with a half life of 30 y leak quickly into an aquifer through a highly permeable ground over a square horizontal area of 2000 m^2. In the aquifer, the groundwater is found to be moving at a speed of 2 m d^{-1} and the radionuclide has a retardation coefficient of 80. The aquifer has a saturated thickness of 50 m and an effective porosity n_e of 0.18. Plot the activity concentration in well water (the well penetrates the entire aquifer) as a function of time after leakage into the aquifer if the well is (a) 300 m directly downgradient from the center of the source, and (b) 400 m meters downgradient and 600 m to the side of the source center.

20. Plot the concentrations for problem 10.19 if the contamination is introduced at a steady rate of 2 Ci y^{-1}.

Chapter 11

Environmental Pathway Modeling

11.1 INTRODUCTION

Radionuclides discharged into the environment can move through the biosphere by many routes that eventually lead to radiation exposure of humans. Radionuclides can migrate through the biosphere on winds and in surface waters and lead directly to externally received doses. Ingestion or inhalation of radionuclides can lead to internally received doses. In the two previous chapters, models for describing how radionuclides are dispersed by atmospheric and hydrospheric mechanisms were presented. In this chapter, some of the important pathways are examined by which radionuclides move through the environment and along the food chain to reach humans.

There are many routes and pathways along which radionuclides discharged by nuclear facilities can travel through the different ecosystems to reach humans. These routes can be quite direct, as in the inhalation of contaminated air, or quite complex, involving many levels of the food chain and different transport mechanisms. Fig. 11.1 shows the three basic ecosystems — marine, aquatic, and terrestrial — and some of the principal pathways through which humans can be subjected to radioactivity emitted from a nuclear facility. While processes that move radionuclides through the atmosphere and hydrosphere tend to reduce radionuclide concentrations, many pathway components or processes that move radionuclides through the food chains to humans cause the radionuclides to become increasingly concentrated. Thus, for example, even though a radionuclide concentration in a lake may be so low as to be of little concern if the lake water were to be drunk by humans, the eating of fish living in the lake may not be acceptable because of the large buildup of radionuclides in the fish.

One of the main tasks for environmental analysts is to predict exposures received by humans from radioactivity reaching them from the many possible exposure pathways. To perform such analyses, not only must the character of the emitted radioactivity be known (e.g., nuclide type, amount, rate of release, chemical form, etc.), but

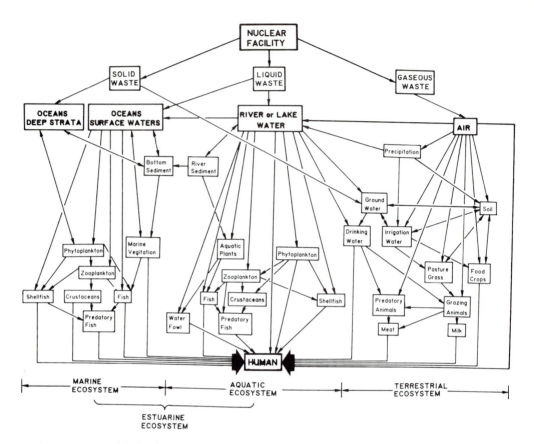

Figure 11.1 Simplified pathways leading to potential exposure in humans through the three basic ecosystems marine (saltwater), aquatic (freshwater), and terrestrial (land). A fourth ecosystem, estuarine (brackish water), has characteristics of both the aquatic and marine ecosystems. [After Peterson (1983).]

also must models be developed to describe how the radionuclides are subsequently absorbed, retained, and passed along by the various compartments of the many possible exposure pathways.

Ideally, models for the various components of a particular pathway should predict the time dependence of the amount of a particular radionuclide in the component, much as the MIRD models of Chap. 8 predict the activity in various body organs or compartments. Transient models have been developed and thoroughly studied for exposure pathways contributing significantly to human exposure. Some of these important transient models will be presented in this chapter. Unfortunately, transient models for many pathway components are still lacking or, at best, are very approximate, with many of the model parameters poorly known. For these cases, simpler steady-state methods based on equilibrium concentration factors have been developed, and these will also be discussed.

11.2 GENERAL DESCRIPTION OF EXPOSURE PATHWAYS

Radionuclides can reach humans from a great many potential exposure pathways in the environment (see Fig. 11.1). The movement of radionuclides along a particular pathway can be envisioned as proceeding through a series of pathway steps (or *trophes* in a food chain pathway). Each step or compartment of a pathway represents some physical entity or process by which radionuclides are received from a donor compartment, accumulated, and eventually passed down the pathway to the next compartment. At the beginning of the pathway is the source of radioactivity released into the environment. At the end of the pathway is the human population of concern whose internal exposure from ingestion of radionuclides or external exposure from radionuclides in the environment is to be estimated.

In most pathway analyses, human exposure is estimated from a set of independent subsystems, each representing a small portion of the general exposure pathway system and each containing a relatively few compartments. An example of such a subsystem is shown in Fig. 11.2. Only those compartments and pathways through which significant amounts of radioactivity can reach the target population for a specified type of release need be considered. A single comprehensive model containing all possible pathways in all three ecosystems of Fig. 11.1 would not only be difficult to develop and verify but would also be computationally inefficient since most of the many possible exposure pathways would make negligible contributions to a particular population group for a particular radioactive release. For example, a release of short-lived radioactivity to a small lake far inland would generally impact only people nearby, and pathways in the marine ecosystem can be neglected.

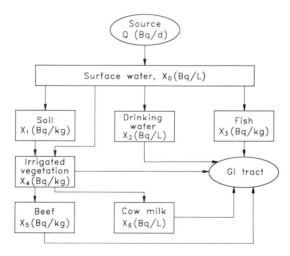

Figure 11.2 An example of a possible compartmentalization for assessing the effect of a release of ^{137}Cs into surface water.

The challenge to the analyst is to develop sufficiently simple subsystem models and to obtain values for the model parameters so that accurate predictions of exposures can be made for a particular problem. The analyst must ensure that all important exposure pathways are incorporated and that unnecessary pathways are omitted.

The advantage of the compartmental approach in the development of exposure pathway models is that each compartment can be treated independently and described by a relatively simple mathematical model. Moreover, most parameters for exposure pathway models must be obtained experimentally, and the use of compartments greatly eases the experimental phase of the model development.

A compartment in an exposure pathway model can have inputs from multiple donor compartments and the output can branch to several receiving compartments. Most pathways are direct, in that radionuclides flow from one compartment to another without reentering a previously visited compartment. Such a serial exposure pathway, one without feedback loops, is shown in Fig. 11.3. Feedback certainly occurs in some pathways, as in the decomposition of dead fish releasing radionuclides back into the aquatic food chain for other fish to absorb; however, the direct pathway of Fig. 11.3 can still be applied if any feedback is embedded inside the individual-compartment models. If a feedback loop is to be shown explicitly, then the following methods must be modified to account for such loops.

Figure 11.3   Schematic of a serial pathway with n compartments in which the output of one compartment is the input for the next.

11.2.1 Use of Compartment Models

A compartment model predicts the output of radioactivity, in whatever physical units are appropriate, in terms of the radioactivity input to the compartment. Some simple compartment models are described by a single algebraic equation. Generally, however, compartment models require multiple coupled equations, both algebraic and differential, involving dependent variables internal to the compartment as well as input and output dependent variables. The internal variables are needed to describe the movement of radioisotopes within various components of the compartment. The task of the modeler is to develop an appropriate set of equations describing the dynamic response of the compartment. Some specific compartment models of importance for radiological assessments are presented later in this chapter. For the moment, a very general description of compartmental dynamics is discussed.

Compartment models for exposure pathways often possess two important properties. First, they are linear; that is, doubling the input doubles the output. Second, they are unchanged by a shift in the reference base for time; that is, the compartments respond the same today as they will tomorrow. If a compartment model has these two properties, then the response of the compartment can be completely described by its *impulse response function* $z(t)$, the output at time $t > 0$ from a compartment which is subjected to a unit instantaneous input at $t = 0$. In this chapter, symbol X is used as a generic symbol for compartment output. The symbol may represent concentration (symbol C in other chapters), total quantities in a compartment, or even quantities per unit area, mass or volume. For an arbitrary input $X_{i-1}(t')$ to the ith compartment, the output $X_i(t)$ at time t is simply the superposition (integral) of responses $z_i(t - t')$ caused by an infinite series of impulse inputs $X_{i-1}(t')dt'$ over all previous times, namely,

$$X_i(t) = \int_{-\infty}^{t} dt' \, X_{i-1}(t')z_i(t - t'). \tag{11.1}$$

This result can be rewritten as

$$X_i(t) = \int_{0}^{\infty} d\tau \, X_{i-1}(t - \tau)z_i(\tau). \tag{11.2}$$

If this result is applied recursively starting from the beginning of the pathway of Fig. 11.3, with $X_o(t) = Q(t)$ the rate of radioactivity release to the environment, the output from the nth compartment is found to be

$$X_n(t) = \int_{0}^{\infty} d\tau_n \, z_n(\tau_n) \int_{0}^{\infty} d\tau_{n-1} \, z_{n-1}(\tau_{n-1}) \cdots \int_{0}^{\infty} d\tau_1 \, z_1(\tau_1)Q(t - T), \tag{11.3}$$

where $T = \sum_{j=1}^{n} \tau_j$. From this general result it is seen that radionuclides tend to become increasingly accumulated in each step of the pathway since the concentration (or amount) leaving the nth compartment at time t is the n-fold integral over all preceding time of the radioactivity leaving all prior compartments.

This formulation of the output from some compartment at time t as given by Eqs. (11.1) or (11.2) indicates that the input is integrated over all previous accumulation times. When the compartment has a finite accumulation time t_a, as when a compartment represents an organism with a finite lifetime or the output represents crops of a certain age, the impulse response function for the compartment, $z_i(\tau)$, must be set to zero for $\tau > t_a$.

11.2.2 A Pathway Example

To illustrate the application of the very general formulation presented above, consider a simple exposure pathway in which radionuclides are discharged at some point in a stream which flows into a small pond in which fish live and which in turn are consumed by humans. For this example, a four-compartment pathway (stream $\to$ pond $\to$ fish $\to$ human) can be used. Each of these compartments is discussed separately in the following subsections.

The Stream Compartment. The first compartment represents stream transport of discharged radionuclides to the pond. The input to this compartment is the rate of radionuclide discharge $Q(t)$ (Bq s^{-1}) and the output $X_1(t)$ is the rate of radionuclide flow (Bq s^{-1}) into the pond. Assume that the pond is sufficiently far downstream from the discharge location that the effluent has become fully mixed in the transverse direction. For this case, an instantaneous (impulse) discharge of Q' (Bq) at time $t = 0$ produces a fully mixed concentration in the stream at a distance x downstream that is given by Eq. (10.32). The total radionuclide flow a distance x below the discharge (assumed to be the lake entrance) is this concentration multiplied by the volumetric flow rate of the stream (i.e., by WHu). Thus, for a unit impulse discharge ($Q' = 1$), the impulse response function for the stream compartment is seen to be

$$z_1(t) = \frac{u}{\sqrt{2\pi}\sigma_x} \exp\left[-\frac{(x - ut)^2}{2\sigma_x^2} - \lambda t\right], \qquad t \geq 0. \tag{11.4}$$

For $t < 0$ the response $z_1(t)$ is zero since there can be no radionuclide flow in the stream before the discharge occurs.

The discharge from the stream compartment for an arbitrary input $Q(t)$ is obtained from substitution of $z_1(t)$ into Eq. (11.2), namely,

$$X_1(t) = \frac{u}{\sqrt{2\pi}\sigma_x} \int_0^\infty d\tau \, Q(t - \tau) \exp\left[-\frac{(x - u\tau)^2}{2\sigma_x^2} - \lambda\tau\right]. \tag{11.5}$$

The integrand is narrowly peaked about $\tau = x/u$ so that this result may be approximated by

$$X_1(t) \cong Q(t - x/u)e^{-\lambda x/u} \left\{\frac{u}{\sqrt{2\pi}\sigma_x} \int_0^\infty d\tau \, \exp\left[-\frac{(x - u\tau)^2}{2\sigma_x^2}\right]\right\}. \tag{11.6}$$

Then, since the integrand of this result is negligibly small for $\tau \leq 0$ (because it is assumed that $x \gg \sigma_x$), the lower limit may be replaced by $-\infty$. In this manner the term in braces evaluates to unity so that

$$X_1(t) \cong Q(t - x/u)e^{-\lambda x/u}. \tag{11.7}$$

This result indicates that the inflow to the pond is just the activity discharged into the stream a time x/u earlier, corrected for radioactive decay during the trip downstream.

The approximations used to reduce Eq. (11.5) to Eq. (11.7) is equivalent to approximating the impulse response of Eq. (11.4) by

$$z_1(t) \cong e^{-\lambda t}\delta(t - x/u). \tag{11.8}$$

The Dirac delta function (see Appendix F, Sec. F-3.3) in this approximation causes the stream compartment to be a *transitory* compartment, since the compartment response is directly proportional to the input, but shifted by the stream travel time.

The Pond Compartment.   The model for the pond in this example is that used in Sec. 10.3. The impulse response function for this system is readily obtained from the general solution of Eq. (10.12). If the initial time t_o is taken as $-\infty$ and the radionuclide input rate $Q(t')$ is taken as the Dirac delta function $\delta(t')$, a unit impulse at $t' = 0$, Eq. (10.12) gives

$$z_2(t) = \frac{1}{V}e^{-\lambda' t} \tag{11.9}$$

where $\lambda' \equiv \lambda + q_o/V$. Here $z_2(t)$ is the concentration in the pond water at time t after a unit activity input into the pond, q_o is the volumetric rate of discharge from the pond, and V is the pond volume.

Thus, for an arbitrary input discharge rate $Q(t)$ into the stream, the radionuclide concentration in the pond at time t is, from Eq. (11.3),

$$X_2(t) = \int_0^\infty d\tau_2 \, z_2(\tau_2) \int_0^\infty d\tau_1 \, z_1(\tau_1) Q(t - \tau_1 - \tau_2), \qquad (11.10)$$

which, on substitution for z_1 and the approximation of Eq. (11.8) for z_2, yields

$$X_2(t) = \frac{e^{-\lambda x/u}}{V} \int_0^\infty d\tau \, e^{-\lambda' \tau} Q(t - \tau - x/u) \qquad (11.11)$$

as an approximation for X_2. If the discharge into the stream begins at $t = 0$ and remains steady at Q_o, then the pond concentration becomes approximately

$$X_2(t) = \frac{Q_o e^{-\lambda x/u}}{V \lambda'} \left[1 - \exp(-\lambda'(t - x/u)) \right], \quad \text{for } t > x/u. \qquad (11.12)$$

For $t < x/u$, $X_2(t) = 0$ since no radionuclides have yet reached the pond. At long times after a steady discharge begins, a steady-state concentration in the pond is reached, namely

$$X_2^\infty = \frac{Q_o e^{-\lambda x/u}}{q_o + \lambda V}. \qquad (11.13)$$

If the discharge is stopped after a time, the pond concentration does not suddenly drop to zero (as in the stream compartment), but rather persists, decreasing exponentially. This behavior is a result of Eq. (11.11) which shows that the pond concentration is the integral of the input. For this reason compartments such as ponds or lakes are sometimes referred to as *integrating* compartments.

The Fish Compartment.   Fish that live in the pond will absorb and accumulate radionuclides into their skeleton and muscle. To determine the concentration of radionuclides in the muscle the following simple model for the fish can be used (Peterson 1983)

$$\frac{dX_3(t)}{dt} = a_i X_2(t) - (\lambda + a_r) X_3(t), \qquad (11.14)$$

where X_2 (Bq cm^{-3}) and X_3 (Bq g^{-1}) are the activity concentrations in water and in muscle, respectively. The constant a_i (cm^3 s^{-1} g^{-1}) is rate of intake of the radionuclide (Bq s^{-1}) per unit mass (g) of fish per unit activity concentration (Bq cm^{-3}) in water, and a_r (s^{-1}) is the rate constant for biological elimination of the radionuclide.

The general solution to this equation (see Appendix F, Sec. F-1.1) can be written as

$$X_3(t) = \int_0^\infty d\tau \, a_i e^{-(\lambda + a_r)\tau} X_2(t - \tau). \qquad (11.15)$$

Upon comparison of this result to Eq. (11.2), it is seen that the impulse response for the fish is

$$z_3(t) = a_i e^{-(\lambda + a_r)t}. \qquad (11.16)$$

The concentration in the fish at time t in terms of the discharge $Q(t)$ in the stream can be obtained by using Eq. (11.3) with $n = 3$, or, more easily, by substituting for $X_2(t)$ given by Eq. (11.11) into Eq. (11.15). For a steady discharge Q_o beginning at $t = 0$, the result is found to be (for $t > x/u$)

$$X_3(t) = \frac{a_i Q_o e^{-\lambda x/u}}{q_o + \lambda V} \left\{ \frac{1}{\lambda + a_r} \left(1 - e^{-(\lambda + a_r)(t - x/u)} \right) \right. \tag{11.17}$$

$$\left. - \frac{e^{-(\lambda + q_o/V)(t - x/u)}}{a_r - q_o/V} \left(1 - e^{-(a_r - q_o/V)(t - x/u)} \right) \right\}.$$

For $t \le x/u$, no radionuclides have reached the fish, and $X_3(t) = 0$. At long times after the discharge has begun, the steady-state concentration in the fish becomes

$$X_3^\infty = \frac{a_i Q_o e^{-\lambda x/u}}{(q_o + \lambda V)(\lambda + a_r)}. \tag{11.18}$$

Exposure to Humans. The output from the fish compartment in turn is ingested by humans. If the consumption rate of fish by a human is denoted by U and, as usual, is assumed to be constant, then the total activity ingested by the human between times t_1 and t_2 is

$$\text{Ingestion} = \int_{t_1}^{t_2} dt \, U \, X_3(t). \tag{11.19}$$

From this ingestion, the resulting committed organ doses to the person can be calculated using the methods described in Chap. 8. Alternately, the ratio of this intake (for a year) to the annual limit of intake (ALI) can be used to calculate the committed effective dose equivalent by multiplying this ratio by the dose used as a basis for the ALI.

11.2.3 Steady-State Analysis

In many analyses of exposure pathways, the radionuclide of concern is released over such a long time and at such a steady rate that the transient nature of the pathway can be neglected. For this equilibrium case, the output from the ith exposure pathway compartment is given by Eq. (11.2) with $X_{i-1}^\infty = 1$, namely

$$CR_i \equiv \int_0^\infty dt \, z_i(t). \tag{11.20}$$

Here CR_i is the steady-state output of compartment i resulting from a steady input of unit strength. The ratio of the equilibrium output to input for a compartment, CR_i, is referred to as the *concentration factor* for the compartment.[1] The concentration factor

---

[1]The ratio of the steady-state output to input for a compartment (or a chain of compartments) is also sometimes referred to as the discrimination factor, accumulation factor, or another name more specific to a given compartment, for example, soil-to-plant transfer factor. In this text, the more general concentration factor or concentration ratio is used.

can also be calculated from the compartment equations by setting all time derivatives to zero and solving the resulting set of algebraic equations. This is usually easier than first finding the impulse response $z_i(t)$ for the compartment and then integrating it to obtain CR_i.

The measurement of concentration factors for various pathway components has been the subject of many investigations, and many of these factors will be presented later in the chapter. Under equilibrium conditions, the concentration factors afford a very simple means of estimating the amount of radioactivity passed along a particular exposure pathway and eliminate the need to develop and use transient models for each pathway compartment. For a steady discharge rate Q_o, the steady-state output from the nth compartment, as given by Eq. (11.3), reduces to

$$X_n^\infty = Q_o \prod_{i=1}^{n} CR_i. \tag{11.21}$$

Thus, if concentration factors are known for each compartment in an exposure pathway model, the amount of radioactivity transferred to humans is obtained by simply multiplying the concentration factors for each pathway compartment and summing over all model pathways.

This steady-state procedure is known as the concentration factor (CF) method (ICRP 1979) and forms the basis of many environmental exposure analyses (USNRC 1977). The far more difficult transient analysis based on the dynamic model equations is sometimes referred to as the system analysis (SA) method (ICRP 1979).

11.3 TERRESTRIAL EXPOSURE PATHWAYS

The exposure pathways that generally are the most important for humans are those in the terrestrial ecosystem. Not only do people live in this system, but the bulk of their food comes from it. While a few terrestrial exposure pathways are transitory, for example, the external exposure caused by airborne radioactivity, the vast majority of pathways are integrating in nature since the exposure continues long after the release of radioactivity to the environment has ceased.

Two types of terrestrial exposure pathway are responsible for most of the population exposure. First, the accumulation of radionuclides on the ground (fallout) leads to direct external exposures. Second, the consumption of vegetables, grains, and animal food products which have become contaminated leads to internal exposures. It is these food pathways resulting in internal exposures that are discussed in this section.

11.3.1 Direct Contamination of Vegetation

Airborne releases of radioactivity can contaminate vegetation that is used as food by humans or as forage for cattle. This contamination arises from two distinct mechanisms: the uptake of radionuclides from contaminated soil by plants, and direct deposition of radionuclides on plants from airborne radionuclides. The direct deposition of radionuclides on vegetation is considered in this section.

Upon reaching the ground, some of the radionuclides in a diffusing gas cloud will be removed from the cloud and adhere to surfaces on the ground including vegetation (see Sec. 9.5.3). After deposition on vegetation, the radionuclides may subsequently be removed by rain, plant weathering, and by radioactive decay. Thus, at some location where the activity concentration in the air is $X_{air}(t)$, the areal activity $X_A(t)$ of a radionuclide species that is on the vegetation growing over a unit horizontal area of the ground is given by

$$\frac{dX_A(t)}{dt} = [f_{veg}v_d + f_w W_v R_p]X_{air}(t) - (\lambda + \lambda_r)X_A(t). \qquad (11.22)$$

Here, f_{veg} and f_w are the fractions of radionuclides removed from the atmospheric plume by dry and wet deposition on the vegetation, respectively, v_d is the dry deposition speed, W_v is the volumetric washout factor, R_p is the precipitation rate, λ is the radioactive decay constant, and λ_r is the rate constant for the subsequent physical removal of radionuclides from the plant surfaces (discussed in detail later).

It is more useful to express the extent of vegetation contamination on a volumetric or mass basis. If Y_{veg} denotes the dry-mass[2] of vegetation per unit area of the ground (kg m^{-2}), then the contamination per unit mass of vegetation X_{veg} (Bq kg^{-1}) is simply X_A/Y_{veg}. Upon division of Eq. (11.22) by Y_{veg}, the contamination per unit mass is given by

$$\frac{dX_{veg}(t)}{dt} = V_{veg}X_{air}(t) - (\lambda + \lambda_r)X_{veg}(t), \qquad (11.23)$$

where V_{veg} is the transfer rate from the air to the vegetation (m^3 kg^{-1} s^{-1}), given by

$$V_{veg} = \frac{f_{veg}v_d + f_w W_v R_p}{Y_{veg}}. \qquad (11.24)$$

The transfer rate V_{veg}, usually without the rainfall deposition component, not only can be calculated from this result but also can be measured readily. It is found that the use of this transfer rate tends to give more reproducible results than the use of the deposition velocity (Peterson 1983). Measured values of this transfer rate for dry conditions are given in Table 11.1 for various forms of iodine on various crops.

Contamination by a Continuous Release. If there is no contamination of the vegetation prior to $t = 0$, the activity per unit mass of vegetation at some later time t is given by the general solution of Eq. (11.23), namely,

$$X_{veg}(t) = e^{-(\lambda+\lambda_r)t} \int_0^t dt'\, V_{veg}X_{air}(t')e^{(\lambda+\lambda_r)t'}. \qquad (11.25)$$

For a constant air concentration of radionuclides beginning at $t = 0$, this solution becomes

---

[2]The water content of vegetation varies widely, depending not only on the type of vegetation but also on climate, collection and storage conditions. For this reason it is preferable to use a dry weight as the mass basis.

TABLE 11.1   Measured Values of Dry Deposition Speeds for Radioiodine.

FORM	VEGETATION	ATMOSPHERIC CONDITIONS	DEPOSITION SPEED v_d (m s^{-1})	TRANSFER RATE V_d (m^3 kg^{-1} s^{-1})
I$_2$	wheat/pasture grass	unstable	$(1.37 \pm 0.69) \times 10^{-2}$	0.12 ± 0.04
	wheat/pasture grass	stable	$(0.31 \pm 0.2) \times 10^{-2}$	0.046 ± 0.03
	dry grass	stable	$(1.03 \pm 0.75) \times 10^{-2}$	0.079 ± 0.057
	pasture grass	stable[a]	$(2.6 \pm 1.3) \times 10^{-2}$	0.33 ± 0.15
	pasture grass	night	$(1.1 \pm 0.37) \times 10^{-2}$	0.019 ± 0.015
	moist grass		$(2.7 \pm 2.2) \times 10^{-2}$	0.144 ± 0.071
	pasture grass	neutral	8×10^{-3}	0.115
	rye grass		$(8.6 \pm 6.8) \times 10^{-3}$	0.057 ± 0.032
	clover		$(2.0 \pm 1.3) \times 10^{-2}$	0.164 ± 0.045
	clover		$(1.2 \pm 0.2) \times 10^{-2}$	0.14 ± 0.03
	clover	stable	3×10^{-3}	0.071
	dandelion	unstable	$(1.1 \pm 0.3) \times 10^{-2}$	0.35 ± 0.20
	dandelion	stable	5×10^{-3}	0.12
CH$_3$I	pasture grass	unstable[a]	1×10^{-4}	0.001
	pasture grass	unstable	1×10^{-6}	1×10^{-5}
	pasture grass	stable	1×10^{-6}	1×10^{-5}

[a] Very high surface roughness.

Source: From various sources as reported by Peterson 1983.

$$X_{veg}(t) = \frac{V_{veg} X_{air}}{\lambda + \lambda_r} \left(1 - e^{-(\lambda + \lambda_r)t} \right). \tag{11.26}$$

Very often this equation is used to represent approximately the equilibrium concentration of vegetation in terms of the average atmospheric concentration, namely, $X_{veg}^{\infty} = V_{veg} X_{air}/(\lambda + \lambda_r)$.

Contamination by an Instantaneous Release.   The ground level airborne concentration at some downwind distance x and cross wind distance y from a point at which an amount Q' of activity is released at time $t = 0$ at an elevation h is found from Eq. (9.30) (after correcting for radioactive decay) to be

$$X_{air}(t) = \frac{Q' e^{-\lambda t}}{(2\pi)^{3/2} \sigma_x(t) \sigma_y(t) \sigma_z(t)} \exp\left[-\frac{(x - ut)^2}{2\sigma_x^2(t)} - \frac{y^2}{2\sigma_y^2(t)} - \frac{h^2}{2\sigma_z^2(t)} \right]. \tag{11.27}$$

Substitution of this result into Eq. (11.25), assumption that the travel time x/u to the point of interest is much less than t and that the downwind distance x is much greater than σ_x, and approximation of the resulting integral with the same method used to reduce Eq. (9.26) to Eq. (9.29) yield the vegetation activity concentration at location (x, y)

$$X_{veg}(t) = \frac{Q' V_{veg} \, e^{-\lambda t} \, e^{-\lambda_r(t - x/u)}}{2\pi \, \sigma_y(x) \sigma_z(x) u} \exp\left[-\frac{y^2}{2\sigma_y^2(x)} - \frac{h^2}{2\sigma_z^2(x)} \right]. \tag{11.28}$$

Loss of Radionuclides by Plants.   In Eq. (11.23), the rate constant λ_r for radionuclide removal from vegetation is often expressed as the sum of constants for three distinct processes: the dilution by new plant growth, removal by weathering, and a loss by plant metabolic processes. Thus, the removal rate constant λ_r can be decomposed as

$$\lambda_r = \lambda_g + \lambda_w + \lambda_p, \tag{11.29}$$

where λ_g is the rate of activity loss per unit mass caused by new plant growth, λ_w is the removal rate by weather (wind and rain), and λ_p is a rate constant for the poorly understood "plant factor" removal process.

Plants in their growth stage double their mass roughly every 20 days (the range from reported measurements on different crops is about 13 to 29 days) (Peterson 1983). This doubling time of 20 days corresponds to an exponential growth constant $\lambda_g = \ln(2)/20 = 0.035$ d^{-1}. Note that this plant-growth rate constant is included in λ_r only when figuring the activity on a per unit mass basis [as in Eq. (11.23)]. In calculating vegetation activity on a per unit ground area basis [as in Eq. (11.22)] it is omitted.

Weathering causes loss of radionuclides from the plant by wind erosion, rain washoff, and possibly by volatilization. For wind and rain removal, λ_w can be expressed as (Peterson 1983)

$$\lambda_w = k_{wind}\bar{u} + k_{rain}R_p, \tag{11.30}$$

where $\bar{u}$ is the mean wind speed, R_p is the amount of rain (mm), and k_{wind} and k_{rain} are empirical constants. The wind erosion constant k_{wind} ranges from (1.2 to 5.9) $\times$ 10^{-5} m^{-1} which corresponds to half-times between 0.14 and 0.65 days for a mean wind speed of 1 m s^{-1} (Peterson 1983). Some measured values of the washoff constant k_{rain} are given in Table 11.2. For a value of $k_{rain} = 0.025$ mm^{-1} and a typical rainfall rate of 2.5 mm d^{-1}, the weather half time is 11 days which is comparable to measured weathering half-times (Peterson 1983).

The mechanisms responsible for the plant-factor rate constant λ_p are not well understood but probably include volatilization as well as actual loss of plant material such as the shedding of waxy cuticles during plant growth. A typical value of λ_p is 0.02 d^{-1} which implies a removal half-time of about 35 days.

From studies of long term retention of radionuclides on vegetation following an initial contaminating event, it is has been found that radioactivity loss occurs with two distinct time constants, one for a half-time of about 5 days and another with a half-time of about 48 days (Dahlam et al. 1969; Krieger and Burmann 1969). The short time constant corresponds to typical weathering half-times, while the longer time constant is related to the "plant factor" removal term. The fraction $R_{veg}(t)$ of activity left on the vegetation at time t after initial plant contamination is referred to as the *plant retention function*, and can be expressed as (Peterson 1983)

$$R_{veg}(t) = e^{-\lambda t}\left(0.70e^{-0.138t} + 0.30e^{-0.0144t}\right), \tag{11.31}$$

where time t is measured in units of days.

With this plant retention function, the activity on the vegetation subjected to a radionuclide deposition rate of $\omega(t)$ (Bq m^{-2} s^{-1}) at some specified location is computed as

TABLE 11.2   Measured Values of the Washoff Coefficient k_{rain}.

RADIONUCLIDE	MATERIAL/CONDITIONS	k_{rain} (mm^{-1})
^{90}Sr	grass (washing)	0.0238
	grass (long-term)	0.009 ± 0.018
^{89}Sr	cabbage	0.0236
^{95}Zr	cabbage	0.0218
Ruthernium chloride	romaine lettuce	0.0625
Nitrosylruthenium chloride	romaine lettuce	0.0428
Nitrosylruthenium tetranitrate	romaine lettuce	0.0914
Nitrosylruthenium hydroxide	romaine lettuce	0.0529
^{131}I	cabbage	0.0256
	grass	0.0020 ± 0.028
^{137}Cs	cabbage	0.0197
	grass (washing)	0.0343
^{144}Ce	cabbage	0.0245
^{238}Pu dioxide	bush beans (4 mm)[a]	$(5.25 \pm 2.75) \times 10^{-3}$
	bush beans (17 mm)	$(1.4 \pm 0.5) \times 10^{-3}$
	sugar beets (17 mm)	0.010 ± 0.0029
^{238}Pu hydrated dioxide	bush beans (4 mm)	$(6.5 \pm 3.5) \times 10^{-3}$
	bush beans (17 mm)	$(2.6 \pm 0.7) \times 10^{-3}$
	sugar beets (17 mm)	$(2.5 \pm 0.6) \times 10^{-3}$
Silicate particles	grass	0.0626
	grain heads	0.0685

[a] Extent of rainfall given in parentheses.

Source: From various references as reported by Peterson 1983.

$$X_{veg}(t) = \frac{1}{Y_{veg}} \int_{-\infty}^{t} dt'\, \omega(t') R_{veg}(t-t') e^{-\lambda(t-t')}, \qquad (11.32)$$

or, equivalently, by

$$X_{veg}(t) = \frac{1}{Y_{veg}} \int_{0}^{\infty} d\tau\, \omega(t-\tau) R_{veg}(\tau) e^{-\lambda\tau}. \qquad (11.33)$$

11.3.2 Radionuclide Buildup in Soil

The accumulation of radioactivity in soil near the surface is of concern since growing vegetation will take up radionuclides through roots and incorporate them into various parts of the plants. Soil contamination thus represents a long-term source of radionuclides that can continue to contaminate human or animal food crops for many years after the initial soil deposition. By contrast, the direct contamination of vegetation discussed in the previous section is usually of concern only for the crops exposed during the deposition.

Soil can become contaminated from the direct deposition of airborne radioactivity, from contaminated rain or irrigation water, from the washoff of radionuclides initially deposited on plants, and from the decay of contaminated vegetation. All soil contamination initially enters the soil from the surface. These radionuclides then migrate into the soil, carried down by rain and irrigation water or mixed mechanically by tillage. The radioactivity available for subsequent uptake by plants is limited to that residing within the plant root zone, for most crops to a depth of 0.15 to 0.2 m. Radionuclides deposited on the ground can be removed from potential plant uptake by being washed away from the surface by surface water *runoff* or by relocation to a depth beyond the root zone brought about by water *infiltration* or by deep tillage.

Calculation of Soil Concentrations.   For calculation of plant uptake at some location, one must first determine the average concentrations $X_{soil}(t)$ of radioactivity (per unit mass of soil) in the root zone. This concentration is given by a simple balance relation between the soil input rate I and loss rates by decay (λX_{soil}) and by migration out of the root zone ($\lambda_s X_{soil}$), namely,

$$\frac{dX_{soil}(t)}{dt} = I(t) - (\lambda + \lambda_s)X_{soil}(t). \tag{11.34}$$

Thus, for a constant input I_o into the root zone beginning at time $t = 0$ and for a constant infiltration loss rate λ_s, the activity concentration in the soil is

$$X_{soil}(t) = \frac{I_o}{\lambda + \lambda_s}\left(1 - e^{-(\lambda+\lambda_s)t}\right). \tag{11.35}$$

Let the rate at which a given type of radionuclide is deposited on a unit area of the surface at time t be denoted by $\omega_s(t)$. A fraction f_r of this surface deposition will be carried away by rain or irrigation water that runs off the field, and the remainder $(1 - f_r)$ will begin to migrate into the soil. Thus, the rate at which radionuclides enter the soil per unit mass of the soil, and averaged over the soil to depth d of the root zone, is

$$I(t) = \frac{(1 - f_r)\omega_s(t)}{\rho d}, \tag{11.36}$$

where ρ is the soil mass density (typically 1.6 to 2.6 g cm^{-3}). If only direct deposition to the ground from airborne radionuclides is considered, then from the dry and wet deposition models of Section 9.5.3, the rate of direct input to the soil becomes

$$I(t) = (1 - f_r)[(1 - f_{veg})v_d + (1 - f_w)W_vR_p]\frac{X_{air}(t)}{\rho d}. \tag{11.37}$$

This result neglects the subsequent input of radionuclides initially deposited on the vegetation but subsequently weathered or washed off the plants onto the ground. To account for plant washoff, Peterson (1983) suggests multiplying this result by the factor $[1 + k_{rain}R_p/(\lambda + k_{rain}R_p)]$. For long-term studies of soils in which the leafy material is allowed to decay back into the soil, the soil gain rate may be approximated by Eq. (11.37) with the plant retention factors f_{veg} and f_w set to zero.

The fraction f_r of the surface deposition that is washed away by the runoff of surface water depends on how quickly the soil can absorbed rain or irrigation water as well as on the slope of the field. The capacity of soil to absorb water is measured by the *soil permeability* μ_{soil} which is defined as the water volume transmitted per unit surface area per unit time. The permeability varies between 2.5 to 25 cm h^{-1} for sandy soils, 1 to 7.5 cm h^{-1} for loam, and 0.025 to 0.5 cm h^{-1} for silt and clay. Thus, if the rainfall (or irrigation rate) R_p is less than μ_{soil}, there is no runoff, that is, $f_r = 0$. When the waterfall exceeds μ_{soil}, Peterson (1983) suggests $f_r = s(1 - \mu_{soil}/R_p)$, where s is the slope of the field.

Finally, the rate constant for radionuclide loss from the root zone by water infiltration, λ_s, is generally very difficult to estimate (see Sec. 10.8). Often this loss is neglected, thereby giving conservative estimates of soil burdens. However, for highly permeable soils with low adsorption capabilities, this neglect of infiltration loss greatly overestimates the long term radionuclide concentrations in soil and food crops.

11.3.3 Resuspension of Radionuclides

Radionuclides once deposited on the ground may become resuspended in the air as a result of wind or other mechanical action such as vehicular or pedestrian traffic. These resuspended airborne radionuclides can present an inhalation hazard to human populations for many years after the initial event that contaminated the ground.

Consider a large ground area which has been instantly and uniformly contaminated at time $t = 0$ by an amount $X_A(0)$ (Bq m^{-2}). Some of these radioactive particles will be resuspended producing an activity concentration in the air of X_{air} (Bq m^{-3}). Under equilibrium conditions, the rate at which radionuclides are resuspended (assumed to be proportional to X_A) must equal the rate at which they are subsequently redeposited on the ground ($v_d X_{air}$). Thus the airborne radionuclide concentration X_{air} above the ground as a result of resuspended radioactive particles is proportional to the surface ground concentration, that is,

$$X_{air} = K_r X_A \tag{11.38}$$

where the proportionality constant K_r is known as the *resuspension factor* and is seen to have dimensions of inverse length.

Measured values of the resuspension factor K_r are found to vary by many orders of magnitude ranging from 10^{-2} to 10^{-9} m^{-1} (Slinn 1978; Peterson 1983). Generally, K_r increases with the wind speed (Anspaugh et al. 1976; Sehmel and Lloyd 1976) and depends on the size of the particles to which the radionuclides are attached. Moreover, the size distribution of the radioactive particles changes with time as large particles are broken up and small particles become attached to larger particles. Finally, weathering of the soil and the migration of the radionuclides below the surface make them less available for resuspension so that K_r is observed to decrease with the time after the initial contamination event.

Thus, the airborne concentration at some time t after the initial ground contamination, in terms of the initial ground concentration, can be expressed as

$$X_{air}(t) = K_r(t) X_A(0) e^{-\lambda t}. \tag{11.39}$$

One empirical expression for $K_r(t)$ in m^{-1}, with t in units of years, is (USNRC 1975)

$$K_r(t) = 10^{-9} + 10^{-5}\, e^{-0.6769t}. \tag{11.40}$$

Another expression, suitable for short-lived as well as long-lived radionuclides, is (Lassey 1980)

$$K_r(t) = 10^{-9} + 10^{-5}\, e^{-0.6769t} + 9 \times 10^{-5}\, e^{-5.776t}. \tag{11.41}$$

11.3.4 Crop Contamination from the Soil

Besides external contamination from the direct deposition of airborne radionuclides on plant surfaces, plants also become radioactive internally by the incorporation of radionuclides into their tissues. The primary mechanism for internal plant contamination is the uptake of radionuclides from the soil through the plant roots. To a lesser extent, soluble airborne radionuclides can be absorbed through the leaves, stems, flowers (*inflorescence*) and fruits.

The calculation of the amount of radionuclide uptake by plants is usually based on a steady-state description. Transient analyses are seldom attempted because of the difficulty in describing the dynamic migration among the different parts of a plant (*translocation*) and the changing metabolic processes which occur over the growing cycle of a plant. A pseudo steady-state approach is usually adequate since most plants contributing to the human exposure pathway do so at a well defined plant age (e.g., when the fruit is ripe) after growing all their life in soil with a constant, equilibrium activity concentration.

The plant activity concentration can thus be readily estimated by multiplying the radionuclide concentration in the soil by an empirical *plant-to-soil concentration ratio* defined as

$$CR = \frac{\text{activity per unit plant mass}}{\text{activity per unit soil mass}}. \tag{11.42}$$

The mass of the soil is usually expressed on a dry weight basis. However, results based on both dry and wet (fresh) weights for the crops are reported. It is important to distinguish between the wet and dry plant measurements since the fresh-weight to dry-weight ratio, FW/DW, can be quite large (see Table 11.3). For a dry-plant weight basis, the plant-to-soil concentration ratio will be denoted by CR, while for a fresh weight basis the symbol B_v will be used. The relation between these two quantities is

$$B_v = CR\frac{DW}{FW}. \tag{11.43}$$

When one uses a published value for the plant-to-soil concentration ratio of a particular plant species, it is important to understand what part of the plant was used for the activity measurements. Usually, only the edible portions are used; however, values based on the whole plant are also found in the literature and these values can be quite different from values for the edible plant parts since radionuclides are often not uniformly distributed throughout a plant.

TABLE 11.3   Fresh-To-Dry-Weight Ratios, FW/DW, for Food Crops and Forage.

CROP	FW/DW	CROP	FW/DW	CROP	FW/DW
Forage (fresh)		*Forage (dry)*		*Grains*	
alfalfa	4.4	alfalfa	1.1	barley	1.08
clover	5.0	oat	1.1	rice	1.14
grass	5.5	soybean	1.1	wheat	1.15
silage	4.2	wheat	1.1	corn	3.8
Leafy Vegetables		*Root Vegetables*		*Fruits*	
asparagus	12.0	beets	7.9	apples	6.7
broccoli	9.1	carrots	8.5	bananas	4.1
brussel		onions	8.6	cherries	5.1
sprouts	6.8	potato	4.5	cucumbers	20
cabbage	13	radish	18	eggplant	13
cauliflower	12	sweet potato	3.4	grapefruit	8.6
celery	16	turnip	11.8	oranges	7.1
kale	8.0	yam	4.0	peaches	9.2
lettuce	20			pears	6.0
rhubarb	19	*Legumes*		pineapples	6.8
spinach	12.6	green beans	10	plums	5.3
turnip		lima beans	3.1	squash	16.7
greens	10	peas	5.9	tomatoes	15

Source: Peterson 1983.

In Table 11.4 values of CR and B_v are shown for several elements. The CR values are concentration factors for dry feed consumed by livestock. The B_v values are for the fresh edible parts of plants consumed directly by humans. These values are averages of many measurements on different plant species grown in different soils and climates. For detailed analyses, plant-specific values should be used. Ng et al. (1982) present extensive tables that summarize concentration ratio measurements for different elements and plant species.

The uptake of radionuclides by plants can be highly variable as can be seen by the large range of CR and B_v values found in Table 11.4 for some elements. Besides depending on the plant species and the radionuclide involved, the concentration ratio is affected by several other factors as discussed in the following subsections (Peterson 1983).

Physical and Chemical Form.   The chemical and physical form of a radionuclide deposited on the soil can greatly influence its subsequent uptake by vegetation. Chemical species that are tightly bound to the soil (see Sec. 10.7) are less available for uptake than are loosely bound radionuclides. The oxidation state and solubility of the radionuclide also influence plant uptake. Often fallout is in the form of small particles, for example, glass or ceramic spheres, and the radionuclides entrapped in these particles are protected from plant uptake.

Plant Species and Translocation.   Legumes that symbiotically interact with nitrogen-fixing bacteria in the roots often have a higher propensity for radionuclide uptake than do other plants. The solubility of the radionuclide in plant liquids affects the degree to which radionuclides can be carried to different parts of a plant. The

TABLE 11.4 Dry (CR) and Fresh (B_v) Plant-To-Soil Concentration Ratios for Several Elements. Mean Values are Averaged over Multiple Crops and Growing Conditions.

ELEMENT	RATIO	NO. OBS.	RANGE	MEAN	NRC
Na	B_v	6	8.2×10^{-4}–2.6×10^{-2}	9.4×10^{-3}	5.2×10^{-2}
	CR	1		7.4×10^{-2}	2.1×10^{-1}
Cr	B_v	5	5.4×10^{-3}–2.2×10^{-2}	8.1×10^{-3}	2.5×10^{-4}
	CR	19	3.6×10^{-3}–2.5×10^{-2}	2.9×10^{-2}	
Mn	B_v	21	6.9×10^{-3}–3.4	3.6×10^{-1}	2.9×10^{-2}
	CR	20	4.2×10^{-2}–20	2.6	1.2×10^{-1}
Fe	B_v	6	2.4×10^{-4}–6.8×10^{-4}	4.4×10^{-4}	6.6×10^{-4}
	CR	16	6.9×10^{-4}–4.7×10^{-2}	1.2×10^{-2}	2.6×10^{-3}
Co	B_v	24	2.2×10^{-3}–2.0×10^{-1}	3.0×10^{-2}	9.4×10^{-3}
	CR	10	8.0×10^{-3}–1.1	2.4×10^{-1}	3.8×10^{-2}
Ni	B_v	10	7.0×10^{-3}–1.5×10^{-1}	3.3×10^{-2}	1.9×10^{-2}
	CR	15	5.7×10^{-3}–5.5×10^{-1}	7.4×10^{-2}	7.6×10^{-2}
Cu	B_v	11	3.0×10^{-3}–8.2×10^{-1}	2.5×10^{-1}	1.2×10^{-1}
	CR	15	5.1×10^{-2}–1.9	6.7×10^{-1}	
Zn	B_v	31	7.2×10^{-3}–1.6	3.7×10^{-1}	4.0×10^{-1}
	CR	21	1.2×10^{-1}–4.4	9.3×10^{-1}	1.6
Rb	CR	2	3.3×10^{-1}–2.0	1.2	
Sr	B_v	109	1.6×10^{-3}–1.7	1.6×10^{-1}	1.7×10^{-2}
	CR	54	1.2×10^{-1}–23	3.5	6.8×10^{-2}
Zr	B_v	23	3.4×10^{-6}–1.8×10^{-2}	3.2×10^{-3}	1.7×10^{-4}
	CR	2	4.4×10^{-2}–1.2×10^{-1}	8.1×10^{-2}	6.8×10^{-4}
Mo	CR	47	2.3×10^{-2}–38	4.5	
Ru	B_v	36	4.8×10^{-5}–1.4×10^{-1}	6.4×10^{-3}	5.0×10^{-2}
	CR	17	1.8×10^{-3}–5.6×10^{-1}	1.4×10^{-1}	2.0×10^{-1}
I	B_v	16	2.0×10^{-4}–1.2×10^{-1}	1.4×10^{-2}	2.0×10^{-2}
	CR	32	1.5×10^{-2}–1.9	3.1×10^{-1}	8.0×10^{-2}
Cs	B_v	144	1.5×10^{-5}–5.9×10^{-2}	1.1×10^{-2}	1.0×10^{-2}
	CR	42	3.8×10^{-3}–5.7×10^{-1}	1.3×10^{-1}	4.0×10^{-2}
Ba	B_v	217	4.9×10^{-5}–3.7×10^{-2}	2.5×10^{-3}	5.0×10^{-3}
	CR	2	1.8×10^{-2}–8.3×10^{-2}	5.1×10^{-2}	2.0×10^{-2}
Ce	B_v	47	4.6×10^{-6}–1.8×10^{-2}	2.4×10^{-3}	2.5×10^{-3}
	CR	13	2.5×10^{-3}–1.9×10^{-1}	4.5×10^{-2}	1.0×10^{-2}
Np	B_v	2	2.8×10^{-2}–1.3×10^{-1}	7.9×10^{-2}	2.5×10^{-3}
	CR	2	3.7×10^{-1}–3.5	1.9	1.0×10^{-2}

Source: Ng et al. 1982; NRC values for B_v from USNRC 1977 and NRC CR values from Ng et al. 1968 as reported by McDowell-Boyer and Baes 1984.

more soluble the radionuclide, the more easily it can reach the fruit and grain from the roots.

Soil Characteristics.   Different types of soils have widely different adsorption capabilities for radioactive ions. Sandy soils have a much smaller retention capacity than do clays. The soil acidity also strongly affects plant uptake. Under alkaline conditions, radionuclides can form insoluble precipitates, thereby making them less available for uptake by plants. As the soil becomes more acidic, radionuclides return to ionic form and are more easily absorbed by the roots. Under highly acid conditions (pH $<$ 5.5) uptake can be up to a few orders of magnitude greater than that under neutral conditions, so high in fact that uptake of trace elements (principally, iron and manganese) can kill the plant (Peterson 1983).

Interference Effects by Similar Elements.   The presence of a naturally occurring (stable) element with similar chemical properties to those of a particular radionuclide species can interfere with radionuclide uptake. For example, potassium in the soil inhibits the uptake of ^{137}Cs while calcium inhibits the uptake of ^{90}Sr. The stable, chemically similar element apparently saturates receptor sites on special molecules used by the plant to carry the radionuclide across its root membranes into the plant. Thus, plants grown in soils deficient in potassium (or calcium) can exhibit much higher than normal uptakes of ^{137}Cs (or ^{90}Sr).

Use of Agricultural Chemicals.   Fertilizers and other agricultural chemicals spread on a field can change the soil characteristics and the subsequent uptake of radionuclides. The use of lime to increase soil pH also decreases plant uptake of many radionuclides. On the other hand, ammonium compounds decrease soil pH and enhance radionuclide uptake. Organic additives (e.g., manure and peat) also change the soil pH, the ion exchange capacity of the soil, and the stable element content of the soil.

Distribution of Radionuclides in the Soil.   The root structures of most crops are densest near the surface although the root patterns depends on the crop, soil properties and water table level. Except in times of draught, more water is usually taken in by the roots near the surface. Since radionuclides mostly enter the plant with the water, the relative uptake of radionuclides generally decreases with depth into the soil. Moreover, radionuclides are rarely uniformly distributed with depth, especially in soils with strong sorption properties. Thus, radionuclide uptake by field crops can be quite different from uptake in plants grown in shallow test beds that have uniform soil properties and radionuclide concentrations and that are often used to measure concentration ratios.

11.3.5 Radionuclide Transfer to Animal Food Products

Livestock that graze on contaminated pastures or that are given contaminated feed in turn transmit some of the radioactivity to humans when the animal food products are consumed. To describe how radionuclides are redistributed and concentrated in various animal products, it is generally necessary to develop a dynamic model describing the movement of radionuclides among the various organs and body compartments of the animal. Such dynamic models are very similar to those used in internal dose estimation for humans (see Chap. 8), and such models have been developed for important exposure pathways. In later sections, two such dynamic models will be discussed. For many environmental dose analyses, a simpler steady-state procedure can be used. In this steady-state method, empirical concentration factors are used to determine radionuclide concentrations in animal food products from the concentrations in the animal feed. To gain a better understanding of the capabilities and limitations of such a steady-state method, the method will be derived from a general dynamic model for animals.

A General Transient Model.　A dynamic model of the retention and elimination of a radionuclide by a particular animal species is usually based on a set of differential equations describing the transient exchange of radioactivity among an appropriate number of model compartments. These compartments may represent actual organs, parts of organs, various body fluids and biophysical systems, and even different physical and chemical forms of the radionuclide. Many such models have been developed and values of the model parameters have been determined by tracer studies (see, for example, Sheppard 1962).

The vast majority of these models are expressed in the form of a set of linear differential equations with constant first-order transfer coefficients (or parameters). With such an n-compartment animal model, the radionuclide concentration in the ith compartment is given by

$$\frac{dX_i(t)}{dt} = \sum_{\substack{j=1 \\ j \neq i}}^{n} \lambda_{ij} X_j(t) - X_i(t) \sum_{j=1}^{n} \lambda_{ji}, \qquad i = 1, \ldots, n \qquad (11.44)$$

where λ_{ij} is a model parameter for transfer from compartment j to compartment i, and $\lambda_{ii} = \lambda$ (the decay constant for the radionuclide). Here we are assuming that the X_i all have the same units (e.g., Bq or Bq cm^{-3}). When, for example X_i and X_j have different units, the single constant λ_{ij} cannot be used to represent both a loss from compartment j and a gain for compartment i. An example transient model for the forage-cow-milk environmental pathway is described later in this chapter. In that model X_i may represent activity per unit area of grass or soil, and X_j total activity in one of the compartments of the cow such as the gut or the thyroid.

The general solution of these coupled equations for an instantaneous ingestion of Q' Bq of a radionuclide at time $t = 0$ is (see Appendix F, Sec. F-1.4)

$$X_i(t) = Q' e^{-\lambda t} \sum_{j=1}^{n} \alpha_{ij} e^{-r_j t} \qquad (11.45)$$

where α_{ij} and r_j are constants determined from the model parameters λ_{ij} and the initial conditions.

For a continuous input $Q(t)$ (Bq s^{-1}) to the animal beginning at time $t = 0$, the radionuclide concentration is obtained by convolution of $Q(t)$ with the result of Eq. (11.45), that is,

$$X_i(t) = \sum_{j=1}^{n} \alpha_{ij} \int_0^t d\tau \, Q(\tau) e^{-(\lambda + r_j)(t - \tau)}. \qquad (11.46)$$

For a constant ingestion rate Q_o, this result reduces to

$$X_i(t) = Q_o \sum_{j=1}^{n} \frac{\alpha_{ij}}{\lambda + r_j} \left[1 - e^{-(\lambda + r_j)t} \right]. \qquad (11.47)$$

Steady-State Results.　To use this expression for calculating the activity concentration in the ith animal compartment (food product), the model parameters must, of course, be known and the constants α_{ij} and r_j determined. Often the model parameters are known only approximately, and finding the constants α_{ij} and r_j is tedious, especially if the model has a large number of compartments. Fortunately, calculation of compartment concentrations can be considerably simplified for steady-state (equilibrium) situations which are often encountered in environmental analyses.

After a sufficiently long time the concentrations in each compartment, as given by Eq. (11.47), reach the constant equilibrium values

$$X_i^\infty = Q_o \sum_{j=1}^{n} \frac{\alpha_{ij}}{\lambda + r_j}. \tag{11.48}$$

The equilibrium value for a compartment that represents a food product (e.g., milk, eggs, or meat) for a unit intake rate is called the *intake-to-food-product transfer factor* F_i, and, for this general model, is given by

$$F_i \equiv \frac{X_i^\infty}{Q_o} = \sum_{j=1}^{n} \frac{\alpha_{ij}}{\lambda + r_j}. \tag{11.49}$$

If the transfer factor for a particular food product is known, then, under equilibrium conditions, the activity concentration in a food product is obtained simply by multiplying the animal's rate of radionuclide intake by the transfer factor.

The transfer factors could be estimated from Eq. (11.49) using values of the model parameters (if they were known). However, for many analyses only the values of F_i are needed, and, consequently, there has been considerable experimental effort to measure directly the transfer factors to milk, eggs, and meat. In Table 11.5 transfer factors for transfer to cow's milk are given, and in Table 11.6 transfer factors for meat and eggs are given. A recent evaluation of the many experimental studies made of transfer factors for important radionuclides is given by Coughtrey (1990). It should be noted that most of these transfer factors were determined by measurements using stable isotopes. The corresponding transfer factor for a radioisotope can be obtained approximately by multiplying the tabulated value by $(1 + \lambda/r)^{-1}$, where r is the effective biological excretion rate into the food product, as discussed in Sec. 11.4.2.

The use of transfer factors in steady-state analyses of the consequences of steady long-term discharges from nuclear facilities, greatly simplifies the human food-chain exposure calculations since specific dynamic models for animals in the food chain are not needed. Thus, many regulatory guidelines specifying procedures for environmental analyses are based on steady-state transfer factors with transient decay corrections made only for delays between different compartments in the exposure pathway (US-NRC 1977).

Short-Term Contamination Analysis.   As an example of the use of animal models for transient conditions consider the case in which a pasture is contaminated at $t = 0$ by an instantaneous release of airborne radioactivity from a nearby nuclear facility. The resulting activity concentration (Bq kg^{-1}) on the surface of the pasture grass $X_{veg}(t)$ is given by Eq. (11.28). If animals grazing on this pasture consume this

TABLE 11.5 Transfer Factors F_m for Intake to Cows' Milk.

ELEMENT	F_m (d L^{-1})	ELEMENT	F_m (d L^{-1})
H^3	1.4×10^{-2}	Sb (SbCl$_3$)	2.0×10^{-5}
C	1.5×10^{-2}	Te	2.0×10^{-4}
Na	3.5×10^{-2}	I	9.9×10^{-3}
P	1.6×10^{-2}	Cs	7.1×10^{-3}
K	7.2×10^{-3}	rare earths	2.0×10^{-5}
Ca	1.1×10^{-2}	Ta (oxalate)	2.8×10^{-6}
Cr	2.0×10^{-3}	W (Na$_2$WO$_4$)	2.9×10^{-4}
Mn	8.4×10^{-5}	Re (NaReO$_4$)	1.3×10^{-3}
Fe	5.9×10^{-5}	Os, Pt, Ir, Au	5.0×10^{-6}
Co	2.0×10^{-3}	Hg (HgCl)	9.7×10^{-6}
Ni	1.0×10^{-2}	Tl (Tl(NO$_3$)$_2$)	1.3×10^{-3}
Cu	1.7×10^{-3}	Pb	2.6×10^{-4}
Zn	1.0×10^{-2}	Bi	5.0×10^{-4}
Br	2.0×10^{-2}	Po	1.4×10^{-2}
Kr	2.0×10^{-2}	Rn	3.0×10^{-2}
Rb	1.2×10^{-2}	Ra	4.5×10^{-4}
Sr	1.4×10^{-3}	Ac	2.0×10^{-5}
Y	2.0×10^{-5}	Th	5.0×10^{-6}
Zr	8.0×10^{-2}	Pa	5.0×10^{-6}
Nb	2.0×10^{-2}	U	6.1×10^{-4}
Mo	1.4×10^{-3}	Np	5.0×10^{-6}
Tc	9.9×10^{-3}	Pu (PuO$_2$)	2.7×10^{-9}
Ru (chloride or NO complex)	6.1×10^{-7}	transuranics	2.0×10^{-5}

Source: Peterson 1983; most values from Ng et al. 1977.

grass at a constant rate of I (kg d^{-1}), then the rate of activity intake $Q(t)$ by each animal is

$$Q(t) = IX_{veg}(0)e^{-(\lambda+\lambda_r)t}. \tag{11.50}$$

Substitution of this intake rate into Eq. (11.46) yields

$$X_i(t) = IX_{veg}(0)e^{-\lambda t} \sum_{j=1}^{n} \frac{\alpha_{ij}}{\lambda_r - r_j} \left(e^{-r_j t} - e^{-\lambda_r t}\right). \tag{11.51}$$

If a person were to consume a fresh product such as milk from these animals at some constant rate, the so-called *usage rate* U (kg d^{-1}), the total activity ingested by that person over all future times would be

$$\text{Total Ingestion} = U \int_0^\infty dt\, X_i(t) = \frac{UIX_{veg}(0)}{\lambda_r + \lambda} \sum_{j=1}^{n} \frac{\alpha_{ij}}{\lambda + r_j}. \tag{11.52}$$

This total ingested activity is needed for internal dose calculations. The total activity consumed by an animal for all times following the release is

$$I_{tot} = I \int_0^\infty dt\, X_{veg}(t) = \frac{IX_{veg}(0)}{\lambda_r + \lambda}. \tag{11.53}$$

TABLE 11.6 Transfer Factors F_f for Meat and Eggs.

ELEMENT	F_f (d kg^{-1}) BEEF	PORK	LAMB	CHICKEN	EGG
Na	8.3×10^{-2}				6.1
P	4.9×10^{-2}				
K	1.8×10^{-2}				1.1
Ca	1.6×10^{-3}			4.4×10^{-2}	0.44
Cr	9.2×10^{-3}				
Mn	5×10^{-4}	3.6×10^{-3}	5.9×10^{-3}	5.1×10^{-2}	0.065
Fe	2.1×10^{-2}	2.6×10^{-2}	7.2×10^{-2}	1.5	1.3
Co	1.2×10^{-2}	1.7×10^{-1}	6.2×10^{-2}		
Ni	2×10^{-3}				
Cu	9.0×10^{-3}	2.2×10^{-2}	3.9×10^{-2}	5.1×10^{-1}	0.49
Zn	9.8×10^{-2}	1.5×10^{-1}	4.1	6.5	2.6
Rb	1.1×10^{-2}				
Sr	8.1×10^{-4}	3.9×10^{-2}	2.2×10^{-3}	3.5×10^{-2}	0.3
Y	1×10^{-3}			1×10^{-2}	2×10^{-3}
Zr	2×10^{-2}				
Nb	2×10^{-3}			2×10^{-3}	3×10^{-3}
Mo	6.8×10^{-3}			5×10^{-2}	0.5
Tc	8.7×10^{-3}				
Ru	2.0×10^{-3}			7×10^{-3}	6×10^{-3}
Sb	9.2×10^{-4}				
Te	1.5×10^{-2}			2.7×10^{-1}	0.8
I	3.6×10^{-3}	3.3×10^{-3}	3.0×10^{-2}		2.8
Cs	2.6×10^{-2}	2.5×10^{-1}	2.9×10^{-1}	4.4	0.49
Ba	9.7×10^{-5}			2×10^{-2}	0.9
Ce	2×10^{-3}			1×10^{-2}	5×10^{-3}
Pr				3×10^{-2}	5×10^{-3}
Nd				9×10^{-2}	3×10^{-4}
Pm				2×10^{-3}	0.02
W	3.7×10^{-2}				
Hg				2.7×10^{-2}	
Pb	4×10^{-4}				
Po	4.5×10^{-3}				
Ra	9×10^{-4}				
Th	2.0×10^{-4}				
U	3.4×10^{-4}	4×10^{-2}		1.2	0.99
PuO$_2$	2×10^{-6}			1.9×10^{-5}	3.3×10^{-5}

Source: Values from Ng et al. 1982 and supplemented by values from Peterson 1983.

From the definition of the intake-to-food-product transfer factor F_i in Eq. (11.49), the total human ingestion is

$$\text{Total Ingestion} = U I_{tot} F_i. \tag{11.54}$$

Finally, this simple accident analysis can be refined to account for depletion of pasture grass by the grazing animals, uptake of radionuclides from the soil by the grass, and radioactive decay between the time the food product leaves the animal and its consumption by people.

Long-Term Contamination Analysis. Of considerable concern in environmental analyses for nuclear facilities is the low-level chronic contamination of pastures and food crops. Consider a facility which begins to release airborne radioactivity at $t = 0$ some of which is deposited on a pasture some distance away. The resulting external contamination $X_{veg}(t)$ on the pasture grass for $t > 0$ is given by Eq. (11.26). The intake rate $Q(t)$ by animals grazing on the pasture is $IX_{veg}(t)$ where I (kg d^{-1}) is the forage rate for an animal. Then from Eq. (11.46), the activity concentration in the ith animal compartment is found to be

$$X_i(t) = \frac{I \, V_{veg} \, X_{air}}{\lambda + \lambda_r} \sum_{j=1}^{n} \alpha_{ij} \left[\frac{1 - e^{-(\lambda + r_j)t}}{\lambda + r_j} - \frac{e^{-(\lambda + r_j)t} - e^{-(\lambda + \lambda_r)t}}{\lambda_r - r_j} \right]. \quad (11.55)$$

After a long time the compartment concentration reaches an equilibrium value which, from the above result, is seen to be proportional to the transfer factor for the compartment, that is,

$$X_i^{\infty} = \frac{I \, V_{veg} \, X_{air}}{\lambda + \lambda_r} \sum_{j=1}^{n} \frac{\alpha_{ij}}{\lambda + r_j} = \frac{I \, V_{veg} \, X_{air}}{\lambda + \lambda_r} F_i, \quad (11.56)$$

or in terms of the equilibrium pasture concentration $X_{veg}^{\infty} = V_{veg} X_{air}/(\lambda + \lambda_r)$ from Eq. (11.26),

$$X_i^{\infty} = X_{veg}^{\infty} I F_i. \quad (11.57)$$

From this result, another interpretation of the transfer factor F_i is seen to be the ratio of the equilibrium food product concentration (per unit animal input rate) to the equilibrium concentration in the pasture vegetation.

Variations of this steady-state analysis to account for plant uptake of radionuclides in the soil, vegetation depletion, seasonal vegetation and grazing patterns, use of contaminated fodder for the animals, and radionuclide decay effects can be made (USNRC 1977; McDowell-Boyer and Baes 1984). However, the key to all these equilibrium models is the use of appropriate intake-to-food-product transfer factors.

11.3.6 The Forage-Cow-Milk Environmental Pathway

Steady-state exposure models are not suitable for some important terrestrial food chains that make large contributions to human exposure. This is especially true for the analysis of chains involving relatively short-lived radionuclides emitted sporadically. One such pathway is the forage-cow-milk pathway which is an exceedingly important route for passing radioiodine to humans, especially infants and children. For this and similar pathways, specific transient models must be used. In this section a transient model specific to the forage-cow-milk pathway is described. In the next section a more general terrestrial food-chain example is discussed.

In Fig. 11.4 a transient model (Garner 1967; Booth et al. 1971) is shown for the transfer of radioiodine from the atmosphere to man via fallout on pasture and uptake by the grazing cow. There are a number of compartments in this model and it is assumed that the rate of transfer of elemental iodine from one compartment to another is described by a first-order rate constant. Let λ_{ij} represent, in general, the fraction of the activity in compartment j transferred to compartment i per unit differential time. Similarly, let $\lambda_j \equiv \sum_i \lambda_{ij}$ represent the total rate of transfer of the element from compartment j. Let λ represent the radiological rate constant for decay of the radioiodine. Table 11.7 lists representative values of the rate constants for the model.

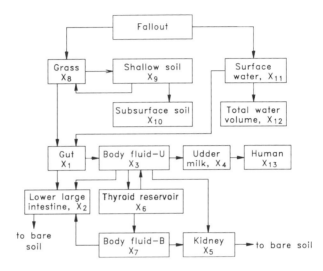

Figure 11.4   Diagram of a model for the movement of radioiodine in the forage-cow-milk food chain. [After Garner (1967), Booth et al. (1971).]

In Fig. 11.4, for $i = 1$ through 7, X_i represents the activity in compartment i. For $i = 8$ through 10, X_i represents the activity in compartment i per unit surface area. Compartments 11 and 12 are not considered in this discussion. The specific activity of radioiodine in milk is given by X_4 divided by the 11.5 liter udder volume.

Transfer from Atmosphere to Grass.   If $X_{air}(t)$ is the concentration of radioiodine in the atmosphere, the rate of transfer to grass is given by $v_g R_g X_{air}(t)$, where v_g is the grass deposition velocity and R_g is the short-term retention factor for iodine on grass. The deposition speed v_g ranges from 0.01 to 0.03 m s^{-1} with a representative value of 0.015 m s^{-1}, and R_g ranges from 0.07 to 0.85 with a representative value of 0.3 (USNRC 1974). The activity per unit area of pasture is given by the solution of

$$\frac{dX_8(t)}{dt} = v_g R_g X_{air}(t) - (\lambda + \lambda_{98} + \frac{V_g}{A_g Y_g})X_8(t) + \lambda_{89}X_9(t). \quad (11.58)$$

The rate constant λ_{98} is also called the *weathering rate constant*, and, for radioactive iodine, the prescribed value for use in radiological-impact analysis is 0.053 d^{-1} (USNRC 1974). The rate of loss of activity from pasture by grazing is not given specif-

TABLE 11.7   Representative Rate Constants and Other Data for Use in
the Forage-Cow-Milk Environmental Pathway Analysis.

PARAMETER	DESCRIPTION	VALUE
Y_g	dry-weight grass density	0.15 kg m^{-2}
V_g	consumption rate of dry grass per cow	10 kg d^{-1}
A_g	pasture area per cow	10^4 m^2
C_{air}	ground-level radionuclide conc.	$\text{Bq m}^{-3} \text{ or Ci m}^{-3}$
v_g	deposition velocity	0.015 m s^{-1}
R_g	short-term retention factor	0.3
λ_{21}	rate constant comp. 1 to comp. 2	0.43 d^{-1}
λ_{31}	rate constant comp. 1 to comp. 3	0.95 d^{-1}
λ_{92}	rate constant comp. 2 to comp. 9	$0.001 \text{ d}^{-1} \text{ m}^{-2}$
λ_{23}	rate constant comp. 3 to comp. 2	0.52 d^{-1}
λ_{43}	rate constant comp. 3 to comp. 4	0.20 d^{-1}
λ_{53}	rate constant comp. 3 to comp. 5	0.37 d^{-1}
λ_{63}	rate constant comp. 3 to comp. 6	0.28 d^{-1}
$\lambda_{13,4}$	rate constant comp. 4 to comp. 13	2.0 d^{-1}
λ_{95}	rate constant comp. 5 to comp. 9	$0.0003 \text{ d}^{-1} \text{ m}^{-2}$
λ_{36}	rate constant comp. 6 to comp. 3	0.07 d^{-1}
λ_{76}	rate constant comp. 6 to comp. 7	0.04 d^{-1}
λ_{27}	rate constant comp. 7 to comp. 2	1.02 d^{-1}
λ_{57}	rate constant comp. 7 to comp. 5	0.03 d^{-1}
λ_{98}	rate constant comp. 8 to comp. 9	0.053 d^{-1}
λ_{89}	rate constant comp. 9 to comp. 8	0.00003 d^{-1}
$\lambda_{10,9}$	rate constant comp. 9 to comp. 10	0.0001 d^{-1}
λ	decay constant for ^{131}I	0.086 d^{-1}
λ	decay constant for ^{133}I	0.81 d^{-1}
	milk capacity of udder	12.5 L
	milk produced per cow	25 L d^{-1}

Source: Drbl 1975.

ically in Table 11.7. Instead, it is calculated as $X_8(t)V_g/Y_g$, the activity per unit time transferred to the grazing animal, divided by A_g, the pasture area over which the animal grazes. Similar equations would have to be written for all other compartments, and then solved simultaneously to yield $X_8(t)$. However, because feedback of activity from shallow soil to pasture is relatively unimportant, it is possible to obtain an approximate value for X_8. From data in Table 11.7, it is apparent that V_g/A_gY_g is much less than λ_{98}; that is, weathering loss is more rapid than grazing loss. Thus, to a first approximation,

$$\frac{dX_8(t)}{dt} \cong v_g R_g X_{air}(t) - (\lambda + \lambda_{98})X_8(t). \qquad (11.59)$$

Under steady-state conditions,

$$X_8 \cong \frac{X_{air} v_g R_g}{\lambda_{98} + \lambda}. \qquad (11.60)$$

Under transient conditions, with a given initial value $X_8(0)$,

$$X_8(t) \cong X_8(0)e^{-(\lambda+\lambda_{98})t}. \tag{11.61}$$

Transfer from Grass to Milk.   The determination of the activity in milk requires simultaneous solution of differential equations of the type of Eq. (11.58) for all compartments of Fig. 11.4 (except here neglecting compartments 11 and 12). The equation for compartment 1, a special case, is given by

$$\frac{dX_1(t)}{dt} = -(\lambda + \lambda_{21} + \lambda_{31})X_1(t) + \frac{V_g}{Y_g}X_8(t). \tag{11.62}$$

Figure 11.5 illustrates the solution of this set of equations, giving the specific activity in milk (Bq L^{-1}) as a function of time after instantaneous surface deposition of 1 Bq m^{-2} of ^{131}I or ^{133}I. Under steady-state conditions, the specific activity in milk is directly proportional to the activity per unit area on pasture. For ^{131}I and ^{133}I respectively, the ratios are approximately 0.2 and 0.08 Bq L^{-1} per Bq m^{-2}, similar to the values 0.2 and 0.09 in (USNRC 1974).

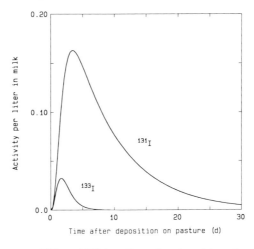

Figure 11.5   Specific activities of ^{131}I and ^{133}I in milk as a function of time after deposition of unit activity per unit area of pasture. [Data are from Drbal (1875).]

Table 11.8 illustrates peak concentrations and time to reach the peak not only for radioiodine in milk but also ^{137}Cs, ^{89}Sr, and ^{90}Sr in meat resulting from instantaneous deposition of radionuclides on pasture. Table 11.9 illustrates steady-state concentrations in milk and meat resulting from continuous fallout of radionuclides on pasture.

Radioiodine Dose Conversion Factors.   For use in assessment of long-term, or annual-average, consequences of routine releases of radioiodine, the U.S. Nuclear Regulatory Commission (USNRC 1974) has provided a set of conversion factors relating annual dose equivalents to annual release rates and annual-average dilution factors for various radioiodine pathways. Specifically, for the release rate Q (Ci y^{-1}),

TABLE 11.8   Predicted Results for a 1 μCi cm^{-3} Source Which Exists for 1 Hour at 1 Meter above a Pasture[a]. In Effect, $Q_8(0) = 15.1$ Ci m^{-2}.

RADIO- NUCLIDE	MAXIMUM CONCENTRATION		TIME FOR MAXIMUM CONCENTRATION	
	MILK (μCi L^{-1})	MEAT (μCi kg^{-1})	MILK (days)	MEAT (days)
^{131}I	25×10^5		3.5	
^{133}I	6×10^5		1.5	
^{137}Cs	25×10^5	31×10^5	7.5	14.0
^{89}Sr	2×10^5	4×10^5	5.0	10.0
^{90}Sr	2×10^5	5×10^5	6.0	11.5

RADIO- NUCLIDE	TOTAL CUMULATIVE INTAKE BY MAN		TIME FOR 90% OF TOTAL CUMULATIVE INTAKE BY MAN	
	MILK (μCI)	MEAT (μCI)	MILK (days)	MEAT (days)
^{131}I	8.1×10^6		16.5	
^{133}I	4.5×10^5		4.0	
^{137}Cs	1.9×10^7	3.0×10^7	43.0	49.0
^{89}Sr	1.3×10^6	3.6×10^6	39.0	46.0
^{90}Sr	1.9×10^6	5.1×10^6	55.0	63.0

[a]Based on $v_g R_g = 0.0042$ m s^{-1} and $Y_g = 0.15$ kg m^{-2}.
[b]Based on consumption of 0.3 L d^{-1} milk and 0.3 kg d^{-1} of beef.
Source: Booth 1971.

TABLE 11.9   Predicted Steady-State Concentrations in Milk and Beef for a Continuous Deposition of 1 μCi m^{-2} d^{-1} on Pasture[a].

RADIONUCLIDE	MILK (μCi L^{-1})	MEAT (μCi kg^{-1})
^{131}I	1.7	
^{133}I	0.1	
^{137}Cs	3.1	7.1
^{89}Sr	0.4	0.8
^{90}Sr	0.4	1.2

[a]Based on $v_g R_g = 0.0042$ m s^{-1} and $Y_g = 0.15$ kg m^{-2}.
Source: Booth 1971.

the dilution factor X_{air}/Q (s m^{-3}), and the correction factor ξ for cloud depletion (see Section 9.5.4), the annual dose equivalent H (mrem y^{-1}) is given by

TABLE 11.10 Concentration-To-Dose Conversion Factors $\mathcal{F}$ {mrem m^3 Ci^{-1} s^{-1}} for Radioiodine Environmental Pathways.

PATHWAY	^{131}I	^{133}I
Infant thyroid dose via cows milk	1.15×10^8	2.12×10^6
Infant thyroid dose via inhalation	4.8×10^5	1.2×10^5
Adult thyroid dose via inhalation	4.0×10^5	9.8×10^4
Adult thyroid dose via leafy vegetables	2.1×10^6	8.4×10^4

Source: USNRC 1974.

$$H = \mathcal{F}\xi Q(X_{air}/Q), \tag{11.63}$$

where values of the factor $\mathcal{F}$ are given in Table 11.10. In evaluating the factors, the following assumptions were made in addition to those previously described. Milk consumption for the infant was assumed to be one liter per day. The annual yield of leafy garden vegetables was assumed to be 2.34 kg m^{-2} and the adult annual consumption of leafy vegetables was assumed to be 18 kg.

11.3.7 A Terrestrial Exposure Pathway Model

A more comprehensive terrestrial pathway model (ICRP 1979) that includes not only the forage-cow-milk pathway but also direct inhalation, contaminated vegetation intake, and contaminated beef intake is shown in Fig. 11.6(a). Fruits, vegetables, and grains grown for human consumption are lumped into two compartments depending upon whether the edible portion is above or below ground. Forage grown for cattle consumption is depicted by the pasture-grass compartment. Cattle are represented by two compartments that yield milk and beef for human consumption. Other animal protein such as pork or poultry are subsumed by the beef compartment. Radionuclide transfers to both adults and children are computed using the lung/GI-tract submodel shown in Fig. 11.6(b). The radionuclide content in each compartment is denoted by X_i where the subscript is the compartment number. For the lung/GI-tract submodel a second subscript is used, namely X_{δ_i} where $\delta = a$ for an adult and $\delta = c$ for a child. The purpose of the model is to compute these compartmental concentrations as a function of time throughout the year.

In Fig. 11.6 transfer coefficients are denoted by λ_{ij} for transfer from compartment j to compartment i. Quantities with a single subscript are either a compartment concentration such as X_i or a specific input to a submodel or compartment. Also, conventionally recognized quantities are shown explicitly, namely, deposition velocities v, ingestion rates E, U, B, and M, inhalation rates I, and lung retention factors Δ_δ.

(a)

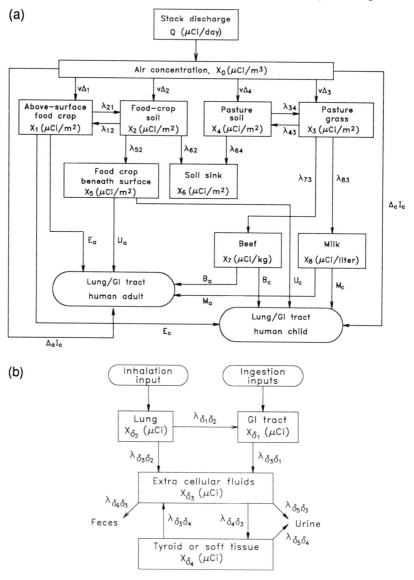

(b)

Figure 11.6 (a) Conceptual compartmentalization of a terrestrial foodchain transient models for use with the system analysis (SA) method. (b) The associated human compartmental model used for both an adult and a child. [From ICRP (1979).]

The differential equations that describe the dynamic behavior of this food-chain model are as follows.

$$\frac{dX_1}{dt} = \Delta_1 v X_o + \lambda_{12} X_2 - [\lambda + \lambda_{21} + \frac{V_h}{A_c Y_c}]X_1 \qquad (11.64)$$

$$\frac{dX_2}{dt} = \Delta_2 v X_o + \lambda_{21} X_1 - [\lambda + \lambda_{12} + \lambda_{52} + \lambda_{62}] X_2 \tag{11.65}$$

$$\frac{dX_3}{dt} = \Delta_3 v X_o + \lambda_{34} X_4 - [\lambda + \lambda_{43} + \frac{V_c}{A_g Y_g}] X_3 \tag{11.66}$$

$$\frac{dX_4}{dt} = \Delta_4 v X_o + \lambda_{43} X_3 - [\lambda + \lambda_{34} + \lambda_{64}] X_4 \tag{11.67}$$

$$\frac{dX_5}{dt} = \lambda_{52} X_2 - [\lambda + \frac{V_h}{A_c Y_c}] X_5 \tag{11.68}$$

$$\frac{dX_6}{dt} = \lambda_{62} X_2 + \lambda_{64} X_4 - \lambda X_6 \tag{11.69}$$

$$\frac{dX_7}{dt} = \lambda_{73} X_3 - [\lambda + \lambda_b] X_7, \text{ with } \lambda_{73} = \frac{f_b}{M_b} \frac{V_c}{Y_g} \tag{11.70}$$

$$\frac{dX_8}{dt} = \lambda_{83} X_3 - [\lambda + \lambda_m] X_8, \text{ with } \lambda_{83} = \frac{f_m}{L} \frac{V_c}{Y_g} \tag{11.71}$$

$$\frac{dX_{\delta_1}}{dt} = E_\delta X_1 + U_\delta X_5 + B_\delta X_7 + M_\delta X_8 + \lambda_{\delta_1 \delta_2} X_{\delta_2} - (\lambda + \lambda_{\delta_3 \delta_1}) X_{\delta_1} \tag{11.72}$$

$$\frac{dX_{\delta_2}}{dt} = \Delta_\delta I_\delta X_o - (\lambda + \lambda_{\delta_1 \delta_2} + \lambda_{\delta_3 \delta_2}) X_{\delta_2} \tag{11.73}$$

$$\frac{dX_{\delta_3}}{dt} = \lambda_{\delta_3 \delta_1} X_{\delta_1} + \lambda_{\delta_3 \delta_2} X_{\delta_2} + \lambda_{\delta_3 \delta_4} X_{\delta_4} - (\lambda + \lambda_{\delta_4 \delta_3} + \lambda_{\delta_5 \delta_3} + \lambda_{\delta_6 \delta_3}) X_{\delta_3} \tag{11.74}$$

$$\frac{dX_{\delta_4}}{dt} = \lambda_{\delta_4 \delta_3} X_{\delta_3} - (\lambda + \lambda_{\delta_3 \delta_4} + \lambda_{\delta_5 \delta_4}) X_{\delta_4} \tag{11.75}$$

Two sets of equations are obtained from Eqs. (11.72) to (11.75) by setting $\delta = a$ for the adult and $\delta = c$ for the child. Thus, this terrestrial model has a total of 16 compartments. Values for the many model parameters are listed in Tables 11.11 to 11.13. It should be noted that these parameter values do not represent officially sanctioned values, but they are realistic for many situations.

As an example of the use of this model, consider a steady release of ^{131}I from a nuclear facility at the rate of 1 Ci per year. For the assumed atmospheric conditions in this example, the release produces an airborne ground level concentration X_o of 0.27 pCi m^{-3} at a pasture 1 km from the facility upon which cattle graze and food and forage crops are grown. The release is begun at the beginning of a 120-day growing season so that during the remainder of the year all radionuclides entering compartments 1 and 3 go directly to compartments 2 and 4, respectively.

Numerical solutions of these equations for the first year are shown in Fig. 11.7(a) for the food-chain compartments and in Fig. 11.7(b) for the compartments of the lung/GI-tract submodel. Integration of these transients over the year has been found to yield total compartmental activities very close to those obtained using a concentration factor or steady-state analysis approach (ICRP 1979). However, this system analysis approach also yields detailed transient information that cannot be obtained by the concentration factor method.

TABLE 11.11 Age Dependent Consumption Parameters for the Terrestrial Food-Chain Model of Fig. 11.6.

PARAMETER	AGE	NOMINAL VALUE	OBSERVED RANGE
Δ_c	child	0.63	
Δ_a	adult	0.63	
I_c	child	$5.7 \text{ m}^3 \text{ d}^{-1}$	$(2.9–11) \text{ m}^3 \text{ d}^{-1}$
I_a	adult	$23 \text{ m}^3 \text{ d}^{-1}$	$(12–46) \text{ m}^3 \text{ d}^{-1}$
E_c	child	$0.03 \text{ kg(wet) d}^{-1a}$	$(0–0.04) \text{ kg(dry) d}^{-1a}$
E_a	adult	$0.10 \text{ kg(wet) d}^{-1a}$	$(0.05–0.16) \text{ kg(dry) d}^{-1a}$
U_c	child	$0.032 \text{ kg(dry) d}^{-1b}$	$(0–0.064) \text{ kg(dry) d}^{-1b}$
U_a	adult	$0.19 \text{ kg(dry) d}^{-1b}$	$(0.1–0.38) \text{ kg(dry) d}^{-1b}$
B_c	child	$0.09 \text{ kg(wet) d}^{-1}$	$(0–0.18) \text{ kg(wet) d}^{-1}$
B_a	adult	$0.28 \text{ kg(wet) d}^{-1}$	$(0.14–0.56) \text{ kg(wet) d}^{-1}$
M_c	child	0.82 L d^{-1}	$(0.41–1.6) \text{ L d}^{-1}$
M_a	adult	0.36 L d^{-1}	$(0.18–0.72) \text{ L d}^{-1}$

[a]To convert to $\text{m}^2 \text{ d}^{-1}$ multiply by 0.7 and to convert to dry mass multiply by 0.07.

[b]To convert to $\text{m}^2 \text{ d}^{-1}$ multiply by a factor of 10.

Source: ICRP 1979.

11.4 AQUATIC AND MARINE EXPOSURE PATHWAYS

Compared to terrestrial food-chain exposure pathways, the analysis of pathways in aquatic (freshwater), marine (saltwater), and estuarine (brackish water) environments is considerably more difficult. There are usually more steps in the food chains leading to humans in aquatic and marine pathways than in terrestrial pathways. Typically, terrestrial pathways consist of only two or three steps (e.g., vegetation → herbivore → human). By contrast, exposure pathways in water ecosystems can be very long with multiple branches (e.g., stream → lake → sediment → algae → zooplankton → crustacean → small fish → predator fish → human). Moreover, predator organisms often consume a variety of different prey from different levels of the food chain, and this food web can change seasonally as the populations and developmental stages of water organisms vary. Also, unlike components of terrestrial pathways, many members of water food chains can move considerable distances and thereby experience large variations in environment and in intake of radionuclides. Two extreme examples are migratory waterfowl and salmon both of which travel many thousands of kilometers in their lifetimes.

The availability of radionuclides in water environments is also usually more difficult to predict than for terrestrial pathways. The physical and chemical form of a radionuclide is usually more variable in a water environment and, consequently, the potential for radionuclide transfer to organisms can vary widely. Adsorption and desorption of radionuclides on suspended and bedded sediment can alter dramatically the dissolved radionuclide concentrations (see Sec. 10.7).

TABLE 11.12 Generally Applied Transfer Parameters for the Terrestrial Food-Chain Model of Fig. 11.6.

PARAMETER	NOMINAL VALUE[a]	OBSERVED RANGE
λ_{12}	6×10^{-9} d^{-1} [0]	$(8 \times 10^{-11}$–$5 \times 10^{-7})$ d^{-1}
λ_{52}	6×10^{-9} d^{-1} [0]	$(8 \times 10^{-11}$–$5 \times 10^{-7})$ d^{-1}
λ_{62}	10^{-4} d^{-1}	$(0$–$10^{-4})$ d^{-1}
V_h/Y_c	1.25 m^2 d^{-1}	$(0.15$–$4)$ m^2 d^{-1}
V_c/Y_g	100 m^2 d^{-1}	$(50$–$160)$ m^2 d^{-1}
f_m/L	0.8×10^{-2} L^{-1}	$(0.4 - 2.2) \times 10^{-2}$ L^{-1}
λ_{34}	8×10^{-3} d^{-1} [0]	$(2 \times 10^{-5}$–$8 \times 10^{-1})$ d^{-1}
λ_{64}	10^{-4} d^{-1}	$(0$–$10^{-4})$ d^{-1}
Δ_1	0.90 [0]	$(0.1$–$1)$
Δ_2	0.10 [1]	$(0$–$0.9)$
Δ_3	0.90 [0]	$(0.1$–$1)$
Δ_4	0.10 [1]	$(0$–$0.9)$
Ac	10^3 m^2	
Ag	10^4 m^2	
λ_b	0.00381 d^{-1}	
λ_m	2.0 d^{-1}	
$\lambda_{\delta_3\delta_1}$	17 d^{-1}	
$\lambda_{\delta_1\delta_2}$	0.88 d^{-1}	
$\lambda_{\delta_3\delta_2}$	1.9 d^{-1}	

[a] Value in brackets denotes value for dormant seasons.

Source: ICRP 1979.

Two other important mechanisms that affect radionuclide transfer to aquatic organisms are the mechanisms of colloid formation and co-precipitation (Peterson 1983).

Colloid Formation.   Some radionuclides tend to adhere or become trapped in colloids which are suspensions of very fine, typically 0.005 to 0.2 μm, particles of insoluble compounds such as amorphous assemblies of hydrated oxides. Such entrapment reconcentrates those dissolved radionuclides that form colloidal suspensions. The colloidal particles are too large to pass through cell membranes but are of the size consumed by many aquatic organisms. Colloidal formation is important for those elements which form hydrated oxides, notably rare earths, iron, cobalt, uranium and thorium.

Co-Precipitation.   If a stable element is dissolved in water with sufficiently high concentration, it may begin to precipitate. If a particular dissolved radionuclide, which is always present in trace quantities, can form a lattice structure similar to that of the precipitating element, then the radionuclide will also become incorporated into the precipitate. For example, radium will co-precipitate with barium sulfate. Under many natural conditions, ferric hydroxide will precipitate, removing by co-precipitation many radionuclides from solution. If the particles of insoluble pre-

TABLE 11.13 Adopted Element or Nuclide Dependent Parameters for the Terrestrial Food-Chain Model of Fig. 11.6.

PARAMETER	NUCLIDE	NOMINAL VALUE[a]	OBSERVED RANGE
f_b/M_b	^{131}I	9×10^{-5} kg^{-1}	$(6.1 \times 10^{-5}\text{--}12 \times 10^{-5})$ kg^{-1}
f_b/M_b	^{137}Cs	4×10^{-5} kg^{-1}	$(6.1 \times 10^{-5}\text{--}12 \times 10^{-5})$ kg^{-1}
λ_{21}	^{131}I	0.14 d^{-1} [0]	$(0.087\text{--}0.23)$ d^{-1}
λ_{21}	^{137}Cs	0.08 d^{-1} [0]	$(0.03\text{--}0.08)$ d^{-1}
λ_{43}	^{131}I	0.14 d^{-1}	$(0.087\text{--}0.23)$ d^{-1}
λ_{43}	^{137}Cs	0.08 d^{-1}	$(0.03\text{--}0.08)$ d^{-1}
v	^{131}I	1.0 cm s^{-1}	$(0.1\text{--}5)$ cm s^{-1}
v	^{137}Cs	0.8 cm s^{-1}	$(0.04\text{--}5)$ cm s^{-1}
λ	^{131}I	8.57×10^{-2} d^{-1}	
λ	^{137}Cs	6.34×10^{-5} d^{-1}	
$\lambda_{\delta_4 \delta_3}$	^{131}I	1.0 d^{-1}	
$\lambda_{\delta_4 \delta_3}$	^{137}Cs	3.1 d^{-1}	
$\lambda_{\delta_5 \delta_3}$	^{131}I	0.060 d^{-1}	
$\lambda_{\delta_5 \delta_3}$	^{137}Cs	0	
$\lambda_{\delta_6 \delta_3}$	^{131}I	0.13 d^{-1}	
$\lambda_{\delta_6 \delta_3}$	^{137}Cs	0.33 d^{-1}	
$\lambda_{\delta_3 \delta_4}$	^{131}I	6.7×10^{-3} d^{-1}	
$\lambda_{\delta_3 \delta_4}$	^{137}Cs	0	
$\lambda_{\delta_5 \delta_4}$	^{131}I	0	
$\lambda_{\delta_5 \delta_4}$	^{137}Cs	0	
$\lambda_{c_5 c_4}$	^{137}Cs	3.5×10^{-2} d^{-1}	
$\lambda_{a_5 a_4}$	^{137}Cs	6.3×10^{-2} d^{-1}	

[a] Value in brackets are for dormant seasons.
Source: ICRP 1979.

cipitates are large enough, they will settle to the bottom making the radioisotopes more available to bottom-dwelling organisms.

11.4.1 Methodology for Aquatic Pathway Analysis

Because of the many different aquatic pathways and the many trophic levels typically found in each pathway, dynamic models for human exposure from aquatic food chains are seldom attempted. Almost always the concentration factor (CF) method is used. In this approach, equilibrium is assumed to exist between the radionuclide in the water and the concentration in all members of the exposure pathway. While this steady-state assumption may sometimes not be valid, estimates obtained with it are always conservative (i.e., over predictive).

To simplify the CR method even further, all compartments or trophes of food chains leading to an aquatic or marine food product used by humans are conceptually collapsed together. The concentration ratios of all compartments in a pathway are multiplied together and the results for the paths contributing to the product are

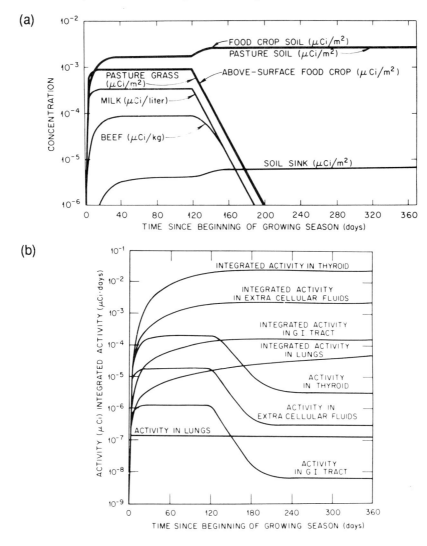

Figure 11.7   (a) Predicted activity concentrations in the food-chain compartments of the model in Fig. 11.6 (a). (b) The corresponding concentrations in the adult lung/GI-tract model compartments of Fig. 11.6(b). Profiles are for the first year following the steady release if 1 Ci y^{-1} of ^{131}I and a corresponding 0.27 pCl m^{-3} concentration in the air. [From ICRP (1979).]

summed to give an overall concentration factor for the food product. Thus, a concentration ratio CR^f, often called the *bioaccumulation factor*, for food product f is defined as the equilibrium ratio of a radionuclide concentration X_f in the food product to its concentration in water X_w, namely,

$$CR^f = \frac{X_f}{X_w}. \tag{11.76}$$

In this manner, the transfer of radioactivity from contaminated water through the many trophic levels of aquatic life to a human food product is described by a single parameter CR^f.

With this aggregated concentration factor, and the assumption of equilibrium conditions, the exposure of humans is readily estimated. Most aquatic or marine exposure models thus calculate the annual human ingestion of radionuclide i, I_i (Bq), from all aquatic food pathways from an equation similar to (USNRC 1977; IAEA 1982; Shor 1984)

$$I_i = \sum_f U^f \, X^f_{wi} \, CR^f_i \, \exp(-\lambda_i t^f), \qquad (11.77)$$

where,

U^f = annual ingestion (usage factor) of aquatic food product f (kg)

X^f_{wi} = average concentration of radionuclide i in water for food f (Bq L^{-1})

CR^f_i = equilibrium concentration ratio for radionuclide i and food product f (L kg^{-1})

λ_i = radioactive decay constant for radionuclide i

t^f = average time between harvesting and ingesting food product f.

Once the total annual ingestion of radionuclide i from all aquatic food is determined, the resulting committed dose to the various body organs can be computed by the methods of Chap. 8. Equivalently, the annual intake can be multiplied by an appropriate ingestion-to-organ-dose conversion factor (USDOE 1988).

11.4.2 Aquatic Concentration Factors

Of critical importance in using the above method for radiological assessments is the availability of accurate concentration ratios. Many experimental measurements of CR^f have been performed for many elements under a wide variety of conditions and for a wide range of aquatic organisms. Since there are many factors that affect the transfer and accumulation of a particular element by organisms in an aquatic food chain, it is not surprising that reported values of CR^f often exhibit a very wide range of values.

In Tables 11.14 and 11.15, concentration ratios and their range of values are given for the edible portion of freshwater and saltwater organisms. These values are the stable element concentrations (per unit wet weight of the edible part of the organism) divided by the average stable element concentration in the water. The single selected value for CR^f in these tables usually lies in the higher part of the range of reported values. Consequently, the selected value should lead to a conservative (over estimated) calculation. Other compilations are given by Peterson (1983), Lowman et al. (1971), and Vanderploeg et al. (1975).

Caution must be exercised when using "generic" concentration factors. If an analysis, based on such factors, indicates that population exposures are near a regulatory limit or design guide, then experimental data should be collected to obtain CR values appropriate to the particular aquatic systems under scrutiny and the analysis redone.

TABLE 11.14 Concentration Factors CR (L kg^{-1} Net Weight) for Three Food Components from Freshwater Environments.

ELEMENT	FISH MUSCLE	RANGE	INVERTEBRATES SOFT PARTS	RANGE	PLANTS	RANGE
Na	20	20–1000	20	20–200	100	
P	1×10^5	$(3–10) \times 10^4$	1×10^5	$(2–10) \times 10^4$	100	
S	800	300–1000	100		100	
Cr	40	40–4000	20	20–2000	40	40–4000
Mn	100	20–500	4×10^4	$(2–100) \times 10^3$	1×10^4	70–10^4
Fe	100	100–1000	1000		1000	
Co	20	10–1000	400	$(2–100) \times 10^2$	1×10^4	20–1×10^4
Ni	100		100		50	
Cu	200	3–800	1000		1000	$(1–2) \times 10^3$
Zn	1000	500–8000	1×10^4	$(4–20) \times 10^3$	1000	$(2–200) \times 10^2$
Sr	5	2–200	300		1000	3–2000
Y	30	10–100	1000		5000	
Zr	3	1–400	7		1000	70–1000
Nb	100	40–400	100		800	20–800
Tc	20	20–1000	5		40	
Ru	10	10–300	300		200	10–200
Ag	2	2–10	800		200	
Sb	1	1–200	10		2000	30–2000
Te	400		6000	$(6–100) \times 10^3$	1000	
I	20	10–800	300	5–300	800	40–800
Cs	400	50–10^4	1000	60–10^4	1000	10–2000
Ba	4	4–200	1000		500	
La	30		200		5000	
Ce	30	10–200	1000		5000	300–5000
Pm	30		1000		5000	
Pb	300	100–300	1000		200	
Bi	20		1×10^5		1×10^5	
Po	50	50–500	2×10^4		2000	
Ra	50	30–100	300		3000	30–3000
Th	30	30–100	500		2000	90–5×10^4
Pa	10		100		1000	
U	10	2–20	100		1000	4–1000
Np	10	10–100	400		1000	
Pu	4	0.4–500	100		400	200–10^4
Am	30	30–100	1000		5000	
Cm	30	30–100	1000		5000	

Source: Shor 1984.

CR Values for Radioactive Isotopes. Most compilations of concentration factors (including those of Tables 11.14 and 11.15) are for stable forms of the elements. The use of stable element CR values is again conservative since the CR for a radioactive isotope of an element will be smaller than the corresponding CR for stable isotopes. To see this, consider the simple one-compartment model of an aquatic organism described by Eq. (11.14). If the activity concentration in the water is held constant (X_2 constant), then in steady-state dX_3/dt vanishes and the equilibrium concentration ratio is seen to be

$$CR \equiv \frac{X_3}{X_2} = \frac{a_i}{\lambda + a_r}, \tag{11.78}$$

TABLE 11.15 Concentration Factors CR (L kg^{-1} Net Weight) for Three Components in Food from the Marine Environment.

ELEMENT	FISH MUSCLE	RANGE	INVERTEBRATES SOFT PARTS	RANGE	PLANTS	RANGE
Na	1	0.1–10	0.2	0.2–20	1	0.1–1
P	3×10^4	$(1-2) \times 10^4$	3×10^4	$(1-3) \times 10$	3000	$(3-100) \times 10^3$
S	2	1–100	2	0.3–5	0.4	0.4–1
Cr	400	70–400	2000	60–2000	2000	$20-3 \times 10^4$
Mn	600	80–3000	1×10^4	$(2-100) \times 10^3$	2×10^4	$(6-20) \times 10^3$
Fe	3000	400–3000	2×10^4	$(6-3000) \times 10^2$	5×10^5	$(7-700) \times 10^2$
Co	100	30–600	1000	$10-10^4$	1000	20–1000
Ni	100	100–500	300	10^2-10^4	300	300–1000
Cu	700	700–1000	2000	$(2-5) \times 10^3$	1000	
Zn	2000	$(3-200) \times 10^2$	10^5	$50-10^5$	1000	80–3000
Sr	2	0.1–2	20	10–50	10	1–100
Y	30	10–300	1000	10–1000	5000	300–5000
Zr	200	1–200	20	$1-2 \times 10^4$	2000	$200-2 \times 10^4$
Nb	100	1–100	100	10–100	1000	200–1000
Tc	10	10–30	50	50–1000	4000	$(4-10) \times 10^3$
Ru	10	0.05–10	1000	10–2000	2000	1000–2000
Ag	3000	1000–3000	3000	$300-10^5$	200	200–1000
Sb	40	40–1000	5	5–1000	2000	100–2000
Te	1000	10–1000	10^5	$10-10^5$	1×10^4	$200-10^4$
I	100	10–300	50	40–100	4000	$(1-10) \times 10^3$
Cs	40	20–200	20	3–100	20	10–50
Ba	10	3–10	100		500	500–1000
La	30		1000		5000	
Ce	30	30–300	1000	100–5000	5000	10–5000
Pm	30	30–100	1000		5000	1000–5000
Pb	300	$300-5 \times 10^5$	1000	$40-2 \times 10^5$	5000	$(1-20) \times 10^3$
Bi	20		10^5	$(1-100) \times 10^3$	1000	
Po	300	300–2000	2×10^4		2000	1000–2000
Ra	50	50–100	100	100–1000	100	
Th	10^4	10^3-10^4	2000	2000–6000	3000	$(1-20) \times 10^3$
Pa	10		10		6	6–100
U	3	1–10	10		70	10–70
Np	10	1–10	10	10–1000	6	6–1000
Pu	3	3–30	100	3–2000	400	400–3000
Am	30	10–30	1000	200–2000	5000	2000–5000
Cm	30	10–30	1000	200–2000	5000	2000–5000

Source: Shor 1984.

where a_i and a_r are the rate constants for intake and for biological elimination, respectively. If the isotope were stable (i.e., $\lambda = 0$), this result shows that the stable concentration ratio CR_s would be larger than that for a radioactive isotope of the same element. Specifically, the relations between the two concentration factors is

$$CR = \frac{CR_s}{1 + \lambda/a_r},$$ (11.79)

which, for very short-lived radionuclides (large λ), makes CR considerably smaller than its corresponding stable-isotope CR_s. On the other hand, long-lived isotopes with half lives much greater than the biological half life behave very much like their stable counterparts.

Nonequilibrium Situations.   Almost all analyses of aquatic exposure pathways are based on the use of equilibrium concentration ratios. However, the assumption of equilibrium is not always valid. The equilibrium concentration ratio given by Eq. (11.78) can be generalized, for the simple organism model, to give a time-dependent concentration ratio $CR(t)$ for situations in which the organism has been in the contaminated water for only a short time t. From the solution of Eq. (11.14), for a constant water concentration X_2 beginning at time $t = 0$, one obtains

$$CR(t) \equiv \frac{X_3(t)}{X_2} = \frac{a_i}{\lambda + a_r} \left[1 - e^{-(\lambda + a_i)t} \right]. \tag{11.80}$$

Only if the contact time t is greater than a few multiples of the effective half life of the radionuclide in the organism, $T_e = \ln 2/(\lambda + a_i)$, can $CR(t)$ be replaced by its equilibrium value.

For accident analyses, the radionuclide concentration in a body of water can decrease rapidly from its initial value. Organisms in such water will consequently be exposed to a variable radionuclide concentration in the water, and it is not clear what value of the water activity should be used when computing organism activities with the CR approach. If the water activity varies sufficiently slowly, then an radionuclide equilibrium between the water and the aquatic organisms and the water activity at the time of interest should be used. However, for rapid water activity transients, an equilibrium will not be reached and organism activities, based on equilibrium CR values and initial water activities, will be very overpredictive.

11.4.3 Factors Affecting the Concentration Factor

From Tables 11.14 and 11.15, it is seen that measured concentration factors for some elements can vary by several orders of magnitude. Most of this variation is a result of natural factors that influence the uptake of radionuclides by aquatic organisms. The manner in which concentration factors are measured can also affect the resulting CR values. In this section some of the more important influences are summarized.

Presence of Chemically Similar Elements.   The concept of the concentration ratio CR assumes that the concentration of a particular element in an aquatic or marine organism is directly proportional to the concentration of that element in the ambient water. For many trace impurities (such as most radionuclides) this is a reasonable assumption. However, it is clearly not true for all elements and all concentration ranges. For example, the amount of calcium needed by a fish to form its skeleton is independent of the calcium concentration in the water, and in waters with either very high or very low calcium concentrations, the amount incorporated into the fish varies only weakly with the water calcium concentration. A fish, through its metabolic processes, can regulate the intake and biological elimination rate of calcium to maintain its required calcium content.

The ability of an aquatic organism to regulate its accumulation of an essential element in varying ambient water concentrations, also affects the amounts of trace elements or radionuclides with chemically similar properties that are accumulated by the organisms. For example, the concentration of calcium in fish is held constant by *homeostatic* control. Thus, if the concentration of calcium in water is high, the concentration ratio of calcium in the fish to water must be low. This leads to a low CR for chemically similar strontium. This effect of calcium concentration in water does not hold for plants and algae since calcium is not an essential element for them. Another important example is the reduced uptake of ^{137}Cs associated with high concentrations of potassium in water.

There is a simple relation of how the CR for a trace radionuclide varies with the water concentration X_s^{water} of a chemically similar, more abundant, stable element whose uptake is regulated homeostatically by an aquatic species. Because of the similar chemical behavior of the stable (subcript s) and radioactive (subscript r) forms, the ratio of the two concentrations in the aquatic organism will equal that in the water, i.e., $X_s/X_r = X_s^{water}/X_r^{water}$. Because of the homeostatic control, the sum of the two concentrations in the organism will be constant (k say) for varying water concentration, i.e. $X_s + X_r = k$. From these two relations it is found for a trace radionuclide (for which $X_r^{water} \ll X_s^{water}$)

$$CR \equiv \frac{X_r}{X_r^{water}} = \frac{k}{X_s^{water} + X_r^{water}} \simeq \frac{k}{X_s^{water}}. \qquad (11.81)$$

From this result it is seen that the concentration factor for these types of radionuclides (ones with stable analogs in homeostasis) is inversely proportional to the concentration in water of the stable element. This result has been verified experimentally for the ^{90}Sr-calcium and ^{137}Cs-potassium pairs (Vanderploeg et al. 1975).

Species.   Concentration factors vary greatly among different species, their habitats, and their positions in the food chain. CR values for different freshwater and saltwater species are given by Peterson (1983). Species which feed on bottom organisms tend to accumulate more of those radionuclides that adsorb strongly to sediment, and hence, collect on the bottom. Species with similar habitats and feeding habits have similar concentration factors.

For some radionuclides there is also a *trophic level* effect. Species which feed further along the food chain from the primary producers (plants) are said to be at a higher trophic level. Because only a portion of most ingested radionuclides are accumulated by a species, the higher up the food web a species is (i.e., the higher the trophic level), the lower is the radionuclide concentration in the species, provided the radionuclide is not in a chemical form needed by the consumer. Thus, plutonium concentration is observed to decrease rapidly as one progresses along the aquatic food chain. By contrast, predator fish are found to have much higher concentrations of ^{137}Cs than smaller fish as a result of muscle protein (in which cesium is readily incorporated) being reconcentrated along the food chain. Since humans are the ultimate predators, this reconcentration effect is of considerable importance.

Measurement Technique.   The concentration factor is defined as the equilibrium ratio of the activity concentration in the organism to that in ambient water. For studies in natural water bodies, the water activity concentration remains relatively unchanged. But in tank studies, the absorption of radionuclides by the organisms decreases the water activity concentration from its initial value. In such studies the final or equilibrium water activity concentration should be used to define the CR.

It is often very hard to determine experimentally the concentration factor for higher trophic level aquatic organisms. Equilibrium should exist not only between radioactivity in the organism and in the water, but also between all members of the food chain used by the organism under scrutiny. This equilibrium condition is difficult to achieve both in tank studies for which it is difficult to simulate the natural food chain and in natural settings for which the mobility of organisms can prevent equilibrium conditions. Measurements made at a single time from a limited population can also be quite unreliable.

The water chemistry and quality can also affect strongly observed concentration factors. Varying water pH and temperatures can alter the uptake of trace elements. Observed CR values for the same radionuclide and aquatic species can be quite different in clear water and in water with a large suspended sediment content.

Finally, the distribution of absorbed radionuclides generally varies throughout an organism. For example ^{90}Sr is more concentrated in the shells of oysters and in the bones of fish than in the soft edible parts. Similarly, the manner in which aquatic food is prepared and consumed by humans can alter the amount of ingested radionuclides. For example, some radionuclides (e.g., ^{90}Y, ^{95}Zr-^{95}Nb, and Pu) are concentrated in the GI tract and kidneys of some aquatic organisms, and, consequently, consumption of the entire organism will lead to more radioactivity ingestion than if only the less contaminated muscle parts are consumed. In using concentration factor tabulations, it is thus very important to determine what part of the organism was used.

11.5 TREATMENT FOR SPECIAL RADIONUCLIDES

In exposure pathway analyses, atmospheric releases of tritium and carbon-14 are usually given special treatments because of the rapidity with which they spread throughout the environment. The basis of these special treatments is the assumption that an equilibrium exists between the concentration of these nuclides in the air (or water) and the human food product produced at a given location.

Tritium. For dose analyses it is usually assumed that all tritium is released from nuclear facilities in the form of tritiated water vapor ^{3}HOH, although other chemical forms (e.g., ^{3}HH or ^{3}HCH$_3$) have been observed. Forms other that ^{3}HOH generally result in lower doses to the exposed population (Till et al. 1980), and so the assumption that all tritium is in the form of ^{3}HOH is conservative.

For atmospheric releases, the concentration of tritium in atmospheric water vapor X_{vap} (Bq L^{-1}) is related to the atmospheric tritium concentration X_{air} (Bq m^{-3}) at some location of interest by

$$X_{vap} = \frac{1000\, X_{air}}{H_a},$$

(11.82)

where H_a (g H$_2$O m^{-3}) is the *absolute humidity* of the air. Because of the dilution of tritium by airborne water vapor, the eventual dose to humans from airborne tritium is inversely proportional to the absolute humidity. A default value of 8 g H_2O m^{-3} is used by USNRC (1977), but the actual value varies geographically and seasonally.

For chronic releases, it is assumed that tritium in atmospheric water vapor, vegetation and surface water are in equilibrium. Thus, if the concentration of tritium (Bq L^{-1}) is known in the atmospheric water vapor (X_{vap}), food products (X_{fn}) and drinking water (X_w) to which an individual is exposed, the mean concentration of tritium in the water ingested and inhaled by Reference Man, X_T (Bq L^{-1}), can be estimated from (NCRP 1979, 1984) as

$$X_T = 0.41X_w + 0.52\sum_n X_{fn}f_n + 0.07X_{vap}, \tag{11.83}$$

where X_{fn} the concentration in the water of food products grown at location n and f_n is the fraction of the total food intake grown at location n. The numerical factor 0.41 is the fraction of total daily water intake (3 L d^{-1}) that is obtained by Reference Man from drinking water, 0.52 the fraction from food, and 0.07 the fraction from inhalation and skin absorption (NCRP 1979). Multiplication of this mean concentration by the annual water intake and by appropriate ingestion-to-organ-dose conversion factors (see Chap. 8, especially Sec. 8.6), then gives the annual committed dose an individual would receive to various body organs.

Carbon-14. Releases of ^{14}C from nuclear facilities are assumed to be converted to $^{14}CO_2$, fixed in vegetation through photosynthesis, and ultimately reach humans through the food chain.[3] The method normally used to calculate human ingestion and doses for chronic ^{14}C releases into the environment assumes an equilibrium relationship between the ^{14}C activity per unit mass of carbon in air at the point of photosynthetic fixation and that in the vegetation or in subsequent trophes in the food chain.

The specific activity of ^{14}C in airborne carbon, A^c_{air} (Bq m^{-3})/(g carbon m^{-3}) is related to the atmospheric concentration of ^{14}C, X^c_{air} (Bq m^{-3}), by

$$X^c_{air} = \rho_c A^c_{air}, \tag{11.84}$$

where ρ_c is the atmospheric mass density of carbon (g m^{-3}). The atmospheric CO_2 concentration varies diurnally and seasonally and can be altered by local industrial or residential sources. A typical value of ρ_c, corresponding to 300 ppm CO_2, is 0.18 g of carbon per m^3 of air.

Since photosynthesis is active only during the day, the atmospheric dispersion of $^{14}CO_2$ from the release point to the location of interest should be calculated using day-time atmospheric conditions typical for the growing season. Although isotopic fractionation occurs during photosynthesis (i.e., different plants can have different $^{14}C/^{12}C$ ratios for the same X^c_{air}, this effect is neglected for population exposure calculations. Thus, the specific activity of the carbon fixed in plants is assumed to the same as that for the air at the same location.

---

[3]For example, Schwibach et al. (1978) report that almost all ^{14}C emitted from boiling water reactors is in the form of $^{14}CO_2$; however, for pressurized reactors, an appreciable fraction can be emitted in the form of ^{14}CO, $^{14}CH_4$, or other hydrocarbons that do not enter the food chain through photosynthesis.

TABLE 11.16 Carbon Content {g(C) per kg} for the Edible Portion of Various Foods. Values are for Unprepared Food and Are Based on Carbon Content of Protein (50%), Carbohydrates (44%), and Fat (76%).

FOOD	CONTENT g(C) kg^{-1}	FOOD	CONTENT g(C) kg^{-1}	FOOD	CONTENT g(C) kg^{-1}
Vegetables and Fruits		*Vegetables and Fruits*		*Grain for Human*	
lettuce	20	peaches	56	*Consumption*	
cabbage	32	pears	76	corn	118
celery	24	plums	62	spring wheat	391
spinach	28	strawberries	44	soybeans	465
broccoli	42	snap beans	47	oats	431
cauliflower	35	summer squash	21	barley	395
brussel sprouts	65	blueberries	74	rye	396
potatoes	95	blackberries	69		
oranges	55	cranberries	57		
sweet cane	118	currants	67	*Miscellaneous*	
sugar cane	438	gooseberries	48	eggs	156
peanuts	574	black raspberries	89	honey	365
carrots	49	asparagus	30		
dry beans	198				
sweet potatoes	137				
cantaloupe	25	*Meat and Poultry*		*Dairy Products*	
dry peas	395	beef	228	whole milk	67
pecans	659	lamb	289	ice cream	196
tomatoes	25	pork	402	cheese	350
grapes	83	turkey	254	cottage cheese	99
apples	70	chicken	156	butter	620

Source: NCRP 1984; by permission of the NCRP.

The ^{14}C in humans comes almost exclusively from the ingestion of vegetation or animal products. Almost none comes from inhalation of atmospheric ^{14}C. Only for accident or short term releases is inhalation an important exposure pathway. Fractionation effects in animals and humans are insignificant compared to those in plants. Thus, under equilibrium conditions, the specific activity of ^{14}C in animals and humans equals that of the food consumed by them, which, in turn equals that in the air where the ^{14}C was first fixed in the precursor vegetation in the food chain. With this equilibrium situation, the average specific activity of ^{14}C in the carbon within the body is (NCRP 1984)

$$A_{body} = \sum_n \frac{G_n}{G} A_{air}^{c,n} \qquad (11.85)$$

where G_n (g carbon y^{-1}) is the annual intake of dietary carbon derived from the nth location, G is the total annual average intake of dietary carbon (about 1.1×10^5 g y^{-1} for a male adult), and $A_{air}^{c,n}$ is the estimated daytime specific activity of ambient airborne carbon during the growing season at location n. The carbon content of various foods is shown in Table 11.16.

Multiplication of this mean specific activity by the annual carbon intake and by appropriate ingestion-to-organ-dose conversion factors (see Chap. 8, especially Sec.

TABLE 11.17   Generic Usage Factors for the Average per Capita Daily Intake of Various Foods by People in the United States at Different Ages.

FOOD	AGE (years)			
	< 1	1–11	12–18	> 18
	mL d^{-1}			
Milk				
fluid	696	542	485	261
milk and milk products (Ca equivalent)	795	606	594	306
	g d^{-1}			
Eggs	17	25	31	41
Meats				
beef	7	38	66	86
pork	4	41	69	76
other and mixtures	34	39	52	70
Poultry	3	18	27	26
Fish				
freshwater fin	—	0.49	0.84	1.48
saltwater fin	—	4.33	7.23	10.68
shellfish (all)	—	0.93	1.45	3.59
Potatoes	6	49	67	69
Vegetables				
leafy, mixtures	2	20	30	50
deep yellow, mixtures	12	7	7	8
legumes, mixtures	12	22	28	25
other, mixtures	50	58	82	99
Fruit				
citrus, tomatoes	23	74	93	99
other, mixtures	112	112	116	87
Grain (flour equivalent)	21	87	113	97

Source: NCRP 1984; by permission of the NCRP.

8.10.2), gives the annual committed dose an individual would receive to various body organs.

11.6 USAGE FACTORS

Evaluation of population exposures from the ingestion of contaminated food and water requires information about the consumption rate of the food and water. These consumption rates are often called *usage factors*. Whenever possible, site-specific usage factors should be used in exposure analyses since diet and amounts of food and water consumed can vary widely geographically, ethnically, seasonally, and with age and sex. For initial calculations, generic usage factors are often used (USNRC 1977). In Table 11.17, an example set of such generic usage factors is given for people in the United States.

Besides the ingestion rate of food, the intake and excretion rates of water and the inhalation rate of air are also important in assessing population doses. These rates were presented earlier in Table 8.2.

REFERENCES

ANSPAUGH, L.R., D.L. PHELPS, N.C. KENNEDY, J.H. SHINN, AND J.M. REICHMAN, "Experimental Studies on the Resuspension of Plutonium and Aged Sources at the Nevada Test Site," in *Atmospheric-Surface Exchange of Gaseous Pollutants*, R.J. Englemann and G.A. Sehmel (coordinators), (Proc. Sympos. Richland, Washington 4-6 Sept. 1974), pp 727–742, U.S. ERDA Report CONF-740921, 1976.

BOOTH, R.S., O.W. BURKE, AND S.V. KAYE, "Dynamics of the Forage-Cow-Milk Pathway for Transfer of Radioactive Iodine, Strontium and Cesium to Man," Proceedings of an ANS Topical Meeting *Nuclear Methods in Environmental Research*, Columbia, Mo., 1971.

COUGHTREY, P.J., *Radioactivity Transfer in Animal Products*, Report EUR 12608EN, Comm. of the European Communities, Luxembourg, 1990.

DAHLMAN, R.C., S.I. AUERBACH, AND P.B. DUNAWAY, "Behavior of ^{137}Cs-Tagged Particles in a Fescue Meadow," in *Environmental Contamination by Radioactive Materials*, pp 153–165, Proc. FAO-IAEA-WHO Seminar, (Vienna, Austria 24-28 March 1969), IAEA, Vienna, 1969.

DRBAL, L.F., Ph.D. Dissertation, Nuclear Engineering Department, Kansas State University, Manhattan, Kans., 1975.

GARNER, R.J., "A Mathematical Analysis of the Transfer of Fission Products to Cow's Milk," *Health Physics 13*, 205, 1967.

IAEA, *Generic Models and Parameters for Assessing the Environmental Transfer of Radionuclides in Predicting Exposures to Critical Groups from Routine Releases*, Safety Series, STI/PUB 611, International Atomic Energy Agency, Vienna, Austria, 1982.

ICRP, *Radionuclide Release into the Environment: Assessment of Doses to Man*, Publication 29, International Commission on Radiological Protection, Pergamon Press, Oxford, *Annals of the ICRP 2* (2), (1979).

KRIEGER, H.L., AND F.J. BURMANN, "Effective Half-lives of ^{85}Sr and ^{134}Cs for a Contaminated Pasture," *Health Physics 17*(4), 565–569 (1969).

LASSEY, K.R., "The Possible Importance of Short Term Exposure to Resuspended Radionuclides," *Health Physics 38*(5), 749–761, (1980).

LOWMAN, F.G., T.R. RICE, AND F.A. RICHARDS, "Accumulation and Redistribution of Radionuclides by Marine Organisms," in *Radioactivity in the Marine Environment*, National Academy of Sciences, Washington, D.C., 1971.

MCDOWELL-BOYER, L.M., AND C.F. BAES, "Terrestrial Food-Chain Transport," in *Models and Parameters for Environmental Radiological Assessments*, C.W. Miller (ed.), DOE/TIC-11468, NTIS, Springfield, Va., 1984.

NCRP, *Tritium in the Environment*, Report No. 62, National Council on Radiation Protection and Measurement, Washington, D.C., 1979.

NCRP, *Radiological Assessment: Predicting the Transport, Bioaccumulation, and Uptake by Man of Radionuclides Released to the Environment*, Report No. 76, National Council on Radiation Protection and Measurement, Bethesda, Md., 1984.

NG, Y.C., C.A. BURTON, S.E. THOMPSON, R.K. TANDY, H.K. KRETNER, AND M.W. PRATT, "Prediction of the Maximum Dosage from the Fallout of Nuclear Devices," in *Handbook for Estimating the Maximum Internal Dose from Radionuclides Released to the Biosphere*, USAEC Report UCRL-50163, Part IV, Lawrence Livermore Lab., Livermore, Ca., 1968.

NG, Y.C., C.S. COLSHER, D.J. QUINN, AND S.E. THOMPSON, *Transfer Coefficients for the Prediction of the Dose to Man Via the Forage-Cow-Milk Pathway from Radionuclides Released to the Biosphere*, Report UCRL-51939, Lawrence Livermore Laboratory, Livermore, Ca., July 1977.

NG, Y.C., S.E. THOMPSON, AND C.S. COLSHER, *Soil-To-Plant Concentration Factors for Radiological Assessments*, NUREG/CR-2975, U.S. Nuclear Regulatory Commission, Washington, D.C., 1982.

NG, Y.C., C.S. COLSHER, AND S.E. THOMPSON, *Transfer Coefficients for Assessing the Dose from Radionuclides in Meat and Eggs*, NUREG/ CR- 2976, U.S. Nuclear Regulatory Commission, Washington, D.C., 1982.

PETERSON, H.T. JR., "Terrestrial and Aquatic Food Chain Pathways," in *Radiological Assessment: A Textbook on Environmental Dose Analysis*, J.E. Hill and H.R. Meyer, eds., NUREG/ CR-3332, U.S. Nuclear Regulatory Commission, Washington, D.C. 1983.

SCHWIBACH, J., H. RIEDEL, AND J. BRETSCHNEIDER, *Investigations into the Emission of Carbon-14 Compounds from Nuclear Facilities*, Report V-3062/78-EN, Comm. of the European Communities, Paris, 1978.

SEHMEL, G.A, AND F.D. LLOYD, "Particle Suspension Rates," in *Atmospheric-Surface Exchange of Gaseous Pollutants*, R.J. Englemann and G.A. Sehmel (coordinators), (Proc. Sympos. Richland, WA 4-6 Sept. 1974), pp 846–855, U.S. ERDA Report CONF-740921, 1976.

SHEPPARD, C.W., *Basic Principles of the Tracer Method*, Wiley, New York (1962).

SLINN, W.G.N., "Parameterization for Resuspension and for Wet and Dry Deposition of Particles and Gases for Use in Radiation Dose Calculations," *Nuclear Safety 19*, 205-219 and 365 (1978).

SHOR, R.W., "Aquatic Food-Chain Transport," in *Models and Parameters for Environmental Radiological Assessments*, C.W. Miller (ed.), DOE/TIC-11468, National Technical Information Service, Springfield, Va., 1984.

TILL, J.E., H.R. MEYER, E.L. ETNIER, E.S. BOMER, R.D. GENTRY, G.G. KILLOUGH, P.S. ROHWER, V.J. TENNEY, AND C.C. TRAVIS, *Tritium — An Analysis of Key Environmental and Dosimetric Questions*, ORNL/TM-6990, Oak Ridge National Lab., Oak Ridge, Tenn., 1980.

USDOE, *Internal Dose Conversion Factors for Calculation of Dose to the Public*, DOE/ EH-0071, U.S. Dept. of Energy, Washington, D.C., 1988.

USNRC, *Calculation of Annual Doses to Man from Routine Releases of Reactor Effluents for the Purpose of Evaluating Compliance with 10 CFR Part 50,* Regulatory Guide 1.109, U.S. Nuclear Regulatory Commission, Washington, D.C., 1977.

USNRC, *Interim Licensing Policy on As Low as Practicable for Gaseous Radioiodine Releases from Light Water Cooled Nuclear Power Reactors,* Regulatory Guide 1.42, U.S. Nuclear Regulatory Commission, Washington, D.C., 1974 (withdrawn 1976).

USNRC, *Reactor Safety Study — An Assessment of Accident Risks in Commercial Nuclear Power Plants*, WASH-1400 (NUREG- 75/014), Appendix VI, U.S. Nuclear Regulatory Commission, Washington, D.C., 1975.

VANDERPLOEG, H.A., D.C. PARZYCK, W.H. WILCOX, J.R. KERCHER, AND S.V. KAYE, *Bioaccumulation Factors for Radionuclides in Freshwater Biota*, ORNL-5002, Oak Ridge National Lab., Oak Ridge, Tenn., 1975.

PROBLEMS

1. Sketch four example exposure pathways in which the compartments of each example are all in the same ecosystem. Sketch four example exposure pathways each of which involves two ecosystems. Sketch two exposure pathways each of which involves all three major ecosystems. Identify each compartment and indicate appropriate units for the output from each.

2. Develop a set of differential equations to describe the transient radionuclide concentration $X_i(t)$ in each of the six compartments of the exposure pathways shown in Fig. 11.2. Assume linear first-order transfer processes (i.e., the rate of transfer from one compartment to another is proportional to the concentration in the source compartment. Clearly define each model parameter. Under steady-state conditions, derive an expression for the rate of radionuclide input to the GI tract.

3. Under what conditions is the approximation of Eq. (11.5) by Eq. (11.7) valid? Develop numerical criteria for the range of validity.

4. With the data given in Chap. 10, estimate the average rate at which airborne molecular radioiodine is (a) removed from the air above a wheat field, (b) deposited on the wheat, and (c) deposited on the soil. The concentration in the air above the wheat field averages 5 pCi m^{-3}, and the average rate of rainfall is 4 mm wk^{-1}.

5. A growing food crop is contaminated by an accidental airborne release of radioactivity with a half life of 10 d. Estimate the fraction of initial radioactivity deposited on the crop that remains after 2 days, 2 weeks, and 2 months.

6. Geese living on a federal reservation consume water plants growing in a lake whose bottom sediment is known to contain ^{137}Cs at a concentration of 2 pCi per gram (dry) of sediment. The concentration ratio for cesium by the plants is 300 (dry-plant basis). (a) If geese consume 100 g of the plants (dry weight) per day, what is the equilibrium activity concentration in the meat of the geese? Assume the transfer factor for geese is similar to that for chicken. (b) If a person were to eat 500 g of meat from one of these geese, what committed effective dose equivalent would be received?

7. Based on the terrestrial exposure pathway model described by Eqs. (11.64) to (11.75), what is the steady-state activity concentration in cow milk (μCi L^{-1}) and in beef (μCi kg^{-1}) during the growing season if the activity concentration in the air over the pasture is constant at 1 pCi m^{-3}?

8. For the forage-cow-milk model presented in Sec. 11.3.6, calculate the intake-to-milk transfer factor.

9. An accident at a nuclear power station caused a release, over a 24-h period, of 15 Ci of ^{131}I and 7.5×10^6 Ci of ^{133}Xe. During and following the accident, generally southerly winds varied between 120° and 200° (directions from which the wind blows) with an average speed of 1.1 m s^{-1}. Atmospheric conditions were slightly unstable (Pasquill category C) and the release elevation was 30 m. The effective exclusion area of the station has a radius of 3.2 km and the nearest dairy farm is 6.4 km directly north of the plant. A city of

population 100,000 is located 17.6 km from the plant at an azimuth of 320°. Aside from the city, the area within 3.2 and 32 km radially from the plant is rural, with an average population density of 31.2 per km². The area between 32 and 80 km of the plant is rural and small-town with an average population density of 105 per km². Estimate the following quantities:

(a) The maximum potential instantaneous dose equivalent rate offsite from ^{133}Xe in mrem h^{-1}.

(b) The maximum potential dose equivalent offsite during the entire course of the accident in mrem.

(c) The ^{131}I concentration in Ci L^{-1}, as a function of time, in milk produced at the nearest dairy farm. Assume the iodine that is deposited over the 24 hour period is deposited instantly. Why is this a conservative assumption?

(d) The maximum potential thyroid committed dose equivalent (in mrem) an infant would receive if it consumed milk from the nearest dairy farm.

10. Consider a simple one-compartment model that describes the concentration of radioactivity, $X(t)$, in an animal food product (e.g., meat, milk or eggs) by the following differential equation

$$\frac{dX(t)}{dt} = \frac{f}{V}Q(t) - (\lambda + r)X(t)$$

where f is the fraction of ingested radionuclides that goes to the food product, V is the volume of the food product, $Q(t)$ is the rate of ingestion of radioactivity (Bq d^{-1}), r is the biological excretion rate of the radioactive element from the food product, and λ is the radioactive decay constant for the radionuclide.

(a) Derive an expression for the activity concentration in the animal food product as a function of time after the animal begins to ingest radioactivity at a constant rate Q_o.

(b) What is the activity concentration in the food product following an instantaneous ingestion of Q_o (Bq) of radioactivity at time $t = 0$.

(c) Use convolution of the result obtained in Part (b) to obtain the result found in Part (a).

(d) Show that the input-to-food-product transfer factor F is given by

$$F = \frac{f}{(r + \lambda)V}.$$

(e) If F_s is the transfer factor for a stable element, derive a multiplicative correction factor to give the transfer factor F for a radioactive isotope of the same element.

11. During the preparation of an environmental impact report for a nuclear power plant, a little-studied freshwater shrimp in found to live in a river near the facility. A related salt-water species of this shrimp is known to have a concentration ratio for ^{137}Cs when the potassium concentration is 410 ppm. Estimate the concentration ratio for ^{137}Cs in the freshwater shrimp if the potassium concentration in the river is 3.2 ppm.

12. A radioisotope with a half life of 150 days is accidentally spilled into a tidal estuary in which measurable activity persists for only 1 week. During the week the average activity concentration in the water is $0.15\ \mu$Ci L^{-1}. For a particular species of fish that lives in this estuary, the biological half life of the radionuclide is 250 days and the concentration ratio is 350. Estimate the maximum activity concentration in the fish.

13. Clams living in a lake with a concentration of ^{60}Co of 2 pCi L^{-1} have a concentration factor of 1500 for this radionuclide. What is the effective committed dose equivalent received each year by an individual who eats 25 kg of these clams each year?

14. Estimate the annual committed effective dose equivalent received from the ^{14}C in the beef consumed by a typical American adult. The ^{14}C/^{12}C ratio in the environment is 1.2 × 10^{-12}.

15. The normal concentration of tritium in surface water on the earth is 0.4 mBq mL^{-1}. What would be the increase in the annual committed dose equivalent per person of a population which receives most of its drinking water from a source with double this normal concentration?

Appendix A

Constants and Conversion Factors

TABLE A.1 Physical Constants.

CONSTANT	SYMBOL	VALUE
Atomic mass unit[a]	u	$1.660\,566 \times 10^{-27}$ kg
Speed of light (in vacuum)	c	$2.997\,925 \times 10^{8}$ m s^{-1}
Electron charge	e	$1.602\,189 \times 10^{-19}$ C
Electron rest mass	m_e	$9.109\,535 \times 10^{-31}$ kg
		$(5.485\,803 \times 10^{-4}$ u)
Proton rest mass	m_p	$1.672\,649 \times 10^{-27}$ kg
		$(1.007\,276$ u)
Neutron rest mass	m_n	$1.674\,954 \times 10^{-27}$ kg
		$(1.008\,665$ u)
Electron radius (classical)	r_e	$2.817\,938 \times 10^{-15}$ m
Planck's constant	h	$6.626\,176 \times 10^{-34}$ J s
Avogadro's constant	N_A	$6.022\,045 \times 10^{23}$ mol^{-1}

[a]Equivalent to 931.502 MeV of energy.

Source: Lederer and Shirley 1978.

TABLE A.2 Radiological Units.

QUANTITY	UNIT	EQUIVALENCE IN OTHER UNITS
Absorbed dose	Gy (gray)	1 J kg^{-1}, 100 rad
Activity	Bq (becquerel)	1 s^{-1}, 2.7027×10^{-11} Ci (curie)
Cross section	m^2	10^{28} b (barn)
Dose equivalent	Sv (sievert)	100 rem
Energy	MeV	1.6021×10^{-6} erg, 1.6021×10^{-13} J
Exposure	R (roentgen)	2.58×10^{-4} C kg^{-1}
Kerma	Gy (gray)	1 J kg^{-1}, 10^4 erg g^{-1}

TABLE A.3   Prefixes for SI Units.

FACTOR	PREFIX	SYMBOL
10^{18}	exa	G
10^{15}	peta	P
10^{12}	tera	T
10^{9}	giga	G
10^{6}	mega	M
10^{3}	kilo	k
10^{2}	hecto	h
10^{1}	deca	da
10^{-1}	deci	d
10^{-2}	centi	c
10^{-3}	milli	m
10^{-6}	micro	μ
10^{-9}	nano	n
10^{-12}	pico	p
10^{-15}	femto	f
10^{-18}	atto	a

TABLE A.4   Conversion Factors.

PROPERTY	UNIT	SI EQUIVALENT	
Length	in.	2.54×10^{-2}	m[a]
	ft	3.048×10^{-1}	m[a]
	mile (int'l)	$1.609\,344 \times 10^{3}$	m[a]
Area	in^2	6.4516×10^{-4}	m^{2}[a]
	ft^2	$9.290\,304 \times 10^{-2}$	m^{2}[a]
	acre	$4.046\,873 \times 10^{3}$	m^2
	square mile (int'l)	$2.589\,988 \times 10^{6}$	m^2
	hectare	$1. \times 10^{4}$	m^2
Volume	oz (U.S. liquid)	$2.957\,353 \times 10^{-5}$	m^3
	in^3	$1.638\,706 \times 10^{-5}$	m^3
	gallon (U.S.)	$3.785\,412 \times 10^{-3}$	m^3
	ft^3	$2.831\,685 \times 10^{-2}$	m^3
Mass	oz (avdp.)	$2.834\,952 \times 10^{-2}$	kg
	lb	$4.535\,924 \times 10^{-1}$	kg
	ton (short)	$9.071\,847 \times 10^{2}$	kg
Force	kg_f	$9.806\,650$	N (newton)[a]
	lb_f	$4.448\,222$	N
	ton	$8.896\,444 \times 10^{3}$	N
Pressure	lb_f/in^2 (psi)	$6.894\,757 \times 10^{3}$	Pa (pascal)
	lb_f/ft^2	$4.788\,026 \times 10^{1}$	Pa
	atm (standard)	$1.013\,250 \times 10^{5}$	Pa[a]
	in. H_2O (@ 4 C)	$2.490\,82 \times 10^{2}$	Pa
	in. Hg (@ 0 C)	$3.386\,39 \times 10^{3}$	Pa
	mm Hg (@ 0 C)	$1.333\,22 \times 10^{2}$	Pa
	bar	$1. \times 10^{5}$	Pa[a]
Energy	eV	$1.602\,19 \times 10^{-19}$	J
	cal	4.184	J[a]
	Btu	$1.054\,350 \times 10^{3}$	J
	kWh	3.6×10^{6}	J[a]
	MWd	8.64×10^{10}	J[a]

[a] Exact conversion factor.

Source: ANSI 1976.

REFERENCES

LEDERER, C. M., AND V. SHIRLEY (eds.), *Table of Isotopes*, 7th ed., Appendix 1., Wiley–Interscience, New York, 1978.

ANSI, American National Standard Institute, *Standard for Metric Practice*, ANSI/ASTM E380–76, New York, 1976.

Appendix B

Decay Characteristics of Selected Radionuclides

This tabulation is based on calculations performed using the EDISTR code[a]. Data were kindly provided by J. C. Ryman and K. F. Eckerman of the Metabolism and Dosimetry Research Group, Health and Safety Research Division, Oak Ridge National Laboratory, Oak Ridge, Tennessee.

For the most part, the tabulation is based on the data used in preparation of the MIRD compendium of radioactive decay data[b]. For those nuclides not included in the MIRD document, tabulations are based on data used in preparation of the ICRP compendium of radioactive decay data[c]. Exceptions are the nuclides ^{16}N and ^{89}Kr, for which the tabulations are based on data of Kocher[d]. Frequencies of annihilation quanta are included.

Individual radiations are listed only if two criteria are met. First, the (maximum) energy must exceed 10 keV. Second, for any one group, e.g., combined gamma and x rays, the product of the energy and the frequency of an individual radiation must exceed 1% of the sum of the products for that group. Group totals include contributions from all radiations in the group, not just those listed. When the sum of the products of those listed is less than 95% of the group total, an asterisk after the group total is used to alert the reader.

Data are listed in order of increasing atomic number. Tables cross-referencing atomic number and elemental designation are provided on the following page.

---

[a]Dillman, L. T., *EDISTR – A Computer Program to Obtain a Nuclear Decay Data Base for Radiation Dosimetry*, Report ORNL/TM-6689, Oak Ridge National Laboratory, Oak Ridge, Tennessee, 1980.

[b]Weber, D. A., K. F. Eckerman, L. T. Dillman, and J. C. Ryman, *MIRD: Radionuclide Data and Decay Schemes*, Society of Nuclear Medicine, New York, 1989.

[c]*Radionuclide Transformations*, Publication 38, International Commission on Radiological Protection, *Annals of the ICRP 11–13,* 1983.

[d]Kocher, D. C., *Radioactive Decay Tables*, Report DOE/TIC 11026, Technical Information Center, U. S. Department of Energy, Washington, D. C., 1981.

Chemical symbol and atomic number for the elements

1	H	26	Fe	51	Sb	76	Os
2	He	27	Co	52	Te	77	Ir
3	Li	28	Ni	53	I	78	Pt
4	Be	29	Cu	54	Xe	79	Au
5	B	30	Zn	55	Cs	80	Hg
6	C	31	Ga	56	Ba	81	Tl
7	N	32	Ge	57	La	82	Pb
8	O	33	As	58	Ce	83	Bi
9	F	34	Se	59	Pr	84	Po
10	Ne	35	Br	60	Nd	85	At
11	Na	36	Kr	61	Pm	86	Rn
12	Mg	37	Rb	62	Sm	87	Fr
13	Al	38	Sr	63	Eu	88	Ra
14	Si	39	Y	64	Gd	89	Ac
15	P	40	Zr	65	Tb	90	Th
16	S	41	Nb	66	Dy	91	Pa
17	Cl	42	Mo	67	Ho	92	U
18	Ar	43	Tc	68	Er	93	Np
19	K	44	Ru	69	Tm	94	Pu
20	Ca	45	Rh	70	Yb	95	Am
21	Sc	46	Pd	71	Lu	96	Cm
22	Ti	47	Ag	72	Hf	97	Bk
23	V	48	Cd	73	Ta	98	Cf
24	Cr	49	In	74	W	99	Es
25	Mn	50	Sn	75	Re	100	Fm

Ac	89	Dy	66	Mn	25	Ru	44
Ag	47	Er	68	Mo	42	S	16
Al	13	Es	99	N	7	Sb	51
Am	95	Eu	63	Na	11	Sc	21
Ar	18	F	9	Nb	41	Se	34
As	33	Fe	26	Nd	60	Si	14
At	85	Fm	100	Ne	10	Sm	62
Au	79	Fr	87	Ni	28	Sn	50
B	5	Ga	31	Np	93	Sr	38
Ba	56	Gd	64	O	8	Ta	73
Be	4	Ge	32	Os	76	Tb	65
Bi	83	H	1	P	15	Tc	43
Bk	97	He	2	Pa	91	Te	52
Br	35	Hf	72	Pb	82	Th	90
C	6	Hg	80	Pd	46	Ti	22
Ca	20	Ho	67	Pm	61	Tl	81
Cd	48	I	53	Po	84	Tm	69
Ce	58	In	49	Pr	59	U	92
Cf	98	Ir	77	Pt	78	V	23
Cl	17	K	19	Pu	94	W	74
Cm	96	Kr	36	Ra	88	Xe	54
Co	27	La	57	Rb	37	Y	39
Cr	24	Li	3	Re	75	Yb	70
Cs	55	Lu	71	Rh	45	Zn	30
Cu	29	Mg	12	Rn	86	Zr	40

^{3}H   12.33 y

Principal Beta Particles

Freq. (%)	E_{avg} (keV)	E_{max} (keV)
100.00	5.67	18.6
keV/decay	5.67	

^{11}C   20.39 min

Positrons

Freq. (%)	E_{avg} (keV)	E_{max} (keV)
99.76	385.6	960.1
keV/decay	384.6	

Principal Gamma and X Rays

Freq. (%)	E (keV)
199.52	511.0
keV/decay	1019.5

^{14}C   5730 y

Principal Beta Particles

Freq. (%)	E_{avg} (keV)	E_{max} (keV)
100.00	49.5	156.5
keV/decay	49.5	

^{13}N   9.965 min

Positrons

Freq. (%)	E_{avg} (keV)	E_{max} (keV)
99.82	491.8	1198.5
keV/decay	490.9	

Principal Conversion and Auger Electrons

keV/decay	4.61×10^{-4} *

Principal Gamma and X Rays

Freq. (%)	E (keV)
199.64	511.0
keV/decay	1020.2

^{16}N   7.13 s

Principal Beta Particles

Freq. (%)	E_{avg} (keV)	E_{max} (keV)
4.9	1461.5	3301.9
68.0	1941.2	4288.3
26.0	4979.2	10419.0
keV/decay	2692.7	

Principal Gamma and X Rays

Freq. (%)	E (keV)
69.0	6129.2
5.0	7115.1
keV/decay	4618.2

^{15}O   122.24 s

Positrons

Freq. (%)	E_{avg} (keV)	E_{max} (keV)
99.89	735.3	1731.8
keV/decay	734.5	

Principal Conversion and Auger Electrons

keV/decay	4.15×10^{-4} *

Principal Gamma and X Rays

Freq. (%)	E (keV)
199.77	511.0
keV/decay	1020.8

^{19}O   26.91 s

Principal Beta Particles

Freq. (%)	E_{avg} (keV)	E_{max} (keV)
56.06	1442.7	3264.8
39.77	2103.5	4621.6
3.83	2216.1	4818.8
keV/decay	1732.7	

Principal Conversion and Auger Electrons

keV/decay	0.431 *

Principal Gamma and X Rays

Freq. (%)	E (keV)
3.45	110.0
90.34	197.0
50.34	1356.0
3.35	1444.0
2.20	1550.0
keV/decay	957.0

^{18}F   109.77 min

Positrons

Freq. (%)	E_{avg} (keV)	E_{max} (keV)
100.00	249.8	633.5
keV/decay	249.8	

Principal Gamma and X Rays

Freq. (%)	E (keV)
200.00	511.0
keV/decay	1022.0

^{22}Na   2.602 y

Positrons

Freq. (%)	E_{avg} (keV)	E_{max} (keV)
89.84	215.4	545.5
keV/decay	194.0	

Principal Conversion and Auger Electrons

keV/decay	0.088 *

Principal Gamma and X Rays

Freq. (%)	E (keV)
179.80	511.0
99.94	1274.5
keV/decay	2192.6

^{24}Na   15.020 h

Principal Beta Particles

Freq. (%)	E_{avg} (keV)	E_{max} (keV)
99.94	553.8	1390.2
keV/decay	553.5	

Principal Conversion and Auger Electrons

keV/decay	0.021 *

Principal Gamma and X Rays

Freq. (%)	E (keV)
100.00	1368.5
99.94	2754.1
keV/decay	4123.1

^{32}P   14.26 d

Principal Beta Particles

Freq. (%)	E_{avg} (keV)	E_{max} (keV)
100.00	694.7	1710.4
keV/decay	694.7	

^{35}S   87.44 d

Principal Beta Particles

Freq. (%)	E_{avg} (keV)	E_{max} (keV)
100.00	48.8	167.5
keV/decay	48.8	

^{36}Cl 3.01×10^5 y

Positrons

keV/decay	0.009 *

Principal Beta Particles

Freq. (%)	E_{avg} (keV)	E_{max} (keV)
98.10	278.8	709.5
keV/decay	273.5	

Principal Conversion and Auger Electrons

keV/decay	0.033 *

Principal Gamma and X Rays

keV/decay	0.155 *

^{38}Cl   37.21 min

Principal Beta Particles

Freq. (%)	E_{avg} (keV)	E_{max} (keV)
32.53	420.1	1107.2
11.41	1181.4	2749.6
56.06	2244.0	4917.2
keV/decay	1529.4	

Principal Conversion and Auger Electrons

keV/decay	0.020 *

Principal Gamma and X Rays

Freq. (%)	E (keV)
32.50	1642.4
44.00	2167.5
keV/decay	1488.4

^{41}Ar   1.827 h

Principal Beta Particles

Freq. (%)	E$_{avg}$ (keV)	E$_{max}$ (keV)
99.17	459.0	1198.1
keV/decay	463.7	

Principal Conversion and Auger Electrons

keV/decay	0.089 *

Principal Gamma and X Rays

Freq. (%)	E (keV)
99.16	1293.6
keV/decay	1283.6

^{38}K   7.636 min

Positrons

Freq. (%)	E$_{avg}$ (keV)	E$_{max}$ (keV)
99.29	1216.5	2733.4
keV/decay	1208.5	

Principal Conversion and Auger Electrons

keV/decay	0.041 *

Principal Gamma and X Rays

Freq. (%)	E (keV)
198.93	511.0
99.80	2167.0
keV/decay	3186.9

^{40}K 1.28×10^9 y

Principal Beta Particles

Freq. (%)	E$_{avg}$ (keV)	E$_{max}$ (keV)
89.30	585.0	1311.6
keV/decay	522.4	

Principal Conversion and Auger Electrons

keV/decay	0.202 *

Principal Gamma and X Rays

Freq. (%)	E (keV)
10.70	1460.7
keV/decay	156.3

^{42}K   12.36 h

Principal Beta Particles

Freq. (%)	E$_{avg}$ (keV)	E$_{max}$ (keV)
17.49	822.0	1996.4
82.07	1563.7	3521.1
keV/decay	1429.5	

Principal Conversion and Auger Electrons

keV/decay	0.013 *

Principal Gamma and X Rays

Freq. (%)	E (keV)
17.90	1524.7
keV/decay	275.9

^{43}K   22.6 h

Principal Beta Particles

Freq. (%)	E$_{avg}$ (keV)	E$_{max}$ (keV)
2.24	136.9	422.4
91.97	297.9	826.7
3.69	469.1	1223.6
1.30	761.5	1817.0
keV/decay	308.8	

Principal Conversion and Auger Electrons

keV/decay	0.292 *

Principal Gamma and X Rays

Freq. (%)	E (keV)
4.11	220.6
87.27	372.8
11.43	396.9
11.03	593.4
80.51	617.5
1.88	1021.8
keV/decay	970.0

^{45}Ca   163.8 d

Principal Beta Particles

Freq. (%)	E$_{avg}$ (keV)	E$_{max}$ (keV)
100.00	77.2	256.9
keV/decay	77.2	

Principal Conversion and Auger Electrons

keV/decay	2.22×10^{-4} *

Principal Gamma and X Rays

keV/decay	1.34×10^{-5}

^{47}Ca   4.536 d

Radioactive Daughter   Half Life
^{47}Sc (100.0%)   3.345 d

Principal Beta Particles

Freq. (%)	E$_{avg}$ (keV)	E$_{max}$ (keV)
80.83	241.0	690.9
18.96	816.7	1988.0
keV/decay	350.5	

Principal Conversion and Auger Electrons

keV/decay	0.069 *

Principal Gamma and X Rays

Freq. (%)	E (keV)
6.51	489.2
6.51	807.9
74.00	1297.1
keV/decay	1047.1

^{46}Sc   83.83 d

Principal Beta Particles

Freq. (%)	E$_{avg}$ (keV)	E$_{max}$ (keV)
100.00	112.0	357.3
keV/decay	112.0	

Principal Conversion and Auger Electrons

keV/decay	0.247 *

Principal Gamma and X Rays

Freq. (%)	E (keV)
99.98	889.3
99.99	1120.5
keV/decay	2009.5

^{47}Sc   3.345 d

Principal Beta Particles

Freq. (%)	E$_{avg}$ (keV)	E$_{max}$ (keV)
68.30	142.7	441.2
31.70	204.1	600.6
keV/decay	162.2	

Principal Conversion and Auger Electrons

keV/decay	0.663 *

more ...

^{47}Sc (continued)

Principal Gamma and X Rays

Freq. (%)	E (keV)
67.90	159.4
keV/decay	108.2

^{51}Cr   27.704 d

Principal Conversion and Auger Electrons

keV/decay	3.87 *

Principal Gamma and X Rays

Freq. (%)	E (keV)
10.08	320.1
keV/decay	33.4

^{54}Mn   312.5 d

Principal Conversion and Auger Electrons

keV/decay	4.22 *

Principal Gamma and X Rays

Freq. (%)	E (keV)
99.98	834.8
keV/decay	836.0

^{52}Fe   8.275 h

Radioactive Daughter   Half Life
^{52m}Mn (100.0%)   21.1 min

Positrons

Freq. (%)	E$_{avg}$ (keV)	E$_{max}$ (keV)
56.00	339.9	803.6
keV/decay	190.4	

Principal Conversion and Auger Electrons

keV/decay	3.15 *

Principal Gamma and X Rays

Freq. (%)	E (keV)
99.23	168.7
112.00	511.0
keV/decay	740.5

^{55}Fe   2.73 y

Principal Conversion and Auger Electrons
keV/decay   4.20 *

Principal Gamma and X Rays
keV/decay   1.69 *

^{59}Fe   496 d

Principal Beta Particles

Freq. (%)	E_{avg} (keV)	E_{max} (keV)
1.29	35.6	130.5
45.26	80.9	273.2
53.17	149.1	465.5
keV/decay	117.5	

Principal Conversion and Auger Electrons
keV/decay   0.252 *

Principal Gamma and X Rays

Freq. (%)	E (keV)
1.02	142.7
3.08	192.3
56.50	1099.3
43.20	1291.6
keV/decay	1188.3

^{57}Co   271.80 d

Principal Conversion and Auger Electrons

Freq. (%)	E (keV)
6.83	13.6
1.84	114.9
1.39	129.4
keV/decay	18.6 *

Principal Gamma and X Rays

Freq. (%)	E (keV)
9.23	14.4
85.95	122.1
10.33	136.5
keV/decay	125.3

^{58}Co   70.916 d

Positrons

Freq. (%)	E_{avg} (keV)	E_{max} (keV)
14.99	201.2	475.0
keV/decay	30.2	

Principal Conversion and Auger Electrons
keV/decay   4.01 *

Principal Gamma and X Rays

Freq. (%)	E (keV)
29.98	511.0
99.45	810.8
keV/decay	975.8

^{60}Co   5.2704 y

Principal Beta Particles

Freq. (%)	E_{avg} (keV)	E_{max} (keV)
99.94	95.8	317.9
keV/decay	96.1	

Principal Conversion and Auger Electrons
keV/decay   0.359 *

Principal Gamma and X Rays

Freq. (%)	E (keV)
99.90	1173.2
99.99	1332.5
keV/decay	2504.5

^{63}Ni   100.1 y

Principal Beta Particles

Freq. (%)	E_{avg} (keV)	E_{max} (keV)
100.00	17.1	65.9
keV/decay	17.1	

^{62}Zn   9.26 h

Radioactive Daughter	Half Life
^{62}Cu (100.0%)	9.74 min

Positrons

Freq. (%)	E_{avg} (keV)	E_{max} (keV)
8.39	258.6	605.0
keV/decay	21.7	

more ...

^{62}Zn   (continued)

Principal Conversion and Auger Electrons

Freq. (%)	E (keV)
14.65	31.9
1.46	39.8
keV/decay	10.9 *

Principal Gamma and X Rays

Freq. (%)	E (keV)
25.19	40.9
2.49	243.4
1.88	247.0
1.34	260.4
2.21	394.0
14.65	507.6
16.78	511.0
15.16	548.3
25.70	596.6
keV/decay	439.0

---

^{65}Zn   243.9 d

Positrons

Freq. (%)	E_{avg} (keV)	E_{max} (keV)
1.42	142.6	328.8
keV/decay	2.02	

Principal Conversion and Auger Electrons

keV/decay	4.79 *

Principal Gamma and X Rays

Freq. (%)	E (keV)
2.83	511.0
50.70	1115.5
keV/decay	583.2

---

^{69m}Zn   13.76 h

Radioactive Daughter

	Half Life
^{69}Zn (100.0%)	55.6 min

Principal Beta Particles

keV/decay	0.096 *

Principal Conversion and Auger Electrons

Freq. (%)	E (keV)
4.53	429.0
keV/decay	22.2 *

more ...

^{69m}Zn   (continued)

Principal Gamma and X Rays

Freq. (%)	E (keV)
94.87	438.6
keV/decay	416.5

---

^{66}Ga   9.49 h

Positrons

Freq. (%)	E_{avg} (keV)	E_{max} (keV)
3.80	397.0	923.7
50.24	1903.4	4153.0
keV/decay	981.0	

Principal Conversion and Auger Electrons

keV/decay	2.31 *

Principal Gamma and X Rays

Freq. (%)	E (keV)
113.07	511.0
6.03	833.5
37.90	1039.2
1.23	1333.1
2.14	1918.6
5.71	2189.9
1.96	2422.8
23.19	2752.0
1.48	3229.2
1.40	3381.3
1.02	3791.6
1.14	4086.3
3.49	4295.9
1.48	4806.6
keV/decay	2454.6 *

---

^{67}Ga   3.261 d

Principal Conversion and Auger Electrons

Freq. (%)	E (keV)
27.77	83.7
2.46	92.1
keV/decay	34.4 *

more ...

^{67}Ga (continued)

Principal Gamma and X Rays

Freq. (%)	E (keV)
2.96	91.3
37.00	93.3
20.40	184.6
2.33	209.0
16.60	300.2
4.64	393.5
keV/decay	154.8

^{68}Ga   68.06 min

Positrons

Freq. (%)	E$_{avg}$ (keV)	E$_{max}$ (keV)
1.12	352.6	821.7
87.92	835.8	1899.1
keV/decay	738.8	

Principal Conversion and Auger Electrons
keV/decay   0.544 *

Principal Gamma and X Rays

Freq. (%)	E (keV)
178.08	511.0
3.00	1077.4
keV/decay	947.4

^{72}Ga   14.10 h

Principal Beta Particles

Freq. (%)	E$_{avg}$ (keV)	E$_{max}$ (keV)
14.97	217.1	650.2
21.65	223.7	667.0
27.64	341.9	956.4
1.87	381.0	1048.5
8.88	569.1	1477.2
2.98	773.9	1927.0
8.48	1054.8	2528.0
10.28	1354.2	3158.0
keV/decay	497.8	

Principal Conversion and Auger Electrons
keV/decay   4.23 *

more ...

^{72}Ga (continued)

Principal Gamma and X Rays

Freq. (%)	E (keV)
5.54	601.0
24.77	630.0
3.20	786.4
2.01	810.2
95.63	834.0
9.88	894.2
1.10	970.5
6.91	1050.7
1.45	1230.9
1.45	1230.9
1.13	1260.1
1.56	1276.8
3.55	1464.0
4.24	1596.7
5.25	1861.1
1.04	2109.5
25.92	2201.7
7.68	2491.0
12.78	2507.8
keV/decay	2724.2

^{72}As   26.0 h

Positrons

Freq. (%)	E$_{avg}$ (keV)	E$_{max}$ (keV)
5.80	821.8	1865.0
64.04	1114.4	2495.0
16.29	1525.7	3329.0
keV/decay	1021.3	

Principal Conversion and Auger Electrons

Freq. (%)	E (keV)
1.43	679.9
keV/decay	12.0 *

Principal Gamma and X Rays

Freq. (%)	E (keV)
175.27	511.0
7.92	629.9
79.50	834.0
1.11	1464.0
keV/decay	1776.4 *

^{74}As   17.76 d

Positrons

Freq. (%)	E_{avg} (keV)	E_{max} (keV)
25.70	407.9	944.3
3.40	700.9	1540.1
keV/decay	128.6	

Principal Beta Particles

Freq. (%)	E_{avg} (keV)	E_{max} (keV)
15.38	242.8	718.3
18.78	530.9	1353.1
keV/decay	137.1	

Principal Conversion and Auger Electrons

keV/decay	2.49 *

Principal Gamma and X Rays

Freq. (%)	E (keV)
1.45	11.0
58.20	511.0
59.22	595.8
15.39	634.8
keV/decay	758.5

^{76}As   26.32 h

Principal Beta Particles

Freq. (%)	E_{avg} (keV)	E_{max} (keV)
1.19	92.8	313.6
1.89	173.8	540.1
2.02	436.3	1181.2
7.60	691.6	1752.8
34.59	996.2	2409.8
50.99	1266.9	2968.9
keV/decay	1063.8	

Principal Conversion and Auger Electrons

keV/decay	0.578 *

Principal Gamma and X Rays

Freq. (%)	E (keV)
44.70	559.1
1.17	563.2
6.08	657.0
1.63	1212.7
3.84	1216.0
1.39	1228.5
keV/decay	430.0 *

^{75}Se   119.770 d

Principal Conversion and Auger Electrons

Freq. (%)	E (keV)
3.02	10.2
2.48	10.4
4.48	10.4
1.05	11.6
5.32	12.7
2.70	84.9
1.55	124.1
keV/decay	14.7 *

Principal Gamma and X Rays

Freq. (%)	E (keV)
16.50	10.5
32.11	10.5
2.41	11.7
4.72	11.7
1.14	66.1
3.48	96.7
17.32	121.1
58.98	136.0
1.47	198.6
59.10	264.7
25.18	279.5
1.34	303.9
11.56	400.7
keV/decay	391.7

^{76}Br   16.2 h

Positrons

Freq. (%)	E_{avg} (keV)	E_{max} (keV)
1.42	333.8	770.0
6.20	373.0	860.3
5.12	425.1	979.4
1.22	549.3	1259.8
2.76	1218.5	2713.9
2.07	1262.7	2807.7
25.41	1529.8	3370.9
5.91	1797.3	3930.0
keV/decay	642.1	

Principal Conversion and Auger Electrons

Freq. (%)	E (keV)
1.16	10.8
1.71	11.0
keV/decay	3.57 *

more ..

^{76}Br   (continued)

Principal Gamma and X Rays

Freq. (%)	E (keV)
6.98	11.2
13.55	11.2
1.04	12.5
2.03	12.5
1.86	472.9
109.53	511.0
74.00	559.1
3.55	563.2
15.91	657.0
4.59	1129.8
1.70	1213.1
8.81	1216.1
2.09	1228.7
2.52	1380.5
2.31	1471.1
14.65	1853.7
1.36	2096.7
2.49	2111.2
4.74	2391.2
1.95	2510.8
5.62	2792.7
7.40	2950.5
1.55	3604.0
keV/decay	2790.4 *

^{77}Br   57.036 h

Positrons
keV/decay	1.14 *

Principal Conversion and Auger Electrons

Freq. (%)	E (keV)
2.58	10.8
2.13	11.0
3.79	11.0
keV/decay	8.36 *

Principal Gamma and X Rays

Freq. (%)	E (keV)
15.48	11.2
30.06	11.2
2.31	12.5
4.51	12.5
1.45	87.6
1.14	161.8
1.25	200.4

more ...

^{77}Br   (continued)

Principal Gamma and X Rays   (continued)

Freq. (%)	E (keV)
23.89	239.0
3.08	249.8
2.37	281.7
4.30	297.2
1.22	303.8
1.62	439.5
1.03	484.6
1.48	511.0
23.17	520.7
1.23	574.6
3.06	578.9
1.62	585.5
1.72	755.3
2.15	817.8
keV/decay	331.4 *

^{82}Br   35.30 h

Principal Beta Particles

Freq. (%)	E$_{avg}$ (keV)	E$_{max}$ (keV)
1.36	76.2	264.6
98.64	137.8	444.3
keV/decay	137.0	

Principal Conversion and Auger Electrons
keV/decay	2.06 *

Principal Gamma and X Rays

Freq. (%)	E (keV)
2.26	221.4
70.78	554.3
1.17	606.3
43.45	619.1
28.49	698.3
83.56	776.5
24.04	827.8
1.27	1007.6
27.20	1044.0
26.52	1317.5
16.32	1474.8
keV/decay	2642.0

^{81m}Kr   13 s

Principal Conversion and Auger Electrons

Freq. (%)	E (keV)
1.10	10.6
3.42	10.8
1.05	12.4
26.86	176.1
2.51	188.5
1.03	188.7
1.04	188.8
keV/decay	59.0

Principal Gamma and X Rays

Freq. (%)	E (keV)
5.07	12.6
9.81	12.7
1.50	14.1
67.10	190.4
keV/decay	130.0

^{85m}Kr   4.480 h

Radioactive Daughter	Half Life
^{85}Kr (21.1%)	10.72 y

Principal Beta Particles

Freq. (%)	E_{avg} (keV)	E_{max} (keV)
79.00	290.3	840.7
keV/decay	229.3	

Principal Conversion and Auger Electrons

Freq. (%)	E (keV)
3.15	136.0
6.00	290.5
keV/decay	26.1 *

Principal Gamma and X Rays

Freq. (%)	E (keV)
1.13	12.6
2.19	12.7
1.18	13.4
75.37	151.2
13.76	304.9
keV/decay	156.7

^{85}Kr   10.72 y

Principal Beta Particles

Freq. (%)	E_{avg} (keV)	E_{max} (keV)
99.56	251.4	687.0
keV/decay	250.5	

Principal Conversion and Auger Electrons
keV/decay   0.016 *

Principal Gamma and X Rays
keV/decay   2.24 *

^{87}Kr   76.3 min

Radioactive Daughter	Half Life
^{87}Rb (100.0%)	4.7×10^{10} y

Principal Beta Particles

Freq. (%)	E_{avg} (keV)	E_{max} (keV)
4.39	326.0	927.4
9.48	499.9	1333.2
5.59	562.1	1473.6
6.88	1293.6	3042.6
40.69	1501.3	3485.4
30.42	1694.0	3888.0
keV/decay	1322.9	

Principal Conversion and Auger Electrons
keV/decay   0.899 *

Principal Gamma and X Rays

Freq. (%)	E (keV)
49.50	402.6
1.91	673.9
7.28	845.4
1.12	1175.4
2.05	1740.5
2.90	2011.9
9.31	2554.8
3.91	2558.1
keV/decay	793.1 *

^{88}Kr   2.84 h

Radioactive Daughter	Half Life
^{88}Rb (100.0%)	17.8 min

Principal Beta Particles

Freq. (%)	E_{avg} (keV)	E_{max} (keV)
2.64	109.5	364.6
66.80	165.4	520.8
9.07	226.5	681.2
1.91	441.0	1198.3
1.31	824.4	2050.7
1.79	1135.9	2716.7
13.96	1233.0	2913.0
keV/decay	358.6	

Principal Conversion and Auger Electrons

Freq. (%)	E (keV)
1.43	11.4
10.75	12.3
1.13	25.4
1.17	181.1
keV/decay	5.70 *

Principal Gamma and X Rays

Freq. (%)	E (keV)
2.40	13.3
4.63	13.4
2.07	27.5
3.10	166.0
25.98	196.3
2.25	362.2
12.97	834.8
1.31	985.8
1.28	1141.3
1.12	1250.7
1.48	1369.5
2.15	1518.4
10.93	1529.8
4.53	2029.8
3.74	2035.4
13.18	2195.8
3.39	2231.8
34.60	2392.1
keV/decay	1954.6 *

^{89}Kr   3.16 min

Radioactive Daughter	Half Life
^{89}Rb (100.0%)	15.44 min

Principal Beta Particles

Freq. (%)	E_{avg} (keV)	E_{max} (keV)
4.00	850.0	2103.9
14.40	970.0	2371.9
5.70	1070.0	2569.1
3.09	1180.0	2810.0
2.53	1260.0	2971.5
10.20	1400.0	3276.2
2.90	1480.0	3439.8
3.60	1580.0	3645.7
1.30	1730.0	3972.5
2.30	1930.0	4384.1
4.40	1940.0	4393.0
1.20	1980.0	4472.5
23.00	2210.0	4970.0
keV/decay	1360.7 *	

Principal Conversion and Auger Electrons

keV/decay	1.3

Principal Gamma and X Rays

Freq. (%)	E (keV)
20.00	220.9
6.64	497.5
5.64	577.0
16.60	585.8
4.20	738.4
5.92	867.1
7.18	904.3
2.92	1107.8
1.66	1116.6
3.06	1324.3
6.88	1472.8
1.32	1501.0
3.32	1530.0
5.12	1533.7
4.38	1693.7
1.04	1903.4
1.56	2012.2
1.74	2866.2
1.04	3140.3
1.04	3361.7
1.34	3532.9
7.12	2181.1
keV/decay	1834.5 *

^{81}Rb

4.576 h

Radioactive Daughter   Half Life
^{81}Kr (100.0%)   2.13×10^5 y

Positrons

Freq. (%)	E$_{avg}$ (keV)	E$_{max}$ (keV)
1.79	252.7	577.4
26.31	446.3	1023.5
keV/decay	123.3	

Principal Conversion and Auger Electrons

Freq. (%)	E (keV)
1.67	10.4
3.64	10.6
2.12	10.7
11.27	10.8
3.23	10.9
2.39	12.2
1.93	12.4
3.45	12.4
25.69	176.0
2.40	188.4
1.00	188.6
keV/decay	60.9 *

Principal Gamma and X Rays

Freq. (%)	E (keV)
16.73	12.6
32.35	12.7
2.54	14.1
4.95	14.1
64.03	190.3
23.20	446.1
3.02	456.7
5.34	510.5
57.05	511.0
2.23	537.6
keV/decay	644.9 *

^{84}Rb

32.87 d

Positrons

Freq. (%)	E$_{avg}$ (keV)	E$_{max}$ (keV)
12.41	338.5	776.7
13.51	756.3	1658.1
keV/decay	144.1	

more ...

^{84}Rb (continue)

Principal Beta Particles

Freq. (%)	E$_{avg}$ (keV)	E$_{max}$ (keV)
4.00	331.2	890.0
keV/decay	13.2	

Principal Conversion and Auger Electrons

Freq. (%)	E (keV)
1.16	10.4
2.52	10.6
1.47	10.7
7.82	10.8
2.24	10.9
1.65	12.2
1.34	12.4
2.40	12.4
keV/decay	4.18 *

Principal Gamma and X Rays

Freq. (%)	E (keV)
11.60	12.6
22.44	12.7
1.76	14.1
3.43	14.1
51.82	511.0
67.87	881.5
keV/decay	886.6

^{86}Rb

18.66 d

Principal Beta Particles

Freq. (%)	E$_{avg}$ (keV)	E$_{max}$ (keV)
8.78	232.6	698.0
91.22	709.4	1774.6
keV/decay	667.5	

Principal Conversion and Auger Electrons
keV/decay   0.045 *

Principal Gamma and X Rays

Freq. (%)	E (keV)
8.78	1076.6
keV/decay	94.5

^{85}Sr   64.84 d

Principal Conversion and Auger Electrons

Freq. (%)	E (keV)
1.57	11.0
3.29	11.2
2.02	11.3
10.21	11.4
3.09	11.5
2.26	12.9
1.82	13.1
3.24	13.1
keV/decay	8.99 *

Principal Gamma and X Rays

Freq. (%)	E (keV)
17.10	13.3
32.98	13.4
2.64	15.0
5.15	15.0
98.30	514.0
keV/decay	513.4

---

^{87m}Sr   2.81 h

Radioactive Daughter
	Half Life
^{87}Rb (0.3%)	4.73×10^{10} y

Principal Conversion and Auger Electrons

Freq. (%)	E (keV)
1.60	12.1
15.02	372.3
1.78	386.2
keV/decay	67.1 *

Principal Gamma and X Rays

Freq. (%)	E (keV)
3.01	14.1
5.79	14.2
82.26	388.4
keV/decay	321.0

---

^{89}Sr   50.5 d

Principal Beta Particles

Freq. (%)	E_{avg} (keV)	E_{max} (keV)
99.99	583.3	1492.1
keV/decay	583.2	

more ...

^{89}Sr (continued)

Principal Conversion and Auger Electrons
keV/decay	7.10×10^{-4} *

Principal Gamma and X Rays
keV/decay	0.085 *

---

^{90}Sr   29.12 y

Radioactive Daughter
	Half Life
^{90}Y (100.0%)	64.0 h

Principal Beta Particles

Freq. (%)	E_{avg} (keV)	E_{max} (keV)
100.00	195.7	546.0
keV/decay	195.7	

---

^{90}Y   64.0 h

Principal Beta Particles

Freq. (%)	E_{avg} (keV)	E_{max} (keV)
99.98	934.8	2284.0
keV/decay	934.6	

Principal Conversion and Auger Electrons
keV/decay	0.280 *

Principal Gamma and X Rays
keV/decay	0.002 *

---

^{91}Y   58.51 d

Principal Beta Particles

Freq. (%)	E_{avg} (keV)	E_{max} (keV)
99.70	603.7	1543.0
keV/decay	602.2	

Principal Conversion and Auger Electrons
keV/decay	0.002 *

Principal Gamma and X Rays
keV/decay	3.61 *

^{95}Zr   64.02 d

Radioactive Daughter   **Half Life**
^{95}Nb   (99.2%)   35.02 d

Principal Beta Particles

Freq. (%)	E_{avg} (keV)	E_{max} (keV)
54.59	109.2	366.4
44.20	120.4	398.9
1.11	327.0	887.4
keV/decay	116.9	

Principal Conversion and Auger Electrons
keV/decay   1.02 *

Principal Gamma and X Rays

Freq. (%)	E (keV)
44.14	724.2
54.50	756.7
keV/decay	732.1

^{95m}Nb   86.6 h

Radioactive Daughter   **Half Life**
^{95}Nb   (94.5%)   35.02 d

Principal Beta Particles

Freq. (%)	E_{avg} (keV)	E_{max} (keV)
2.31	334.9	957.2
3.22	437.7	1161.3
keV/decay	21.9	

Principal Conversion and Auger Electrons

Freq. (%)	E (keV)
1.45	13.7
1.04	13.8
4.42	14.0
1.58	14.1
1.14	15.9
1.61	16.2
54.39	216.7
7.41	233.0
1.23	233.2
2.23	233.3
2.00	235.3
keV/decay	152.6

more ...

^{95m}Nb   (continued)

Principal Gamma and X Rays

Freq. (%)	E (keV)
11.71	16.5
22.35	16.6
1.99	18.6
3.73	18.6
2.24	204.1
24.07	235.7
keV/decay	68.7

^{95}Nb   35.02 d

Principal Beta Particles

Freq. (%)	E_{avg} (keV)	E_{max} (keV)
99.95	43.3	159.7
keV/decay	43.5	

Principal Conversion and Auger Electrons
keV/decay   1.09 *

Principal Gamma and X Rays

Freq. (%)	E (keV)
99.79	765.8
keV/decay	764.3

^{99}Mo   65.94 h

Radioactive Daughter   **Half Life**
^{99m}Tc   (88.6%)   6.01 h

Principal Beta Particles

Freq. (%)	E_{avg} (keV)	E_{max} (keV)
16.39	133.0	436.3
1.13	289.6	847.9
81.86	442.7	1214.3
keV/decay	389.2	

Principal Conversion and Auger Electrons

Freq. (%)	E (keV)
3.44	19.5
keV/decay	3.51 *

more ...

^{99}Mo (continued)

Principal Gamma and X Rays

Freq. (%)	E (keV)
1.06	18.3
2.01	18.4
1.05	40.6
4.52	140.5
6.08	181.1
1.15	366.4
12.13	739.6
4.34	778.0
keV/decay	149.2

^{99m}Tc   6.01 h

Radioactive Daughter	Half Life
^{99}Tc (100.0%)	2.11×10^5 y

Principal Beta Particles

keV/decay	0.004 *

Principal Conversion and Auger Electrons

Freq. (%)	E (keV)
8.84	119.5
keV/decay	16.1 *

Principal Gamma and X Rays

Freq. (%)	E (keV)
2.10	18.3
3.99	18.4
89.06	140.5
keV/decay	126.5

^{103}Ru   39.26 d

Radioactive Daughter	Half Life
^{103m}Rh (99.7%)	56.11 min

Principal Beta Particles

Freq. (%)	E$_{avg}$ (keV)	E$_{max}$ (keV)
6.61	30.8	116.1
92.18	64.2	229.4
keV/decay	63.8	

Principal Conversion and Auger Electrons

keV/decay	2.77 *

more ...

^{103}Ru (continued)

Principal Gamma and X Rays

Freq. (%)	E (keV)
90.90	497.1
5.73	610.3
keV/decay	494.7

^{106}Ru   368.2 d

Radioactive Daughter	Half Life
^{106}Rh (100.0%)	29.9 s

Principal Beta Particles

Freq. (%)	E$_{avg}$ (keV)	E$_{max}$ (keV)
100.00	10.0	39.4
keV/decay	10.0	

^{103m}Rh   56.114 min

Principal Conversion and Auger Electrons

Freq. (%)	E (keV)
9.54	16.5
29.34	36.6
41.40	36.8
14.39	39.3
2.48	39.8
keV/decay	37.1 *

Principal Gamma and X Rays

Freq. (%)	E (keV)
2.20	20.1
4.17	20.2
keV/decay	1.71 *

^{106}Rh   29.9 s

Principal Beta Particles

Freq. (%)	E$_{avg}$ (keV)	E$_{max}$ (keV)
1.70	779.2	1978.1
9.69	976.1	2406.4
8.39	1266.5	3028.2
78.84	1508.0	3540.0
keV/decay	1412.3	

Principal Conversion and Auger Electrons

keV/decay	0.815 *

more ...

106Rh (continued)

Principal Gamma and X Rays

Freq. (%)	E (keV)
20.60	511.8
9.81	621.8
1.46	1050.1
keV/decay	204.9 *

110mAg   249.9 d

Radioactive Daughter

	Half Life
110Ag (1.3%)	24.6 s

Principal Beta Particles

Freq. (%)	E$_{avg}$ (keV)	E$_{max}$ (keV)
67.49	21.8	83.9
30.59	165.5	530.7
keV/decay	65.7	

Principal Conversion and Auger Electrons

keV/decay	6.54 *

Principal Gamma and X Rays

Freq. (%)	E (keV)
3.66	446.8
2.78	620.3
94.74	657.7
10.72	677.6
6.49	687.0
16.74	706.7
4.66	744.3
22.36	763.9
7.32	818.0
72.86	884.7
34.32	937.5
24.35	1384.3
3.99	1475.8
13.11	1505.0
1.18	1562.3
keV/decay	2750.5

110Ag   24.6 s

Principal Beta Particles

Freq. (%)	E$_{avg}$ (keV)	E$_{max}$ (keV)
4.40	893.9	2235.1
95.21	1199.1	2892.8
keV/decay	1181.4	

more ...

110Ag (continued)

Principal Conversion and Auger Electrons

keV/decay	0.110 *

Principal Gamma and X Rays

Freq. (%)	E (keV)
4.50	657.7
keV/decay	30.6

109Cd   462.9 d

Principal Conversion and Auger Electrons

Freq. (%)	E (keV)
1.29	17.8
2.04	18.1
1.72	18.3
5.89	18.5
2.40	18.7
1.87	21.1
1.46	21.4
2.52	21.6
41.77	62.5
3.26	84.2
18.86	84.5
21.93	84.7
8.95	87.5
1.61	88.0
keV/decay	82.6 *

Principal Gamma and X Rays

Freq. (%)	E (keV)
28.47	22.0
53.66	22.2
4.95	24.9
9.66	24.9
2.71	25.5
3.61	88.0
keV/decay	26.0

^{111}In   2.83 d

Principal Conversion and Auger Electrons

Freq. (%)	E (keV)
1.01	18.6
1.57	18.9
1.34	19.0
4.48	19.3
1.85	19.5
1.46	22.0
1.13	22.3
1.96	22.5
8.41	144.6
5.04	218.6
keV/decay	34.7 *

Principal Gamma and X Rays

Freq. (%)	E (keV)
23.61	23.0
44.38	23.2
4.15	26.1
8.08	26.1
2.35	26.6
90.24	171.3
94.00	245.4
keV/decay	404.9

^{113m}In   1.658 h

Principal Conversion and Auger Electrons

Freq. (%)	E (keV)
1.21	20.1
28.80	363.7
4.09	387.5
1.13	391.0
keV/decay	134.0 *

Principal Gamma and X Rays

Freq. (%)	E (keV)
6.99	24.0
13.12	24.2
1.24	27.2
2.40	27.3
64.23	391.7
keV/decay	257.7

^{115m}In   4.486 h

Radioactive Daughter	Half Life
^{115}In (96.3%)	4.36 h

Principal Beta Particles

Freq. (%)	E$_{avg}$ (keV)	E$_{max}$ (keV)
4.95	279.3	831.2
keV/decay	13.9	

Principal Conversion and Auger Electrons

Freq. (%)	E (keV)
1.65	20.1
39.29	308.3
5.78	332.0
1.51	332.5
1.69	335.6
keV/decay	158.5

Principal Gamma and X Rays

Freq. (%)	E (keV)
9.54	24.0
17.90	24.2
1.69	27.2
3.28	27.3
45.37	336.2
keV/decay	161.1

^{125}Sb   2.77 y

Radioactive Daughter	Half Life
^{125m}Te (23.1%)	119.7 d

Principal Beta Particles

Freq. (%)	E$_{avg}$ (keV)	E$_{max}$ (keV)
13.54	25.0	95.4
5.72	33.1	124.8
18.15	34.8	130.8
1.50	67.6	241.8
40.32	87.0	303.5
7.22	134.6	445.9
13.54	215.5	622.2
keV/decay	86.6	

Principal Conversion and Auger Electrons

Freq. (%)	E (keV)
1.80	22.7
6.17	30.5
1.35	34.6
keV/decay	13.6 *

more ...

^{125}Sb (continued)

Principal Gamma and X Rays

Freq. (%)	E (keV)
13.43	27.2
25.01	27.5
2.43	30.9
4.73	31.0
1.45	31.7
4.31	35.5
6.70	176.3
1.50	380.5
29.50	427.9
10.33	463.4
17.64	600.6
4.84	606.7
11.33	636.0
1.72	671.5
keV/decay	430.8

^{123}I   13.2 h

Radioactive Daughter	Half Life
^{123}Te (100.0%)	1×10^{13} y

Principal Conversion and Auger Electrons

Freq. (%)	E (keV)
1.24	22.1
1.10	22.4
3.30	22.7
1.42	23.0
1.19	26.0
1.57	26.6
13.62	127.2
1.61	154.0
keV/decay	28.4 *

Principal Gamma and X Rays

Freq. (%)	E (keV)
24.65	27.2
45.90	27.5
4.46	30.9
8.68	31.0
2.66	31.7
83.30	159.0
1.39	529.0
keV/decay	172.7 *

^{124}I   4.18 d

Positrons

Freq. (%)	E$_{avg}$ (keV)	E$_{max}$ (keV)
11.29	685.9	1532.3
11.29	973.6	2135.0
keV/decay	188.4	

Principal Conversion and Auger Electrons

Freq. (%)	E (keV)
2.20	22.7
1.05	26.6
keV/decay	7.14 *

Principal Gamma and X Rays

Freq. (%)	E (keV)
16.47	27.2
30.67	27.5
2.98	30.9
5.80	31.0
1.78	31.7
45.73	511.0
60.50	602.7
9.98	722.8
1.43	1325.5
1.66	1376.0
2.99	1509.5
10.41	1691.0
keV/decay	1083.2 *

^{125}I   60.14 d

Principal Conversion and Auger Electrons

Freq. (%)	E (keV)
1.37	21.8
2.00	22.1
1.78	22.4
5.31	22.7
2.28	23.0
1.92	26.0
1.50	26.3
2.54	26.6
9.49	30.6
2.13	34.7
keV/decay	19.5 *

more ...

^{125}I (continued)

Principal Gamma and X Rays

Freq. (%)	E (keV)
39.71	27.2
73.95	27.5
7.19	30.9
13.98	31.0
4.29	31.7
6.65	35.5
keV/decay	42.0

^{129}I 1.57×10^7 y

Principal Beta Particles

Freq. (%)	E_{avg} (keV)	E_{max} (keV)
100.00	50.3	154.4
keV/decay	50.3	

Principal Conversion and Auger Electrons

Freq. (%)	E (keV)
2.27	24.5
1.13	28.8
9.64	32.1
2.14	36.7
keV/decay	13.0 *

Principal Gamma and X Rays

Freq. (%)	E (keV)
19.96	29.5
37.02	29.8
3.65	33.6
7.11	33.6
2.37	34.4
7.50	37.6
keV/decay	24.5

^{131}I 8.04 d

Radioactive Daughter	Half Life
^{131m}Xe (1.1%)	11.84 d

Principal Beta Particles

Freq. (%)	E_{avg} (keV)	E_{max} (keV)
2.13	69.4	247.9
7.36	96.6	333.8
89.41	191.5	606.3
keV/decay	181.7	

more ...

^{131}I (continued)

Principal Conversion and Auger Electrons

Freq. (%)	E (keV)
3.63	45.6
1.55	329.9
keV/decay	10.1 *

Principal Gamma and X Rays

Freq. (%)	E (keV)
1.40	29.5
2.59	29.8
2.62	80.2
6.06	284.3
81.24	364.5
7.27	637.0
1.80	722.9
keV/decay	381.6

^{133}I 20.8 h

Radioactive Daughter	Half Life
^{133}Xe (97.1%)	5.25 d

Principal Beta Particles

Freq. (%)	E_{avg} (keV)	E_{max} (keV)
1.24	109.8	373.8
3.74	139.7	461.8
3.12	161.5	523.6
4.15	298.5	884.7
1.81	351.6	1016.2
83.52	440.6	1230.1
1.07	573.2	1526.8
keV/decay	406.4	

Principal Conversion and Auger Electrons

keV/decay	4.28 *

Principal Gamma and X Rays

Freq. (%)	E (keV)
1.81	510.5
86.31	529.9
1.49	706.6
1.23	856.3
4.47	875.3
1.49	1236.4
2.33	1298.2
keV/decay	607.2 *

^{135}I 6.61 h

Radioactive Daughter   Half Life
^{135}Xe (83.5%)   9.09 h

Principal Beta Particles

Freq. (%)	E_{avg} (keV)	E_{max} (keV)
1.04	86.4	302.4
1.41	103.1	353.8
4.79	137.5	455.5
7.42	145.4	478.0
1.59	195.9	618.1
1.11	213.4	665.1
8.00	242.9	742.7
8.81	312.5	919.8
21.77	358.4	1032.9
8.10	405.1	1145.7
7.70	450.4	1253.4
24.10	534.9	1450.6
1.11	591.1	1579.5
keV/decay	363.9	

Principal Conversion and Auger Electrons
keV/decay   3.19 *

Principal Gamma and X Rays

Freq. (%)	E (keV)
1.75	220.5
3.09	288.5
3.52	417.6
7.13	546.6
6.67	836.8
1.20	972.6
7.93	1038.8
1.60	1101.6
3.61	1124.0
22.53	1131.5
28.63	1260.4
8.65	1457.6
1.07	1502.8
1.29	1566.4
9.53	1678.0
4.09	1706.5
7.70	1791.2
keV/decay	1576.3 *

^{127}Xe 36.4 d

Principal Conversion and Auger Electrons

Freq. (%)	E (keV)
1.18	23.0
1.06	23.3
3.08	23.6
1.33	23.9
3.93	24.4
1.15	27.1
1.50	27.7
1.54	112.1
3.65	139.0
6.62	169.7
keV/decay	32.4 *

Principal Gamma and X Rays

Freq. (%)	E (keV)
24.98	28.3
46.43	28.6
4.55	32.2
8.82	32.3
2.83	33.1
1.23	57.6
4.29	145.3
25.54	172.1
68.30	202.9
17.21	375.0
keV/decay	280.1

^{133m}Xe 2.188 d

Radioactive Daughter   Half Life
^{133}Xe (100.0%)   5.25 d

Principal Conversion and Auger Electrons

Freq. (%)	E (keV)
1.83	24.5
63.51	198.6
11.87	227.7
2.56	228.1
6.26	228.4
4.57	232.2
1.23	233.2
keV/decay	192.4

more ...

^{133m}Xe (continued)

Principal Gamma and X Rays

Freq. (%)	E (keV)
16.05	29.5
29.78	29.8
2.94	33.6
5.72	33.6
1.91	34.4
9.99	233.2
keV/decay	40.8

^{133}Xe   5.245 d

Principal Beta Particles

Freq. (%)	E_{avg} (keV)	E_{max} (keV)
99.33	100.5	346.0
keV/decay	100.3	

Principal Conversion and Auger Electrons

Freq. (%)	E (keV)
1.44	25.5
53.54	45.0
6.52	75.3
1.45	80.0
keV/decay	35.4 *

Principal Gamma and X Rays

Freq. (%)	E (keV)
13.71	30.6
25.34	31.0
2.52	34.9
4.89	35.0
1.70	35.8
37.42	81.0
keV/decay	46.1

^{135m}Xe   15.29 min

Radioactive Daughter

	Half Life
^{135}Xe (100.0%)	9.09 h

Principal Beta Particles

keV/decay	0.011 *

Principal Conversion and Auger Electrons

Freq. (%)	E (keV)
15.59	492.0
2.26	521.1
keV/decay	97.6 *

more ...

^{135m}Xe (continued)

Principal Gamma and X Rays

Freq. (%)	E (keV)
3.94	29.5
7.31	29.8
1.40	33.6
80.66	526.6
keV/decay	429.0

^{135}Xe   9.09 h

Radioactive Daughter

	Half Life
^{135}Cs (100.0%)	2.3×10^6 y

Principal Beta Particles

Freq. (%)	E_{avg} (keV)	E_{max} (keV)
3.12	170.7	549.8
96.00	307.4	908.2
keV/decay	302.1	

Principal Conversion and Auger Electrons

Freq. (%)	E (keV)
5.64	213.8
keV/decay	15.1 *

Principal Gamma and X Rays

Freq. (%)	E (keV)
1.46	30.6
2.70	31.0
90.13	249.8
2.90	608.2
keV/decay	248.6

^{138}Xe   14.17 min

Radioactive Daughter

	Half Life
^{138}Cs (100.0%)	32.20 min

Principal Beta Particles

Freq. (%)	E_{avg} (keV)	E_{max} (keV)
3.06	150.2	492.4
• 9.46	176.9	566.9
32.55	266.0	803.3
20.01	948.9	2379.8
13.34	966.2	2417.9
5.57	1036.7	2571.6
4.98	1139.5	2819.2
8.96	1148.5	2814.3
keV/decay	652.3	

more ...

^{138}Xe   (continued)

Principal Conversion and Auger Electrons

Freq. (%)	E (keV)
2.56	10.9
1.74	117.8
1.80	222.3
keV/decay	20.3 *

Principal Gamma and X Rays

Freq. (%)	E (keV)
1.15	30.6
2.12	31.0
5.95	153.7
3.50	242.6
31.50	258.3
6.30	396.4
2.17	401.4
20.32	434.5
1.47	1114.3
16.73	1768.3
1.42	1850.9
5.35	2004.7
12.25	2015.8
1.44	2079.2
2.29	2252.3
keV/decay	1125.0 *

^{129}Cs   32.06 h

Principal Conversion and Auger Electrons

Freq. (%)	E (keV)
1.33	23.9
1.19	24.2
3.38	24.5
1.47	24.9
1.29	28.1
1.68	28.8
3.81	34.1
keV/decay	17.8 *

more ...

^{129}Cs   (continued)

Principal Gamma and X Rays

Freq. (%)	E (keV)
29.74	29.5
55.18	29.8
5.45	33.6
10.60	33.6
3.53	34.4
2.99	39.6
1.33	278.6
2.46	318.2
30.80	371.9
22.45	411.5
3.42	548.9
keV/decay	281.2

^{130}Cs   29.9 min

Positrons

Freq. (%)	E_{avg} (keV)	E_{max} (keV)
44.02	893.0	1988.0
keV/decay	397.2	

Principal Conversion and Auger Electrons

Freq. (%)	E (keV)
1.36	24.5
keV/decay	3.89 *

Principal Gamma and X Rays

Freq. (%)	E (keV)
11.95	29.5
22.16	29.8
2.19	33.6
4.26	33.6
1.42	34.4
89.29	511.0
4.10	536.1
keV/decay	516.7

^{131}Cs   9.69 d

Principal Conversion and Auger Electrons

Freq. (%)	E (keV)
2.39	24.5
1.04	24.9
1.19	28.8
keV/decay	6.60 *

more ...

^{131}Cs (continued)

Principal Gamma and X Rays

Freq. (%)	E (keV)
21.02	29.5
38.99	29.8
3.85	33.6
7.49	33.6
2.50	34.4
keV/decay	22.8

^{134m}Cs　　　　2.91 h

Radioactive Daughter	Half Life
^{134}Cs (100.0%)	2.062 y

Principal Conversion and Auger Electrons

Freq. (%)	E (keV)
16.03	10.2
3.92	11.2
35.18	91.5
3.01	121.8
19.60	122.1
18.33	122.5
9.13	126.5
2.30	127.5
keV/decay	112.4 *

Principal Gamma and X Rays

Freq. (%)	E (keV)
9.03	30.6
16.68	31.0
1.66	34.9
3.22	35.0
1.12	35.8
12.70	127.5
keV/decay	27.1

^{134}Cs　　　　2.062 y

Principal Beta Particles

Freq. (%)	E_{avg} (keV)	E_{max} (keV)
27.28	23.1	88.6
2.47	123.4	415.2
69.80	210.1	657.9
keV/decay	157.4	

Principal Conversion and Auger Electrons

keV/decay	6.83 *

more ...

^{134}Cs (continued)

Principal Gamma and X Rays

Freq. (%)	E (keV)
1.46	475.4
8.38	563.2
15.43	569.3
97.56	604.7
85.44	795.8
8.73	801.9
1.80	1167.9
3.04	1365.2
keV/decay	1555.3

^{137}Cs　　　　30.0 y

Radioactive Daughter	Half Life
^{137m}Ba (94.6%)	2.552 min

Principal Beta Particles

Freq. (%)	E_{avg} (keV)	E_{max} (keV)
94.43	173.4	511.5
5.57	424.6	1173.2
keV/decay	187.4	

^{131}Ba　　　　11.8 d

Radioactive Daughter	Half Life
^{131}Cs (100.0%)	9.69 d

Positrons

keV/decay	6.76×10^{-4} *

Principal Conversion and Auger Electrons

Freq. (%)	E (keV)
1.17	24.8
1.05	25.1
2.92	25.5
1.27	25.8
1.14	29.2
1.48	29.9
1.17	42.8
18.13	87.8
1.69	118.1
2.08	118.4
2.23	118.8
1.28	122.8
1.83	180.1
keV/decay	45.8 *

more ...

^{131}Ba   (continued)

Principal Gamma and X Rays

Freq. (%)	E (keV)
27.72	30.6
51.24	31.0
5.10	34.9
9.89	35.0
3.43	35.8
29.05	123.8
2.19	133.6
19.90	216.1
2.41	239.6
2.81	249.4
13.33	373.3
1.29	404.0
1.89	486.5
43.78	496.3
1.23	585.0
1.57	620.0
1.19	1047.6
keV/decay	458.9

^{133m}Ba   38.9 h

Radioactive Daughter	Half Life
^{133}Ba (100.0%)	10.74 y

Principal Conversion and Auger Electrons

Freq. (%)	E (keV)
15.85	11.2
5.36	12.3
1.47	26.4
59.24	238.6
10.87	270.1
2.31	270.5
4.88	270.8
4.04	275.0
1.16	276.1
keV/decay	221.3 *

Principal Gamma and X Rays

Freq. (%)	E (keV)
1.39	12.3
15.18	31.8
27.95	32.2
2.80	36.3
5.42	36.4
1.96	37.3
17.51	276.1
keV/decay	66.8

^{135m}Ba   28.7 h

Principal Conversion and Auger Electrons

Freq. (%)	E (keV)
1.50	26.4
60.16	230.8
11.17	262.2
2.39	262.6
5.22	263.0
4.21	267.2
1.21	268.2
keV/decay	208.1

Principal Gamma and X Rays

Freq. (%)	E (keV)
15.41	31.8
28.39	32.2
2.84	36.3
5.51	36.4
1.99	37.3
15.64	268.2
keV/decay	60.1

^{137m}Ba   2.552 min

Principal Conversion and Auger Electrons

Freq. (%)	E (keV)
8.32	624.2
1.19	655.7
keV/decay	65.1 *

Principal Gamma and X Rays

Freq. (%)	E (keV)
2.13	31.8
3.92	32.2
89.78	661.6
keV/decay	596.5

^{140}Ba — 12.74 d

Radioactive Daughter	Half Life
^{140}La (100.0%)	40.27 h

Principal Beta Particles

Freq. (%)	E_{avg} (keV)	E_{max} (keV)
24.60	136.5	453.9
9.82	176.6	567.4
3.80	305.6	872.4
38.86	339.6	991.2
22.92	356.9	1005.0
keV/decay	276.3	

Principal Conversion and Auger Electrons

Freq. (%)	E (keV)
10.82	12.7
2.57	13.9
53.82	23.7
5.62	24.1
2.45	24.5
12.80	28.8
3.45	30.0
1.48	123.7
keV/decay	36.5 *

Principal Gamma and X Rays

Freq. (%)	E (keV)
1.22	13.9
13.78	30.0
1.01	33.4
6.21	162.6
4.30	304.9
3.15	423.7
1.93	437.6
24.39	537.3
keV/decay	182.7

^{140}La — 40.272 h

Principal Beta Particles

Freq. (%)	E_{avg} (keV)	E_{max} (keV)
11.12	440.8	1238.2
5.71	443.2	1243.9
5.61	465.0	1295.7
44.87	487.1	1347.7
5.11	514.4	1411.8
20.83	629.2	1676.5
4.81	845.8	2163.5
keV/decay	525.8	

more ...

^{140}La (continued)

Principal Conversion and Auger Electrons

keV/decay	9.33 *

Principal Gamma and X Rays

Freq. (%)	E (keV)
1.05	34.7
20.74	328.8
2.99	432.5
45.94	487.0
4.41	751.8
23.64	815.8
5.59	867.8
2.68	919.6
7.05	925.2
95.40	1596.5
3.43	2521.7
keV/decay	2315.9

^{141}Ce — 32.50 d

Principal Beta Particles

Freq. (%)	E_{avg} (keV)	E_{max} (keV)
70.20	130.0	435.9
29.80	181.1	581.3
keV/decay	145.2	

Principal Conversion and Auger Electrons

Freq. (%)	E (keV)
18.71	103.5
2.35	138.6
keV/decay	25.5 *

Principal Gamma and X Rays

Freq. (%)	E (keV)
4.86	35.6
8.87	36.0
1.76	40.7
48.20	145.4
keV/decay	76.5

^{144}Ce   284.3 d

Radioactive Daughter   Half Life
^{144}Pr   (98.6%)   17.28 min

Principal Beta Particles

Freq. (%)	E_{avg} (keV)	E_{max} (keV)
19.58	49.4	181.9
4.60	65.3	235.3
75.82	90.2	315.4
keV/decay	81.1	

Principal Conversion and Auger Electrons

Freq. (%)	E (keV)
3.48	38.1
5.33	91.5
keV/decay	11.1 *

Principal Gamma and X Rays

Freq. (%)	E (keV)
2.96	35.6
5.40	36.0
1.07	40.7
1.64	80.1
10.80	133.5
keV/decay	20.7 *

^{144}Pr   17.28 min

Principal Beta Particles

Freq. (%)	E_{avg} (keV)	E_{max} (keV)
1.08	266.9	811.3
1.17	894.7	2300.5
97.74	1221.7	2997.0
keV/decay	1207.5	

Principal Conversion and Auger Electrons
keV/decay   0.056 *

Principal Gamma and X Rays

Freq. (%)	E (keV)
1.48	696.5
keV/decay	31.9 *

^{147}Pm   2.6234 y

Radioactive Daughter   Half Life
^{147}Sm   (100.0%) 6.9×10^9 y

Principal Beta Particles

Freq. (%)	E_{avg} (keV)	E_{max} (keV)
99.99	62.0	224.7
keV/decay	62.0	

Principal Conversion and Auger Electrons
keV/decay   0.003 *

Principal Gamma and X Rays
keV/decay   0.004 *

^{169}Yb   32.022 d

Principal Conversion and Auger Electrons

Freq. (%)	E (keV)
7.14	10.6
1.89	18.9
8.38	34.2
1.26	39.5
2.09	40.9
1.12	47.3
1.35	48.8
34.85	50.4
3.90	53.0
1.43	53.5
1.79	54.5
1.33	58.8
1.58	61.2
6.07	71.1
1.15	83.5
4.78	99.7
1.26	107.9
10.30	117.8
2.44	120.9
2.14	121.9
1.25	128.6
12.79	138.6
1.39	167.1
1.74	187.8
keV/decay	123.2 *

more ...

^{169}Yb (continued)

Principal Gamma and X Rays

Freq. (%)	E (keV)
53.00	49.8
93.73	50.7
9.94	57.3
19.17	57.5
8.20	59.1
43.74	63.1
2.66	93.6
17.37	109.8
1.88	118.2
11.11	130.5
21.45	177.2
34.94	198.0
1.90	261.1
10.80	307.7
keV/decay	311.4

^{198}Au   2.696 d

Principal Beta Particles

Freq. (%)	E_{avg} (keV)	E_{max} (keV)
1.30	79.4	284.8
98.68	314.5	960.7
keV/decay	311.5	

Principal Conversion and Auger Electrons

Freq. (%)	E (keV)
2.88	328.7
keV/decay	15.5 *

Principal Gamma and X Rays

Freq. (%)	E (keV)
1.38	70.8
95.51	411.8
1.06	675.9
keV/decay	404.4

^{197}Hg   64.1 h

Principal Conversion and Auger Electrons

Freq. (%)	E (keV)
8.88	11.1
3.57	11.2
2.93	11.8
1.42	13.0
31.98	63.0
14.91	63.6
10.91	65.4
14.03	74.6
4.39	77.3
keV/decay	66.5 *

Principal Gamma and X Rays

Freq. (%)	E (keV)
1.81	11.2
12.84	11.4
4.39	11.6
2.31	11.6
2.57	13.4
20.47	67.0
34.96	68.8
18.00	77.3
4.14	77.6
7.94	78.0
3.27	80.2
keV/decay	70.3 *

^{203}Hg   46.612 d

Principal Beta Particles

Freq. (%)	E_{avg} (keV)	E_{max} (keV)
100.00	57.8	212.6
keV/decay	57.8	

Principal Conversion and Auger Electrons

Freq. (%)	E (keV)
13.80	193.7
2.07	263.8
1.05	264.5
keV/decay	41.2 *

Principal Gamma and X Rays

Freq. (%)	E (keV)
2.17	10.3
1.62	12.2
3.86	70.8
6.53	72.9
1.49	82.6
81.46	279.2
keV/decay	238.0

^{201}Tl   73.1 h

Principal Conversion and Auger Electrons

Freq. (%)	E (keV)
7.09	11.5
3.23	11.5
1.55	12.1
1.14	13.4
7.30	15.8
6.28	17.4
1.90	27.8
1.63	29.3
7.47	52.2
15.44	84.3
1.14	120.5
2.35	152.6
keV/decay	43.4 *

Principal Gamma and X Rays

Freq. (%)	E (keV)
1.21	11.6
10.55	11.8
4.19	11.9
1.51	12.0
2.15	13.8
27.15	68.9
46.18	70.8
5.48	79.8
10.53	80.3
4.43	82.6
2.65	135.3
10.00	167.4
keV/decay	93.5

^{204}Tl   3.779 y

Principal Beta Particles

Freq. (%)	E_{avg} (keV)	E_{max} (keV)
97.42	243.9	763.4
keV/decay	237.6	

Principal Conversion and Auger Electrons

keV/decay	0.229 *

Principal Gamma and X Rays

keV/decay	1.13 *

^{208}Tl   3.053 min

Principal Beta Particles

Freq. (%)	E_{avg} (keV)	E_{max} (keV)
3.11	340.9	1033.0
24.66	439.4	1285.6
21.94	533.2	1518.9
49.01	647.2	1796.3
keV/decay	557.1	

Principal Conversion and Auger Electrons

Freq. (%)	E (keV)
1.27	10.1
2.84	189.4
1.97	422.8
1.28	495.2
keV/decay	53.9 *

Principal Gamma and X Rays

Freq. (%)	E (keV)
1.23	10.6
2.23	72.8
3.76	75.0
6.31	277.4
22.61	510.8
84.48	583.2
1.81	763.1
12.42	860.6
99.16	2614.5
keV/decay	3360.2

^{210}Pb   22.3 y

Radioactive Daughter	Half Life
^{210}Bi (100.0%)	5.013 d

Alpha Particles

keV/decay	7.07×10^{-5} *

Principal Beta Particles

Freq. (%)	E_{avg} (keV)	E_{max} (keV)
80.00	4.14	16.5
20.00	16.1	63.0
keV/decay	6.54	

more ...

^{210}Pb (continue)

Principal Conversion and Auger Electrons

Freq. (%)	E (keV)
8.25	10.3
2.94	10.6
1.42	12.5
2.39	12.6
2.14	13.3
51.90	30.1
5.43	30.8
13.60	43.3
4.60	46.5
keV/decay	31.4 *

Principal Gamma and X Rays

Freq. (%)	E (keV)
8.37	10.8
2.13	12.7
2.09	13.0
4.11	13.0
2.46	13.2
4.05	46.5
keV/decay	4.81 *

^{212}Pb   10.64 h

Radioactive Daughter

	Half Life
^{212}Bi (100.0%)	50.55 min

Principal Beta Particles

Freq. (%)	E_{avg} (keV)	E_{max} (keV)
5.08	41.9	157.5
82.98	94.4	334.2
11.77	172.6	572.8
keV/decay	101.0	

Principal Conversion and Auger Electrons

Freq. (%)	E (keV)
5.76	10.3
2.19	12.6
3.42	24.7
32.32	148.1
1.29	209.6
5.04	222.2
1.32	235.5
keV/decay	74.4 *

more ...

^{212}Pb (continued)

Principal Gamma and X Rays

Freq. (%)	E (keV)
5.84	10.8
1.46	13.0
3.75	13.0
10.42	74.8
17.51	77.1
2.08	86.8
3.99	87.3
1.84	89.9
43.65	238.6
3.34	300.1
keV/decay	145.2

^{214}Pb   26.8 min

Radioactive Daughter

	Half Life
^{214}Bi (100.0%)	19.9 min

Principal Beta Particles

Freq. (%)	E_{avg} (keV)	E_{max} (keV)
2.52	49.7	184.9
48.10	207.2	672.1
42.12	227.4	728.8
6.29	336.7	1024.0
keV/decay	219.2	

Principal Conversion and Auger Electrons

Freq. (%)	E (keV)
5.06	10.3
1.79	12.6
9.50	36.8
2.48	50.0
5.32	151.4
7.43	204.6
9.48	261.4
1.16	278.8
1.47	335.5
keV/decay	74.0 *

more ...

^{214}Pb (continued)

Principal Gamma and X Rays

Freq. (%)	E (keV)
5.13	10.8
1.28	13.0
3.08	13.0
1.10	53.2
6.39	74.8
10.73	77.1
1.28	86.8
2.45	87.3
1.13	89.9
7.46	241.9
19.17	295.2
37.06	351.9
1.09	785.9
keV/decay	249.8 *

^{210}Bi   5.013 d

Radioactive Daughter

	Half Life
^{210}Po (100.0%)	138.7 d

Alpha Particles

keV/decay	0.006 *

Principal Beta Particles

Freq. (%)	E$_{avg}$ (keV)	E$_{max}$ (keV)
100.00	388.9	1161.4
keV/decay	388.9	

^{212}Bi   60.55 min

Radioactive Daughter

	Half Life
^{212}Po (64.1%)	0.298 μs

Alpha Particles

Freq. (%)	E (keV)
25.24	6050.9
9.64	6090.0
keV/decay	2174.7

Principal Beta Particles

Freq. (%)	E$_{avg}$ (keV)	E$_{max}$ (keV)
1.89	190.5	625.4
1.46	228.6	733.3
4.49	530.6	1518.8
55.30	831.5	2246.0
keV/decay	491.9	

more ...

^{212}Bi (continued)

Principal Conversion and Auger Electrons

Freq. (%)	E (keV)
18.25	24.5
1.87	25.2
4.73	36.9
1.57	39.9
keV/decay	11.7 *

Principal Gamma and X Rays

Freq. (%)	E (keV)
2.82	10.3
1.50	12.2
1.09	39.9
6.65	727.2
1.11	785.4
1.51	1620.6
keV/decay	106.3 *

^{214}Bi   19.9 min

Radioactive Daughter

	Half Life
^{214}Po (100.0%)	164.3 μs

Alpha Particles

keV/decay	1.15 *

Principal Beta Particles

Freq. (%)	E$_{avg}$ (keV)	E$_{max}$ (keV)
1.04	248.1	787.5
2.78	260.8	822.3
5.48	352.0	1065.9
4.31	385.0	1151.4
2.48	424.5	1252.7
1.49	427.0	1259.2
1.18	433.4	1275.4
1.57	474.8	1379.7
8.28	491.9	1422.6
17.60	525.2	1505.5
17.90	539.3	1540.4
3.32	615.2	1726.6
1.01	667.9	1854.5
7.52	683.6	1892.3
17.70	1268.6	3270.0
keV/decay	637.6	

Principal Conversion and Auger Electrons

keV/decay	21.7 *

more ...

^{214}Bi (continued)

Principal Gamma and X Rays

Freq. (%)	E (keV)
46.09	609.3
1.56	665.5
4.88	768.4
1.23	806.2
3.16	934.1
15.04	1120.3
1.69	1155.2
5.92	1238.1
1.47	1281.0
4.02	1377.6
1.39	1401.5
2.48	1408.0
2.19	1509.2
1.15	1661.3
3.05	1729.6
15.92	1764.5
2.12	1847.4
1.21	2118.5
4.99	2204.1
1.55	2447.7
keV/decay	1508.2 *

^{210}Po   138.376 d

Alpha Particles

Freq. (%)	E (keV)
100.00	5304.6
keV/decay	5304.5

Principal Conversion and Auger Electrons
keV/decay	9.34×10^{-5} *

Principal Gamma and X Rays
keV/decay	0.010 *

^{212}Po   0.298 μs

Alpha Particles

Freq. (%)	E (keV)
100.00	8784.3
keV/decay	8784.3

^{214}Po   164.3 μs

Radioactive Daughter   Half Life
^{210}Pb (100.0%)   22.3 y

Alpha Particles

Freq. (%)	E (keV)
99.99	7687.2
keV/decay	7687.1

Principal Conversion and Auger Electrons
keV/decay	8.21×10^{-4} *

Principal Gamma and X Rays
keV/decay	0.083 *

^{218}Po   3.05 min

Radioactive Daughter   Half Life
^{214}Pb (100.0%)   26.8 min

Alpha Particles

Freq. (%)	E (keV)
99.98	6002.6
keV/decay	6001.4

Principal Beta Particles
keV/decay	0.014 *

^{220}Rn   55.6 s

Radioactive Daughter   Half Life
^{216}Po (100.0%)   0.145 s

Alpha Particles

Freq. (%)	E (keV)
99.93	6288.4
keV/decay	6288.0

Principal Conversion and Auger Electrons
keV/decay	0.009 *

Principal Gamma and X Rays
keV/decay	0.386 *

^{222}Rn   3.8235 d

Radioactive Daughter   Half Life
^{218}Po (100.0%)   3.05 min

Alpha Particles

Freq. (%)	E (keV)
99.92	5489.7
keV/decay	5489.3

Principal Conversion and Auger Electrons
keV/decay   0.011 *

Principal Gamma and X Rays
keV/decay   0.399 *

^{224}Ra   3.66 d

Radioactive Daughter   Half Life
^{220}Rn (100.0%)   55.6 s

Alpha Particles

Freq. (%)	E (keV)
4.90	5448.9
95.08	5685.6
keV/decay	5673.9

Principal Conversion and Auger Electrons
keV/decay   2.22 *

Principal Gamma and X Rays

Freq. (%)	E (keV)
3.90	241.0
keV/decay	9.90 *

^{226}Ra   1600 y

Radioactive Daughter   Half Life
^{222}Rn (100.0%)   3.824 d

Alpha Particles

Freq. (%)	E (keV)
5.55	4601.8
94.44	4784.5
keV/decay	4774.4

Principal Conversion and Auger Electrons
keV/decay   3.59 *

more ...

^{226}Ra (continued)

Principal Gamma and X Rays

Freq. (%)	E (keV)
3.28	186.0
keV/decay	6.75 *

^{241}Am   432.2 y

Alpha Particles

Freq. (%)	E (keV)
1.40	5388.4
12.80	5443.1
85.19	5485.8
keV/decay	5479.1

Principal Conversion and Auger Electrons

Freq. (%)	E (keV)
22.30	10.6
11.59	10.8
3.64	11.6
16.04	13.1
7.94	13.4
2.29	14.2
20.61	14.6
2.19	15.6
2.71	16.0
11.87	16.1
1.38	16.9
5.49	17.1
1.72	17.9
2.74	21.0
3.50	21.8
3.68	21.9
2.79	25.8
1.44	26.3
12.01	27.7
4.38	28.8
4.76	32.2
1.73	33.2
3.24	34.1
11.06	37.1
10.06	37.9
1.24	38.5
2.40	39.0
10.10	41.9
7.72	55.1
2.71	59.5
keV/decay	58.6 *

more ...

^{241}Am (continued)

Principal Gamma and X Rays

Freq. (%)	E (keV)
1.38	11.9
2.75	13.8
24.59	13.9
6.53	16.8
3.19	17.1
1.41	17.5
20.02	17.8
2.82	18.0
4.86	20.8
1.33	21.1
1.22	21.3
1.03	21.5
2.41	26.3
35.90	59.5
keV/decay	34.1

Gamma-Ray Interaction Coefficients

TABLE C.1 Mass Interaction, Energy Transfer, and Energy Absorption Coefficients, in Units of $cm^2 g^{-1}$, for Various Compounds and Mixtures.

E (MeV)	AIR μ/ρ	AIR μ_{tr}/ρ	AIR μ_{en}/ρ	WATER μ/ρ	WATER μ_{tr}/ρ	WATER μ_{en}/ρ	ICRU TISSUE[a] μ/ρ	ICRU TISSUE[a] μ_{tr}/ρ	ICRU TISSUE[a] μ_{en}/ρ	LUCITE μ/ρ	LUCITE μ_{tr}/ρ	LUCITE μ_{en}/ρ	POLYETHYLENE μ/ρ	POLYETHYLENE μ_{tr}/ρ	POLYETHYLENE μ_{en}/ρ
0.010	4.9100	4.5330	4.5330	5.0660	4.6840	4.6840	4.6860	4.3150	4.3150	3.1510	2.8200	2.8210	1.9280	1.6210	1.6200
0.015	1.5220	1.2420	1.2420	1.5680	1.2690	1.2690	1.4590	1.1670	1.1670	1.0230	0.7539	0.7535	0.6874	0.4250	0.4250
0.020	0.7334	0.4943	0.4942	0.7613	0.5019	0.5016	0.7157	0.4610	0.4609	0.5358	0.2969	0.2968	0.4056	0.1672	0.1672
0.030	0.3398	0.1395	0.1395	0.3612	0.1411	0.1411	0.3468	0.1300	0.1299	0.2927	0.0857	0.0857	0.2631	0.0514	0.0514
0.040	0.2429	0.0625	0.0625	0.2629	0.0637	0.0637	0.2557	0.0593	0.0593	0.2310	0.0417	0.0417	0.2245	0.0288	0.0288
0.050	0.2053	0.0382	0.0382	0.2245	0.0396	0.0396	0.2200	0.0373	0.0373	0.2055	0.0287	0.0287	0.2070	0.0229	0.0229
0.060	0.1861	0.0290	0.0289	0.2046	0.0305	0.0305	0.2012	0.0292	0.0292	0.1914	0.0242	0.0242	0.1962	0.0215	0.0215
0.080	0.1658	0.0236	0.0236	0.1833	0.0255	0.0255	0.1809	0.0249	0.0249	0.1748	0.0227	0.0227	0.1820	0.0224	0.0224
0.10	0.1539	0.0231	0.0231	0.1706	0.0253	0.0253	0.1686	0.0249	0.0248	0.1639	0.0235	0.0235	0.1718	0.0241	0.0241
0.15	0.1356	0.0249	0.0249	0.1505	0.0276	0.0276	0.1490	0.0273	0.0273	0.1456	0.0266	0.0266	0.1534	0.0279	0.0279
0.20	0.1234	0.0267	0.0267	0.1370	0.0297	0.0297	0.1357	0.0294	0.0294	0.1328	0.0287	0.0287	0.1401	0.0303	0.0303
0.30	0.1068	0.0288	0.0287	0.1187	0.0320	0.0320	0.1175	0.0316	0.0316	0.1152	0.0310	0.0310	0.1216	0.0328	0.0327
0.40	0.0955	0.0295	0.0295	0.1061	0.0328	0.0328	0.1051	0.0325	0.0325	0.1031	0.0319	0.0318	0.1089	0.0337	0.0336
0.50	0.0871	0.0297	0.0297	0.0969	0.0330	0.0330	0.0959	0.0327	0.0327	0.0941	0.0321	0.0320	0.0994	0.0339	0.0339
0.60	0.0806	0.0296	0.0295	0.0896	0.0329	0.0328	0.0887	0.0326	0.0326	0.0870	0.0320	0.0319	0.0920	0.0338	0.0337
0.80	0.0707	0.0289	0.0288	0.0786	0.0321	0.0320	0.0779	0.0318	0.0318	0.0764	0.0312	0.0312	0.0808	0.0330	0.0329
1.0	0.0636	0.0280	0.0279	0.0707	0.0311	0.0310	0.0700	0.0308	0.0308	0.0687	0.0302	0.0301	0.0726	0.0320	0.0319
1.5	0.0518	0.0256	0.0255	0.0575	0.0284	0.0283	0.0570	0.0282	0.0282	0.0559	0.0276	0.0275	0.0591	0.0292	0.0291
2.0	0.0445	0.0236	0.0235	0.0494	0.0262	0.0260	0.0489	0.0260	0.0259	0.0480	0.0255	0.0253	0.0507	0.0269	0.0267
3.0	0.0358	0.0208	0.0206	0.0397	0.0230	0.0228	0.0393	0.0228	0.0228	0.0384	0.0223	0.0221	0.0405	0.0234	0.0233
4.0	0.0308	0.0190	0.0187	0.0340	0.0209	0.0206	0.0337	0.0207	0.0207	0.0329	0.0202	0.0199	0.0344	0.0211	0.0209
5.0	0.0275	0.0177	0.0174	0.0303	0.0195	0.0191	0.0300	0.0192	0.0192	0.0292	0.0187	0.0184	0.0304	0.0194	0.0192
6.0	0.0251	0.0168	0.0164	0.0276	0.0183	0.0180	0.0273	0.0181	0.0181	0.0265	0.0176	0.0172	0.0275	0.0182	0.0179
8.0	0.0221	0.0156	0.0152	0.0242	0.0170	0.0165	0.0239	0.0167	0.0167	0.0231	0.0161	0.0157	0.0237	0.0165	0.0161
10.0	0.0205	0.0151	0.0145	0.0222	0.0163	0.0157	0.0219	0.0160	0.0160	0.0211	0.0153	0.0148	0.0215	0.0155	0.0151

[a] Composition, by weight percent: H 10.1, C 11.1, N 2.6, O 76.2.

Source: Data Library DLC-139 (1988) distributed by the Radiation Shielding Information Center, Oak Ridge National Laboratory, and based on Report UCRL-50400, Vol. 6, Rev. 1, National Technical Information Service, Springfield, Va., 1975.

TABLE C.1 (cont'd).

E (MeV)	POLYSTYRENE μ/ρ	μ_{tr}/ρ	μ_{en}/ρ	LiF (TLD) μ/ρ	μ_{tr}/ρ	μ_{en}/ρ	SODIUM IODIDE μ/ρ	μ_{tr}/ρ	μ_{en}/ρ	CONCRETE[b] μ/ρ	μ_{tr}/ρ	μ_{en}/ρ
0.010	2.0480	1.7450	1.7460	6.0760	5.7130	5.7140	140.1000	133.1000	133.1000	26.6300	25.0800	25.0700
0.015	0.7115	0.4571	0.4570	1.8330	1.5620	1.5630	46.9000	44.2500	44.1700	8.1810	7.5690	7.5700
0.020	0.4084	0.1791	0.1791	0.8480	0.6183	0.6181	21.5600	20.0300	19.9900	3.5620	3.1740	3.1740
0.030	0.2558	0.0540	0.0540	0.3613	0.1712	0.1712	7.2460	6.4450	6.4250	1.1750	0.9175	0.9177
0.040	0.2151	0.0292	0.0292	0.2440	0.0740	0.0740	19.0200	8.4150	8.3850	0.5939	0.3823	0.3823
0.050	0.1970	0.0226	0.0226	0.1998	0.0430	0.0430	10.6400	5.7790	5.7520	0.3853	0.1979	0.1980
0.060	0.1861	0.0209	0.0208	0.1780	0.0310	0.0310	6.5590	3.9730	3.9530	0.2909	0.1194	0.1195
0.080	0.1722	0.0213	0.0213	0.1560	0.0235	0.0235	3.0420	2.0460	2.0320	0.2110	0.0605	0.0605
0.10	0.1623	0.0228	0.0228	0.1440	0.0222	0.0222	1.6880	1.1770	1.1680	0.1779	0.0416	0.0416
0.15	0.1447	0.0263	0.0263	0.1261	0.0233	0.0233	0.6164	0.4183	0.4139	0.1435	0.0303	0.0303
0.20	0.1322	0.0286	0.0286	0.1145	0.0249	0.0248	0.3312	0.2055	0.2029	0.1271	0.0290	0.0290
0.30	0.1147	0.0309	0.0309	0.0990	0.0267	0.0266	0.1669	0.0860	0.0846	0.1082	0.0295	0.0295
0.40	0.1027	0.0318	0.0317	0.0885	0.0274	0.0273	0.1176	0.0542	0.0532	0.0963	0.0299	0.0299
0.50	0.0937	0.0320	0.0319	0.0808	0.0275	0.0275	0.0951	0.0418	0.0410	0.0877	0.0299	0.0299
0.60	0.0867	0.0319	0.0318	0.0747	0.0274	0.0274	0.0822	0.0358	0.0350	0.0810	0.0297	0.0298
0.80	0.0762	0.0311	0.0311	0.0656	0.0268	0.0267	0.0674	0.0301	0.0293	0.0710	0.0290	0.0290
1.0	0.0685	0.0301	0.0300	0.0589	0.0259	0.0258	0.0586	0.0272	0.0264	0.0638	0.0281	0.0281
1.5	0.0557	0.0275	0.0274	0.0480	0.0237	0.0236	0.0469	0.0237	0.0226	0.0520	0.0257	0.0257
2.0	0.0478	0.0254	0.0252	0.0412	0.0219	0.0217	0.0414	0.0222	0.0210	0.0448	0.0238	0.0237
3.0	0.0382	0.0221	0.0220	0.0332	0.0193	0.0190	0.0367	0.0219	0.0203	0.0366	0.0212	0.0213
4.0	0.0326	0.0200	0.0198	0.0286	0.0176	0.0173	0.0351	0.0227	0.0209	0.0319	0.0198	0.0198
5.0	0.0289	0.0185	0.0182	0.0256	0.0165	0.0161	0.0347	0.0240	0.0218	0.0290	0.0189	0.0189
6.0	0.0262	0.0173	0.0170	0.0233	0.0156	0.0152	0.0343	0.0250	0.0223	0.0268	0.0181	0.0181
8.0	0.0227	0.0158	0.0154	0.0206	0.0145	0.0140	0.0351	0.0275	0.0239	0.0243	0.0175	0.0175
10.0	0.0206	0.0149	0.0145	0.0190	0.0140	0.0135	0.0369	0.0303	0.0258	0.0231	0.0174	0.0174

[b]Composition, by weight percent: H 0.56, O 49.83, Na 1.71, Mg 0.24, Al 4.56, Si 31.58, S 0.12, K 1.92, Ca 8.26, Fe 1.22.

Source: Data Library DLC-139 (1988) distributed by the Radiation Shielding Information Center, Oak Ridge National Laboratory, and based on Report UCRL-50400, Vol. 6, Rev. 1, National Technical Information Service, Springfield, Va., 1975.

TABLE C.2 Mass Interaction, Energy Transfer, and Energy Absorption Coefficients, in Units of $cm^2 g^{-1}$, for Various Elements.

E (MeV)	HYDROGEN μ/ρ	μ_{tr}/ρ	μ_{en}/ρ	HELIUM μ/ρ	μ_{tr}/ρ	μ_{en}/ρ	LITHIUM μ/ρ	μ_{tr}/ρ	μ_{en}/ρ	BERYLLIUM μ/ρ	μ_{tr}/ρ	μ_{en}/ρ	BORON μ/ρ	μ_{tr}/ρ	μ_{en}/ρ
0.010	0.3853	0.0098	0.0098	0.2410	0.0258	0.0258	0.3216	0.1203	0.1203	0.5848	0.3628	0.3627	1.1460	0.8963	0.8963
0.015	0.3764	0.0110	0.0110	0.2071	0.0104	0.0104	0.2112	0.0324	0.0324	0.2866	0.0931	0.0931	0.4439	0.2304	0.2304
0.020	0.3694	0.0135	0.0135	0.1952	0.0086	0.0086	0.1827	0.0158	0.0158	0.2164	0.0384	0.0384	0.2840	0.0905	0.0905
0.030	0.3570	0.0186	0.0186	0.1836	0.0098	0.0098	0.1635	0.0104	0.0104	0.1767	0.0162	0.0162	0.2013	0.0297	0.0297
0.040	0.3458	0.0231	0.0231	0.1762	0.0118	0.0118	0.1548	0.0109	0.0109	0.1630	0.0133	0.0133	0.1773	0.0187	0.0187
0.050	0.3355	0.0271	0.0271	0.1703	0.0137	0.0137	0.1487	0.0122	0.0122	0.1550	0.0135	0.0135	0.1655	0.0163	0.0163
0.060	0.3260	0.0305	0.0305	0.1651	0.0154	0.0154	0.1437	0.0135	0.0135	0.1490	0.0144	0.0144	0.1578	0.0162	0.0162
0.080	0.3091	0.0362	0.0362	0.1561	0.0182	0.0182	0.1356	0.0159	0.0159	0.1399	0.0165	0.0165	0.1470	0.0176	0.0176
0.10	0.2944	0.0406	0.0406	0.1486	0.0205	0.0205	0.1289	0.0178	0.0178	0.1327	0.0183	0.0183	0.1390	0.0193	0.0193
0.15	0.2651	0.0481	0.0481	0.1336	0.0242	0.0242	0.1158	0.0210	0.0210	0.1190	0.0216	0.0216	0.1243	0.0225	0.0225
0.20	0.2429	0.0525	0.0525	0.1224	0.0265	0.0265	0.1060	0.0229	0.0229	0.1089	0.0235	0.0235	0.1136	0.0245	0.0245
0.30	0.2112	0.0569	0.0569	0.1064	0.0287	0.0287	0.0921	0.0248	0.0248	0.0946	0.0255	0.0255	0.0986	0.0266	0.0266
0.40	0.1893	0.0586	0.0586	0.0953	0.0295	0.0295	0.0825	0.0255	0.0255	0.0847	0.0262	0.0262	0.0883	0.0273	0.0273
0.50	0.1728	0.0590	0.0590	0.0871	0.0297	0.0297	0.0753	0.0257	0.0257	0.0774	0.0264	0.0264	0.0806	0.0275	0.0275
0.60	0.1599	0.0588	0.0587	0.0805	0.0296	0.0296	0.0697	0.0256	0.0256	0.0716	0.0263	0.0263	0.0746	0.0274	0.0274
0.80	0.1405	0.0574	0.0574	0.0708	0.0289	0.0289	0.0612	0.0250	0.0250	0.0629	0.0257	0.0257	0.0655	0.0268	0.0267
1.0	0.1263	0.0556	0.0555	0.0636	0.0280	0.0280	0.0550	0.0242	0.0242	0.0565	0.0249	0.0248	0.0589	0.0259	0.0259
1.5	0.1027	0.0508	0.0507	0.0517	0.0256	0.0256	0.0448	0.0221	0.0221	0.0460	0.0227	0.0227	0.0479	0.0237	0.0236
2.0	0.0877	0.0466	0.0465	0.0442	0.0235	0.0234	0.0383	0.0203	0.0203	0.0394	0.0209	0.0209	0.0411	0.0218	0.0217
3.0	0.0692	0.0400	0.0399	0.0350	0.0202	0.0202	0.0304	0.0176	0.0176	0.0314	0.0182	0.0181	0.0329	0.0190	0.0189
4.0	0.0581	0.0353	0.0352	0.0295	0.0180	0.0179	0.0257	0.0157	0.0156	0.0266	0.0163	0.0163	0.0280	0.0171	0.0170
5.0	0.0505	0.0318	0.0317	0.0258	0.0163	0.0162	0.0226	0.0143	0.0143	0.0235	0.0150	0.0149	0.0248	0.0158	0.0157
6.0	0.0449	0.0291	0.0290	0.0230	0.0150	0.0149	0.0203	0.0133	0.0132	0.0212	0.0139	0.0139	0.0224	0.0148	0.0146
8.0	0.0374	0.0253	0.0251	0.0193	0.0132	0.0131	0.0172	0.0118	0.0117	0.0181	0.0125	0.0125	0.0194	0.0135	0.0132
10.0	0.0325	0.0227	0.0225	0.0170	0.0120	0.0119	0.0153	0.0109	0.0107	0.0163	0.0117	0.0116	0.0176	0.0127	0.0124

Source: Data Library DLC-139 (1988) distributed by the Radiation Shielding Information Center, Oak Ridge National Laboratory, and based on Report UCRL-50400, Vol. 6, Rev. 1, National Technical Information Service, Springfield, Va., 1975.

TABLE C.2 (cont'd).

E (MeV)	CARBON μ/ρ	CARBON μtr/ρ	CARBON μen/ρ	NITROGEN μ/ρ	NITROGEN μtr/ρ	NITROGEN μen/ρ	OXYGEN μ/ρ	OXYGEN μtr/ρ	OXYGEN μen/ρ	SODIUM μ/ρ	SODIUM μtr/ρ	SODIUM μen/ρ	MAGNESIUM μ/ρ	MAGNESIUM μtr/ρ	MAGNESIUM μen/ρ
0.010	2.1870	1.8910	1.8910	3.6600	3.3270	3.3270	5.6560	5.2730	5.2730	15.6400	15.0600	15.0600	21.2300	20.5500	20.5500
0.015	0.7396	0.4945	0.4945	1.1490	0.8840	0.8839	1.7180	1.4280	1.4280	4.5840	4.2050	4.2050	6.2360	5.8080	5.8070
0.020	0.4117	0.1930	0.1930	0.5771	0.3457	0.3456	0.8107	0.5634	0.5633	1.9760	1.6790	1.6780	2.6600	2.3310	2.3310
0.030	0.2473	0.0569	0.0569	0.2944	0.0975	0.0974	0.3617	0.1565	0.1565	0.6858	0.4593	0.4592	0.8839	0.6396	0.6395
0.040	0.2041	0.0297	0.0297	0.2240	0.0455	0.0455	0.2524	0.0688	0.0688	0.3818	0.1878	0.1877	0.4663	0.2597	0.2597
0.050	0.1854	0.0222	0.0222	0.1957	0.0298	0.0298	0.2105	0.0412	0.0412	0.2728	0.0985	0.0985	0.3180	0.1338	0.1338
0.060	0.1744	0.0200	0.0200	0.1806	0.0242	0.0242	0.1893	0.0305	0.0305	0.2227	0.0620	0.0620	0.2508	0.0818	0.0818
0.080	0.1607	0.0200	0.0200	0.1635	0.0217	0.0217	0.1674	0.0242	0.0234	0.1783	0.0361	0.0361	0.1928	0.0444	0.0444
0.10	0.1512	0.0213	0.0213	0.1528	0.0221	0.0221	0.1550	0.0234	0.0234	0.1580	0.0287	0.0287	0.1677	0.0331	0.0331
0.15	0.1346	0.0245	0.0245	0.1353	0.0247	0.0247	0.1361	0.0251	0.0250	0.1335	0.0258	0.0257	0.1392	0.0275	0.0275
0.20	0.1229	0.0266	0.0265	0.1233	0.0267	0.0267	0.1237	0.0268	0.0268	0.1199	0.0264	0.0264	0.1244	0.0276	0.0276
0.30	0.1066	0.0287	0.0287	0.1068	0.0288	0.0287	0.1070	0.0288	0.0288	0.1030	0.0278	0.0278	0.1065	0.0288	0.0288
0.40	0.0954	0.0295	0.0295	0.0955	0.0295	0.0295	0.0957	0.0296	0.0295	0.0919	0.0284	0.0284	0.0949	0.0293	0.0293
0.50	0.0871	0.0297	0.0297	0.0872	0.0297	0.0297	0.0873	0.0298	0.0297	0.0837	0.0285	0.0285	0.0864	0.0295	0.0294
0.60	0.0806	0.0296	0.0295	0.0806	0.0296	0.0295	0.0807	0.0296	0.0296	0.0774	0.0284	0.0283	0.0799	0.0293	0.0292
0.80	0.0708	0.0289	0.0289	0.0708	0.0289	0.0288	0.0709	0.0289	0.0289	0.0679	0.0277	0.0276	0.0701	0.0286	0.0285
1.0	0.0636	0.0280	0.0279	0.0636	0.0280	0.0279	0.0637	0.0280	0.0279	0.0610	0.0268	0.0267	0.0629	0.0277	0.0276
1.5	0.0518	0.0256	0.0255	0.0518	0.0256	0.0255	0.0518	0.0256	0.0255	0.0497	0.0245	0.0244	0.0513	0.0253	0.0252
2.0	0.0444	0.0236	0.0234	0.0445	0.0236	0.0235	0.0446	0.0236	0.0235	0.0428	0.0227	0.0225	0.0442	0.0235	0.0233
3.0	0.0356	0.0206	0.0204	0.0358	0.0207	0.0205	0.0360	0.0209	0.0206	0.0349	0.0203	0.0200	0.0361	0.0210	0.0207
4.0	0.0305	0.0187	0.0185	0.0308	0.0189	0.0187	0.0310	0.0191	0.0188	0.0304	0.0188	0.0185	0.0316	0.0196	0.0192
5.0	0.0271	0.0174	0.0171	0.0274	0.0176	0.0173	0.0278	0.0179	0.0176	0.0276	0.0179	0.0175	0.0288	0.0187	0.0183
6.0	0.0246	0.0163	0.0160	0.0250	0.0167	0.0163	0.0254	0.0170	0.0166	0.0255	0.0172	0.0167	0.0267	0.0181	0.0175
8.0	0.0214	0.0150	0.0146	0.0220	0.0154	0.0150	0.0225	0.0159	0.0154	0.0230	0.0165	0.0159	0.0242	0.0175	0.0168
10.0	0.0196	0.0143	0.0138	0.0203	0.0149	0.0143	0.0209	0.0155	0.0148	0.0218	0.0164	0.0156	0.0231	0.0175	0.0166

Source: Data Library DLC-139 (1988) distributed by the Radiation Shielding Information Center, Oak Ridge National Laboratory, and based on Report UCRL-50400, Vol. 6, Rev. 1, National Technical Information Service, Springfield, Va., 1975.

TABLE C.2 (cont'd).

E (MeV)	ALUMINUM μ/ρ	ALUMINUM μ_{tr}/ρ	ALUMINUM μ_{en}/ρ	SILICON μ/ρ	SILICON μ_{tr}/ρ	SILICON μ_{en}/ρ	PHOSPHORUS μ/ρ	PHOSPHORUS μ_{tr}/ρ	PHOSPHORUS μ_{en}/ρ	SULFUR μ/ρ	SULFUR μ_{tr}/ρ	SULFUR μ_{en}/ρ	CHLORINE μ/ρ	CHLORINE μ_{tr}/ρ	CHLORINE μ_{en}/ρ
0.010	26.3500	25.5700	25.5700	34.3900	33.4100	33.4100	41.0200	39.8200	39.8200	49.3000	47.7000	47.7000	58.9200	56.7900	56.7800
0.015	7.8730	7.4110	7.4100	10.2500	9.7170	9.7160	12.3400	11.7500	11.7500	15.0200	14.3200	14.3200	18.1600	17.3400	17.3300
0.020	3.3750	3.0310	3.0300	4.3540	3.9710	3.9700	5.2450	4.8390	4.8380	6.4420	5.9810	5.9800	7.7880	7.2860	7.2840
0.030	1.1010	0.8514	0.8510	1.3770	1.1070	1.1070	1.6360	1.3600	1.3600	2.0210	1.7210	1.7210	2.4020	2.0950	2.0950
0.040	0.5571	0.3489	0.3486	0.6721	0.4496	0.4495	0.7763	0.5525	0.5524	0.9493	0.7101	0.7099	1.0990	0.8597	0.8594
0.050	0.3632	0.1790	0.1788	0.4231	0.2277	0.2277	0.4735	0.2787	0.2786	0.5677	0.3613	0.3612	0.6375	0.4332	0.4331
0.060	0.2757	0.1076	0.1075	0.3121	0.1347	0.1347	0.3391	0.1632	0.1632	0.3970	0.2118	0.2117	0.4332	0.2510	0.2509
0.080	0.2015	0.0547	0.0547	0.2197	0.0658	0.0658	0.2285	0.0770	0.0770	0.2566	0.0981	0.0981	0.2675	0.1129	0.1128
0.10	0.1706	0.0381	0.0380	0.1823	0.0439	0.0439	0.1850	0.0491	0.0491	0.2018	0.0603	0.0603	0.2045	0.0670	0.0670
0.15	0.1381	0.0285	0.0285	0.1447	0.0308	0.0308	0.1431	0.0317	0.0317	0.1510	0.0356	0.0356	0.1483	0.0367	0.0367
0.20	0.1225	0.0277	0.0276	0.1276	0.0291	0.0291	0.1251	0.0291	0.0291	0.1305	0.0312	0.0312	0.1268	0.0310	0.0310
0.30	0.1043	0.0283	0.0282	0.1082	0.0294	0.0294	0.1055	0.0288	0.0288	0.1093	0.0301	0.0301	0.1055	0.0292	0.0292
0.40	0.0928	0.0287	0.0287	0.0962	0.0298	0.0298	0.0936	0.0290	0.0290	0.0967	0.0301	0.0301	0.0932	0.0291	0.0290
0.50	0.0845	0.0288	0.0287	0.0875	0.0298	0.0298	0.0851	0.0290	0.0290	0.0878	0.0300	0.0300	0.0845	0.0289	0.0289
0.60	0.0780	0.0286	0.0285	0.0808	0.0296	0.0296	0.0785	0.0288	0.0288	0.0810	0.0298	0.0297	0.0780	0.0287	0.0286
0.80	0.0684	0.0279	0.0278	0.0708	0.0289	0.0288	0.0688	0.0281	0.0280	0.0710	0.0290	0.0289	0.0683	0.0279	0.0278
1.0	0.0615	0.0270	0.0268	0.0636	0.0280	0.0279	0.0618	0.0272	0.0270	0.0637	0.0280	0.0279	0.0613	0.0269	0.0268
1.5	0.0501	0.0247	0.0245	0.0518	0.0256	0.0254	0.0504	0.0249	0.0247	0.0519	0.0256	0.0254	0.0499	0.0246	0.0245
2.0	0.0432	0.0229	0.0226	0.0448	0.0237	0.0235	0.0436	0.0231	0.0229	0.0450	0.0238	0.0236	0.0433	0.0229	0.0227
3.0	0.0354	0.0206	0.0202	0.0368	0.0214	0.0211	0.0359	0.0209	0.0206	0.0372	0.0217	0.0213	0.0359	0.0209	0.0205
4.0	0.0311	0.0193	0.0188	0.0324	0.0202	0.0197	0.0317	0.0198	0.0193	0.0329	0.0205	0.0201	0.0319	0.0199	0.0194
5.0	0.0284	0.0185	0.0179	0.0297	0.0194	0.0189	0.0292	0.0192	0.0186	0.0304	0.0200	0.0194	0.0295	0.0195	0.0188
6.0	0.0264	0.0180	0.0172	0.0277	0.0189	0.0182	0.0273	0.0187	0.0180	0.0285	0.0196	0.0188	0.0278	0.0191	0.0183
8.0	0.0242	0.0175	0.0166	0.0255	0.0185	0.0176	0.0253	0.0184	0.0175	0.0265	0.0194	0.0184	0.0260	0.0191	0.0180
10.0	0.0232	0.0176	0.0165	0.0246	0.0188	0.0176	0.0245	0.0188	0.0176	0.0259	0.0199	0.0185	0.0255	0.0197	0.0182

Source: Data Library DLC-139 (1988) distributed by the Radiation Shielding Information Center, Oak Ridge National Laboratory, and based on Report UCRL-50400, Vol. 6, Rev. 1, National Technical Information Service, Springfield, Va., 1975.

TABLE C.2 (cont'd).

E (MeV)	POTASSIUM μ/ρ	μ_{tr}/ρ	μ_{en}/ρ	CALCIUM μ/ρ	μ_{tr}/ρ	μ_{en}/ρ	TITANIUM μ/ρ	μ_{tr}/ρ	μ_{en}/ρ	VANADIUM μ/ρ	μ_{tr}/ρ	μ_{en}/ρ	CHROMIUM μ/ρ	μ_{tr}/ρ	μ_{en}/ρ
0.010	80.8700	76.6600	76.6600	93.3600	87.3800	87.3700	111.700	100.900	100.900	122.000	107.500	107.500	139.300	119.300	119.300
0.015	25.2300	23.9200	23.9100	29.4000	27.6800	27.6700	35.8300	33.0800	33.0700	38.9000	35.3700	35.3600	44.9400	40.1500	40.1400
0.020	10.8500	10.1700	10.1600	12.8000	11.9600	11.9500	15.7200	14.5400	14.5300	17.0600	15.6200	15.6100	19.8300	17.9500	17.9400
0.030	3.3170	2.9570	2.9560	3.9810	3.5750	3.5720	4.8880	4.4150	4.4120	5.3200	4.7900	4.7860	6.2190	5.5830	5.5780
0.040	1.4830	1.2180	1.2170	1.7910	1.5030	1.5010	2.1740	1.8700	1.8680	2.3690	2.0450	2.0430	2.7670	2.4010	2.3990
0.050	0.8337	0.6131	0.6128	1.0040	0.7689	0.7677	1.1950	0.9586	0.9575	1.3000	1.0550	1.0540	1.5120	1.2450	1.2430
0.060	0.5473	0.3536	0.3534	0.6519	0.4474	0.4466	0.7578	0.5572	0.5564	0.8211	0.6164	0.6155	0.9470	0.7279	0.7269
0.080	0.3169	0.1550	0.1550	0.3662	0.1971	0.1966	0.4046	0.2427	0.2422	0.4323	0.2695	0.2690	0.4888	0.3179	0.3173
0.10	0.2309	0.0883	0.0883	0.2590	0.1107	0.1105	0.2735	0.1332	0.1329	0.2877	0.1476	0.1473	0.3188	0.1730	0.1727
0.15	0.1576	0.0431	0.0431	0.1689	0.0505	0.0504	0.1664	0.0558	0.0557	0.1698	0.0602	0.0600	0.1813	0.0682	0.0680
0.20	0.1318	0.0339	0.0339	0.1385	0.0375	0.0374	0.1324	0.0385	0.0384	0.1330	0.0402	0.0401	0.1394	0.0440	0.0439
0.30	0.1080	0.0304	0.0303	0.1119	0.0320	0.0319	0.1047	0.0307	0.0306	0.1039	0.0309	0.0308	0.1073	0.0325	0.0324
0.40	0.0950	0.0298	0.0298	0.0980	0.0309	0.0308	0.0910	0.0290	0.0290	0.0899	0.0289	0.0288	0.0924	0.0299	0.0298
0.50	0.0860	0.0295	0.0294	0.0886	0.0304	0.0303	0.0820	0.0283	0.0282	0.0808	0.0280	0.0279	0.0829	0.0289	0.0287
0.60	0.0792	0.0291	0.0291	0.0815	0.0300	0.0298	0.0753	0.0278	0.0277	0.0742	0.0275	0.0273	0.0760	0.0282	0.0281
0.80	0.0693	0.0283	0.0282	0.0712	0.0291	0.0288	0.0657	0.0269	0.0267	0.0646	0.0265	0.0263	0.0662	0.0271	0.0269
1.0	0.0622	0.0273	0.0272	0.0639	0.0281	0.0278	0.0589	0.0259	0.0257	0.0579	0.0255	0.0253	0.0593	0.0261	0.0259
1.5	0.0507	0.0250	0.0248	0.0521	0.0257	0.0253	0.0480	0.0237	0.0234	0.0472	0.0233	0.0230	0.0483	0.0238	0.0236
2.0	0.0440	0.0233	0.0230	0.0452	0.0240	0.0234	0.0418	0.0221	0.0218	0.0411	0.0218	0.0214	0.0421	0.0223	0.0220
3.0	0.0367	0.0214	0.0210	0.0378	0.0221	0.0214	0.0351	0.0205	0.0200	0.0347	0.0203	0.0198	0.0356	0.0208	0.0203
4.0	0.0328	0.0205	0.0200	0.0340	0.0213	0.0205	0.0317	0.0200	0.0193	0.0314	0.0198	0.0191	0.0324	0.0204	0.0197
5.0	0.0306	0.0203	0.0195	0.0318	0.0211	0.0201	0.0298	0.0199	0.0190	0.0296	0.0198	0.0189	0.0306	0.0204	0.0196
6.0	0.0289	0.0200	0.0191	0.0301	0.0209	0.0197	0.0284	0.0198	0.0188	0.0283	0.0198	0.0188	0.0293	0.0205	0.0195
8.0	0.0273	0.0202	0.0189	0.0286	0.0212	0.0197	0.0272	0.0203	0.0190	0.0272	0.0204	0.0190	0.0283	0.0213	0.0198
10.0	0.0270	0.0210	0.0193	0.0283	0.0222	0.0202	0.0272	0.0214	0.0196	0.0273	0.0216	0.0198	0.0285	0.0226	0.0207

Source: Data Library DLC-139 (1988) distributed by the Radiation Shielding Information Center, Oak Ridge National Laboratory, and based on Report UCRL-50400, Vol. 6, Rev. 1, National Technical Information Service, Springfield, Va., 1975.

TABLE C.2 (cont'd).

E (MeV)	MANGANESE μ/ρ	MANGANESE μ_{tr}/ρ	MANGANESE μ_{en}/ρ	IRON μ/ρ	IRON μ_{tr}/ρ	IRON μ_{en}/ρ	NICKEL μ/ρ	NICKEL μ_{tr}/ρ	NICKEL μ_{en}/ρ	COPPER μ/ρ	COPPER μ_{tr}/ρ	COPPER μ_{en}/ρ	ZINC μ/ρ	ZINC μ_{tr}/ρ	ZINC μ_{en}/ρ
0.010	150.900	124.900	124.900	172.800	137.700	137.600	210.600	151.900	151.900	219.200	148.700	148.700	243.400	153.800	153.800
.015	48.8700	42.7800	42.7700	57.6300	49.3300	49.3200	71.7400	57.8700	57.8600	74.7600	58.2100	58.1800	83.9300	62.7800	62.7500
0.020	21.6200	19.3000	19.2900	25.7200	22.6200	22.6100	32.3300	27.3000	27.2900	33.8600	27.9000	27.8700	38.1100	30.5600	30.5400
0.030	6.7950	6.0630	6.0590	8.0760	7.1690	7.1660	10.2100	8.8720	8.8680	10.8000	9.2590	9.2450	12.1400	10.2500	10.2400
0.040	3.0210	2.6210	2.6190	3.5550	3.0920	3.0900	4.4870	3.8710	3.8690	4.7790	4.0970	4.0890	5.3470	4.5490	4.5420
0.050	1.6450	1.3630	1.3610	1.9100	1.5970	1.5960	2.3960	2.0110	2.0100	2.5610	2.1510	2.1460	2.8490	2.3870	2.3830
0.060	1.0260	0.7998	0.7986	1.1760	0.9310	0.9303	1.4600	1.1750	1.1740	1.5620	1.2660	1.2630	1.7260	1.4040	1.4000
0.080	0.5227	0.3496	0.3490	0.5836	0.4010	0.4006	0.7067	0.5057	0.5052	0.7519	0.5506	0.5486	0.8201	0.6082	0.6065
0.10	0.3360	0.1897	0.1893	0.3672	0.2144	0.2142	0.4325	0.2686	0.2683	0.4549	0.2934	0.2922	0.4898	0.3227	0.3216
0.15	0.1858	0.0733	0.0731	0.1964	0.0801	0.0800	0.2186	0.0968	0.0967	0.2226	0.1044	0.1039	0.2335	0.1132	0.1127
0.20	0.1406	0.0460	0.0459	0.1465	0.0491	0.0490	0.1578	0.0567	0.0566	0.1570	0.0594	0.0591	0.1622	0.0633	0.0630
0.30	0.1069	0.0329	0.0328	0.1102	0.0342	0.0342	0.1155	0.0371	0.0370	0.1125	0.0371	0.0369	0.1146	0.0385	0.0383
0.40	0.0916	0.0299	0.0298	0.0941	0.0308	0.0308	0.0977	0.0325	0.0324	0.0944	0.0318	0.0316	0.0956	0.0326	0.0324
0.50	0.0821	0.0287	0.0286	0.0842	0.0295	0.0294	0.0870	0.0307	0.0306	0.0837	0.0298	0.0295	0.0846	0.0303	0.0301
0.60	0.0751	0.0279	0.0278	0.0770	0.0287	0.0286	0.0794	0.0297	0.0296	0.0763	0.0287	0.0283	0.0770	0.0290	0.0288
0.80	0.0653	0.0268	0.0266	0.0669	0.0275	0.0273	0.0688	0.0283	0.0281	0.0660	0.0272	0.0268	0.0665	0.0275	0.0272
1.0	0.0585	0.0258	0.0256	0.0599	0.0264	0.0262	0.0615	0.0271	0.0269	0.0589	0.0260	0.0256	0.0594	0.0262	0.0259
1.5	0.0477	0.0235	0.0233	0.0488	0.0241	0.0238	0.0501	0.0247	0.0244	0.0480	0.0237	0.0231	0.0483	0.0239	0.0234
2.0	0.0416	0.0220	0.0217	0.0426	0.0226	0.0222	0.0438	0.0232	0.0228	0.0420	0.0223	0.0215	0.0424	0.0224	0.0219
3.0	0.0352	0.0206	0.0201	0.0362	0.0212	0.0206	0.0374	0.0220	0.0213	0.0360	0.0211	0.0201	0.0363	0.0213	0.0206
4.0	0.0321	0.0203	0.0196	0.0331	0.0209	0.0201	0.0344	0.0218	0.0209	0.0332	0.0211	0.0198	0.0336	0.0214	0.0203
5.0	0.0304	0.0204	0.0195	0.0314	0.0211	0.0200	0.0329	0.0221	0.0209	0.0317	0.0214	0.0199	0.0322	0.0218	0.0205
6.0	0.0292	0.0205	0.0195	0.0302	0.0213	0.0199	0.0317	0.0224	0.0209	0.0307	0.0217	0.0200	0.0312	0.0222	0.0205
8.0	0.0283	0.0214	0.0199	0.0295	0.0223	0.0203	0.0311	0.0236	0.0214	0.0302	0.0230	0.0207	0.0309	0.0235	0.0213
10.0	0.0287	0.0227	0.0209	0.0299	0.0238	0.0211	0.0318	0.0254	0.0223	0.0310	0.0248	0.0218	0.0317	0.0254	0.0225

Source: Data Library DLC-139 (1988) distributed by the Radiation Shielding Information Center, Oak Ridge National Laboratory, and based on Report UCRL-50400, Vol. 6, Rev. 1, National Technical Information Service, Springfield, Va., 1975.

TABLE C.3   Mass Interaction, Energy Transfer, and Energy Absorption Coefficients, in Units of $cm^2 g^{-1}$, for Zirconium and Tungsten.

E	ZIRCONIUM			E	TUNGSTEN		
(MeV)	μ/ρ	μ_{tr}/ρ	μ_{en}/ρ	(MeV)	μ/ρ	μ_{tr}/ρ	μ_{en}/ρ
0.0100	75.16	72.57	72.57	0.0100	97.93	93.34	93.30
0.0125	40.41	38.50	38.49	0.0102	93.06	88.60	88.40
0.0150	24.46	22.96	22.95	0.0102	236.50	184.40	184.00
0.0175	15.92	14.69	14.68	0.0115	172.70	138.20	137.80
0.0180	14.73	13.54	13.54	0.0115	241.90	194.80	194.30
0.0180	92.84	40.71	40.69	0.0121	214.40	174.10	173.70
				0.0121	248.00	201.90	201.40
0.0200	71.58	35.26	35.24	0.0200	66.95	58.08	57.85
0.0300	25.05	16.28	16.26	0.0300	22.93	20.17	20.06
0.0400	11.46	8.263	8.250	0.0400	10.67	9.304	9.239
0.0500	6.177	4.669	4.660	0.0500	5.901	5.052	5.010
0.0600	3.721	2.869	2.862	0.0695	2.487	2.019	1.997
0.0700	2.428	1.878	1.873	0.0695	11.010	3.194	3.159
0.0800	1.692	1.299	1.295	0.0800	7.779	2.925	2.889
0.1000	0.9444	0.697	0.6946	0.1000	4.430	2.150	2.119
0.1500	0.3709	0.2319	0.2308	0.1500	1.577	0.9627	0.9435
0.2000	0.2203	0.1152	0.1146	0.2000	0.7779	0.5025	0.4902
0.3000	0.1310	0.0545	0.0542	0.3000	0.3181	0.2010	0.1946
0.4000	0.1016	0.0394	0.0391	0.4000	0.1881	0.1114	0.1072
0.5000	0.0867	0.0336	0.0333	0.5000	0.1344	0.0753	0.0720
0.6000	0.0774	0.0307	0.0305	0.6000	0.1067	0.0575	0.0548
0.7000	0.0706	0.0290	0.0287	0.7000	0.0902	0.0476	0.0452
0.8000	0.0656	0.0278	0.0275	0.8000	0.0790	0.0412	0.0389
1.0000	0.0580	0.0260	0.0257	1.0000	0.0652	0.0340	0.0319
1.2500	0.0514	0.0245	0.0241	1.2500	0.0550	0.0292	0.0271
2.0000	0.0414	0.0220	0.0215	2.0000	0.0437	0.0242	0.0219
3.0000	0.0364	0.0215	0.0207	3.0000	0.0402	0.0246	0.0218
4.0000	0.0345	0.0222	0.0211	4.0000	0.0399	0.0267	0.0231
5.0000	0.0338	0.0232	0.0218	5.0000	0.0407	0.0291	0.0247
6.0000	0.0333	0.0240	0.0223	6.0000	0.0412	0.0309	0.0257
7.0000	0.0334	0.0250	0.0230	7.0000	0.0421	0.0329	0.0269
8.0000	0.0338	0.0262	0.0238	8.0000	0.0434	0.0349	0.0280
10.0000	0.0354	0.0289	0.0256	10.0000	0.0466	0.0392	0.0303

Source: Data Library DLC-139 (1988) distributed by the Radiation Shielding Information Center, Oak Ridge National Laboratory, and based on Report UCRL-50400, Vol. 6, Rev. 1, National Technical Information Service, Springfield, Va., 1975.

TABLE C.4 Mass Interaction, Energy Transfer, and Energy Absorption Coefficients, in Units of $cm^2 \, g^{-1}$, for Lead and Uranium.

E	LEAD			E	URANIUM		
(MeV)	μ/ρ	μ_{tr}/ρ	μ_{en}/ρ	(MeV)	μ/ρ	μ_{tr}/ρ	μ_{en}/ρ
0.0100	132.70	126.80	126.70	0.0100	176.20	167.00	166.50
0.0130	67.68	63.52	63.44	0.0125	100.60	94.48	94.14
0.0130	161.40	114.20	114.10				
0.0152	108.70	80.52	80.35	0.0150	63.95	59.39	59.13
0.0152	151.90	113.50	113.30				
0.0159	136.00	102.80	102.60	0.0172	45.77	42.04	41.84
0.0159	157.20	119.20	119.00	0.0172	100.40	62.04	61.75
0.0200	85.91	68.56	68.32	0.0200	69.10	45.78	45.53
				0.0210	61.66	41.59	41.36
				0.0210	85.92	58.65	58.32
				0.0218	78.21	54.15	53.84
				0.0218	90.35	62.82	62.46
0.0300	29.73	24.93	24.79	0.0300	40.45	30.81	30.58
0.0400	14.00	11.87	11.78	0.0400	19.56	15.60	15.46
0.0500	7.811	6.594	6.534	0.0500	11.10	9.020	8.919
0.0600	4.863	4.054	4.011	0.0600	6.976	5.697	5.623
0.0700	3.268	2.677	2.645	0.0700	4.704	3.832	3.775
0.0800	2.331	1.869	1.844	0.0800	3.358	2.714	2.669
0.0880	1.835	1.444	1.422				
0.0880	7.483	2.118	2.086				
0.1000	5.461	1.974	1.942	0.1000	1.920	1.511	1.482
				0.1156	1.342	1.029	1.007
				0.1156	4.994	1.410	1.380
0.1500	1.994	1.075	1.0510	0.1500	2.586	1.106	1.077
0.2000	0.9913	0.6027	0.5859	0.2000	1.278	0.6863	0.6640
0.3000	0.3996	0.2540	0.2447	0.3000	0.5095	0.3130	0.2996
0.4000	0.2294	0.1419	0.1356	0.4000	0.2870	0.1798	0.1705
0.5000	0.1588	0.0945	0.0898	0.5000	0.1943	0.1205	0.1133
0.6000	0.1226	0.0707	0.0668	0.6000	0.1467	0.0895	0.0836
0.7000	0.1014	0.0572	0.0538	0.7000	0.1189	0.0716	0.0665
0.8000	0.0872	0.0485	0.0453	0.8000	0.1005	0.0597	0.0552
1.0000	0.0701	0.0386	0.0358	1.0000	0.0787	0.0462	0.0422
1.2500	0.0581	0.0322	0.0296	1.2500	0.0634	0.0372	0.0336
1.5000	0.0518	0.0287	0.0261	1.5000	0.0557	0.0323	0.0289
2.0000	0.0453	0.0257	0.0230	2.0000	0.0479	0.0280	0.0247
3.0000	0.0417	0.0259	0.0225	3.0000	0.0439	0.0278	0.0243
4.0000	0.0415	0.0281	0.0238	4.0000	0.0436	0.0299	0.0260
5.0000	0.0423	0.0305	0.0253	5.0000	0.0445	0.0324	0.0280
6.0000	0.0429	0.0324	0.0263	6.0000	0.0450	0.0344	0.0295
7.0000	0.0440	0.0345	0.0274	7.0000	0.0461	0.0365	0.0312
8.0000	0.0454	0.0367	0.0285	8.0000	0.0475	0.0387	0.0328
10.0000	0.0488	0.0412	0.0308	10.0000	0.0509	0.0432	0.0362

Source: Data Library DLC-139 (1988) distributed by the Radiation Shielding Information Center, Oak Ridge National Laboratory, and based on Report UCRL-50400, Vol. 6, Rev. 1, National Technical Information Service, Springfield, Va., 1975.

TABLE C.5 Composition and Mass Density of Selected Radiologic Compounds and Mixtures. More Extensive Compilations for Various Body Tissues are Given in ICRU Report 44 (1989).

MATERIAL	(Z/A)	DENSITY (g/cm^3)	COMPOSITION (Z OF CONSTITUENT: WEIGHT FRACTION)			
Adipose tissue (ICRP-23)	0.55846	0.9200	1:0.119477	6:0.637240	7:0.007970	8:0.232333
			11:0.000500	12:0.000020	15:0.000160	16:0.000730
			17:0.001190	19:0.000320	20:0.000020	26:0.000020
			30:0.000020			
Air (dry)	0.49919	1.293(-3)	6:0.000124	7:0.755267	8:0.231781	18:0.012827
Bone, compact (ICRU-10b)	0.53010	1.850	1:0.063984	6:0.278000	7:0.027000	8:0.410016
			12:0.002000	15:0.070000	16:0.002000	20:0.147000
Bone, cortical (ICRP-23)	0.52129	1.850	1:0.047234	6:0.144330	7:0.041990	8:0.446096
			12:0.002200	15:0.104970	16:0.003150	20:0.209930
			30:0.000100			
Concrete, ordinary	0.50029	2.35	1:0.0056	8:0.4983	11:0.0171	12:0.0024
			13:0.0456	14:0.3158	16:0.0012	19:0.0192
			20:0.0826	26:0.0122		
Lithium floride	0.46265	2.635	3:0.267585	9:0.732415		
Lucite,Plexiglas,Perspex	0.53936	1.190	1:0.080538	6:0.599848	8:0.319614	
Muscle, skeletal (ICRP-23)	0.54937	1.040	1:0.100637	6:0.107830	7:0.027680	8:0.754773
			11:0.000750	12:0.000190	15:0.001800	16:0.002410
			17:0.000790	19:0.003020	20:0.000030	26:0.000040
			30:0.000050			
Muscle, striated (ICRU-10b)	0.55004	1.040	1:0.101997	6:0.123000	7:0.035000	8:0.729003
			11:0.000800	12:0.000200	15:0.002000	16:0.005000
			19:0.003000			
Polyethylene	0.57032	0.940	1:0.143711	6:0.856289		
Polystyrene	0.53767	1.060	1:0.077418	6:0.922582		
Sodium Iodide	0.42697	3.67	11:0.153374	53:0.846626		
Soil, (composite average)	0.49753	variable	8:0.5090	13:0.0500	14:0.3920	19:0.0141
			20:0.0129	26:0.0220		
Tissue, generic (ICRU-33)	0.54966	1.0	1:0.1010	6:0.1110	7:0.026	8:0.7620
Water (liquid)	0.55508	1.000	1:0.111894	8:0.888106		
Water (vapor)	0.55508	7.562(-4)	1:0.111894	8:0.888106		

Sources: ICRU Report 37 (1984); Chilton, Shultis and Faw (1987).

TABLE C.6 Mass Densities of Selected Elements at Standard Temperature and Pressure.

ELEMENT	Z	A	DENSITY (g/cm^3)	Element	Z	A	DENSITY (g/cm^3)
Hydrogen	1	1.00797	8.99(-5)[a]	Potassium	19	39.102	0.86
Helium	2	4.0026	1.786(-4)[a]	Calcium	20	40.08	1.55
Lithium	3	6.939	0.53	Titanium	22	47.90	4.51
Beryllium	4	9.0122	1.85	Vanadium	23	50.942	6.1
Boron	5	10.811	2.34	Chromium	24	51.996	7.19
Carbon	6	12.0112	2.26	Manganese	25	54.938	7.43
Nitrogen	7	14.0067	1.250(-3)[a]	Iron	26	55.847	7.86
Oxygen	8	15.9994	1.429(-3)[a]	Nickel	28	58.71	8.9
Sodium	11	22.9898	0.97	Copper	29	63.54	8.96
Magnesium	12	24.312	1.74	Zinc	30	65.37	7.14
Aluminum	13	26.9815	2.70	Zirconium	40	91.22	6.49
Silicon	14	28.086	2.33	Tungsten	74	183.85	19.3
Phosphorous	15	30.9738	1.82	Lead	82	207.19	11.34
Sulfur	16	32.064	2.07	Bismuth	83	208.980	9.80
Chlorine	17	35.453	3.214(-3)[a]	Uranium	92	238.03	19.05

[a] Density is for gaseous phase in molecular form.

Gamma Ray Yields and Cross Sections for Neutron Capture

TABLE D.1   Capture Cross Sections and the Number Capture Gamma Rays Produced in Common Elements with Natural Isotopic Abundance. The Thermal Capture Cross Sections are for 2200 m s^{-1} (0.025-eV) Neutrons in Units of the Barn (10^{-24} cm^2). Listed are the Numbers of Gamma Rays Produced per Capture in Each of 11 Energy Groups.

		ENERGY GROUPS (MeV)										
	σ_c (b)	0–1	1–2	2–3	3–4	4–5	5–6	6–7	7–8	8–9	9–10	10–11
H	3.32E−01	0.0000	0.0000	1.0000	0.0000	0.0000	0.0000	0.0000	0.0000	0.0000	0.0000	0.0000
Li	3.63E−02	0.1242	0.0491	0.8933	0.0000	0.0000	0.0000	0.0107	0.0402	0.0000	0.0000	0.0000
Be	9.20E−03	0.2641	0.0000	0.2356	0.4530	0.0000	0.0175	0.6375	0.0000	0.0000	0.0000	0.0000
B	1.03E−01	0.0000	0.0000	0.0000	0.0000	1.1014	0.0000	0.3950	0.4785	0.0000	0.0000	0.0000
C	3.37E−03	0.0000	0.2953	0.0000	0.3210	0.6764	0.0000	0.0000	0.0000	0.0000	0.0000	0.0000
N	7.47E−02	0.2632	0.3716	0.2450	0.2819	0.1578	0.8031	0.2064	0.1019	0.0397	0.0222	0.1412
O	2.70E−04	1.0000	0.8200	0.8200	0.1800	0.0000	0.0000	0.0000	0.0000	0.0000	0.0000	0.0000
Na	4.00E−01	0.9266	0.2046	0.7264	0.6537	0.0323	0.0633	0.2244	0.0000	0.0000	0.0000	0.0000
Mg	6.30E−02	0.5963	0.6876	0.6404	0.9584	0.0662	0.1078	0.1156	0.0372	0.0474	0.0075	0.0000
Al	2.30E−01	0.2751	0.0855	0.3225	0.3057	0.3499	0.0863	0.0778	0.3235	0.0000	0.0000	0.0000
Si	1.60E−01	0.1233	0.3403	0.3362	0.8046	0.6467	0.0634	0.1635	0.0903	0.0254	0.0000	0.0040
P	1.80E−01	0.4065	0.5411	0.5212	0.5446	0.2689	0.1290	0.1809	0.0788	0.0000	0.0000	0.0000
S	5.20E−01	0.7555	0.0000	0.7716	0.3642	0.1794	0.6348	0.0000	0.0391	0.0266	0.0000	0.0000
Cl	3.32E+01	0.4818	0.7536	0.4286	0.2597	0.1992	0.1516	0.3858	0.1966	0.0291	0.0000	0.0000
K	2.10E+00	0.7491	0.5656	0.7044	0.4329	0.2654	0.3638	0.0344	0.0670	0.0000	0.0000	0.0000
Ca	4.30E−01	0.2400	0.9349	0.5187	0.1712	0.2303	0.1256	0.4383	0.0216	0.0000	0.0000	0.0000
Ti	6.10E+00	0.3213	0.9772	0.0832	0.1221	0.1187	0.0283	0.6089	0.0109	0.0043	0.0022	0.0003
V	5.04E+00	0.3837	0.2486	0.1335	0.0591	0.0877	0.3158	0.3947	0.1972	0.0000	0.0000	0.0000
Cr	3.10E+00	0.4051	0.1608	0.2067	0.0921	0.0421	0.1103	0.1189	0.2461	0.3766	0.1097	0.0000
Mn	1.33E+01	0.7128	0.1242	0.3838	0.2199	0.2049	0.2981	0.0949	0.3257	0.0000	0.0000	0.0000
Fe	2.55E+00	0.2781	0.2383	0.1018	0.1328	0.1137	0.1097	0.1045	0.5865	0.0087	0.0415	0.0011
Co	3.72E+01	0.9375	0.1737	0.0794	0.0920	0.1121	0.2991	0.2893	0.0980	0.0000	0.0000	0.0000
Ni	4.43E+00	0.2616	0.0658	0.0604	0.0364	0.0371	0.0746	0.1703	0.1404	0.5898	0.0000	0.0000
Cu	3.79E+00	0.8176	0.0602	0.0458	0.0588	0.0917	0.1018	0.1621	0.6488	0.0000	0.0000	0.0000
Zn	1.10E+00	0.1598	0.3949	0.0943	0.0420	0.0596	0.0985	0.1200	0.1429	0.0071	0.0109	0.0000
Ge	2.30E+00	0.9587	0.1459	0.0360	0.4219	0.0576	0.0697	0.1096	0.0266	0.0113	0.0000	0.0000
Br	6.80E+00	0.6347	0.0666	0.0101	0.0030	0.0044	0.0187	0.0318	0.0227	0.0000	0.0000	0.0000
Zr	1.85E−01	0.8081	0.3048	0.2119	0.1361	0.0847	0.0820	0.1745	0.0042	0.0082	0.0000	0.0000
Mo	2.65E+00	0.8097	0.2000	0.0816	0.0416	0.0590	0.0542	0.0611	0.0074	0.0054	0.0000	0.0000
Ag	6.36E+01	0.6831	0.0166	0.0105	0.0102	0.0312	0.0877	0.0266	0.0112	0.0000	0.0000	0.0000
Cd	2.45E+03	1.0399	0.2239	0.1895	0.0736	0.0410	0.0957	0.0129	0.0073	0.0036	0.0025	0.0000
In	1.94E+02	0.3362	0.3534	0.1365	0.0311	0.0381	0.0328	0.0029	0.0000	0.0000	0.0000	0.0000
Sn	6.30E−01	0.1411	0.3611	0.1246	0.0715	0.0307	0.0335	0.0168	0.0035	0.0000	0.0034	0.0000
Ba	1.20E+00	0.3751	0.3971	0.2427	0.0910	0.3210	0.0848	0.0332	0.0088	0.0034	0.0065	0.0000
Ta	2.11E+01	0.4396	0.0090	0.0010	0.0292	0.0771	0.0421	0.0054	0.0000	0.0000	0.0000	0.0000
W	1.85E+01	0.7986	0.0557	0.0520	0.0887	0.1072	0.0987	0.0725	0.0067	0.0000	0.0000	0.0000
Hg	3.76E+02	0.9153	0.3383	0.2106	0.1322	0.2744	0.3008	0.0698	0.0000	0.0000	0.0000	0.0000
Pb	1.70E−01	0.0000	0.0000	0.0000	0.0000	0.0000	0.0000	0.0504	0.9406	0.0000	0.0000	0.0000
Bi	3.30E−02	0.0000	0.0000	0.0000	0.0000	1.1170	0.0000	0.0000	0.0000	0.0000	0.0000	0.0000

Source: Lone, M.A., R.A. Leavitt, and D.A. Harrison. 1981. Prompt Gamma Rays from Thermal Neutron Capture. Atomic Data and Nuclear Data Tables 26(6): 511–559. [Data files released as Package DLC-140 THERMGAM by the Radiation Shielding Information Center, Oak Ridge National Laboratory, Oak Ridge, Tenn.]

TABLE D.2 Capture Cross Sections and Energies of Capture Gamma Rays Produced in Common Elements with Natural Isotopic Abundance. The Thermal Capture Cross Sections are for 2200 m s^{-1} (0.025-eV) Neutrons in Units of the Barn (10^{-24} cm^2). Listed are the Total Energies Carried by Gamma Rays Produced per Neutron Capture in Each of 11 Energy Groups.

		ENERGY GROUPS (MeV)										
	σ_c (b)	0–1	1–2	2–3	3–4	4–5	5–6	6–7	7–8	8–9	9–10	10–11
H	3.32E−01	0.0000	0.0000	2.2233	0.0000	0.0000	0.0000	0.0000	0.0000	0.0000	0.0000	0.0000
Li	3.63E−02	0.1088	0.0517	1.8156	0.0000	0.0000	0.0000	0.0724	0.2913	0.0000	0.0000	0.0000
Be	9.20E−03	0.2241	0.0000	0.6110	1.5343	0.0000	0.1042	4.3410	0.0000	0.0000	0.0000	0.0000
B	1.03E−01	0.0000	0.0000	0.0000	0.0000	4.9848	0.0000	2.6699	3.3519	0.0000	0.0000	0.0000
C	3.37E−03	0.0000	0.3726	0.0000	1.1825	3.3450	0.0000	0.0000	0.0000	0.0000	0.0000	0.0000
N	7.47E−02	0.1218	0.6868	0.5715	1.0192	0.7115	4.3196	1.3115	0.7451	0.3303	0.2052	1.5291
O	2.70E−04	0.8709	0.8921	1.7913	0.5888	0.0000	0.0000	0.0000	0.0000	0.0000	0.0000	0.0000
Na	4.00E−01	0.5507	0.3508	1.7780	2.3518	0.1395	0.3509	1.4343	0.0000	0.0000	0.0000	0.0000
Mg	6.30E−02	0.3395	1.1726	1.7368	3.4943	0.2945	0.5869	0.7379	0.2742	0.3865	0.0696	0.0000
Al	2.30E−01	0.1395	0.1406	0.8275	1.0620	1.5670	0.4633	0.4954	2.4946	0.0000	0.0000	0.0000
Si	1.60E−01	0.0495	0.5001	0.7370	2.8495	3.1848	0.3273	1.0513	0.6517	0.2151	0.0000	0.0424
P	1.80E−01	0.2242	0.8311	1.2151	1.9395	1.2219	0.7049	1.2086	0.5902	0.0000	0.0000	0.0000
S	5.20E−01	0.6355	0.0000	1.9928	1.1973	0.8496	3.4312	0.0000	0.3050	0.2298	0.0000	0.0000
Cl	3.32E+01	0.3195	1.2437	1.1040	0.8857	0.9134	0.8450	2.4523	1.4894	0.2496	0.0000	0.0000
K	2.10E+00	0.5389	0.8437	1.6580	1.5228	1.1536	1.9911	0.2380	0.5206	0.0000	0.0000	0.0000
Ca	4.30E−01	0.1562	1.7510	1.1616	0.6146	1.0383	0.7075	2.8153	0.1564	0.0000	0.0000	0.0000
Ti	6.10E+00	0.1365	1.4180	0.2071	0.4232	0.5734	0.1573	3.9971	0.0787	0.0353	0.0204	0.0032
V	5.04E+00	0.2327	0.3971	0.3086	0.2104	0.3956	1.7377	2.6051	1.4211	0.0000	0.0000	0.0000
Cr	3.10E+00	0.3237	0.2902	0.4860	0.3200	0.1905	0.6247	0.7736	1.8743	3.3047	1.0663	0.0000
Mn	1.33E+01	0.1279	0.2214	0.9016	0.7660	0.9404	1.5850	0.6261	2.3343	0.0000	0.0000	0.0000
Fe	2.55E+00	0.1384	0.3670	0.2547	0.4607	0.5035	0.6402	0.6352	4.4584	0.0765	0.3859	0.0111
Co	3.72E+01	0.3744	0.2959	0.2001	0.3309	0.5075	1.6866	1.9356	0.7130	0.0000	0.0000	0.0000
Ni	4.43E+00	0.1294	0.1049	0.1537	0.1264	0.1703	0.4183	1.1363	1.0831	5.1953	0.0000	0.0000
Cu	3.79E+00	0.3230	0.0929	0.1193	0.2099	0.4119	0.5420	1.0671	4.9839	0.0000	0.0000	0.0000
Zn	1.10E+00	0.0975	0.5201	0.2252	0.1451	0.2673	0.5483	0.8040	1.0960	0.0590	0.0994	0.0000
Ge	2.30E+00	0.5385	0.2056	0.0829	1.2958	0.2648	0.3816	0.7025	0.1940	0.0948	0.0000	0.0000
Br	6.80E+00	0.2521	0.0831	0.0248	0.0104	0.0197	0.1045	0.2019	0.1666	0.0000	0.0000	0.0000
Zr	1.85E−01	0.5529	0.4767	0.5284	0.4711	0.3800	0.4264	1.1022	0.0323	0.0708	0.0000	0.0000
Mo	2.65E+00	0.5839	0.2804	0.2014	0.1482	0.2695	0.3015	0.4073	0.0558	0.0452	0.0000	0.0000
Ag	6.36E+01	0.1913	0.0175	0.0235	0.0373	0.1417	0.4819	0.1665	0.0802	0.0000	0.0000	0.0000
Cd	2.45E+03	0.6229	0.3222	0.4758	0.2464	0.1853	0.5267	0.0839	0.0564	0.0305	0.0226	0.0000
In	1.94E+02	0.1493	0.4477	0.2973	0.1078	0.1753	0.1765	0.0182	0.0000	0.0000	0.0317	0.0000
Sn	6.30E−01	0.0733	0.4790	0.2997	0.2456	0.1414	0.1809	0.1078	0.0253	0.0000	0.0317	0.0000
Ta	2.11E+01	0.1817	0.0144	0.0029	0.1088	0.3492	0.2309	0.0327	0.0000	0.0000	0.0000	0.0000
W	1.85E+01	0.2324	0.0785	0.1360	0.3129	0.4732	0.5196	0.4483	0.0495	0.0000	0.0000	0.0000
Au	9.88E+01	0.1580	0.1830	0.4255	0.0295	0.6278	0.9170	1.0132	0.0000	0.0000	0.0000	0.0000
Hg	3.76E+02	0.3720	0.5297	0.5016	0.4438	1.2917	1.7014	0.4486	0.0000	0.0000	0.0000	0.0000
Pb	1.70E−01	0.0000	0.0000	0.0000	0.0000	0.0000	0.0000	0.3395	6.9301	0.0000	0.0000	0.0000
Bi	3.30E−02	0.0000	0.0000	0.0000	0.0000	4.5998	0.0000	0.0000	0.0000	0.0000	0.0000	0.0000

Source: Lone, M.A., R.A. Leavitt, and D.A. Harrison. 1981. Prompt Gamma Rays from Thermal Neutron Capture. Atomic Data and Nuclear Data Tables 26(6): 511–559. [Data files released as Package DLC-140 THERMGAM by the Radiation Shielding Information Center, Oak Ridge National Laboratory, Oak Ridge, Tenn.]

Appendix E

Gamma-Ray Buildup Factors

TABLE E.1 Gamma-Ray Buildup Factors for Air-Kerma Response to an Isotropic Point Source in an Infinite Air Medium. Also Given is the Total Mass Interaction Coefficient Used to Calculate the Mean-Free-Path Length.

MEAN FREE PATHS	ENERGY (MeV)										
	0.04	0.06	0.08	0.1	0.2	0.5	1	2	5	10	15
0	1.00	1.00	1.00	1.00	1.00	1.00	1.00	1.00	1.00	1.00	1.00
0.5	2.20	2.58	2.52	2.35	1.90	1.60	1.47	1.38	1.29	1.20	1.15
1	3.38	4.76	4.83	4.46	3.28	2.44	2.08	1.83	1.57	1.37	1.28
2	5.85	10.8	12.0	11.4	7.74	4.84	3.60	2.81	2.09	1.68	1.49
3	8.47	18.9	22.9	22.5	15.00	8.21	5.46	3.86	2.60	1.97	1.70
4	11.2	29.1	37.9	38.4	25.6	12.6	7.60	4.96	3.11	2.26	1.90
5	14.1	41.5	57.4	59.9	40.0	17.9	10.0	6.13	3.61	2.54	2.11
6	17.0	56.1	82	87.8	58.9	24.2	12.7	7.35	4.12	2.82	2.30
7	20.1	73.2	112	123	82.8	31.6	15.6	8.61	4.62	3.10	2.50
8	23.3	92.7	148	166	112	40.1	18.8	9.92	5.12	3.37	2.70
10	30.0	140	242	282	192	60.6	25.8	12.6	6.13	3.92	3.08
15	49.0	316	636	800	545	134	47.0	20.0	8.63	5.25	4.03
20	71.4	596	1350	1810	1220	241	72.8	27.9	11.1	6.55	4.96
25	97.2	1010	2540	3570	2360	385	103	36.2	13.6	7.84	5.87
30	126	1600	4390	6430	4150	567	136	45.0	16.1	9.11	6.75
35	159	2410	7140	10600	6770	788	173	54.0	18.5	10.4	7.58
40	195	3480	11100	15700	10500	1050	212	63.2	21.0	11.6	8.31
μ/ρ (cm^2/g)	.2486	.1875	.1662	.1541	.1234	.08712	.06358	.04447	.02751	.02045	.01810

Source: Extracted from American National Standard ANSI/ANS-6.4.3-1991 with permission of the publisher, the American Nuclear Society.

624

TABLE E.2 Gamma-Ray Buildup Factors for Air-Kerma Response to an Isotropic Point Source in an Infinite Water Medium. Data are for Air Kerma Response, but Apply within a Few Percent to Water Kerma Response as Well. Also Given is the Total Mass Interaction Coefficient Used to Calculate the Mean-Free-Path Length.

MEAN FREE PATHS	ENERGY (MeV)										
	0.04	0.06	0.08	0.1	0.2	0.5	1	2	5	10	15
0	1.00	1.00	1.00	1.00	1.00	1.00	1.00	1.00	1.00	1.00	1.00
0.5	2.22	2.62	2.55	2.37	1.92	1.60	1.47	1.38	1.28	1.20	1.15
1	3.47	4.90	4.95	4.55	3.42	2.44	2.08	1.83	1.56	1.37	1.28
2	6.18	11.4	12.6	11.8	8.31	4.88	3.62	2.81	2.08	1.68	1.49
3	9.14	20.4	24.5	23.8	16.0	8.35	5.50	3.87	2.58	1.97	1.70
4	12.3	32.0	41.1	41.3	27.0	12.8	7.68	4.98	3.08	2.25	1.90
5	15.7	46.4	63.3	65.2	42.2	18.4	10.1	6.15	3.58	2.53	2.10
6	19.2	63.6	91.4	96.7	62.5	25.0	12.8	7.38	4.08	2.80	2.30
7	22.9	83.9	126	137	88.5	32.7	15.8	8.65	4.58	3.07	2.49
8	26.8	107	169	187	121	41.5	19.0	9.97	5.07	3.34	2.68
10	35.1	165	281	321	208	62.9	26.1	12.7	6.05	3.86	3.05
15	59.2	386	762	938	600	139	47.7	20.1	8.49	5.14	3.96
20	88.5	751	1660	2170	1350	252	74.0	28.0	10.9	6.38	4.84
25	123	1310	3200	4360	2670	403	104	36.5	13.3	7.59	5.69
30	163	2110	5630	7970	4810	594	139	45.2	15.7	9.96	6.51
35	208	3240	9300	13500	8170	828	177	54.4	18.0	11.2	7.26
40	259	4740	14600	21100	13300	1110	218	63.7	20.4	11.3	7.91
μ/ρ (cm^2/g)	.2683	.2058	.1836	.1707	.1370	.09687	.07072	.04941	.03031	.02219	.01941

Source: Extracted from American National Standard ANSI/ANS-6.4.3-1991 with permission of the publisher, the American Nuclear Society.

TABLE E.3 Gamma-Ray Buildup Factors for Air-Kerma Response to an Isotropic Point Source in an Infinite Concrete Medium. Also Given is the Total Mass Interaction Coefficient Used to Calculate the Mean-Free-Path Length.

MEAN FREE PATHS	ENERGY (MeV)										
	0.04	0.06	0.08	0.1	0.2	0.5	1	2	5	10	15
0	1.00	1.00	1.00	1.00	1.00	1.00	1.00	1.00	1.00	1.00	1.00
0.5	1.30	1.68	1.84	1.89	1.78	1.57	1.45	1.37	1.27	1.19	1.15
1	1.46	2.15	2.58	2.78	2.72	2.27	1.98	1.77	1.53	1.35	1.26
2	1.69	2.89	3.96	4.63	5.05	4.03	3.24	2.65	2.04	1.64	1.46
3	1.87	3.54	5.31	6.63	8.00	6.26	4.72	3.60	2.53	1.93	1.66
4	2.01	4.17	6.69	8.8	11.6	8.97	6.42	4.61	3.03	2.22	1.86
5	2.14	4.77	8.09	11.1	15.9	12.2	8.33	5.68	3.54	2.51	2.07
6	2.25	5.34	9.52	13.6	20.9	15.9	10.4	6.80	4.05	2.80	2.28
7	2.35	5.90	11.0	16.3	26.7	20.2	12.7	7.97	4.57	3.10	2.50
8	2.45	6.44	12.5	19.2	33.4	25.0	15.2	9.18	5.09	3.40	2.71
10	2.62	7.52	15.7	25.6	49.6	36.4	20.7	11.7	6.15	4.01	3.16

(Continued)

TABLE E.3 (cont'd).

MEAN FREE PATHS	ENERGY (MeV)										
	0.04	0.06	0.08	0.1	0.2	0.5	1	2	5	10	15
15	2.98	10.2	24.3	44.9	109	75.6	37.2	18.6	8.85	5.57	4.34
20	3.27	12.7	33.8	69.1	201	131	57.1	26.0	11.6	7.19	5.59
25	3.51	15.2	44.3	97.9	331	203	80.1	33.9	14.4	8.86	6.91
30	3.73	18.2	55.4	131	507	292	106	42.2	17.3	10.6	8.27
35	3.91	21.9	66.8	170	734	399	134	50.9	20.5	12.3	9.63
40	4.03	26.5	78.1	214	1020	523	164	59.8	24.8	14.5	10.9
μ/ρ (cm^2/g)	.6122	.2957	.2125	.1783	.1270	.08768	.06382	.04482	.02895	.02311	.02153

Source: Extracted from American National Standard ANSI/ANS-6.4.3-1991 with permission of the publisher, the American Nuclear Society.

TABLE E.4   Gamma-Ray Buildup Factors for Air-Kerma Response to an Isotropic Point Source in an Infinite Lead Medium. Also Given is the Total Mass Interaction Coefficient Used to Calculate the Mean-Free-Path Length.

MEAN FREE PATHS	ENERGY (MeV)											
	0.04	0.88	0.89	0.1	0.12	0.14	0.5	1	2	5	10	15
0	1.00	1.00	1.00	1.00	1.00	1.00	1.00	1.00	1.00	1.00	1.00	1.00
0.5	1.01	1.05	1.59	1.51	1.38	1.28	1.14	1.20	1.21	1.25	1.28	1.31
1	1.01	1.07	2.24	2.04	1.70	1.48	1.24	1.38	1.40	1.41	1.51	1.63
2	1.02	1.10	4.12	3.39	2.36	1.77	1.39	1.68	1.76	1.71	2.01	2.34
3	1.02	1.11	7.66	5.60	3.13	1.98	1.52	1.95	2.14	2.05	2.63	3.34
4	1.02	1.13	14.9	9.59	4.19	2.10	1.62	2.19	2.52	2.44	3.42	4.78
5	1.03	1.14	30.0	17.0	5.71	2.40	1.71	2.43	2.91	2.88	4.45	6.83
6	1.03	1.15	61.1	30.6	7.90	2.62	1.80	2.66	3.32	3.38	5.73	9.70
7	1.03	1.16	119	54.9	11.2	2.87	1.88	2.89	3.74	3.93	7.37	13.7
8	1.03	1.17	229	94.7	15.8	3.18	1.95	3.10	4.17	4.56	9.44	19.4
10	1.03	1.18	875	294	32.1	3.99	2.10	3.51	5.07	6.03	15.4	38.7
15	1.04	1.21	2.73E4	5.80E3	235	7.47	2.39	4.45	7.44	11.4	50.8	208
20	1.04	1.23	9.51E5	1.33E5	2.14E3	17.1	2.64	5.27	9.98	19.9	61	1070
25	1.04	1.25	3.57E7	3.34E6	2.11E4	45.0	2.85	5.98	12.6	32.9	495	5330
30	1.05	1.27	1.40E9	8.77E7	2.15E5	128	3.02	6.64	15.4	52.2	1470	2.57E4
35	1.05	1.28	5.70E10	2.36E9	2.22E6	380	3.18	7.23	18.2	79.9	4280	1.21E5
40	1.05	1.29	2.36E12	6.43E10	2.33E7	1.60E3	3.31	7.79	21.0	119	12200	5.59E5
μ/ρ (cm^2/g)	13.49	1.668	7.229	5.355	3.368	2.284	.1515	.06843	.04536	.04256	.04965	.05653

Source: Extracted from American National Standard ANSI/ANS-6.4.3-1991 with permission of the publisher, the American Nuclear Society.

TABLE E.5 Parameters for the Taylor Form of the Exposure Buildup Factor for Gamma Photons from a Point Isotropic Source in Ordinary Concrete, to 40 Mean Free Path Lengths.

SOURCE ENERGY (MeV)	PARAMETER A_1	α_1	σ_2	MAXIMUM PERCENT DEVIATION
0.04	2.33	−0.0147	0.317	4.5
0.06	5.29	−0.0414	0.210	5.3
0.08	18.3	−0.0382	0.0469	4.7
0.1	73.8	−0.0394	−0.0145	6.0
0.2	144	−0.0741	−0.0598	20.2
0.5	62.0	−0.0688	−0.0424	22.2
1.0	97.0	−0.0396	−0.0271	15.2
2.0	38.7	−0.0250	−0.00227	7.0
5.0	10.42	−0.0244	0.0269	1.5
10.0	5.10	−0.0269	0.0450	2.3
15.0	4.04	−0.0267	0.0393	2.7

Note: $B(E_o, \mu r) = A_1 e^{-\alpha_1 \mu r} + A_2 e^{-\alpha_2 \mu r}$, where $A_2 = 1 - A_1$.

Source: Chilton, *Nucl. Sci. Eng. 64*, 799–800 (1977); by permission; Copyright 1977, American Nuclear Society, La-Grange Park, Ill.

TABLE E.6 Parameters for the Berger Form of the Air Kerma Buildup Factor in Five Attenuating Media.

ENERGY (MeV)	AIR a	b	WATER a	b	CONCRETE a	b	IRON a	b	LEAD a	b
0.015	0.08	−0.034	0.09	−0.036	0.01	−0.029	0.00	0.000	0.00	0.000
0.020	0.23	−0.032	0.26	−0.032	0.03	−0.041	0.02	−0.032	0.00	0.000
0.030	0.93	−0.009	1.01	−0.006	0.10	−0.036	0.01	−0.036	0.00	0.000
0.040	2.40	0.018	2.58	0.024	0.26	−0.035	0.02	−0.032	0.01	−0.066
0.050	4.05	0.050	4.36	0.057	0.52	−0.026	0.04	−0.034	0.01	−0.046
0.060	5.27	0.075	5.59	0.082	0.78	−0.008	0.07	−0.039	0.01	−0.028
0.080	6.11	0.102	6.47	0.108	1.42	0.007	0.14	−0.034	0.02	−0.029
0.100	5.93	0.113	6.11	0.120	1.83	0.028	0.24	−0.030	0.20	0.479
0.150	4.70	0.121	4.88	0.125	2.19	0.054	0.52	−0.015	0.21	−0.075
0.200	3.94	0.113	4.13	0.118	2.20	0.065	0.77	0.004	0.08	−0.054
0.300	3.10	0.094	3.18	0.096	2.03	0.067	1.06	0.022	0.08	−0.040
0.400	2.61	0.079	2.67	0.080	1.87	0.061	1.15	0.033	0.11	−0.033
0.500	2.29	0.067	2.32	0.068	1.73	0.055	1.16	0.036	0.15	−0.028
0.600	2.05	0.058	2.07	0.059	1.60	0.049	1.14	0.036	0.19	−0.024
0.800	1.71	0.045	1.74	0.045	1.41	0.040	1.09	0.032	0.25	−0.019
1.000	1.50	0.035	1.50	0.036	1.27	0.032	1.03	0.028	0.30	−0.015

(Continued)

TABLE E.6 (cont'd).

ENERGY (MeV)	AIR a	AIR b	WATER a	WATER b	CONCRETE a	CONCRETE b	IRON a	IRON b	LEAD a	LEAD b
1.500	1.16	0.021	1.16	0.021	1.02	0.021	0.88	0.020	0.36	−0.007
2.000	0.97	0.013	0.97	0.013	0.89	0.014	0.76	0.018	0.38	0.004
3.000	0.75	0.005	0.74	0.005	0.71	0.007	0.66	0.014	0.37	0.019
4.000	0.61	0.001	0.62	0.000	0.59	0.004	0.56	0.015	0.31	0.038
5.000	0.53	−0.002	0.52	0.002	0.49	0.004	0.49	0.017	0.24	0.062
6.000	0.47	−0.004	0.47	−0.005	0.45	0.002	0.42	0.021	0.19	0.082
8.000	0.37	−0.004	0.38	−0.006	0.36	0.001	0.33	0.028	0.11	0.125
10.000	0.31	−0.004	0.31	−0.005	0.30	0.003	0.25	0.039	0.07	0.161
15.000	0.23	−0.006	0.23	−0.008	0.21	0.004	0.15	0.066	0.00	0.000

Sources:

Chilton, Shultis and Faw, *Principles of Radiation Shielding*, Prentice–Hall 1984.

Chilton, Eisenhauer, and Simmons, *Nucl. Sci. Eng. 73*, 97–107 (1980).

Chilton, *Nucl. Sci. Eng. 69*, 436–438 (1979).

TABLE E.7   Coefficients for the Geometric-Progression Form of the Gamma-Ray Buildup Factor.

E (MeV)	WATER KERMA/WATER MEDIUM b	c	a	ξ	d	AIR KERMA/WATER MEDIUM b	c	a	ξ	d
0.015	1.188	0.464	0.172	14.00	−0.0829	1.182	0.463	0.175	14.23	−0.0908
0.020	1.449	0.532	0.152	14.61	−0.0764	1.427	0.549	0.143	14.86	−0.0707
0.030	2.411	0.741	0.084	14.62	−0.0452	2.335	0.736	0.087	13.28	−0.0419
0.040	3.587	1.114	−0.018	12.48	0.0013	3.477	1.117	−0.019	11.67	0.0026
0.050	4.554	1.457	−0.084	13.69	0.0341	4.461	1.457	−0.084	13.62	0.0341
0.060	5.018	1.735	−0.127	13.70	0.0676	4.983	1.730	−0.126	13.64	0.0561
0.080	5.030	2.054	−0.167	13.84	0.0763	5.059	2.059	−0.168	13.67	0.0770
0.100	4.627	2.207	−0.184	13.27	0.0799	4.663	2.221	−0.186	13.33	0.0826
0.150	3.888	2.206	−0.180	14.27	0.0738	3.897	2.242	−0.185	14.19	0.0777
0.200	3.462	2.132	−0.173	14.51	0.0750	3.478	2.154	−0.176	14.50	0.0774
0.300	2.897	2.008	−0.162	14.18	0.0641	2.920	2.022	−0.164	14.21	0.0655
0.400	2.646	1.874	−0.148	14.16	0.0591	2.660	1.882	−0.149	14.24	0.0595
0.500	2.499	1.749	−0.132	14.36	0.0517	2.500	1.766	−0.135	14.33	0.0546
0.600	2.383	1.662	−0.121	14.19	0.0482	2.377	1.679	−0.124	14.23	0.0503
0.800	2.223	1.524	−0.101	14.31	0.0403	2.212	1.544	−0.105	14.36	0.0437
1.000	2.106	1.436	−0.088	14.19	0.0367	2.103	1.441	−0.089	14.22	0.0378
1.500	1.948	1.265	−0.057	14.98	0.0245	1.939	1.269	−0.058	14.52	0.0246
2.000	1.843	1.169	−0.038	14.22	0.0157	1.839	1.173	−0.039	14.07	0.0161
3.000	1.716	1.050	−0.011	13.63	0.0027	1.710	1.056	−0.013	11.82	0.0047
4.000	1.633	0.979	0.007	14.23	−0.0060	1.621	0.989	0.004	13.45	−0.0041
5.000	1.571	0.928	0.022	13.20	−0.0157	1.554	0.939	0.018	13.55	−0.0122
6.000	1.521	0.893	0.033	11.92	−0.0208	1.507	0.903	0.029	16.13	−0.0272
8.000	1.432	0.873	0.038	11.56	−0.0204	1.422	0.879	0.035	13.36	−0.0191
10.000	1.378	0.849	0.045	14.34	−0.0280	1.362	0.859	0.042	13.37	−0.0247
15.000	1.280	0.829	0.052	14.85	−0.0367	1.267	0.843	0.047	15.08	−0.0336

Source: Extracted from American National Standard ANSI/ANS−6.4.3−1991, *Gamma-Ray Attenuation Coefficients and Buildup Factors for Engineering Materials*, with permission of the publisher, the American Nuclear Society (1991); data also available from Data Library DLC−129/ANS643, issued by the Radiation Shielding Information Center, Oak Ridge National Laboratory, Oak Ridge, Tenn., 1988.

TABLE E.8 Coefficients for the Geometric-Progression Form of the Gamma-Ray Buildup Factor.

E (MeV)	AIR KERMA/AIR MEDIUM					AIR KERMA/CONCRETE MEDIUM				
	b	c	a	ξ	d	b	c	a	ξ	d
0.015	1.170	0.459	0.175	13.73	−0.0862	1.029	0.364	0.240	14.12	−0.1704
0.020	1.407	0.512	0.161	14.40	−0.0819	1.067	0.389	0.214	12.68	−0.1126
0.030	2.292	0.693	0.102	13.34	−0.0484	1.212	0.421	0.201	14.12	−0.1079
0.040	3.390	1.052	−0.004	19.76	−0.0068	1.455	0.493	0.171	14.53	−0.0925
0.050	4.322	1.383	−0.071	13.51	0.0270	1.737	0.628	0.115	15.82	−0.0600
0.060	4.837	1.653	−0.115	13.66	0.0511	2.125	0.664	0.118	11.90	−0.0615
0.080	4.929	1.983	−0.159	13.74	0.0730	2.557	0.895	0.042	14.37	−0.0413
0.100	4.580	2.146	−0.178	12.83	0.0759	2.766	1.069	0.001	12.64	−0.0251
0.150	3.894	2.148	−0.173	14.46	0.0698	2.824	1.315	−0.049	8.66	−0.0048
0.200	3.345	2.147	−0.176	14.08	0.0719	2.716	1.430	−0.070	18.52	0.0108
0.300	2.887	1.990	−0.160	14.13	0.0633	2.522	1.492	−0.082	16.59	0.0161
0.400	2.635	1.860	−0.146	14.24	0.0583	2.372	1.494	−0.085	15.96	0.0194
0.500	2.496	1.736	−0.130	14.32	0.0505	2.271	1.466	−0.082	16.25	0.0195
0.600	2.371	1.656	−0.120	14.27	0.0472	2.192	1.434	−0.078	17.02	0.0199
0.800	2.207	1.532	−0.103	14.12	0.0425	2.066	1.386	−0.073	15.07	0.0202
1.000	2.102	1.428	−0.086	14.35	0.0344	1.982	1.332	−0.065	15.38	0.0193
1.500	1.939	1.265	−0.057	14.24	0.0232	1.848	1.227	−0.047	16.41	0.0160
2.000	1.835	1.173	−0.039	14.07	0.0161	1.775	1.154	−0.033	14.35	0.0100
3.000	1.712	1.051	−0.011	13.67	0.0024	1.671	1.054	−0.010	10.47	−0.0008
4.000	1.627	0.983	0.006	13.51	−0.0051	1.597	0.988	0.008	12.53	−0.0115
5.000	1.558	0.943	0.017	13.82	−0.0117	1.527	0.951	0.020	9.99	−0.0184
6.000	1.505	0.915	0.025	16.37	−0.0231	1.478	0.940	0.021	13.11	−0.0163
8.000	1.418	0.891	0.032	12.06	−0.0167	1.395	0.917	0.028	13.45	−0.0213
10.000	1.358	0.875	0.037	14.01	−0.0226	1.334	0.901	0.035	12.56	−0.0267
15.000	1.267	0.844	0.048	14.55	−0.0344	1.260	0.823	0.065	14.28	−0.0581

Source: Extracted from American National Standard ANSI/ANS-6.4.3-1991, *Gamma-Ray Attenuation Coefficients and Buildup Factors for Engineering Materials*, with permission of the publisher, the American Nuclear Society (1991); data also available from Data Library DLC-129/ANS643, issued by the Radiation Shielding Information Center, Oak Ridge National Laboratory, Oak Ridge, Tenn., 1988.

TABLE E.9 Coefficients for the Geometric-Progression Form of the Gamma-Ray Buildup Factor.

E (MeV)	AIR KERMA/IRON MEDIUM					AIR KERMA/LEAD MEDIUM				
	b	c	a	ξ	d	b	c	a	ξ	d
0.015	1.004	1.561	−0.554	5.60	0.3524					
0.020	1.012	0.130	0.620	11.39	−0.6162					
0.030	1.028	0.374	0.190	29.34	−0.3170	1.007	0.322	0.246	13.67	−0.1030
0.040	1.058	0.336	0.248	11.65	−0.1188	1.014	0.317	0.245	14.95	−0.0867
0.050	1.099	0.366	0.232	14.01	−0.1354	1.023	0.312	0.252	14.17	−0.1005
0.060	1.148	0.405	0.208	14.17	−0.1142	1.033	0.320	0.260	13.89	−0.1223
0.080	1.267	0.470	0.180	14.48	−0.0974	1.058	0.362	0.233	13.91	−0.1127
0.088						1.067	0.382	0.220	14.14	−0.1048
0.089						2.368	1.580	0.075	12.44	−0.0635
0.090						2.187	1.693	0.050	18.21	−0.0415
0.100	1.389	0.557	0.144	14.11	−0.0791	1.930	1.499	0.061	29.65	−0.1162
0.110						1.821	1.196	0.102	16.64	−0.0756
0.120						1.644	0.970	0.136	16.10	−0.1135
0.130						1.540	0.718	0.194	15.69	−0.1685
0.140						1.472	0.479	0.273	16.50	−0.2153
0.150	1.660	0.743	0.079	14.12	−0.0476	1.402	0.352	0.269	17.09	−0.0247
0.160						1.334	0.329	0.145	11.38	−0.0643
0.200	1.839	0.911	0.034	13.23	−0.0334	1.201	0.158	0.426	14.12	−0.1873
0.300	1.973	1.095	−0.009	11.86	−0.0183	1.148	0.422	0.203	13.49	−0.1013
0.400	1.992	1.187	−0.027	10.72	−0.0140	1.187	0.562	0.137	14.19	−0.0706
0.500	1.967	1.240	−0.039	8.34	−0.0074	1.233	0.634	0.109	14.20	−0.0556
0.600	1.947	1.247	−0.040	8.20	−0.0096	1.269	0.685	0.089	13.78	−0.0440
0.800	1.906	1.233	−0.038	7.93	−0.0110	1.329	0.759	0.065	13.69	−0.0317
1.000	1.841	1.250	−0.048	19.49	0.0140	1.367	0.811	0.051	13.67	−0.0283
1.500	1.750	1.197	−0.040	15.90	0.0110	1.369	0.942	0.020	14.65	−0.0207
2.000	1.712	1.123	−0.021	7.97	−0.0057	1.384	0.980	0.014	13.51	−0.0216
3.000	1.627	1.059	−0.005	11.99	−0.0132	1.367	1.006	0.017	13.33	−0.0377
4.000	1.553	1.026	0.005	12.93	−0.0191	1.337	1.009	0.024	14.15	−0.0455
5.000	1.483	1.009	0.012	13.12	−0.0258	1.360	0.957	0.049	14.04	−0.0683
6.000	1.442	0.980	0.023	13.37	−0.0355	1.363	0.965	0.054	14.21	−0.0715
8.000	1.354	0.974	0.029	13.65	−0.0424	1.441	0.994	0.061	14.18	−0.0800
10.000	1.297	0.949	0.042	13.97	−0.0561	1.464	1.148	0.032	14.08	−0.0554
15.000	1.199	0.957	0.049	14.37	−0.0594	1.573	1.337	0.016	13.54	−0.0463

Source: Extracted from American National Standard ANSI/ANS-6.4.3-1991, *Gamma-Ray Attenuation Coefficients and Buildup Factors for Engineering Materials*, with permission of the publisher, the American Nuclear Society (1991); data also available from Data Library DLC-129/ANS643, issued by the Radiation Shielding Information Center, Oak Ridge National Laboratory, Oak Ridge, Tenn., 1988.

Appendix F

Assorted Mathematical Tidbits

F1 FIRST-ORDER, LINEAR, ORDINARY DIFFERENTIAL EQUATIONS

The most frequently encountered type of equation used in models for radiological assessments is that involving only the first derivative of some unknown function of a single variable, $y(t)$, and a multiple of that function. In this appendix, the solution of such first-order, linear, ordinary differential equations is summarized.

F1.1 Single Equation with Constant Coefficient

This equation can be written as

$$\frac{dy(t)}{dt} + \lambda y(t) = f(t) \tag{F.1}$$

where λ is some constant, and the inhomogeneous term, $f(t)$, is a specified function. To obtain the solution $y(t)$ subject to an initial condition $y(0) = y_o$, multiply through by the integrating factor $e^{\lambda t}$ to obtain

$$\frac{d}{dt}\left[e^{\lambda t}y(t)\right] = e^{\lambda t}f(t). \tag{F.2}$$

Upon integrating this result over t from 0 to t and using the initial condition, one obtains the solution

$$y(t) = y_o e^{-\lambda t} + e^{-\lambda t}\int_0^t dt' f(t')\, e^{\lambda t'}. \tag{F.3}$$

F1.2 Single Equation with Variable Coefficient

If λ in Eq. (F.1) is a function of t, the integrating factor needed to make the left hand side of Eq. (F.1) an exact differential is

$$\text{Integrating Factor} = \exp\left[\int_0^t dt''\lambda(t'')\right]. \tag{F.4}$$

631

Multiplication of Eq. (F.1) through by this integrating factor, integration of the result from 0 to t, and rearrangement yields

$$y(t) = \exp\left[-\int_0^t dt'' \lambda(t'')\right]\left\{y_o + \int_0^t dt' f(t') \exp\left[\int_0^{t'} dt'' \lambda(t'')\right]\right\}. \quad \text{(F.5)}$$

F1.3 Coupled Equations for a Decay Chain

In many analyses, one has to calculate the activities of daughter nuclides in a radioactive decay chain

$$N_1 \xrightarrow{\lambda_1} N_2 \xrightarrow{\lambda_2} N_3 \xrightarrow{\lambda_3} \cdots \xrightarrow{\lambda_{i-1}} N_i \xrightarrow{\lambda_i} \cdots \xrightarrow{\lambda_{n-1}} N_n$$

The equation governing the concentration (or amount) of species i, here denoted by $y_i(t)$ with a radioactive decay constant λ_i, is

$$\frac{dy_i(t)}{dt} + \lambda_i\, y_i(t) = \lambda_{i-1}\, y_{i-1}(t), \qquad i = 2, 3, \ldots, n. \quad \text{(F.6)}$$

For $i = 1$, the parent, the right hand side is zero and, from Eq. (F.3), one has

$$y_1(t) = y_1(0)e^{-\lambda_1 t}. \quad \text{(F.7)}$$

If it is assumed that at time $t = 0$ only the parent is present, that is, $y_i(0) = 0$ for $i > 1$, then it is possible to obtain an explicit solution for the coupled equations.

First consider $y_2(t)$. Equation (F.6) with $i = 2$ has the same form as Eq. (F.1) with

$$f(t) = y_1(0)\lambda_1 e^{\lambda_1 t}. \quad \text{(F.8)}$$

Substitution of this expression for $f(t)$ into Eq. (F.3) and subsequent reduction yield

$$y_2(t) = y_1(0)\lambda_1\left[\frac{e^{-\lambda_1 t}}{\lambda_2 - \lambda_1} + \frac{e^{-\lambda_2 t}}{\lambda_1 - \lambda_2}\right]. \quad \text{(F.9)}$$

Similarly, Eq. (F.6) for $i = 3$ has the form of Eq. (F.1), where $f(t) = \lambda_2 y_2(t)$. Substitution of the explicit result for $f(t)$ into Eq. (F.3) then yields

$$y_3(t) = y_1(0)\lambda_1\lambda_2\left[\frac{e^{-\lambda_1 t}}{(\lambda_3 - \lambda_1)(\lambda_2 - \lambda_1)} + \frac{e^{-\lambda_2 t}}{(\lambda_1 - \lambda_2)(\lambda_3 - \lambda_2)} + \frac{e^{-\lambda_3 t}}{(\lambda_1 - \lambda_3)(\lambda_2 - \lambda_3)}\right]$$
$$\text{(F.10)}$$

or, more compactly,

$$y_3(t) = y_1(0)B_2 \sum_{j=1}^{3} C_{j3}\, e^{-\lambda_j t} \quad \text{(F.11)}$$

where

$$B_i \equiv \prod_{k=1}^{i} \lambda_k \quad \text{and} \quad C_{ji} \equiv \left\{\prod_{\substack{k=1 \\ k \neq j}}^{i} (\lambda_k - \lambda_j)\right\}^{-1}. \quad \text{(F.12)}$$

One could continue in this manner to obtain $y_4(t)$, $y_5(t)$, and so on. From the pattern developed, one might suppose

$$y_{i-1}(t) = y_1(0) \, B_{i-2} \sum_{j=1}^{i-1} C_{j,i-1} e^{-\lambda_j t}. \tag{F.13}$$

Then Eq. (F.6) for $y_i(t)$ has the form of Eq. (F.1) whose solution, Eq. (F.3), yields

$$y_i(t) = y_1(0) \, \lambda_{i-1} \, B_{i-2} \, e^{-\lambda_i t} \int_0^t dt' \, e^{\lambda_i t'} \sum_{j=1}^{i-1} C_{j,i-1} \, e^{-\lambda_j t'} \tag{F.14}$$

or, on integrating,

$$y_i(t) = y_1(0) \, B_{i-1} \left[\sum_{j=1}^{i-1} C_{ji} e^{-\lambda_j t} - e^{-\lambda_i t} \sum_{j=1}^{i-1} C_{ji} \right]. \tag{F.15}$$

One can show that

$$\sum_{j=1}^{i-1} C_{ji} = -C_{ii} \tag{F.16}$$

so that the solution for $y_i(t)$ becomes

$$y_i(t) = y_1(0) \, B_{i-1} \sum_{j=1}^{i} C_{ji} \, e^{-\lambda_j t}. \tag{F.17}$$

This is the same result as was assumed in Eq. (F.13) with i replaced by $i - 1$. Thus, by induction, Eq. (F.17) is the desired solution.

F1.4 Coupled Linear Homogeneous Equations

In many models for the distribution of a particular radionuclide among several body organs or among different environmental regions, one encounters a set of first-order linear differential equations of the form

$$
\begin{aligned}
\frac{dy_1(t)}{dt} &= a_{11} y_1(t) + a_{12} y_2(t) + \cdots + a_{1n} y_n(t) \\[2mm]
\frac{dy_2(t)}{dt} &= a_{21} y_1(t) + a_{22} y_2(t) + \cdots + a_{2n} y_n(t) \\[2mm]
&\ \ \vdots \\[2mm]
\frac{dy_n(t)}{dt} &= a_{n1} y_1(t) + a_{n2} y_2(t) + \cdots + a_{nn} y_n(t)
\end{aligned}
\tag{F.18}
$$

where a_{ij} is the (constant) coefficient for transfer from organ (region) j to organ (region) i. One seeks the solution of these equations, subject to some specified initial conditions $y_1(0)$, $y_2(0)$, ..., $y_n(0)$. This set of equations can be written more compactly using matrix notation as

$$\frac{d\mathbf{y}(t)}{dt} = \mathbf{A}\mathbf{y}(t) \tag{F.19}$$

where the matrix $\mathbf{A}$ has elements a_{ij}, and vector $\mathbf{y}(t)$ has elements $y_i(t)$.

To find the general solution of Eq. (F.19), begin by looking for a solution of the form

$$\mathbf{y}(t) = \mathbf{u}\,e^{\lambda t} \tag{F.20}$$

where the vector $\mathbf{u}$ and the constant λ are to be determined. Substitution into Eq. (F.19) yields the following set of linear algebraic equations for $\mathbf{u}$

$$[\mathbf{A} - \lambda\mathbf{I}]\mathbf{u} = 0 \tag{F.21}$$

where $\mathbf{I}$ is the identity matrix (i.e., $I_{ij} = 0$, $i \neq j$; $I_{ii} = 1$). This set of equations has a nontrivial (i.e., nonzero) solution for $\mathbf{u}$ only if

$$\det|\mathbf{A} - \lambda\mathbf{I}| = 0. \tag{F.22}$$

From this condition (which is a polynomial in λ of degree n), one obtains n (usually distinct) acceptable values of λ. These values are called the *eigenvalues* of $\mathbf{A}$ and are denoted by λ_k, $k = 1, \ldots, n$. For each λ_k, Eq. (F.21) then yields a nontrivial solution $\mathbf{u}_k$ (a so-called *eigenvector* of $\mathbf{A}$). Thus, for the case that all the λ_k are different, the most general solution of Eq. (F.18) or (F.19) will be a linear combination of these individual solutions, namely

$$\mathbf{y}(t) = \sum_{k=1}^{n} \alpha_k \mathbf{u}_k e^{\lambda_k t}. \tag{F.23}$$

The constants α_k are determined from the initial condition $\mathbf{y}(0)$. From Eq. (F.23), these constants are seen to be given by the algebraic equations

$$\mathbf{y}(0) = \sum_{k=1}^{n} \alpha_k \mathbf{u}_k, \tag{F.24}$$

which can be rewritten in matrix form as

$$\mathbf{y}(0) = \mathbf{S}\,\alpha \tag{F.25}$$

where $\mathbf{S}$ is a matrix whose kth column is the eigenvector $\mathbf{u}_k$ and α is a vector with components α_i. This set of linear equations is solved by any standard method for α in terms of the known $\mathbf{y}(0)$ vector. Eq. (F.23) then gives the desired solution.

F2 THE LOG-NORMAL DISTRIBUTION

Measured data such as occupational exposures and indoor radon concentrations are frequently found empirically to be well described by the log-normal distribution function. Suppose that x is the quantity being measured (e.g., exposure), and, from physical principles must necessarily be nonnegative. The measured values of x will be found to form a distribution $p(x)$ where $p(x)\,dx$ is the probability a measured value will fall within the range dx about x.

If $p(x)$ is described by a log-normal distribution, then $y \equiv \ln x$ has a normal distribution $n(y)$, that is,

$$n(y) = \frac{1}{\sqrt{2\pi\sigma^2}} \exp\left[\frac{-(y-\mu)^2}{2\sigma^2}\right], \quad -\infty < y < \infty \tag{F.26}$$

in which the parameters μ and σ^2 are, respectively, the mean and the variance of the y-distribution. Since $n(y)$ is symmetric about the mean μ, μ is also the median of the distribution (i.e., on the average, half the y values will be greater than μ and half will be less).

Since $p(x)$ and $n(y)$ describe the same phenomenon, the probability of obtaining values in corresponding dy and dx intervals must be equal, that is, $p(x)\,dx = n(y)\,dy$. Thus, it is seen that the log-normal distribution $p(x)$ is given by

$$p(x) = n(y)\frac{dy}{dx} = \frac{1}{x\sqrt{2\pi\sigma^2}} \exp\{-[\ln x - \mu]^2/(2\sigma^2)\}, \quad 0 < x < \infty. \tag{F.27}$$

For this log-normal distribution, both the geometric mean and the median value of x are $\exp(\mu)$ and the average (mean) value is $\exp(\mu + \sigma^2/2)$. Thus, knowledge of the median x_m and the mean x_{av} fully describe the log-normal distribution (i.e., from the observed mean and median, values for μ and σ^2 of the approximating log-normal distribution can be estimated). Explicitly one has

$$\mu = \ln x_m \tag{F.28}$$

and

$$\sigma^2 = 2[\ln x_{av} - \mu] = 2[\ln x_{av} - \ln x_m]. \tag{F.29}$$

For many analyses, it is often useful to express the measured distribution in terms of the *cumulative probability distribution* $P(x)$ which is the probability (or frequency) that a measured value will be less than or equal to x. For the log-normal distribution, this cumulative distribution is given by

$$P(x) \equiv \int_0^x dx'\, p(x') = \frac{1}{\sqrt{2\pi}} \int_{-\infty}^{[\ln x - \mu]/\sigma} dt\, e^{-t^2/2}. \tag{F.30}$$

Values of the integral on the right-hand side of Eq. (F.30) are available in standard mathematics tables as the *standard normal distribution* function. Sample values are as follows:

$[\ln x - \mu]/\sigma$	$P(x)$
0.0	0.50000
0.5	0.69146
1.0	0.84134
2.0	0.97725
3.0	0.99865
4.0	0.99997

Since the standard normal is symmetric about $y = 0$,

$$P\left(\frac{-[\ln x - \mu]}{\sigma}\right) = 1 - P\left(\frac{[\ln x - \mu]}{\sigma}\right). \tag{F.31}$$

Alternatively, the cumulative log-normal distribution can be expressed in terms of the error function $\mathrm{erf}(x)$ (see the next Sec.) as

$$P(x) = \frac{1}{2}\left\{1 + \mathrm{erf}\left(\frac{\ln x - \mu}{\sqrt{2}\sigma}\right)\right\}. \tag{F.32}$$

If one has a set of data $\{x_i\}$ which is suspected to be distributed according to a log-normal distribution, a plot of $\ln x_i$ versus the frequency with which x_i or smaller values occur in the sample will yield a straight line if the data are truly log-normal and if the x-axis (frequency) is plotted in equal units of $[\ln x - \mu]/\sigma$. An example for measured indoor radon concentrations is shown in Fig. 4.11. In such plots a logarithmic scale is used for the y-axis and the x-axis is given in units of $P(x)$ on an equal $[\ln x - \mu]/\sigma$ scale. From such plots, a straight line fit to the data will have the slope σ for the approximating log-normal distribution and the median is readily found from the intersection of the straight line at the $P(x) = 0.5$ ordinate. From this median value, Eq. (F.28) then yields μ for the approximating log-normal distribution.

F3 EVALUATION OF SPECIAL FUNCTIONS

Several special functions appear repeatedly in many of the formulas derived in this book. While tabulated values of these special functions are sufficient for hand calculations, an accurate analytic approximation would be more useful for computer calculations. Rational approximations are not known for all the special functions appearing in this textbook [e.g., the Sievert integral of Eq. (6.26)]; however, the following approximations for two commonly encountered functions are presented.

F3.1 The Error Function

The error function $\text{erf}(x)$ is defined for real x as

$$\text{erf}(x) = \frac{2}{\sqrt{\pi}} \int_0^x dt\, e^{-t^2}, \tag{F.33}$$

and has the symmetry property $\text{erf}(-x) = -\text{erf}(x)$.

Many rational approximations are available (see Ambramowitz and Stegun, 1964), one of which, for $0 \le x < \infty$, is

$$\text{erf}(x) = 1 - (a_1 t + a_2 t^2 + a_3 t^3 + a_4 t^4 + a_5 t^5)\, e^{-x^2} + \epsilon(x) \tag{F.34}$$

where $t = (1+px)^{-1}$. The error $|\epsilon(x)| < 1.5 \times 10^{-7}$, and the approximation constants are

$$p = .3275911 \quad\quad a_1 = .254829592 \quad\quad a_2 = -.284496736$$
$$a_3 = 1.421413741 \quad\quad a_4 = -1.453152027 \quad\quad a_5 = 1.061405429$$

A closely related function is the complimentary error function $\text{erfc}(x)$ defined as

$$\text{erfc}(x) = \frac{2}{\sqrt{\pi}} \int_x^\infty dt\, e^{-t^2} = 1 - \text{erf}(x). \tag{F.35}$$

F3.2 The Exponential Integral

The exponential integral function of order n, $E_n(x)$, is defined for real positive x as

$$E_n(x) = x^{n-1} \int_x^\infty dt \frac{e^{-t}}{t^n} = \int_1^\infty du \frac{e^{-ux}}{u^n} = \int_0^1 dv\, v^{n-2}\, e^{-x/v}. \tag{F.36}$$

The first order exponential integral for $0 \le x \le 1$ can be approximated with an error $|\epsilon(x)| < 2 \times 10^{-7}$ by (Ambramowitz and Stegun, 1964)

$$E_1(x) = -\ln x + a_0 + a_1 x + a_2 x^2 + a_3 x^3 + a_4 x^4 + a_5 x^5 + \epsilon(x), \tag{F.37}$$

where

$$a_0 = -.57721\,566 \quad\quad a_1 = .99999\,193 \quad\quad a_2 = -.24991\,055$$
$$a_3 = .05519\,968 \quad\quad a_4 = -.00976\,004 \quad\quad a_5 = .00107\,857$$

For $1 \le x < \infty$ with $|\epsilon(x)| < 5 \times 10^{-5}$

$$xe^x E_1(x) = \frac{x^2 + a_1 x + a_2}{x^2 + b_1 x + b_2} + \epsilon(x) \tag{F.38}$$

where $a_1 = 2.334733$, $a_2 = .250621$, $b_1 = 3.330657$, and $b_2 = 1.681534$.

The higher order exponential integral functions can readily be evaluated from the recurrence relation

$$E_{n+1}(x) = \frac{1}{n}[e^{-x} - x E_n(x)], \quad n = 1, 2, 3, \ldots \tag{F.39}$$

F4 THE DIRAC DELTA FUNCTION

A very useful but unusual function is the Dirac delta function $\delta(x - x_o)$ which vanishes at all real x values except at $x = x_o$ where it is infinite. Moreover, this function with zero width and support at only one point x_o has unit area. One can represent this function as a limiting case of the function $f_\epsilon(x - x_o)$ defined as

$$f_\epsilon(x - x_o) = \begin{cases} \dfrac{1}{\epsilon}, & x_o - \dfrac{\epsilon}{2} \le x \le x_o + \dfrac{\epsilon}{2} \\ 0, & \text{otherwise.} \end{cases} \tag{F.40}$$

The Dirac delta function is then defined as

$$\delta(x - x_o) = \lim_{\epsilon \to 0} f_\epsilon(x - x_o). \tag{F.41}$$

Alternate representations of this function are

$$\delta(x - x_o) = \frac{1}{\pi} \lim_{\lambda \to 0} \frac{\sin \lambda(x - x_o)}{(x - x_o)} \tag{F.42}$$

or

$$\delta(x - x_o) = \frac{1}{\pi} \lim_{\epsilon \to 0} \frac{\epsilon}{(x - x_o)^2 - \epsilon^2}. \tag{F.43}$$

This delta function has the following properties:

$$\delta(x) = \delta(-x) \tag{F.44}$$

$$\delta(ax) = \frac{1}{|a|} \delta(x) \tag{F.45}$$

$$x\delta(x) = 0 \tag{F.46}$$

$$\delta[g(x)] = \sum_n \frac{1}{|g_n'(x_n)|} \delta(x - x_n) \tag{F.47}$$

where $g(x)$ is a function that has zeros at x_n [i.e., $g(x_n) = 0$] and whose first derivative does not vanish at these zeros [i.e., $g_n'(x_n) \ne 0$].

These properties really have meaning only when the delta function is inside an integral. For example, the most widely used relation involving the delta function is

$$\int_a^b dx \, f(x) \, \delta(x - x_o) = \begin{cases} f(x_o), & \text{if } a < x_o < b \\ 0, & \text{otherwise} \end{cases} \tag{F.48}$$

where $f(x)$ is any function.

REFERENCES

Abramowitz, M., and I. R. Stegun, *Handbook of Mathematical Functions*, Applied Mathematics Series—55, National Bureau of Standards, Washington, D.C., 1964.

Appendix G

ICRP Data for Internal Dosimetry

TABLE G.1 Masses of Organs and Tissues of Reference Man, Based on ICRP Publication 23 and Used in the ICRP Method.

ORGAN OR TISSUE	MASS (g)[a]	ORGAN OR TISSUE	MASS (g)
Adipose tissue		Prostate	16
Subcutaneous	7500	Salivary glands	85
Other separable	5000	Skeleton	
Adrenals (2)	14	Cortical bone	4000
Blood	3690	Trabecular bone	1000
Connective tissue	1600	Red marrow	1500
Central nervous system	1430	Yellow marrow	1500
Cerebrospinal fluid	120	Cartilage	1100
Eyes (2)	15	Periarticular material	900
Gall bladder & bile	72	Skin	
GI tract	1200	Epidermis	100
Contents	1005	Dermis	2500
Heart	330	Spleen	180
Kidneys (2)	310	Teeth	46
Larynx	28	Testes (2)	35
Liver	1800	Thymus	20
Lungs (2)	1000	Thyroid	20
Lymph nodes	250	Tongue	70
Miscellaneous		Tonsils (2)	4
(blood vessels, hair, etc.)	3276.1	Trachea	10
Muscle	28000	Ureters	16
Pancreas	100	Urethra	10
Parathyroids (4)	0.12	Urinary bladder	45
Pineal	0.18	Urine	102
Pituitary	0.6	TOTAL BODY	70000

[a] Total mass of both adrenals, both eyes, and so on.

Source: ICRP 1975.

TABLE G.2   Fractional Transfer of Ingested and Inhaled Elements to the Body Fluids.

Z	f_1	INGESTION FORM	INHALATION CLASS	Z	f_1	INGESTION FORM	INHALATION CLASS
Be 4	5E−03[a]	all	W,Y	I 53	1E+00	all	D
F 9	1E+00	all	D,W,Y	Cs 55	1E+00	all	D
Na 11	1E+00	all	D	Ba 56	1E−01	all	D
Mg 12	5E−01	all	D,W	La 57	1E−03	all	D,W
Al 13	1E−02	all	D,W	Ce 58	3E−04	all	W,Y
Si 14	1E−02	all	D,W,Y	Pr 59	3E−04	all	W,Y
P 15	8E−01	all	D,W	Nd 60	3E−04	all	W,Y
S 16	8E−01	inorganic	D,W	Pm 61	3E−04	all	W,Y
S 16	1E−01	elemental		Sm 62	3E−04	all	W
Cl 17	1E+00	all	D,W	Eu 63	1E−03	all	W
K 19	1E+00	all	D,W	Gd 64	3E−04	all	D,W
Ca 20	3E−01	all	W	Tb 65	3E−04	all	W
Sc 21	1E−04	all	Y	Dy 66	3E−04	all	W
Ti 22	1E−02	all	D,W,Y	Ho 67	3E−04	all	W
V 23	1E−02	all	D,W	Er 68	3E−04	all	W
Cr 24	1E−02	trivalent		Tm 69	3E−04	all	W
Cr 24	1E−01	hexavalent	D,W,Y	Yb 70	3E−04	all	W,Y
Mn 25	1E−01	all	D,W	Lu 71	3E−04	all	W,Y
Fe 26	1E−01	all	D,W	Hf 72	2E−03	all	D,W
Co 27	3E−01	orga. complex		Ta 73	1E−03	all	W,Y
Co 27	5E−02	inorg. tracer	D,W	W 74	1E−02	tungstic acid	
Ni 28	5E−02	all	D,W	W 74	3E−01	all other	D,W
Cu 29	5E−01	all	D,W,Y	Re 75	8E−01	all	D,W
Zn 30	5E−01	all	Y	Os 76	1E−02	all	D,W,Y
Ga 31	1E−03	all	D,W	Ir 77	1E−02	all	D,W,Y
Ge 32	1E+00	all	D,W	Pt 78	1E−02	all	D
As 33	5E−01	all	W	Au 79	1E−01	all	D,W,Y
Se 34	5E−02	elemental		Hg 80	2E−02	inorganic	D,W
Se 34	8E−01	all other	D,W	Tl 81	1E+00	all	D
Br 35	1E+00	all	D,W	Pb 82	2E−01	all	D
Rb 37	1E+00	all	D	Bi 83	5E−02	all	D,W
Sr 38	3E−01	soluble salts	D	Po 84	1E−01	all	D,W
Sr 38	1E−02	SrTiO$_3$	Y	At 85	1E+00	all	D,W
Y 39	1E−04	all	W,Y	Fr 87	1E+00	all	D
Zr 40	2E−03	all	D,W,Y	Ra 88	2E−01	all	W
Nb 41	1E−02	all	D,W,Y	Ac 89	1E−03	all	D,W,Y
Mo 42	5E−02	MoS$_2$	Y	Th 90	2E−04	all	W,Y
Mo 42	8E−01	all other	D	Pa 91	1E−03	all	W,Y
Tc 43	8E−01	all	D,W	U 92	2E−03	rel. insol.	Y
Ru 44	5E−02	all	D,W,Y	U 92	5E−02	water sol.	D,W
Rh 45	5E−02	all	D,W,Y	Np 93	1E−02	all	W
Pd 46	5E−03	all	D,W,Y	Pu 94	1E−04	all other	W
Ag 47	5E−02	all	D,W,Y	Pu 94	1E−05	oxide/hydroxide	Y
Cd 48	5E−02	inorganic	D,W,Y	Am 95	5E−04	all	W
In 49	2E−02	all	D,W	Cm 96	5E−04	all	W
Sn 50	2E−02	all	D,W	Bk 97	5E−04	all	W
Sb 51	1E−02	all other	W	Cf 98	5E−04	all	W,Y
Sb 51	1E−01	tartar emetic	D	Es 99	5E−04	all	W
Te 52	2E−01	all	D,W	Fm 100	5E−04	all	W
				Md 108	5E−04	all	W

[a] Read as 5×10^{-3}, and so on.

Source: ICRP 1979.

TABLE G.3   Distribution of Elements from the Body Fluid Transfer Compartment and Retention in Source Organs, Based on ICRP Publication 30.

Z	ORGAN OR TISSUE	f_2	f_2'	f_2''	BIOLOGICAL HALF LIVES FOR f_2 FRACTIONS (days)		
Be 4	mineral bone	0.4			1500		
Be 4	all other	0.16	0.04		15	1500	
F 9	skeleton	1			∞		
Na 11	skeleton	0.297	0.003		10	500	
Na 11	all other	0.7			10		
Mg 12	mineral bone	0.4			100		
Mg 12	all other	0.4			100		
Al 13	mineral bone	0.3			100		
Al 13	all other	0.7			100		
Si 14	all	0.4	0.6		5	100	
P 15	soft tissue	0.4			19		
P 15	intracellular fluids	0.15			2		
P 15	mineral bone	0.3			∞		
S 16	total body	0.15	0.05		20	2000	
Cl 17	total body	1			10		
K 19	total body	1			30		
Cr 24	bone	0.05			1000		
Cr 24	all other	0.4	0.25		6	80	
Sc 21	skeleton	0.04	0.36		5	1500	
Sc 21	liver	0.03	0.27		5	1500	
Sc 21	spleen	0.01	0.09		5	1500	
Sc 21	all other	0.02	0.18		5	1500	
Ti 22	all	1			600		
V 23	mineral bone	0.25			10000		
V 23	all other	0.05			10000		
Mn 25	liver	0.1	0.15		4	40	
Mn 25	all other	0.2	0.2		4	40	
Mn 25	bone	0.35			40		
Fe 26	all other	0.907			2000		
Fe 26	liver	0.08			2000		
Fe 26	spleen	0.013			2000		
Co 27	liver	0.03	0.01	0.01	6	60	800
Co 27	all other	0.27	0.09	0.09	6	60	800
Ni 28	kidneys	0.02			0.2		
Ni 28	all other	0.3			1200		
Cu 29	all other	0.794			40		
Cu 29	liver	0.1			40		
Cu 29	pancreas	0.006			40		
Cu 29	brain	0.1			40		
Zn 30	all other	0.24	0.56		20	400	
Zn 30	skeleton	0.2			400		
Ga 31	mineral bone	0.09	0.21		1	50	
Ga 31	liver	0.027	0.063		1	50	
Ga 31	spleen	0.003	0.007		1	50	
Ga 31	all other	0.18	0.42		1	50	
Ge 32	kidneys	0.5			0.02		
Ge 32	all other	0.5			1		
As 33	liver	0.028	0.042		1	10	
As 33	kidneys	0.006	0.009		1	10	
As 33	spleen	0.002	0.003		1	10	

(Continued)

TABLE G.3 (cont'd).

Z	ORGAN OR TISSUE	f_2	f_2'	f_2''	BIOLOGICAL HALF LIVES FOR f_2 FRACTIONS (days)		
As 33	all other	0.224	0.336		1	10	
Se 34	liver	0.015	0.06	0.075	3	30	150
Se 34	kidneys	0.005	0.02	0.025	3	30	150
Se 34	spleen	0.001	0.004	0.005	3	30	150
Se 34	pancreas	0.0005	0.002	0.0025	3	30	150
Se 34	all other	0.0785	0.314	0.3925	3	30	150
Br 35	total body	1			10		
Rb 37	all other	0.75			44		
Rb 37	skeleton	0.25			44		
Y 39	skeleton	0.5			105		
Y 39	liver	0.15			105		
Y 39	all other	0.1			105		
Zr 40	bone	0.5			8000		
Zr 40	all other	0.5			7		
Nb 41	spleen	0.005	0.005		6	200	
Nb 41	all other	0.13	0.13		6	200	
Nb 41	mineral bone	0.355	0.355		6	200	
Nb 41	kidneys	0.009	0.009		6	200	
Nb 41	testes	0.001	0.001		6	200	
Mo 42	kidneys	0.005	0.045		1	50	
Mo 42	mineral bone	0.015	0.135		1	50	
Mo 42	liver	0.03	0.27		1	50	
Mo 42	all other	0.05	0.45		1	50	
Tc 43	liver	0.0225	0.006	0.0015	1.6	3.7	22
Tc 43	stomach wall	0.075	0.02	0.005	1.6	3.7	22
Tc 43	thyroid	0.04			0.5		
Tc 43	all other	0.6225	0.166	0.0415	1.6	3.7	22
Ru 44	total body	0.35	0.3	0.2	8	35	1000
Rh 45	total body	0.35	0.3	0.2	8	35	1000
Pd 46	liver	0.45			15		
Pd 46	kidneys	0.15			15		
Pd 46	mineral bone	0.07			15		
Pd 46	all other	0.03			15		
Ag 47	liver	0.08	0.72		3.5	50	
Ag 47	all other	0.02	0.18		3.5	50	
Cd 48	liver	0.3			9131		
Cd 48	all other	0.4			9131		
Cd 48	kidneys	0.3			9131		
In 49	kidneys	0.07			∞		
In 49	all other	0.42			∞		
In 49	red bone marrow	0.3			∞		
In 49	liver	0.2			∞		
In 49	spleen	0.01			∞		
Sn 50	mineral bone	0.07	0.07	0.21	4	25	400
Sn 50	all other	0.03	0.03	0.09	4	25	400
Sb 51	mineral bone	0.19	0.01		5	100	
Sb 51	liver	0.095	0.005		5	100	
Sb 51	all other	0.475	0.025		5	100	
Te 52	all others	0.25			20		
Te 52	mineral bone	0.25			5000		
I 53	thyroid	0.3			80		

(Continued)

TABLE G.3 (cont'd).

Z	ORGAN OR TISSUE	f_2	f_2'	f_2''	BIOLOGICAL HALF LIVES FOR f_2 FRACTIONS (days)	
Cs 55	total body	0.1	0.9		2	110
Ba 56	(see text)					
La 57	liver	0.6			3500	
La 57	mineral bone	0.2			3500	
La 57	all other	0.2			3500	
Ce 58	all other	0.15			3500	
Ce 58	bone	0.2			3500	
Ce 58	spleen	0.05			3500	
Ce 58	liver	0.6			3500	
Pr 59	liver	0.6			3500	
Pr 59	mineral bone	0.25			3500	
Pr 59	kidneys	0.05			10	
Nd 60	liver	0.45			3500	
Nd 60	mineral bone	0.45			3500	
Pm 61	liver	0.45			3500	
Pm 61	mineral bone	0.45			3500	
Sm 62	liver	0.45			3500	
Sm 62	mineral bone	0.45			3500	
Eu 63	liver	0.4			3500	
Eu 63	mineral bone	0.4			3500	
Eu 63	kidneys	0.06			3500	
Gd 64	mineral bone	0.45			3500	
Gd 64	liver	0.3			3500	
Gd 64	kidneys	0.03			3500	
Tb 65	mineral bone	0.5			3500	
Tb 65	liver	0.25			3500	
Tb 65	kidneys	0.05			10	
Dy 66	mineral bone	0.6			3500	
Dy 66	liver	0.1			3500	
Dy 66	kidneys	0.02			10	
Ho 67	mineral bone	0.4			3500	
Ho 67	liver	0.4			3500	
Ho 67	pancreas	0.05			3500	
Er 68	mineral bone	0.6			3500	
Er 68	liver	0.05			3500	
Er 68	all other	0.1			3500	
Tm 69	mineral bone	0.65			3500	
Tm 69	liver	0.04			3500	
Tm 69	all other	0.1			3500	
Yb 70	mineral bone	0.5			3500	
Yb 70	liver	0.03			3500	
Yb 70	kidneys	0.02			10	
Yb 70	spleen	0.005			3500	
Lu 71	mineral bone	0.6			3500	
Lu 71	liver	0.02			3500	
Lu 71	kidneys	0.005			10	
Hf 72	mineral bone	0.5			8000	
Hf 72	all other	0.5			7	
Ta 73	kidneys	0.03	0.03		4	100
Ta 73	mineral bone	0.3			100	
Ta 73	all other	0.32	0.32		4	100

(Continued)

TABLE G.3 (cont'd).

Z	ORGAN OR TISSUE	f_2	f_2'	f_2''	BIOLOGICAL HALF LIVES FOR f_2 FRACTIONS (days)		
W 74	mineral bone	0.005	0.0025	0.0175	5	100	1000
W 74	kidneys	0.007	0.003		5	100	
W 74	liver	0.007	0.003		5	100	
W 74	spleen	0.0035	0.0015		5	100	
Re 75	thyroid	0.04			0.5		
Re 75	stomach wall	0.075	0.02	0.005	1.6	3.7	22
Re 75	liver	0.0225	0.006	0.0015	1.6	3.7	22
Re 75	all other	0.6225	0.166	0.0415	1.6	3.7	22
Os 76	liver	0.04	0.16		8	200	
Os 76	kidneys	0.008	0.032		8	200	
Os 76	spleen	0.004	0.016		8	200	
Os 76	all other	0.108	0.432		8	200	
Ir 77	liver	0.04	0.16		8	200	
Ir 77	kidneys	0.008	0.032		8	200	
Ir 77	spleen	0.004	0.016		8	200	
Ir 77	all other	0.108	0.432		8	200	
Pt 78	kidneys	0.095	0.005		8	200	
Pt 78	liver	0.095	0.005		8	200	
Pt 78	spleen	0.0095	0.0005		8	200	
Pt 78	adrenals	0.0009	0.0001		8	200	
Pt 78	all other	0.5595	0.0294		8	200	
Au 79	all	1			3		
Hg 80	kidneys	0.076	0.004		40	10000	
Hg 80	all other	0.874	0.046		40	10000	
Tl 81	kidneys	0.03			10		
Tl 81	all other	0.97			10		
Pb 82	skeleton	0.33	0.11	0.11	12	180	10000
Pb 82	liver	0.2	0.045	0.005	12	180	10000
Pb 82	kidneys	0.016	0.0036	0.0004	12	180	10000
Pb 82	all other	0.144	0.0324	0.0036	12	180	10000
Bi 83	kidneys	0.24	0.16		0.6	5	
Bi 83	all other	0.18	0.12		0.6	5	
Po 84	liver	0.1			50		
Po 84	all other	0.7			50		
Po 84	spleen	0.1			50		
Po 84	kidneys	0.1			50		
At 85	all	1			10		
Fr 87	all	1			0.0		
Ac[a] 89	mineral bone	0.45			36525		
Ac[a] 89	liver	0.45			14610		
Ac[a] 89	testes	0.0003			∞		
Ac[a] 89	ovaries	0.0001			∞		
Th 90	liver	0.04			700		
Th 90	all other	0.16			700		
Th 90	bone	0.7			8000		
Pa 91	mineral bone	0.4			36525		
Pa 91	liver	0.105	0.045		10	60	
Pa 91	kidneys	0.004	0.016		10	60	
U 92	kidneys	0.12	0.0005		6	1500	
U 92	all other	0.12	0.0005		6	1500	
U 92	mineral bone	0.2	0.023		20	5000	

[a] Includes transuranium actinides.

Source: ICRP 1979.

TABLE G.4 Fraction AF of Photon Energy Released in Source Organ and Absorbed in Target Organ, Based on ICRP Publication 23.

Source = Target

	PHOTON ENERGY (MeV)					
SOURCE	0.010	0.015	0.020	0.030	0.050	0.100
Bladder[a]	3.83E−02[c]	6.31E−02	6.45E−02	4.43E−01	2.02E−02	1.15E−02
Stomach[a]	4.82E−02	1.14E−01	1.34E−01	1.03E−01	4.92E−02	2.82E−02
Small intestine[a]	7.70E−01	6.87E−01	5.82E−01	3.86E−01	1.94E−01	1.18E−01
Upper large intestine[a]	7.59E−02	1.67E−01	1.80E−01	1.16E−01	5.45E−02	3.18E−02
Lower large intestine[a]	8.53E−02	1.81E−01	1.86E−01	1.12E−01	4.75E−02	2.82E−02
Kidneys	9.32E−01	7.78E−01	5.79E−01	2.93E−01	1.12E−01	6.67E−02
Liver	9.75E−01	9.02E−01	8.06E−01	5.40E−01	2.76E−01	1.66E−01
Lungs	8.16E−01	6.57E−01	4.71E−01	2.30E−01	8.98E−02	5.04E−02
Other tissue (muscle)	8.72E−01	8.19E−01	7.33E−01	5.54E−01	3.44E−01	2.45E−01
Ovaries	8.01E−01	4.89E−01	2.70E−01	9.51E−02	2.96E−02	1.84E−02
Pancreas	8.86E−01	6.57E−01	4.31E−01	1.91E−01	6.63E−02	3.93E−02
Bone surface[b]	9.90E−01	9.42E−01	8.84E−01	7.18E−01	4.31E−01	1.90E−01
Red marrow[b]	1.41E−01	1.35E−01	1.26E−01	1.03E−01	6.15E−02	2.72E−02
Skin	5.80E−01	2.91E−01	1.58E−01	6.34E−02	2.28E−02	1.51E−02
Spleen	9.47E−01	8.11E−01	6.25E−01	3.24E−01	1.26E−01	7.33E−02
Testes	9.02E−01	6.90E−01	4.56E−01	1.98E−01	6.57E−02	3.97E−02
Thyroid	8.41E−01	5.74E−01	3.55E−01	1.45E−01	4.74E−02	2.82E−02
Total body	9.93E−01	9.65E−01	9.23E−01	7.97E−01	5.68E−01	3.78E−01

	PHOTON ENERGY (MeV)					
SOURCE	0.200	0.500	1.000	1.500	2.000	4.000
Bladder[a]	1.11E−02	1.15E−02	1.00E−02	9.34E−03	8.88E−03	7.08E−03
Stomach[a]	2.67E−02	2.69E−02	2.48E−02	2.22E−02	2.12E−02	1.73E−02
Small intestine[a]	1.11E−01	1.11E−01	9.95E−02	9.43E−02	8.56E−02	7.16E−02
Upper large intestine[a]	3.07E−02	3.07E−02	2.91E−02	2.59E−02	2.47E−02	1.95E−02
Lower large intestine[a]	2.78E−02	2.83E−02	2.59E−02	2.43E−02	2.22E−02	1.73E−02
Kidneys	6.79E−02	7.16E−02	6.42E−02	6.08E−02	5.48E−02	4.63E−02
Liver	1.60E−01	1.61E−01	1.47E−01	1.36E−01	1.25E−01	1.02E−01
Lungs	5.00E−02	5.00E−02	4.55E−02	4.32E−02	3.92E−02	3.08E−02
Other tissue (muscle)	2.42E−01	2.52E−01	2.41E−01	2.24E−01	2.10E−01	1.76E−01
Ovaries	2.05E−02	2.17E−02	2.00E−02	1.89E−02	1.75E−02	1.42E−02
Pancreas	4.15E−02	4.40E−02	4.06E−02	3.82E−02	3.45E−02	2.76E−02
Bone surface[b]	1.35E−01	1.25E−01	1.17E−01	1.06E−01	9.86E−02	8.18E−02
Red marrow[b]	1.94E−02	1.79E−02	1.67E−02	1.52E−02	1.41E−02	1.17E−02
Skin	1.71E−02	1.96E−02	1.96E−02	1.81E−02	1.70E−02	1.35E−02
Spleen	7.52E−02	7.81E−02	7.13E−02	6.56E−02	6.18E−02	4.75E−02
Testes	4.34E−02	4.56E−02	4.30E−02	3.86E−02	3.64E−02	2.84E−02
Thyroid	3.04E−02	3.25E−02	3.02E−02	2.84E−02	2.57E−02	2.06E−02
Total body	3.46E−01	3.47E−01	3.30E−01	3.08E−01	2.91E−01	2.43E−01

(Continued)

TABLE G.4 (cont'd).

Target: Lungs

SOURCE	PHOTON ENERGY (MeV)					
	0.010	0.015	0.020	0.030	0.050	0.100
Bladder	4.25E−26[a]	6.38E−26	5.54E−14	7.64E−08	1.31E−05	8.88E−05
Stomach	1.76E−04	2.64E−04	2.60E−03	7.90E−03	9.60E−03	6.91E−03
Small intestine	4.44E−15	6.66E−15	9.64E−09	6.30E−05	4.61E−04	7.92E−04
Upper large intestine	3.97E−15	5.94E−15	1.18E−08	7.15E−05	6.13E−04	9.95E−04
Lower large intestine	2.61E−14	3.92E−17	4.15E−10	2.91E−06	1.15E−04	2.87E−04
Kidneys	3.57E−05	5.35E−05	7.14E−05	1.18E−03	3.27E−03	3.37E−03
Liver	7.75E−05	3.19E−03	9.45E−03	1.64E−03	1.45E−02	9.91E−03
Lungs	8.16E−01	6.57E−01	4.71E−01	2.30E−01	8.98E−02	5.04E−02
Other tissue (muscle)	3.89E−03	6.81E−03	9.43E−03	9.99E−03	7.04E−03	4.87E−03
Ovaries	5.11E−23	7.67E−23	1.02E−11	9.25E−07	8.57E−05	2.06E−04
Pancreas	2.75E−06	4.12E−06	1.16E−03	8.86E−03	1.29E−02	1.02E−02
Skeleton	1.69E−04	2.54E−04	1.04E−03	3.20E−03	4.42E−03	3.63E−03
Skin	8.03E−05	1.21E−04	7.44E−04	2.54E−03	2.94E−03	2.17E−03
Spleen	6.49E−04	9.75E−04	4.97E−03	1.25E−02	1.23E−02	8.96E−03
Testes	2.24E−33	3.36E−33	1.83E−17	1.65E−09	1.49E−06	1.23E−05
Thyroid	1.52E−10	2.28E−09	3.30E−05	1.23E−03	3.89E−03	3.67E−03
Total body	1.32E−02	1.34E−02	1.43E−02	1.26E−02	8.39E−03	5.40E−03

SOURCE	PHOTON ENERGY (MeV)					
	0.200	0.500	1.000	1.500	2.000	4.000
Bladder	9.32E−05	1.65E−04	2.83E−04	3.07E−04	3.31E−04	4.61E−04
Stomach	5.96E−03	6.39E−03	5.71E−03	5.50E−03	4.91E−03	4.28E−03
Small intestine	9.01E−04	1.16E−03	1.17E−03	1.07E−03	1.17E−03	1.13E−03
Upper large intestine	9.82E−04	1.17E−03	1.38E−03	1.43E−03	1.28E−03	1.13E−03
Lower large intestine	3.14E−04	3.49E−04	5.48E−04	5.48E−04	7.80E−04	5.77E−04
Kidneys	2.99E−03	3.28E−03	3.30E−03	2.96E−03	2.80E−03	2.72E−03
Liver	8.83E−03	8.22E−03	7.89E−03	7.71E−03	6.95E−03	5.59E−03
Lungs	5.00E−02	5.00E−02	4.55E−02	4.32E−02	3.92E−02	3.08E−02
Other tissue (muscle)	4.61E−03	4.62E−03	4.31E−03	4.04E−03	3.81E−03	3.17E−03
Ovaries	2.56E−04	3.88E−04	5.39E−04	5.61E−04	5.85E−04	7.00E−04
Pancreas	8.91E−03	8.20E−03	7.42E−03	6.76E−03	6.53E−03	5.19E−03
Skeleton	3.50E−03	3.38E−03	3.29E−03	3.03E−03	3.09E−03	2.30E−03
Skin	2.27E−03	2.34E−03	2.37E−03	2.33E−03	2.23E−03	1.84E−03
Spleen	7.90E−03	7.55E−03	6.86E−03	6.20E−03	6.37E−03	5.22E−03
Testes	4.38E−05	4.94E−05	1.01E−04	1.60E−04	1.20E−04	1.68E−04
Thyroid	3.38E−03	3.67E−03	3.83E−03	3.35E−03	3.17E−03	3.22E−03
Total body	4.97E−03	5.35E−03	4.58E−03	4.05E−03	3.85E−03	3.45E−03

(Continued)

[a] Read as 4.25×10^{-26}, and so on.

TABLE G.4 (cont'd).

Target: Thyroid

SOURCE	PHOTON ENERGY (MeV)					
	0.010	0.015	0.020	0.030	0.050	0.100
Bladder	1.22E−23[a]	1.82E−23	2.43E−23	5.61E−13	3.04E−09	7.94E−08
Stomach	1.11E−24	1.67E−24	4.45E−14	1.12E−08	1.14E−06	5.08E−06
Small intestine	8.51E−34	1.27E−33	2.18E−18	1.01E−10	6.82E−08	7.53E−07
Upper large intestine	1.03E−33	1.55E−33	3.12E−18	1.34E−10	8.13E−08	8.43E−07
Lower large intestine	2.06E−36	3.10E−36	7.37E−20	1.42E−11	2.47E−07	2.47E−07
Kidneys	1.60E−28	2.39E−28	7.59E−16	1.82E−09	4.06E−07	2.74E−06
Liver	2.53E−23	3.80E−23	2.20E−13	2.37E−08	1.73E−06	7.45E−06
Lungs	7.94E−11	1.19E−10	4.49E−07	1.59E−05	4.51E−05	7.55E−05
Other tissue (muscle)	6.57E−05	1.70E−04	2.33E−04	2.27E−04	1.42E−04	9.43E−05
Ovaries	4.51E−22	6.76E−22	9.04E−22	4.90E−12	1.17E−08	2.14E−07
Pancreas	2.37E−25	3.55E−25	4.08E−14	1.53E−08	1.57E−06	7.41E−06
Skeleton	2.59E−06	3.88E−06	5.17E−06	3.80E−05	5.70E−05	5.41E−05
Skin	2.45E−05	3.68E−05	4.90E−05	1.03E−04	6.53E−05	4.90E−05
Spleen	1.23E−25	1.85E−25	1.82E−14	8.37E−09	1.01E−06	5.29E−06
Testes	4.90E−27	7.35E−27	9.80E−27	1.43E−14	3.29E−10	1.55E−08
Thyroid	8.41E−01	5.74E−01	3.55E−01	1.45E−01	4.74E−02	2.82E−02
Total body	3.86E−04	2.55E−04	3.02E−04	1.70E−04	1.46E−04	6.00E−05

SOURCE	PHOTON ENERGY (MeV)					
	0.200	0.500	1.000	1.500	2.000	4.000
Bladder	2.61E−07	6.41E−07	1.11E−06	1.44E−06	1.68E−06	2.06E−06
Stomach	8.53E−06	1.03E−05	1.14E−05	1.17E−05	1.18E−05	1.11E−05
Small intestine	1.63E−06	2.70E−06	3.63E−06	4.12E−06	4.43E−06	4.70E−06
Upper large intestine	1.79E−06	2.90E−06	3.82E−06	4.33E−06	4.65E−06	4.88E−06
Lower large intestine	6.25E−07	1.23E−06	1.83E−06	2.21E−06	2.49E−06	2.84E−06
Kidneys	4.74E−06	6.41E−06	7.57E−06	8.06E−06	8.31E−06	8.13E−06
Liver	1.61E−05	1.24E−05	1.33E−05	1.35E−05	1.35E−05	1.25E−05
Lungs	5.82E−05	8.13E−05	9.13E−05	5.39E−05	5.88E−05	5.17E−05
Other tissue (muscle)	9.07E−05	9.33E−05	8.72E−05	8.21E−05	7.72E−05	6.41E−05
Ovaries	5.90E−07	1.22E−06	1.88E−06	2.29E−06	2.57E−06	2.96E−06
Pancreas	1.08E−05	1.27E−05	1.80E−06	1.39E−05	1.39E−05	1.30E−05
Skeleton	6.49E−05	7.57E−05	3.14E−05	7.49E−05	3.57E−06	3.74E−05
Skin	5.35E−05	5.66E−05	4.70E−05	4.53E−05	5.53E−05	4.27E−05
Spleen	8.15E−06	1.00E−05	1.11E−05	1.14E−05	1.15E−05	1.09E−05
Testes	6.92E−08	2.29E−07	4.82E−07	6.88E−07	8.53E−07	1.18E−06
Thyroid	3.04E−02	3.25E−02	3.02E−02	2.84E−02	2.57E−02	2.06E−02
Total body	8.53E−05	9.68E−05	1.03E−04	7.90E−05	8.08E−05	9.62E−05

(Continued)

[a] Read as 1.22×10^{-23}, and so on.

TABLE G.4 (cont'd).

Target: Testes

SOURCE	PHOTON ENERGY (MeV)					
	0.010	0.015	0.020	0.030	0.050	0.100
Bladder	6.64E−08[a]	9.98E−08	5.16E−05	5.71E−04	1.02E−03	6.97E−04
Stomach	1.27E−24	1.90E−24	7.01E−14	1.93E−08	2.04E−06	5.86E−06
Small intestine	1.02E−16	1.53E−16	6.23E−10	1.78E−06	1.81E−05	4.08E−05
Upper large intestine	4.56E−16	6.86E−16	9.46E−10	1.78E−06	1.88E−05	3.46E−05
Lower large intestine	1.73E−08	2.60E−08	8.20E−06	1.64E−04	3.19E−04	2.33E−04
Kidneys	1.63E−25	2.44E−25	3.16E−14	1.61E−08	1.95E−06	1.02E−05
Liver	8.05E−25	1.21E−24	3.48E−14	1.03E−08	1.27E−06	7.05E−06
Lungs	3.16E−34	4.75E−34	2.14E−18	2.06E−10	1.87E−07	7.12E−07
Other tissue (muscle)	1.85E−06	2.67E−05	9.87E−05	2.12E−04	2.05E−04	1.59E−04
Pancreas	3.58E−28	5.38E−28	1.81E−15	4.34E−09	9.28E−07	6.08E−06
Skeleton	6.72E−10	1.01E−09	8.24E−07	3.08E−05	6.42E−05	9.61E−05
Skin	3.07E−05	2.16E−04	3.19E−04	3.51E−04	1.80E−04	1.13E−04
Spleen	4.53E−28	6.79E−28	1.71E−15	3.61E−09	7.75E−07	5.27E−06
Testes	9.02E−01	6.90E−01	4.56E−01	1.98E−01	6.57E−02	3.97E−02
Thyroid	9.20E−27	1.38E−26	1.84E−26	2.72E−14	6.20E−10	2.92E−08
Total body	5.75E−04	4.71E−04	3.47E−04	3.30E−04	2.29E−04	1.62E−04

SOURCE	PHOTON ENERGY (MeV)					
	0.200	0.500	1.000	1.500	2.000	4.000
Bladder	5.90E−04	6.49E−04	5.82E−04	4.75E−04	5.19E−04	4.49E−04
Stomach	8.94E−06	2.99E−06	2.12E−05	2.19E−05	2.21E−05	2.08E−05
Small intestine	4.75E−05	4.60E−05	6.64E−05	7.12E−05	7.87E−05	4.38E−05
Upper large intestine	4.27E−05	6.60E−05	7.12E−05	4.82E−05	6.46E−05	5.60E−05
Lower large intestine	2.69E−04	2.65E−04	1.80E−04	1.79E−04	2.26E−04	1.45E−04
Kidneys	1.57E−05	1.92E−05	2.14E−05	2.20E−05	2.22E−05	2.10E−05
Liver	1.13E−05	1.45E−05	3.25E−05	1.74E−05	1.78E−05	1.71E−05
Lungs	1.71E−06	3.13E−06	4.53E−06	5.31E−06	5.86E−06	6.46E−06
Other tissue (muscle)	1.54E−04	1.58E−04	1.51E−04	1.47E−04	1.70E−04	1.19E−04
Pancreas	1.02E−05	7.98E−06	1.59E−05	1.67E−05	1.71E−05	1.66E−05
Skeleton	8.27E−05	7.35E−05	8.31E−05	8.79E−05	1.44E−04	7.90E−05
Skin	1.42E−04	1.06E−04	1.15E−04	1.14E−04	1.10E−04	1.19E−04
Spleen	9.09E−06	1.22E−05	1.44E−05	1.54E−05	1.58E−05	1.55E−05
Testes	4.34E−02	4.56E−02	4.30E−02	3.86E−02	3.64E−02	2.84E−02
Thyroid	1.31E−07	4.34E−07	9.13E−07	1.30E−06	1.62E−06	2.23E−06
Total body	1.42E−04	2.41E−04	2.12E−04	1.82E−04	1.70E−04	1.60E−04

(Continued)

[a] Read as 6.64×10^{-8}, and so on.

TABLE G.4 (cont'd).

Target: Ovaries

SOURCE	PHOTON ENERGY (MeV)					
	0.010	0.015	0.020	0.030	0.050	0.100
Bladder	5.15E−08[a]	7.73E−08	1.34E−05	2.30E−04	2.96E−04	2.33E−04
Stomach	1.65E−13	2.46E−13	8.68E−09	2.39E−06	1.95E−05	1.76E−05
Small intestine	1.14E−04	1.71E−04	3.66E−04	5.18E−04	5.38E−04	3.50E−04
Upper large intestine	1.66E−04	2.50E−04	4.41E−04	6.13E−04	4.62E−04	3.75E−04
Lower large intestine	3.39E−04	5.09E−04	1.34E−03	1.46E−03	8.77E−04	5.17E−04
Kidneys	3.04E−13	4.57E−13	1.46E−08	3.98E−06	3.07E−05	3.42E−05
Liver	8.12E−14	1.22E−13	4.59E−09	1.35E−06	1.25E−05	1.35E−05
Lungs	4.34E−23	6.52E−23	2.88E−13	2.59E−08	1.90E−06	2.50E−06
Other tissue (muscle)	3.40E−05	7.70E−05	9.84E−05	1.02E−04	8.27E−05	6.26E−05
Ovaries	8.01E−01	4.89E−01	2.70E−01	9.51E−02	2.96E−02	1.84E−02
Pancreas	4.37E−17	6.55E−17	2.54E−10	6.61E−07	1.07E−05	1.51E−05
Skeleton	1.75E−07	2.63E−07	7.47E−06	3.18E−05	2.65E−05	2.32E−05
Skin	7.12E−11	1.07E−10	1.60E−07	6.35E−06	1.54E−05	1.01E−05
Spleen	8.12E−16	1.22E−15	6.57E−10	7.28E−07	9.84E−06	9.43E−06
Thyroid	1.93E−22	2.89E−22	3.85E−22	2.07E−12	4.94E−09	9.01E−08
Total body	1.37E−04	2.04E−04	1.17E−04	1.02E−04	6.75E−05	5.80E−05

SOURCE	PHOTON ENERGY (MeV)					
	0.200	0.500	1.000	1.500	2.000	4.000
Bladder	2.25E−04	1.79E−04	1.12E−04	1.91E−04	1.37E−04	1.22E−04
Stomach	1.27E−05	1.41E−05	4.21E−05	2.50E−05	2.37E−05	2.00E−05
Small intestine	3.10E−04	2.48E−04	2.80E−04	1.91E−04	1.76E−04	2.32E−04
Upper large intestine	3.39E−04	3.63E−04	1.91E−04	1.55E−04	2.75E−04	1.72E−04
Lower large intestine	5.90E−04	4.48E−04	4.55E−04	4.15E−04	3.83E−04	2.90E−04
Kidneys	3.67E−05	3.27E−05	4.49E−05	3.17E−05	2.97E−05	1.46E−07
Liver	1.49E−05	5.40E−06	2.06E−05	2.84E−05	1.84E−05	1.59E−05
Lungs	3.65E−06	4.37E−06	4.75E−06	4.85E−06	4.91E−06	4.61E−06
Other tissue (muscle)	5.81E−05	5.66E−05	5.29E−05	4.96E−05	4.67E−05	3.93E−05
Ovaries	2.05E−02	2.17E−02	2.00E−02	1.89E−02	1.75E−02	1.42E−02
Pancreas	9.43E−06	1.06E−05	2.89E−05	1.96E−05	1.86E−05	1.60E−05
Skeleton	2.03E−05	3.10E−05	2.61E−05	2.31E−05	2.27E−05	1.84E−05
Skin	1.46E−05	8.44E−06	1.74E−05	1.45E−05	1.74E−05	1.01E−05
Spleen	1.70E−05	3.06E−05	1.41E−05	1.79E−05	2.86E−06	8.20E−07
Thyroid	2.48E−07	5.14E−07	7.96E−07	9.68E−07	1.08E−06	1.25E−06
Total body	5.38E−05	4.83E−05	3.47E−05	5.43E−05	5.08E−05	5.35E−05

(Continued)

[a] Read as 5.15×10^{-8}, and so on.

TABLE G.4 (cont'd).

Target: Red Marrow

SOURCE	PHOTON ENERGY (MeV)					
	0.010	0.015	0.020	0.030	0.050	0.100
Bladder	4.26E−05[a]	6.38E−05	8.51E−05	5.82E−03	1.85E−02	1.50E−02
Stomach	6.00E−05	9.00E−05	1.00E−03	5.48E−03	1.28E−02	1.07E−02
Small intestine	1.36E−03	4.46E−03	1.11E−02	3.12E−02	4.59E−02	2.88E−02
Upper large intestine	1.04E−03	1.56E−03	8.60E−03	2.90E−02	3.93E−02	2.49E−02
Lower large intestine	2.72E−04	8.55E−03	3.56E−02	7.86E−02	7.02E−02	3.42E−02
Kidneys	8.99E−04	1.35E−03	8.58E−03	3.02E−02	4.14E−02	2.54E−02
Liver	9.05E−05	7.91E−04	2.94E−03	8.52E−03	1.40E−02	1.07E−02
Lungs	4.22E−05	1.13E−03	5.96E−03	1.80E−02	2.19E−02	1.25E−02
Other tissue (muscle)	3.02E−03	6.90E−03	1.31E−02	2.39E−02	2.34E−02	1.32E−02
Ovaries	1.71E−04	2.57E−04	8.78E−03	4.92E−02	6.77E−02	3.75E−02
Pancreas	1.18E−04	1.77E−04	2.19E−03	1.37E−02	2.67E−02	1.91E−02
Skeleton	1.41E−01	1.33E−01	1.25E−01	9.68E−02	5.66E−02	2.48E−02
Skin	2.42E−03	5.85E−03	8.96E−03	1.20E−02	1.11E−02	6.03E−03
Spleen	1.06E−05	4.53E−04	2.61E−03	9.11E−03	1.58E−02	1.12E−02
Testes	4.86E−06	7.29E−06	9.74E−06	9.69E−04	4.83E−03	4.68E−03
Thyroid	4.02E−06	6.03E−06	4.71E−04	4.11E−03	9.69E−03	7.31E−03
Total body	2.34E−02	2.54E−02	2.93E−02	3.44E−02	2.85E−02	1.52E−02

SOURCE	PHOTON ENERGY (MeV)					
	0.200	0.500	1.000	1.500	2.000	4.000
Bladder	9.15E−03	6.50E−03	5.58E−03	5.21E−03	5.04E−03	4.61E−03
Stomach	6.96E−03	5.42E−03	4.77E−03	4.35E−03	4.17E−03	3.99E−03
Small intestine	1.74E−02	1.29E−02	1.17E−02	1.06E−02	1.02E−02	8.28E−03
Upper large intestine	1.49E−02	1.12E−02	9.60E−03	9.08E−03	8.79E−03	7.28E−03
Lower large intestine	2.04E−02	1.61E−02	1.45E−02	1.32E−02	1.27E−02	1.02E−02
Kidneys	1.58E−02	1.29E−02	1.17E−02	1.11E−02	1.02E−02	8.78E−03
Liver	6.96E−03	5.58E−03	4.82E−03	4.89E−03	4.76E−03	3.93E−03
Lungs	7.82E−03	6.53E−03	6.03E−03	5.54E−03	5.55E−03	4.53E−03
Other tissue (muscle)	8.37E−03	7.13E−03	6.39E−03	5.78E−03	5.93E−03	4.47E−03
Ovaries	2.15E−02	1.59E−02	1.40E−02	1.28E−02	1.21E−02	9.48E−03
Pancreas	1.16E−02	8.96E−03	8.01E−03	7.71E−03	7.14E−03	5.81E−03
Skeleton	1.73E−02	1.59E−02	1.52E−02	1.38E−02	1.26E−02	1.06E−02
Skin	4.28E−03	4.08E−03	3.95E−03	3.89E−03	3.48E−03	3.09E−03
Spleen	7.56E−03	5.85E−03	5.46E−03	5.03E−03	4.65E−03	4.23E−03
Testes	3.38E−03	2.72E−03	2.54E−03	2.27E−03	2.42E−03	2.12E−03
Thyroid	4.91E−03	4.34E−03	3.86E−03	3.84E−03	3.71E−03	3.17E−03
Total body	9.86E−03	8.52E−03	7.76E−03	7.07E−03	6.98E−03	5.46E−03

(Continued)

[a] Read as 4.26×10^{-5}, and so on.

TABLE G.4 (cont'd).

Target: Stomach

SOURCE	PHOTON ENERGY (MeV)					
	0.010	0.015	0.020	0.030	0.050	0.100
Bladder	4.25E−16[a]	6.38E−16	1.80E−09	4.92E−06	6.80E−05	1.58E−04
Stomach	4.82E−02	1.14E−01	1.34E−01	1.03E−01	4.92E−02	2.82E−02
Small intestine	8.54E−05	2.70E−04	1.23E−03	2.13E−03	2.66E−03	2.18E−03
Upper large intestine	3.17E−04	4.74E−04	1.71E−03	3.33E−03	3.08E−03	2.24E−03
Lower large intestine	1.88E−06	2.82E−06	1.29E−04	7.55E−04	1.20E−03	1.04E−03
Kidneys	2.03E−07	3.05E−07	8.18E−05	1.20E−03	2.39E−03	2.21E−03
Liver	5.78E−05	8.67E−05	1.16E−04	6.12E−04	1.34E−03	1.06E−03
Lungs	8.06E−05	1.21E−04	5.18E−04	1.21E−03	1.42E−03	1.08E−03
Other tissue (muscle)	5.97E−06	1.73E−04	5.88E−04	1.15E−03	1.10E−03	6.90E−04
Ovaries	1.21E−11	1.82E−11	3.11E−07	7.83E−05	3.60E−04	4.86E−04
Pancreas	3.39E−04	5.87E−03	1.65E−02	2.36E−02	1.70E−02	1.01E−02
Skeleton	2.17E−05	3.27E−05	4.35E−05	1.59E−04	2.60E−04	3.17E−04
Skin	4.11E−05	6.17E−05	8.22E−05	3.03E−04	3.87E−04	3.09E−04
Spleen	1.29E−04	1.94E−04	2.01E−03	8.00E−03	8.79E−03	6.14E−03
Testes	2.09E−23	3.12E−23	5.60E−13	9.99E−08	9.27E−06	1.29E−05
Thyroid	3.11E−23	4.67E−23	6.75E−13	1.09E−07	3.02E−05	2.85E−05
Total body	2.07E−03	2.12E−03	2.19E−03	1.71E−03	1.29E−03	8.64E−04

SOURCE	PHOTON ENERGY (MeV)					
	0.200	0.500	1.000	1.500	2.000	4.000
Bladder	1.53E−04	1.65E−04	2.79E−04	1.95E−04	2.75E−04	2.30E−04
Stomach	2.67E−02	2.69E−02	2.48E−02	2.22E−02	2.12E−02	1.73E−02
Small intestine	1.97E−03	1.74E−03	1.71E−03	1.62E−03	1.48E−03	1.18E−03
Upper large intestine	1.98E−03	1.86E−03	1.74E−03	1.42E−03	1.52E−03	1.16E−03
Lower large intestine	9.51E−04	9.35E−04	7.47E−04	7.10E−04	8.24E−04	7.38E−04
Kidneys	1.80E−03	1.74E−03	1.61E−03	1.27E−03	1.50E−03	1.22E−03
Liver	1.04E−03	9.75E−04	9.66E−04	9.00E−04	9.17E−04	7.76E−04
Lungs	9.92E−04	9.48E−04	9.51E−04	9.21E−04	9.68E−04	4.58E−04
Other tissue (muscle)	7.44E−04	7.26E−04	6.90E−04	6.57E−04	6.15E−04	5.18E−04
Ovaries	4.23E−04	4.35E−04	3.72E−04	4.88E−04	4.68E−04	4.70E−04
Pancreas	9.81E−03	8.76E−03	8.10E−03	7.44E−03	6.81E−03	6.03E−03
Skeleton	3.11E−04	3.03E−04	2.64E−04	2.85E−04	2.58E−04	2.22E−04
Skin	2.96E−04	3.36E−04	3.35E−04	3.06E−04	2.91E−04	2.42E−04
Spleen	5.04E−03	5.08E−03	4.40E−03	4.71E−03	3.98E−03	3.09E−03
Testes	2.57E−05	6.11E−05	4.07E−05	6.09E−05	7.20E−05	6.78E−05
Thyroid	2.17E−05	6.66E−05	6.93E−05	9.63E−05	5.55E−05	7.80E−05
Total body	1.01E−03	9.21E−04	8.22E−04	6.14E−04	6.90E−04	5.97E−04

(Continued)

[a] Read as 4.25×10^{-16}, and so on.

TABLE G.4 (cont'd).

Target: Small Intestine

SOURCE	PHOTON ENERGY (MeV)					
	0.010	0.015	0.020	0.030	0.050	0.100
Bladder	1.58E−07[a]	2.37E−07	1.83E−04	5.31E−03	1.29E−02	1.27E−02
Stomach	2.07E−04	3.11E−04	1.28E−03	6.99E−03	1.27E−02	1.12E−02
Small intestine	7.70E−01	6.87E−01	5.82E−01	3.86E−01	1.94E−01	1.18E−01
Upper large intestine	5.68E−03	5.26E−02	1.24E−01	1.56E−01	1.07E−01	6.75E−02
Lower large intestine	2.52E−03	2.89E−02	6.39E−02	8.12E−02	5.64E−02	3.80E−02
Kidneys	6.77E−06	1.02E−05	1.00E−03	6.74E−03	1.28E−02	1.19E−02
Liver	1.68E−04	2.53E−04	8.01E−04	3.07E−03	6.65E−03	6.57E−03
Lungs	1.72E−14	2.57E−14	3.04E−08	4.86E−05	3.74E−04	7.39E−04
Other tissue (muscle)	7.24E−04	2.23E−03	4.41E−03	7.60E−03	7.79E−03	6.19E−03
Ovaries	2.98E−03	2.32E−02	6.15E−02	9.83E−02	7.51E−02	4.99E−02
Pancreas	2.69E−08	4.05E−08	4.17E−05	1.94E−03	6.65E−03	7.39E−03
Skeleton	1.58E−04	2.37E−04	5.21E−04	1.76E−03	3.21E−03	2.92E−03
Skin	6.71E−05	1.01E−04	1.34E−04	8.99E−04	1.79E−03	1.76E−03
Spleen	1.84E−07	2.76E−07	9.77E−05	1.55E−03	5.28E−03	5.82E−03
Testes	2.78E−15	4.17E−15	1.68E−08	4.85E−05	8.36E−04	1.30E−03
Thyroid	4.76E−32	7.14E−32	1.12E−16	5.47E−09	3.63E−06	2.05E−05
Total body	1.50E−02	1.51E−02	1.45E−02	1.39E−02	1.03E−02	7.86E−03

SOURCE	PHOTON ENERGY (MeV)					
	0.200	0.500	1.000	1.500	2.000	4.000
Bladder	1.07E−02	9.33E−03	9.19E−03	8.50E−03	8.76E−03	6.92E−03
Stomach	9.64E−03	9.44E−03	8.49E−03	7.66E−03	7.97E−03	6.64E−03
Small intestine	1.11E−01	1.11E−01	9.95E−02	9.43E−02	8.56E−02	6.95E−02
Upper large intestine	6.13E−02	5.83E−02	5.25E−02	4.72E−02	4.38E−02	3.42E−02
Lower large intestine	3.45E−02	3.19E−02	2.95E−02	2.75E−02	2.61E−02	2.07E−02
Kidneys	1.06E−02	9.80E−03	9.38E−03	9.27E−03	8.37E−03	6.93E−03
Liver	6.25E−03	5.66E−03	5.37E−03	5.30E−03	4.83E−03	4.33E−03
Lungs	7.43E−04	9.95E−04	1.12E−03	1.08E−03	1.21E−03	1.15E−03
Other tissue (muscle)	5.73E−03	5.54E−03	5.20E−03	4.92E−03	4.61E−03	3.92E−03
Ovaries	4.36E−02	4.23E−02	3.71E−02	3.47E−02	3.28E−02	2.61E−02
Pancreas	6.94E−03	6.30E−03	6.14E−03	5.84E−03	5.48E−03	4.70E−03
Skeleton	2.84E−03	2.88E−03	2.90E−03	2.45E−03	2.44E−03	2.06E−03
Skin	1.80E−03	1.98E−03	2.09E−03	2.01E−03	1.87E−03	1.62E−03
Spleen	5.28E−03	4.96E−03	4.75E−03	4.86E−03	4.26E−03	3.69E−03
Testes	1.56E−03	1.92E−03	1.85E−03	1.85E−03	1.93E−03	1.64E−03
Thyroid	6.03E−05	2.57E−05	1.44E−04	1.99E−04	2.00E−04	2.12E−04
Total body	7.29E−03	7.08E−03	6.56E−03	6.08E−03	5.62E−03	5.01E−03

(Continued)

[a] Read as 1.58×10^{-7}, and so on.

TABLE G.4 (cont'd).

Target: Upper Large Intestine

	PHOTON ENERGY (MeV)					
SOURCE	0.010	0.015	0.020	0.030	0.050	0.100
Bladder	3.57E−08[a]	5.37E−08	2.13E−05	6.35E−04	1.80E−03	1.74E−03
Stomach	1.69E−04	2.53E−04	1.19E−03	3.39E−03	3.97E−03	2.93E−03
Small intestine	9.41E−02	9.36E−02	8.26E−02	5.73E−02	2.97E−02	1.81E−02
Upper large intestine	7.59E−02	1.67E−01	1.80E−01	1.16E−01	5.45E−02	3.18E−02
Lower large intestine	2.15E−03	4.64E−03	6.60E−03	6.42E−03	4.62E−03	3.36E−03
Kidneys	1.56E−07	2.34E−07	7.73E−05	1.31E−03	2.45E−03	2.36E−03
Liver	3.95E−05	5.94E−05	3.57E−04	1.79E−03	2.47E−03	2.11E−03
Lungs	5.29E−15	7.94E−15	9.45E−09	1.61E−05	1.07E−04	1.75E−04
Other tissue (muscle)	9.15E−04	1.01E−03	1.30E−03	1.72E−03	1.62E−03	1.28E−03
Ovaries	1.87E−03	1.41E−02	2.26E−02	2.15E−02	1.33E−02	8.69E−03
Pancreas	8.11E−09	1.22E−08	1.27E−05	5.83E−04	1.73E−03	1.74E−03
Skeleton	4.43E−05	6.63E−05	1.35E−04	2.88E−04	5.37E−04	5.35E−04
Skin	2.63E−05	3.95E−05	5.27E−05	2.49E−04	4.08E−04	3.82E−04
Spleen	2.47E−08	3.70E−08	1.40E−05	4.56E−04	1.24E−03	1.30E−03
Testes	2.82E−15	4.24E−15	5.75E−09	9.99E−06	1.40E−04	2.22E−04
Thyroid	1.95E−32	2.93E−32	3.93E−17	1.48E−09	8.78E−07	9.09E−06
Total body	3.07E−03	3.16E−03	2.80E−03	2.84E−03	2.03E−03	1.54E−03

	PHOTON ENERGY (MeV)					
SOURCE	0.200	0.500	1.000	1.500	2.000	4.000
Bladder	1.75E−03	1.71E−03	1.29E−03	1.29E−03	1.21E−03	1.11E−03
Stomach	2.63E−03	2.36E−03	2.01E−03	2.45E−03	1.76E−03	1.65E−03
Small intestine	1.72E−02	1.64E−02	1.51E−02	1.40E−02	1.32E−02	1.04E−02
Upper large intestine	3.07E−02	3.07E−02	2.91E−02	2.59E−02	2.47E−02	1.95E−02
Lower large intestine	3.14E−03	3.03E−03	2.61E−03	2.82E−03	2.26E−03	2.06E−03
Kidneys	2.15E−03	2.07E−03	1.92E−03	1.93E−03	1.52E−03	1.31E−03
Liver	1.82E−03	1.85E−03	1.61E−03	1.36E−03	1.45E−03	1.09E−03
Lungs	1.79E−04	2.76E−04	2.88E−04	3.03E−04	2.86E−04	3.66E−04
Other tissue (muscle)	1.20E−03	1.17E−03	1.11E−03	1.05E−03	9.78E−04	8.30E−04
Ovaries	8.34E−03	7.98E−03	7.02E−03	6.67E−03	6.21E−03	4.62E−03
Pancreas	1.66E−03	1.55E−03	1.50E−03	1.41E−03	1.24E−03	1.05E−03
Skeleton	5.58E−04	5.02E−04	5.45E−04	4.87E−04	4.85E−04	3.70E−04
Skin	3.47E−04	4.18E−04	4.70E−04	3.72E−04	3.89E−04	3.01E−04
Spleen	9.11E−04	1.01E−03	9.57E−04	9.68E−04	9.72E−04	6.33E−04
Testes	2.78E−04	2.30E−04	2.93E−04	2.42E−04	4.81E−04	2.97E−04
Thyroid	8.42E−06	7.59E−06	3.16E−05	4.79E−05	3.87E−05	8.19E−05
Total body	1.32E−03	1.41E−03	1.26E−03	1.12E−03	1.08E−03	8.84E−04

(Continued)

[a] Read as 3.57×10^{-8}, and so on.

TABLE G.4 (cont'd).

Target: Lower Large Intestine

SOURCE	PHOTON ENERGY (MeV)					
	0.010	0.015	0.020	0.030	0.050	0.100
Bladder	5.47E−05[a]	8.19E−05	1.25E−03	5.62E−03	6.66E−03	4.82E−03
Stomach	2.74E−07	4.10E−07	4.48E−05	2.93E−04	7.65E−04	7.10E−04
Small intestine	4.50E−03	8.19E−03	1.09E−02	1.01E−02	6.85E−03	4.51E−03
Upper large intestine	3.95E−04	5.92E−04	1.82E−03	2.90E−03	2.42E−03	2.08E−03
Lower large intestine	8.85E−02	1.81E−01	1.86E−01	1.12E−01	4.75E−02	2.82E−02
Kidneys	1.33E−09	1.98E−09	1.86E−06	1.06E−04	2.90E−04	3.98E−04
Liver	4.37E−15	6.56E−15	3.23E−09	4.24E−06	5.34E−05	1.40E−04
Lungs	1.00E−17	1.50E−17	1.81E−10	1.17E−06	1.33E−05	2.85E−05
Other tissue (muscle)	1.09E−05	2.16E−04	6.59E−04	1.27E−03	1.32E−03	1.04E−03
Ovaries	1.59E−03	1.30E−02	2.34E−02	2.32E−02	6.85E−03	8.99E−03
Pancreas	2.26E−12	3.39E−12	1.41E−07	4.93E−05	2.77E−04	3.55E−04
Skeleton	1.04E−04	1.55E−04	2.94E−04	5.54E−04	6.56E−04	6.35E−04
Skin	9.15E−06	1.37E−05	1.82E−05	1.45E−04	2.64E−04	2.94E−04
Spleen	2.91E−10	4.37E−10	9.25E−07	7.28E−05	3.22E−04	3.76E−04
Testes	2.08E−07	3.12E−07	1.26E−04	1.10E−03	2.10E−03	1.70E−03
Thyroid	1.33E−35	2.00E−35	4.40E−19	9.06E−11	1.06E−07	1.60E−06
Total body	2.34E−03	1.94E−03	2.19E−03	2.02E−03	1.66E−03	1.07E−03

SOURCE	PHOTON ENERGY (MeV)					
	0.200	0.500	1.000	1.500	2.000	4.000
Bladder	4.02E−03	3.90E−03	3.34E−03	3.46E−03	3.23E−03	2.18E−03
Stomach	7.44E−04	6.96E−04	7.25E−04	7.22E−04	6.37E−04	4.42E−04
Small intestine	4.03E−03	3.71E−03	3.50E−03	3.25E−03	3.28E−03	2.61E−03
Upper large intestine	1.70E−03	1.65E−03	1.73E−03	1.38E−03	1.30E−03	9.58E−04
Lower large intestine	2.78E−02	2.83E−02	2.59E−02	2.43E−02	2.22E−02	1.73E−02
Kidneys	4.88E−04	4.85E−04	4.54E−04	5.42E−04	4.69E−04	4.03E−04
Liver	1.37E−04	1.79E−04	1.42E−04	1.98E−04	1.52E−04	2.24E−04
Lungs	6.48E−05	4.30E−05	5.14E−05	3.41E−05	3.74E−05	4.30E−05
Other tissue (muscle)	9.62E−04	9.34E−04	8.77E−04	8.29E−04	7.82E−04	6.54E−04
Ovaries	8.11E−03	7.81E−03	6.80E−03	6.43E−03	6.14E−03	5.39E−03
Pancreas	3.30E−04	2.94E−04	2.96E−04	3.73E−04	2.46E−04	3.06E−04
Skeleton	6.11E−04	5.18E−04	5.73E−04	4.85E−04	4.35E−04	4.06E−04
Skin	2.85E−04	3.33E−04	3.60E−04	3.06E−04	3.33E−04	2.32E−04
Spleen	3.63E−04	3.86E−04	3.30E−04	3.38E−04	3.55E−04	2.90E−04
Testes	1.59E−03	1.57E−03	1.16E−03	1.28E−03	1.17E−03	1.18E−03
Thyroid	4.14E−06	8.35E−06	1.29E−05	1.59E−05	1.78E−05	2.08E−05
Total body	9.47E−04	1.09E−03	1.02E−03	9.42E−04	1.14E−03	6.45E−04

(Continued)

[a] Read as 5.47×10^{-5}, and so on.

TABLE G.4 (cont'd).

Target: Liver

SOURCE	PHOTON ENERGY (MeV)					
	0.010	0.015	0.020	0.030	0.050	0.100
Bladder	4.72E−16[a]	7.08E−16	4.29E−09	3.71E−05	4.92E−04	1.00E−03
Stomach	3.87E−04	5.83E−04	7.76E−04	7.22E−03	1.50E−02	1.46E−02
Small intestine	1.38E−04	6.37E−04	2.41E−03	8.31E−03	1.39E−02	1.30E−02
Upper large intestine	2.86E−04	4.29E−04	3.49E−03	1.34E−02	2.14E−02	1.81E−02
Lower large intestine	5.29E−14	7.93E−14	4.24E−08	5.88E−05	8.33E−04	1.80E−03
Kidneys	8.02E−04	1.20E−03	8.47E−03	2.79E−02	3.53E−02	2.77E−02
Liver	9.70E−01	8.98E−01	7.86E−01	5.38E−01	2.75E−01	1.65E−01
Lungs	1.30E−04	4.13E−03	1.31E−02	2.53E−02	2.33E−02	1.72E−02
Other tissue (muscle)	1.11E−03	3.48E−03	6.43E−03	9.99E−03	9.50E−03	7.40E−03
Ovaries	1.90E−11	2.86E−11	1.04E−06	4.38E−04	2.43E−03	3.62E−03
Pancreas	2.79E−04	4.16E−04	6.53E−03	2.97E−02	4.05E−02	3.19E−02
Skeleton	2.24E−04	3.38E−04	8.67E−04	3.02E−03	4.29E−03	4.62E−03
Skin	8.63E−05	1.29E−04	8.80E−04	3.22E−03	4.20E−03	3.64E−03
Spleen	1.30E−09	1.95E−09	7.69E−06	1.08E−03	5.30E−03	6.84E−03
Testes	4.67E−23	7.00E−23	1.74E−12	4.96E−07	4.94E−05	1.75E−04
Thyroid	2.28E−21	3.44E−21	2.10E−11	2.15E−06	3.49E−04	5.52E−04
Total body	2.61E−02	2.57E−02	2.57E−02	2.23E−02	1.63E−02	1.13E−02

SOURCE	PHOTON ENERGY (MeV)					
	0.200	0.500	1.000	1.500	2.000	4.000
Bladder	1.39E−03	1.77E−03	1.85E−03	1.74E−03	2.15E−03	1.88E−03
Stomach	1.27E−02	1.27E−02	1.15E−02	1.11E−02	1.12E−02	8.85E−03
Small intestine	1.20E−02	1.12E−02	1.11E−02	1.03E−02	9.92E−03	7.95E−03
Upper large intestine	1.69E−02	1.53E−02	1.47E−02	1.39E−02	1.31E−02	1.14E−02
Lower large intestine	1.85E−03	2.15E−03	2.33E−03	2.72E−03	2.24E−03	2.14E−03
Kidneys	2.46E−02	2.50E−02	2.21E−02	2.08E−02	1.99E−02	1.59E−02
Liver	1.60E−01	1.60E−01	1.46E−01	1.35E−01	1.24E−01	1.01E−01
Lungs	1.59E−02	1.48E−02	1.40E−02	1.30E−02	1.18E−02	1.05E−02
Other tissue (muscle)	6.91E−03	6.97E−03	6.68E−03	6.26E−03	5.97E−03	5.07E−03
Ovaries	3.82E−03	3.82E−03	3.93E−03	3.87E−03	4.16E−03	3.19E−03
Pancreas	2.82E−02	2.55E−02	2.33E−02	2.19E−02	1.95E−02	1.72E−02
Skeleton	4.33E−03	4.42E−03	4.36E−03	4.16E−03	4.53E−03	3.71E−03
Skin	3.66E−03	4.24E−03	4.07E−03	3.98E−03	4.00E−03	3.13E−03
Spleen	6.64E−03	6.68E−03	6.39E−03	6.24E−03	5.92E−03	5.10E−03
Testes	2.68E−04	3.20E−04	6.53E−04	7.53E−04	7.37E−04	7.19E−04
Thyroid	7.75E−04	9.94E−04	1.18E−03	1.32E−03	1.59E−03	1.40E−03
Total body	1.05E−02	1.08E−02	1.07E−02	9.09E−03	8.58E−03	7.75E−03

(Continued)

[a] Read as 4.72×10^{-16}, and so on.

TABLE G.4 (cont'd).

Target: Kidneys

SOURCE	PHOTON ENERGY (MeV)					
	0.010	0.015	0.020	0.030	0.050	0.100
Bladder	4.32E−17[a]	6.48E−17	1.12E−09	6.99E−06	1.15E−04	2.83E−04
Stomach	1.09E−07	1.63E−07	1.18E−04	2.32E−03	4.69E−03	3.86E−03
Small intestine	1.40E−06	2.10E−06	2.39E−04	1.97E−03	3.64E−03	3.69E−03
Upper large intestine	9.03E−08	1.35E−07	1.01E−04	1.64E−03	3.55E−03	3.24E−03
Lower large intestine	1.20E−09	1.81E−09	3.72E−06	2.53E−04	7.36E−04	8.86E−04
Kidneys	9.32E−01	7.78E−01	5.79E−01	2.93E−01	1.12E−01	6.67E−02
Liver	2.38E−04	3.58E−04	1.28E−03	4.46E−03	5.54E−03	4.49E−03
Lungs	1.82E−08	2.74E−08	1.65E−05	3.52E−04	8.43E−04	9.23E−04
Other tissue (muscle)	4.03E−04	1.26E−03	2.10E−03	2.50E−03	1.93E−03	1.45E−03
Ovaries	1.05E−11	1.58E−11	5.40E−07	1.55E−04	7.50E−04	1.00E−03
Pancreas	6.42E−06	9.63E−06	9.17E−04	7.10E−03	9.57E−03	7.53E−03
Skeleton	8.58E−05	1.29E−04	1.72E−04	5.14E−04	9.26E−04	8.55E−04
Skin	5.17E−05	7.72E−05	3.35E−04	7.98E−04	8.26E−04	5.91E−04
Spleen	1.99E−03	2.98E−03	1.03E−02	1.89E−02	1.51E−02	1.03E−02
Testes	1.33E−24	1.99E−24	2.56E−13	1.23E−07	1.08E−05	3.29E−05
Thyroid	2.52E−27	3.78E−27	1.13E−14	2.64E−08	5.85E−06	3.41E−05
Total body	3.98E−03	4.12E−03	3.64E−03	3.01E−03	2.33E−03	1.70E−03

SOURCE	PHOTON ENERGY (MeV)					
	0.200	0.500	1.000	1.500	2.000	4.000
Bladder	2.84E−04	3.78E−04	3.32E−04	4.80E−04	3.58E−04	3.89E−04
Stomach	3.58E−03	3.21E−03	3.01E−03	2.79E−03	2.55E−03	2.11E−03
Small intestine	3.07E−03	3.10E−03	2.93E−03	2.46E−03	2.44E−03	1.78E−03
Upper large intestine	2.79E−03	2.65E−03	2.67E−03	2.44E−03	2.08E−03	1.78E−03
Lower large intestine	9.85E−04	8.07E−04	8.75E−04	8.46E−04	1.01E−03	7.10E−04
Kidneys	6.79E−02	7.16E−02	6.42E−02	6.08E−02	5.48E−02	4.63E−02
Liver	3.86E−03	3.66E−03	3.35E−03	3.24E−03	3.12E−03	2.34E−03
Lungs	8.92E−04	1.03E−03	9.14E−04	1.03E−03	6.56E−04	8.01E−04
Other tissue (muscle)	1.41E−03	1.40E−03	1.34E−03	1.27E−03	1.20E−03	1.01E−03
Ovaries	9.85E−04	1.11E−03	1.03E−03	9.43E−04	1.03E−03	8.46E−04
Pancreas	6.45E−03	6.05E−03	5.51E−03	5.17E−03	4.63E−03	4.17E−03
Skeleton	9.34E−04	8.86E−04	1.14E−03	8.97E−04	6.90E−04	6.59E−04
Skin	6.50E−04	7.36E−04	7.38E−04	6.76E−04	6.53E−04	5.62E−04
Spleen	8.92E−03	8.32E−03	7.36E−03	7.24E−03	6.62E−03	5.08E−03
Testes	6.82E−05	9.97E−05	1.44E−04	1.60E−04	1.15E−04	2.06E−04
Thyroid	3.98E−05	4.80E−05	1.17E−04	1.69E−04	1.24E−04	8.78E−05
Total body	1.73E−03	1.72E−03	1.68E−03	1.39E−03	1.33E−03	1.12E−03

[a] Read as 4.32×10^{-17}, and so on.

Source: ICRP 1975.

[a] Target is wall, source is content.

[b] Source is either cortical or trabecular bone.

[c] Read as 3.83×10^{-2}, and so on.

Source: ICRP 1975.

TABLE G.5 Specific Committed Dose Equivalents (Sv Bq^{-1}) for Selected Ingested Radionuclides and Their Daughters, Based on ICRP Publication 30. RM = Red Marrow, BS = Bone Surface, ST = Stomach, SI = Small Intestine, ULI = Upper Large Intestine, LLI = Lower Large Intestine.

A	f_1	Gonads	Breast	RM	Lung	Thyroid	BS	ST	SI	ULI	LLI	Liver	Kidneys	Other	(Weight)[b]
Na 24	1E+00	3.4E−10[a]	2.7E−10	3.7E−10	2.6E−10		4.7E−10	1.2E−09			3.4E−10			4.4E−10	0.18
P 32	8E−01	6.5E−10	6.5E−10	8.1E−09			7.9E−09				7.2E−09				
K 40	1E+00	5.1E−09	4.9E−09	4.9E−09	4.9E−09	4.9E−09	4.9E−09	5.5E−09	5.0E−09	3.0E−09	5.0E−09			5.4E−09	0.12
Cr 51	1E−01	4.0E−11	4.7E−11	1.2E−11					4.7E−11	1.1E−10	2.5E−10				
Cr 51	1E−02	4.0E−11								1.1E−10	2.7E−10				
Mn 54	1E−01	9.5E−10	2.8E−10	4.9E−10	2.3E−10			9.0E−10	9.8E−10	1.4E−09	2.2E−09	1.0E−09		5.0E−10	0.06
Mn 56	1E−01	8.5E−11							1.1E−09	1.4E−09	5.4E−10				
Fe 55	1E−01	1.1E−10	1.0E−10	1.1E−10	1.0E−10				1.2E−09	1.7E−10	3.0E−10	3.4E−10		5.6E−10	0.06
Fe 59	1E−01	1.7E−09	7.4E−10	8.5E−10	6.4E−10				2.1E−09	3.9E−09	8.4E−09	1.5E−09		1.8E−09	0.06
Co 58	5E−02	1.0E−09	1.8E−10	2.6E−10					1.1E−09	1.9E−09	4.0E−10			4.8E−10	0.06
Co 60	5E−02	3.2E−09	1.1E−09	1.3E−09	8.7E−10				3.6E−10	5.7E−09	1.1E−08	2.3E−09		2.1E−09	0.06
Ni 63	5E−02	8.5E−11	8.5E−11	8.5E−11	8.5E−11			1.0E−10	1.3E−10	3.6E−10	9.2E−10				
Ni 65	5E−02	2.4E−11						6.2E−10	7.3E−10	9.3E−10	3.7E−10				
Cu 64	5E−01	4.8E−11						1.7E−10	2.1E−10	6.1E−10	7.5E−10				
Zn 65	5E−01	3.5E−09	3.3E−09	4.5E−09	3.1E−09	3.2E−09	4.5E−09		4.3E−09	4.2E−09	5.0E−09			4.8E−09	0.12
Zn 69	5E−01								1.1E−10	6.1E−11					
Br 83	1E+00	7.3E−12						2.1E−10							
Br 84	1E+00							3.0E−10							
Rb 88	1E+00							6.8E−10							
Rb 89	1E+00							7.3E−10							
Sr 89	3E−01			3.2E−09			4.8E−09	3.6E−10		7.3E−09	2.1E−08				
Sr 90	3E−01			1.9E−07			4.2E−07								
Y 90	1E−04									1.3E−08	3.2E−08				
Y 91m	1E−04	6.9E−12						4.9E−11	3.1E−11	3.1E−11	2.4E−11			1.1E−11	0.06
Y 91	1E−04									1.0E−08	3.0E−08				
Y 92	1E−04							1.4E−09	2.0E−09	3.3E−09	1.7E−09				
Y 93	1E−04							1.3E−09	2.5E−09	7.8E−09	8.8E−09				
Zr 95	2E−03	8.1E−10							1.1E−09	3.0E−09	7.8E−09				
Zr 97	2E−03	6.2E−10							3.4E−09	1.2E−08	1.8E−08				
Nb 95	1E−02	8.0E−10							9.1E−10	1.8E−08	4.0E−09				
Mo 99	8E−01	2.2E−10	1.8E−10	5.3E−10	1.9E−10		7.7E−10	6.7E−10			3.1E−09	2.7E−09	2.5E−09		
Tc 99m	8E−01	9.7E−12	3.6E−12	6.3E−12		8.5E−11		7.2E−11	2.2E−11	1.4E−09	2.5E−11			1.1E−11	0.06
Tc 101	8E−01							1.5E−10	2.3E−11	3.7E−11					

[a] Read as 3.4 × 10^{-10}.

[b] Factor to be applied to other-organ dose equivalent when calculating the effective dose equivalent.

Source: ICRP 1979.

TABLE G.5 (cont'd).

A	f_1	Gonads	Breast	RM	Lung	Thyroid	BS	ST	SI	ULI	LLI	Liver	Kidneys	Other	(Weight)[b]
Ru 103	5E−02	5.7E−10[a]						5.0E−10	8.5E−10	2.5E−09	6.5E−09				
Ru 105	5E−02	9.7E−11							7.9E−10	1.6E−09	1.3E−09				
Ru 106	5E−02									2.5E−08	7.1E−08				
Ag 110m	5E−02	3.0E−09	7.5E−10	9.4E−10	8.3E−10				3.5E−09	5.9E−09	1.1E−08	8.6E−09		1.6E−09	0.06
Te 125m	2E−01			1.2E−09			1.3E−08			1.7E−09	4.7E−09				
Te 127m	2E−01			5.4E−09			2.1E−08			3.1E−09	1.1E−08				
Te 127	2E−01							2.4E−10	3.9E−10	1.2E−09	1.3E−09				
Te 129m	2E−01			3.5E−09			8.0E−09			8.4E−09	2.5E−08				
Te 129	2E−01							4.0E−10	2.7E−10	1.9E−10					
Te 131m	2E−01	7.3E−10				4.3E−08				4.6E−09	8.2E−09				
Te 131	2E−01					4.2E−09		6.3E−10	5.5E−10	5.3E−10					
Te 132	2E−01					5.9E−08					3.8E−09				
I 125	1E+00					3.4E−07									
I 130	1E+00					3.9E−08									
I 131	1E+00					4.8E−07									
I 133	1E+00					9.1E−08									
I 134	1E+00					6.2E−10									
I 135	1E+00					1.8E−08		5.5E−10							
Cs 134	1E+00	2.1E−08	1.7E−08	1.9E−08	1.8E−08	1.8E−08	1.7E−08		2.2E−08		2.2E−08			2.3E−08	0.18
Cs 136	1E+00	3.0E+00	2.6E−09	3.0E−09	2.6E−09	2.7E−09	2.7E−09		3.4E−09		3.4E−09			3.8E−09	0.18
Cs 137	1E+00	1.4E−08	1.2E−08	1.3E−08	1.3E−08	1.3E−08	1.3E−08		1.4E−08	1.4E−08	1.4E−08			1.5E−08	0.12
Cs 138	1E+00							7.0E−10							
Ba 139	1E−01							6.9E−10	5.5E−10	4.4E−10	9.9E−11				
Ba 140	1E−01	1.0E−09						3.9E−10	1.9E−10	7.7E−09	2.6E−08				
Ba 141	1E−01								3.0E−09	2.2E−10	1.2E−10				
La 140	1E−03	1.3E−09						8.5E−10	3.0E−09	9.1E−10	1.7E−08				
La 142	1E−03	7.0E−11							7.8E−10	7.2E−10	1.9E−10				
Ce 141	3E−04								3.0E−09	3.0E−09	8.6E−09				
Ce 143	3E−04								1.4E−09	5.7E−10	1.2E−08				
Ce 144	3E−04									2.2E−08	6.6E−08				
Pr 143	3E−04									5.1E−09	1.5E−08				
Pr 144	3E−04							4.1E−10							
Nd 147	3E−04								9.6E−11	4.6E−09	1.3E−08				
W 187	3E−01	1.9E−10							7.5E−10	2.6E−09	4.4E−09				
Np 239	1E−02								8.7E−10	3.8E−09	8.6E−09				

[a]Read as 5.7×10^{-10}.

[b]Factor to be applied to other-organ dose equivalent when calculating the effective dose equivalent.

Source: ICRP 1979.

TABLE G.6 Specific Committed Dose Equivalents (Sv Bq^{-1}) for Selected Inhaled Radionuclides and Their Daughters, Based on ICRP Publication 30. RM = Red Marrow, BS = Bone Surface, ST = Stomach, SI = Small Intestine, LLI = Lower Large Intestine.

A	Class	Gonads	Breast	RM	Lung	Thyroid	BS	ST	SI	ULI	LLI	Liver	Kidneys	Other	(Weight)[b]
Na 24	D	1.8E−10[a]	1.6E−10	2.1E−10	1.2E−09			3.1E−10							
P 32	W			4.2E−09	2.6E−08										
P 32	D	4.8E−10	4.8E−10	6.0E−09	2.5E−09		5.8E−09				1.5E−09			3.4E−09	0.12
K 40	D	3.2E−09	3.1E−09	3.1E−09	4.7E−09	3.1E−09	3.1E−09	3.2E−09	3.2E−09	3.8E−11	3.2E−09			2.5E−11	0.12
Cr 51	D	2.7E−11	1.9E−11	2.7E−11	3.8E−10		2.7E−11				5.9E−11				
Cr 51	W	2.2E−11			3.8E−10			3.0E−11	3.0E−11		1.1E−10				
Cr 51	Y				5.3E−10						1.2E−10				
Mn 54	D	8.9E−10	9.1E−10	1.7E−09	1.2E−09		2.6E−09			1.3E−09		4.6E−09		1.8E−09	0.18
Mn 54	W	7.1E−10	8.6E−10	1.1E−09	6.7E−09							2.5E−09		1.8E−09	0.18
Mn 56	D	2.1E−11			4.4E−10			1.4E−10	1.6E−10						
Mn 56	W				5.4E−10					2.0E−10					
Fe 55	D	5.2E−10	5.1E−10	5.2E−10	5.2E−10					5.2E−10	5.4E−10	1.7E−09		1.7E−09	0.12
Fe 55	W	1.8E−10	1.7E−10	1.8E−10	1.1E−09						2.8E−10	5.8E−10		9.5E−10	0.06
Fe 59	W	1.4E−09	1.3E−09		1.4E−08						4.5E−09			2.9E−09	0.06
Fe 59	D	3.3E−09	3.0E−09	3.2E−09	3.5E−09	2.9E−09	2.9E−09			4.1E−09	4.8E−09	7.1E−09		6.5E−09	0.12
Co 58	D	6.5E−10			7.9E−09										
Co 58	Y				1.6E−08						2.0E−09				
Co 60	Y				3.4E−07										
Co 60	W	4.0E−09	4.2E−09	4.2E−09	3.6E−08						8.2E−09	9.2E−09		8.0E−09	0.06
Ni 63	W	2.5E−10	2.5E−10		3.1E−09						6.7E−10				
Ni 63	D	8.2E−10	8.2E−10	8.2E−10	8.7E−10	8.2E−10	8.2E−10	8.3E−10	8.3E−10	8.7E−10	9.5E−10				
Ni 65	W				3.8E−10								8.2E−10		
Ni 65	D				3.1E−10			9.2E−11	1.1E−10	1.3E−10					
Cu 64	D	1.6E−11			2.0E−10					1.0E−10	1.2E−10				
Cu 64	W				3.4E−10				4.2E−11	1.3E−10	1.6E−10				
Cu 64	Y				3.5E−10					1.5E−10	1.9E−10				
Zn 65	Y	2.0E−09	3.1E−09	3.6E−09	2.1E−08									5.9E−09	0.18
Zn 69	Y				8.0E−11										
Br 83	D				1.5E−10			4.3E−11							
Br 83	W				1.8E−10										
Br 84	W				1.7E−10										
Br 84	D				1.6E−10			7.4E−11							

[a] Read as 1.8×10^{-10}.
[b] Factor to be applied to other-organ dose equivalent when calculating the effective dose equivalent.

Source: ICRP 1979.

TABLE G.6 (cont'd).

A	Class	Gonads	Breast	RM	Lung	Thyroid	BS	ST	SI	ULI	LLI	Liver	Kidneys	Other	(Weight)[b]
Rb 88	D				1.5E−10[a]			6.2E−11							
Rb 89	D				6.8E−11			3.0E−11							
Sr 89	D	4.2E−10		5.6E−09	2.2E−09		8.4E−09			1.5E−09	3.6E−09				
Sr 89	Y				8.4E−08										
Sr 90	D			3.3E−07			7.3E−07								
Sr 90	Y				2.9E−06										
Y 90	W				8.9E−09					4.6E−09	1.1E−08				
Y 90	Y				9.3E−09					5.3E−09	1.3E−08				
Y 91m	Y				7.0E−11										
Y 91m	W				4.2E−11										
Y 91	W			5.6E−09	5.3E−08						8.6E−12				
Y 91	Y				9.9E−08						1.4E−08				
Y 92	W				1.2E−09					3.2E−10					
Y 92	Y				1.2E−09					3.9E−10					
Y 93	W				2.5E−09				5.7E−10	1.8E−09	2.0E−09				
Y 93	Y				2.4E−09					1.5E−09	1.6E−09				
Zr 95	W			3.2E−09	1.9E−08		2.2E−08								
Zr 95	Y				4.1E−08						4.2E−09				
Zr 95	D	1.9E−09		1.3E−08			1.0E−07		8.9E−10	2.9E−09	4.3E−09				
Zr 97	W				4.0E−09										
Zr 97	D	1.8E−10		5.0E−10	2.1E−09				6.1E−10	1.9E−09	2.8E−09				
Zr 97	Y				4.1E−09				1.0E−09	3.5E−09	5.1E−09				
Nb 95	W	4.8E−10		6.7E−10	5.5E−09		2.4E−09				1.9E−09				
Nb 95	Y	4.3E−10			8.3E−09						1.9E−09				

[a] Read as 1.5×10^{-10}.
[b] Factor to be applied to other-organ dose equivalent when calculating the effective dose equivalent.
Source: ICRP 1979.

660

TABLE G.6 (cont'd).

A	Class	Gonads	Breast	RM	Lung	Thyroid	BS	ST	SI	ULI	LLI	Liver	Kidneys	Other	(Weight)[b]
Mo 99	Y				4.3E−09[a]				2.3E−09	5.5E−09					
Mo 99	D	1.3E−10	1.3E−10	3.7E−10	1.2E−09		5.4E−10			3.3E−10	5.8E−10	1.9E−09	1.7E−09		
Tc 99m	W	1.7E−12			3.1E−11	2.1E−11		1.5E−11							
Tc 99m	D	2.8E−12	2.1E−12	3.4E−12	2.3E−11	5.0E−11		2.9E−11	4.8E−12	7.0E−12	5.0E−12			5.0E−12	0.06
Tc 101	W				3.0E−11										
Tc 101	D				2.8E−11			1.4E−11							
Ru 103	D	7.3E−10	6.1E−10	6.7E−10	9.9E−09		6.2E−10	7.2E−10	8.4E−10	1.1E−09	3.1E−09			8.1E−10	0.06
Ru 103	W				1.0E−08						1.7E−09				
Ru 103	Y				1.6E−08						3.1E−09				
Ru 105	D	2.7E−11			3.7E−10			8.9E−11	1.3E−10	2.5E−10	2.1E−10				
Ru 105	W				5.4E−10					2.3E−10	2.6E−10				
Ru 105	Y				5.7E−10			1.2E−10	1.2E−10	2.8E−10	3.1E−10				
Ru 106	W				2.1E−07									1.4E−08	0.06
Ru 106	Y				1.0E−06									6.9E−09	0.12
Ru 106	D	1.4E−08	1.4E−08	1.4E−08	1.4E−08	1.4E−08	1.4E−08	1.4E−08	1.5E−08	1.7E−08	2.5E−08	2.6E−08			
Ag 110m	W	2.3E−09	2.9E−09		3.2E−08										
Ag 110m	Y				1.2E−07										
Ag 110m	D	3.3E−09	4.1E−09		8.1E−09					8.4E−09		8.1E−08		1.4E−08	0.18
Te 129	D				1.3E−10			5.1E−11	3.6E−11						
Te 125m	W			1.1E−09	1.0E−08		1.2E−08				2.2E−09				
Te 125m	D			3.0E−09			3.2E−08								
Te 127m	D			1.4E−08			5.2E−08								
Te 127	D				2.8E−10				6.4E−11	1.8E−10	1.9E−10				
Te 127	W				4.3E−10					2.1E−10	2.3E−10				
Te 127m	W			5.4E−09	3.3E−08		2.0E−08				1.1E−08				
Te 129m	W				4.0E−08						4.2E−09				
Te 129m	D			8.8E−09	2.2E−08		2.0E−08								
Te 129	W				1.5E−10										
Te 131	W				3.0E−10	2.7E−09									
Te 131m	D				9.3E−10	3.3E−08									
Te 131m	W				2.2E−09	3.6E−08					2.4E−09				
Te 131	D				2.5E−10	2.6E−09									

[a] Read as 4.3×10^{-9}.

[b] Factor to be applied to other-organ dose equivalent when calculating the effective dose equivalent.

Source: ICRP 1979.

TABLE G.6 (cont'd).

A	Class	Gonads	Breast	RM	Lung	Thyroid	BS	ST	SI	ULI	LLI	Liver	Kidneys	Other	(Weight)[b]
Te 132	W				1.7E−09[a]	6.3E−08									
Te 132	D					5.9E−08									
I 125	D					2.2E−07									
I 130	D				6.0E−10	2.0E−08									
I 131	D					2.9E−07									
I 133	D					4.9E−08									
I 134	D				1.4E−10	2.9E−10									
I 135	D				4.4E−10	8.5E−09		7.1E−11							
Cs 134	D	1.3E−08	1.1E−08	1.2E−08	1.2E−08	1.1E−08	1.1E−08		1.4E−08		1.4E−08			1.5E−08	0.18
Cs 136	D	1.9E−09	1.7E−09	1.9E−09	2.3E−09	1.7E−09	1.7E−09		2.1E−09		2.1E−09			2.4E−09	0.18
Cs 137	D	8.8E−09	7.8E−09	8.3E−09	8.8E−09	7.9E−09	7.9E−09		9.1E−09	9.0E−09	9.1E−09			9.5E−09	0.12
Cs 138	D				1.6E−10			7.7E−11	7.4E−11	5.8E−11					
Ba 139	D				2.5E−10			9.1E−11	5.3E−10	1.5E−09	4.4E−09				
Ba 140	D	4.3E−10	2.9E−10	1.3E−09	1.7E−09		2.4E−09								
Ba 141	D				1.2E−10			3.9E−11	2.3E−11	3.1E−11					
La 140	D	3.6E−10	2.0E−10	4.6E−10	1.7E−09				6.5E−10	1.6E−09	2.8E−09	3.5E−09		4.4E−10	0.06
La 140	W	4.5E−10			4.2E−09				9.7E−10	2.9E−09	5.5E−09				
La 142	D	1.6E−11			3.0E−10			1.2E−10	1.1E−10	1.0E−10					
La 142	W				3.5E−10										
Ce 141	W				1.1E−08						3.8E−09	3.5E−09		2.8E−09	0.06
Ce 141	Y				1.7E−08					1.8E−09	4.1E−09				
Ce 143	W				3.5E−09						3.7E−09				
Ce 143	Y				3.9E−09					2.1E−09	4.3E−09				
Ce 144	Y				7.9E−07										
Ce 144	W			2.7E−08	1.8E−07							2.5E−07		2.1E−07	0.06
Pr 143	Y				1.3E−08						6.8E−09				
Pr 143	W				1.1E−08						6.1E−09				
Pr 144	W				8.8E−11										
Pr 144	Y				9.4E−11										
Nd 147	Y				1.1E−08					1.9E−09	5.9E−09				
Nd 147	W	2.9E−11			8.4E−09					3.9E−10	5.2E−09				
W 187	D				6.0E−10						6.6E−10				
Np 239	W				2.4E−09		1.4E−09			1.3E−09	2.9E−09				

[a] Read as 1.7 × 10⁻⁹.

[b] Factor to be applied to other-organ dose equivalent when calculating the effective dose equivalent.

Source: ICRP 1979.

662

TABLE G.7 Masses and Effective Dimensions of Organs of the Standard Man, Based on ICRP Publication 2. Note: This Table is Included for Historical Purposes. It is Not to be Used with the ICRP-30 and ICRP-60 Methods.

ORGAN	MASS (g)[c]	EFFECTIVE DIMENSION (cm)
Total body[a]	70000	30
Muscle[b]	30000	30
Skin and subcutaneous tissue	6100	0.1
Fat	10000	20
Skeleton		
without bone marrow	7000	5
red marrow	1500	
yellow marrow	1500	
Blood	5400	
Gastrointestinal tract	2000	30
Contents of GI tract		
lower large intestine	150	5
stomach	250	10
small Intestine	1100	30
upper large intestine	35	5
Liver	1700	10
Brain	1500	15
Lungs (2)	1000	10
Lymphoid tissue	700	
Kidneys (2)	300	7
Heart	300	7
Spleen	150	7
Urinary bladder	150	
Pancreas	70	5
Salivary glands (6)	50	
Testes (2)	40	3
Spinal cord	30	1
Eyes (2)	30	0.25
Thyroid gland	20	3
Teeth	20	
Prostate gland	20	3
Adrenal glands or suparenal (2)	20	3
Thymus	20	
Ovaries (2)	8	3
Hypophysis (Pituitary)	0.6	0.5
Pineal Gland	0.2	0.04
Parathyroids (4)	0.15	0.06
Miscellaneous (blood vessels, cartilage, nerves, etc.)	390	

[a] Does not include contents of the gastrointestinal tract.

[b] The mass of the skin alone is taken to be 2000 g.

[c] Masses of both lungs, both kidneys, and so on.

Source: ICRP 1960.

TABLE G.8 Fraction AF of Photon Energy Released in Source Organ and Absorbed in Target Organ, Based on ICRP Publication 2. Note: This Table is Included for Historical Purposes. It is Not to be Used with the ICRP-30 and ICRP-60 Methods.

Source = Target

TARGET	PHOTON ENERGY (MeV) 0.010	0.015	0.020	0.030	0.050	0.100
Stomach	1.00E+00[a]	1.00E+00	9.94E−01	7.56E−01	3.22E−01	2.23E−01
Small intestine	1.00E+00	1.00E+00	1.00E+00	9.85E−01	6.89E−01	5.30E−01
Upper large intestine	1.00E+00	9.98E−01	9.20E−01	5.06E−01	1.77E−01	1.18E−01
Lower large intestine	1.00E+00	9.98E−01	9.20E−01	5.06E−01	1.77E−01	1.18E−01
Kidneys	1.00E+00	1.00E+00	9.71E−01	6.27E−01	2.38E−01	1.62E−01
Liver	1.00E+00	1.00E+00	9.94E−01	7.56E−01	3.22E−01	2.23E−01
Lungs	1.00E+00	1.00E+00	9.94E−01	7.56E−01	3.22E−01	2.23E−01
Other tissue (muscle)	1.00E+00	1.00E+00	1.00E+00	9.85E−01	6.89E−01	5.30E−01
Ovaries	1.00E+00	9.79E−01	7.81E−01	3.45E−01	1.10E−01	7.28E−02
Pancreas	1.00E+00	9.98E−01	9.20E−01	5.06E−01	1.77E−01	1.18E−01
Skeleton (endosteal cells)	1.00E+00	9.98E−01	9.20E−01	5.06E−01	1.77E−01	1.18E−01
Skin	3.76E−01	1.21E−01	4.93E−02	1.40E−02	3.88E−03	2.52E−03
Spleen	1.00E+00	1.00E+00	9.71E−01	6.27E−01	2.38E−01	1.62E−01
Testes	1.00E+00	9.79E−01	7.81E−01	3.45E−01	1.10E−01	7.28E−02
Thyroid	1.00E+00	9.79E−01	7.81E−01	3.45E−01	1.10E−01	7.28E−02
Total Body	1.00E+00	1.00E+00	1.00E+00	9.85E−01	6.89E−01	5.30E−01

	PHOTON ENERGY (MeV) 0.200	0.500	1.000	1.500	2.000	4.000
Stomach	2.57E−01	2.81E−01	2.67E−01	2.48E−01	2.32E−01	1.93E−01
Small intestine	5.90E−01	6.28E−01	6.07E−01	5.75E−01	5.47E−01	4.74E−01
Upper large intestine	1.38E−01	1.52E−01	1.44E−01	1.33E−01	1.24E−01	1.01E−01
Lower large intestine	1.38E−01	1.52E−01	1.44E−01	1.33E−01	1.24E−01	1.01E−01
Kidneys	1.88E−01	2.06E−01	1.96E−01	1.81E−01	1.69E−01	1.39E−01
Liver	2.57E−01	2.81E−01	2.67E−01	2.48E−01	2.32E−01	1.93E−01
Lungs	2.57E−01	2.81E−01	2.67E−01	2.48E−01	2.32E−01	1.93E−01
Other tissue (muscle)	5.90E−01	6.28E−01	6.07E−01	5.75E−01	5.47E−01	4.74E−01
Ovaries	8.52E−02	9.43E−02	8.91E−02	8.19E−02	7.61E−02	6.22E−02
Pancreas	1.38E−01	1.52E−01	1.44E−01	1.33E−01	1.24E−01	1.01E−01
Skeleton (endosteal cells)	1.38E−01	1.52E−01	1.44E−01	1.33E−01	1.24E−01	1.01E−01
Skin	2.97E−03	3.29E−03	3.11E−03	2.85E−03	2.64E−03	2.14E−03
Spleen	1.88E−01	2.06E−01	1.96E−01	1.81E−01	1.69E−01	1.39E−01
Testes	8.52E−02	9.43E−02	8.91E−02	8.19E−02	7.61E−02	6.22E−02
Thyroid	8.52E−02	9.43E−02	8.91E−02	8.19E−02	7.61E−02	6.22E−02
Total Body	5.90E−01	6.28E−01	6.07E−01	5.75E−01	5.47E−01	4.74E−01

[a] Read as 1.00×10^0, and so on.

Source: ICRP 1960.

REFERENCES

ICRP, Publication 2, *Recommendations of the International Commission on Radiological Protection Report of Committee II on Permissible Dose for Internal Radiation (1959)*, Publication 2, International Commission on Radiological Protection, Pergamon Press, Oxford. *Health Physics 3* (1960).

ICRP, *Report of the Task Group on Reference Man. A Report Prepared by a Task Group of Committee 2 of ICRP*, Publication 23, International Commission on Radiological Protection, Pergamon Press, Oxford, 1975.

ICRP, *Limits for Intakes of Radionuclides by Workers*, Publication 30, Parts 1–3, including addenda and supplements, International Commission on Radiological Protection, Pergamon Press, Oxford, *Annals of the ICRP 2* (3–4), 1979, *3* (1–4), 1979, *4* (3–4), 1980, *6* (2–3), 1981, *7* (1–3), 1982, and *8* (1–4), 1982.

Fission Product Yields for Fast and Thermal Fission of ^{235}U and ^{239}Pu

TABLE H.1 Fission Yields for ^{235}U and ^{239}Pu Induced by Thermal Neutrons Characteristic of Light Water Reactors and by Fast Neutrons Characterized by the Fission Spectrum.

NUCLIDE	U-235 FISSION THERMAL	U-235 FISSION FAST	PU-239 FISSION THERMAL	PU-239 FISSION FAST	NUCLIDE	U-235 FISSION THERMAL	U-235 FISSION FAST	PU-239 FISSION THERMAL	PU-239 FISSION FAST
1 H 3	0.0108	0.0200	0.0164	0.0230	37 Rb 88	0.0000	0.1260	0.0000	0.0000
30 Zn 72	0.0000	0.0000	0.0001	0.0000	35 Br 89	4.5900	0.0000	1.6400	0.0000
31 Ga 73	0.0001	0.0000	0.0002	0.0000	36 Kr 89	0.0000	4.4800	0.0000	1.6400
31 Ga 74	0.0003	0.0014	0.0008	0.0000	37 Rb 89	0.0000	0.3890	0.0000	0.2050
31 Ga 75	0.0008	0.0175	0.0008	0.0000	38 Sr 89	0.2000	0.0000	0.0710	0.0000
33 As 75	0.0000	0.0000	0.0000	0.0000	36 Kr 90	5.0000	4.1800	1.9800	1.5400
31 Ga 76	0.0025	0.0272	0.0030	0.0000	37 Rb 90	0.0000	0.8760	0.0000	0.5960
32 Ge 76	0.0000	0.0019	0.0000	0.0000	38 Sr 90	0.7700	0.0389	0.2300	0.0308
32 Ge 77m	0.0067	0.0409	0.0081	0.0175	36 Kr 91	3.4500	3.5000	1.4400	1.1300
32 Ge 77	0.0016	0.0058	0.0019	0.0021	37 Rb 91	1.9800	1.6500	0.8300	1.3400
32 Ge 78	0.0200	0.0720	0.0250	0.0401	38 Sr 91	0.3800	0.1950	0.1600	0.2050
33 As 78	0.0000	0.0019	0.0000	0.0000	40 Zr 91	0.0300	0.0000	0.1800	0.0000
33 As 79	0.0560	0.1170	0.0400	0.0719	36 Kr 92	1.8700	2.7200	0.9730	0.5650
35 Br 79	0.0000	0.0000	0.0000	0.0000	37 Rb 92	3.4300	2.3400	1.7900	1.9000
33 As 80	0.1000	0.1850	0.0700	0.1130	38 Sr 92	0.0000	0.5840	0.0000	0.7190
35 Br 80m	0.0000	0.0000	0.0000	0.0000	39 Y 92	0.7300	0.0000	0.3800	0.1030
33 As 81	0.0084	0.2920	0.0070	0.1750	36 Kr 93	0.4800	1.7500	0.3000	0.1850
34 Se 81	0.1320	0.0097	0.1100	0.0103	37 Rb 93	5.6200	2.8200	3.4600	1.8700
34 Se 82	0.3200	0.4180	0.2000	0.2670	38 Sr 93	0.0000	1.2600	0.0000	1.6400
35 Br 82m	0.0000	0.0000	0.0000	0.0000	39 Y 93	0.0000	0.0973	0.0000	0.2050
35 Br 82	0.0000	0.0000	0.0000	0.0010	40 Zr 93	0.3500	0.0000	0.2100	0.0000
34 Se 83m	0.2900	0.0000	0.0480	0.0000	36 Kr 94	0.1000	0.8470	0.0700	0.0411
34 Se 83	0.2200	0.5640	0.0360	0.3800	37 Rb 94	2.4000	2.6600	1.6800	1.1300
35 Br 83	0.0000	0.0341	0.0000	0.0103	38 Sr 94	2.9000	2.1400	2.0300	2.8400
36 Kr 83	0.0340	0.0000	0.2060	0.0000	39 Y 94	0.0000	0.3890	0.0000	0.7190
34 Se 84	0.9200	1.0100	0.4300	0.4930	40 Zr 94	1.0000	0.0000	0.7000	0.0000
35 Br 84m	0.0190	0.0292	0.0089	0.0411	37 Rb 95	1.5500	2.6300	1.2400	0.4420
36 Kr 84	0.0610	0.0000	0.0290	0.0000	38 Sr 95	3.0000	2.8200	2.4100	2.9500
34 Se 85	1.1000	1.1600	0.4560	0.5340	39 Y 95	1.6500	1.0700	1.3200	1.9500
35 Br 85	0.0000	0.2820	0.0000	0.1640	40 Zr 95	0.0000	0.0000	0.0000	0.1030
36 Kr 85m	0.1430	0.0000	0.0590	0.0000	42 Mo 95	0.0700	0.0000	0.0560	0.0000
36 Kr 85	0.0570	0.0097	0.0240	0.0103	39 Y 96	6.3300	5.9400	5.1700	4.8300
35 Br 86	2.0200	0.0000	0.7600	0.0000	40 Zr 96	0.0000	0.2920	0.0000	0.5140
36 Kr 86	0.0000	1.8800	0.0000	0.9250	41 Nb 96	0.0006	0.0000	0.0036	0.0000
37 Rb 86m	0.0000	0.0000	0.0000	0.0000	39 Y 97	5.9000	5.9400	5.5000	3.8000
35 Br 87	2.4900	0.0000	0.9200	0.0000	40 Zr 97	0.0000	0.6520	0.0000	1.5400
36 Kr 87	0.0000	2.5900	0.0000	1.1700	41 Nb 97	0.0000	0.0000	0.0000	0.1030
37 Rb 87	0.0000	0.0000	0.0000	0.0103	42 Mo 97	0.1900	0.0000	0.1500	0.0000
35 Br 88	3.5700	0.0000	1.4200	0.0000	40 Zr 98	5.7200	5.5500	5.6900	5.3400
36 Kr 88	0.0000	3.4100	0.0000	1.4400	41 Nb 98m	0.0640	0.3890	0.2000	0.3080

(Continued)

TABLE H.1 (cont'd).

NUCLIDE	U-235 FISSION THERMAL	U-235 FISSION FAST	PU-239 FISSION THERMAL	PU-239 FISSION FAST	NUCLIDE	U-235 FISSION THERMAL	U-235 FISSION FAST	PU-239 FISSION THERMAL	PU-239 FISSION FAST
41 Nb 99	6.0600	5.7400	6.1000	6.0400	49 In 121	0.0000	0.0049	0.0000	0.0103
41 Nb 100	6.3000	5.3500	7.1000	5.9600	50 Sn 121	0.0000	0.0010	0.0000	0.0000
42 Mo 100	0.0000	0.0000	0.0000	0.2050	49 In 122	0.0160	0.0000	0.0450	0.0000
41 Nb 101	5.0000	4.4800	5.9100	5.4500	50 Sn 122	0.0000	0.0603	0.0000	0.0997
42 Mo 101	0.0000	0.7790	0.0000	0.7190	49 In 123m	0.0160	0.0419	0.0500	0.0515
43 Tc 101	0.0000	0.0584	0.0000	0.0000	49 In 123	0.0013	0.0419	0.0050	0.0515
42 Mo 102	4.1000	4.3800	5.9900	6.0600	50 Sn 123m	0.0000	0.0078	0.0000	0.0103
43 Tc 102	0.0000	0.1460	0.0000	0.1030	49 In 124	0.0220	0.0000	0.0700	0.0000
42 Mo 103	3.0000	3.1100	5.6700	5.7500	50 Sn 124	0.0000	0.1260	0.0000	0.1340
43 Tc 103	0.0000	0.2920	0.0000	0.2570	51 Sb 124	0.0000	0.0013	0.0000	0.0000
42 Mo 104	1.8000	1.8400	5.9300	5.0300	50 Sn 125m	0.0080	0.0000	0.0440	0.0000
43 Tc 104	0.0000	0.4280	0.0000	0.7190	50 Sn 125	0.0130	0.0759	0.0710	0.1640
44 Ru 104	0.0000	0.0195	0.0000	0.0000	51 Sb 125	0.0000	0.1280	0.0000	0.0103
42 Mo 105	0.9000	1.2600	5.3000	3.8000	50 Sn 126	0.0440	0.2920	0.2000	0.2770
43 Tc 105	0.0000	0.1950	0.0000	1.5400	51 Sb 126	0.0000	0.0389	0.0000	0.0103
44 Ru 105	0.0000	0.0000	0.0000	0.1030	50 Sn 127m	0.0000	0.1660	0.0000	0.1900
43 Tc 106	0.3800	0.8950	4.5700	4.6200	50 Sn 127	0.1300	0.1660	0.3900	0.1900
44 Ru 106	0.0000	0.0487	0.0000	0.2050	51 Sb 127	0.0000	0.1170	0.0000	0.0617
44 Ru 107	0.1900	0.3500	3.5000	3.8000	52 Te 127	0.0000	0.0097	0.0000	0.0000
44 Ru 108	0.0650	0.2340	2.5000	2.7800	50 Sn 128	0.3700	0.5060	1.1000	0.4930
45 Rh 109	0.0300	0.1420	1.4000	1.6800	51 Sb 128	0.0390	0.1650	0.1100	0.2260
45 Rh 110	0.0200	0.1070	0.5000	0.8430	52 Te 128	0.0000	0.0195	0.0000	0.0103
46 Pd 111m	0.0002	0.0448	0.0023	0.3800	53 I 128	0.0000	0.0000	0.0001	0.0000
46 Pd 111	0.0190	0.0000	0.2300	0.0103	50 Sn 129m	1.0000	0.3500	1.7000	0.2800
46 Pd 112	0.0100	0.0389	0.1200	0.1560	50 Sn 129	0.0000	0.3500	0.0000	0.2800
46 Pd 113	0.0314	0.0333	0.0700	0.0925	51 Sb 129	0.0000	0.4670	0.0000	0.5750
46 Pd 114	0.0120	0.0329	0.0520	0.0997	52 Te 129	0.0000	0.0973	0.0000	0.1030
47 Ag 114	0.0000	0.0004	0.0000	0.0134	54 Xe 129	0.0004	0.0000	0.0006	0.0000
46 Pd 115	0.0104	0.0302	0.0410	0.0894	50 Sn 130	2.0000	0.8470	2.6000	0.5650
47 Ag 115	0.0000	0.0029	0.0000	0.0082	51 Sb 130m	0.0000	0.9050	0.0000	1.4900
47 Ag 116	0.0105	0.0349	0.0380	0.0904	52 Te 130	0.0000	0.2920	0.0000	0.3080
48 Cd 116	0.0000	0.0001	0.0000	0.0000	53 I 130	0.0005	0.0058	0.0006	0.0093
47 Ag 117	0.0110	0.0350	0.0390	0.0863	50 Sn 131	2.9300	1.2100	3.7800	0.3290
48 Cd 117	0.0000	0.0019	0.0000	0.0031	51 Sb 131	0.0000	1.3600	0.0000	1.9300
48 Cd 118	0.0110	0.0390	0.0390	0.0883	52 Te 131	0.0000	0.4870	0.0000	1.1300
48 Cd 119m	0.0060	0.0185	0.0200	0.0436	53 I 131	0.0000	0.0097	0.0000	0.1030
48 Cd 119	0.0060	0.0185	0.0200	0.0436	50 Sn 132	4.2400	0.9730	5.1000	0.1130
49 In 119	0.0000	0.0000	0.0000	0.0010	51 Sb 132	0.0000	2.2400	0.0000	1.5300
48 Cd 120	0.0130	0.0360	0.0400	0.0884	52 Te 132	0.0000	1.5600	0.0000	1.9500
49 In 120	0.0000	0.0019	0.0000	0.0031	53 I 132	0.1400	0.0000	0.1600	0.2670
48 Cd 121	0.0150	0.0341	0.0440	0.0843	51 Sb 133	3.8700	3.3100	4.0400	1.2400

(Continued)

TABLE H.1 (cont'd).

	U-235 FISSION		PU-239 FISSION			U-235 FISSION		PU-239 FISSION	
NUCLIDE	THERMAL	FAST	THERMAL	FAST	NUCLIDE	THERMAL	FAST	THERMAL	FAST
52 Te 133m	2.2400	0.0000	2.3400	0.0000	58 Ce 145	3.9800	3.7400	3.1300	3.1100
52 Te 133	0.0000	3.2000	0.0000	3.4900	59 Pr 145	0.0000	0.0097	0.0000	0.0205
53 I 133	0.5000	0.0000	0.1530	1.1300	58 Ce 146	3.0700	2.8200	2.6000	2.5100
51 Sb 134	6.9000	1.5600	6.3900	0.5340	59 Pr 146	0.0000	0.0973	0.0000	0.0925
52 Te 134	0.0000	3.2100	0.0000	3.5800	58 Ce 147	2.3600	2.0400	2.0700	1.7500
53 I 134	0.9000	1.8500	0.8300	2.3600	59 Pr 147	0.0000	0.1950	0.0000	0.3080
54 Xe 134	0.2600	0.2820	0.2400	0.1030	58 Ce 148	1.7100	1.3600	1.7300	1.1300
52 Te 135	6.0900	3.3100	5.7000	2.6700	59 Pr 148	0.0000	0.3210	0.0000	0.5860
53 I 135	0.0000	2.5300	0.0000	3.8000	60 Nd 148	0.0000	0.0195	0.0000	0.0411
54 Xe 135	0.3200	0.4870	1.4700	0.5140	59 Pr 149	1.1300	0.0000	1.3200	0.0000
55 Cs 135	0.0000	0.0389	0.0000	0.0000	60 Nd 149	0.0000	1.1200	0.0000	1.4400
53 I 136	3.1000	0.0000	2.1000	0.0000	61 Pm 149	0.0000	0.0097	0.0000	0.0000
54 Xe 136	3.3600	5.7700	4.5300	7.0900	60 Nd 150	0.6700	0.8100	1.0100	1.0400
55 Cs 136	0.0068	0.3600	0.1100	0.1540	61 Pm 150	0.0000	0.0370	0.0000	0.0068
53 I 137	6.0000	0.0000	6.4700	0.0000	60 Nd 151	0.4400	0.3800	0.8000	0.8530
54 Xe 137	0.0000	5.2600	0.0000	6.3700	61 Pm 151	0.0000	0.0467	0.0000	0.0617
55 Cs 137	0.1500	0.8270	0.1600	0.3080	61 Pm 152	0.2810	0.2980	0.6200	0.7190
53 I 138	5.7400	0.0000	6.3100	0.0000	62 Sm 152	0.0000	0.0029	0.0000	0.0000
54 Xe 138	0.0000	4.6700	0.0000	5.1400	61 Pm 153	0.1500	0.0000	0.3700	0.0000
55 Cs 138	0.0000	1.6100	0.0000	1.1300	62 Sm 153	0.0190	0.2040	0.0470	0.4930
56 Ba 138	0.0000	0.1460	0.0000	0.0000	61 Pm 154	0.0770	0.0866	0.2900	0.3080
53 I 139	5.4000	0.0000	4.8400	0.0000	62 Sm 154	0.0000	0.0088	0.0000	0.0617
54 Xe 139	0.0000	3.2100	0.0000	3.0800	63 Eu 154	0.0000	0.0000	0.0000	0.0010
55 Cs 139	1.0700	2.2400	0.9590	2.1600	62 Sm 155	0.0330	0.0691	0.2300	0.2570
56 Ba 139	0.0800	0.4870	0.0700	0.2050	63 Eu 155	0.0000	0.0000	0.0000	0.0103
54 Xe 140	3.8000	2.0400	3.3200	1.4400	62 Sm 156	0.0140	0.0234	0.1100	0.1640
55 Cs 140	2.2000	2.5300	1.9300	3.0800	63 Eu 156	0.0000	0.0010	0.0000	0.0205
56 Ba 140	0.3500	0.9730	0.3100	0.7190	62 Sm 157	0.0078	0.0146	0.0800	0.0853
58 Ce 140	0.0900	0.0000	0.0790	0.0000	63 Eu 157	0.0000	0.0019	0.0000	0.0380
54 Xe 141	1.3300	1.1700	1.1800	0.4110	63 Eu 158	0.0020	0.0080	0.0400	0.0688
55 Cs 141	3.2700	2.7700	2.3100	3.5600	64 Gd 158	0.0000	0.0002	0.0000	0.0051
56 Ba 141	1.7000	1.8500	1.5100	1.4400	63 Eu 159	0.0011	0.0033	0.0210	0.0360
57 La 141	0.1000	0.2920	0.0890	0.1030	64 Gd 159	0.0000	0.0000	0.0000	0.0103
54 Xe 142	0.3500	0.5060	0.2900	0.1130	63 Eu 160	0.0003	0.0000	0.0098	0.0134
55 Cs 142	3.1100	2.2200	2.5900	1.9400	64 Gd 160	0.0000	0.0002	0.0000	0.0123
56 Ba 142	2.2400	2.3400	1.8700	2.9800	65 Tb 160	0.0000	0.0000	0.0000	0.0010
57 La 142	0.3100	0.6030	0.2600	0.5140	64 Gd 161	0.0001	0.0004	0.0039	0.0123
54 Xe 143	0.0510	0.1750	0.0380	0.0134	65 Tb 161	0.0000	0.0000	0.0000	0.0021
55 Cs 143	1.8000	1.3800	1.3600	0.6850	64 Gd 162	0.0000	0.0000	0.0020	0.0000
56 Ba 143	2.9200	2.6300	2.2100	2.6100	65 Tb 163m	0.0000	0.0000	0.0004	0.0000
57 La 143	0.9600	1.3600	0.9500	1.1700	65 Tb 163	0.0000	0.0000	0.0004	0.0000
58 Ce 143	0.0000	0.0973	0.0000	0.0514	65 Tb 164	0.0000	0.0000	0.0003	0.0000
57 La 144	5.6200	4.6700	3.9300	3.4200	66 Dy 165m	0.0000	0.0000	0.0001	0.0000
58 Ce 144	0.0000	0.3410	0.0000	0.2050	66 Dy 166	0.0000	0.0000	0.0001	0.0000

Source: Ryman, J.C., "Origen-S Data Libraries," Vol. 3, Section M6 in Scale-3: *A Modular Code System for Performing Standardized Computer Analyses for Licensing Evaluations*, Report NUREG/CR-0200 (ORNL/NUREG/CSD-2), Oak Ridge National Laboratory, Oak Ridge, Tenn., 1984. [Released by the Radiation Shielding Information Center, Oak Ridge National Laboratory, as Code Package CCC-466.]

Index